SOIL PROCESSES AND THE CARBON CYCLE

Edited by

**Rattan Lal
John M. Kimble
Ronald F. Follett
Bobby A. Stewart**

Taylor & Francis
Taylor & Francis Group

Boca Raton   London   New York   Singapore

A CRC title, part of the Taylor & Francis imprint, a member of the
Taylor & Francis Group, the academic division of T&F Informa plc.

FIRST INDIAN REPRINT, 2014

Library of Congress Cataloging-in-Publication Data

Soil processes and the carbon cycle / edited by R. Lal . . . [et al.]
  p. cm.
  Papers from a symposium entitled, "Carbon Sequestration in Soils" held at Ohio State University, July 1996.
  Includes bibliographical references and index.
  ISBN 0-8493-7441-3 (alk. paper)
  1. Soils- -Carbon content- -Congresses. 2. Carbon sequestration- -Congresses. 3. Carbon cycle (Biogeochemistry)- -Congresses.
  I. Lal, R. II. Series: Advances in soil science (Boca Raton, Fla.)
S592.6.C35S65 1997
631.4'17—dc21   97-29442

This book contains information obtained from authentic and highly regarded sources. Reprinted material is quoted with permission, and sources are indicated. A wide variety of references are listed. Reasonable efforts have been made to publish reliable data and information, but the author and the publisher cannot assume responsibility for the validity of all materials or for the consequences of their use.

Neither this book nor any part may be reproduced or transmitted in any form or by any means, electronic or mechanical, including photocopying, microfilming, and recording, or by any information storage or retrieval system, without prior permission in writing from the publisher.

All rights reserved. Authorization to photocopy items for internal or personal use, or the personal or internal use of specific clients, may be granted by CRC Press LLC, provided that $.50 per page photocopied is paid directly to Copyright Clearance Center, 222 Rosewood Drive, Danvers, MA 01923 USA. The fee code for users of the Transactional Reporting Service is ISBN 0-8493-7441-3/00/$0.00+$.50. The fee is subject to change without notice. For organizations that have been granted a photocopy license by the CCC, a separate system of payment has been arranged.

The consent of CRC Press LLC does not extend to copying for general distribution, for promotion, for creating new works, or for resale. Specific permission must be obtained in writing from CRC Press LLC for such copying.

Direct all inquiries to CRC Press LLC, 2000 N.W. Corporate Blvd., Boca Raton, Florida 33431.

Trademark Notice: Product or corporate names may be trademarks or registered trademarks, and are used only for identification and explanation, without intent to infringe.

Visit the CRC Press Web site at www.crcpress.com

© 1998 by CRC Press LLC

No claim to original U.S. Government works
International Standard Book Number 0-8493-7441-3
Library of Congress Card Number 97-29442

Printed and bound in India by Replika Press Pvt. Ltd.

FOR SALE IN SOUTH ASIA ONLY

PREFACE

The pedosphere is the interface between the atmosphere, biosphere, hydrosphere, and geosphere; it plays a major role in the overall global carbon cycle, and in the production of food and fiber around the world. Because of human manipulation of the pedosphere, its interaction with the other spheres, and the attendant impacts on ecosystem productivity and environment quality, a symposium entitled, "Carbon Sequestration in Soils" was held at The Ohio State University in July 1996. This volume and two others contain the papers covering the information presented at the symposium.

These volumes are the state-of-the-art compendium on this topical issue of global significance. They point out "Knowledge Gaps" and researchable issues related to "Soils and the Carbon Cycle", pedospheric processes and their interactions with other natural spheres, and relevant strategic and policy considerations. Information concerning soil organic carbon pool in different ecosystems, the impact of land use and management on this pool, relationships of soil organic matter to soil structure, soil quality, and to mechanisms governing carbon sequestration in soil, and other themes are presented in a comprehensive manner with the objectives of identifying and developing policy and management options. However, the information presented also points out that much more knowledge is needed to improve our future understanding of these fundamental processes that govern the dynamics of soil organic carbon and their accompanying effects on the entire fabric of life on earth. Papers dealing with the global C cycle and pools in different ecoregions are compiled in the first volume, those dealing with C sequestration in soil in relation to management and land use in the second volume, and those dealing with site-specific issues in the third volume. The editorial committee has provided introductory and concluding chapters to highlight the issues and summarize the salient features. A total of 34 chapters are included in volume 1, 41 in volume 2, and 18 in volume 3.

The organization of the symposium and publication of these volumes were made possible by cooperation and funding of the USDA - Natural Resources Conservation Service, the Agricultural Research Service, the Forest Service, Soil Science Society of America, and The Ohio State University. The editors thank all authors for their outstanding efforts to document and present their information on the current understanding of soil processes and the carbon cycle in a timely fashion. Their efforts have contributed to enhancing the overall understanding of pedospheric processes and how to better use soils as a sink for carbon while also managing soils to minimize pedosphere contributions of carbon dioxide and other greenhouse gases to the atmosphere. These efforts have advanced the frontiers of soil science and improved the understanding of the pedosphere into the broader scientific arena of linking soils to the global carbon cycle, soil productivity, and environment quality.

Thanks are also due to the staff of CRC for their timely efforts in publishing this information on time to make it available to the overall scientific community. The field tour was successfully organized by Dr. Ed Redmond and scientists from the NRCS office in Ohio. In addition, valuable contributions were made by numerous colleagues, graduate students, university staff, and staff of the Fawcette Center. We especially thank Ms. Lynn Everett for her efforts in organizing the conference and for handling the flow of chapters to and from the authors throughout the review process. Her tireless efforts, good humor, and good nature are greatly appreciated. We also offer special thanks to Ms. Brenda Swank for her help in preparing this material and for her assistance in all aspects of the symposium. Mrs. Maria Lemon of the "Editors" helped in editing several chapters. The efforts of many others were also very important in getting this relevant and important scientific information out in a timely manner. Financial support received from NRCS, ARS, FS, SSSA, The Ohio State University, and others is gratefully acknowledged.

The Editorial Committee

About the Editors:

Dr. R. Lal is a Professor of Soil Science in the School of Natural Resources at The Ohio State University. Prior to joining Ohio State in 1987, he served as a soil scientist for 18 years at The International Institute of Tropical Agriculture, Ibadan, Nigeria. Prof. Lal is a fellow of the Soil Science Society of America, American Society of Agronomy, The Third World Academy of Sciences, American Association for Advancement of Sciences, and the Soil and Water Conservation Society. He is recipient of the International Soil Science Award, the Soil Science Applied Research Award of the Soil Science Society of America, and the International Agronomy Award of the American Society of Agronomy. He is past President of the World Association of the Soil and Water Conservation and the International Soil Tillage Research Organization.

Dr. John Kimble is a Research Soil Scientist at the USDA Natural Resources Conservation Service National Soil Survey Laboratory in Lincoln, Nebraska. Dr. Kimble manages the Global Change project of the Natural Resources Conservation Service, and has worked more than 15 years with the US Agency for International Development projects dealing with soils-related problems in more than 40 developing countries. He is a member of the American Society of Agronomy, the Soil Science Society of America, the International Soil Science Society, and the International Humic Substances Society.

Dr. R.F. Follett is Supervisory Soil Scientist, USDA-ARS, Soil Plant Nutrient Research Unit, Fort Collins, CO. He previously served 10 years as a National Program Leader with ARS headquarters in Beltsville, MD. Dr. Follett is a Fellow of the Soil Science Society of America, American Society of Agronomy, and the Soil and Water Conservation Society. He was twice awarded the USDA Distinguished Service Award (USDA's highest award). Dr. Follett organized and wrote the ARS Strategic Plans for both "Ground-Water Quality Protection–Biogeochemical Dynamics." Dr. Follett has been a lead editor for several books and a guest editor for the Journal of Cantaminant Hydrology. His scientific publications include topics about nutrient management for forage production, soil-N and -C cycling, ground-water quality protection, global climate change, agroecosystems, soil and crop management systems, soil erosion and crop productivity, plant mineral nutrition, animal nutrition, irrigation, and drainage.

Dr. B.A. Stewart is a Distinguished Professor of Soil Science, and Director of the Dryland Agriculture Institute at West Texas A&M University. Prior to joining West Texas A&M University in 1993, he was Director of the USDA Conservation and Production Research Laboratory, Bushland, Texas. Dr. Stewart is past president of the Soil Science Society of America, and was a member of the 1990-93 Committee on Long Range Soil and Water Policy, National Research Council, National Academy of Sciences. He is a Fellow of the Soil Science Society of America, American Society of Agronomy, Soil and Water Conservation Society, a recipient of the USDA Superior Service Award, and a recipient of the Hugh Hammond Bennett Award of the Soil and Water Conservation Society.

Contributors

E. Amezquita, Centro Internacional de Agricultura Tropical, Tropical Lowlands Program, Apartado Aereo 6713, Cali, Colombia.

D.A. Angers, Research Center, Agriculture and Agri-Food Canada, 2560 Hochelaga Blvd., Ste. Foy, Quebec G1V 2J3, Canada.

M.A. Arshad, Northern Alberta Research Centre, Agriculture and Agri-Food Canada, Beaverlodge, Alberta, T0H 0C0, Canada.

R.M. Bajracharya, The Ohio State University, School of Natural Resources, 2021 Coffey Road, Columbus, OH, 43210, U.S.A.

J.A. Baldock, CSIRO Division of Soils, Glen Osmond, S.A. 5064, Australia.

T.O. Barnwell, USEPA, Environmental Research Laboratory, Athens, GA 30613, U.S.A.

J.G. Bockheim, University of Wisconsin, Department of Soil Science, 1525 Observatory Drive, Madison, WI 53706-1299, U.S.A.

C.A. Cambardella, USDA-ARS, National Soil Tilth Laboratory, 2150 Pammel Dr., Ames, IA 50011-0001, U.S.A.

C.A. Campbell, Agriculture Canada Research Station, P.O. Box 1030, Swift Current, Saskatchewan S9H 3X2, Canada.

C.C. Cerri, Universidade de Sao Paulo, Centro de Energia Nuclear na Agricultura, Campus Luiz de Queiroz, Avenida Centenario 303, Caixa Postal 96, Cep 13400-970, Piracicaba SP, Brazil.

R.J. Chandler, University of Alaska-Fairbanks, Palmer Research Center, 533 E. Firewood Av., Palmer, AK 99645

Z. Qi Chen, University of Alberta, Department of Renewable Resources, Earth Sciences Building, Room 442, Edmonton, Alberta T6G 2E3, Canada.

C. Chenu, INRA, Unite de Science du Sol, 78026 Versailles, France.

R.V. Chinnaswamy, AQUA TERRA Consultants, 2672 Bayshore Parkway, Suite 1001, Mountain View, CA 94043, U.S.A.

C.E. Clapp, University of Minnesota, Department of Soil, Water, and Climate; 1991 Upper Buford Circle, St. Paul, MN 55108, U.S.A.

J.F.L. de Moraes, Instituto Agronômico de Campinas, Caixa Postal 28, 13020-902, Campinas, SP, Brazil.

H. Dinel, 960 Carling Avenue, K.W. Neatby Building, Agriculture and Agri-Food Canada, Ottawa, Ontario K1A 0C6, Canada.

R.F. Dodson, Dynamac Corporation, USEPA National Health and Environmental Effects Laboratory, Western Ecology Division, 200 SW 35th St., Corvallis, OR 97333.

A.S. Donigian, Jr., AQUA TERRA Consultants, 2672 Bayshore Parkway, Suite 1001, Mountain View, CA 94043, U.S.A.

W. Eisner, Byrd Polar Research Center, The Ohio State University, 108 Scott Hall, 1090 Carmack Rd., Columbus, OH 43210-1002, U.S.A.

B.H. Ellert, Agriculture and Agri-Food Canada, Box 3000, Lethbridge, Alberta T1J 4B1, Canada.

E.T. Elliott, Colorado State University, Natural Resource Ecology Laboratory, Office of Ecosystem Research and Management, Fort Collins, CO 80523, U.S.A.

L.R. Everett, Byrd Polar Research Center, The Ohio State University, 108 Scott Hall, 1090 Carmack Rd., Columbus, OH 43210-1002, U.S.A.

B.J. Feigl, Centro de Energia Nuclear na Agricultura, Avenida Centenório, 303, Caixa Postal 96, CEP 13416000, Piracicaba, SP, Brazil.

M.J. Fisher, Centro Internacional de Agricultura Tropical, Tropical Lowlands Program, Apartado Aereo 6713, Cali, Colombia.

R.F. Follett, USDA-ARS, Soil Plant Nutrient Research, P.O. Box E, Ft. Collins, CO 80522-0470, U.S.A.

J.R. Fortner, USDA-NRCS, National Soil Survey Center, Federal Bldg. Room 152, 100 Centennial Mall N., Lincoln, NE 68508-3866, U.S.A.

D.S. Gamble, 960 Carling Avenue, K.W. Neatby Building, Agriculture and Agri-Food Canada, Ottawa, Ontario K1A 06C, Canada.

A. Gennadiyev, Moscow State University, Department of Geography, Moscow 119899, Russia.

S.A. Glaum, USDA-NRCS, National Soil Survey Center, Federal Bldg. Room 152, 100 Centennial Mall N., Lincoln, NE 68508-3866, U.S.A.

J.A. Golchin, The University of Adelaide, Department of Soil Science, Waite Agricultural Research Institute, Glen Osmond, S.A. 5064, Australia.

R.F. Grant, University of Alberta, Department of Renewable Resources, Edmonton, Alberta T6G 2E3, Canada.

D.J. Greenland, Low Wood, The Street South Stoke, Reading RG8 0JS, U.K.

E.G. Gregorich, Agriculture and Agri-Food Canada, K.W. Neatby Building, CEF, Ottawa, Ontario K1A 0C5, Canada.

R. B. Grossman, USDA-NRCS, National Soil Survey Center, Federal Bldg. Room 152, 100 Centennial Mall N.,Lincoln, NE 68508-3866, U.S.A.

D.H. Harms,USDA-NRCS, National Soil Survey Center, Federal Bldg. Room 152, 100 Centennial Mall N., Lincoln, NE 68508-3866, U.S.A.

K.G. Harrison, Duke University Phytotron, Box 90340, Durham, NC 27708-0340, U.S.A.

S.L. Hartung, USDA-NRCS, National Soil Survey Center, Federal Bldg. Room 152, 100 Centennial Mall N., Lincoln, NE 68508-3866, U.S.A.

J.E. Herrick, USDA-ARS Jornada Exp. Range, Box 30003, NMSU, Dept. 3JER, Las Cruces, NM 88003-0003, U.S.A.

Frank L. Himes, The Ohio State University, School of Natural Resources, 2021 Coffey Road, Columbus, OH 43210, U.S.A.

R.C. Izaurralde, University of Alberta, 442 Earth Sciences Building, Department of Renewable Resources, Edmonton, Alberta T6G 2E3, Canada.

D.C. Jans-Hammermeister, Agrium Biologicals, 402-15 Innovation Blvd., Saskatoon, SK S7N 2X8, Canada.

H. Henry Janzen, Agriculture and Agri-Food Canada, Box 3000, Lethbridge, Alberta T1J 4B1, Canada.

Julie D. Jastrow, DOE/Argonne National Laboratory, 9700 South Cass Ave., Argonne, IL 60439, U.S.A.

N.G. Juma, University of Alberta, Department of Renewable Resources, Room 442, Earth Sciences Building, Edmonton, Alberta T6G 2E3, Canada.

D.L. Karlen, USDA-ARS, National Soil Tilth Laboratory, 2150 Pammel Drive, Ame, IA 50011, U.S.A.

B.D. Kay, University of Guelph, Department of Land Resource Science, Guelph, Ontario N1G 2W1, Canada.

J.S. Kern, Dynamac Corporation, USEPA National Health and Environmental Effects Laboratory, Western Ecology Division, 200 SW 35th Street, Corvallis, OR 97333, U.S.A.

K. Killian, Colorado State University, Natural Resource Ecology Laboratory, B266, Ft. Collins, CO 80523, U.S.A.

J. Kimble, USDA-NRCS, National Soil Survey Center, Federal Building, Room 152, 100 Centennial Mall North, Lincoln, NE 68508-3866, U.S.A.

D.S. Kimes, NASA/Goddard Space Flight Center, Code 923 Biospheric Sciences Branch, Greenbelt, MA 20771, U.S.A.

E.-M. Klimanek, Umweltforschungszentrum Leipzig-Halle GmbH, Sektion Bodenforschung, Hallesche Str. 44, D-06246 Bad Lauchstadt, Germany.

Dmitri Ye Konyushkov, V.V. Docuchaev Soils Institute, Apt. 79, 82Pyzheuskiiy Lane, Moscow 109017, Russia.

M. Karschens, UFZ Centre for Environmental Research Liepzig-Halle, Department of Soil Science, Hallesche Strabe 44, 06246 Bad Lauchstadt, Germany.

M.S. Kuzila, USDA-NRCS, National Soil Survey Center, Federal Bldg., Room 152, 100 Centennial Mall N., Lincoln, NE 68508-3866, U.S.A.

B. Lacelle, Agric. Canada Eastern Cereal & Oilseed Research Centre Branch, Agriculture and Agri-Food Canada, K.W. Neatby Bldg. C.E.F., Ottawa, Ontario K1A 0C6, Canada.

R. Lal, The Ohio State University, School of Natural Resources, 2021 Coffey Road, Columbus, OH 43210, U.S.A.

E. Levine, NASA/Goddard Space Flight Center, Code 923, Biospheric Sciences Branch, Greenbelt, MD 20771, U.S.A.

D.R. Linden, USDA-ARS, University of Minnesota, Soil and Water Management Research Unit, 1530 Cleveland Ave. N., 439 Borlaug Hall, St. Paul, MN 55108-1004, U.S.A.

E. Lioubimtseva, Faculte des Sciences Agronomiques, Department des Sciences du Milieu et Amenagement du Territoire, Croix du Sud, 2 Bte 9, 1348 Louvain-la-Neuve, Belgium.

W.M. Loya, c/o C.L. Ping, University of Alaska-Fairbanks, Palmer Research Center, 533 E. Firewood Av., Palmer, AK 99645, U.S.A.

Ronald L. Malcolm, U.S. Geological Survey, Denver, CO, U.S.A.

S.S. Malhi, Research Station, Agriculture and Agri-Food Canada, P.O. Box 1240, Highway 6 South Melfort, SK S0E 1A0, Canada.

M.J. Mausbach, USDA-NRCS, 2150 Pammel Dr., Ames, IA 50011-0001, U.S.A.

W.B. McGill, University of Alberta, Department of Renewable Resources, Edmonton, Alberta T6J 2M1, Canada.

B.F. McQuaid, USDA-NRCS, North Carolina State University, 1509 Varsity Dr., Raleigh, NC 27606-2083, U.S.A.

J.M. Melillo, The Ecosystems Center, Marine Biological Laboratory, Woods Hole, MA 02543, U.S.A.

S. Mercik, Warsaw Agricultural University, Faculty of Agriculture, Department of Agricultural Chemistry, Warsaw, Poland.

G.J. Michaelson, University of Alaska-Fairbanks, Palmer Research Center, 533 E. Fireweed Ave., Palmer, AK 99645, U.S.A.

R.M. Miller, Environmental Research Division, Argonne National Laboratory, Argonne, IL 60439, U.S.A.

M. Molina-Ayala, University of Alberta, Department of Renewable Resources, Edmonton, Alberta T6J 2M1, Canada.

C.M. Monreal, Agriculture Canada, K.W. Neatby Building, Room 3042, 960 Carlig Avenue, Ottawa, Ontario K1A 0C6, Canada.

Tim R. Moore, McGill University, Department of Geography, 805 Sherbrooke St. W., Montreal H3A 2K6, Canada.

C. Neill, Marine Biological Laboratory, The Ecosystems Center, Woods Hole, MA 02543, U.S.A.

M. Nyborg, University of Alberta, Department of Renewable Resources, Edmonton, Alberta T6J 2M1, Canada.

J.M. Oades, The University of Adelaide, Faculty of Agricultural and Natural Resource Sciences, Waite Campus, Private Bag No. 1, Glen Osmond, SA 5064, Australia.

Gail L. Olson, INEL, 215 N. Blvd., Box 16258, Idaho Falls, ID 83415-2213, U.S.A.

A.S. Patwardhan, Minnesota Pollution Control Agency, 520 Lafayette Road, St. Paul, MN 55155-4194, U.S.A.

K. Paustian, Colorado State University, Natural Resources Ecology Laboratory, Fort Collins, CO 80523-1499, U.S.A.

S. Pawluk, University of Alberta, Department of Renewable Resources, Room 442, Earth Sciences Building, Edmonton, Alberta T6G 2E3, Canada.

E.M. Pfeiffer, University of Hamburg, Institute of Soil Science, Allendeplatz 2, 20146 Hamburg, Germany.

Marisa C. Piccolo, Centro de Energia Nuclear na Agricultura, Avenida Centenrio, 303 Caixa Postal 96, CEP 13416000, Piracicaba, SP, Brazil.

C.L. Ping, University of Alaska-Fairbanks, Palmer Research Center, 533 E. Firewood Av., Palmer, AK 99645, U.S.A.

H.H. Rogers, USDA-ARS, National Soil Dynamics Laboratory, P.O. Box 3439, Auburn, AL 36831-3439, U.S.A.

H.-W. Scharpenseel, University of Hamburg, Institute of Soil Science, Allendeplatz 2, 20146 Hamburg, Germany.

M. Schnitzer, Agriculture and Agri-Food Canada, 960 Carling Avenue, K.W. Neatby Building, Ottawa, Ontario K1A 0C6, Canada.

C. Seybold, Oregon State University, Soil Science Department, ALS Building Room 3017, Corvallis, OR 97331-8507, U.S.A.

E.D. Solberg, Alberta Agriculture Food and Rural Development, Agronomy Unit, Edmonton, Alberta T6H 5T6, Canada.

P.A. Steudler, Marine Biological Laboratory, The Ecosystems Center, Woods Hole, MA 02543.

C. Tarnocai, Eastern Cereal and Oilseed Research Centre, Agriculture Canada, K.W. Neatby Building, 960 Carling Avenue, Ottawa K1A 0C6, Canada.

W. Trujillo, The Ohio State University, School of Natural Resources, 2021 Coffey Road, Columbus, OH 43210, U.S.A.

D.P. Turner, Oregon State University, Forest Science Department, Corvallis, OR 97331, U.S.A.

D.A. Walker, INSTAAR, University of Colorado, Campus Box 450, Boulder, CO 80309, U.S.A.

M.M. Wander, University of Illinois, Natural Resources and Environmental Sciences, Urbana, IL 61801, U.S.A.

Annett Weigel, Umweltforschungszentrum Leipzig-Halle GmbH Sektion Boden-forschung, Hallesche Str. 44, D-06246, Bad Lauchstadt, Germany.

Contents

SOIL PROCESSES AND C CYCLES

Chapter 1. Pedospheric Processes and the Carbon Cycle 1
R. Lal, J. Kimble, and R. Follett

CARBON POOLS IN DIFFERENT BIOMES

Chapter 2. Stocks and Dynamics of Soil Carbon Following Deforestation for Pasture in Rondônia 9
Christopher Neill, Carlos Cerri, Jerry M. Melillo, Brigitte J. Feigl, Paul A. Steudler, Jener F.L. Moraes, and Marisa C. Piccolo

Chapter 3. Spatial Patterns in Soil Organic Carbon Pool Size in the Northwestern United States 29
Jeffrey S. Kern, David P. Turner, and R.F. Dodson

Chapter 4. Organic Carbon in Deep Alluvium in Southeast Nebraska and Northeast Kansas 45
R.B. Grossman, D.H. Harms, M.S. Kuzila, S.A. Glaum, S.L. Hartung, and J.R. Fortner

Chapter 5. Soil Carbon Dynamics in Canadian Agroecosystems 57
H.H. Janzen, C.A. Campbell, E.G. Gregorich, and B.H. Ellert

Chapter 6. The Amount of Organic Carbon in Various Soil Orders and Ecological Provinces in Canada 81
D. Tarnocai

Chapter 7. Canada's Soil Organic Carbon Database 93
B. Lacelle

Chapter 8. Rate of Humus (Organic Carbon) Accumulation in Soils of Different Ecosystems 103
A. Gennadiyev

Chapter 9. Land Use and Soil Management Effects on Soil Organic Carbon Dynamics on Alfisols in Western Nigeria 109
R. Lal

Chapter 10. Arctic Paleoecology and Soil Processes: Developing New Perspectives for Understanding Global Change 127
Wendy R. Eisner

Chapter 11. Soil Carbon Distribution in Nonacidic and Acidic Tundra of Arctic Alaska 143
J.G. Bockheim, D.A. Walker, and L.R. Everett

Chapter 12. Characteristics of Soil Organic Matter in Arctic Ecosystems of Alaska .. 157
C.L. Ping, G.J. Michaelson, W.M. Loya, R.J. Chandler, and R.L. Malcolm

SOIL STRUCTURE AND OTHER PHYSICAL PROCESSES FOR C SEQUESTRATION

Chapter 13. Soil Structure and Organic Carbon: a Review 169
B.D. Kay

Chapter 14. Dynamics of Soil Aggregation and C Sequestration 199
Denis A. Angers and Claire Chenu

Chapter 15. Soil Aggregate Stabilization and Carbon Sequestration: Feedbacks Through Organo-Mineral Associations 207
J.D. Jastrow and R.M. Miller

Chapter 16. Impact of Variations in Granular Structures on C Sequestration in Two Alberta Mollisols 225
Z. Chen, S. Pawluk, and N.G. Juma

Chapter 17. A Model Linking Organic Matter Decomposition, Chemistry, and Aggregate Dynamics 245
J.A. Golchin, J.A. Baldock, and J.M. Oades

Chapter 18. Soil Organic Carbon Dynamics and Land Use in the Colombian Savannas: Aggregate Size Distribution 267
W. Trujillo, E. Amezquita, M.J. Fisher, and R. Lal

SOIL CHEMICAL PROCESSES

Chapter 19. Dissolved Organic Carbon: Sources, Sinks, and Fluxes and Role in the Soil Carbon Cycle 281
T.R. Moore

Chapter 20. Geochemical History of Carbon on the Planet: Implications for Soil Carbon Studies 293
D. Ye. Konyushkov

Chapter 21. Nitrogen, Sulfur, and Phosphorus and the Sequestering of Carbon 315
F.L. Himes

SOIL BIOLOGICAL PROPERTIES

Chapter 22. Management of Soil C by Manipulation of Microbial Metabolism: Daily vs. Pulsed C Additions 321
D.C. Jans-Hammermeister, W.B. McGill, and R.C. Izaurralde

Chapter 23. Investigations to the Carbon and Nitrogen Dynamics of Different Long-Term Experiments by Means of Biological Soil Properties 335
A. Weigel, E.-M. Klimanek, M. Körschens, and St. Mercik

Chapter 24. Effect of Corn and Soybean Residues on Earthworm Cast Carbon Content and Natural Abundance Isotope Signature 345
Dennis R. Linden and C.Edward Clapp

SOIL EROSION AND C DYNAMICS

Chapter 25. Soil Organic Carbon Distribution in Aggregates and Primary Particle Fractions as Influenced by Erosion Phases and Landscape Position 353
R.M. Bajracharya, R. Lal, and J.M. Kimble

Chapter 26. Carbon Storage in Eroded Soils after Five Years of Reclamation Techniques 369
R.C. Izaurralde, M. Nyborg, E.D. Solberg, H.H. Janzen, M.A. Arshad, S.S. Malhi, and M. Molina-Ayala

SOIL QUALITY AND C SEQUESTRATION

Chapter 27. Quantification of Soil Quality 387
C.A. Seybold, M.J. Mausbach, D.L. Karlen, and H.H. Rogers

Chapter 28. Relationships Between Soil Organic Carbon and Soil Quality in Cropped and Rangeland Soils: The Importance of Distribution, Composition, and Soil Biological Activity 405
Jeffrey E. Herrick and Michelle M. Wander

Chapter 29. Soil Quality Indices of Piedmont Sites under Different Management Systems 427
Betty F. McQuaid and Gail L. Olson

Chapter 30. Impact of Carbon Sequestration on Functional Indicators of Soil Quality as Influenced by Management in Sustainable Agriculture 435
C.M. Monreal, H. Dinel, M. Schnitzer, D.S. Gamble, and V.O. Biederbeck

MODELLING C DYNAMICS

Chapter 31. Modeling Soil Carbon in Relation to Management and Climate Change in Some Agroecosystems in the Central North America 459
Keith Paustian, Edward T. Elliott, and Kendrick Killian

Chapter 32. Predicting Soil Carbon in Mollisols Using Neural Networks 473
Elissa R. Levine and Daniel Kimes

Chapter 33. A Retrospective Modeling Assessment of Historical Changes in Soil Carbon and Impacts of Agricultural Development in Central U.S.A., 1900 to 1990 485
A.S. Patwardhan, A.S. Donigian Jr., R.V. Chinnaswamy, and T.O. Barnwell

Chapter 34. Modeling Soil Carbon and Agricultural Practices in the Central U.S.: An Update of Preliminary Study Results 499
A.S. Donigian, A.S. Patwardhan, R.V. Chinnaswamy, and T.O. Barnwell

Chapter 35. Experimental Verification of Simulated Soil Organic Matter Pools 519
Cynthia A. Cambardella

Chapter 36. Modeling Tillage and Surface Residue Effects on C Storage under Ambient vs. Elevated CO_2 and Temperature in *Ecosys* 527
R.F. Grant, R.C. Izaurralde, M. Nyborg, S.S. Malhi, E.D. Solberg, and D. Jans Hammermeister

METHODS OF SOC DETERMINATION

Chapter 37. Using Bulk Radiocarbon Measurements to Estimate Soil Organic Matter Turnover Times 549
K.G. Harrison

Chapter 38. Impacts of Climatic Change on Carbon Storage Variations in African and Asian Deserts 561
E. Lioubimtseva

Chapter 39. Carbon Turnover in Different Climates and Environments 577
H.W. Scharpenseel and E.M. Pfeiffer

IMPACT OF CLIMATE ON C DYNAMICS

Chapter 40. Carbon Sequestration in Soil: Knowledge Gaps Indicated by the Symposium Presentations 591
D.J. Greenland

Chapter 41. Knowledge Gaps and Researchable Priorities 595
R. Lal, J. Kimble, and R. Follett

Index 605

CHAPTER 1

Pedospheric Processes and the Carbon Cycle

R. Lal, J. Kimble, and R.F. Follett

I. Introduction

The atmospheric CO_2 concentration was about 250 ppmv from 900 to 1200 AD, and about 280 ppmv from 1300 to 1800 AD. In 1994, the concentration was 358 ppmv (IPCC, 1995, Figure 1). The major increase is atmospheric CO_2 (about 80 ppmv) has occurred since the 1850s. This increase is attributed to two principal human activities, land use changes (1.6 ± 1.0 Pg C yr^{-1}) and the fossil fuel combustion (5.5 ± 0.5 Pg C yr^{-1}). Annual increase in the atmospheric CO_2 due to these two activities is estimated at 1.5 ppmv, 3.3 ± 0.2 Pg C yr^{-1}, or 1.2 μL L^{-1} of CO_2 (Post et al., 1990). With oceanic uptake of 2.0 ± 0.8 Pg C yr^{-1} and forest regrowth in the northern hemisphere estimated to account for 0.5 ± 0.5 Pg C yr^{-1}, the unknown or missing terrestrial sink is estimated at 1.3 ± 1.5 Pg C yr^{-1} (IPCC, 1995). Since the unknown sink is not ocean (Kern, 1992) and is likely within the terrestrial ecosystems (Tans et al., 1990), the question about the role of soil in C sequestration is very relevant. Another debatable issue is how much of the 80 ppmv increase in atmospheric CO_2 concentration since 1850 came from soils, and what is the potential of world soils to sequester C? Some estimates show that between 400 and 800 Tg y^{-1} of C could be sequestered globally in agricultural soils through judious management (IPCC, 1995).

The modern global carbon cycle of principal C pools and exchanges between them (Figure 2) shows two significant fluxes: (i) between atmosphere and the land plants (120 Pg yr^{-1}), and (ii) between atmosphere and the ocean (105 to 107 Pg yr^{-1}). The exchange between the atmosphere and the land plant includes soil-related efflux of 60 Pg C yr^{-1}. World soils, therefore, play an important role in the global carbon cycle. Thus, important and strategic objectives of selecting soil management options are to decrease the efflux of CO_2 from soil to the atmosphere, and increase the total soil organic carbon (SOC) pool at the expense of the atmospheric C pool (Figure 2). How much of the missing terrestrial sink (Tans et al., 1990; Ciais et al. 1995) that can be attributed to soils and soil-related processes is not known.

II. Interactive Pedospheric Processes

The pedosphere lies at the interface between the lithosphere and the atmosphere. It is a one to two meter deep layer (and likely to be deeper in the tropics) on top of the Earth's crust. The pedosphere supports all biotic activity within the terrestrial ecosystems and interacts with the atmosphere, lithosphere, biosphere, and hydrosphere (Figure 3). These interactions influence the biogeochemical cycles of principal nutrient elements (e.g., N, P, K, S) and H_2O. Interactive processes with the atmosphere lead to gaseous and energy exchanges between the soil and the atmosphere. Mechanisms

ISBN 0-8493-7441-3
©1997 by CRC Press LLC

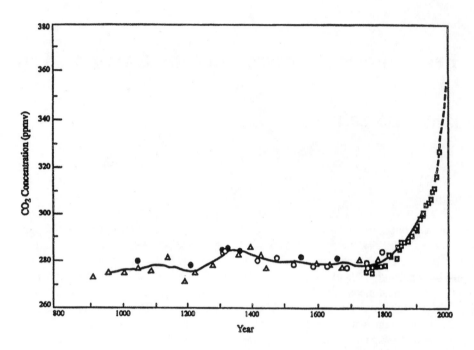

Figure 1. Changes in atmospheric CO_2 concentration. (Adapted from IPCC, 1995.)

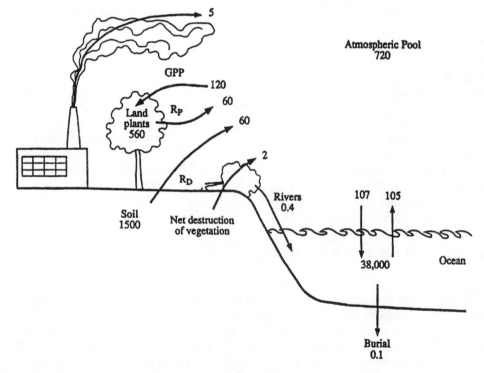

Figure 2. The global carbon cycle, showing the major annual transfers between land, sea, and atmosphere, expressed as 10^{15} g C/yr. (Modified from Schlesinger, 1991.)

of interaction between the lithosphere and pedosphere include leaching of nutrients and new soil formation due to weathering. Elemental cycling and pedoturbation (due to the activity of soil fauna) are interactive processes between the pedosphere and the biosphere. Water exchange between soil and atmosphere plays a principal role in the local, regional, and global hydrological cycle.

In addition to interactive linkages with the pedosphere, there are several crucial processes linking all five predominant spheres (Figure 3). The hydrosphere interacts with the lithosphere through runoff, seepage, and ground water recharge. The hydrosphere interacts with atmosphere through the precipitation-evaporation cycle. Photosynthesis and respiration are the predominant processes linking the atmosphere and the biosphere. The lithosphere is linked with the biosphere through weathering of parent materials and elemental cycling by biota.

Interactive processes that play a major role in the global carbon cycle are those between the pedosphere, the atmosphere, and the biosphere.

III. Pedospheric Processes and the Global Carbon Cycle

There are two types of carbon pools in the pedosphere, e.g., soil organic carbon (SOC) and soil inorganic carbon (SIC). The current SOC pool in the world soils is estimated at 1500 Pg (Eswaran et al., 1995; Figure 2; Table 1). The SOC pool is about 2.1 times that of the atmosphere pool and about 2.7 times that of the biotic pool comprising land plants. Estimates of the SIC pool are more tentative than those of SOC pool, but may be about 12% more than those of the SOC pool (Schlesinger, 1991; Grossman et al., 1995). Most of the SIC pool, as carbonates, lies in soils of the semi-arid regions.

The pedosphere has played a significant role in influencing the gaseous composition of the atmosphere, especially since 1850. However, the magnitude of total contribution to the atmospheric pool, and past and current rates of C flux between the pedosphere and the atmosphere, are not known. One method to estimate C release from world soils to the atmosphere is to compute a C budget for each soil order as influenced by changes in land use. This method is based on the assumption that the antecedent and final C pools (both SOC and SIC) are known.

Little, if any, is known about the dynamics of the SIC pool in relation to land use. However, soil scientists are beginning to understand the dynamics of SOC and factors affecting it. Predominant pedospheric processes that affect SOC dynamics may be grouped into two categories: (i) SOC enhancing and, (ii) SOC degrading processes (Figure 4). Processes that enhance SOC content are plant biomass production, humification, aggregation, and sediment deposition. Processes that degrade SOC content are soil erosion, leaching, and soil organic matter decomposition. It is the net balance between these SOC aggrading and degrading processes, as influenced by land use and anthropogenic factors, that determines the net SOC pool of the pedosphere.

IV. Mechanisms and Potential of C Sequestration in the Pedosphere

An increase in SOC, through C sequestration into the pedosphere, has two notable positive effects. First is the enhancement of soil quality, and second is the improvements in the soil's environmental regulatory capacity (Figure 5). These positive effects form the basis of any strategy for sustainable management of soil and water resources. The soil quality effects of SOC are related to several strongly interacting edaphological factors including soil structure, rooting depth and solum properties, available water capacity or least limiting water range (Thomasson, 1978; Letey, 1985; de Silva et al., 1995), soil biodiversity, and elemental cycling and nutrient reserves. The environmental regulatory effects of SOC are due to its impact on water quality and gaseous composition of the atmosphere. The water quality effect is strongly related to soil structure, its resistance to climatic erosivity or forces of

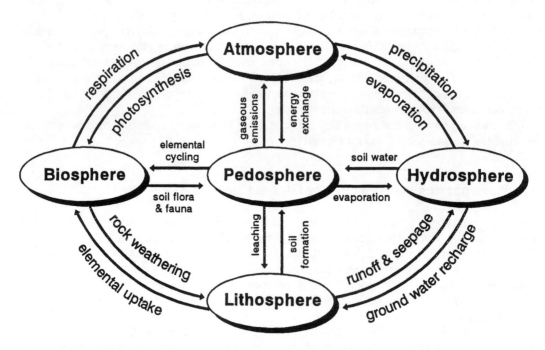

Figure 3. Interactive processes linking pedosphere with atmosphere, biosphere, hydrosphere, and lithosphere.

Table 1. Estimates of soil organic and inorganic carbon pools in world soils

Soils	Carbon pools to 1-m depth	
	Organic	Inorganic
Ultisols	101	0
Andisols	69	1
Aridisols	110	1044
Oxisols	150	0
Inceptisols	267	258
Alfisols	136	127
Mollisols	72	139
Vertisols	38	25
Spodosols	98	0
Entisols	106	117
Histosols	390	0
Miscellaneous	18	0
Total	1555	1738

(Modified from Eswaran et al., 1995.)

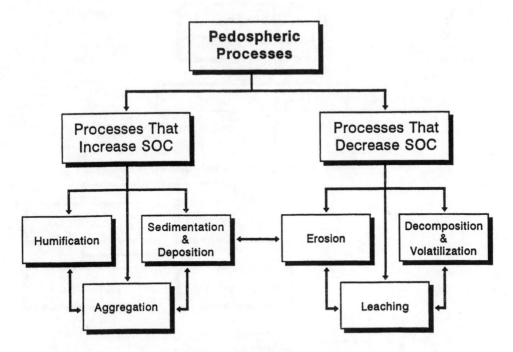

Figure 4. Principal pedospheric processes affecting soil organic carbon content.

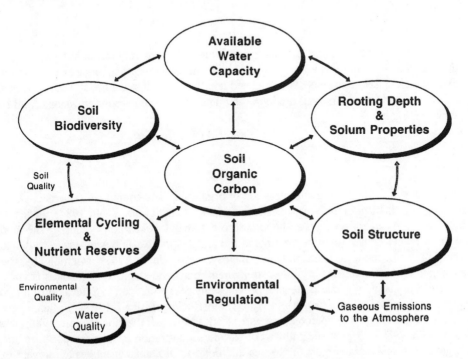

Figure 5. Soil organic carbon impact on soil and environmental quality.

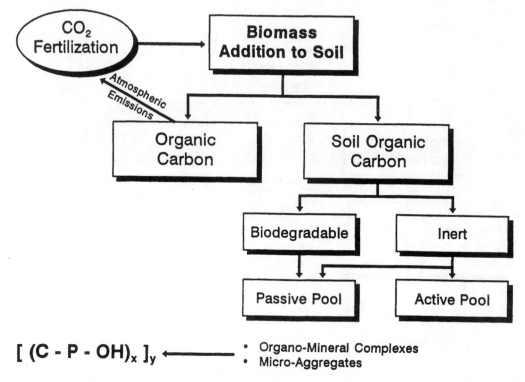

Figure 6. A pathway of carbon sequestration in soil through formation of organo-mineral complexes and stable aggregates.

wind and water, and transmission of water and solutes through the soil solum. The gaseous composition of the atmosphere is affected by emissions of radiatively-active gases from soil to the atmosphere, i.e., CO_2, CH_4, Nox, etc. The SOC and its effect on soil structure, soil moisture regime, elemental cycling, and transformations are important determinants of gaseous emissions from soil to the atmosphere (Figure 4).

A. Aggregation

Improvement in soil structure through formation of organo-mineral complexes (Harris et al., 1966) is an important mechanism of carbon sequestration in soils (Chaney and Swift, 1984; Oades, 1988; Emerson, 1995). Aggregation plays an important role in the global carbon cycling. "The union of mineral and organic matter to form the organo-mineral complexes is a synthesis as vital to the continuance of life as, and less understood than, photosynthesis" (Jacks, 1963). Microbial byproducts form an important cementing material that strengthen bonds and stabilize aggregates (Lynch and Bragg, 1985). Formation of stable aggregates provides physical protection to SOC against microbial decomposition (Powlson, 1980; Oades and Waters, 1991; Hassink et al., 1993). In fact, aggregate disruption and exposure of C to microbial processes leads to C mineralization (Elliott, 1986; Martel, 1994; Powlson, 1980). A possible pathway of carbon sequestration in soils through formation of stable aggregates is outlined in Figure 6. Regular additions of substantial quantities of biomass to soil leads to enhancement of the quantity and quality of humus and the formation of organo-mineral complexes (Monnier, 1965; Harris et al., 1966) that are stabilized by microbial byproducts (Lynch and Bragg, 1985). Plant debris is often the nucleus of a macro-aggregate (Oades and Waters, 1991; Golchin et al., 1994; 1995 Buyanovsky et al., 1994).

B. Biogeochemical Processes

There is a need to assess the magnitude of C translocated into the sub-soil as DOC (Schlesinger, 1984). Considered over a long time, this process could account for a large sink of C. Weathering of carbonates and silicate minerals is another process that deserves attention (Berner, 1990). The role and magnitude of these biochemical processes in soil are not known.

IV. Conclusions

The pedosphere plays a major role in the global carbon cycle, through several interactive mechanisms with the atmosphere and the biosphere. The magnitude of the pedosphere's contribution of carbon to the atmospheric pool is not known. There is a conspicuous lack of information on the magnitude and dynamics of the SIC pool, and although some data on the SIC pool in the pedosphere exist, the information is very sketchy, vague and unreliable.

Estimates of C released from world soils to the atmosphere range from 40 to 50 Pg; however, these estimates are tentative and difficult to verify with the present database. There is a strong need to study and understand interactive mechanisms and processes between the pedosphere on the one hand and atmosphere and the biosphere on the other. It is the knowledge of these interactions that can also lead to the identification of mechanisms and the formulation of strategies for carbon sequestration in the pedosphere.

Three principal processes of C sequestration in the pedosphere are humification, aggregation, and sedimentation. Each of these processes and their relations to each other need to be clearly understood as do the effects of SOC and SIC leaching and the chemical transformations of dissolved carbon.

References

Berner, R.A. 1990. Atmospheric CO_2 levels over Phanerozoic time. *Science* 249:1382-1386.
Buyanovsky, G.A., M. Ashlam, and G.H. Wagner. 1994. Carbon turnover in soil physical fractions. *Soil Sci. Soc. Am. J.* 58:1167-1173.
Chaney, K. and R.S. Swift. 1984. The influence of organic matter on aggregate stability in some British soils. *J. Soil Sci.* 35:223-230.
Ciais, P., P.O. Tans, M. Trolier, J.W.C. White, and R.J. Francey. 1995. A large Northern Hemisphere Terrestrial CO_2 sink indicated by the $^{13}C/^{13}C$ ratio of atmospheric CO_2. *Science* 269:1098-1102.
da Silva, A.P., B.D. Kay, and E. Perfect. 1994. Characterization of the least limiting water range of soils. *Soil Sci. Soc. Amer. J.* 58:1775-1781.
Elliott, E.T. 1986. Aggregate structure and carbon, nitrogen and phosphorus in native and cultivated soils. *Soil Sci. Soc. Am. J.* 50:627-633.
Emerson, W.W. 1995. Water retention, organic C and soil texture. *Aust. J. Soil Res.* 33:241-251.
Eswaran, H., E. Van den Berg, P. Reich, and J. Kimble. 1995. Global soil carbon resources. p. 27-43. In: R. Lal, J. Kimble, E. Levine, and B.A. Stewart (eds.), *Soils and Global Change*. CRC/Lewis Publishers, Boca Raton, FL.
Golchin, A., J.M. Oades, J.O. Skjemstad, and P. Clarke. 1994. Soil structure and carbon cycling. *Aust. J. Soil Res.* 32:1043-1068.

Golchin, A., P. Clarke, J.M. Oades, and J.O. Skjemstad. 1995. The effects of cultivation on the composition of organic matter and structural stability of soils. *Aust. J. Soil Res.* 33:975-993.

Grossman, R.B., R.J. Ahrens, L.H. Gile, C.E. Montoya, and O.A. Chadwick. 1995. A real evaluation of carbonate carbon in a desert area of southern New Mexico. p. 81-91. In: R. Lal, J. Kimble, E. Levine and B.A. Stewart (eds.), *Soils and Global Change*. CRC/Lewis Publishers, Boca Raton, FL.

Harris, R.F., G. Chesters, and O.N. Allen. 1966. Dynamics of soil aggregation. *Adv. Agron.* 18:107-169.

Hassink, J., L.A. Bouwman, K.B. Zwart, J. Bloem, and L. Brussard. 1993. Relationships between soil texture, physical protection of organic matter, soil biota, and C and N mineralization in grassland soils. *Soil and Tillage Res.* 19:77-87.

IPCC 1995. Technical Summary. Inter-Governmental Panel on Climate Change, WMO, Geneva, Switzerland, 44 pp.

Jacks, G.V. 1963. The biological nature of soil productivity. *Soils and Ferts.* 26:147-150.

Letey, J. 1985. Relationship between soil physical properties and crop production. *Adv. Soil Sci.* 1:277-294.

Lynch, J.M. and E. Bragg. 1985. Microorganisms and soil aggregate stability. *Adv. Soil Sci.* 2:133-171.

Martel, J. 1994. Influence du Travail du Sol Sur La Distribution et La Qualité De La Matière Organique D'un Sol Argileux. Mémoire M.Sc. Université Laval, Quebec. 81 pp.

Monnier, G. 1965. Action des matières organiques sur la stabilité structurale des sols. *Annales Agronomiques* 16:327-400.

Oades, J.M. 1988. The retention of organic matter in soils. *Biogeochemistry* 5:35-70.

Oades, J.M. and A.G. Waters. 1991. Aggregate hierarchy in soils. *Aust. J. Soil Res.*

Post, W.M., T.H. Peng, W.R. Emanuel, A.W. King, V.H. Dale, and D.L. De Angelis. 1990. The global carbon cycle. *American Scientist* 78:310-326.

Powlson, D.S. 1980. The effects of grinding on microbial and non-microbial organic matter in soil. *J. Soil Sci.* 31:77-85.

Schlesinger, W.H. 1984. The world carbon pool in soil organic matter: a source of atmospheric CO_2. p. 111-124. In: G.M. Woodewell (ed.), *The Role of Terrestrial Vegetation in the Global Carbon Cycle: Measurement by Remote Sensing*. SCOPE 23, J. Wiley & Sons, U.K.

Schlesinger, W.H. 1991. *Biogeochemistry: An Analysis of Global Change*. Academic Press, San Diego, CA.

Tans, P.P., I.Y. Fung, and T. Takahashi. 1990. Observational constraints on the global atmospheric CO_2 budget. *Science* 247:1431-1438.

Thomasson, A.J. 1978. Towards an objective classification of soil structure. *J. Soil Sci.* 29:38-46.

CHAPTER 2

Stocks and Dynamics of Soil Carbon Following Deforestation for Pasture in Rondônia

Christopher Neill, Carlos C. Cerri, Jerry M. Melillo, Brigitte J. Feigl,
Paul A. Steudler, Jener F.L. Moraes, and Marisa C. Piccolo

I. Introduction

Soils play an important role in the carbon cycle of the earth because they are a major location of C storage in terrestrial ecosystems. At the global scale, the upper 1 m of mineral soils contains 1300-1500 Gt C, more than twice the C stored in terrestrial plant biomass (Deevy, 1970: Post et al., 1982; Schlesinger, 1986). Human conversion of natural ecosystems to agricultural use can have an important influence on the fate of this stored C. Soils in natural ecosystems converted to cultivation show large changes in C concentrations, C stocks, and associated properties such as bulk density and soil structure (Schlesinger, 1986). Losses of soil C from a wide variety of soils under cultivation are in the range of 20-30% of the C originally present (Mann, 1986; Schlesinger, 1986; Davidson and Ackerman, 1993). Most of these losses occur within the first 20 years or sooner (Mann, 1986; Davidson and Ackerman, 1993). Grazing also influences soil C balance and soil properties, but these effects can differ from those of cultivation (Detwiler, 1986; Burke et al., 1989; Hassink and Neeteson, 1991; Naeth et al., 1990) and these changes are not as well studied. The losses or gains of soil C that occur following land use change, large amounts of C stored in soils, and large areas of land currently being altered for agriculture make soils potentially significant sources or sinks of C to the atmosphere (Houghton et al., 1985; Detwiler and Hall, 1988).

While much of our understanding of how land use conversion influences soil C stocks and dynamics comes from the temperate zone, the most rapid land use conversions today are taking place in the tropics. In recent decades, tropical forests and savannas have replaced temperate forests and grasslands as the regions of most rapid land cover conversion (Turner et al., 1990; Ojima et al., 1994; Melillo, 1996). The total C stored per unit area in tropical soils is generally similar to that found in temperate soils (Sanchez and Logan, 1992). Globally, Eswaran et al. (1993) estimate that 32% of the organic C stored in soils is found in the tropics. The speed at which these conversions are now taking place and the large amount of C in tropical soils make understanding of changes to soil C storage in the tropics essential to an understanding of global soil-atmosphere C exchanges.

The Brazilian Amazon Basin is the largest region of intact tropical forest in the world. Its deforestation rate of approximately 15,000 km^2 y^{-1} (INPE, 1992; Skole and Tucker, 1993) is also the world's highest. In this large and important region, approximately 70% of cleared lands are used for cattle pasture (Serrão, 1992; Fearnside, 1993; Skole et al., 1994). Despite the predominance of this land cover conversion, little information currently exists about how soil organic matter stocks and cycling patterns change after pastures are created on cleared moist tropical forest lands. Clearing for crop cultivation typically reduces tropical soil C concentrations and stocks (Seubert et al., 1977; Aina, 1979; Sanchez et al., 1983; Lal, 1985; Vitorello et al., 1989; Detwiler, 1986; Bonde et al., 1992a). In contrast, the widely practiced pasture agriculture attempts to maintain contiguous grass cover and may have very different effects on soil C stocks.

Based on what we know from previous studies, the direction of changes to soil C stocks under pasture can vary. Some studies report lower soil C stocks in pastures compared with the original forest (Detwiler, 1986; Veldkamp, 1994; Reiners et al., 1994). In other reports from forests and savannas, inputs of C from roots of pasture grasses result in increased soil C stocks following pasture formation (Lugo et al., 1986; Choné et al., 1991; Fisher et al., 1994). In several locations, increases or decreases may occur, depending on pasture management (Hecht, 1982; Buschbacher et al., 1988; Eden et al., 1990; Trumbore et al., 1995). A better understanding of the direction, magnitude, and controls on these changes based on additional field research would aid predictions of the effects of land use change in the Amazon Basin on soil C stocks. It could also potentially help to improve management to prevent soil C losses from areas currently in pasture.

The gain or loss of soil C in planted pastures depends on both the rate of decay of organic C derived from the former forest vegetation and the contribution of C from grass roots and litter. Because pasture creation results in a change from C_3 forest vegetation to C_4 pasture grasses, the origin of soil C in pastures can be traced using ^{13}C (Cerri et al., 1985; Balesdent et al., 1987; Vitorello et al., 1989). This allows determination of the amount of soil C derived from the original forest vegetation and soil C derived from pasture grasses at a given time since pasture creation. Studies from Hawaii (Townsend et al., 1995), Costa Rica (Veldkamp, 1994), Mexico (García-Oliva et al., 1994), and the Brazilian Amazon (Choné et al., 1991, Trumbore et al., 1995) all suggest that more than 50% of the forest-derived C in surface soils of pastures on converted tropical forests turns over in 10 to 30 years. This high rate of surface soil turnover and C stocks in the top 50 cm of tropical Ultisols and Oxisols of 8-10 kg C m^{-2} (Kimble et al., 1990) suggest a high potential for significant changes to C storage.

Organic matter quality may also be an important control on rates of soil C cycling. The conversion of forest to pasture may lead to substantial changes in soil organic matter quality as C derived from pasture grasses replaces C derived from the original forest vegetation, because grasses contain a lower concentration of lignin than woody vegetation (Parton et al., 1987). Understanding organic matter quality is also important because organic C is a substrate for microbial reactions that control soil biogeochemical cycles of nitrogen (N) and phosphorus (P) and soil-atmosphere exchanges of trace gases. Currently we know little about how soil organic matter quality changes after pasture formation.

II. Chronosequence Studies in Rondônia

In this paper, we summarize our findings of changes to soils following forest clearing and pasture creation at Fazenda (Ranch) Nova Vida in the western Amazon Basin state of Rondônia (Figure 1). Rondônia has been a region of very active forest clearing (Fearnside and Salati, 1985; Skole and Tucker, 1993). These studies were conducted in chronosequences that consisted of forests and pastures of different ages. The following factors were considered following deforestation and pasture planting: 1) changes to soil chemical and physical characteristics, 2) changes to soil C concentrations and stocks, 3) changes to the origin of soil C in different organic fractions, and 4) changes to the quality of soil organic carbon.

The chronosequences at Nova Vida consist of native forests and pastures between 3 and 81 years old (Table 1). They represent one of the best known and longest sequences in the Amazon Basin. They were also converted directly from forest without an intermediate use for annual crop cultivation. This makes them particularly valuable for evaluating the effects of continuous pasture use alone, without the confounding factor of other brief cropping phases that is common in the Amazon and

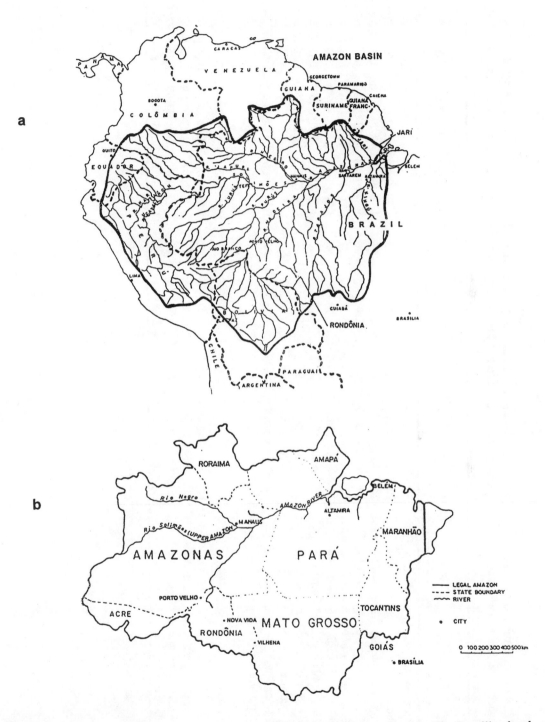

Figure 1. Map showing the extent of the Amazon Basin in South America (**a**). The Brazilian legal Amazon region and the location of Rondônia and Fazenda Nova Vida in the southwestern portion of the Basin (**b**).

Table 1. Land use and soil characteristics in the top 10 cm measured in two chronosequences at Fazenda Nova Vida. Asterisk indicates where bulk density, carbon and nitrogen in pastures were significantly different than in the reference forest (Bonferroni t-test, $p<0.05$). Standard deviations in parentheses

Land use	Pasture vegetation	Clay %	Soil texture class	pH H_2O	ECEC $Cmol(+)kg^{-1}$	Base saturation %	Density $Mg\ kg^{-1}$	Carbon $g\ kg^{-1}$	Nitrogen $g\ kg^{-1}$
Chronosequence 1 (Kandiudult)									
Forest		22	sandy loam	4.78	6.09	45	1.29 (0.07)	12.81 (5.06)	1.09 (0.31)
3-year-old pasture	*Brachiaria brizantha*	24	sandy loam	6.60	6.90	93	1.36 (0.10)	12.49 (4.83)	0.90 (0.30)
5-year-old pasture	*B. brizantha*	29	sandy clay loam	7.18	8.41	96	1.37 (0.08)	17.53 (11.99)	1.33 (0.86)
9-year-old pasture	*Panicum maximum*	25	sandy loam	6.05	5.32	91	1.51 (0.10)*	15.50 (5.67)	1.16 (0.38)
13-year-old pasture	*P. maximum*	22	sandy loam	6.06	3.79	86	1.39 (0.08)	14.95 (9.32)	0.97 (0.46)
20-year-old pasture	*B. brizantha*	23	sandy clay loam	5.67	4.57	58	1.31 (0.09)	16.40 (4.31)	1.13 (0.31)
41-year-old pasture	*P. maximum*	25	sandy clay loam	5.74	4.75	73	1.36 (0.12)	18.25 (3.66)	1.30 (0.24)
81-year-old pasture	*B. brizantha*	15	loamy sand	5.86	4.74	82	1.33 (0.06)	20.84 (6.12)	1.60 (0.65)
Chronosequence 2 (Paleudult)									
Forest		11	loamy sand	4.92	4.59	72	1.22 (0.09)	11.44 (4.91)	0.67 (0.36)
3-year-old pasture	*B. brizantha*	23	sandy loam	5.12	6.26	73	1.30 (0.07)	14.77 (6.19)	1.09 (0.38)
5-year-old pasture	*B. brizantha*	13	loamy sand	6.51	7.06	96	1.30 (0.09)	17.38 (11.98)	1.16 (0.72)
20-year-old pasture	*P. maximum*	19	loamy sand	6.10	3.28	87	1.28 (0.07)	17.95 (11.55)	0.88 (0.37)

which complicates many pasture studies. Sites in the Nova Vida sequences were within 5 km of each other and in areas of similar topography (Neill et al., 1995; Moraes et al., 1996). Soils were classified as Podzolico Vermellho Amarelo (Red Yellow Podzolic) in the Brazilian classification scheme and as Ultisols (kandiuldults and paleudults) in the U.S. soil taxonomy (Moraes et al., 1995; Moraes et al., 1996) and had similar clay content and texture. Ultisols cover more than 100 million ha in the Brazilian Amazon Basin (Richter and Babbar, 1991; Moraes et al., 1995) and more than 700 million ha worldwide in the tropics (Sanchez and Logan, 1992). The climate of Rondônia is moist tropical. Annual rainfall at Nova Vida averages 2.2 m and a dry season with little rain occurs from June through September. All pastures in the chronosequences were established in a similar manner by slashing, burning after the first rains in September and October, and planting pasture grasses in December or January. All pastures had similar management. They were actively grazed at an average annual rate of approximately one animal ha^{-1}. Neither mechanized practices nor chemical fertilizers were used. Pastures were burned every 4-10 years to control weeds. Selective logging in the forests removed about 3-4 trees ha^{-1}.

Soil samples, excluding the litter layer, were obtained at each site from five pits to a depth of 50 cm. Samples were further prepared by air drying and sieving (2 mm) to remove stones and root fragments. Soil bulk density, texture, pH, effective cation exchange capacity (ECEC), and C and N concentrations were measured by using techniques described in Anderson and Ingram, 1989; EMBRAPA, 1979; and Moraes et al., 1996. For example, particle size fractions were determined by hydrometer after dispersion in a mixer with hexametaphosphate and digestion of organic matter with H_2O_2. Soil pH was measured in water (2.5:1) on air-dried soil. Potassium, calcium, and magnesium were extracted with ammonium acetate at pH 7 and analyzed by emission spectrophotometry. Aluminum was extracted with unbuffered KCl and analyzed by emission spectrophotometry. Total C and N were analyzed by combustion on a Perkin-Elmer 2400 Elemental Analyzer. Bulk density was measured in the field with volumetric steel rings.

Results of these measures from the top 10 cm where changes following clearing were greatest are presented in this report. Stocks of C were calculated to 50 cm and corrected based on sampling of a total soil mass in the pastures that was equal to the soil mass to 50 cm in the original forest (Davidson and Ackerman, 1993; Veldkamp, 1994). The top 50 cm of the profile was the depth that contained the pool of soil C most likely to be altered by land use during the 81 years represented by our chronosequence. Soil samples were analyzed for $\delta^{13}C$ on a Micromass 602E or a Finnigan Delta-S mass spectrometer. The ratio of $^{13}C/^{12}C$ of the sample was expressed in δ values in parts per mil relative to the PDB standard. Soil organic C fractions were separated physically by floating, sieving, and centrifugation (Feigl et al., 1995). Lignocellulose ratio, or LCI, was determined from sequential digestions which resulted in acid-soluble (cellulose) and acid-insoluble (lignin) fractions (Feigl et al., 1995). Respiration rates were determined from laboratory incubation of bulk soil (Neill et al., 1996). Some analyses were not performed on every site from both chronosequences.

III. Chemical and Physical Properties

A. pH and Cation Exchange Capacity

Forest conversion to pasture had a large and predictable effect on the chemical properties of surface soil (Table 1). These changes are important because they can be linked to higher pasture grass production and potentially greater plant-derived C inputs to surface soils. Soil pH in the top 10 cm increased by 1 to 2 units within 3-5 years. The pH remained elevated by approximately 1 unit in the 41- and 81-year-old pastures. Effective cation exchange capacity showed a similar increase in young pastures, but it declined to levels near or below that in the original forest by 20 years. Concentrations of Ca^{2+}, Mg^{2+}, and K^+ showed a similar pattern. Base saturation increased in young pastures and decreased somewhat in older pastures, driven by declines in exchangeable base cations and increases in exchangeable Al^{3+}.

The short-term increases to soil pH, exchangeable cations, and ECEC at Nova Vida parallel changes observed following the forest cutting and burning typical of tropical slash-and-burn agriculture (Popenoe, 1957; Nye and Greenland, 1964; Seubert et al., 1977; Ewel et al., 1981; Brinkmann and Nascimento, 1973; Sanchez et al., 1983; Jordan, 1985). Because the process of land clearing and burning for pasture creation are similar, increases of pH and CEC are common following tropical forest clearing for pasture (Falesi, 1976; Hecht, 1982; Eden, 1990; Bonde et al., 1992b; Luizão et al., 1992). Soil pH in pastures tends to remain elevated for more than 20 years after clearing, similar to the pattern at Nova Vida (Falesi, 1976; Hecht, 1982; Eden et al., 1990), but this increase is not universal (Choné et al., 1991; Veldkamp, 1994) and it does not occur where soils are heavily disturbed by mechanical clearing (Buschbacher et al., 1988). Soil CEC and the concentrations of individual base cations can also remain elevated for many years (Falesi, 1976; Hecht, 1982; Serrão et al., 1979). This contrasts with the declines in base cations and ECEC to near or slightly below forest levels that we observed in older pastures at Nova Vida. ECEC and base saturation in the forest in the relatively sandy soils at Nova Vida were much higher than measured on the more acidic and higher clay Oxisols where comparable chronosequences containing similar old pastures exist. Maintenance of high soil pH and CEC in older pastures does not necessarily reflect continued high soil fertility if critical nutrients, particularly P, are lacking (Fearnside, 1980; Serrão et al., 1979).

B. Bulk Density

Accurate measures of changes to soil bulk density are critical to accurate determination of soil C stocks after forest clearing. At Nova Vida, bulk density increased under pasture, but in most cases these increases were relatively modest and in the range of about 10% (Table 1). The relatively small magnitude of changes and some spatial variability in densities resulted in significantly higher densities than the original forest in only one pasture (13 years old). Several studies report increases in soil bulk density in tropical pastures relative to the original forest (Hecht, 1982; Eden et al., 1990; Luizão et al., 1992; Veldkamp, 1994; Reiners et al., 1994). This increased density reflects the collapse of soil aggregates and is a widespread problem in tropical soils (Cassell and Lal, 1992). Higher densities in pastures are associated with changes in pore space distribution and decreased water infiltration and porosity (Chauvel et al., 1991; Reiners et al., 1994), but there are few data on infiltration for pasture soils. Relatively sandy Oxisols and Ultisols, such as those at Nova Vida, may be less prone to severe compaction. Desjardins et al. (1994) reported relatively no increase in density in a 10-year-old pasture on a similar kandiudult with 13-20% clay from Capitão Poço in Pará. We have also recorded relatively small increases in density from sandy Ultisols from Ouro Preto, Rondônia, compared with greater increases in density on Oxisols with higher clay contents (unpublished data). The use of heavy machinery for land clearing amplifies bulk density increases for both slash-and-burn agriculture and for pasture formation on a variety of soils (Seubert et al., 1977; Buschbacher et al., 1988).

IV. Carbon Concentration and Stocks

At Nova Vida, soil C concentrations in the top 10 cm of pasture soils were generally greater than concentrations in the original forest; however, C concentrations within each site were variable and these differences were not statistically significant. While C concentrations ranged from 12.49 to 20.84 g C kg^{-1}, the lowest C concentrations in both chronosequences occurred in the forest or the youngest pasture and the highest concentrations occurred in the oldest pastures. Nitrogen concentrations were

Table 2. Soil carbon and nitrogen stocks in the top 50 cm measured at Fazenda Nova Vida; asterisks indicate where stocks in pastures were significantly different from stocks in the reference forest (Bonferroni t-test, p<0.05); stocks were corrected for changes to soil bulk density; standard deviations in parentheses; accumulation rates were calculated as the total difference in carbon stock divided by the number of years in pasture

Land use	Carbon stock	Total carbon gain	Mean carbon accumulation rate
	kg m^{-2}		kg m^{-2} y^{-1}
Chronosequence 1			
Forest	4.33 (0.72)		
3-year-old pasture	4.61 (0.66)	0.28	0.093
5-year-old pasture	5.85 (2.09)	1.52	0.304
9-year-old pasture	5.26 (0.85)	1.21	0.134
13-year-old pasture	4.64 (1.58)	1.07	0.082
20-year-old pasture	5.28 (0.64)	1.22	0.061
41-year-old pasture	6.56 (0.73)*	2.27	0.055
81-year-old pasture	6.12 (0.88)*	1.79	0.022
Chronosequence 2			
Forest	3.69 (0.72)		
3-year-old pasture	5.44 (0.80)*	1.75	0.583
5-year-old pasture	4.75 (1.67)	1.06	0.212
20-year-old pasture	4.82 (2.37)	1.13	0.056

even more variable. Eight of ten pastures had higher N concentrations than the original forest, but N concentrations were not significantly different.

Carbon stocks in the top 50 cm were variable but increased with time under pasture. All the pastures sampled had higher C stocks than the forest (Table 2). In one chronosequence, the oldest 41- and 81-year-old pastures had significantly higher C stocks than the forest. In the other chronosequence, only stocks in the 3-year-old pasture were significantly greater than stocks in the forest. The total C gained in pasture soils was 0.28-2.27 kg m^{-2}. Except for the 3-year-old pasture in Chronosequence 2, the highest gains were in the oldest pastures (Table 2). Rates of C accumulation were highest in young 3-5 year-old pastures (0.21-0.58 kg C m^{-2} y^{-1}), but these rates declined sharply with increasing pasture age (Table 2) and the potential for further C accumulation in older pastures was much less.

Increased soil C concentrations in surface horizons are a common consequence of pasture formation from cleared moist tropical forest in the Amazon Basin (Eden et al., 1990; Choné et al., 1991; Bonde et al., 1992b; Luizão et al., 1992; Trumbore et al., 1995). Other studies report soil C concentration declines (Detwiler, 1986; Veldkamp, 1994), undetectable changes, or increases in some pastures and losses in others (Hecht, 1982; Lugo et al., 1986; Buschbacher et al., 1988; Desjardins et al., 1994; Reiners et al., 1994). Patterns in change of C stocks are similar, although not all studies that report soil C concentrations report C stocks per unit area or the bulk density values that allow determination of C stocks. Eden et al. (1990), Choné et al. (1991), Bonde et al. (1992b) and Luizão et al. (1992) all reported increased C stocks under pasture or a combination of higher densities and higher C concentrations that result in higher stocks. All of these sites were on typical Amazon Basin Ultisols or Oxisols, although clay contents varied. From Costa Rica, Veldkamp (1994) reported high C losses and Reiners et al. (1994) found changes to pasture C stocks were variable. These pastures were created from forests on relatively fertile volcanic soil high in C. All of these studies suggest that a value for the loss of surface soil C from tropical soil under pasture of 11.5% used in land use change C accounting models (Houghton et al., 1990) is probably too high for the Amazon Basin.

We currently do not have a good understanding of the role of soil type or original vegetation structure on the direction and magnitude of changes to soil C stocks under pasture. Results from other chronosequences spanning Oxisols and Ultisols ranging from 12-75% clay in Rondônia (unpublished) show C accumulation in the top 50 cm to be quite variable and not clearly related to soil type or soil clay content. Gains of soil C in the top meter of 0.29-1.47 kg m^{-2} y^{-1} have been measured following planting pasture grasses on cleared tropical savannas (Fisher et al., 1994). These rates are higher but not completely unlike the rates we measured at Nova Vida. Evidence from older pastures at Nova Vida suggests that these high rates are not likely to continue. In a seasonally dry area at Chamelas, Mexico, García-Oliva et al. (1994) found generally similar C stocks in pastures ranging to 11 years.

Pasture management may play a particularly important role in C accumulation or loss. Trumbore et al. (1995) reported soil C losses in degraded pasture but soil C gains from reclaimed fertilized pasture in Pará. Degraded pastures with little grass cover probably will be less likely to accumulate soil C because inputs to soil organic C from pasture roots will be diminished, but this might not be true in more vigorously regrowing secondary forests. Greater grazing intensity and soil damage during pasture clearing, including by heavy machinery, would all likely cause soil C loss rather than gain. Similar processes that influence the magnitude of annual soil organic matter inputs also regulate the accumulation of soil C in soils of North American grasslands (Burke et al., 1991). There are few long-term studies that have examined the effect of pasture grass species, fertilization, grazing intensity, and burning frequency on pasture soil properties and soil C balance in the Amazon. The potential for soil C gains under pasture differs from the pattern for cultivated tropical soils where declines in soil C stocks are the almost universal pattern (Brams, 1971; Ayanaba et al., 1976; Juo and Lal, 1977; Sanchez et al., 1983; Vitorello et al., 1989; Bonde et al., 1992a). Future human-induced clearing or climate changes that alter the distribution of cultivated vs. pasture lands in the tropics are likely to have a major effect on regional soil C balance.

All of these studies of changes to soil C stocks are based upon surface soils (0-1 m). But in many tropical regions like the Amazon Basin, soils are often very deep, with accumulation of weathered material over bedrock extending to many meters (Richter and Babbar, 1991). Nepstad et al. (1994) and Trumbore et al. (1995) found that C stored in deep soils of the Brazilian Amazon is significant and that the total soil C inventory below 1 m exceeds that in the top 1 m. They found that 13% of deep soil C, or approximately 3 kg m^{-2}, could be traced to recent inputs from tree roots and that some portion of this pool potentially could be lost when forest clearing removes deep-rooted trees. Nepstad et al. (1994) identified areas covering 36% of the Brazilian Legal Amazon where trees remained evergreen despite severe soil moisture deficits in the top 1 m indicating the presence of deeply rooted trees that could lead to C storage in deep soil. Root biomass and its role in C balance of deep soils remain little studied.

V. Carbon Origin

Total C in the top 10 cm in the Nova Vida chronosequence increased from 1.61 kg m^{-2} in the forest to 2.74 kg m^{-2} in the 81-year-old pasture (Table 3). There was a transition in the origin of this C with increasing time since forest clearing from C$_3$ plants associated with the original forest to C$_4$ pasture grasses (Figure 2). The rate of gain of C derived from pasture grasses exceeded rate of loss of C derived from the original forest vegetation. In the 81-year-old pasture, 92% of soil C in the top 10 cm originated from pasture grasses. The half life calculated from an exponential decay of the C derived from the original forest in the top 10 cm was 40 years, but C turnover varied among sites in the chronosequence. For example, the 20-year-old pasture had 37% of the original forest C remaining but the 41-year-old pasture had 84% remaining. The rate of gain of pasture-derived C also varied.

Table 3. Origin of carbon in the top 10 cm in different soil organic matter fractions measured at Fazenda Nova Vida; percent C derived from pasture vegetation (% Cdp) was determined from linear mixing equations using the δ¹³C value of each fraction in the forest and a δ¹³C value of -13 ‰ for pasture as endpoints; soil organic fractions were separated physically; abbreviations are: OF—floatable organic fraction, OMF—organo-mineral fraction separated by sieving, OMF1—organo-mineral fraction separated by centrifugation, OMF2—organo-mineral fraction separated by centrifugation and precipitation; respired CO₂ was captured in laboratory incubations of bulk soil

	Total C kg m⁻²	Bulk soil δ¹³C	Bulk soil % Cdp	OF (2000-200 μm) δ¹³C	OF % Cdp	OMF (200-50 μm) δ¹³C	OMF % Cdp	OMF 2 (<50 μm) δ¹³C	OMF 2 (<50 μm) % Cdp	OMF 2 (<50 μm) δ¹³C	OMF 2 (<50 μm) % Cdp	Respired CO₂ δ¹³C	Respired CO₂ % Cdp
Forest	1.61	-27.4	0	-28.6	0	-29.5	0	-27.3	0	-27.7	0	-26.5	0
3-year-old pasture	1.68	-26.8	5	-28.5	1	-27.5	13	-27.3	0	-26.1	12	-17.1	69
5-year-old pasture	2.28	-22.6	27	-22.9	40	-24.5	33	-24.1	25	-24.6	23	-17.6	65
9-year-old pasture	2.31	-21.9	42	-18.8	69	-24.0	36	-24.4	22	-24.0	28	-17.5	66
13-year-old pasture	2.06	-21.7	44	-19.9	62	-23.2	42	-23.0	33	-23.1	34	-16.8	71
20-year-old pasture	2.12	-18.4	69	-16.5	85	-19.1	68	-19.2	62	-19.7	60	-14.1	91
41-year-old pasture	2.47	-21.2	45										
81-year-old pasture	2.74	-15.4	92	-15.1	93	-15.3	93	-15.9	88	-16.7	82	-12.9	100

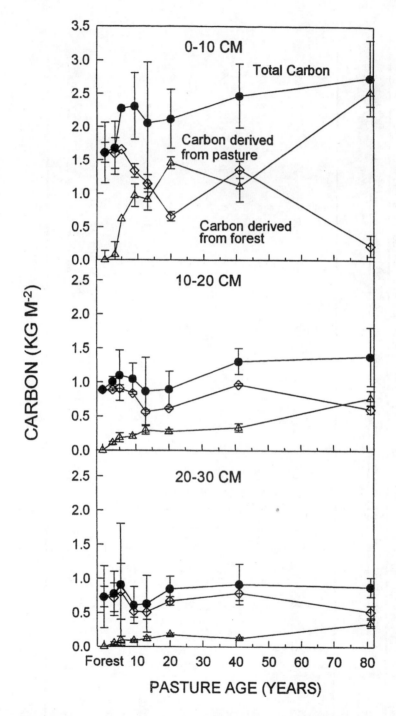

Figure 2. Changes of total soil C: (•) soil C derived from the original forest vegetation; (◊) soil C derived from pasture vegetation; (Δ) from 0-10, 10-20, and 20-30 cm depths at Fazenda Nova Vida. The rate of decay of forest-derived C and the rate of replacement of the total soil C stock with pasture-derived C decreased with depth. Error bars are ± 1 SD and are not shown when they are smaller than the symbol.

Pastures 9, 13, and 20 years old all contained 42-45% pasture-derived C (Table 3). Total C stocks decreased with depth, and the rate of decay of forest-derived C and the rate of replacement of the original forest-derived C with pasture-derived C both declined at 10-20 cm and 20-30 cm depths (Figure 2). The half life of forest-derived C was 159 years at 10-20 cm and 199 years at 20-30 cm. Pasture-derived C in the 81-year-old pasture made up 56% of the total C stock at 10-20 cm and 40% at 20-30 cm.

Results from the few additional studies of soil $\delta^{13}C$ values and soil C turnover under pasture in the tropics suggest that replacement and augmentation of forest-derived C with pasture grass-derived C can be similarly rapid. Choné et al. (1991) found that pasture-derived C comprised 30% of the soil C stock at 0-3 cm in a pasture near Manaus after 2 years and 68% of the soil C stock after 8 years. Forty-three percent of the original forest-derived C remained after 8 years. Replacement at 3-10 cm and 10-20 cm was slower, with pasture-derived C accounting for 9% of C at 10-20 cm after 2 years and 27% after 8 years (Choné et al., 1991). Desjardins et al. (1994) found a similar pattern in Pará, where pasture-derived C made up 52% of total C stock at 0-10 cm in a 10-year-old pasture and 45% of the original forest-derived C remained. In a dry forest, the pattern was similar with pasture-derived C comprising about 40% of total C stock at 0-6 cm and about 52% of the original forest-derived C remaining after 11 years (García-Oliva et al., 1994). Trumbore et al. (1995) found that only about 16-21% of total soil C at 0-10 cm was pasture-derived in both degraded and managed 23-year-old pastures in Pará. A high proportion of C_3 weeds may account for relatively low $\delta^{13}C$ values and low amount of C derived from pasture grasses. Significant inputs of C derived from C_3 weeds in pastures cannot be separated from the original forest C using $\delta^{13}C$ techniques.

In different organic matter size fractions from Nova Vida, soil $\delta^{13}C$ values showed a similar pattern of replacement of forest-derived C with pasture-derived C, but the rates of replacement differed among the fractions (Table 3). Compared with the proportion of C derived from pasture in bulk soil, grass C made up a larger proportion of the C in the OF (2000-200 μm - floatable organic fraction) and the OMF (200-50 μm - organo-mineral fraction separated by sieving) fractions in 5- to 20-year-old pastures, and a lower proportion of C in the smaller OMF1 (<50 μm - organo-mineral fraction separated by centrifugation) and OMF2 (<50 μm - organo-mineral fraction separated by centrifugation and precipitation) fractions. Respired CO_2 turned over the fastest of any fraction (Table 3). A change from the forest $\delta^{13}C$ value in respired CO_2 of -26.5‰ to approximately -17‰ in the 3- to 13-year-old pastures indicated that the shift in the origin of the substrate used for soil microbial activity rapidly switched from forest-derived C to pasture-derived C soon after pasture establishment. Pasture-derived C made up 69% of respired CO_2 in the 3-year-old pastures. The $\delta^{13}C$ value of -12.9‰ in CO_2 respired from soil in the 81-year-old pasture was very close to widely reported values of -13‰ for C_4 plant material (Cerri et al.,1985; Balesdent et al., 1987) and indicated respired CO_2 contained 100% pasture-derived C.

Large differences among size fractions existed in the rates of decay of the original forest-derived C (Table 4). These differences did not follow a simple pattern of higher turnover in the larger size fractions. Turnover was most rapid in the respired CO_2 and in the OF (2000-200 μm) and OMF1 (<50 μm) fractions. Turnover was slowest in the OMF (200-50 μm) and in the OMF2 (<50 μm) fractions. The half life of original forest C varied from a low of 4 years for respired CO_2 to a high of 510 years for the OMF2 (<50 μm) fractions (Table 4).

In a comparison of soil C turnover in different soil organic fractions after clearing of forest for sugar cane in southern Brazil, Bonde et al. (1992a) found that 51% of the original forest-derived C remained in the clay sized fraction (<2 μm) of a 50-year-old sugar cane field but only 6% remained in the fine sand (63-200 μm) and coarse sand (200-200 μm) fractions. Inputs of crop-derived C in the <50 μm fractions at Nova Vida were much greater, as might be expected in a pasture system which does not disturb soil structure and results in much higher C inputs to soils from grasses. In sugar cane, inputs from pasture grasses in most fractions were small and $\delta^{13}C$ values in all particle size fractions were about -23‰.

Table 4. Carbon derived from the original forest remaining in the top 10 cm in different soil organic matter size fractions at Fazenda Nova Vida; half life of carbon pools was derived from an exponential decay function for each fraction; total carbon in the size fractions does equal exactly the total carbon in bulk soil because a small amount of carbon contained in the 2000-200 μm mineral fraction was lost during separation

	Bulk soil	*OF (2000-200 μm)	OMF (200-50 μm)	OMF1 (<50 μm)	OMF2 (<50 μm)	Respired CO_2
			—kg C m^{-2}—			mg C g^{-1} C d^{-1}
Forest	1.61	0.48	0.11	0.75	0.21	1.00
3-year-old pasture	1.60	0.26	0.23	0.67	0.29	0.39
5-year-old pasture	1.66	0.36	0.17	0.28	0.29	0.34
9-year-old pasture	1.34	0.20	0.22	0.24	0.24	0.34
13-year-old pasture	1.36	0.14	0.17	0.29	0.28	0.28
20-year-old pasture	0.66	0.06	0.09	0.14	0.15	0.10
41-year-old pasture	1.36					0.21
81-year-old pasture	0.22	0.03	0.01	0.15	0.23	0.00
Half life of C (years)	40	7	98	7	510	4

*OF—floatable organic fraction, OMF—organo-mineral fraction separated by sieving, OMF1—organo-mineral fraction separated by centrifugation, OMF2—organo-mineral fraction separated by centrifugation and precipitation.

Examination of the dynamics of C loss and accumulation in soil size fractions at Nova Vida yields two important insights. First, although losses of forest C from different size fractions occurs at markedly different rates, C in all fractions does turn over in less than the 1000-year time frame commonly specified as the definition of "passive" soil C in models of soil C dynamics, such as CENTURY (Parton et al., 1987; Townsend et al., 1995). In the soils at Nova Vida, the slowest turnover size fractions contained a relatively small portion of the total soil C in surface soils (total C in the top 10 cm contained in the slower turnover OMF (200-50 μm) and OMF2 (<50 μm) pools was 0.32 kg C m^{-2} out of 1.61 kg m^{-2} at that depth), but even these fractions were subject to substantial loss over the 81 years of land use change represented by the chronosequence. This conclusion would support Trumbore's claim (1993) that tropical soils contain a much smaller proportion of total C in slower turnover pools. Overall changes to the total soil C pool are moderated by slower C turnover at deeper depths.

A second point is that pasture-derived C is quickly incorporated into soil organic matter in the smallest size fractions. Pasture-derived C made up about 25% of total C in the top 10 cm after only 5 years. This same pattern was reported by Desjardins et al. (1994), where pasture-derived C comprised 40-46% of total C in the clay (0-0.2 μm) fraction after 10 years. Clearly, each SOM size fraction contains a range of organic compounds, with presumably different turnover times (Tiessen and Stewart, 1983). It was difficult to determine if the mix of compounds and stability of recently-formed pasture-derived C are comparable to those in the same fractions from the original forest. This is an important unanswered question because pastures may undergo future land use conversions and the rate of decay of the now relatively large amount of grass-derived soil C in pastures will be an important determinant of overall soil C balance.

VI. Carbon Quality

The C:N ratios of bulk soil ranged from 10.8 to 16.0 but showed no clear pattern with increasing pasture age (Table 5). These C:N ratios were close to worldwide average values for agricultural soils (Jenny 1980, Stevenson 1982). C:N ratios were highest in the OF (2000-200 μm) and OMF1 (<50 μm) fractions and lowest in the OMF (200-50 μm) and OMF2 (<50 μm) fractions. This pattern followed closely the pattern of loss of original forest-derived C and increase of pasture-derived C in the fractions. Fractions with higher C:N ratios had higher C turnover. The C:N ratios and turnover rates indicated a higher proportion of fresher, more labile C in the OF (2000-200 μm) and OMF1 (<50 μm) fractions and more humified material in the OMF (200-50 μm) and OMF2 (<50 μm) fractions. There were no indications based on C:N ratios that C quality differed between the forest and any of the pastures (Table 5).

The approximately equal respiration rates of forest and pasture soils also supported the conclusion that there were only very minor changes in the availability of labile C between forest and pasture soils and among pasture ages (Table 5). Approximately equal proportions of C were respired from forest and pasture soils even though the majority of respired CO_2 in all of the pastures was grass-derived. Higher values of LCI for bulk soil in the forest and lower values in the 81-year-old pasture did suggest a small increase in the lability of C in the older pastures. Among the size fractions, the OF (2000-200 μm) fraction had the lowest LCI values (0.58-0.70) and the OMF2 (200-50 μm) fraction had the highest LCI values (0.85-0.91) (Feigl et al., 1995). These are again consistent with the gradient from a higher relative rate of C turnover in the OF (2000-200 μm) fraction to a low rate of C turnover in the OMF2 (200-50 μm) fraction. LCI values in both forest and pasture soils, in the range of 0.7-0.8, were all quite high and indicated highly processed SOM, especially when compared with values for forest soils from the temperate zone, which typically have LCI values in the range of 0.5-0.7 (Melillo et al., 1989). LCI values of 0.7 to 0.9 from both forest and pastures were more typical of cultivated agricultural soils in the temperate zone. These represent the stage at which organic matter inputs to

Table 5. Indices of the quality of total soil carbon in the top 10 cm measured in bulk soil and in soil organic matter size fractions at Fazenda Nova Vida

	Bulk soil	*OF (200-200 μm)	OMF (200-50 μm)	OMF1 (<50 μm)	OMF2 (<50 μm)	LCI ratio**	Respiration rate
			C:N				mg CO_2-C g^{-1} C d^{-1}
Forest	13.3	24.2	4.8	10.9	11.1	0.85	1.00
3-year-old pasture	10.8	15.3	10.5	9.8	8.0	0.84	1.24
5-year-old pasture	14.3	22.1	13.0	13.6	10.0	0.80	0.98
9-year-old pasture	13.3	24.9	9.8	21.1	9.3	0.82	0.99
13-year-old pasture	13.6	19.6	13.6	10.9	10.4	0.79	0.96
20-year-old pasture	16.0	25.8	18.0	12.6	10.3	0.79	1.03
81-year-old pasture	14.2	22.6	15.4	15.6	10.7	0.74	0.90

*OF—floatable organic fraction, OMF—organo-mineral fraction separated by sieving, OMF1—organo-mineral fraction separated by centrifugation, OMF2—organo-mineral fraction separated by centrifugation and precipitation.
**Ratio of lignin/(lignin + cellulose).

soils have been reworked to the extent that they have reached a relatively constant chemical quality. Like soil C turnover in the size fractions, C quality and humification indicated by C:N ratios and LCI in this chronosequence did not follow strictly from organic matter fraction size. Some of the C in the <50 μm OF2 fraction had a higher C:N ratio, a lower LCI, and turned over at approximately the same rate as C in the much larger 2000-200 μm OF fraction.

VII. Some Considerations in Chronosequence Studies

While the general pattern of the decay of forest-derived C, the accumulation of grass-derived C and how they contribute to total soil C was clear along the Nova Vida sequence of pasture ages, spatial variability in rates of loss and accumulation and pre-existing differences among sites both contributed to the overall variability along the chronosequence. For example, the 41-year-old pasture showed lower rates of C turnover than the 20-year-old pasture. Some of this variation undoubtedly represented spatial variability in the rates, loss and accumulation in the forest- and pasture-derived pools. This could have been caused by differences in factors such as soil moisture or pasture grass cover that could influence soil organic matter losses or inputs. Another possibility is pre-existing differences in the total amount of soil C between sites at the time of clearing. Such differences, for example, may have been responsible for the small increase in the forest-derived C in the 5-year-old pasture compared with the original forest. It is impossible to separate this affect from spatial differences in decay or accumulation rates from other factors such as inputs of forest-derived C from the decay of incompletely burned wood during the first 5 years in pasture. Calculations of turnover times along chronosequences are also very sensitive to the number of sites the sequence contains (Bernoux et al. personal communication). This occurs because even the most complete chronosequences contain a relatively small number of ages.

Chronosequence studies are a sensible approach to deriving inferences about the effects over time of changes to land use. This is especially true in regions such as Rondônia where rapid land colonization, high turnover of land tenure, and often poor records of land use make chronological studies of soils following deforestation difficult. Chronosequence studies cannot, however, completely replace long-term experimental chronological studies where randomization can control for pre-existing differences between sites and most aspects of post-clearing site history and land management can be controlled.

VIII. Global Implications

The increases in soil C in the top 50 cm at Nova Vida represent net storage of up to about 2 kg C m^{-2} over approximately 80 years. This represents the best long-term mean accumulation rate we can determine from the chronosequence. Total future gains in older pastures are limited by a lower rate of annual C sequestration. Based on the two chronosequences at Nova Vida, the ultimate potential for net soil C storage above that contained in the top 50 cm of the original forest soil is 2-3 kg C m^{-2}; if we make the assumption that this C accumulation levels off at approximately 100 years, the average annual accumulation rate is in the range of 0.02-0.03 kg m^{-2} y^{-1} over this time. These are large additions to the soil C pool and they are significant even when compared with total aboveground C in forest biomass of 14.2-17.9 kg m^{-2}, such as reported for Rondônia by Kauffman et al. (1995). This biomass represents a reasonable upper bound on total aboveground ecosystem C losses from deforestation. During slash-burning, combustion losses represent 27-48% of total aboveground C (Kauffman et al., 1995; de Graças, 1997). It is evident that almost all of the aboveground, unburned, coarse woody debris decays and is lost over approximately 10-20 years after clearing. As a result, almost 100% of the C in forest aboveground biomass is lost within 20 years of clearing. These data on soil C stocks in pastures suggest that soil C sequestration in pastures at Nova Vida reclaimed approximately 10-20% of this loss.

Recent results from measurements of net C exchange in intact tropical forest in Rondônia show that the forest is a net sink for approximately 0.1 kg C m^{-2} y^{-1} (Grace et al., 1995), presumably accumulated in growing vegetation. This sink is 3-5 times larger on an areal basis than the Nova Vida pasture C sink. If this is correct, then the relative balance between C lost by forest clearing and C accumulated in pasture soils for Rondônia must consider not only the losses from burned and decomposing forest plant biomass (14.2-17.9 kg C m^{-2}) but also the loss of the forest as a potential C sink, equivalent to 10 kg C m^{-2} over 100 years.

Several additional pieces of information are required to assess the importance of these changes of soil C storage to total ecosystem C stocks at the scale of the Amazon Basin. Although similar relatively low clay Ultisols are widespread, information from 6 other chronosequences in Rondônia on a diversity of soil types ranging from 12-75% clay indicate that these patterns of C accumulation in pastures are not universal and that soil type is not a predictor of soil C accumulation (unpublished data). It is hypothesized that aspects of land history and land management, such as grazing intensity and burning frequency, can influence soil C accumulation, but most of the current information on the effect of pasture management is anecdotal, and we currently have little information from chronosequence or chronological studies to evaluate the effects of different management practices. The Nova Vida chronosequences represent land planted directly to pasture after clearing. It is assumed this is the standard pattern of clearing on large ranches in the region, but we currently have little information on how much land was converted to pasture in this manner compared with other clearing patterns, such as planting annual crops for 1-2 years or rotating other crops or fallows before planting for pasture, as is common on small farms. Another factor in total soil C balance is the importance of losses of deep soil C that follows forest removal as indicated by Nepstad et al. (1994) in Pará. This has been evaluated in only one location. A substantial dry season occurs in Rondônia and many other parts of the Amazon Basin and could lead to dependence on deep roots and C losses from deep soil after clearing.

Finally, future trajectories of land use and land management have the potential to alter the pattern of soil C accumulation that we observed in pasture soils. The legacy of more than two decades of high rates of forest clearing in the Amazon is an enormous area of pasture that is now 10-20 years old. This aging pasture is poised to play a dominant role in the land cover change of the Basin in the near future. The length of time land remains in pasture can influence soil C storage because the largest changes to stocks occur during the first 5 years after clearing. Continued cycles of pasture abandonment and reclearing have the potential to change patterns of C accumulation and loss, either by degrading C stocks during burning or by augmenting stocks by resetting the clock on soil C accumulation. In addition, if large areas of aging pasture are abandoned to second growth forests, C accumulations in aboveground plant biomass may dwarf any changes to soil stocks. Fertilization with P, now uncommon throughout most of Amazonia but increasing in the eastern part of the Basin (Nepstad et al., 1991; Serrão, 1992), could lead to important changes. So could a transition to cultivated crops, although these areas currently are small in comparison to areas of pasture (Smith et al., 1995). In southern Brazil and in the central savanna region, soybean production increased dramatically between 1970 and 1990 (Skole et al., 1994). A similar change in Amazonia would set in motion a series of dramatic new changes to regional soil C stocks and dynamics.

References

Aina, P. O. 1979. Soil changes resulting from long-term management practices in western Nigeria. *Soil Sci. Soc. Am. J.* 43:173-177.

Anderson, J. M. and J. S. I. Ingram (eds.). 1989. *Tropical Soil Biology and Fertility: A Handbook of Methods.* Wallingford, Oxon. 171 pp.

Ayanaba, A., S. B. Tuckwell, and D. S. Jenkinson. 1976. The effect of clearing and cropping on the organic reserves and biomass of tropical forest soils. *Soil Biol. Biochem.* 8:519-525.

Balesdent, J., A. Mariotti, and B. Guillet. 1987. Natural ^{13}C abundance as a tracer for studies of soil organic matter dynamics. *Soil Biol. Biochem.* 19:25-30.

Bonde, T. A., B. T. Christensen, and C. C. Cerri. 1992a. Dynamics of soil organic matter as reflected by natural ^{13}C abundance in particle size fractions of forested and cultivated Oxisols. *Soil Biol. Biochem.* 24:275-277.

Bonde, T. A., T. Rosswall and R. L. Victoria. 1992b. *The Dynamics of Soil Organic Matter and Soil Microbial Biomass Following Clearfelling and Cropping of a Tropical Rainforest in the Central Amazon.* Linköping Studies in Arts and Science 63. Linköping, Sweden. p. 1-19.

Brams, E. A. 1971. Continuous cultivation of west African soils: organic matter diminution and effects of applied lime and phosphorus. *Plant Soil* 35:401-414.

Brinkmann, W. L. F. and J .C. de Nascimento. 1973. The effect of slash and burn agriculture on plant nutrients in the Tertiary region of Central Amazonia. *Turrialba* 23:284-290.

Burke, I. C., C. M. Yonker, W. J. Parton, C.V. Cole, K. Flach, and D. S. Schimel. 1989. Texture, climate and cultivation effects on soil organic matter content in U. S. grassland soils. *Soil Sci. Soc. Am. J.* 53:800-805.

Burke, I. C., T. G .F. Kittel, W. K. Lauenroth, P. Snook, C. M. Yonker, and W. J. Parton. 1991. Regional analysis of the central great plains. *BioScience* 41:685-692.

Buschbacher, R., C. Uhl, and E. A. S. Serrão. 1988. Abandoned pastures in eastern Amazonia. II. Nutrient stocks in the soil and vegetation. *J. Ecol.* 76:682-699.

Cassell, D. K. and R. Lal. 1992. Soil properties of the tropics: common beliefs and management restraints. p. 61-89. In: R. Lal and P. A. Sanchez (eds.), Myths and Science of Soils of the Tropics. Soil Science Society of America, Madison, WI. SSSA Special Publ. No. 29.

Cerri, C., C. Feller, J. Balesdent, R. Victoria, and A. Plenecassagne. 1985. Application du traçage isotopique naturel in ^{13}C à l'étude de la dynamique de la matière organique dans les sols. *C.R. Acad. Sc. Paris, Ser. 2,* 300:423-428.

Chauvel, A., M. Grimaldi and D. Tessier. 1991. Changes in soil pore-space distribution following deforestation and revegetation: an example from the central Amazon Basin, Brazil. *For. Ecol. Manage.* 38:259-271.

Choné, T., F. Andreux, J. C. Correa, B. Volkoff, and C.C. Cerri. 1991. Changes in organic matter in an oxisol from the central Amazonian forest during eight years as pasture, determined by ^{13}C composition. p. 307-405. In: J. Berthelin (ed.), *Diversity of Environmental Biogeochemistry.* Elsevier, N.Y. 537 pp.

Davidson, E. A. and I.L. Ackerman. 1993. Changes in soil carbon inventories following cultivation of previously untilled soils. *Biogeochem.* 20:161-193.

Deevy, E. S. 1970. Mineral cycles. *Sci. Am.* 223:148-158.

Desjardins, T., F. Andreux, B. Volkoff, and C.C. Cerri. 1994. Organic carbon and ^{13}C contents in soils and soil size-fractions, and their changes due to deforestation and pasture installation in eastern Amazonia. *Geoderma* 61:103-118.

Detwiler, R.P. 1986. Land use change and the global carbon cycle: the role of tropical soils. *Biogeochem.* 2:67-93.

Detwiler, R.P. and C.A.S. Hall. 1988. Tropical forests and the global carbon cycle. *Science* 239:42-47.

Eden, M.J., D. F. M. McGregor, and N. A. Q. Viera. 1990. III. Pasture development on cleared forest land in northern Amazonia. *Geogr. J.* 156:283-296.

EMBRAPA 1979. Manual de métodos de análise do solo. Empresa Brasileira de Pesquisa Agropecuária, Serviço Nacional de Levantamento e Conservação de Solo. Rio de Janeiro.

Eswaran, H., E. van der Berg, and P. Reich. 1993. Organic carbon in soils of the world. *Soil Sci. Soc. Am. J.* 57:192-194.

Ewel, J., C. Berish, B. Brown, N. Price, and J. Raich. 1981. Slash and burn impacts on a Costa Rican wet forest site. *Ecology* 62:816-829.

Falesi, I.C. 1976. Ecosistema de pastagem cultivada na Amazonia Brasileira. Centro de Pesquisa Agropecuária do Trópico Umido. Empresa Brasileira de Pesquisa Agropecuária. Boletim Técnico no. 1, Belem. 193 pp.

Fearnside, P.M. 1980. The effects of cattle pasture on soil fertility in the Brazilian Amazon: consequences for beef production sustainability. *Trop. Ecol.* 21:125-137.

Fearnside, P.M. 1993. Deforestation in Brazilian Amazonia: the effect of population and land tenure. *Ambio* 22:537-545.

Fearnside, P.M. and E. Salati. 1985. Explosive deforestation in Rondônia, Brazil. *Environ. Cons.* 12:355-356.

Feigl, B. J., J. M. Melillo, and C. C. Cerri. 1995. Changes in the origin and quality of soil organic matter after pasture introduction in Rondônia (Brazil). *Plant Soil* 175:21-29.

Fisher, M.J., I.M. Rao, M.A. Ayarza, C.E. Lascano, J.I. Sanz, R.J. Thomas, and R.R. Vera. 1994. Carbon storage by introduced deep-rooted grasses in the South American savannas. *Nature* 371:236-238.

García-Oliva, F., I. Casar, P. Morales, and J. M. Mass. 1994. Forest-to-pasture conversion influences on soil organic carbon dynamics in a tropical deciduous forest. *Oecologia* 99:392-396.

Graças, P.M. de. A. 1997. Conteúdo de carbono da biomassa florestal na Amazônia e alterações apôs a quema. M. S. Thesis, Escola Superior de Agricultura Luiz de Queiroz, University of São Paulo, Piracicaba, Brazil.

Grace, J., J. Lloyd, J. McIntyre, A.C. Miranda, P. Meir, H.S. Miranda, C. Nobre, J. Moncrieff, J. Massheder, Y. Malhi, I. Wright, and J. Gash. 1995. Carbon dioxide uptake by undisturbed tropical rain forest in southwest Amazonia, 1992 to 1993. *Science* 270:778-780.

Hassink, J. and J.J. Neeteson. 1991. Effect of grassland management on the amounts of soil organic N and C. *Neth. J. Agric. Sci.* 39:225-236.

Hecht, S.B. 1982. Deforestation in the Amazon Basin: magnitude, dynamics and soil resource effects. *Stud. Third World Soc.* 13:61-101.

Houghton, R.A., D.L. Skole, and D.S. Lefkowitz. 1990. Changes in the landscape of Latin America between 1850 and 1985. II. Net release of CO_2 to the atmosphere. *For. Ecol. Manage.* 38:173-199.

Houghton, R.A., R.D. Boone, J.M. Melillo, C.A. Palm, G.M. Woodwell, N. Myers, B. Moore, and D. L. Skole. 1985. Net flux of carbon dioxide from tropical forests in 1980. *Nature* 316:617-620.

INPE (Instituto Nacional de Pesquisas Espaciais). 1992. Deforestation in Brazilian Amazonia. Summary of Basin deforestation. São Jose dos Campos, SP, Brazil.

Jenny, H. 1980. The Soil Resource. Springer, NY 377 pp.

Jordan, C.F. 1985. *Nutrient Cycling in Tropical Forest Ecosystems*. Wiley, NY 190 pp.

Juo, A.S.R. and R. Lal. 1977. The effect of fallow and continuous cultivation on the chemical and physical properties of an alfisol in western Nigeria. *Plant Soil* 47:567-584.

Kauffman, J.B., D.L. Cummings, D.E. Ward, and R. Babbit. 1995. Fire in the Brazilian Amazon: biomass, nutrient pools and losses in slashed primary forests. *Oecologia* 140:397-408.

Kimble, J.M., H. Eswaran, and T. Cook. 1990. Organic carbon on a volume basis in tropical and temperate soils. *Proc. Int. Cong. Soil Sci.* 14:248-253.

Lal, R. 1985. Mechanized tillage systems effects on properties of a tropical alfisol in a watershed cropped to maize. *Soil Tillage Res.* 6:149-162.

Lugo, A.E., M. J. Sanchez, and S. Brown. 1986. Land use and organic carbon content of some subtropical soils. *Plant Soil* 96:185-196.

Luizão, R.C.C., T.A. Bonde, and T. Rosswall. 1992. Seasonal variation of soil microbial biomass–the effect of clearfelling a tropical rainforest and establishment of pasture in the central Amazon. *Soil Biol. Biochem.* 24:805-813.

Mann, L.K. 1986. Changes in soil carbon storage after cultivation. *Soil Sci.* 142:279-288.

Melillo, J. M. 1996. Tropical deforestation and the global carbon budget. *Ann Rev. Energy Environ.* 21:293-310.

Melillo, J.M., J.D. Aber, A.E. Linkins, A. Ricca, B. Fry, and K.J. Nadelhoffer. 1989. Carbon and nitrogen dynamics along the decay continuum: plant litter to soil organic matter. p. 53-62. In: M. Clarholm and L. Bergström (eds.), *Ecology of Arable Land.* Kluwer Academic Publ., NY 295 pp.

Moraes, J.F.L., B. Volkoff, C.C. Cerri, and M. Bernoux. 1996. Soil properties under Amazon forest and changes due to pasture installation in Rondônia, Brazil. *Geoderma* 70:63-81.

Moraes, J.L., C.C. Cerri, J.M. Melillo, D. Kicklighter, C. Neill, D. L. Skole, and P. A. Steudler. 1995. Soil carbon stocks of the Brazilian Amazon basin. *Soil Sci. Soc. Am. J.* 59:244-247.

Naeth, M.A., D.J. Pluth, D.S. Chanasyk, A.W. Bailey, and A.W. Fedkenheuer. 1990. Soil compacting impacts of grazing in mixed prairie and fescue grassland ecosystems of Alberta. *Can. J. Soil Sci.* 70:157-167.

Neill, C., J.M. Melillo, P.A. Steudler, J.L. Moraes, and C. C. Cerri. 1996. Forest- and pasture-derived carbon contributions to carbon stocks and microbial respiration of tropical pasture soils. *Oecologia* 107:113-119.

Neill, C., M.C. Piccolo, P.A. Steudler, J.M. Melillo, B. Feigl, and C.C. Cerri. 1995. Nitrogen dynamics in soils of forests and active pastures in the western Brazilian Amazon Basin. *Soil Biol. Biochem.* 27:1167-1175.

Nepstad, D.C., C. Uhl, and A.E.S. Serrão. 1991. Recuperation of a degraded Amazonian landscape: forest recovery and agricultural restoration. *Ambio* 20:248-255.

Nepstad, D. C., C. R. de Carvalho, E. A. Davidson, P. H. Jipp, P. A. Lefebvre, G. H. Negreiros, E. D. da Silva, T. A. Stone, S. E. Trumbore, and S. Vieira. 1994. The deep-soil link between water and carbon cycles of Amazonian forests and pastures. *Nature* 372:666-669.

Nye, P.H. and D.J. Greenland. 1964. Changes in the soil after clearing tropical forest. *Plant Soil* 21:101-112.

Ojima, D.S., K.A. Galvin, and B.L. Turner, II. 1994. The global impact of land-use change. *BioScience* 44:300-304.

Parton, W.J., D.S. Schimel, C.V. Cole, and D.S. Ojima. 1987. Analysis of factors controlling soil organic matter levels in Great Plains grasslands. *Soil Sci. Soc. Am. J.* 51:1173-1179.

Popenoe, H. 1957. The influence of the shifting cultivation cycle on soil properties in Central America. Proc. *Ninth Pacific Sci. Congr.* 7:72-77.

Post, W.M., W.R. Emanuel, P.J. Zinke, and A.G. Stangenberger. 1982. Soil carbon pools and world life zones. *Nature* 298:156-159.

Reiners, W.A., A.F. Bouwman, W.F.J. Parsons, and M. Keller. 1994. Tropical rain forest conversion to pasture: changes in vegetation and soil properties. *Ecol. Appl.* 4:363-377.

Richter, D.D. and L.I. Babbar. 1991. Soil diversity in the tropics. *Adv. Ecol. Res.* 21:315-389.

Sanchez, P. A. and T. J. Logan. 1992. Myths and science about the chemistry and fertility of soils in the tropics. p. 38-59. In: R. Lal and P. A. Sanchez (eds.), Myths and Science of Soils in the Tropics. Soil Science Society of America, Madison, WI. SSSA Special Publ. No. 29. 185 pp.

Sanchez, P.A., J.H. Villachia, and D.E. Bandy. 1983. Soil fertility dynamics after clearing a tropical rainforest in Peru. *Soil Sci. Soc. Am. J.* 47:1171-1178.

Schlesinger, W.H. 1986. Changes in soil carbon storage and associated properties with disturbance and recovery. p. 194-220. In: J. R. Trabalka and D. E. Reichle (eds.), *The Changing Carbon Cycle.* Springer, NY 592 pp.

Serrão, E. A. S. 1992. Alternative models for sustainable cattle ranching on already deforested lands in the Amazon. *An. Acad. Bras. Ci. 64 (Suppl. 1):* 97-104.

Serrão, E.A.S., I.C. Falesi, J.B. da Veiga, and J.F.T. Neto. 1979. Productivity of cultivated pastures on low fertility soil of the Amazon Basin. p. 195-225 In: P.A. Sanchez and L. E. Tergas (eds.), *Pasture Production in Acid Soils of the Tropics*. Centro Internacional de Agricultura Tropical, Cali, Colombia. 488 pp.

Seubert, C.E., P.A. Sanchez, and C. Valverde. 1977. Effects of land clearing methods on soil properties of an ultisol and crop performance in the Amazon jungle of Peru. *Trop. Agric. (Trin.)* 54:307-321.

Skole, D.S. and C. Tucker. 1993. Tropical deforestation and habitat fragmentation in the Amazon: satellite data from 1978 to 1988. *Science* 260:1904-1910.

Skole, D.S., W.H. Chomentowski, W.A. Salas, and A.D. Nobre. 1994. Physical and human dimensions of deforestation in Amazonia. *BioScience* 44:314-328.

Smith, N.J.H., E.A.S. Serrão, P.T. Alvim, and I.C. Falesi. 1995. Amazonia: Resiliency and Dynamism of its Land and its People. United Nations Univ. Press, NY 253 pp.

Stevenson, F.J. 1982. Biochemistry of the formation of humic substances. p. 195-200. In: F.J. Stevenson (ed.), *Humus Chemistry: Genesis, Composition, Reactions*. Wiley, NY 443 pp.

Tiessen, H. and J.W.B. Stewart. 1983. Particle-size fractions and their use in studies of soil organic matter: II. Cultivation effects on organic matter composition in size fractions. *Soil Sci. Soc. Am. J.* 47:509-514.

Townsend, A.B., P.M. Vitousek, and S.E. Trumbore. 1995. Soil organic matter dynamics along gradients in temperature and land use on the island of Hawaii. *Ecology* 76:721-733.

Trumbore, S.E. 1993. Comparison of carbon dynamics in tropical and temperate soils using radiocarbon measurements. *Global Biogeochem. Cycles* 7:275-290.

Trumbore, S.E., E.A. Davidson, P.B. de Camargo, D.C. Nepstad, and L.A. Martinelli. 1995. Belowground cycling of carbon in forests and pastures of eastern Amazonia. *Global Biogeochem. Cycles* 9:515-528.

Turner, B.-L., W.C. Clark, R.W. Kates, J.F. Richards, J.T. Mathews, and W. B. Meyer (eds.). 1990. *The Earth as Transformed by Human Action*. Cambridge Univ. Press, NY.

Veldkamp, E. 1994. Organic carbon turnover in three tropical soils under pasture after deforestation. *Soil Sci. Soc. Am. J.* 58:175-180.

Vitorello, V.A., C.C. Cerri, F. Andreux, C. Feller, and R.L. Victória. 1989. Organic matter and natural carbon-13 distribution in forested and cultivated Oxisols. *Soil Sci. Soc. Am. J.* 53:773-778.

CHAPTER 3

Spatial Patterns of Soil Organic Carbon Pool Size in the Northwestern United States

Jeffrey S. Kern, David P. Turner, and Rusty F. Dodson

I. Introduction

Spatially-distributed estimates of current soil organic carbon (SOC) pools and flux are important requirements for understanding the role of soil in the global carbon (C) cycle and for assessing potential biospheric responses to climatic change or variation (Schimel et al., 1994). Much interest has been focused on SOC because soil may be the largest terrestrial pool of C (Post et al., 1990). Because of limitations in available input data, spatially-distributed pool and flux estimates have often been generated at relatively coarse spatial resolutions, typically for grid cells greater than 50-km on a side (e.g., VEMAP Members, 1995). In areas of complex topography, however, the major environmental gradients potentially useful for model testing and validation tend to be lost at such coarse spatial resolutions. Landuse patterns, which strongly influence SOC pools and flux, also vary at relatively fine spatial resolutions. Soil data are becoming increasingly available at finer spatial resolutions and offer the opportunity to explore possible region-specific patterns in climate and soil relationships. In this chapter, we describe an approach to characterizing SOC pools at a 4-km spatial resolution over the Pacific Northwest region of the United States.

Spatial patterns in SOC pools have previously been examined in several regions of the U.S.A. Franzmeir et al. (1985) studied SOC in the northcentral U.S.A. using soil survey data and expert judgement. Parton et al. (1987; 1989) modeled SOC accumulation using the CENTURY model in the Great Plains grasslands using soil temperature, soil moisture, particle size analysis, plant lignin content, and N inputs. Burke et al. (1989) used soil characterization and climate data to conclude that soil organic matter in the Central Plains Grasslands increases with precipitation and clay content, and decreases with temperature. Nichols (1984) found that SOC in Mollisols of the southern Great Plains is related to clay content and, to a lesser degree, annual precipitation. Sims and Nielsen (1986) reported that elevation and precipitation are correlated to SOC in frigid and cryic soils of Montana, but texture is not. Grigal and Ohmann (1992) concluded that SOC in upland forests of the Lake States varies significantly with forest type, and that forest type, stand age, soil available water holding capacity (AWHC), actual evapotranspiration (AET), and clay content explained some of the SOC variation. Homann et al. (1995) found that SOC in forested western Oregon increases with mean annual temperature (MAT), mean annual precipitation (MAP), AET, clay content, and AWHC but decreases with slope. Davidson and Lefebvre (1993) compared several approaches to estimate SOC for the state of Maine and concluded that SOC content was correlated with soil drainage and not clay

ISBN 0-8493-7441-3
©1997 by CRC Press LLC

content. These studies generally suggest that the relative importance of the major factors influencing SOC accumulation varies among regions.

Efforts have also been made to assess and map SOC distribution at national to global scales. Spatial patterns of SOC in the contiguous U.S.A. were estimated by Kern (1994) by aggregating soil characterization data by ecosystem complex, soil great group, and by UN soil map of the world soil unit. Tarnocai and Ballard (1994) characterized the spatial patterns of SOC in Canada by linking the Soil Landscape of Canada map data with SOC values calculated or estimated from soil characterization data. Bliss et al. (1995) provide a detailed description of how to use spatial soil data for the U.S. to develop SOC inventories. On a global scale, Schlesinger (1984) used soil characterization data extrapolated by ecosystem type to estimate SOC pools. Post et al. (1982) extrapolated SOC estimated from a large soil characterization database (Zinke et al., 1984) using a map of ecosystem complexes. Eswaran et al. (1993) updated the work of Kimble et al. (1990) and used soil characterization data extrapolated at the soil suborder level. Schimel et al. (1994) used the CENTURY model to estimate global SOC. These broader scale analyses indicate some of the more robust relationships of SOC pool size to environmental variables, and help provide perspective on the regional studies.

The Pacific Northwest region is particularly attractive for investigating the relationship of SOC pool size to environmental factors because of its strong climatic gradients and variation in soil texture. The annual precipitation drops from 4500 mm yr^{-1} along the Pacific Coast to less than 230 mm yr^{-1} in the Great Basin east of the Cascade Mountains (Kittel et al., 1995). Associated vegetation changes from dense coniferous forests to semi-desert shrubland (Franklin and Dyrness, 1990). Relatively coarse textured soils have developed from volcanic ash deposits in the Cascades, high clay allophanic soils are found along the coast, and sandy soils are found in the drier eastern portion of the region (Natural Resources Conservation Service [NRCS], 1994). These combinations of climate and soil texture have resulted in a large range in the current SOC pool size. The intent here is to examine the strength of the relationships among SOC pool size and particular environmental variables.

II. Methods and Materials

Soil organic C at 0- to 50-cm and 50- to 100-cm depths were estimated using a soil characterization database aggregated by soil taxonomic unit. These estimates were spatially extrapolated using the spatial distribution of soil taxonomic units, rock fragment content, and soil depth from a soil geographic database. The relationships of SOC pool size to soil physical characteristics and climate were studied by using geographic databases of AWHC, MAP, MAT, AET, and particle size distribution.

A. Soil Organic C by Great Group

Soil organic C was estimated at the great group level of Soil Taxonomy (Soil Survey Staff 1975; 1994) because Kern (1994) found that it was at this level and finer of soil classification that factors important to SOC accumulation become most apparent. The U.S. National Soil Characterization Database (NSCD) was used to calculate SOC at the great group level. The NSCD is maintained by U.S. Department of Agriculture (USDA) NRCS (formerly the Soil Conservation Service [SCS]) and consists of soil characterization data from their laboratories at Lincoln, NE and Riverside, CA, as well as samples analyzed by the USDA Agricultural Research Service (ARS) in Beltsville, MD. Soil organic C is reported as percentage weight determined by wet combustion with $Cr_2O_7^{-2}$, and bulk density (ρ_b) was determined using the clod method (Soil Survey Staff, 1984). Bulk density measured at -33 kPa matric potential water content was used because it more closely approximates field-moist conditions than that measured at oven dry moisture content. In cases where oven-dry ρ_b only was available, it was adjusted to approximate that measured at -33 kPa matric potential using the equation developed by Kern (1994).

The May 1994 version of the NSCD used contains analytical data for more than 20,000 pedons from the U.S.A. and 1100 from outside the U.S.A.. The database contained 115,396 samples from 18,202 pedons or sites in the contiguous U.S.A. The database was checked for incorrectly coded taxonomy and horizon depths. Pedons with unknown soil taxonomy codes were deleted. Missing taxonomic codes were obtained by merging a file of soil series names and classification, provided by the NRCS (John Vrana, personal communication). Pedons with missing classification were deleted from the database. Soil depths were checked to verify the sequence of horizons and, where problems were obvious, they were corrected. Missing ρ_b data were obtained by using data from the NSCD to regress sand, clay, SOC, and SOC squared to ρ_b (similar to the work of Manrique and Jones [1991]), at the great group, suborder, and order level. These regressions were first applied to great groups. For those great groups where no regression equations could be calculated, the suborder level regression equations were used, and, if ρ_b data were still missing, then regression equations developed at the order level were used.

Soil organic C was calculated using the pedon data for Idaho, Oregon, and Washington states. This reduced the sample size but it presumably increased the likelihood that the data are more representative of the region. If a very small sample size (< 2 pedons) was obtained for a great group, then the results were compared with analyses using data from the contiguous U.S.A. If the mean for the national dataset and the region value were similar, the value for the northwest U.S. was used. If data were unavailable, or unacceptable from the northwest U.S.A., then data were used from the national dataset.

Soil organic C by great group was estimated by calculating the depth weighted average SOC in the 0- to 50-cm and 50- to 100-cm depth increments for each pedon that had data for at least 30-cm in the increment. The average SOC for each great group was calculated as the mean of all the pedons with the same classification. The percent coefficient of variation (CV) of the great group level SOC estimates was calculated as 100 times the standard deviation divided by the mean.

B. Spatial Extrapolation of Great Group SOC

The most detailed, consistent source of geographic information about soils in Idaho, Montana, and Oregon (and the entire U.S.A.) is the State Soil Geographic (STATSGO) database (NRCS, 1994). The STATSGO database was created "to be used primarily for regional, multi-state, state, and river basin resource planning, management and monitoring". It was compiled at a map scale of 1:250,000, with the minimum map delineations of 625 ha and no more than 400 map delineations for each map sheet (NRCS, 1994). The data were obtained from the NRCS on a CD-ROM dated October 1994. Each mapped area can contain up to 21 components for which there are accompanying data about soil properties and land use interpretations (NRCS, 1994). The spatial resolution of the minimum map delineations are equivalent to square cells of 2.5-km size. The STATSGO polygon map data were converted to raster cells at 4-km spatial resolution.

The soil component information was merged with the SOC database by great group. The SOC for each component was calculated adjusting for volumetric rock fragment content and soil depth. Rock fragment content was calculated as the average of the lower and upper values from the STATSGO layer attribute dataset. Rock fragment content is an important attribute for many applications, but the data are confusing because they are reported in the STATSGO database as a weight percent of rock fragments greater than 7.62-cm, and as a weight percent of the material that is less than 7.62-cm and greater than 2.00-mm that are converted to a volume basis using ρ_b and rock density. The rock fragment content of the layers described in the STATSGO database were averaged using depth as a weight to makes estimates for 0- to 50-cm and 50- to 100-cm increments. Soil depth was defined as the bottom of the deepest layer described for each component. The component SOC and CV of the SOC were area-weighted averaged for each map unit.

C. Factors Affecting SOC Pools

The relationship of SOC pools to various climatic and soil-texture-related variables was examined using multiple linear regression (SAS Institute, 1990) with all variables averaged at the level of the soil map polygons. The STATSGO data (NRCS, 1994) were used to make area-weighted averages of clay and sand. Clay content was estimated as the mean of the lower and upper values of each layer described by the STATSGO database. Sand content of the STATSGO layers was estimated as the summation of the mean of various sand-sized fractions given. The average clay and sand content of the STATSGO layers were averaged using depth as a weight for the 0- to 50-cm and 50- to 100-cm depth increments. Silt content of the increments is 100 minus the sum of clay and sand content. Sand and clay content were used with the Saxton et al. (1986) model to estimate water retention at -33 kPa and -1500 kPa soil water pressures. Available soil water holding capacity was estimated as the difference between water retention at these water pressures, adjusted for rock fragment content and soil depth. The Saxton et al. (1986) model has the advantage that it does not require SOM or SOC data. Kern (1995) found that a model similar to Saxton et al. (1986) that did use SOM data produced the same result for AWHC with various soil textures because the water retention at both -33 kPa and -1500 kPa soil water pressure were nearly equally affected.

For temperature, the Vegetation/Ecosystem Modeling and Analysis (VEMAP) climate data set (Kittel et al., 1995) provided an initial daily-time-step climatology based on 30-year average climate at a spatial resolution of 0.5 x 0.5 degree of latitude and longitude (grid cells about 50-km on a side in the temperate zone). For this study, a 2.5 minute (~4 km) temperature database was developed from the VEMAP data. VEMAP mean monthly minimum and maximum temperatures were assigned to the mid-points of the VEMAP grid cells with their associated elevation on the VEMAP Digital Elevation Model (DEM). These temperatures were then adjusted to their sea level equivalents using the assumption of locally adiabatic conditions (Byers, 1974). An inverse distance squared interpolation was subsequently made to the mid-points of the 4-km resolution grid and all values were adjusted to their elevation on a 4-km DEM (Defense Mapping Agency, 1981) by inverting the conversion procedure. Mean monthly precipitation at the 4-km resolution was based on interpolations of 30-year average meteorological station data using the PRISM model (C. Daly, Oregon State University, personal communication). PRISM is a moving window regression model which accounts for effects of elevation and aspect on precipitation (Daly et al., 1994). Annual evapotranspiration was the sum of the daily evaporation and transpiration as estimated by the Forest-BGC model (Running and Coughlan, 1988) when driven with our basic climate data. Approaches for development of spatially-distributed surfaces for climate and land cover characteristics to drive Forest-BGC are described elsewhere (Turner and Marks, 1996; Turner et al., 1996).

III. Results

A. Soil Organic C by Great Group

After duplicate samples for the same depths were eliminated and after the data were proofread, there were 72,851 horizons (12,207 pedons or sites) that had SOC data in the contiguous U.S.A. The states of Idaho, Oregon, and Washington had a total of 7049 horizons (2693, 2024, 2332, respectively) for a total of 1306 pedons or sites (485, 375, 446, respectively). Soil organic C was calculated for Idaho, Oregon, and Washington and the results for Fluvaquents and Humaquepts were deleted because, although they are expected to have large SOC content (Kern, 1994), they were extremely large (27.5 and 50.0 kg C m^{-2} to 1 m depth). The great groups that had acceptable data for Idaho, Oregon, and Washington accounted for 95% of the land area. The great group SOC for 3.5% of the land area was obtained from the pedon data for all the contiguous U.S.A. The remaining 1.5% of the soil area was assigned SOC based on the most similar great groups (Soil Survey Staff, 1994). In the discussion below, generalizations are given about the meaning of taxonomic groups for people not familiar with Soil Taxonomy and for a more complete understanding of these complex groups one should refer to Soil Taxonomy (Soil Survey Staff, 1975) and the updated Keys to Soil Taxonomy (Soil Survey Staff, 1994).

The results for SOC by great group are presented in Table 1 sorted by soil order and subgroup with decreasing SOC. The great groups that had extremely large amounts of SOC (> 30.0 kg C m^{-2} to 1 m depth) were Endoaquands, Fluvudands, Melanuands, Hydraquents, and Andaqepts. All of these great groups, with the exception of the very poorly drained soils with little development (Hydraquents), are soils formed in volcanic materials (Andisol soil order). The great groups with the smallest amounts of SOC (< 4.5 kg C m^{-2} to 1 m depth) were Vitritorrands, Udipsamments, Halaquepts, Siderqauods, and Haploxererts. These diverse great groups include Andisols with aridic moisture regimes (Vitritorrands), sandy soils with udic moisture regimes and a low degree of development (Udipsamments), salt affected wet soils with moderate degree of development (Halaquepts), soils with horizons of aluminum and iron complexed with SOC but with a larger ratio of iron to SOC than is typical (Siderquods), and soils with expanding clays and xeric moisture regimes (Haploxererts).

Poorly drained (Aqualfs) and cold (Boralfs) tended to have the largest SOC for the Alfisols (soils with a clay accumulation in the subsoil and high base saturation). Some of the Andisol great groups (Endoaquands, Hapludands, Fulvicryands, Fulvudands, and Melanudands) had very large (> 20.0 kg C m^{-2} to 1 m depth) total SOC, but there was considerable variation within some subgroups. Endoaquands are very poorly drained Andisols, Hapludands are Andisols with udic moisture regimes and no other special features, and the other Andisols with very large SOC content have classifications that indicate high amounts of SOC with their first syllable (Fulv, Fulvi, and Melan). Great groups of the Aridisol soil order (soils with aridic moisture regimes) found in the northwest U.S. had consistently low total SOC. The Entisols (soils with a minimal degree of soil development) with the largest SOC pools were the wet soils (the Aquent subgroup). An exception to cold soils having large amounts of SOC was Cryorthents with 7.2 kg C m^{-2} to 1 m depth. Sandy Entisols (Psamments subgroup) had the lowest SOC of the Entisols. Poorly drained Inceptisols (soils with a moderate degree of development) tended to have large amounts of SOC as did warm Inceptisols (Tropepts) and Umbrepts which, by definition, have large amounts of organic matter.

There was a large range (5.5 to 20.0 kg C m^{-2}) of SOC content to 1 m depth for great groups within suborders of Mollisols. Mollisols are soils with dark surface layers indicating moderate to high amounts of SOC. Soil organic C for Spodosols (soils with horizons of accumulated iron and aluminum complexed with SOC) also tended to be large except for Sideraquods which, as noted above, have a larger ratio of iron to SOC than is typical for Spodosols. Great groups in the Humult suborder of Ultisols (soils with clay accumulations in the subsoil and low base saturation) had relatively large amounts of SOC as one would expect from the name (Hum- denoting high amounts of organic matter). Vertisols (soils with high amounts of expanding clays) had from very small to moderate amounts of SOC. The CVs of the SOC estimates in the 0- to 50-cm depth increment tended to be in the 30 to 50% range with the extreme value (132% for Ustochrepts) being twice that of the next highest CVs (Table 1). The CVs of the SOC estimates for the 50- to 100-cm depth increment, that are not presented, were either similar to or slightly larger than the CVs for the 0- to 50-cm depth layer. The SOC content by great group calculated from pedons throughout the contiguous U.S. had similar trends but tended to have slightly lower SOC content compared to pedons from Idaho, Oregon, and Washington.

Table 1. Soil organic C (SOC, kg C m^{-2}) and area by great group for Idaho, Oregon, and Washington

	Source[1]	Total SOC[2]	0- to 50-cm depth			
			SOC	CV[3]	n[4]	%[5]
Albaqualfs	NW	10.7	8.9	41	6	83
Endoqualfs	US	7.5	5.8	47	34	77
Glossaqualfs	US	4.9	3.7	32	17	76
Ochraqualfs	US	9.3	7.3	59	112	78
Umbraqualfs	NW	14.3	13.0	-	1	91
Cryboralfs	NW	9.6	6.9	57	4	72
Eutroboralfs	NW	12.1	8.1	-	1	67
Fragiboralfs	NW	8.2	6.1	11	4	74
Glossoboralfs	NW	10.2	8.6	36	3	84
Paleoborolls	US	7.8	5.5	30	4	71
Glossudalfs	US	4.9	3.8	45	11	78
Hapludalfs	US	7.3	5.6	42	717	77
Durixeralfs	NW	7.6	5.8	40	5	76
Fragixeralfs	NW	10.2	8.2	22	6	80
Haploxeralfs	NW	8.6	6.7	45	41	78
Natrixeralfs	NW	5.7	4.5	32	4	79
Palexeralfs	NW	13.3	10.8	64	10	81
Rhodoxeralfs	US	12.4	9.0	-	1	73
Aquands	ES	32.4	10.6	-	1	33
Cryaquands	NW	7.5	6.3	37	4	84
Endoaquands	NW	32.4	10.6	-	1	33
Haplaquands	ES	32.4	10.6	-	1	33
Vitriaquands	NW	4.3	3.4	15	2	79
Fulvicryands	NW	21.9	16.4	1	2	75
Haplocryands	NW	18.5	14.0	37	9	76
Vitricryands	NW	8.0	6.2	33	26	78
Vitritorrands	US	1.7	1.6	68	2	94
Fulvudands	NW	37.5	28.0	34	15	75
Hapludands	NW	25.2	18.0	25	11	71
Melanudands	NW	49.1	32.1	52	2	65
Udivtrands	NW	9.5	7.2	51	15	76
Haploxerands	NW	18.8	13.1	24	3	70
Vitrixerands	NW	11.0	8.6	48	45	78
Durargids	NW	7.3	4.7	46	14	64
Haplargids	NW	5.4	4.1	33	16	76
Nadurargids	US	5.6	3.7	62	8	66
Natragrids	NW	7.6	5.1	26	8	67
Paleargids	NW	6.6	4.6	41	8	70
Camborthids	ES	6.1	4.1	48	20	67
Calciortids	NW	8.7	6.0	44	16	69
Camborthids	NW	6.1	4.1	48	20	67
Durothids	NW	7.1	5.0	50	15	70
Salorthids	NW	6.4	4.5	-	1	70
Haplaquents	ES	5.7	3.7	32	5	65
Fluvaqents	US	11.4	7.8	59	87	68
Haplaquents	US	5.7	3.7	32	5	65

Table 1. continued--

	Source[1]	Total SOC[2]	0- to 50-cm depth SOC	CV[3]	n[4]	%[5]
Hydraquents	US	32.1	17.5	27	5	55
Psammaquents	US	8.6	7.6	45	12	88
Udorthents	ES	9.7	6.4	57	67	66
Udifluvents	ES	8.4	5.4	51	65	64
Torrifluvents	NW	9.9	6.0	-	1	61
Tropofluvents	ES	8.4	5.4	51	65	64
Udifluvents	US	8.4	5.4	51	65	64
Xerofluvents	US	7.6	4.8	54	30	63
Cryothents	NW	7.2	6.7	-	1	93
Torriorthents	NW	4.5	3.1	26	7	69
Udorthents	US	9.7	6.4	57	67	66
Xerorthents	NW	9.6	8.4	80	2	88
Cryopsamments	ES	4.2	3.4	49	65	81
Torripsamments	NW	2.7	1.8	37	8	67
Tropopsamments	ES	4.2	3.4	49	65	81
Udipsamments	US	4.2	3.4	49	65	81
Xeropsamments	NW	7.8	6.4	27	3	82
Andepts	ES	10.0	6.9	30	3	69
Cryandepts	NW	10.0	6.9	30	3	69
Durandepts	ES	10.0	6.9	30	3	69
Dystrandepts	ES	10.0	6.9	30	3	69
Vitrandepts	NW	10.0	6.4	47	9	64
Andaquepts	US	32.4	18.5	22	2	57
Cryaquepts	NW	20.5	17.9	64	2	87
Endoaquepts	US	12.4	9.2	41	12	74
Fragiaquepts	NW	6.7	5.6	26	2	84
Halaquepts	NW	4.3	2.8	66	9	65
Haplaquepts	NW	13.6	9.6	40	4	71
Humaquepts	US	28.5	18.7	63	21	66
Placaquepts	ES	13.6	9.6	40	4	71
Tropaquepts	NW	14.5	12.2	-	1	84
Cryochrepts	NW	8.9	6.7	28	15	75
Durochrepts	NW	8.3	6.8	43	5	82
Dystrochrepts	NW	16.7	13.3	26	5	80
Eutrochrepts	NW	14.1	10.5	47	5	74
Fragiochrepts	NW	12.3	10.8	6	3	88
Ustochrepts	US	9.2	6.8	132	110	74
Xerochrepts	NW	8.6	6.6	56	48	77
Dystropepts	NW	28.9	17.3	-	1	60
Humitropepts	NW	26.4	18.1	21	11	69
Cryumbrepts	US	19.5	14.4	33	20	74
Fragiumbrepts	NW	14.0	11.8	11	2	84
Haplumbrepts	NW	23.1	16.5	27	14	71
Xerumbrepts	NW	10.4	7.7	31	4	74
Argialbolls	NW	13.4	11.0	40	14	82
Argiaquolls	US	13.8	10.5	41	87	76

Table 1. continued--

	Source[1]	Total SOC[2]	0- to 50-cm depth			
			SOC	CV[3]	n[4]	%[5]
Calciaquolls	NW	23.3	19.1	15	4	82
Cryaquolls	NW	16.3	14.6	10	2	90
Duraquolls	NW	6.5	4.8	-	1	74
Endoaquolls	NW	12.7	10.8	15	2	85
Haplaquolls	NW	24.3	20.2	36	2	83
Natraquolls	US	10.3	7.7	36	17	75
Argiborolls	US	11.5	8.4	40	179	73
Cryoborolls	NW	14.1	10.3	41	40	73
Haplaborolls	US	11.6	8.5	42	272	73
Paleborolls	NW	20.6	15.2	23	15	74
Rendolls	US	12.9	11.0	11	4	85
Argiudolls	NW	5.1	2.5	0	2	49
Haplaudolls	US	12.0	9.1	47	426	76
Argiustolls	US	9.9	7.0	35	450	71
Calciustolls	US	12.7	9.5	54	59	75
Haplustolls	US	10.3	7.4	49	325	72
Argixerolls	NW	11.7	8.4	54	171	72
Calcixerolls	NW	15.9	11.5	44	13	72
Durixerolls	NW	8.5	5.7	33	37	67
Haploxerolls	NW	10	6.9	59	182	69
Natrixerolls	NW	11.6	9.1	49	7	78
Palexerolls	NW	10.5	8.1	37	10	77
Sideraquods	NW	3.3	2.6	-	1	79
Tropaquods	NW	21.4	17.3	-	1	81
Cryods	ES	25.3	17.0	55	7	67
Haplocryods	NW	25.3	17.0	55	7	67
Humods	ES	17.3	14.3	-	1	83
Haplohumods	US	17.3	14.3	-	1	83
Cryorthods	NW	12.9	11.2	68	9	87
Haplorthods	NW	16.6	11.8	35	18	71
Umbraquults	US	27	22.8	51	7	84
Haplohumults	NW	15.4	11.6	30	14	75
Palehumults	NW	17.4	13.6	27	19	78
Tropohumults	US	11.9	7.9	15	2	66
Hapluduluts	NW	18.7	15.9	8	2	85
Rhodudults	NW	10.9	9.4	4	2	86
Haplustults	US	6.4	4.7	62	6	73
Haploxerults	NW	10.6	8.5	18	5	80
Pelluderts	US	9.3	6.5	35	9	70
Calcixererts	ES	6.5	3.9	41	5	60
Chromoxererts	NW	6.5	3.9	41	5	60
Durixererts	ES	6.5	3.9	41	5	60
Haploxererts	US	3.8	2.7	42	4	71
Pelloxererts	US	11.5	7.2	33	19	63

Source[1]: NW = ID, OR, WA; US = contiguous U.S.A., ES = estimated; Total SOC[2] = kg C m^{-2} to 100-cm depth; CV[3] = % coefficient of variation; n[4] = number of pedons; %[5] = % of total SOC to 100-cm depth.

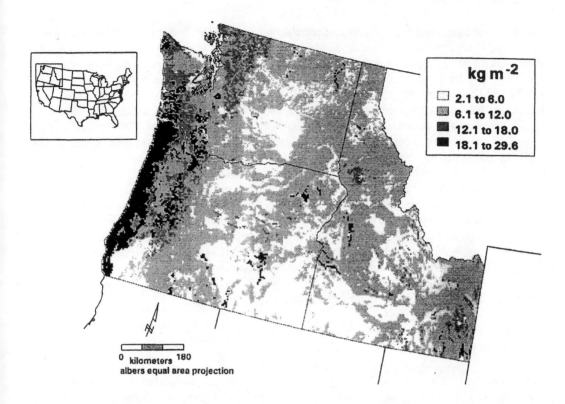

Figure 1. Generalized spatial patterns of average soil organic C at 0 to 100 cm for Idaho, Oregon, and Washington.

B. Spatial Extrapolation of Great Group SOC

The largest SOC pools to 1-m depth (Figure 1) are in the coastal ranges of western Oregon and southwestern Washington (12 to 30 kg C m^{-2}). Note that in the display (Figure 1), considerable complexity in the digital data is omitted by the need to aggregate to four classes for publication in gray scale. The Olympic Peninsula did not have high levels of SOC to 1-m depth (6 to 12 kg C m^{-2}) because of rock fragments and some shallow soils. The Cascade Range of Oregon and southern Washington had high levels of SOC (12 to 30 kg C m^{-2}) but the North Cascades has lower SOC (12 to 18 kg C m^{-2}) because of rock fragment content and soil depth. Scattered small mountain ranges coincided with isolated areas of high SOC content. The drier portions of the study area, portions of east central and southwest Idaho, southeast and northcentral Oregon, and south central Washington, tended to have SOC of less than 6 kg C m^{-2} to 1-m depth. The area-weighted average CVs of the map unit component SOC were mostly 30 to 70% and the CVs of the mean SOC by map unit were larger (mostly 40 to 120%). The relatively large CVs for the mean map unit SOC estimates are because, at this moderately detailed scale of mapping, a wide range of soils can be described within a map unit.

C. Factors Affecting SOC Pools

Actual evapotranspiration had the strongest correlation with SOC in the 0- to 100-cm depth and 0- to 50-cm depth increments ($R^2 = 0.43$ and 0.46) among the single variables (Table 2). Mean annual precipitation was the next most strongly correlated variable in the same depth increments ($R^2 = 0.37$ and 0.41). Available soil water holding capacity, the next mostly strongly correlated single factor ($R^2 = 0.29$ and 0.15), was the only variable with $R^2 > 0.30$ ($R^2 = 0.50$) in the 50- to 100-cm depth increment. Mean annual temperature regressed against SOC resulted in no correlation.

The most strongly correlated sets of two variables vs. SOC were MAP and AWHC, and AET and AWHC. The results were nearly the same for both sets of variables with the strongest correlations in the 0- to 100-cm depth interval ($R^2 = 0.63$ and 0.61) and the 50- to 100-cm depth interval ($R^2 = 0.64$ and 0.63). The correlations for MAP and AWHC, and AET and AWHC in the 50- to 100-cm depth interval were slightly weaker ($R^2 = 0.58$ and 0.55). A surface modeled from the data used for the regression using a bivariate interpolation of AET and AWHC vs. SOC in the 0- to 100-cm depth interval is shown in Figure 2. According to this modeled response surface, increasing AWHC with AET below 700 mm caused modest increases in SOC but, above that point, SOC increased sharply with increasing AWHC. Exploratory analyses were conducted to determine if using more factors, factors squared, or cross-products of factors increased the correlations, but they did not. The data were grouped by the dominant vegetation type (forest, shrub, grassland) and/or by land resource region, but that did not result in larger correlation coefficients.

IV. Discussion

The results for SOC by taxonomic unit were generally consistent with earlier studies. At the order level, the SOC for great groups with the Entisol order found here was similar to Davidson and Lefebvre (1993) for Maine using STATSGO data (except that the Udorthent SOC was much lower here). Inceptisol and Spodosol great group SOC also compared well with Davidson and Lefebvre (1993) except that Humaquept SOC was twice as large in Maine. Franzmeier et al. (1985), working with pedons from the northcentral U.S., reported results as mixtures of suborders and the ranges of values for great groups presented here were comparable except that great groups of sandy Entisols (Psamment suborders) were generally higher here.

The regression analyses of factors that might affect SOC pool size indicated significant relationships in many cases, as has been found in other studies. However, differences between regional studies are also evident and may give some evidence about the relative importance of the different soil forming factors in specific regions. Burke et al. (1989) obtained R^2s of 0.51 for range soils and 0.54 for cultivated soils in the U.S. Central Plains Grasslands using MAT, MAP, silt, and clay (with various terms squared or multiplied). Those results compare well with the R^2s of 0.55 to 0.64 for MAP and AWHC, and AET and AWHC vs. SOC obtained here based on spatially distributed SOC and climate. Grigal and Ohmann (1992) report an R^2 of 0.57 when they regressed forest type, forest age, MAP, AWHC, AET, and clay against SOC pool size for 845 pedons in the north central U.S. Homann et al. (1995) also obtained a modest R^2 (0.42) when they regressed AET and clay against SOC at 0- to 100-cm for pedon data from western Oregon. Nichols (1984) found that clay content vs. SOC resulted in a R^2 of 0.86 and silt resulted in a R^2 of 0.56 for surface horizons of Mollisols in the Southern Great Plains. We found R^2s of 0 for clay and silt but we found a similar R^2 for MAP (0.45 vs. our 0.41 for 0- to 50-cm depth) and for MAT (0.15 vs. our 0.08 for 0- to 50-cm depth).

Knowledge of relationships among SOC pools and climatic variables could give an indication of changes in pool size which might be expected over a specified region in response to climatic change. However, the generally weak relationships of climatic variables to SOC pools suggests caution in using an equilibrium approach to predicting changes in SOC storage as a function of climate change

Table 2. Results of regression analyses of factors that affect soil organic C (SOC) pools

Variable	0- to 100-cm depth				0- to 50-cm depth				50- to 100-cm depth			
	Coeff.[a]	Inter.[b]	P[c]	R^2	Coeff.[a]	Inter.[b]	P[c]	R^2	Coeff.[a]	Inter.[b]	P[c]	R^2
AET[d]	0.011	3.710	0.0001	0.43	0.009	3.024	0.0001	0.46	0.003	0.681	0.0001	0.26
MAP[e]	0.035	5.260	0.0001	0.37	0.028	4.153	0.0001	0.41	0.007	1.100	0.0001	0.18
AWHC[f]	0.500	2.755	0.0001	0.29	0.650	2.561	0.0001	0.15	0.299	0.251	0.0001	0.50
MAT[g]	0.163	7.625	0.0001	0.08	0.013	6.053	0.0001	0.08	0.003	1.572	0.0001	0.04
MAP	0.043	0.139	0.0001	0.63	0.287	-0.086	0.0001	0.58	0.006	-0.196	0.0001	0.64
AWHC	0.480		0.0001			0.693	0.0001		0.286		0.0001	
AET	0.010	-0.007	0.0001	0.61	0.008	0.078	0.0001	0.55	0.002	-0.296	0.0001	0.63
AWHC	0.395		0.0001		0.519		0.0001		0.263		0.0001	

[a]Coeff. = regression coefficient; [b]Inter. = intercept; [c]P = significance; [d]AET = actual evapotranspiration; [e]MAP = mean annual precipitation; [f]AWHC = available soil water holding capacity; [g]MAT = mean annual temperature.

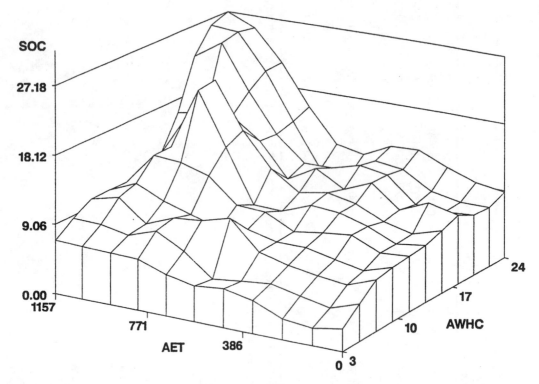

Figure 2. Actual evapotranspiration (AET, mm) and available soil water holding capacity (AWHC) vs. soil organic C (SOC, kg m^{-2}) at 0- to 100-cm depth.

projections (Dixon and Turner, 1991; Smith and Shugart, 1993). That the sign of the slope relating SOC pool size to temperature changes from negative for pedons from the Great Plains (Burke et al. 1989) to positive for pedons from the Pacific Northwest (this study, Homann et al., 1995) is of particular note. The application of process-based models, which account for environmental influences on the uptake of C via plant photosynthesis and its release via decomposition, is an important complimentary approach to assessing effects of potential climate change on SOC pool size. The results of efforts such as this study to use existing data for mapping current SOC pools will be useful in initializing, calibrating, and validating the spatially-distributed process model simulations.

Besides yielding information on SOC pool size, soil taxonomic data may provide information on other variables relevant to modeling of soil C dynamics. Models of SOC dynamics based on only one pool are generally not able to generate results consistent with the observed incorporation rates of bomb ^{14}C into the surface soil (Harrison et al., 1993). SOC is clearly a mixture of materials with very different turnover times and a process-based approach with multiple fractions, as in the CENTURY model (Parton et al., 1987), is desirable for assessing the transient response of soils to climate change. One of the critical variables needed for such models, but difficult to estimate over large areas, is the initial fractionation into slow and fast pools. Trumbore (1993) compared density fractionation of Oxisols and Ultisols and found much higher proportions of high density, slow turnover material in the Oxisols. She noted the possibility of using soil taxonomic information as a way of differentiating patterns in SOM dynamics. Thus, in combination with more efforts to chemically fractionate SOC into slow and fast turnover pools, the STATSGO database taxonomic information may ultimately contribute to procedures for distributing this parameter.

VI. Conclusions

Analysis of spatial patterns in SOM pools and flux under current conditions will aid in understanding climatic regulation of SOC pools and potential biospheric responses to climate change. Information is increasingly available on factors such as climate and land cover at spatial resolutions on the order of 1- to 10-km, and associated information on soil properties is also needed. The STATSGO database, which details spatial distributions of soil taxonomic types, along with the national pedon database, which indicates trends in SOC pools size among different soil taxonomic types, provides an opportunity to map SOC pools at a relatively fine spatial resolution. Examination of the relationships among environmental variables and SOC pools for a map of the Pacific Northwest using this approach indicates similarities with other regional studies but also some significant differences. The generally weak relationships of SOC pools to spatial variation in climate suggests caution in taking an equilibrium approach to assessing potential effects of climatic change on soils. Additional information related to soil taxonomy, which would be particularly useful for the purposes of modeling SOM dynamics, includes the fractionation into slow and fast turnover pools.

Acknowledgments

The information in this document has been funded by the U.S. Environmental Protection Agency under contracts 68-C6-0005 to Dynamac Corportation and 68-C4-0019 to ManTech Environmental Technology Inc. It has been subjected to Agency review and approved for publication. Mention of trade names or commercial products does not constitute endorsement or recommendation for use. Dr. Elissa Levine of the NASA Goddard Flight Center provided many useful comments for improving this paper.

References

Byers, H.R. 1974. *General Meteorology*. Fourth Ed., McGraw-Hill, New York, 461 pp.

Bliss, N.B., S.W. Waltman, and G.W. Petersen. 1995. Preparing a soil carbon inventory for the United States, pp. 275-295. In: R. Lal, J. Kimble, E. Levine, and B.A. Stewart (eds.), *Soils and Global Change*, CRC Press Inc., Boca Raton, FL.

Burke, I.C., C.M. Yonker, W.J. Parton, C.V. Cole, K. Flach, and D.S. Schimel. 1989. Texture, climate, and cultivation effects on soil organic matter content in U.S. grassland soils. *Soil Sci. Soc. Am. J.* 53:800-805.

Daly, C., R. P. Neilson, and D. L. Phillips. 1994. A statistical-topographic model for mapping climatological precipitation over mountainous terrain. *Appl. Meteor.* 33:140–158.

Davidson, E.A. and P.A. Lefebvre. 1993. Estimating regional carbon stocks and spatially covarying edaphic factors using soil maps at three scales. *Biogeochem.* 22:107-131.

Defense Mapping Agency. 1981. *Product Specifications for Digital Terrain Elevation Data*. 2nd Ed. Defense Mapping Agency, Aerospace Center, St. Louis Airforce Station, St. Louis, MO. 23 p.

Dixon, R. and Turner, D.P. 1991. The global carbon cycle and climate change: responses and feedbacks from below-ground systems. *Environmental Pollution* 73:245-262.

Eswaran, H., E. Van Den Berg, and P. Reich. 1993. Organic carbon in soils of the world. *Soil Sci. Soc. Am. J.* 57:192-194.

Franklin, J.F. and C.T. Dyrness. 1990. *Natural Vegetation of Oregon and Washington*. Oregon State University Press, Corvallis, OR.

Franzmeier, D.P., G.D. Lemme, and R.J. Miles. 1985. Organic carbon in soils of north Central United States. *Soil Sci. Soc. Am. J.* 49:702-708.

Grigal, D.F. and L.F. Ohmann. 1992. Carbon storage in upland forests of the Lake States. *Soil Sci. Soc. Am. J.* 56:935-943.

Harrison, K., W. Broecker, and G. Bonani. 1993. A strategy for estimating the impact of CO_2 fertilization of soil carbon storage. *Global Biogeochem. Cycles* 7:69–80.

Homann, P.S., P. Sollins, H.N. Chappell, and A.G. Stangenberger. 1995. Soil organic carbon in a mountainous, forested region: Relations to site characteristics. *Soil Sci. Soc. Am. J.* 59:1468-1475.

Kern, J.S. 1994. Spatial patterns of soil organic carbon in the contiguous United States. *Soil Sci. Soc. Am. J.* 58:439-455.

Kern, J.S. 1995. Evaluation of soil water retention models based on basic soil physical properties. *Soil Sci. Soc. Am. J.* 59:1134-1141

Kimble, J.M., H. Eswaran, and T. Cook. 1990. Organic carbon on a volume basis in tropical and temperate soils. p. V248-V253. In: *Trans. Int. Congr. Soil Sci.* 14th, Kyoto, Japan.

Kittel, T.G.F., N.A. Rosenbloom, T.H. Painter, D.S. Schimel, and VEMAP Modeling Participants. 1995. The VEMAP integrated database for modeling United States ecosystem/vegetation sensitivity to climate change. *J. Biogeography* 22: 857-862.

Manrique, L.A. and C.A. Jones. 1991. Bulk density of soils in relation to soil physical and chemical properties. *Soil Sci. Soc. Am. J.* 55:476-481.

Natural Resources Conservation Service (NRCS). 1994. *State Soil Geographic (STATSGO) Data Base: Data use information*. USDA-NRCS National Soil Survey Center Miscellaneous Publication No. 1492, Lincoln, NE.

Nichols, J.D. 1984. Relation of organic carbon to soil properties and climate in the southern Great Plains. *Soil Sci. Soc. Am. J.* 48:1382-1384.

Parton, W.J., D.S. Schimel, C.V. Cole, and D.S. Ojima. 1987. Analysis of factors controlling soil organic matter levels in Great Plains grasslands. *Soil Sci. Soc. Am. J.* 51:1173-1179.

Parton, W.J., C.V. Cole, J.W.B. Stewart, D.S. Ojima, and D.S. Schimel. 1989. Simulating regional patterns of soil C, N, and P dynamics in US central grasslands region. p. 99-108. In: M. Clarholm and L. Bergstrom (eds.), *Ecology of Arable Lands*. Kluwer Academic Publ., Dordrecht, The Netherlands.

Post, W.M., W.R. Emanuel, P.J. Zinke, and A.G. Stangenberger. 1982. Soil carbon pools and world life zones. *Nature* 298:156-159.

Post, W.M., T.H. Peng, W.R. Emanuel, A.W. King, V.H. Dale, and D.L. DeAngelis. 1990. The global carbon cycle. *Am. Sci.* 78:310-326.

Running, S.W. and J.C. Coughlan. 1988. A general model of forest ecosystem processes for regional applications: I. Hydrologic balance, canopy gas exchange and primary production processes. *Ecol. Model.* 42:125-154.

SAS Institute. 1990. *SAS/STAT User's Guide, Version 6, Fourth Edition, Volume 2*. SAS Institute, Cary, NC, 795 pp.

Saxton, K.E., W.J. Rawls, J.S. Romberger, and R.I. Papendick. 1986. Estimating generalized soil-water characteristics from texture. *Soil Sci. Soc. Am. J.* 50:1031-1036.

Schimel, D.S., B.H. Braswell, E.A. Holland, R. McKeown, D.S. Ojima, T.H. Painter, W.J. Parton, and A.R. Townsend. 1994. Climatic, edaphic, and biotic controls over storage and turnover of carbon in soils. *Global Biogeochem.* 8:279-293.

Schlesinger, W.H. 1984. Soil organic matter: a source of atmospheric CO_2. p. 111-127. In: G.W. Woodwell (ed.), *The Role of Terrestrial Vegetation in the Global Carbon Cycle*. SCOPE 23. Wiley and Sons, NY.

Sims, Z.R. and G.A. Nielsen. 1986. Organic carbon in Montana soils as related to clay content and climate. *Soil Sci. Soc. Am. J.* 50:1269-1271.

Smith, T.M. and H.H. Shugart. 1993. The potential response of global terrestrial carbon storage to a climate change. *Water, Air, and Soil Pollution* 70:629-642.

Soil Survey Staff. 1975. *Soil Taxonomy: a Basic System of Soil Classification for Making and Interpreting Soil Surveys.* Agric. Handb. 436. U.S. Gov. Print. Office, Washington, D.C.

Soil Survey Staff. 1984. *Soil Survey Laboratory Methods and Procedures for Collecting Soil Samples.* USDA-SCS, Soil Survey Investigations Report no. 1, U.S. Gov. Print. Office, Washington, D.C.

Soil Survey Staff. 1994. *Keys to Soil Taxonomy, 6th edition*, USDA-SCS, U.S. Gov. Print. Office, Washington, D.C.

Tarnocai, C. and M. Ballard. 1994. Organic carbon in Canadian Soils, pp. 31-45. in: R. Lal, J.M. Kimble, and E. Levine (eds.), *Soil Processes and Greenhouse Effect*. U.S. Department of Agriculture, Soil Conservation Service, National Soil Survey Center, Lincoln, NE.

Trumbore, S. E. 1993. Comparison of carbon dynamics in tropical and temperate soils using radiocarbon measurements. *Global Biogeochem. Cycles* 7:257–290.

Turner, D.P. and D.G. Marks. 1996. Application of togographically distributed models of energy, water, and carbon balance over the Columbia River Basin: A framework for simulating potential climate change effects at the regional scale. p. 215-233. In: S.J. Ghan, W.T. Pennell, K.L. Peterson, E. Rykiel, M.J. Scott, and L.W. Vail (eds.), *Regional Impacts of Global Climate Change: Assessing Change and Response at the Scales that Matter*. Battelle Press, Columbus Ohio.

Turner, D. P., Dodson, R., and D. Marks. 1996. Comparison of alternative spatial resolutions in the application of a spatially distributed biogeochemical model over complex terrain. *Ecol. Mod.* 90:53-67.

VEMAP Members. 1995. Vegetation/ecosystem modeling and analysis project: Comparing biogeography and biogeochemical models in a continental-scale study of terrestrial ecosystem responses to climate change and CO_2 doubling. *Global Biogeochem. Cycles* 9:407-437.

Zinke, P.J., A.G. Stangenberger, W.M. Post, W.R. Emanuel, and J.S. Olson. 1984. *Worldwide Organic Soil Carbon and Nitrogen Data.* ORNL/TM-8857. Oak Ridge Natl. Lab., Oak Ridge, TN.

CHAPTER 4

Organic Carbon in Deep Alluvium in Southeast Nebraska and Northeast Kansas

R.B. Grossman, D.S. Harms, M.S. Kuzila, S.A. Glaum,
S.L. Hartung, and J.R. Fortner

I. Introduction

The U.S. national data base for organic carbon in soils and associated sediments has gaps and uncertainties. The contribution by O horizons of forest soils is incompletely documented. Standing biomass in the forest and the contribution by large tree roots is generally not included in agricultural sampling. Further, sampling of soils in some areas that are not farmed has very low density or is nonexistent. Despite these uncertainties, rather impressive documentation is available on the amount of organic carbon in soils largely originating from studies for agricultural purposes. The documentation has been summarized recently by Kern (1994).

In this chapter we address one of the uncertainties in areal organic carbon estimates: The importance of the amount of organic carbon that is not measured in the standard depth of sampling for agricultural purposes for soils formed in deep alluvium. The study area is the Nemaha River Valley located in and considered representative of Major Land Resource Area 106 (MLRA 106), which is located in southeast Nebraska and Northeast Kansas (Figure 1). MLRA 106 is one of about 150 areas into which the U.S. is divided on the basis of geology, climate, and agriculture (Soils Cons. Ser., 1981). The area of MLRA 106 is 2.8×10^6 ha.

II. Major Land Resource Region 106

Southeast Nebraska was inhabited by Pawnee Indians in 1804 at the time of the Lewis and Clark expedition through the land obtained by the Louisiana Purchase. In 1854, President Pierce signed a treaty with the Indians allowing agricultural settlement of the lands bordering the west side of the Missouri River. By 1856, settlers were establishing claims 65 hectares in size, and spreading throughout the uplands. Settlers came mostly from neighboring states to the east. Some came from New England and Europe. Agriculture was the main economy as settlers plowed the prairie and planted corn, wheat, potatoes, fruits, and garden crops. The area experienced a boom in the early 1800s, but by the end of the century consolidation of farm units had begun and the population growth rate had declined.

Geologic materials in MLRA 106 are generally unconsolidated deposits of Pleistocene and Holocene age overlying bedrock formations of Cretaceous, Permian, and Pennsylvanian age.

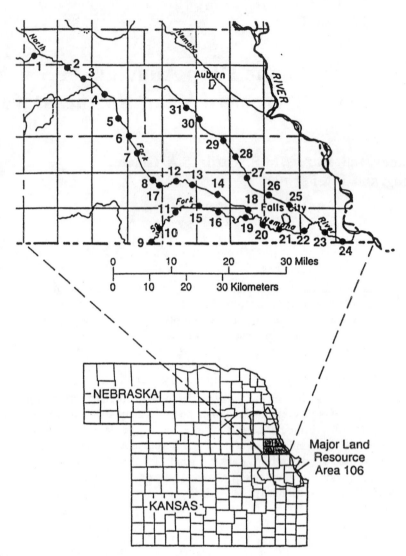

Figure 1. Location of MLRA 106 and sample sites in streams of the Nemaha River Valley.

Erosion and deposition have produced three prominent topographic features: uplands, terraces, and bottomlands. The uplands are mantled by loess which generally thins as distance from the Missouri River increases. In some areas, Illinoian-age sediments, ranging from sandy to clayey in texture, underlie the loess. Till underlies these sediments and has been exposed in many areas by erosion. The till is a heterogeneous mixture of clay, sand, gravel, cobbles, and some boulders. Bedrock, generally limestone and sandstone, underlies the till and exposures are common throughout the area.

Elevation of the MLRA is 300-500 m. Average annual precipitation is 750-925 mm and average annual temperature is about 13°C. Soil textures are mostly silt loam, silty clay loam, silty clay, and clay. Sandy textures are uncommon. Most of the soils are Udolls with a mesic temperature and a udic moisture regime. The most common soils have argillic horizons. Upland soils have gentle through moderately steep slopes. Native vegetation consists mainly of mid and tall grasses. Between one-half and three-fourths of the area is cropland. Pasture and range comprises another one-fourth. Grain sorghum, soybeans, corn, wheat, and alfalfa are important crops.

A. The Nemaha River Valley

Major portions of the current Nemaha River valley (Figure 1) likely follow pre-Pleistocene paleo-valleys that traverse southeastern Nebraska. These paleo-valleys were exhumed during a period of erosion and subsequently filled with alluvium. The alluvium in the Nemaha Valley (Figure 1) averages about 10 to 15 meters in thickness and generally overlies bedrock. It consists mostly of silty and clayey sediment eroded from the loess- and till-derived soils of the adjacent uplands. Some thin lenses of sand are found within the silty and clayey sediment. The thickest layers of sand and gravel are found overlying the bedrock. This sand and gravel is probably the coarse remains of the till that had eroded into the valleys from the uplands. It is thought that much of this alluvium is late-Pleistocene or Holocene in age because there is evidence that major episodes of erosion have occurred in southeastern Nebraska during this period. Since farming operations began in the late 1850s, severe upland and stream bank erosion has occurred. It is possible that a sizeable portion of the alluvium has been deposited since that time, based on evidence of "modern" artifacts found in stream banks in the area.

III. Procedures

Thirty-one sites were sampled (Figure 1). The objective was to place the sites so that the sampling density was relatively uniform along the major stream courses. The sites were located at five to eight kilometer intervals along the north and south forks of the Big Nemaha River and Muddy Creek, a major tributary of the Big Nemaha River. A second group of sites located in northeast Kansas is under study. The Kansas sites have been evaluated geomorphically, which should improve the predictions. Sites were selected to stratify, to an extent, on the basis of the major soils that had been mapped. The drilling was done under very wet conditions and most fields could not be entered. A number of sites were, therefore, placed in the middle of unimproved roads. The sites on abandoned roads have not been tilled since road construction early in this century. The top of the soil column was established at the base of the road fill material. Cores of 5 cm diameter were taken with a hollow stem auger and split-tube core barrel with a Central Mining Equipment Company Model 75 drill rig[1]. Cores were extracted from the land surface (or base of the road fill) to the underlying sand and gravel. Samples could not be extracted from the sand and gravel, which was generally saturated. The cores were described using standard pedological descriptive techniques (Soil Survey Division Staff, 1993). The average length of the core field segment was 60 cm. Cores were placed in cardboard sleeves and then placed into core boxes for transport to the laboratory. The core segments were evaluated in the laboratory and additional divisions were made as needed.

Organic carbon in the upper 20 cm below the fill is 4.7 kg m^{-2} for the 11 sites under roads and 3.8kg m^{-2} for the other sites. We may assume that the 0.9 kg m^{-2} greater value for the sites under roads is due to the reduced tillage. As will be discussed, the mean amount of organic carbon 1.5m to the base of the cores is 40.0kg m^{-2}. The 0.9kg m^{-2} is small compared to the 40.0kg m^{-2} and therefore the possibility of a correction was not pursued.

Each horizon was processed to prepare the <2mm sample. Particle size analysis, organic carbon by wet combustion, and 15 bar water retention were run on all samples (Soil Survey Laboratory Staff

---

[1]Company given for convenience of reader and does not imply endorsement.

1996). Forty sections of the 5 cm diameter cores were cut into 70-90 mm long segments. The outer 10 mm was removed to reduce the effect of compaction that resulted from the drilling operation. Bulk density at 1/3 bar and at oven dryness and water retention at 1/3 bar were determined on these core segments (Soil Survey Lab Staff, 1996). The objective was to be able to calculate saturated hydraulic conductivity by the method of Ahuja et al. (1989).

On each layer, the kg m^{-2} of organic carbon was calculated using the equation:

$$OC \text{ kg m}^{-2} = \frac{L \times D \times OC}{10}$$

where L is the thickness in centimeters, D is the bulk density at or near field capacity, and OC is the organic carbon percent. Rock fragments were absent except for a very few core segments near the underlying sand and gravel. No corrections were made for the >2mm, which, at most, was small. For most core segments, bulk densities were not measured. The following protocol was employed to estimate bulk density:

> 50 percent clay - 1.30 g/cc
20-50% clay and < 15% sand (2-0.1mm) -1.35 g/cc
< 20% clay and < 15% sand (2-0.1mm) -1.45 g/cc
< 20% clay and ≥ 15% sand (2-0.1mm) -1.55 g/cc

IV. Calculation of the Amount of Organic Carbon in MLRA 106

Evaluation of the amount of organic carbon in MLRA 106 involved a number of steps and considerations. Information that pertains is in Table 1. Numerical estimates were made by application of an MLRA 106 map unit legend prepared in 1990. This legend contains the area of each map unit. (For those not acquainted with the systematics of the National Cooperative Soil Survey consult the appendix in Grossman et al., 1995). These map units are grouped under the soil series that names the map unit. The soil series, not the map unit, were used for evaluation of the amount of organic carbon. Assignments by map unit would have been better; however, that would require a familiarity with the soils that was not readily available. The procedure, therefore, differs from that by Grossman et al. (1992) and Grossman et al. (1995) in which assignments to map units were made by locally knowledgeable soil scientists.

Pedon data were drawn from the data base of the National Soil Survey Laboratory, Lincoln, Nebraska. Pedons were chosen to represent MLRA 106 and came from both within and outside of the MLRA. No judgments were made about the applicability of the pedon. Some of the pedons had been classified formally after sampling and the requisite documentation completed (NCSS Soils form S-8). Others had not been evaluated formally. In some instances pedons were used that had been sampled as a particular series, but upon evaluation were found to not represent the series and were identified as "series not designated." This type of data was used only if there were few or no other pedons applicable to the soil series. Laboratory data were unavailable for some soil series; however, these series represented a small portion of the MLRA (see item 1, Table 1). For these series, the mean organic carbon for the area that had laboratory data was assigned. Another small areal percentage of the MLRA (item 2, Table 1) consisted of Miscellaneous Land Types (Gravel Pits, for example). Organic carbon values were assigned to Miscellaneous Land Types on the basis of general knowledge.

Table 1. Information pertaining to calculation and evaluation of areal organic carbon for MLRA 106

1. Proportion of map units without sampled pedons	3.1 percent
2. Proportion of miscellaneous land types	2.1 percent
3. Proportion of MLRA	
a. Series ≥25,000 ha aggregate area	76.0 percent
b. Series <25,000 ha aggregate area	24.0 percent
4. Number of series employed	77
a. ≥25,000 ha aggregate area	20
b. <25,000 ha aggregate area	57
5. Number of pedons	161
a. Series ≥25,000 ha	85
b. Series <25,000 ha	76
6. Proportion of pedons with Ap horizons	
a. All series	79.0 percent
b. Series ≥25,000 ha aggregate area	73.0 percent
7. Decade pedons sampled	
a. 1950-1960	3.0 percent
b. 1960-1970	9.0 percent
c. 1970-1980	19.0 percent
d. 1980-1990	35.0 percent
e. 1990s	24.0 percent

Bulk density data were not available for many pedons from which organic carbon data was obtained. In those instances, estimates were made from pedons of the same series on which bulk density had been measured. If such data was unavailable, the bulk density was estimated from the texture and the moist rupture resistance. The moist rupture resistance was generally obtained from the typifying pedon for the soil survey area where the pedon was sampled.

Pedons were checked for the presence or absence of an Ap horizon (item 6, Table 1). An Ap indicates that the area had been plowed at one time. It does not mean that the pedon is currently cultivated. Pedons with Ap horizons should generally have lower organic carbon than noncultivated counterparts. Noncultivated pedons are more common in soil series that are less suitable for cultivation because of slope, depth to bedrock, chemistry, or other properties. The organic carbon data were obtained from pedons that mostly were all sampled before the introduction of no-till (item 7, Table 1), and would not reflect the expected increase in organic carbon.

Twenty soil series of 25,000 ha extent make up 76 percent of the MLRA. The other 57 series of <25,000 ha extent make up the remaining 24 percent (item 4, Table 1). As shown in item 7, Table 1, about 60 percent of the pedons have been sampled since 1980. None of the pedons were sampled before 1950.

V. Characteristics of the Cores

A total of 396 core segments were delimited from 1.5 m below the ground surface to the base of the cores. The average length of the segments was 60 cm. About 80 percent of the core length is silt loam, silty clay loam, or silty clay (Table 2).

Table 2. Proportion of the total core length below a depth of 1.5 m for the various texture classes

Texture class	Length proportion[a] (%)
Coarse sand	0.9
Sand	1.5
Loamy coarse sand	0.2
Loamy sand	1.6
Loamy fine sand	0.8
Coarse sandy loam	0.4
Sandy loam	0.6
Very fine sandy loam	0.8
Fine sandy loam	1.2
Loam	10.3
Sandy clay loam	0.2
Clay loam	0.9
Silt loam	27.1
Silty clay loam	32.3
Silty clay	19.7
Clay	1.5

[a]The percentages pertain to the proportion of the aggregate length that has the properties stipulated.

Table 3. Proportion of the total core length below a depth of 1.5 m for various classes of organic carbon

Organic carbon (%)	Length proportion[a] (%)
< 0.1	5.3
0.1 - 0.2	14.4
0.2 - 0.4	48.1
0.4 - 0.6	16.4
0.6 - 0.8	7.3
0.8 - 1.0	3.5
≥ 1.0	5.0

[a]The percentages pertain to the proportion of the aggregate length that has the properties stipulated.

The organic carbon percentages generally are modest. Nearly 70 percent of the aggregate core length has organic carbon values below 0.40 percent (Table 3). Only 5 percent has values exceeding 1.0 percent. The maximum value is 1.93 percent. Organic carbon tends to increase as the clay percentage rises (Table 4).

Thirty-two bulk density determinations showed that bulk density varied with clay percentage (Table 5). None of the samples had appreciable amounts of sand (2-0.1 mm). Sand content is part of the basis for assigning a bulk density of 1.35 g/cc to sections with 20 to 50 percent clay and 1.30 g/cc if >50 percent clay.

Saturated hydraulic conductivity was calculated from the air-filled porosity at 1/3 bar water retention of the clods, which was obtained by subtracting the volume of water at 1/3 bar from the calculated total porosity (Ahuja et al., 1989). The average saturated hyraulic conductivity for 10 silt

Table 4. Relationship between organic carbon and clay percentage

Clay (%)	Organic carbon (%)
< 10	0.16
10 - 20	0.31
20 - 30	0.39
30 - 40	0.42
40 - 50	0.58
≥ 50	0.62

Table 5. Relationship between bulk density and clay percentage

Clay (%)	Number horizons	Bulk density (g mm^3)
20 - 30	17	1.36
30 - 40	6	1.42
40 - 50	7	1.35
≥ 50	2	1.29

loam sections was 0.33 cm/hr, for 10 silty clay loam sections was 0.046 cm/hr, and for 8 silty clay sections it was 0.076 cm/hr.

VI. Results

The mean organic carbon below 1.5 m for the 31 cores is 40.0 kg m^{-2} (Table 6). The amount of organic carbon below 1.5 m in alluvium as a percentage of the total for MLRA 106 is 39.0 percent (Table 7). These figures lack precision because the cores are highly variable with an SD of 23 kg m^{-2} and a range from 7 to 106 kg m^{-2}.

The average amount of organic carbon weighted by area for the Argiudolls of the MLRA is 13.5 kg m^{-2} (Table 8). This compares to 11.5 kg m^{-2} given by Kern (1994) for Argiudolls. The Argiudoll great group predominates the areas >25,000 ha. Organic carbon values in Table 8 for the Argiudoll series ranges from 9.7 to 15.8 kg m^{-2}.

VII. Conclusions

The Nemaha River Valley of southeast Nebraska is considered to typify the larger stream valleys of MLRA 106. The area of MLRA 106 is 2.8 x 10^6 ha and the amount of organic carbon is estimated to be 429 Tg. For 31 cores from the Nemaha River Valley, the mean amount of organic carbon below the common depth of agricultural sampling, taken as 1.5 m, is 40 kg m^{-2}. Alluvial soils are 15.1 percent of MLRA 106. The amount of organic carbon in the alluvium below 1.5 m is estimated to be 167 Tg, which is 39 percent of the total organic carbon estimate for the MLRA.

Table 6. Organic carbon (kg m^{-2}) to the base of the cores, from the soil surface to 1.5 m, and from 1.5 m to the base of the cores

Core number	Total depth[a] (cm)	Organic carbon[a]		
		Complete	Upper 1.5 m	1.5 m to base
		—————kg m^{-1}—————		
1	762	38.4	23.8	14.6
2	1021	49.8	20.8	29.0
3	335	34.2	21.2	13.0
4	402	35.4	23.9	11.5
5	688	51.2	15.7	35.5
6	1024	56.3	16.6	39.7
7	930	35.1	10.3	24.8
8	868	37.6	21.4	16.2
9	409	39.6	32.9	6.7
10	961	62.8	21.8	41.0
11	1116	56.8	29.2	27.6
12	1131	72.4	17.7	54.7
13	970	73.6	10.9	62.7
14	792	39.9	16.9	23.0
15	1234	117.8	26.8	91.0
16	1209	124.2	17.8	106.4
17	762	27.8	17.4	10.4
18	899	65.6	20.1	45.5
19	879	60.5	21.5	39.0
20	838	57.9	27.3	30.6
21	885	74.5	24.8	49.7
22	1054	73.3	23.6	49.7
23	930	79.7	17.3	62.4
24	399	35.2	15.8	19.4
25	840	78.6	23.4	55.2
26	868	43.9	17.4	26.5
27	930	88.5	21.7	66.8
28	790	64.6	16.9	47.7
29	884	79.3	21.6	57.7
30	735	49.1	28.3	20.8
31	1212	86.1	23.3	62.8
Mean		60.9	20.9	40.0

[a]Measured from beneath fill, if present.

Table 7. Computation of the organic carbon below 1.5 m in alluvium as a percentage of the total for the MLRA

1. Amount of organic carbon in MLRA 106
    a. Area - 2,766,626 ha
    b. Areal concentration inclusive of miscellaneous areas - 15.5 kg m^{-2}
    c. Product (a) and (b) = 428.8 Tg (10^{12})

2. Amount of organic carbon in alluvial part of MLRA 106
    a. Areal proportion alluvium of MLRA - 15.1 percent
    b. Areal concentration organic carbon 1.5 m to base of cores - 40.0 kg m^{-2}
    c. Amount of organic carbon 1.5 m to base of alluvium
    $$\frac{1a \times 2a \times 2b}{100} = 167.1 \text{ Tg}$$

3. Percentage of organic carbon below 1.5 m in alluvium as a percent of the total for the MLRA
    (2c / 1c) x 100 = 39.0 percent

The study prompts these observations:
- The proportion of the organic carbon in alluvium below 1.5 m is a significant part of the total organic carbon in MLRA 106.
- It is not known whether comparable deep pools of organic carbon occur in other areas. It is possible that the amount of organic carbon below the depth of agricultural sampling is considerable over large areas.
- This deep organic carbon is relatively old and not subject to rapid exchange with the atmosphere (Houghton, 1995). A drop in the water table levels presumably would increase the rate of oxidation.
- This pool of organic carbon is near the surface water in adjacent streams and may be partially saturated by groundwater. The pool may have a significant effect on the quality of both surface and groundwater. The organic carbon may retain potential contaminants such as pesticides before they reach water. On the other hand, decomposition of organic carbon within this pool could be a source of contaminants to water.

Table 8. Series with ≥25,000 ha area arranged in descending order of area, and with pedon numbers and kg m^{-2} of organic carbon to approximately 1.5 m

Series, Classification	Area ha x 10^3	Number pedons Total	Number pedons Cultivated	Organic C kg m^{-2}
Pawnee	441	4	1	13.3
Fine, Aquertic Argiudolls				
Wymore	407	6	6	13.6
Fine, Aquertic Argiudolls				
Sharpsburg	301	9	5	15.8
Fine, Typic Argiudolls				
Marshall	150	7	7	15.5
Fine-silty, Typic Hapludolls				
Kennebec	146	2	2	26.8
Fine-silty, Cumulic Hapludolls				
Shelby	112	3	0	12.2
Fine-loamy, Typic Argiudolls				
Grundy	105	2	2	14.0
Fine, Aquertic Argiudolls				
Crete	90	16	15	13.5
Fine, Pachic Argiudolls				
Judson	83	9	9	23.4
Fine-silty, Cumulic Hapludolls				
Morrill	70	2	2	10.8
Fine-loamy, Typic Argiudolls				
Burchard	59	5	1	15.8
Fine-loamy, Typic Argiudolls				
Steinauer	39	4	7	9.7
Fine-loamy, Typic Udorthents				
Vinland	35	0	0	10.0
Loamy, shallow, Typic Hapludolls				
Mayberry	32	2	1	13.1
Fine, Aquertic Argiudolls				
Zook	28	1	1	25.0
Fine, Cumulic Vertic Endoaquolls				

Acknowledgments

Dr. Carolyn Olson suggested the study. G.A. Borchers, F.V. Belohlavy, A.T. Labenz, and S.A. Scheinost were instrumental in completion of the drilling.

References

Ahuja, L.R., D.K. Cassel, R.R. Bruce, and B.B. Bannes. 1989. Evaluation of the spatial distribution of hydraulic conductivity using effective porosity data. *Soil Sci. Soc. Am J.* 48:404-411.

Grossman, R.B., R.J. Ahrens, L.A. Gile, C.E. Montoya, and O. Chadwick. 1995. Areal evaluation of organic and carbonate carbon in a desert area of southern New Mexico. p. 81-91. In: R. Lal, J. Kimble, E. Levine, and B.A. Stewart (eds.), *Soils and Global Change*. Lewis Pub., Boca Raton, FL.

Grossman, R.B., E.C. Benham, J.R. Fortner, S.W. Waltman, J.M. Kimble, and C.E. Branham. 1992. A demonstration of the use of soil survey information to obtain areal estimates of organic carbon. ASPRS/ACSM/RX 92. Tech. Papers Vol. 4. p. 457-465. In: Am. Soc. Photogrammerty and Remote Sensing and Am. Cong. Surveying and Mapping. Bethesda, MD.

Houghton, R.A. 1995. Changes in the storage of terrestrial carbon since 1850. p. 45-66. In: R. Lal, J. Kimble, E. Levine and B.A. Stewart (eds.), *Soils and Global Change*. Lewis Pub., Boca Raton, FL.

Kern, J.S. 1994. Spatial patterns of soil organic carbon in the contiguous United States. *Soil Sci. Soc. Am. J.* 58:439-455.

Lal, R. 1995. Global soil erosion by water and carbon dynamics. p. 131-142. In: R. Lal, J. Kimble, E. Levine, and B.A. Stewart. (eds.), *Soils and Global Change*. Lewis Pub., Boca Raton, FL.

Soil Conservation Staff. 1981. Land Resource Regions and Major Land Resource Areas of the United States. USDA Hnbk. 296. Washington, D.C.

Soil Survey Division Staff. 1993. Soil Survey Manual. USDA Hnbk 18. Washington, D.C.

Soil Survey Laboratory Staff. 1996. Soil Survey Laboratory Methods Manual. Soil Survey Invest. Report 42. Version 3.0., USDA, NRCS, Lincoln, NE.

CHAPTER 5

Soil Carbon Dynamics in Canadian Agroecosystems

H.H. Janzen, C.A. Campbell, E.G. Gregorich, and B.H. Ellert

1. Introduction

Agricultural soils in Canada occupy 68 million ha, of which about two thirds (representing 5% of total land area) is devoted to arable cropland or seeded pasture (Acton, 1995). Most of the agricultural lands occur in three ecozones, the Prairies of western Canada, the Mixed Wood Plains of central Canada, and the Appalachian highlands of eastern Canada (Figure 1). Significant agricultural production also occurs in various smaller regions throughout the country, though predominantly along the southern fringe.

Soils in the various agroecosystems are continually evolving in response to management practices imposed on them (Janzen et al., 1997a). The most rapid changes occurred upon the initial conversion to arable agriculture, but soils continue to change as management practices evolve. In the future, soils may also change in response to external perturbations such as global changes in temperature, precipitation, and CO_2 concentration.

Of the various soil attributes subject to change, soil organic carbon (SOC) deserves particular attention. The loss of SOC threatens the future productivity of soils. Furthermore, SOC is a large storehouse of carbon, and changes in the size of this reservoir have important implications for atmospheric CO_2, the chief contributor to global warming. Although agricultural soils represent a relatively small proportion of C in Canadian landscapes, their C reserves are perhaps most sensitive to human practices.

Our objective is to review changes in SOC in Canadian agroecosystems, with particular emphasis on the effects of various management options. More specifically, we hope to determine the current trajectory of soil C dynamics, describe some of the rate-determining mechanisms, and estimate the potential of various agroecosystems for removing atmospheric CO_2. Rather than attempt a comprehensive review of all pertinent data, we propose to provide a broad conceptual framework, illustrated with examples selected from various studies reported within the last decade.

II. Overview of Carbon Cycle in Canadian Agroecosystems

Agricultural systems vary widely across the country, largely because of climatic diversity. Production of high-value fruits and vegetables is primarily limited to localized areas in southern British Columbia, Ontario, Quebec, and the Maritime provinces. The comparatively warm, humid conditions of southern Ontario and Quebec (mean annual temperature ≈ 3 to 9°C) also allow the production of crops with

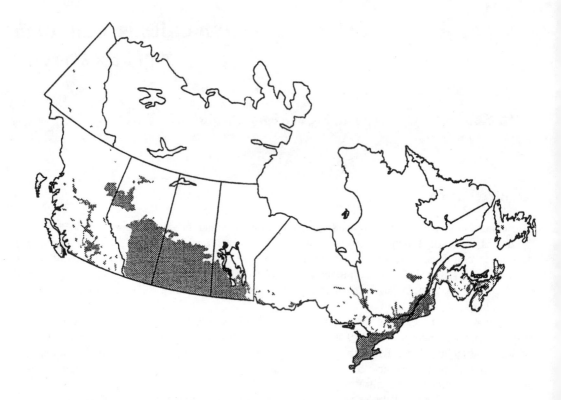

Figure 1. Distribution of agricultural lands in Canada. (From Acton 1995.)

relatively long growing seasons such as corn and soybeans. The prairies of western Canada, by comparison, are cool (mean annual temperature ≈ 0 to 5° C), so that much of the arable cropland is devoted to small-grain cereals and other short-season annual crops. Because of aridity in the southern prairies, a large proportion of arable land is left unseeded (summer fallowed) at regular intervals to replenish soil moisture.

The cultivated or 'improved' cropland of Canada contains about 3 Pg C in the 0-30 cm layer (Dumanski et al., 1997). Largest reserves occur in the Chernozemic soils of western Canada, both because of the large area they occupy and their relatively high C density (Figure 2).

The current C balance of any agroecosystem is a function of the net inputs from photosynthesis and the export of C either as CO_2 from respiration or in the form of harvested C (Figure 3). Fluxes of C tend to be higher in the more humid environments (Figure 3a) than in semi-arid regions (Figure 3b). For example, net primary production in corn in southern Ontario may be more than 2 times that under wheat in western Canada. These higher C inputs, however, are associated with higher respiration rates, so that net dynamics of stored C may be comparable among regions.

In any ecosystem, the change in soil C is a function of the difference between C added in plant residues and the C lost via decomposition processes (assuming no redistribution of C via erosion or organic C removal or addition). Over time, the rates of input and loss tend to converge (i.e., ratio of C input/decomposition = 1), so that amount of stored C is relatively static or near steady state. A

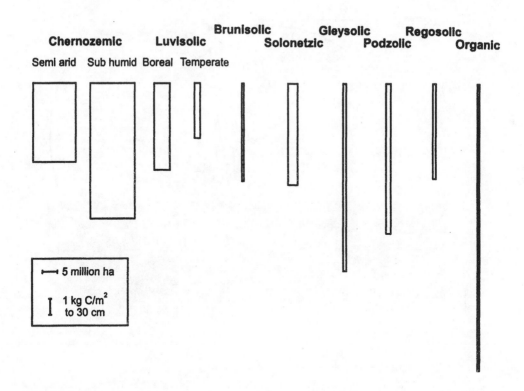

Figure 2. Relative carbon contents of cultivated soils in various Soil Orders. Approximate equivalents in U.S. taxonomic system are as follows: Chernozemic: Boroll; Luvisolic: Boralf and Udalf; Brunisolic: Inceptisol; Solonetzic: Mollisol and Alfisol (Natric great groups); Gleysolic: Aqusuborders; Podzolic: Spodosol; Regosolic: Entisol; Organic: Histosol. (From Canada Soil Survey Committee, 1978.)

change in management, however, can disrupt that condition, resulting in a temporary gain or loss of stored C until the ratio of C input/decomposition again approaches 1 (Figure 4).

III. Current Trajectory of Soil Carbon

An assessment of current SOC trajectory requires an understanding of previous changes that influence the present status. In this section, we first provide a review of historical trends in SOC and then try to describe the current SOC trajectory.

A. Historical Losses of Soil Organic Carbon

Most of the arable lands in Canada were first cultivated within the last century. The conversion to arable agriculture almost invariably resulted in a loss of SOC. Recent estimates suggest that surface soil layers typically lost about 25% of the C present prior to cultivation (Table 1). For example, in an extensive comparison of cultivated and uncultivated soils in Alberta, McGill et al. (1988) reported

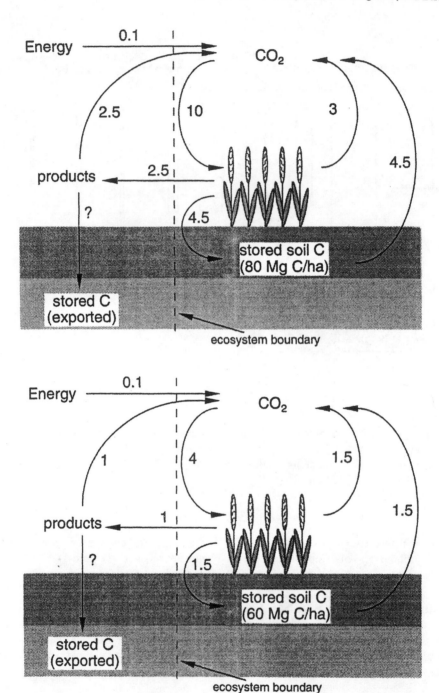

Figure 3. Approximate annual carbon fluxes in a corn system under humid conditions in southern Ontario (a) and in a wheat system under semi-arid conditions in western Canada (b). Actual values will vary widely among years, cropping practices, and growing conditions. Fluxes are shown in units of Mg C ha^{-1} yr^{-1}. Stored soil C refers to C stored in the surface 30 cm layer. The flows assume complete conversion of harvested products to CO_2 via respiration (e.g., human or animal respiration), though a small portion of this C may accumulate in other ecosystems (e.g., as sewage sludge).

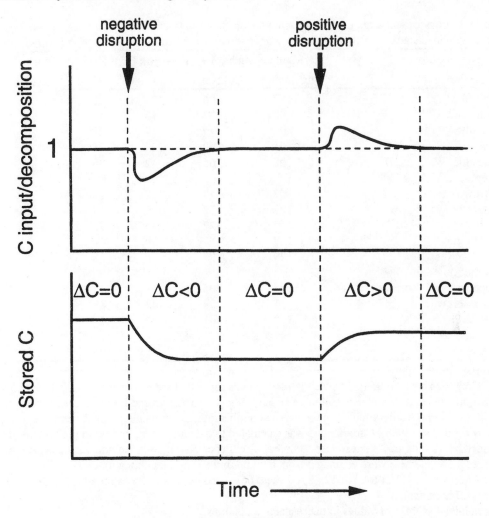

Figure 4. Conceptual view of the impact of negative and positive disruptions induced by a change in management on the C input/C loss ratio and on stored C. Similar illustrations have been presented by Johnson (1995).

losses of organic carbon from the A horizon ranging from 15 to 30%, depending on soil zone. A recent survey of 66 paired profiles at 12 sites in Saskatchewan (Anderson, 1995) indicated that the SOC content of cultivated sites was, on average, about 18 Mg C ha^{-1} less than that of native sites, suggesting a loss of about 17%. In many of these soils, erosion losses were small, based on ^{137}Cs analyses, indicating that most of the loss occurred via mineralization. Recent analysis of paired sites in eastern Canada also suggested losses of SOC from surface soil layers in the range of about 15 to 30% (Ellert and Gregorich, 1996; Gregorich et al., 1995a). Most of these estimates are appreciably lower than values often cited previously, largely because earlier measurements were biased by changes in bulk density and dilution of SOC in surface layer during plowing (McGill et al., 1988). Earlier estimates did not account for variation in soil mass between arable and uncultivated sites (Ellert and Bettany, 1995), nor did they adequately account for redistribution of SOC via erosion.

The loss of SOC upon conversion to arable agriculture is traditionally attributed to the physical effect of tillage, which can disrupt soil aggregates and expose previously inaccessible materials to

Table 1. Loss of soil organic C upon conversion of various grassland or forest soils to arable agriculture in Canada, as estimated from comparison between cultivated and non-cultivated sites

Soil order	Soil C (Mg ha^{-1})		Soil C loss		Source[a]
	Uncultivated	Cultivated	Mg C ha^{-1}	%	
Gleysolic	133	106	27	20	1
Brunisolic	98	83	15	15	1
Podzolic	93	74	19	20	1
Luvisolic (E. Canada)	64	40	24	38	1
Luvisolic (W. Canada)	108	66	42	39	2
Chernozemic					
Brown	67	44	23	34	2
Dark brown	107	78	29	27	2
Black	162	127	35	22	2
Dark gray	127	97	30	24	2
(Peace River)	129	93	36	28	2
Chernozemic					
Brown	105	108	-3	-3	3
Dark brown	101	58	43	43	3
Black	115	101	14	12	3
Mean[b]			26	25	

[a]Sources are as follows:

1. Gregorich et al. (1995a). 22 sites sampled; values presented are means for each soil order. Data were corrected to an 'equivalent soil mass' of 3500 Mg soil ha^{-1} (Ellert and Bettany, 1995). Numerical columns 1 and 4 were presented in the original publication; the remaining values were calculated.

2. McGill et al. (1988). Carbon mass was calculated to a depth corresponding to the bottom of the B horizon. A total of 72 sites were sampled. The 'Peace River' soils include various soil orders. Values presented here were estimated from a graph in the original publication.

3. Anderson (1995). A total of 12 paired sites (66 paired pedons) were sampled to depth of solum (A and B horizons).

[b]Unweighted average of values in the respective columns.

rapid decomposition. Although this process is undoubtedly a factor, the SOC loss upon cultivation probably reflects multiple effects. Indeed, the primary reason for C loss may be the enhanced removal of C from agroecosystems; since the intent of agriculture is to trap C in a marketable form, C inputs in agricultural systems are usually lower than those in native systems (e.g., Voroney et al., 1981). Agricultural systems may also enhance opportunities for decomposition by elimination of constraints within natural systems. Thus, conversion of arid soils from grassland to annual crops allows the accumulation of soil moisture between crops, thereby favoring decomposition (Janzen et al., 1997a). Finally, the composition of plant litter entering the soil in arable agroecosystems may be very different from that in native systems. For example, Gregorich et al. (1996b) found that forest leaves had a higher content of high molecular weight compounds than did maize leaves. In view of all these factors, loss of SOC upon conversion to arable agriculture should be viewed as the broad result of replacing one ecosystem with another, rather than the simple effect of cultivation *per se*.

Much of the SOC loss observed upon cultivation occurs in the 'young' or 'labile' organic matter fractions. For example, cultivation of forest soils in eastern Canada resulted in the disproportinate loss of light fraction C, a SOC pool believed to be largely comprised of partially decomposed litter (Gregorich et al., 1995a). Carbon-dating analyses have provided a direct measure of the loss of

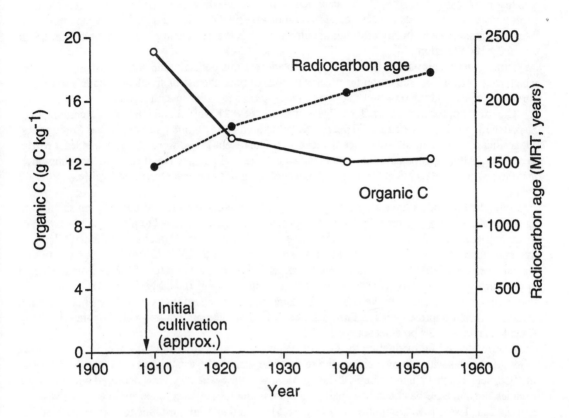

Figure 5. Change in the concentration and mean residence time of organic C during the decades after cultivation of a soil in southern Alberta, based on analysis of archived soil samples (adapted from Ellert, unpublished).

'young' organic matter upon cultivation (Anderson, 1995). For example, Ellert (unpublished) observed that the mean residence time of SOC increased with time following cultivation, particularly in a system with regular fallow (Figure 5). Because SOC loss occurs predominantly at the expense of decomposable or 'young' SOC pools, highest rates of SOC decline are typically observed shortly after initial cultivation (Monreal and Janzen, 1993; Tiessen and Stewart, 1983).

Based on recent surveys, Canadian agricultural soils have been a significant source of CO_2 in the past. For example, the loss of 25% of the SOC in grassland soils of western Canada would have resulted in the emission of about 1.3 Pg of C into the atmosphere (Ellert 1993, unpublished estimate). Anderson (1995) estimated the net efflux of CO_2-C from western Canadian agriculture over the last century to be about 0.5 to 1 Pg. A loss of 1 Pg, from about 80% of the area of agricultural soils in Canada, amounts to about 8 years of CO_2 production from fossil fuel combustion in Canada at the current rate.

B. Current Rate of SOC Loss

The historical losses of C from Canadian agricultural soils have now been widely established. A more pressing and pertinent question is whether these soils are still a net source of CO_2. Given that highest rates of emission occur shortly after initial cultivation, has the decline in stored C abated after about a century of cultivation?

Three general approaches have been used to address this question: the analysis of adjacent sites which have been under cultivation for varying periods of time, the use of simulation models to predict current changes, and the repeated sampling of long-term sites over several decades.

The first approach was used by Tiessen et al. (1982) who found that the C content of a Saskatchewan soil cultivated for 90 years was less than that of a similar soil cultivated for 60 years, suggesting continued losses over recent decades. The recent losses were attributed, at least partly, to erosion. Although this approach offers valuable historical perspective, it rests on the assumption that the two sites had identical C contents at the beginning of the measurement interval (e.g., after 60 years).

Simulation models have also been used to predict the current rates of net CO_2 emission from agricultural soils. For example, Smith et al. (1997) used the Century model to predict soil C dynamics in the agricultural soils of Canada. According to their estimates, agricultural soils have lost, on average, about 22% of their C content (0-30 cm) from 1910 to 1990, a value comparable to that presented earlier (Table 1). The model predicted, furthermore, that net rates of C loss from Canadian agricultural soils in 1990 were close to zero, averaging about 0.04 Mg C ha^{-1} yr^{-1}. Given the uncertainties associated with the use of models across diverse regions, and the possible offsetting effects of recent improvements in farm practices, these predictions suggest that agricultural soils in Canada may no longer be a net source of CO_2.

A third method for evaluating the current trajectory of soil C dynamics is the repeated sampling of long-term sites with a known cropping history. One such study, established at Lethbridge, Alberta in 1910, suggests that much of the SOC loss in various fallow-wheat systems occurred within several decades and that SOC has been reasonably constant thereafter, at least in the surface layer (Monreal and Janzen, 1993; Ellert, unpublished) (Figure 5). Another study, established on a forest soil at Breton, Alberta in 1930, showed either no appreciable change or an increase in SOC over time in various forage systems typical of those used in the area (R.C. Izaurralde, personal communication). Other studies of shorter duration (e.g., Nyborg et al., 1995; Liang and MacKenzie, 1992; Campbell et al., 1995) also suggest that SOC content is often maintained or even increased at least under good management practices.

From these observations, we infer that the decline in SOC resulting from the conversion of soils to arable agriculture has abated to low rates in many Canadian soils. This conclusion, in agreement with other recent reports (Anderson, 1995; Ellert and Gregorich, 1996), is consistent with the view that most SOC loss upon conversion of a soil to arable agriculture occurs within 20 years (Bowman et al., 1990).

A corollary observation is that current trajectories in soil C dynamics are largely a function of present management practices. Thus, we submit, the effects of present or future cropping practices usually overshadow any lingering effects of the initial cultivation of these soils.

IV. Management Practices That Promote C Gain

A gain in SOC can be prompted by adopting a new management practice that increases C inputs relative to C loss. The management change must elicit at least one of the following effects: increased primary production, increased proportion of the primary production returned to the soil, or suppressed rate of decomposition.

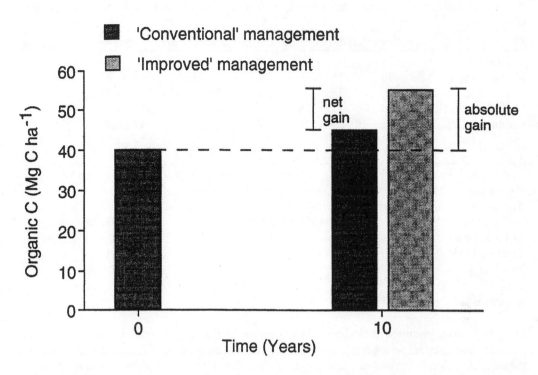

Figure 6. Alternative methods of measuring the change in soil organic carbon (SOC) in response to a new management practice. One way, which estimates the absolute gain, is to measure SOC at the same site over time after adoption of the new practice. Another approach, which estimates the net gain, is to measure the difference in SOC between the new system and a conventional system maintained alongside. The 'net gain' may be smaller or larger than the absolute gain, depending on the temporal change of the 'conventional system'.

Recent interest in C cycling has spawned a flurry of studies seeking to measure SOC changes in response to adoption of an 'improved' management practice. Evaluation of C gains in response to a management change can be performed in several ways (Figure 6):
1. Measurement of changes in SOC by repeated analysis over time after the adoption of the new practice
2. Quantification of the difference in SOC between the 'new' practice (e.g., no-till system) and 'control' treatment (e.g., conventional tillage system) which is assumed to represent the status quo. If these studies are of sufficient duration for SOC to approach steady state, then the difference can be assumed to represent the eventual gain in SOC upon adoption of the 'new' practice.

In this review, we present results from studies presenting both types of data, though most are from the latter approach.

The various 'C-sequestering' practices identified can be categorized into the following strategies: intensified cropping systems, adoption of no-tillage agriculture, improved crop nutrition, organic amendment, and more extensive use of perennial crops. Although the effects of many of these practices are interactive, we present here a brief review of recent studies measuring the individual impacts of these practices on SOC.

Table 2. Partial list of potential gains in soil C in various crop rotations relative to those in fallow-wheat systems

Site	Years	Depth (cm)	ΔC^a (Mg ha^{-1})	Treatment[b]	Baseline[b]	Reference[c]
Melfort, Sk.	31	15	3.0	W (N,P)	FW (N,P)	1
Indian Head, Sk.	30	15	3.6	W (N,P)	FW (N,P)	2
	30	15	3.4	FWWHHH	FW (N,P)	
Swift Current, Sk.	24	15	3.4	W (N,P)	FW (N,P)	3
Lethbridge, Ab.	41	30	2.1	W (unfert.)	FW (unfert.)	4
	41	30	5.9	FWWHHH	FW (unfert.)	

[a] ΔC = (SOC in 'treatment') - (SOC in 'baseline'). Differences in SOC may not all be statistically significant; readers are referred to the references cited for detailed information.
[b] Designations are as follows; W = spring wheat, F = fallow, H = hay (each letter represents one phase of the rotation); N,P = nitrogen and phosphorus fertilizers applied; unfert. = no fertilizer applied.
[c] Sources are as follows - 1: Campbell et al., 1991c; 2: Campbell et al., 1991b, 1997; 3: Campbell and Zentner, 1993; 4: Bremer et al., 1994.

A. Intensified Cropping Systems

Cropping systems that increase the frequency or duration of plant growth favor higher SOC contents. In Canada, the best opportunity for intensification of cropping systems may be the elimination of summer fallow, a practice whereby land is left unplanted for an entire season, usually to replenish soil moisture. Although its use has gradually diminished in recent years, the current area devoted to summer fallow annually is about 6 million ha (Dumanski et al., 1997), most of it in the semi-arid and arid regions of western Canada. Typically, land is fallowed once every 2 to 4 years, in rotation with annual grains or oilseeds.

Numerous studies have compared SOC content in cropping systems with and without summer fallow. After several decades, C contents of soil under continuous wheat (i.e., without fallow) are generally about 2 to 5 Mg ha^{-1} higher than those in alternate fallow-wheat systems (Table 2). Consequently, conversion of land from systems with frequent summer fallow to continuous cropping would be expected to gain a similar amount. Although some gains in SOC might be anticipated by partial reduction of fallowing frequency (e.g., including fallow every three years instead of every two years), several studies in western Canada demonstrate little benefit without complete elimination of fallow (Campbell et al., 1991b; Campbell and Zentner, 1993; Bremer et al., 1995).

B. Reduction in Tillage Intensity

A large number of studies have been initiated recently across Canada to investigate the potential benefits of reduced tillage intensity for SOC content (Arshad et al., 1990; Carter et al., 1988; Angers et al., 1993a,b, 1997; Carter, 1991a, 1992; Grant and Lafond, 1994; Campbell et al., 1995a, 1996a,b; Franzluebbers and Arshad, 1996a,b,c). Most of these studies have shown that the adoption of no-till or minimum tillage practices result in redistribution of SOC in the soil profile, with an accumulation of SOC near the surface. Furthermore, adoption of reduced tillage usually results in an increase in labile SOC fractions near the surface, including microbial biomass, light fraction C, and carbohydrates.

Although a reduction in tillage intensity usually results in higher SOC concentrations in surface layers, its impact on total SOC in the profile is not as consistent (Table 3). A number of studies have shown significant accumulations, amounting to as much as 7.5 Mg C ha^{-1} after about 10 years. Other

Table 3. Partial list of potential gains in soil C in no-till or reduced tillage treatments relative to conventionally-tilled treatments. Additional examples from eastern Canada, including more recent data from several sites presented here, were recently reported by Angers et al., 1997

Site	Years	Depth (cm)	ΔC^a (Mg ha^{-1})	Treatment^b	Baseline^b	Reference^c
1. Stewart Valley, Sk.	11	15	3.9	No-till (W)	Tilled (W)	1
2. Cantur, Sk.	11	15	0	No-till (W)	Tilled (W)	2
3. Swift Current, Sk.	12	15	1.6	No-till (W)	Tilled (W)	1,3
	12	15	0	No-till (FW)	Tilled (FW)	3
4. Breton, Ab.	11	15	7.5	No-till	Tilled	4
5. Ellerslie, Ab.	11	15	0	No-till	Tilled	4
6. Lethbridge, Ab.	16	15	0	No-till (FW)	Tilled (FW)	5
7. Lethbridge, Ab.	8	15	2.0	No-till (W)	Tilled (W)	5
8. St-Lambert, Que.	11	24	0	Min-till	Plowed	6
9. Ste. Anne de Bellevue, Que (SL)	7	20	0	No-till	Tilled	7
10. Ste. Anne de Bellevue, Que (C)	7	20	0	No-till	Tilled	7
11. Charlottetown, PEI	3	~25	-1.7	Direct seeded	Plowed	8
12. Charlottetown, PEI	4	~25	3.0	Direct seeded	Plowed	8
13. N. Alberta/B.C.	7	20	0	No-till	Tilled	9
(four sites)	16	20	0	No-till	Tilled	9
	4	20	7.2	No-till	Tilled	9
	6	20	0	No-till	Tilled	9

$^a\Delta C$ = (SOC in 'treatment') - (SOC in 'baseline'). A value of '0' indicates no significant difference; in some cases, the statistical significance was uncertain - readers are referred to the original publication for detailed informantion.
^{b}W and FW denote continuous wheat and fallow-wheat, respectively.
^{c}Sources are as follows - 1: Campbell et al., 1996b; 2: Campbell et al., 1996a; 3: Campbell et al., 1995; 4: Nyborg et al., 1995; 5: Larney et al., 1997; 6: Angers et al., 1993a, 1995; 7: O'Halloran, 1993; 8: Angers and Carter, 1996; 9: Franzluebbers and Arshad, 1996a,b.
^{d}Revised from value presented in reference 1.
Notes (by reference): 3 - values are means for three sampling times (1986, 1990, 1994); 4 - values shown are for treatments with straw retained and N fertilizer added; 5 (FW system) - value shown is for a comparison of a stubble-mulch and no-till treatment (no-till had significantly higher SOC than intensively-tilled systems); 7 - the study showed no effect of tillage on SOC concentration in the 0-10 and 10-20 cm soil layers (bulk density and Mg C ha^{-1} values were not reported); 8 - statistical significance was not reported.

studies, however, show no measurable benefits of reduced tillage when the entire profile was considered. Indeed Angers and Carter (1996) proposed that, in cool-humid conditions, tillage may sometimes even favor long-term SOC retention via its mixing effect. They concluded that adoption of reduced tillage in humid, cool soils would affect primarily the distribution of SOC in the profile, unless C inputs were increased.

The variability in SOC response to reduced tillage among studies may be attributable, in part, to the inherent difficulties of quantifying small changes in large pools of SOC (e.g., detecting a change of 2 Mg C ha^{-1} in a soil containing 80 Mg C ha^{-1}). The inconsistencies, however, may also indicate that the effect of tillage interacts with that of other factors. For example, benefits of reduced tillage presumably depend on the initial SOC status of the particular site (Janzen et al., 1997b). Soil texture has been implicated as another important factor; Campbell et al., (1996b) demonstrated a strong positive link between soil clay content and SOC gain upon adoption of no-till systems. Also, the potential SOC gain may depend on prior tillage practices; a soil previously subject to very intensive tillage (e.g., moldboard plowing) may show greater SOC gain upon adoption of no-tillage than a soil which had been under minimum tillage (e.g., stubble-mulch). Finally, the benefit of reduced tillage may depend on the cropping system; for example, frequent summer fallow may minimize the SOC gains associated with no-tillage (Larney et al., 1997). Because of these numerous interactions, adoption of reduced tillage, on its own, may not necessarily elicit SOC gain.

C. Improved Crop Nutrition

Alleviation of nutrient deficiencies by the addition of nutritive amendments can enhance crop residue inputs and, thus, SOC contents (e.g., Campbell et al., 1991a; Liang and McKenzie, 1992; Campbell and Zentner, 1996). Such increases typically amount to several Mg C ha^{-1} after one or more decades (Campbell et al., 1991b; Campbell and Zentner, 1993; Nyborg et al., 1995; Gregorich et al., 1996a). In soils with high indigenous fertility, however, SOC response to nutrient application may be negligible (e.g., Campbell et al., 1991c; Nyborg et al., 1995).

Much of the increase in SOC usually occurs in labile fractions like light fraction SOC (Janzen et al., 1992; Biederbeck et al., 1994; Gregorich et al., 1997). For example, Gregorich et al. (1996a) observed that light fraction organic matter accounted for a significant part of the SOC gain in response to fertilization (Figure 7). Furthermore, the gain in light-fraction C was largely derived from recent residue additions, as determined by ^{13}C techniques.

Earlier studies suggested that application of N might accelerate decomposition of SOC. In a recent study using ^{13}C techniques, however, application of N fertilizer was shown to have no measurable effect on the breakdown of indigenous SOC (Gregorich et al., 1996a, 1997).

This brief review of Canadian studies provides strong evidence that fertilized soils generally have higher SOC contents than unfertilized soils. Many soils in Canada are already fertilized at rates approaching optimum, however, and the differences observed between fertilized and unfertilized systems probably overestimate potential gains in many soils.

D. Organic Amendments

Application of organic materials can also elevate SOC. For example, regular applications of livestock manure can induce substantial increases over the course of a few years (Sommerfeldt et al., 1988; N'dayegamiye, 1990; Angers and N'dayegamiye, 1991). Similar responses have been observed to the addition of other organic amendments like wood residues (N'dayegamiye and Angers, 1993). Although organic amendments usually have positive net effect on SOC, one recent study suggests that application of high amounts of soluble C (e.g., extracts from composted manure) can accelerate indigenous SOC decomposition (Liang et al., 1996).

Not all of the SOC gains attributed to organic amendments result in net removal of atmospheric CO_2. In cases where the organic C is imported from elsewhere (e.g., manure from another agro-ecosystem) a portion of the apparent SOC gains occurs simply through transfer or redistribution of C. Only to the extent that the amendment enhances primary production, by improving soil structure or fertility, can the increase in SOC be considered a true C gain.

The handling of crop residues also has an impact on net C gains. For example, removal of straw or stover can result in significant loss of SOC (Nyborg et al., 1995; Campbell et al., 1991a,1997). The net effect of that practice on atmospheric C budgets, however, depends on the disposition of the residues; for example, if they are used as bedding for livestock, then much of the C may be returned to the soil as manure.

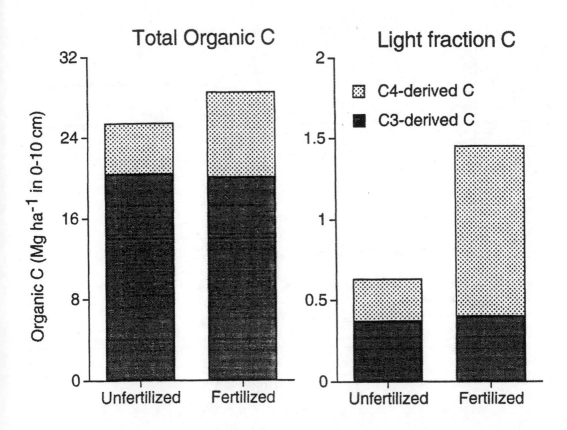

Figure 7. Amounts of total organic C and C_4-C in the whole soil and light fraction (LF) C in the surface 0-10 cm of fertilized and unfertilized soil after 32 years under corn. (From Gregorich et al., 1996a.)

E. Increased Use of Perennial Crops

The use of cropping systems that include perennial forage legumes or grasses can preserve or enhance SOC storage in soils. For example, Angers (1992) reported significant gains in SOC concentration of soil within 5 years of establishing alfalfa in eastern Canada (Figure 8). Similar beneficial effects have been observed in western Canada (Bremer et al., 1994; Campbell et al., 1991b).

Perhaps the most effective method of inducing SOC gains is the reversion of lands to permanent grassland. Dormaar and Smoliak (1985) observed progressive increases in SOC of soils which had been cultivated and then abandoned to grassland. Recent efforts to re-seed grasses onto marginal agricultural lands may also promote SOC storage, though few comprehensive studies of this effect have been completed in Canada. Bremer et al. (1994) observed apparent accumulation of SOC 8 years after seeding native grasses to land which had previously been in various fallow-wheat systems. The accumulation of SOC upon reversion to grassland depends on adequate availability of other nutrients, notably nitrogen. In the absence of fertilizer N or reserves of inorganic N in the soil profile, the accumulation of SOC may be dependent on relatively small inputs of N from biological N_2 fixation or atmospheric deposition.

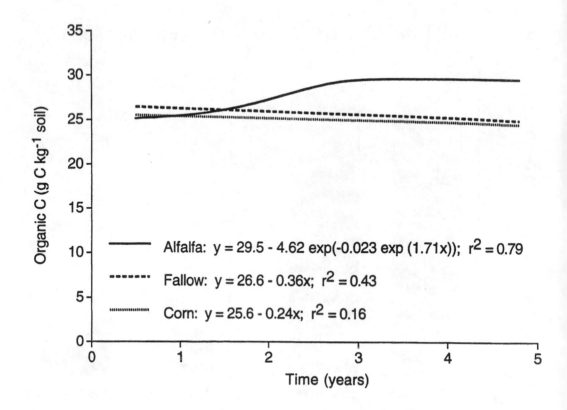

Figure 8. Changes in carbon content of a clay soil (5-15 cm) at La Pocatiere, Quebec in response to adoption of various cropping systems. Curves were derived by regression analysis. (Adapted from Angers, 1992.)

The preceding overview demonstrates appreciable complexity in SOC response to soil management practices. The diversity in response, presumably, reflects interactive effects of various management practices, as well as their interaction with other factors like initial SOC status, soil type, and climate. Because of this complexity, the estimation of SOC gains based on assessment of cropping practices is probably a grave oversimplification. A preferred method may be to identify the mechanisms of SOC gain, determine the conditions which promote SOC accrual, and finally select management systems that favor those conditions. In the next section, therefore, we present briefly several mechanisms that may enhance SOC in Canadian agricultural soils.

V. Mechanisms of SOC Gain

Soil organic matter is composed of a heterogeneous mixture of chemical structures, often in association with soil minerals. These diverse forms can be broadly categorized into three groups:
1. **Plant litter**, consisting of photosynthetically-assimilated materials, minimally affected by decomposition, with cellular structures still recognizable. This fraction may originate from the

residues and roots of crops grown at the site, or from organic amendments imported from elsewhere.

2. **Inert SOC**, comprised of decomposition products which, because of chemical configuration or association with soil minerals, are essentially inaccessible to agents of biological decay (Hsieh 1992, 1993). Carbon in this fraction typically has a turnover time of more than 1000 years (Campbell et al., 1967; Harrison et al., 1993; Scharpenseel and Becker-Heidmann, 1994), and is largely unaffected by management practices imposed on the soil.

3. **Dynamic SOC**, consisting of photosynthetically-reduced C in various stages of transition from plant litter to CO_2 (or inert SOC). This fraction includes any SOC which is inherently decomposable. By definition, therefore, it includes C in faunal or microbial biomass, most of which originated in plant litter. Various names have been assigned to describe portions of this fraction, including 'light fraction' OM (Gregorich and Janzen, 1996), particulate OM (Cambardella and Elliott, 1992), macro-organic matter (Gregorich and Ellert, 1993), mineralizable C (Campbell, 1978), coarse OM (Tiessen et al., 1994), and OM in macroaggregates (Buyanovsky et al., 1994; Angers and Giroux, 1996). These fractions typically have turnover times ranging from a few years to several decades. For example, light fraction OC typically has a half-life of about 10 years (e.g., Gregorich et al., 1995c, 1996b; Ellert, unpublished data).

A. Changes in Inert SOC

From the standpoint of atmospheric CO_2 removal, the ideal reservoir for C gains is the inert SOC. By definition, however, the size of this pool changes only very slowly. Furthermore, the rate and magnitude of its change may likely be influenced more by inherent soil and climatic conditions than by agronomic practices. Consequently, the potential for adopting management practices that favor significant inert SOC gains over time scales of several years or decades may be limited. While the importance of inert C accumulation is highly significant from the perspective of soil genesis, its rate is likely to be too slow to materially offset atmospheric CO_2 concentration.

B. Changes in Dynamic SOC

A more probable repository for C gains in the short term is the dynamic SOC pool. As indicated earlier, this pool includes C in transition between plant litter and CO_2 (and inert C). Assuming that the rate of inert SOC formation is negligible over the course of several decades, then dynamic C can be viewed simply as an intermediate in the following reaction:

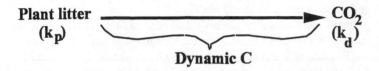

$$\text{Plant litter} \xrightarrow{\quad (k_p) \quad} \underbrace{\qquad\qquad}_{\text{Dynamic C}} \xrightarrow{\quad (k_d) \quad} CO_2$$

The size of the dynamic SOC pool, therefore, depends on the relative rate of two processes: rate of plant litter C input (k_p), and rate of CO_2 formation (k_d).

The rate of plant litter input in agroecosystems is closely related to crop yield. Numerous studies have shown strong correlations between crop residue inputs and SOC contents (e.g., Campbell and Zentner 1993; Biederbeck et al., 1994; Nyborg et al., 1995; Gregorich et al., 1996a). Many of the SOC gains in response to improved management practices can be directly linked to higher yields arising from better crop nutrition, more efficient water utilization, and higher yielding crops. In part, the variable response of SOC to a given management change depends on whether the new practice elicits a yield response. For example, under semiarid conditions of western Canada, adoption of no-tillage can maintain or enhance crop yields (Lafond et al., 1992) because of greater moisture retention, thereby favoring higher SOC (Campbell et al., 1995). Under humid conditions like those in eastern Canada, however, reduced tillage may have little yield advantage and therefore elicit only limited gains in SOC (Angers et al., 1995; Anger and Carter 1996).

Plant litter input is determined not only by the crop yield, but also by the proportion returned to the soil after harvest. For example, production of corn for silage, where most of the above-ground portion is harvested, returns few residues, resulting in loss of SOC (Angers et al., 1995). Higher return of plant litter can also explain the benefits to SOC of straw retention (Nyborg et al., 1995) and use of perennial forages, which usually have higher proportions of plant C below-ground.

But amount of plant litter alone cannot explain all of the management effects on SOC storage. At least as important as the amount of C added to the soil is the rate at which it decomposes to CO_2. Any practice that suppresses the rate of decomposition lengthens the turnover time of dynamic SOC, thereby increasing its content in the soil. Suppression of decomposition rate can be achieved through one of two general mechanisms: 1. suppression of biological activity, and 2. physical protection.

Because decomposition is a biological process, any practice which reduces moisture content, temperature, or aeration will favor the accumulation of dynamic SOC. For example, frequent use of summer fallow results in SOC loss, in part because it creates moisture and temperature conditions conducive to biological activity (Janzen et al., 1992; Bremer et al., 1995; Janzen et al., 1997a). Similarly, application of fertilizer can retard decomposition by enhancing plant growth and desiccating the soil (Paustian et al., 1992). Placement of plant residues may affect decomposition by altering the physical environment. Under arid conditions, for example, no-till practices may slow decomposition by retaining residues on the soil surface where they remain desiccated.

The suppression of decomposition, by adoption of practices that retard biological activity, results in the accumulation of dynamic SOC which has little inherent stability against further breakdown. Consequently, it remains highly susceptible to decomposition should that practice be discontinued. For example, adoption of a fallow-wheat system in a semiarid environment resulted in rapid depletion of light fraction C within a decade, relative to that under continuous cropping (Bremer et al., 1995).

Reducing biological activity, however, may not be the only way of suppressing decomposition. Accumulation of SOC may also be favored by practices which encourage the formation of aggregates that limit accessibility to decomposition (Gregorich et al., 1989; Angers and Carter, 1996; Gregorich et al., 1997; Carter and Gregorich, 1996). For example, reduction in tillage intensity can increase soil aggregation and the amount of SOC stored within aggregates (Carter 1992; Angers et al., 1992, 1993c; Franzluebbers and Arshad, 1996a). Aggregation is also favored by reduction in fallow frequency (Campbell et al., 1993a,b) and by the use of perennial forages (Campbell et al., 1993b), though different species may have variable effects (Carter et al., 1994). Similar effects have been observed with extractable carbohydrate concentration (Angers and Mehuys, 1989; Angers et al., 1993c), which is highly correlated to aggregation (Baldock et al., 1987).

The C stored in aggregates represents a temporary storehouse of SOC, less susceptible to rapid depletion than 'free' dynamic SOC, but still subject to gradual turnover. Using ^{13}C analyses, Gregorich et al. (1997) found that the half-life of 'protected' light fraction OC was about 2-fold that of 'free' light fraction OC. Some findings have suggested that SOC stored within microaggregates is more effectively 'sequestered' than SOC within macroaggregates (Gregorich et al., 1989; Carter, 1996). Others, however, have shown little apparent difference in SOC breakdown among aggregates of various sizes (Gregorich et al., 1994). For example, Angers et al. (1995), using ^{13}C analyses, found after 11 years of silage corn that SOC from previous C3 crops within microaggregates decomposed as quickly as that within macroaggregates. The disruption of aggregates is one of the mechanisms proposed for lower SOC under tilled systems than in no-tillage systems.

C. Dynamic SOC as Early Indicator of SOC Gain

The dynamic SOC fractions invariably show much greater and more rapid response than total SOC to management practices. For example, differences in light fraction C among various cropping systems may vary by as much as 2- or 3-fold, compared to differences of only about 1.3-fold for total SOC concentration (Bremer et al., 1994; Biederbeck et al., 1994; Janzen et al., 1992). Similarly, aggregation, carbohydrate composition, potential C mineralization rate, and microbial biomass content are highly sensitive to changes in cropping, often within a few years of the change (Angers and Mehuys, 1990; Angers et al., 1993c; Carter 1986, 1991b; Campbell et al., 1992a, 1992b,1996a). Consequently, these fractions have been proposed as early indicators of SOC change (Gregorich et al., 1994).

The reliability of these indicators, however, depends on the mechanism of dynamic SOC accumulation. If it occurs because of higher C inputs, then it is reasonable to assume that it will eventually also be reflected in higher inert and total SOC content. If, however, the change in dynamic SOC occurs in response to a suppression of decomposition rate, then it simply represents a gain in transient SOC. In the latter case, the dynamic SOC is not an *indicator* of SOC change; it *is* the SOC change.

VI. Limits to SOC Gain

Under relatively constant conditions, SOC eventually approaches a steady-state concentration at which the rate of C input is balanced by CO_2 loss via respiration. A gain in SOC is prompted by adoption of a practice that disrupts this steady-state, suppressing decomposition relative to C input (Figure 4). With the accumulation of SOC, however, rate of decomposition will eventually again converge upon C input, at which time the SOC approaches a new steady-state. Thus a gain in SOC in response to a new management practice can occur only during the transition from one steady-state to another, and is therefore of finite magnitude and duration.

Losses of SOC upon adoption of a degradative cropping system are finite and approach zero after several decades (e.g., Bremer et al., 1995). The duration of SOC gains in response to improved practices, however, has not been firmly established. Campbell et al. (1995) showed that SOC upon adoption of improved management in an arid soil approached a maximum after only six years. Angers (1992), from a study under humid conditions, also suggested a new SOC plateau within 5 years of seeding a perennial forage. These studies support the view that most of the SOC gains in response to adoption of improved practices may occur within several years, and that gains may subside within a decade. Another study, evaluating SOC gains in response to forages in a soil with low initial SOC (Izaurralde, personal communication), suggested that maximum SOC was achieved only after several decades, perhaps in part because of liming and progressive increases in fertilization. The duration of potential SOC gains, therefore, may vary widely among sites and deserves further investigation.

The potential SOC gain may also vary widely among agroecosystems in Canada, depending on a range of factors. Among the most important, perhaps, is the potential primary production (C input), as dictated by climatic constraints. Thus, soils in areas with severe constraints to productivity (e.g., aridity) may have limited potential for SOC gains.

A second variable is the SOC status prior to the adoption of a new management practice. If the SOC, and, in particular, the dynamic SOC, is already at a high level for its environment, then there may be limited potential for further gains. For example, when a soil that had been in perennial forages was subsequently used for silage corn production under various treatments, it showed SOC losses in all treatments because of its very high initial SOC content (Angers et al., 1995). Similarly, on a soil that had been in manured, mixed rotations for several decades, adoption of continuous cropping, otherwise considered to be a C-conserving practice, only just maintained SOC over subsequent decades (Bremer et al., 1995). In estimating potential C gains from adoption of a new practice, therefore, the previous management system may be just as important a variable as the proposed new system. In theory, the soils which have the greatest potential for SOC gain are those which have been most depleted in dynamic SOC content.

A third factor that determines potential SOC gains is the soil's capacity for 'protecting' recent C inputs, whether directly by association with minerals or within aggregates, or indirectly by suppressing biological activity. For example, recent studies have proposed that potential SOC gains may be related to clay content; thus Campbell et al. (1996b) found that adoption of reduced tillage at three sites was directly related to soil clay content. Similarly, the results of Voroney et al. (1981) suggest that the interaction between SOC and clay particles might be of greater significance in protecting organic matter than the nature of residues.

VII. Potential SOC Gains in Canadian Agroecosystems

Given the complexity of SOC dynamics, potential gains over large regions can be only crudely estimated. For example, recent estimates suggest that potential SOC gains in western Canadian soils under suboptimal cropping practices may amount to about 3 Mg C ha^{-1} over one or more decades (Tables 2 and 3). Assuming that this amount could be gained in all cultivated Chernozemic soils in Canada (27 million ha), the increase in SOC would amount to about 81 Tg.

Gains in SOC, however, can be expected only in areas where improvements in soil management are feasible. For example, only those soils currently under systems with frequent fallow can gain maximum incremental SOC from conversion to continuous cropping practices. Dumanski et al. (1997) estimated that partial conversion of summer fallow into cereal cropland in western Canada would result in a total gain of 10 Tg C in the 0-30 cm soil layer after 15 years. Conversion of the same area of fallow to hay, according to their estimates, would enhance SOC content (0-30 cm) by about 47 Tg after 25 years.

Changes in SOC cannot be divorced entirely from other fluxes within the C cycle. Most Canadian agroecosystems depend on external energy inputs for manufacture of inputs (e.g., fertilizers), tractive power, and transportation. The emission of CO_2 from this energy expenditure also merits inclusion in the overall C cycle of a given agroecosystem. Annual CO_2 emission from energy use in Canadian agroecosystems is often about 0.1 Mg C ha^{-1}yr^{-1} (e.g., Coxworth et al., 1995). Thus, such a system will be a net sink for CO_2 until annual SOC gains fall below 0.1 Mg C ha^{-1}yr^{-1}.

Based on our general overview, we submit that soils in Canadian agroecosystems are no longer a source of CO_2 and have the potential to be a sink for CO_2 during the transition to more C-conserving practices. The magnitude of this SOC gain may more than offset current CO_2 emissions from Canadian agriculture. Once the soils approach a new steady state, and rates of SOC accumulation diminish; however, these agroecosystems may again become a small source of CO_2 equivalent to the emission of CO_2 from energy inputs.

VIII. Future SOC Dynamics

Soil OC contents tend to approach a steady state content if maintained under constant management and environmental conditions. But management systems are continually evolving, and other variables like climate may also be changing over time. Consequently, the projected steady-state OC content of a given soil may be perpetually revised.

One variable that may influence future SOC content is crop yield, which has shown consistent increase in recent decades because of improved cultivars, enhanced nutrient application, and more

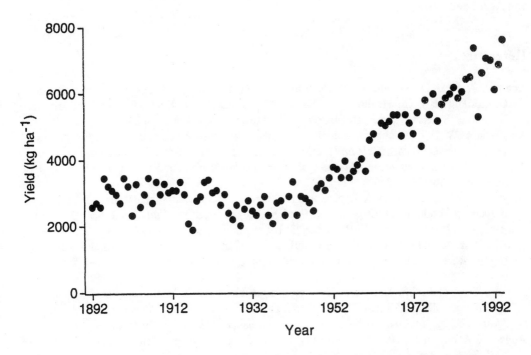

Figure 9. Long-term corn yield trend in Ontario, Canada. (Adapted from Tollenaar et al., 1995.)

intensive management practices (e.g., Figure 9; Janzen 1995). Further yield increases may favor higher SOC contents to the extent that they furnish higher residue C inputs to the soil.

Soil OC is affected not only by agronomic practices but also by various global changes. Projected changes whose influence on future SOC dynamics also deserve attention include changes to climate, atmospheric CO_2 concentration, ultraviolet-B intensity, and N deposition patterns.

IX. Conclusions

Historically, Canadian agricultural soils have been a significant source of atmospheric CO_2. Approximately 25% of the SOC originally present in the surface layer was lost to the atmosphere upon conversion to arable agriculture. Most agriculture soils in Canada have now been cultivated for a sufficient duration that any lingering effects of the initial cultivation are probably overshadowed by the impact of current management options. With the adoption of improved management practices, these soils have some potential for SOC gains, though such gains may be of finite magnitude and duration.

Acknowledgments

We acknowledge financial contributions from the Greenhouse Gas Research Initiative of Agriculture and Agri-Food Canada.

References

Acton, D.A. 1995. Development and effects of farming in Canada. p. 11-18. In: D.F. Acton and L.J. Gregorich (eds.), The Health of Our Soils. Publication 1906/E, Centre for Land and Biological Resources Research, Research Branch, Agriculture and Agri-Food Canada.

Anderson, D.W. 1995. Decomposition of organic matter and carbon emissions from soils. p. 165-175 In: R. Lal et al., (eds.), *Soils and Global Change*. Lewis Publishers, Boca Raton, FL.

Angers, D.A. 1992. Changes in soil aggregation and organic carbon under corn and alfalfa. *Soil Sci. Soc. Am. J.* 56:1244-1249.

Angers, D.A., N. Bissonnette, A. Légère, and N. Samson. 1993b. Microbial and biochemical changes induced by rotation and tillage in a soil under barley production. *Can. J. Soil Sci.* 73:39-50.

Angers, D.A., M.A. Bolinder, M.R. Carter, E.G. Gregorich, C.F. Drury, B.C. Liang, R.P. Voroney, R.R. Simard, R.G. Donald, R.P. Beyaert, and J. Martel. 1997. Impact of tillage practices on organic carbon and nitrogen storage in cool, humid soils of eastern Canada. *Soil Till. Res.* (in press).

Angers, D.A. and M.R. Carter. 1996. Aggregation and organic matter storage in cool, humid agricultural soils. p. 193-211 In: M.R. Carter and B.A. Stewart (eds.), *Structure and Soil Organic Matter Storage in Agricultural Soils*. Lewis Publishers. CRC Press, Boca Raton, FL.

Angers, D.A. and M. Giroux. 1996. Recently deposited organic matter in soil water-stable aggregates. *Soil Sci. Soc. Am. J.* 60:1547-1551.

Angers, D.A. and G.R. Mehuys. 1989. Effects of cropping on carbohydrate content and water-stable aggregation of a clay soil. *Can. J. Soil Sci.* 69:373-380.

Angers, D.A. and G.R. Mehuys. 1990. Barley and alfalfa cropping effects on carbohydrate contents of a clay soil and its size fractions. *Soil Biol. Biochem.* 22:285-288.

Angers, D.A. and A. N'dayegamiye. 1991. Effects of manure application on carbon, nitrogen, and carbohydrate contents of a silt loam and its particle-size fractions. *Biol. Fertil. Soil* 11:79-82.

Angers, D.A., A. N'dayegamiye, and D. Côté. 1993a. Tillage-induced differences in organic matter of particle-size fractions and microbial biomass. *Soil Sci. Soc. Am. J.* 57:512-516.

Angers, D.A., A. Pesant, and J. Vigneux. 1992. Early cropping-induced changes in soil aggregation, organic matter, and microbial biomass. *Soil Sci. Soc. Amer. J.* 56:115-119.

Angers, D.A., N. Samson, and A. Légère. 1993c. Early changes in water-stable aggregation induced by rotation and tillage in a soil under barley production. *Can. J. Soil Sci.* 73:51-59.

Angers, D.A., R.P. Voroney, and D. Côté. 1995. Dynamics of soil organic matter and corn residues affected by tillage practices. *Soil Sci. Soc. Am. J.* 59:1311-1315.

Arshad M.A., M. Schnitzer, D.A. Angers, and J.A. Ripmeester. 1990. Effects of till vs. no-till on the quality of soil organic matter. *Soil Biol. Biochem.* 22:595-599.

Baldock, J.A., B.D. Kay, and M. Schnitzer. 1987. Influence of cropping treatments on the monosaccharide content of the hydrolysates of a soil and its aggregate fractions. *Can. J. Soil Sci.* 67:489-499.

Biederbeck, V.O., H.H. Janzen, C.A. Campbell, and R.P. Zentner. 1994. Labile soil organic matter as influenced by cropping practices in an arid environment. *Soil Biol. Biochem.* 26:1647-1656.

Bowman, R.A., J.D. Reeder, and R.W. Lober. 1990. Changes in soil properties in a central plains rangeland soil after 3, 20, and 60 years of cultivation. *Soil Sci.* 150:851-857.

Bremer, E., B.H. Ellert, and H.H. Janzen. 1995. Total and light-fraction carbon dynamics for four decades after cropping changes. *Soil Sci. Soc. Amer. J.* 59:1398-1403.

Bremer, E., H.H. Janzen, and A.M. Johnston. 1994. Sensitivity of total, light fraction and mineralizable organic matter to management practices in a Lethbridge soil. *Can. J. Soil Sci.* 74:131-138.

Buyanovsky, G.A., M. Aslam, and G.H. Wagner. 1994. Carbon turnover in soil physical fractions. *Soil Sci. Soc. Amer. J.* 58:1167-1173.

Cambardella, C.A. and E.T. Elliott. 1992. Particulate soil organic-matter changes across a grassland cultivation sequence. *Soil Sci. Soc. Am. J.* 56:777-783.

Campbell, C.A. 1978. Soil organic carbon, nitrogen and fertility. p. 173-271. In: M. Schnitzer and S.U. Khan (ed.), *Soil Organic Matter*. Developments in Soil Science 8. Elsevier Scientific Publ. Co., Amsterdam.

Campbell, C.A., V.O. Biederbeck, R.P Zentner, and G.P. LaFond. 1991b. Effect of crop rotations and cultural practices on soil organic matter, microbial biomass and respiration in a thin Black Chernozem. *Can. J. Soil Sci.* 71:363-376.

Campbell, C.A., K.E. Bowren, M. Schnitzer, R.P. Zentner, and L. Townley-Smith. 1991c. Effect of crop rotations and fertilization on soil organic matter and some biochemical properties of a thick Black Chernozem. *Can. J. Soil Sci.* 71:377-387.

Campbell, C.A., S.A. Brandt, V.O. Biederbeck, R.P. Zentner, and M. Schnitzer. 1992a. Effect of crop rotations and rotation phase on characteristics of soil organic matter in a Dark Brown Chernozemic soil. *Can. J. Soil Sci.* 72:403-416.

Campbell, C.A., D. Curtin, S. Brandt, and R.P. Zentner. 1993a. Soil aggregation as influenced by cultural practices in Saskatchewan: II. Brown and Dark Brown Chernozemic Soils. *Can. J. Soil Sci.* 73:597-612.

Campbell, C.A., G.P. Lafond, R.P. Zentner, and V.O. Biederbeck. 1991. Influence of fertiliser and straw baling on soil organic matter in a thin Black Chernozem in western Canada. *Soil Biol. Biochem.* 23:443-446.

Campbell, C.A., B.G. McConkey, R.P. Zentner, F.B. Dyck, F. Selles, and D. Curtin. 1995. Carbon sequestration in a Brown Chernozem as affected by tillage and rotation. *Can. J. Soil Sci.* 75:449-458.

Campbell, C.A., G.P. Lafond, A.P. Moulin, L. Townley-Smith, and R.P. Zentner. 1997. Crop production and soil organic matter in long-term crop rotations in the sub-humid northern Great Plains of Canada. p. 297-315. In: E.A. Paul et al. (eds.), *Soil Organic Matter in Temperate Agroecosystems: Long-Term Experiments of North America*. CRC Press, Boca Raton, FL.

Campbell, C.A., B.G. McConkey, R.P. Zentner, F.B. Dyck, F. Selles, and D. Curtin. 1996b. Long-term effects of tillage and crop rotations on soil organic C and total N in a clay soil in southwestern Saskatchewan. *Can. J. Soil Sci.* 76:395-401.

Campbell, C.A., B.G. McConkey, R.P. Zentner, F. Selles, and D. Curtin. 1996a. Tillage and crop rotation effects on soil organic C and N in a coarse-textured Typic Haploboroll in southwestern Saskatchewan. *Soil Till. Res.* 37:3-14.

Campbell, C.A., A.P. Moulin, K.E. Bowren, H.H. Janzen, L. Townley-Smith, and V.O. Biederbeck. 1992b. Effect of crop rotations on microbial biomass, specific respiratory activity and mineralizable nitrogen in a Black Chernozemic soil. *Can. J. Soil Sci.* 72:417-427.

Campbell, C.A., A.P. Moulin, D. Curtin, and L. Townley-Smith. 1993b. Soil aggregation as influenced by cultural practices in Saskatchewan: I. Black Chernozemic soils. *Can. J. Soil Sci.* 73:579-595.

Campbell, C.A., E.A. Paul, D.A. Rennie, and K.J. McCallum. 1967. Applicability of the carbon-dating method of analysis to soil humus studies. *Soil Sci.* 104:217-224.

Campbell, C.A. and R.P. Zentner. 1993. Soil organic matter as influenced by crop rotations and fertilization. *Soil Sci. Soc. Amer. J.* 57:1034-1040.

Campbell, C.A. and R.P. Zentner. 1996. Response of soil organic matter to crop management. Technical Bulletin 1996-4E. Agriculture and Agri-Food Canada, Research Branch, Swift Current, Saskatchewan.

Canada Soil Survey Committee. 1978. The Canadian system of soil classification. Research Branch, Canada Department of Agriculture. Publication 1646.

Carter, M.R. 1986. Microbial biomass as an index for tillage-induced changes in soil biological properties. *Soil Till. Res.* 7:29-40.

Carter, M.R. 1991a. Evaluation of shallow tillage for spring cereals on a fine sandy loam. 2. Soil physical, chemical and biological properties. *Soil Till. Res.* 21:37-52.

Carter, M.R. 1991b. The influence of tillage on the proportion of organic carbon and nitrogen in the microbial biomass of medium-textured soils in a humid climate. *Biol. Fertil. Soils* 11:135-139.

Carter, M.R. 1992. Influence of reduced tillage systems on organic matter, microbial biomass, macro-aggregate distribution and structural stability of the surface soil in a humid climate. *Soil Till. Res.* 23:361-372.

Carter, M.R. 1996. Analysis of soil organic matter storage in agroecosystems. p. 3-11 In: M.R. Carter and B.A. Stewart (eds.), *Structure and Soil Organic Matter Storage in Agricultural Soils*. Lewis Publishers. CRC Press, Boca Raton, FL.

Carter, M.R., D.A. Angers, and H.T. Kunelius. 1994. Soil structural form and stability, and organic matter under cool-season perennial grasses. *Soil Sci. Soc. Am. J.* 58:1194-1199.

Carter, M.R. and E.G. Gregorich. 1996. Methods to characterize and quantify organic matter storage in soil fractions and aggregates. p. 449-466 In: M.R. Carter and B.A. Stewart (eds.), *Structure and Soil Organic Matter Storage in Agricultural Soils*. Lewis Publishers. CRC Press, Boca Raton, FL.

Carter, M.R., H.W. Johnston, and J. Kimpinski. 1988. Direct drilling and soil loosening for spring cereals on a fine sandy loam in Atlantic Canada. *Soil Till. Res.* 12:365-384.

Coxworth, E., M.H. Entz, S. Henry, K.C. Bamford, A. Schoofs, P.D. Ominski, P. Leduc, and G. Burton. 1995. Study of the effects of cropping and tillage systems on the carbon dioxide released by manufactured inputs to western Canadian agriculture: Identification of methods to reduce carbon dioxide emissions. Unpublished report, submitted to Agriculture and Agri-Food Canada.

Dormaar, J.F. and S. Smoliak. 1985. Recovery of vegetative cover and soil organic matter during revegetation of abandoned farmland in a semiarid climate. *J. Range Management* 38:487-491.

Dumanski, J., R.L. Desjardins, C. Tarnocai, C. Monreal, E.G. Gregorich, C.A. Campbell, and V. Kirkwood. 1997. Possibilities for future carbon sequestration in Canadian agriculture in relation to land use changes. Climate Change (in press).

Ellert, B.H. and J.R. Bettany. 1995. Calculation of organic matter and nutrients stored in soils under contrasting management regimes. *Can. J. Soil Sci.* 75:529-538.

Ellert, B.H. and E.G. Gregorich. 1996. Storage of carbon, nitrogen and phosphorus in cultivated and adjacent forested soils of Ontario. *Soil Sci.* 161:587-603.

Franzluebbers, A.J. and M.A. Arshad. 1996a. Water-stable aggregation and organic matter in four soils under conventional and zero tillage. *Can. J. Soil Sci.* 76:387-393.

Franzluebbers, A.J. and M.A. Arshad. 1996b. Soil organic matter pools during early adoption of conservation tillage in northwestern Canada. *Soil Sci. Soc. Amer. J.* 60:1422-1427.

Franzluebbers, A.J. and M.A. Arshad. 1996c. Soil organic matter pools with conventional and zero tillage in a cold, semiarid climate. *Soil Till. Res.* 39:1-11.

Grant, C.A. and G.P. Lafond. 1994. The effects of tillage systems and crop rotations on soil chemical properties of a Black Chernozemic soil. *Can. J. Soil Sci.* 74:301-306.

Gregorich, E.G., M.R. Carter, D.A. Angers, C.M. Monreal, and B.H. Ellert. 1994. Towards a minimum data set to assess soil organic matter quality in agricultural soils. *Can. J. Soil Sci.* 74:367-385.

Gregorich, E.G., C.F. Drury, B.H. Ellert, and B.C. Liang. 1997. Fertilization effects on physically protected light fraction organic matter. *Soil Sci.. Soc. Amer. J.* 61: (in press).

Gregorich, E.G. and B.H. Ellert. 1993. Light fraction and macroorganic matter in mineral soils. p. 397-407 In: M.R. Carter (ed.), *Soil Sampling and Methods of Analysis*. CRC Press, Bocan Raton, FL.

Gregorich, E.G., B.H. Ellert, D.A. Angers, and M.R. Carter. 1995a. Management-induced changes in the quantity and composition of organic matter in soils of eastern Canada. 1996. p. 273-283 In: M.A. Beran (ed.), Carbon Sequestration in the Biosphere, NATO ASI Series I, Vol. 33. Springer-Verlag, Berlin.

Gregorich, E.G., B.H. Ellert, C.F. Drury, and B.C. Liang. 1996a. Fertilization effects on soil organic matter turnover and corn residue C storage. *Soil Sci. Soc. Am. J.* 60:472-476.

Gregorich, E.G., B.H. Ellert, and C.M. Monreal. 1995c. Turnover of soil organic matter and storage of corn residue carbon estimated from natural ^{13}C abundance. *Can. J. Soil Sci.* 75:161-167.

Gregorich, E.G. and H.H. Janzen. 1996. Storage of soil carbon in the light fraction and macroorganic matter. p. 167-190 In: M.R. Carter and B.A. Stewart (eds.), *Structure and Soil Organic Matter Storage in Agricultural Soils*. Lewis Publishers. CRC Press, Boca Raton, FL.

Gregorich, E.G., R.G. Kachanoski, and R.P. Voroney. 1989. Carbon mineralization in soil size fractions after various amounts of aggregate disruption. *J. Soil Sci.* 40:649-659.

Gregorich, E.G., C.M. Monreal, M. Schnitzer, and H.-R. Schulten. 1996b. Transformations of plant residues into soil organic matter: Chemical characterization of plant tissue, isolated soil fractions, and whole soils. *Soil Sci.* 161:680-693.

Harrison, K.G., Broecker, W.S., and Bonani, G. 1993. The effect of changing land use on soil radiocarbon. *Science* 262:725-726.

Hsieh, Y.-P. 1992. Pool size and mean age of stable soil organic carbon in cropland. *Soil Sci. Soc. Am. J.* 56:460-464.

Hsieh, Y.-P. 1993. Radiocarbon signatures of turnover rates in active soil organic carbon pools. *Soil Sci. Soc. Am. J.* 57:1020-1022.

Janzen, H.H. 1995. The role of long-term sites in agroecological research: A case study. *Can. J. Soil Sci.* 75:123-133.

Janzen, H.H., C.A. Campbell, S.A. Brandt, G.P. Lafond, and L. Townley-Smith. 1992. Light fraction organic matter in soils from long-term crop rotations. *Soil Sci. Soc. Amer. J.* 56:1799-1806.

Janzen, H.H., C.A. Campbell, B.H. Ellert, and E. Bremer. 1997a. Soil organic matter dynamics and their relationship to soil quality. In: E.G. Gregorich and M.R. Carter (eds.), *Soil Quality for Crop Production and Ecosystem Health*. Elsevier Science Publ. Amsterdam, The Netherland. (submitted).

Janzen, H.H., C.A. Campbell, R.C. Izaurralde, B.H. Ellert, N. Juma, W.B. McGill, and R.P. Zentner. 1996b. North American Agricultural Soil Organic Matter Site Network: The Canadian Prairie. *Soil Till. Res.* (submitted).

Johnson, M.G. 1995. The role of soil management in sequestering soil carbon. p. 351-363. In: R. Lal et al. (eds.), *Soil Management and Greenhouse Effect*. Lewis Publishers, Boca Raton, FL.

Lafond, G.P., H. Loeppky, and D.A. Derksen. 1992. Effect of tillage system and crop rotations on soil water conservation, seedling establishment and crop yield. *Can. J. Plant Sci.* 72:103-115.

Larney, F.J., E. Bremer, H.H. Janzen, A.M. Johnston, and C.W. Lindwall. 1997. Changes in total, mineralizable and light fraction organic matter with cropping and tillage intensities in a semi-arid region. *Soil Till. Res.* (submitted).

Liang, B.C., E.G. Gregorich, M. Schnitzer, and R.P. Voroney. 1996. Carbon mineralization in soils of different textures as affected by water-soluble organic carbon extracted from composted dairy manure. *Biol. Fertil. Soils* 21:10-16.

Liang, B.C. and A.F. MacKenzie. 1992. Changes in soil organic carbon and nitrogen after six years of corn production. *Soil Sci.* 153:307-313.

McGill, W.B., J.F. Dormaar, and E. Reinl-Dwyer. 1988. New perspectives on soil organic matter quality, quantity, and dynamics on the Canadian Prairies. In: *Land Degradation and Conservation Tillage*, pp. 30-48, Proc. Annual Can. Soc. Soil Sci. Meeting, Calgary, Alberta.

Monreal, C. and H.H. Janzen. 1993. Soil organic carbon dynamics after eighty years of cropping a Dark Brown Chernozem. *Can. J. Soil Sci.* 73:133-136.

N'dayegamiye, A. 1990. Effets à long terme d'apports de fumier solide de bovins sur l'évolution des caracteristiques chimiques du sol et de la production de mais-ensilage. *Can. J. Plant Sci.* 70:767-775.

N'dayegamiye, A. and D.A. Angers. 1993. Organic matter characteristics and water-stable aggregation of a sandy loam soil after 9 years of wood-residue applications. *Can. J. Soil Sci.* 73:115-122.

Nyborg, M., E.D. Solberg, S.S. Malhi, and R.C. Izaurralde. 1995. Fertilizer N, crop residue, and tillage alter soil C and N content in a decade. p. 93-99 In: R. Lal et al. (eds.), *Soil Management and Greenhouse Effect*. Lewis Publishers, Boca Raton, FL.

O'Halloran, I.P. 1993. Effect of tillage and fertilization on inorganic and organic soil phosphorus. *Can. J. Soil Sci.* 73:359-369.

Paustian, K., W.J. Parton, and J. Persson. 1992. Modeling soil organic matter in organic-amended and nitrogen-fertilized long-term plots. *Soil Sci. Soc. Am. J.* 56:476-488.

Scharpenseel, H.W. and P. Becker-Heidmann. 1994. Sustainable land use in the light of resiliency/elasticity to soil organic matter fluctuations. p. 249-264 In: D.J. Greenland and I. Szabolcs (eds.), *Soil Resilience and Sustainable Land Use*. CAB International.

Smith, W.N., P. Rochette, C. Monreal, R.L. Desjardins, E. Pattey, and A. Jaques. 1997. Estimated net carbon dioxide balance from Canadian agroecosystems - 1990. *Can. J. Soil Sci.* 77: (in press).

Sommerfeldt, T.G., C. Chang, and T. Entz, 1988. Long-term annual manure applications increase soil organic matter and nitrogen, and decrease carbon to nitrogen ratio. *Soil Sci. Soc. Am. J.* 52:1668-1672.

Tiessen, H., E. Cuevas, and P. Chacon. 1994. The role of soil organic matter in sustaining soil fertility. *Nature* 371:783-785.

Tiessen, H. and J.W.B. Stewart. 1983. Particle-size fractions and their use in studies of soil organic matter: II. Cultivation effects on organic matter composition in size fractions. *Soil Sci. Soc. Amer. J.* 47:509-514.

Tiessen, H., J.W.B. Stewart, and J.R. Bettany. 1982. Cultivation effects on the amounts and concentration of carbon, nitrogen, and phosphorus in grassland soils. *Agron. J.* 74:831-835.

Tollenaar, M., S.P. Nissanka, A. Aguilera, and L. Dwyer. 1995. Improving stress tolerance: the key to increased corn yields. Agricultural Research in Ontario (September 1995): 2-7.

Voroney, R.P., J.A. Van Veen, and E.A. Paul. 1981. Organic C dynamics in grassland soils. 2. Model validation and simulation of the long-term effects of cultivation and rainfall erosion. *Can. J. Soil Sci.* 61:211-224.

CHAPTER 6

The Amount of Organic Carbon in Various Soil Orders and Ecological Provinces in Canada

C. Tarnocai

I. Introduction

Approximately three times more carbon occurs in soils than in terrestrial vegetation (Schlesinger, 1986). Post et al. (1982) estimated that 27% of this soil carbon occurs in tundra and boreal forest ecosystems. Since a large part of Canada lies within these regions, a significant portion of the world's soil carbon occurs in Canadian territory. There is little published data concerning the amounts of organic carbon in various Canadian soils, but some values are provided for cultivated and uncultivated pedons of Chernozemic and Luvisolic soils (Anderson et al., 1985; Gregorich and Anderson, 1985), for highly cryoturbated pedons (Turbic Cryosols) (Kimble et al., 1993), and for Cryosols (Tarnocai and Smith, 1992). Some information concerning amounts of soil carbon in the various ecoclimatic provinces is available from other researchers, including values for various global life zones (Post et al., 1982), and for the ecological provinces in Canada (Kurz et al., 1992), which also includes an estimate of 76.4 Gt (gigatonne; 1 Gt=10^9 tonnes=10^{12} kilograms=10^{15} grams) for the mass of organic carbon in Canadian forest soils. Most of these estimates use a relatively small number of pedons to represent large areas. For example, estimates for the Arctic and Boreal areas of the entire world (Post et al., 1982) use a total of 308 pedons, with only 48 pedons being from Arctic areas and 260 from Boreal areas.

In order to obtain a more accurate estimate of the amount of organic carbon in Canadian soils, the Soil Carbon Project (Soil Carbon Data Base Working Group, 1993), which used a geographical information system (GIS), was initiated in late 1991. The database produced as a result of this project was used to generate the amount of organic carbon in various soils. This database provided a means of determining the relationships between amounts of soil carbon and the attributes included in the database, including soil drainage, mode of deposition of the soil parent material, broad vegetation cover and land use. It was designed so that carbon values for the upper 30 cm of both mineral and organic soils could be determined separately. It should be noted that the surface organic horizons of mineral soils (designated LFH in the Canadian system and O in the American system) are included in this soil layer (surface layer). This surface soil layer, which is the one most likely to interact with the atmosphere, is the layer that is most sensitive to environmental change.

The first attempt to calculate the organic carbon content of Canadian soils, using this incomplete Soil Organic Carbon Database, was carried out by Tarnocai and Ballard (1994). Shortly thereafter, using an advanced, but still incomplete version of this database, a second estimate was made by Tarnocai (1994). Since the database had not been completed for all areas of Canada at that time, both of these works used selected typical segments of the area, representing about 20% of the soil area of

Canada, and these carbon values were extrapolated to the entire country. The results derived from these two studies are, therefore, considered to be very preliminary.

During the past two years the Soil Organic Carbon Database has been updated and the quality checked (Lacelle, 1996). This paper gives the calculated total and surface amounts of organic soil carbon in Canadian soils, determined for various soil orders and ecoclimatic provinces using this latest version of the Soil Organic Carbon Database.

II. Analysis of Soil Carbon Data

Carbon values in the Soil Organic Carbon Database are derived from all soil areas throughout the entire area of Canada, ranging from the Temperate to the High Arctic ecoclimatic provinces (Figure 1). The Soil Organic Carbon Database consists of a digital cover in ARC/INFO format. It contains three attribute tables: a carbon polygon attribute table (CARBON.PAT), a carbon component table (CARBON.CMP), and a carbon layer table (CARBON.LYR). A detailed description of the database, including the structure, all of the attributes, and the methods of calculating the carbon values, is given by Lacelle (1996) in these proceedings.

The soil carbon masses and contents are initially determined on the basis of the individual soil components. Although in this paper the total and surface carbon contents and masses are given for each soil order and ecoclimatic province in Canada, it is possible to combine these values in various ways in order to obtain soil carbon values for many different purposes, limited only by the information contained in the database. Examples of such database products are soil carbon maps or tables in which the amount of organic carbon is expressed on the basis of map polygons, soil orders or great groups, parent materials, ecological divisions, etc.

A. Definitions

Total carbon content, expressed as kilograms per square meter ($kg\ m^{-2}$), is a measure of the average amount of organic carbon in either a soil component or a soil landscape. The total carbon content is thus the amount of carbon in a one meter square column of soil. If the soil column is one meter deep, the total carbon content is analogous to carbon density.

Total carbon mass, expressed as kilograms (kg) or gigatonnes (Gt) of carbon, refers to the total mass of organic carbon in each polygon. The carbon mass is determined by multiplying the total carbon content by the area of a particular soil order in a polygon or the land area of the polygon.

Surface carbon content, expressed as kilograms per square meter ($kg\ m^{-2}$), is a measure of the average amount of organic carbon within the top 30 cm layer of soil. It is calculated in a manner similar to that used to determine the total carbon content, except that this value refers only to the 0 to 30 cm depth of the soil.

Surface carbon mass, expressed as kilograms (kg) or gigatonnes (Gt) of carbon, refers to the mass of organic carbon within the top 30 cm (0 to 30 cm depth) layer of a soil. It is calculated in a manner similar to that used to determine the total carbon mass, except that this value refers only to the 0 to 30 cm depth of the soil.

B. Analysis of Carbon Data

The organic carbon percentages given in the database were determined using the Walkley and Black and the induction furnace laboratory methods (Sheldrick, 1984). For mineral soils, in most cases, carbon content is calculated for a depth of one meter, but for mineral soils with lithic contact (shallow

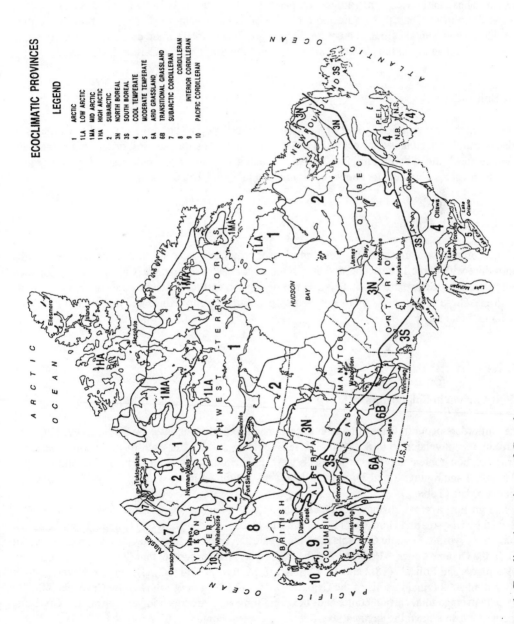

Figure 1. Ecoclimatic provinces of Canada.

soils over bedrock) it is calculated for the depth to the contact (if less than one meter). For organic soils the carbon mass is calculated for the total depth of the peat deposit.

The total carbon content of each soil layer is calculated by combining the attributes (percent organic carbon, bulk density, and layer thickness) stored in the records of the soil carbon layer table (CARBON.LYR). The carbon values thus calculated for each of these soil layers are then combined and adjusted by the amounts of coarse fragments to obtain the total carbon content of each soil component. Each of these values is adjusted according to the proportion of the component in the polygon. These carbon contents are stored permanently in the CARBON.CMP table of the Soil Organic Carbon Database. The adjusted total carbon content values for each component are then summed to obtain a total carbon content for the polygon. These total carbon contents of each polygon are stored permanently in the CARBON.PAT table of the Soil Organic Carbon Database.

1. By Soil Order

Carbon contents are grouped by soil order, adjusted to the proportion of the soil order within a soil landscape, and then summed for each soil order. Carbon masses for each soil order within each soil landscape are obtained by multiplying the carbon contents by the areas of the respective soil orders. The carbon values obtained are shown in Figures 2 and 3, and Tables 1 and 2.

2. By Ecoclimatic Province

Carbon contents are grouped by ecoclimatic subprovince, adjusted to the proportion of the subprovince within a soil landscape, and then summed for each subprovince. Carbon masses for each ecoclimatic subprovince are calculated by multiplying the carbon content of the subprovince by its land area. These carbon masses are summed by ecoclimatic subprovince and reported in Figures 4 and 5 and Table 3.

III. Results

A. Soil Carbon in Soil Orders

Nine soil orders occur in Canada (Agriculture Canada Expert Committee on Soil Survey, 1987; for approximate conversion to U.S. taxonomy, see Tables 1 and 2). For each of these soil orders, the average carbon content and the carbon mass were calculated for both the surface layer and the total soil (Table 1 and Figures 2 and 3). Similar data were calculated in more detail for the Cryosolic and Organic orders (Table 2).

The greatest average surface carbon content occurs in soils of the Organic Order (18.7 kg m^{-2}), followed by Gleysols and Cryosols (11.7 kg m^{-2} and 11.3 kg m^{-2}, respectively) (Figure 2 and Table 1). Within the Cryosols, Organic Cryosols have the greatest average surface soil carbon content (16.7 kg m^{-2}). In the Organic Order, Mesisols have the highest surface carbon content (21.7 kg m^{-2}), followed by Humisols and Folisols (18.2 and 16.7 kg m^{-2}, respectively) (Table 2). The greatest average total carbon content is found in soils of the Organic Order (133.7 kg m^{-2}), followed by Cryosols (40.6 kg m^{-2}). The average total carbon contents of the remaining soils are much lower. Within the Cryosols, Organic Cryosols have the greatest average total carbon content (107.2 kg m^{-2}) (Table 2). In the Organic Order, Mesisols have the highest total carbon content (150.0 kg m^{-2}), followed by Fibrisols and Humisols (111.8 and 109.1 kg m^{-2}, respectively) (Table 2).

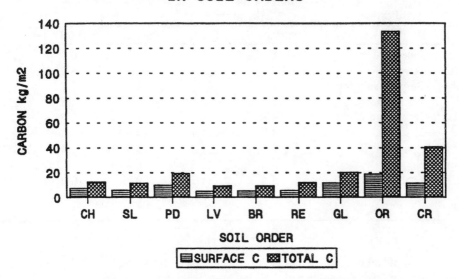

Figure 2. Soil carbon contents in the various soil orders. The codes used for the soil orders are: CH-Chernozem, SL-Solonetz, PD-Podzol, LV-Luvisol, BR-Brunisol, RE-Regosol, GL-Gleysol, OR-Organic, and CR-Cryosol.

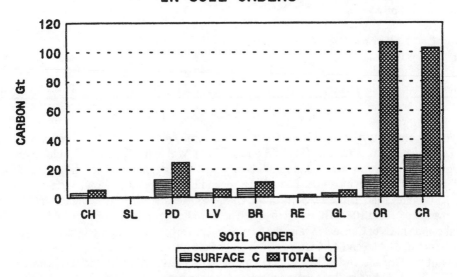

Figure 3. Soil carbon masses in the various soil orders. The codes used for the soil orders are: CH-Chernozem, SL-Solonetz, PD-Podzol, LV-Luvisol, BR-Brunisol, RE-Regosol, GL-Gleysol, OR-Organic, and CR-Cryosol.

Table 1. Amount of soil organic carbon in various soil orders in Canada and correlation of Canadian and U.S. soil classification terminology

Soil classification		Soil carbon content (kg m^{-2})		Soil carbon mass (Gt)		Area (10^3 km^2)
Canada	U.S.	Surface	Total	Surface	Total	
Brunisol	Inceptisol	5.2	9.3	6.1	10.9	1170
Chernozem	Boroll	7.4	12.4	3.2	5.4	434
Cryosol	Pergelic subgroups	11.3	40.6	28.7	102.7	2530
Gleysol	Aqu- suborders	11.7	20.0	2.7	4.6	230
Luvisol	Boralf and Udalf	4.9	9.3	3.0	5.7	611
Organic	Histosol	18.7	133.7	14.9	106.3	795
Podzol	Spodosol	9.9	19.3	12.5	24.4	1267
Regosol	Entisol	5.6	11.8	0.8	1.7	144
Solonetz	Mollisol and Alfisol	5.8	11.5	0.3	0.6	52
All Soils				72.2	262.3	7233

Table 2. Amount of soil organic carbon in the great groups of the Organic and Cryosolic orders in Canada and correlation of Canadian and U.S. soil classification terminology

Soil classification		Soil carbon content (kg m^{-2})		Soil carbon mass (Gt)		Area (10^3 km^2)
Canada	U.S.	Surface	Total	Surface	Total	
Organic	*Histosol*					
Fibrisol	Fibrist	13.9	111.8	4.0	32.2	288
Mesisol	Hemist	21.7	150.0	10.5	72.6	484
Humisol	Saprist	18.2	109.1	0.2	1.2	11
Folisol	Folist	16.7	25.0	0.2	0.3	12
Cryosol	*Pergelic Subg.*					
Turbic	Pergelic	10.4	27.2	18.5	48.5	1786
Static	Pergelic	9.3	22.2	2.8	6.7	302
Organic	Pergelic Histosol	16.7	107.2	7.4	47.4	442

Cryosols contribute by far the largest surface soil carbon mass (28.7 Gt), approximately 40% of the surface carbon mass of Canada (Figure 3 and Table 1). Within the Cryosols, Turbic Cryosols have the largest surface mass (18.5 Gt), approximately 26% of the total surface carbon mass found in Canada. Soils of the Organic Order contain the second largest amount of total surface carbon mass (14.9 Gt) (Figure 3 and Table 1). Soils in the Organic Order and Cryosols contain the largest total carbon mass (106.3 and 102.7 Gt, respectively). These two soil orders contain approximately 80% of the total carbon mass of Canada. Within the Cryosols, the Turbic Cryosols and Organic Cryosols each contain about 18% (48.5 and 47.4 Gt, respectively) of the total carbon mass of Canada, while within the soils of the Organic Order, Mesisols contain about 28% (72.6 Gt) of the total carbon mass of Canada (Table 2). Together, these three soil groups contain nearly two-thirds of the total carbon mass of Canada.

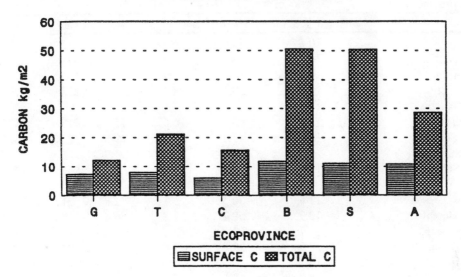

Figure 4. Soil carbon contents in the various ecoclimatic provinces. The codes used for the ecoclimatic provinces are: G-Grassland, T-Temperate, C-Cordilleran, B-Boreal, S-Subarctic, and A-Arctic.

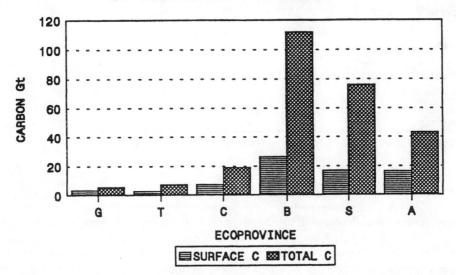

Figure 5. Soil carbon masses in the various ecoclimatic provinces. The codes used for the ecoclimatic provinces are: G-Grassland, T-Temperate, C-Cordilleran, B-Boreal, S-Subarctic, and A-Arctic.

Table 3. Amount of soil organic carbon in the various ecoclimatic provinces and subprovinces in Canada

Ecoclimatic provinces and subprovinces	Soil carbon content (kg m^{-2})		Soil carbon mass (Gt)		Area (10^3 km^2)	
	Surface	Total	Surface	Total	Land[a]	Soil
Arctic	10.8	28.6	16.2	43.0	2375	1506
High Arctic	10.2	19.7	2.2	6.2	681	315
Mid Arctic	14.3	34.3	5.7	13.7	476	399
Low Arctic	10.5	29.2	8.3	23.1	1218	792
Subarctic	11.0	50.4	16.6	75.7	1712	1503
High Subarctic	10.5	47.5	6.5	29.3	725	617
Mid Subarctic	9.7	20.4	0.9	1.9	126	93
Low Subarctic	11.6	56.1	9.2	44.5	861	793
Boreal	11.8	50.5	26.1	111.8	2521	2212
High Boreal	12.3	59.2	11.8	56.8	1085	959
Mid Boreal	12.7	52.2	10.5	43.1	933	825
Low Boreal	8.9	27.8	3.8	11.9	503	428
Cordilleran	5.9	15.7	7.2	19.0	1440	1211
Subarctic Cordilleran	5.5	10.0	0.6	1.1	137	110
Cordilleran	5.7	15.0	4.6	12.1	913	806
Interior Cordilleran	4.1	10.3	0.6	1.5	149	145
Pacific Cordilleran	9.3	28.7	1.4	4.3	241	150
Temperate	7.9	21.1	2.7	7.2	370	341
Cool Temperate	8.3	22.4	2.6	7.0	341	313
Moderate Temperate	3.6	7.1	0.1	0.2	29	28
Grassland	7.2	12.2	3.3	5.6	460	460
Transitional Grassland	9.7	15.4	1.9	3.0	195	195
Subhumid Grassland	7.1	14.3	0.1	0.2	14	14
Arid Grassland	5.2	9.6	1.3	2.4	251	251
All Provinces			72.2	262.3	8879	7233

[a]Includes nonsoil areas such as rockland, glacier ice, and urban land.

B. Soil Carbon in Ecoclimatic Provinces

Canada is divided into eighteen ecoclimatic subprovinces in six ecoclimatic provinces (Figure 1 and Table 3) (Ecoregions Working Group, 1989). The surface and total carbon contents and masses found in the soils of these ecoclimatic provinces are presented in Figures 4 and 5 and Table 3. The greatest surface and total carbon contents, at the province level, are found in the Boreal Ecoclimatic Province (11.8 kg m^{-2} and 50.5 kg m^{-2}, respectively), while the second highest are found in the Subarctic Ecoclimatic Province (11.0 kg m^{-2} and 50.4 kg m^{-2}, respectively). In addition, the highest total carbon content (59.2 kg m^{-2}) and the third highest surface carbon content (12.3 kg m^{-2}), at the subprovince level, occur in the High Boreal Ecoclimatic Subprovince. The highest surface carbon content, at the subprovince level, however, occurs in the Mid Arctic Ecoclimatic Subprovince (14.3 kg m^{-2}), while the second highest total carbon content (56.1 kg m^{-2}) occurs in the Low Subarctic Ecoclimatic Subprovince.

The largest surface carbon masses occur in the Boreal Ecoclimatic Province (26.1 Gt), followed by the Subarctic (16.6 Gt) and Arctic (16.2 Gt) ecoclimatic provinces. Approximately 82% of the surface carbon mass of Canada is found in these three ecoclimatic provinces. Soils in these three ecoclimatic provinces also contain the highest total soil carbon masses, with 111.8 Gt in the Boreal, 75.7 Gt in the Subarctic, and 43.0 Gt in the Arctic. Together, these three ecoclimatic provinces contain 230.5 Gt of organic carbon, approximately 88% of the total carbon mass of Canada, with 43% in the Boreal, 29% in the Subarctic, and 16% in the Arctic ecoclimatic provinces.

IV. Discussion

A large portion of the organic carbon found in Canadian soils occurs at mid and high latitudes (northward from the southern limit of the Boreal forest). Two soil orders, Cryosols and Organic soils, contain high amounts of total carbon and are also the dominant soils in these regions, covering large areas. In these two soil orders, the surface carbon contents are also high (Gleysols, however, have a slightly higher surface carbon content than the Cryosols), suggesting that large amounts of carbon could be directly affected if environmental change occurs. The low pH and nutrient status and the anaerobic conditions in soils of the Organic Order contribute to carbon fixing. For Cryosolic soils, Organic and Turbic Cryosols are the major contributors of soil carbon because of both their properties and the large areas they cover. In addition to low pH, low nutrient status and anaerobic conditions, cold soil temperatures make Organic Cryosols an even more effective carbon sink (per unit area) than other Cryosols (Table 2). The mechanism responsible for the large amount of carbon in Turbic Cryosols (mineral soils) is cryoturbation, which continuously translocates organic material from the surface to the lower soil horizons. Since the depth of the active layer in Turbic Cryosols fluctuates with changes in the permafrost conditions, the near-surface permafrost can contain significant amounts of carbon.

Gleysolic and Podzolic soils also have relatively high carbon content. These soils usually have a peaty or high-organic-content surface horizon. Because of their wetness, the rate of decomposition of organic matter in Gleysols is slow and these soils also act as a carbon sink. High carbon content Podzolic soils are concentrated in Pacific coastal areas. Because of the high biomass production of these areas, these soils are generally associated with a thick organic surface horizon.

The carbon layer data in the carbon layer table (CARBON.LYR) were compiled from field samples collected over the last two decades. The information does not represent the carbon content of Canadian soils at any single point in time, but it is based on measured data, not estimates. Since the data were collected over a period of time, however, the carbon data cannot be used directly to measure changes in organic carbon content during these two decades. Process-based models can, however, incorporate data from this database to predict changes in carbon content resulting from climate change, management practices, or other factors.

The total soil carbon content of Black Chernozemic pedons has been estimated to be 4.6 to 16.4 kg m^{-2} and for Luvisolic pedons, 5.3 to 7.8 kg m^{-2} (Anderson et al., 1985; Gregorich and Anderson, 1985). Results obtained during this study, however, indicate that the average total soil carbon content for Chernozemic soils is 12.4 kg m^{-2} and for Luvisolic soils, 9.3 kg m^{-2}. The total soil carbon content of Turbic Cryosols has variously been found to be 3.2 to 100.9 kg m^{-2} (Kimble et al., 1993) and 17.7 to 82.2 kg m^{-2} (Tarnocai and Smith, 1992). Values of 3.9 to 5.4 kg m^{-2} for Static Cryosols and 86.3 to 176.6 kg m^{-2} for Organic Cryosols have also been found (Tarnocai and Smith, 1992). Although Cryosol studies report a wide variation in soil carbon content values for these soils, the average soil

Table 4. Comparison of total carbon contents for four ecoclimatic provinces

Ecoclimatic province	Total carbon content (kg m^{-2})	
	This study	Post et al. (1982)
Arctic	28.6	21.8
Subarctic	50.4	10.2
Boreal	50.5	11.6 - 19.3
Grassland	12.2	13.3

carbon content values obtained in other studies (Kimble et al., 1993; Tarnocai and Smith, 1992) are within the ranges found in this study, except for the Static Cryosols.

A comparison of data concerning average soil carbon contents in the various ecoclimatic provinces (Post et al., 1982) and data calculated during this project is presented in Table 4. The values presented in this paper are higher than those presented in Post et al. (1982), except for the Grassland Ecoclimatic Province, where Post et al. (1982) has a slightly higher value (13.3 kg m^{-2}) than was found in this study (12.2 kg m^{-2}). The average carbon contents presented in this paper for the Arctic, Subarctic, and Boreal ecoclimatic provinces include the soil carbon contributed by organic soils (peatlands), which are very common in all three of these ecoclimatic provinces. It is not clear, however, whether the study by Post et al. (1982) incorporated organic soil (peatland) data.

Little previous data is available for comparing the total amount of carbon of all Canadian soils. Kurz et al. (1992) found that forest soils in Canada contain approximately 76.4 Gt of carbon, but this estimate does not include the carbon contained in peatlands. Gorham (1988) estimated that Canadian peatlands contain 135 Gt of carbon. This is comparable to the 153.7 Gt of carbon found for the peatland soils (soils of the Organic Order and Organic Cryosols) in this study.

V. Summary

1. The Canadian Soil Organic Carbon Database was set up to determine the amount of organic carbon in the various soil orders and ecoclimatic provinces of Canada. This database (containing major land and soil components for each of the more than 15,000 polygons) is in the ARC/INFO GIS format.

2. The Soil Organic Carbon Database includes those soil attributes (soil horizon designation, thickness, texture, bulk density, percent organic carbon, and coarse fragments) used to calculate the mass of soil carbon in the soils occurring within a polygon.

3. Average soil carbon content for the nine soil orders ranged from 4.9 to 18.7 kg m^{-2} (surface) and from 9.3 to 133.7 kg m^{-2} (total).

4. Soils of the Organic and Cryosolic orders (Histosols and Pergelic subgroups) had the highest total carbon contents (133.7 and 40.6 kg m^{-2}, respectively).

5. An estimate of the mass of organic carbon in Canadian soils produced values of 72.2 Gt (surface) and 262.3 Gt (total). Soils of the Cryosolic and Organic orders contain approximately 80% of the total organic carbon mass of Canada.

6. The Boreal, Subarctic, and Arctic ecoclimatic provinces had the largest total average soil carbon contents at the province level (50.5, 50.4, and 28.6 kg m^{-2}, respectively).

7. The Boreal, Subarctic, and Arctic ecoclimatic provinces had the largest total soil carbon masses (111.8, 75.7, and 43.0 kg m^{-2}, respectively), containing approximately 88% of Canada's total soil carbon mass.

Acknowledgments

Many soil scientists from Soil Survey Units across Canada participated in compiling the soil landscape and soil carbon databases. Special thanks are due to Stephen Zoltai of the Canadian Forestry Service, who provided some of the soil carbon data for northern Canada.

References

Agriculture Canada Expert Committee on Soil Survey. 1987. *The Canadian System of Soil Classification.* (2nd ed.) Research Branch, Agriculture Canada, Ottawa, Canada. Publication 1646, 164 pp.

Anderson, D.W., E.G. Gregorich, and G.E. Verity. 1985. Erosion and cultivation effects on the loss of organic matter from Prairie soils. *Proc. Soils and Crops Workshop*, Extension Division, University of Saskatchewan, Saskatoon, Sask.

Ecoregions Working Group. 1989. *Ecoclimatic Regions of Canada, First Approximation.* Ecoregions Working Group of the Canada Committee on Ecological Land Classification. Ecological Land Classification Series, No. 23, Sustainable Development Branch, Canadian Wildlife Service, Conservation and Protection, Environment Canada, Ottawa, Canada. 119 pp. and map.

Gorham, E. 1988. Canada's peatlands: their importance for the global carbon cycle and possible effects of "greenhouse" climatic warming. *Transact. Royal Soc. Can.* V, 3:21–23.

Gregorich, E.G. and D.W. Anderson. 1985. Effects of cultivation and erosion on soils of four toposequences in the Canadian Prairies. *Geoderma* 36:343–354.

Kimble, J.M., C. Tarnocai, C.L. Ping, R. Ahrens, C.A.S. Smith, J. Moore, and W. Lynn. 1993. Determination of the amount of carbon in highly cryoturbated soils. p. 277–291. Proc. of the Joint Russian–American Seminar on Cryopedology and Global Change, November 15–16, 1992, Pushchino, Russia.

Kurz, W.A., M.J. Apps, T.M. Webb, and P.J. McNamee. 1992. *The Carbon Budget of the Canadian Forest Sector: Phase I.* Forestry Canada, Northwest Region, Northern Forestry Centre, Edmonton, Alberta. Information Report NOR–X–326, 92 pp.

Lacelle, B. 1996. Canada's Soil Organic Carbon Database. (this volume).

Post, W.M., W.R. Emanuel, P.J. Zinke, and G. Stangenberger. 1982. Soil carbon pools and world life zones. *Nature* 298:156–159.

Schlesinger, W.H. 1986. Changes in soil carbon storage and associated properties with disturbance and recovery. Chapter 11, p. 194–220. In: J.R. Trabalka and D.E. Reichle (eds.) *The Changing Carbon Cycle–A Global Analysis.* Springer–Verlag, New York, NY.

Sheldrick, B.H. (ed.). 1984. *Analytical Methods Manual.* Land Resource Research Institute, Agriculture Canada, Ottawa, Canada.

Soil Carbon Data Base Working Group. 1993. *Soil Carbon Data for Canadian Soils.* Centre for Land and Biological Resources Research, Research Branch, Agriculture Canada, Ottawa. 137 pp. and maps.

Tarnocai, C. 1994. Amount of organic carbon in Canadian soils. p. 67–82. In: *Transactions of the 15th World Congress of Soil Science*, Volume 6a, Commission V, Acapulco, Mexico.

Tarnocai, C. and M. Ballard. 1994. Organic carbon in Canadian soils. p. 31–45. In: L. Rattan, J. Kimble and E. Levine (eds.), *Soil Processes and Greenhouse Effect*. USDA, Soil Conservation Service, National Soil Survey Center, Lincoln, NE.

Tarnocai, C. and C.A.S. Smith. 1992. The formation and properties of soils in the permafrost regions of Canada. p. 21–42. Proc. of the 1st International Conference on Cryopedology, Pushchino, Russia.

CHAPTER 7

Canada's Soil Organic Carbon Database

B. Lacelle

I. Introduction

The Canadian Soil Carbon Project was initiated in 1991 to determine the amount of organic carbon in all Canadian soils. The goal was to establish a uniform database for soil carbon data for the whole of Canada. The laboratory attribute data methodology was evaluated and standardized for this project. This enables users to confidently compare the data and draw conclusions. To be useful to researchers, the design of the database also had to acknowledge the relationships between the soil and the carbon content. A digital database was designed in ARC/INFO GIS format which included a spatial network and incorporated the attributes relevant to calculating organic carbon. The Soil Landscapes of Canada Database, part of the National Soils Database (NSDB), and maintained by the Eastern Cereal and Oilseed Research Centre of Agriculture Canada provided an excellent spatial framework. The SLC database contained generalized soil landscape polygons at 1:1 million scale along with the attributes of the soil landscapes. Some of the SLC database was incomplete. The mapsheets for northern British Columbia and Quebec were not in digital form and no attribute files for the Northwest Territories, Labrador, Quebec, and the northern regions of Ontario and British Columbia existed. These areas were digitized and verified according to NSDB requirements and the attribute files were compiled. The current attribute files were redesigned to allow for more complete descriptions of the landscape. A data table consisting of layer data pertinent for calculating a carbon density was obtained mostly from the NSDB Soil Layer Files (provincial files describing the layer characteristics of the major soils). Computerized routines were developed to generalize the data into the Carbon Layer Table structure. The data was evaluated and completed by regional staff. In some areas (Yukon, Northwest Territories and northern Ontario, Quebec and Labrador), staff compiled the data from detailed point data providing a modal or typical pedon for each soil name and modifier. Carbon calculations were performed on the data and assessed. One of the weak areas was the mapsheet boundaries. Inconsistencies were apparent both in the linework and the attribute data. A concentrated effort took place to join the 28 mapsheets into a seamless database. All lines were reevaluated and the attribute data was correlated using a set of computerized routines. This new SLC database formed the base for the latest version of the Canadian Soil Carbon Database. The Canadian Soil Carbon Database was further upgraded by information provided by the Canadian Peatlands Project. The peatlands team identified areas where peat was missing from the SLC database. The areas of concern were verified and the data were updated. Also, estimates of percentages of land and water components were reaccessed and adjusted. Previous soil carbon publications were based on the production and findings of an earlier version of the database, Ballard et al. (1993), Tarnocai (1994). This paper describes the latest version of the Canadian Soil Carbon Database. A paper using this later version of database describing the findings of calculated organic carbon values for the various soil orders and ecoclimatic provinces is given by Tarnocai (1996).

II. Materials and Methods

For the regions of Southern Ontario, Southern Quebec, Southern British Columbia, Saskatchewan, Alberta, Manitoba, and the maritime provinces except Labrador maps were compiled from existing larger scale, soil survey source maps and their reports. Major drainage pattern, major physiographic features, and large uniform soil landscape areas were delineated. A change in any one differentiating property class on the source map resulted in a seperate generalized polygon. For the regions of Labrador, Yukon, the Northwest Territories, and the northern areas of British Columbia, Ontario, and Quebec, maps were compiled from 1:1 million scale, cloud-free, high quality, black and white LANDSAT mosaics (controlled). These were interpreted manually with the aid of panchromatic photographs, colour LANDSAT imagery, and other relevant information. Features most readily available and consistently observable on LANDSAT imagery (regional landforms, surficial materials, local surface forms, water bodies, wetlands, vegetation, and patterned ground were identified and grouped into polygons. Data from existing ground truth sites were reviewed and compiled to the attribute file. Detailed information relating to terrain and vegetation were collected during ground stops. Both types of mapping resulted in the following criteria. The smallest mappable area was selected as 100 km^2 or 1 cm x 1 cm on a 1:1 million scale map. Most polygons were larger with a few exceptions. Small islands with a minimum map size of 0.25 cm^2 (1:1 million) were coded as a polygon. Where small islands were situated close together in a group, a line was drawn around them to form a polygon. The water portion of such a polygon was estimated and is coded as a component of the landscape.

III. Soil Organic Carbon Database

Canada's Soil Organic Carbon Database consists of a digital cover in ARC/INFO format compiled at 1:1 million scale. The database is also composed of 3 attribute tables, a carbon polygon attribute table (CARBON.PAT), a carbon component table (CARBON.CMP), and a carbon layer table (CARBON.LYR). Organic material above the mineral soil is contained in the database. The data of each attribute file can be linked together, but care must be taken to ensure the proper relationships occur. Figure 1 refers to the spatial extent of the database and a zoomed-in area to demonstrate one spatial polygon, i.e., soil landscape number 460210. This soil landscape number is also located in each of the attribute tables and is used as a link to find all the data associated with that polygon.

A. Digital Cover

The digital cover, stored as CARBON, consists of over 15,000 polygons. It is stored in a lambert conformal projection in meters. The central meridian is at -91 degrees, 52 minutes. The standard parallels are located at 49 degrees and 77 degrees with an origin of 0,0.

B. Carbon Polygon Attribute Table (CARBON.PAT)

The Carbon Polygon Attribute Table (CARBON.PAT) links the digital cover and to the other attribute tables (i.e., key = Soil Landscape Number). It consists of one record/polygon. Any variables that describe the polygon as a whole find their home in this table. Each polygon has a unique soil landscape number, except for 100 percent water bodies (last 4 numbers of soil landscape number = 5000) and 100 percent urban areas (last 4 numbers of soil landscape number = 5001). See Table 1 for listing of attributes of CARBON.PAT.

Canada's Soil Organic Carbon Database

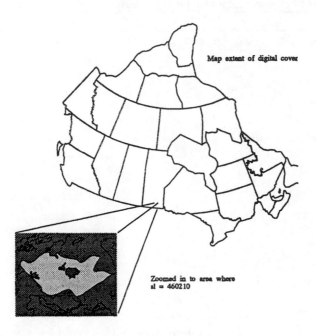

Figure 1. Spatial extent of the Soil Organic Carbon Database of Canada and zoomed-in area showing more detail spatial polygon.

Table 1. File structure of Carbon Polygon Attribute Table (CARBON.PAT)

Attribute name	Width	Type
Soil landscape number[a]	6	Integer
Reliability index	1	Character
Ecoclimatic District	10	Character
Land area of the polygon (km^2)	8 (1 decimal place)	Numeric
Soil area of the polygon (km^2)	8 (1 decimal place)	Numeric
Water area of the polygon (km^2)	8 (1 decimal place)	Numeric
Rock area of the polygon (km^2)	8 (1 decimal place)	Numeric
Glacier area of the polygon (km^2)	8 (1 decimal place)	Numeric
Total carbon content of the polygon (kg/m^2)	6 (2 decimal places)	Numeric
Surface carbon content of polygon (kg/m^2)	6 (2 decimal places)	Numeric
Total carbon mass (10^6 kg)	10 (1 decimal place)	Numeric
Surface carbon mass (10^6 kg)	10 (1 decimal place)	Numeric

[a]Uniquely identifies one spatial polygon.

Table 2. File structure of carbon component table (CARBON.CMP)

Attribute name	Width	Type
Soil landscape number[a]	6	Integer
Component number[a]	2	Integer
Percent of polygon	3	Integer
Kind of material	2	Character
Vegetation	2	Character
Parent material mode of deposition	2	Character
Coarse fragments	1	Character
Rooting depth (cm)	3	Character
Soil development	1	Character
Soil drainage	1	Character
Calcareousness	1	Character
Local surface form	3	Character
Slope	1	Character
Soil code	3	Character
Soil modifier	3	Character
Total carbon content of component (kg/m^2)	6 (2 decimal places)	Numeric
Surface carbon content of component (kg/m^2)	6 (2 decimal places)	Numeric

[a]Together uniquely identifies one component of one polygon.

C. Carbon Component Table (CARBON.CMP)

This table describes each soil and/or nonsoil component found in each spatial polygon. Each component is allocated to be a percentage of the polygon. The total of all components for each polygon must equal 100 percent. Other attributes related to the soil/nonsoil components are listed in Table 2.

D. Carbon Layer Table

The Carbon Layer Table (CARBON.LYR) contains pertinent information for calculating a soil's carbon density for each layer of each soil. It refers back to CARBON.CMP by the soil landscape number and the component number. There are up to three layer records for all mineral and organic soils listed in the carbon component table. Organic carbon was determined using the induction furnace laboratory method. The variables described as texture type, bulk density type, and organic carbon type indicate whether the data are measured or estimated. Estimated values were evaluated by calculating the mean values of like measured data (i.e., same horizon designation, texture, etc). See Table 3 for listing of attributes of CARBON.LYR.

Table 3. File structure of carbon layer table (CARBON.LYR)

Attribute name	Width	Type
Soil landscape number[a]	6	Integer
Component number[a]	2	Integer
Layer number[a]	1	Integer
Layer designation	3	Character
Thickness of layer (cm)	3	Integer
Soil texture	5	Character
Soil texture type	2	Character
Bulk density (g/m^3)	4 (1 decimal places)	Numeric
Bulk density type	2	Character
Organic carbon (%)	4 (2 decimal places)	Numeric
Organic carbon type	2	Character

[a]Together uniquely identifies one layer of one component of one polygon.

IV. Results

Please note that organic material above the mineral soil are included in these calculations of soil carbon.

A. Definitions of Carbon Variables in the Carbon Component File

TOTAL CARBON CONTENT OF THE COMPONENT (kg/m^2) is a measure of the average amount of carbon in a particular soil within a soil landscape polygon.
SURFACE CARBON CONTENT OF THE COMPONENT (kg/m^2) is a measure of the average amount of carbon of the active layer (top 30 cm) for a particluar soil within a soil landscape polygon.

B. Definitions of Carbon Variables in the Carbon Polygon Attribute Table

TOTAL CARBON CONTENT OF THE POLYGON (kg/m^2) is a measure of the average amount of carbon found in the land area of a soil landscape polygon.
SURFACE CARBON CONTENT OF THE POLYGON (kg/m^2) is a measure of the average amount of carbon of the active layer (top 30 cm) found in the land area of a soil landscape polygon.
TOTAL CARBON MASS (kg x 10^6) refers to the total weight of organic carbon for a soil landscape polygon. This item is caluated by multiplying the total carbon content of the polygon by the land area.
SURFACE CARBON MASS (kg x 10^5) refers to the total weight of organic carbon found in the active layer (top 30 cm) for a soil landscape polygon. This item is calculated by multiplying the surface carbon content by the land area.

C. Calculation of Carbon Variables

To calculate these variables, first the carbon content is calculated for each layer of the Carbon Layer File. The following formula was used:

Carbon Content (kg/m^2) = Thick (cm) x Bulk Density (g/m^3) x Organic Carbon (%)/10

Surface carbon is calculated for the first 30 cm while total carbon content is calculated to the depth of soil. For mineral soils, in most cases, the carbon content was calculated for a depth of one meter, but for mineral soils with lithic contact (shallow soils over bedrock), it is calculated for the depth to the contact (if < one meter). For organic soils, the carbon content was calculated for the total depth of the peat deposit. The content for each layer are summed for each component; then they are multiplied by a factor relating to the amount of coarse fragments in the soil. These values refer to TOTAL CARBON CONTENT OF THE COMPONENT and SURFACE CARBON CONTENT OF THE COMPONENT described in section A. Calculating the carbon content for each polygon is done by multiplying the carbon contents found in the Carbon Component Table by the percentage of the polygon and summing up the values for each polygon. To obtain the carbon content related to the land area, the values are divided by the percentage of land of the polygon.

D. Organic Carbon For Canadian Soils

The data in this Soil Organic Carbon Database has been used to assess the carbon mass by soil orders and ecoclimatic provinces and subprovinces, Tarnocai (1997). Average carbon contents for the various soil orders range from 9.3 kg/m^2 to 133.7 kg/m^2 for total carbon content and 4.9 kg/m^2 to 18.7 kg/m^2 for surface carbon content. The soil carbon mass ranged from 0.6 Gt to 106.3 Gt for total carbon and 0.3 Gt to 28.7 Gt for surface carbon. Canada has a total carbon mass of approximately 262.3 Gt and a surface carbon mass of approximately 72.8 Gt. Figure 2 is a generalized map of the total carbon content for Canada. Figure 3 is a generalized map of the surface carbon content for Canada.

V. Summary

The Canadian Soil Organic Carbon Database is capable of combining variables and analyzing the data for many different applications. There are over 15,000 soil landscape polygons in the database that link to three attribute tables, which contain information describing the soil landscape and the carbon content. A lot of care was taken during the development of the database to ensure consistency and integrity. It can be used confidently by policy makers to evaluate the state of carbon in our soils and identify areas of concern. The Soil Organic Carbon Database is being used as a basis for calculating the amount of carbon in various soils and ecoclimatic provinces and subprovinces, Tarnocai (1997). It has been imported into models which estimate the rate of carbon change in agricultural soils. Also, in collaboration with the U.S. and Mexico, a North American Carbon Map and associated database is being compiled.

Acknowledgments

Many soil scientists from Canadian Soil Survey Units across Canada participated in compiling the soil landscape and soil carbon data. Special thanks to Michael Ballard for setting up the initial process for calculating the carbon variables. Also, special thanks to Steve Zoltai of the Canadian Forestry Service, who provided some of the soil data for northern Canada. Thanks to the Canadian Peatlands Project team for informing us of some missing data within the Soil Landscapes of Canada Database. Special thanks to Charles Tarnocai.

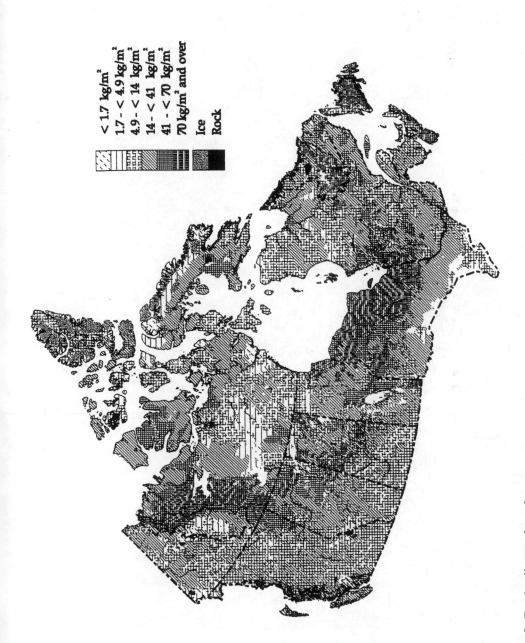

Figure 2. Total soil organic carbon content.

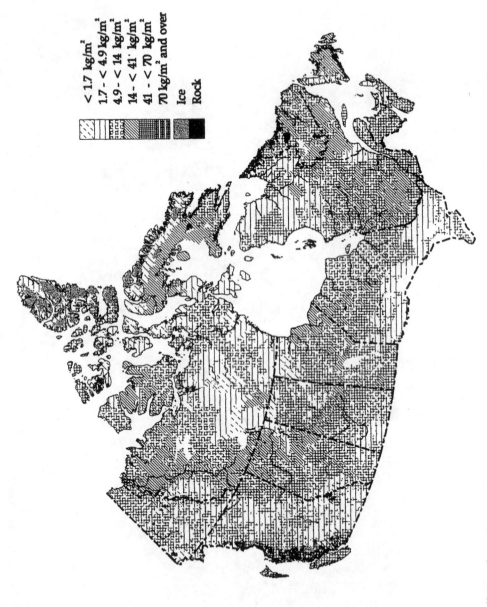

Figure 3. Surface soil organic carbon content.

References

Ballard, M., I. Jarvis, and C. Tarnocai. 1993. Analysis of the Canadian National Soil Organic Carbon Data Base. p.650-661 in: *Proceedings of the Canadian Conference on GIS-1993.* March 23-25, 1993, Ottawa, Canada. CLBRR Contrib. No.93-12.

Tarnocai, C. 1994. Amount of Organic Carbon in Canadian Soils. p.67-74 In: *Proceedings of the 15th World Congress of Soil Science,* Acapulco, Mexico. July 10-16, 1994.

Tarnocai, C. 1997. The Amount of Organic Carbon in Various Soil Orders and Ecological Provinces in Canada. (this volume).

CHAPTER 8

Rate of Humus (Organic Carbon) Accumulation in Soils of Different Ecosystems

Alexander Gennadiyev

I. Introduction

The rate of organic carbon accumulation in soils and of the stability of humus substances remains a disputable problem. On one hand, humus is regarded as a conservative matter which preserves its properties from the zero-moment of soil formation, when it appeared simultaneously with the soil, to the present (Rubilin and Kozyrev, 1974). On the other hand, humus is supposed to be very mobile, subject to quick renewal (Ponomareva and Plotnikova, 1980).

I.V. Tiurin assumed that humus is a specific "very stable system" and its properties and composition do not change in case of soil burial (Tiurin and Tiurina, 1940). In contrast, according to Gerasimov (1971), humus, in such situations, has "very few chances to survive" since it lacks any reliable defense against microbial decomposition. Radiocarbon data of different humus fractions in the upper soil horizon of recent soils range from 500 to 1–2000 years. However, data show that a shallow humus horizon evolves in some dozen years on loose sediments, and that its humus composition is very close to that of normal background soils.

Evaluation of the humus (organic carbon) accumulation rate has been made using the chronosequence approach; i.e., soils of different ages in similar topographic and lithological conditions have been studied (Stevens and Walker, 1970; Gennadiyev, 1978). Soil chronosequences were investigated in the forest-meadow subalpine zone of the Elbrus region, in Caspian semideserts, and in Stavropol' Uplandsteppe. Soils study objects were chosen in such a way that the age of their parent material was known, either for natural substrates (drifts of mountain glaciers) or for man-made ones (archeological mounds).

II. Interaction Between Processes of Organic Carbon Storage, Humification, and Mineralization at Different Stages of High-Mountain Forest and Meadow Soils Development

Case studies have been done in the following soil sequences: a) mountainous-meadow – profile 24 (weakly developed soil, 100 years old), profile 21 (weakly developed soil, 100 years old), profile 22 (moderately developed soil, 300 years old), profile 53 (fully developed soil, about 1000 years old); b) mountainous forest-meadow soils – profile 17 (weakly developed soil, 100 years), profile 18 (moderately developed soil, 500 years), profile 35 (fully developed soil, 1000 years).

When comparing these chronosequences, the following regularities become evident: in older soils, the total carbon content increases, from 0.5–1.5% in weakly developed soils to 11–12% in topsoils of fully developed soils. The latter display broader $C_{h.a.}/C_{f.a.}$ ratios; however, some different trends occur in both soil chronosequences.

Within the meadow sequence, the most conspicuous increment of total carbon content occurs between the stages of weakly developed to moderately developed soil (profiles 24, 21, and 22, respectively). Moreover, carbon content in the topsoil of the latter profile is the same as in a mature soil - 11%. The humic acid proportion in topsoils increases in this sequence: the $C_{h.a.}/C_{f.a.}$ equals 0.24 – 0.40 – 0.53 – 0.59.

In this way, the organic matter balance stabilized in the course of mountainous-meadow soil evolution during several hundreds years from the moment of the start of pedogenesis (in topsoils, first). At the initial stages of profile development, the rate of annual input of humified plant residues is far ahead of the humus mineralization rate, due to low soil biochemical activity.

The next stage of mountainous-meadow soil formation corresponding to the transition to a fully developed soil is characterized by a weak but strongly accelerated increase in humus accumulation. The organic matter balance equilibrate because of "expenses"; the part of neoformed mobile humus substances removed from the solum is greater. Simultaneously, intensification of humus mineralization processes occurs enhanced by the gradual growth of the soil biological activity. This activity involves mostly those humus ingredients that have weak bonds with the soil solid phase.

The dynamics of these parameters differs slightly in the forest-meadow chronosequence. The humus content increases gradually within the series: weakly developed soil (17) - moderately developed soil (18) - fully developed soil (35). The humus content equals 1 – 4 – 12%, respectively, while the $C_{h.a.}/C_{f.a.}$ ratio does not follow this sequence. Thus, the humus development rate is broader in the fully developed soil (0.71) than in the weakly developed one (0.57), and the highest rate is in the moderately developed soil (0.98).

Micromorphological observations demonstrated a strong zoogenic reworking of soil material in forest-meadow high-mountainous soil, and this may be the reason for the drastic difference in organic matter storage between forest-meadow and meadow soils of moderate and strong profile development. In contrast, the relative concentration of humic acids proves to be the highest in forest-meadow soils. "An explosion" of invertebrate activity at early stages of forest-meadow soils' evolution promotes elevated microbial activity. Since fulvic acids are the first to be mineralized, the humic acids' proportion (due to residual accumulation) increases considerably.

III. Accumulation of Organic Carbon and Changes in Time of the Humus Status of Chernozems on the Stavropol' Upland

Chernozemic formation on anthropogenic substrates of different ages (2000 and 3500 years) hardly differ in terms of total organic carbon content, while the quality of their humus is different. In older soils the $C_{h.a.}/C_{f.a.}$ ratio increases from 1.7–1.8 to 1.9–2.5. It is worth remembering here that the neoformed humus in soils on stable substrates several hundreds years old, and lacking any initial organic matter, didn't yet reach 1 for their $C_{h.a.}/C_{f.a.}$ ratio. This trend lets us assume that the participation of humates in chernozems' humus is a function of time, and it becomes more prominent in the course of soil maturing at least during the first millennia.

Humus of soils buried under funeral mounds is much poorer in humates in comparison with humus of the most mature recent soils on the mound. The quality of buried soils' humus changes with the depth of anthropogenic deposits above the soil. Thus, in the central part of a Bronze Age burial mound with the embankment 3m deep, the $C_{h.a.}/C_{f.a.}$ exceeded 1. Towards the periphery of the mound, where the depth of overlying sediment equaled 70-100m, the $C_{h.a.}/C_{f.a.}$ decreased to 0.6–0.4. It grew again to 1, when the overlying sediments were only 30–40cm deep. This phenomenon derives from either the difference in the preservation rate of paleosoils, or from their involvement into current pedogenesis. The two peripheral profiles prove to be almost completely integrated with the recent soil, but the buried soils coincide in the first case with the lower horizons of the recent chernozem, and in the second case with the intermediate horizon; hence, the humus type is humate-fulvic and vice versa.

In addition, however, the humus of the soil buried under 3m deep anthropogenic sediment is somewhat transformed by the neoformed organic matter, i.e., diluted by fulvic acids. We presume, however, that the main reason for the lower humates content ($C_{h.a.}/C_{f.a.}$ is 1.1 versus 1.5-2.0 in the background carbonate chernozems) is its initial humus composition, corresponding to the paleo-geographic situation 3500 years ago in this area. This time span is known as the final part of a rather hot and dry episode in the middle part of the Middle Holocene. The Bronze Age paleosols might have been similar to modern southern chernozems, so they have an arrow $C_{h.a.}/C_{f.a.}$ ratio.

The type of data lets us evaluate the real rates of pedotransformations operative in anthropogenic substrates, which were initially heterogeneous in terms of humus composition. They resemble the modeling of chernozemic humus horizons "fate," when "inserted" by ancient pedogenesis into the profile of a present-day chernozem.

IV. Stages of Organic Carbon Storage and Humus Profile Formation in Soils of a Topogenic Semidesertic Microcatena

Investigations of soils' ingredients of microcatenas of different ages (burial mounds and anthropogenic depressions) in the Caspian semidesert revealed high rates of topogenic differentiation of soil humus status and development of humus profile.

Neoformed soils on 500-year-old substrates have much in common with the background soil profiles in terms of their humus status. More than 1% of fulvic humus are recorded in microelevations soils, while in those of microdepression there are 3–5% of humate and fulvate-humate humus. The depth of the humus horizon reaches 20–40 cm in neoformed soils of microdepressions, and a trend to a certain increase as optical density of humic acids occurs. A regular horizontal pattern of humus enriched zones may occur within 500-year-old soil catenas. At the same time, a certain inversion may appear, caused by the occurrence, in paleosols, of humus horizons fragments property initially in the background. The vertical humus pattern is as follows: the down profile decrease of humus content to the 0.5m depth (1–1.2% to 0.6–0.8%) is replaced by a weak increase to 0.7–0.9% in 60–80cm layer, followed by a new decrease with depth.

The $C_{h.a.}/C_{f.a.}$ ratio becomes wider in the zone of paleosol-embankment interface, due, probably, to specific humification processes in the inherited horizon, as well as to higher moisture here.

An almost complete maturation of the humus profile happens in 2000-year-old soils. In microelevations, soils with any traces of the inversion mentioned earlier disappear. In soils of microdepressions, the depth of the humus horizon (humus content being above 1%) is considerable - up to 60 cm, which is sometimes more than in the background meadow-chestnut soils.

Four-thousand-year-old soils on microelevations are similar to the 2000-year-old soils. Humus profile distribution is regressive-accumulative or regular-accumulative, as in background soils. The humus type is fulvic or humate-fulvic in topsoils and has a low or medium optical density. Occasionally, maximum $C_{h.a.}/C_{f.a.}$ ratio values are observed at a depth of about 50 cm (1.4–2.8); they may be explained by the presence of residual humic acids tightly bound to the mineral soil constituents. These acids might have been formed in the period of the system "soil/embankment" consolidation.

Some microdepression soils are very rich in humus, as the background meadow-chestnut soils are.

Table 1. Approximate rate of humus horizon formation

Soils	Humus horizon formation rate (mm/year)
Aquolls, Udolls	0.80 – 1.00
Histosols, Aqualfs.	0.50 – 0.80
Ochrepts, Aquic Cryoboralfs	0.45 – 0.50
Argiborolls, Haplborolls	0.40 – 0.45
Eutroboralfs, Calciborolls	0.35 – 0.40
Rendolls	0.30 – 0.35
Vermiustolls, Calciustolls, Glossudalfs	0.20 – 0.30
Spodosols, Boralfs	0.10 – 0.20
Natriboralfs, Camborthids	<10

V. Conclusion

The evaluation of humus (organic carbon) accumulation rate, substantiated by this author, derives from pedochronological concepts for natural soils (Gennadiyev, 1990), with some corrections for arable soils. The humus (organic carbon) accumulation rate may be calculated showing the real characteristic times of humus profiles formation - in other words, the time interval required for the development of a mature humus profile with its integrated set of properties. Experimental data-comprehensive studies of various soils on natural or man-made surfaces of a known age let us obtain a sequence of characteristic time values for the humus profiles (Bockheim, 1980; Buol et al., 1989; Gennadiyev, 1990; Stevens and Walker, 1970; Vreeken, 1975).

The substrates, with ages suitable for these purposes include: chernozems and dark chestnut soils > 2500–3000 years; light-chestnut soils and solonetzes 1000–2000 years; sod-podzolic, grey and brown forest soils 800–1000 years; rendzinas, peaty-gley, alpine meadow, and meadow-chestnut soils approximately 500–800 years. The vertical depth of humus profiles was assumed to comprise the totals of A0, A1 horizons thickness, and, if the humus content in the transitional AB horizon was high enough, half of the horizon's depth was taken into account for calculations.

Pedochronological experimental data (Table 1) let us develop an approximate scheme of humus (organic carbon) accumulation rates for some soils of Russia and the U.S.A. (Buol et al., 1989; Chandler, 1942; Fisher, 1983; Franzmeier and Whiteside, 1963; Gennadiyev, 1978, 1979, 1982, 1990; Parsons et al., 1962, 1970; Parsons and Herriman, 1976; Sondheim et al., 1981; Tarnocai, 1982).

Numerous opinions exist regarding the trends and rate of man-influenced humus profile formation. The research reported here demonstrated a higher rate compared to natural rate due to considerable changes in hydrothermic regimes, to fertilizers application, regular tillage, etc. All these effects contribute to more active pedogenetic processes in arable soils by involving larger volumes of soil material in active biological processes, reworking by pedofauna in particular, oxidation, mechanical regrouping, chemical interactions, and other processes.

References

Bockheim, I. G. 1980. Solution and use of chronofunction in Studying Soil development. *Geoderma*. 24:71-85.

Buol, S. W., F. D. Hole, and R. J. McCraken. 1989. *Soil Genesis and Classification*. Iowa State University Press/Ames. 446 pp.

Chandler, R. J. 1942. The time required for podzol profile formation as evidenced by the Mendenhall glacial deposits near Juneau, Alaska. *Proc. Soil Soc. Am.* 7:454-459.

Fisher, P. F. 1983. Pedogenesis within the archaelogical landscape at South Lodge camp, Wiltshire, England. *Geoderma.* 29:93-105.

Franzmeier, D. P. and E. P. Whiteside. 1963. A chronosequence of Podzols in Northern Michigan. Quart. *Bull. Mich. State Univ. Agr. Exp. Stat.* 46:37-57.

Gennadiyev, A.N. 1978. Study of soil formation by the chronosequence method. *Soviet Soil Science.* 10:707-716.

Gennadiyev, A.N. 1979. Soil formation under meadow and forest vegetation in the Alpine region of the Central Caucasus. *Soviet Soil Science.* 11:142-150.

Gennadiyev, A.N. 1982. Chronological model of differentiation of soils as a function of the man-made microrelief in the Caspian Region. *Soviet Soil Science.* 14:27-36.

Gennadiyev, A.N. 1990. *Pochvi i Vremia: Modeli Razvitia (Soils and Time: Models of Development).* Moscow University Press. 240 pp.

Gerasimov, I.P. 1971. Soil formation and evolution of soil cover. *Pochvovedenie (Soil Science).* 7:147–165.

Parsons, R.B., C.A. Balster, and A.O. Ness. 1970. Soil development and geomorphic surfaces, Willamette Valley, Oregon. *Soil Sci. Soc. Am. J.* 34:485-491.

Parsons, R.B. and R.C. Herriman. 1976. Geomorphic surfaces and soil development in the Upper Roque River Valley, Oregon. *Soil Sci. Soc. Am. J.* 40:933-938.

Parsons, R.B., W.H. Schotes, and F.F. Riecken. 1962. Soil of Indian mounds in northern eastern Iowa as benchmarks for studies of soil genesis. *Soil Sci. Soc. Am. Proc.* 26:491-496.

Ponomareva, V.V. and T.A. Plotnikova. 1980. *Gumus i Pochvoobrazovanie (Humus and Soil Formation).* Leningrad Nauka Press. 222 pp.

Rubilin, E.V. and M.G. Kozyreva. 1974. On age of Russian chernozem. *Pochvovedenie (Soil Science).* 7:16–23.

Soundheim, M.W., G.A. Singleton, and L.M. Lavkulich. 1981. Numerical analysis of a chronosequence including the development of a chronofunction. *Soil Sci. Soc. Am. J.* 45:558-563.

Stevens, P.R. and T.W.Walker. 1970. The chronosequence concept and soil formation. *Quart. Rev. Biol.* 45:333-350.

Tarnocai, C. 1982. Soil and terrain development on the York Factory peninsula, Hudson Bay lowland. *Natur. Can.* 109:511-522.

Tiurin, I.V. and E.I. Tiurina. 1940. On humus composition in fossil soils. *Pochvovedenie (Soil Science)* 2:10–22.

Vreeken, W.J. 1975. Principal kinds of chronosequences and their significance in soil history. *J. Soil Sci.* 24:378-394.

CHAPTER 9

Land Use and Soil Management Effects on Soil Organic Matter Dynamics on Alfisols in Western Nigeria

R. Lal

I. Introduction

Tropical ecosystems play a major role in global biogeochemical cycles. Humid tropics are characterized by abundant annual rainfall ranging between 1000 mm and 2500 mm (but can be as high as 8000 mm), consistently high temperatures (mean annual > 22° C), high relative humidity throughout the year, and highly diverse rainforest vegetation (Lugo and Brown, 1991). There are three principal types of lowland tropical rainforests (TRF): dry, moist, and wet. Dry TRFs have mean annual rainfall of 1000 to 1500 mm, and a pronounced dry season of 4 to 5 months. Regions with moist TRF vegetation have mean annual rainfall of 1500 to 2000 mm and no more than 4 months with rainfall of < 200 mm, and wet TRFs have mean annual rainfall of 2000 to 2500 mm and no more than 2 months with rainfall < 200 mm.

Predominant soils of the humid tropics are Oxisols, Ultisols, Inceptisols, Entisols, Alfisols, Histosols, and Spodosols (NRC, 1993). Oxisols and Ultisols are the most abundant soils of moist and wet ecoregions. These soils are deeply weathered, well-drained, and generally have favorable physical properties. Chemically, these soils are acidic, low in inherent fertility, and often contain high concentrations of Al^{+3} and low concentrations of Ca^{+2} and Mg^{+2} in the sub-soil. Alfisols are predominant soils of dry TRFs. These soils are relatively less weathered than Oxisols and Ultisols, have a pH of about 6 to 7, have weak soil structure, and are prone to compaction and erosion. Inceptisols and Entisols, relatively young soils, have high inherent soil fertility. Poorly drained (Aquepts) or hydromorphic soils are fertile and agriculturally important soils of the region. Tropical soils constitute a large proportion of the total SOC pool in world soils (Kimble et al., 1990; Eswaran et al., 1993).

Tropical rainforests, through photosynthesis, evapotranspiration, decomposition of biomass, and nutrient recycling, play a crucial role in the global carbon cycle. These ecosystems influence the regional climate through their effects on convection currents, rainfall regime, albedo and the regional water balance, and global climate through their effects on biogeophysical cycles of C, N, H_2O and other elements. The interaction that this ecosystem has with the global carbon cycle has a profound effect on global atmospheric composition and the mean global temperature. The TRF in the Amazon Basin has a notable effect on global C and H_2O cycling (Dale et al., 1993; Grace et al., 1996; Jordan, 1989; Kruijt, 1996). Patterns of nutrient and C cycling are very specific for the TRF ecosystem (Medina and Cuevas, 1989, 1996), and soil respiration can be an important factor in evaluating the

SOC budget (Meir et al., 1996; Miranda et al., 1996). The carbon pools in the biomass (total, plant, above ground, below ground) and soils beneath the tropical rainforest are a major component of the terrestrial carbon pool. Deforestation and subsequent decomposition of carbon contained in the biomass pools and soil can alter the global C cycle by increasing the atmospheric and decreasing the terrestrial C pool (Houghton, 1990, 1995; Salati and Vose, 1984).

Whereas a considerable amount of literature exists for deforestation and its actual and perceived impact on cycles of C, H_2O, and other elements, there is a relative paucity of such information about soils of west Africa (Lal, 1987). Yet, because of high demographic pressure and need to increase agricultural production, vast areas of rainforest and wooded savannas have been converted to agricultural land use since 1950. The impact of change in land use on soil organic carbon (SOC) content is not known for a wide range of soils and environment of the region.

The objective of this chapter is to summarize data from several long-term deforestation and soil management experiments conducted at the International Institute of Tropical Agriculture (IITA). These experiments were conducted at Ibadan (Alfisols, rainfall 1250 mm per annum) and Okomu (Ultisols, rainfall 2200 mm per annum). Results of soil analyses for SOC content as affected by deforestation, tillage, cropping systems, mulching, and agroforestry are summarized in this chapter.

II. Agricultural Activities in the Tropics and the Greenhouse Effect

Land use and agricultural activities in the tropics, in general, and in the TRF ecosystems, in particular, play an important role in the global C cycling. Principal land uses in the TRF of the world and their relative contribution to total gaseous emissions include pastures (54.3%), arable land use (37.0%), deforestation (7.8%), and cultivation of organic/peat soils (0.9%) (Lal and Logan, 1995). Total C emissions from historical land use activities in the tropics is estimated at about 45 Pg. The current annual rate of C emission from tropical ecosystems is about 0.5 Pg yr^{-1}. The relative contribution of different land use activities to current emissions is estimated at 37.1% from annual burning of grasslands, 30.6% for deforestation, 18.2% for pastures, 12.4% for arable land use, and 1.2% for shifting cultivation (Lal and Logan, 1995).

III. Soil Organic Carbon (SOC) Pool

The SOC content plays an important role in enhancing soil fertility (Tiessen et al., 1994a, b), and in sustainable management of tropical agroecosystems (Sanchez et al., 1989). Similar to vegetation and soils, the SOC pool of the TRF ecosystem is highly variable. Soils of the wet lowland TRF regions usually contain more SOC than those of the moist or dry ecoregions. The SOC pool under dry TRF for some upland soils in west Africa range from 4 to 10 kg/m^2 to 1-m depth (Table 1). Kang and Juo (1986) observed that an Oxisol from Itaituba, Brazil, contained about 3 times more SOC in the surface 0 to 10 cm layer than an Oxisol from Zaire or an Alfisol from Nigeria. Landscape position, through its effect on soil depth and moisture regime, can also have a drastic effect on SOC and associated N and P contents. Differences in SOC content are primarily due to the total biomass pool, especially with regard to the litter fall and the root biomass. The above ground biomass production, rate of litter fall and root biomass also vary widely depending on soil and other edaphological factors. The relative distribution of C in soil vs. the above ground biomass may be drastically different in temperate compared with the tropical ecosystems. Some researchers believe that the SOC content of the soils of the humid tropics compare favorably with those of the temperate zone soils. Sanchez et al. (1982) compared data from 61 randomly chosen profiles from the tropics with 41 from the temperate region for soils classified as Oxisols, Mollisols, Alfisols, and Ultisols. They observed no significant differences in total C and C:N ratio (to 1-m depth) between soils from the tropics and those from

Table 1. Soil organic carbon profile to 1-meter depth under secondary forest regrowth in some upland soils of western Nigeria

Depth (cm)	Iwo	Egbeda	Ibadan	Apomu	Mean
			—kg/m^2—		
0-10	1.8	1.9	2.0	2.7	2.1 ± 0.4
10-20	2.0	1.5	2.0	0.4	1.5 ± 0.8
20-30	1.5	0.9	1.2	0.4	1.0 ± 0.8
30-40	0.9	0.9	0.6	0.3	0.7 ± 0.3
40-50	0.6	0.7	0.5	0.1	0.5 ± 0.3
50-60	0.4	0.4	0.4	0.02	0.3 ± 0.2
60-70	0.4	0.3	0.4	0.02	0.3 ± 0.2
70-80	0.6	0.4	0.04	0.02	0.3 ± 0.2
80-90	0.7	0.4	0.47	0.02	0.3 ± 0.2
90-100	0.7	0.2	0.4	0.02	0.3 ± 0.2
Total	9.6	7.6	8.3	4.0	7.4 ± 0.2

These values are computed from SOC content and bulk density of the fine earth fraction for each horizon, and extrapolation among horizons for respective depth increments.
(Recalculated from Moormann et al., 1975.)

temperate regions. However, as in temperate regions, estimates of C pool in soils of the tropics, vary widely depending on parent material, climate, vegetation, and terrain characteristics. In the Pacific region, Kira and Shidei (1967) observed that surface soil under temperate forest contained a much larger proportion of total organic carbon than that in tropical forest ecosystems.

IV. Land Use Effects on Carbon Cycling

The SOC pool depends on rates of renewal (source) and loss (removal) of carbon from the soil. The primary source of the soil carbon is through biomass produced and its incorporation into the soil. The principal loss of carbon is through respiration, decomposition, erosion, and leaching. Land use and management affect both the magnitude of production and rate of removal of SOC. These principal land use activities affecting the SOC pools, and rates of accretion and losses are described below.

A. Deforestation

Deforestation affects SOC content through its influence on the quantity and quality of biomass produced and incorporated into the soil. Soon after deforestation, there may be little or no effect on SOC content (Lal and Cummings, 1979; Table 2). Hernani et al. (1987) also observed no effect of land clearing on SOC content soon after deforestation of a Yellow Latosol of the Ribeira Valley, Sao Paulo, Brazil. Deforestation by inappropriate methods, however, can lead to removal of topsoil by a dozer blade to windrows thereby exposing the subsoil containing low SOC content. Mechanized clearing can also lead to soil compaction, reduction in infiltration and increase in runoff, and soil erosion that reduce SOC content.

Tillage and intensive cropping following deforestation usually decrease SOC by 25 to 60% for the top 0 to 5 cm depth and 15 to 30% for 5 to 15 cm depth (Cunningham, 1963; Seubert et al., 1977; Sanchez and Salinas, 1981). Aina (1979) reported that SOC content in the 0 to 10 cm depth under 15-

Table 2. Effects of land clearing methods on soil organic carbon (SOC) and total soil nitrogen (TSN) contents of 0 to 10 cm depth

Clearing method	SOC		TSN		C:N ratio	
	I	F	I	F	I	F
			% by weight			
Mechanical	2.00	2.10	0.340	0.261	5.9	8.0
Slash and burn	2.05	2.34	0.353	0.412	5.8	5.7
Slash	2.18	2.41	0.356	0.361	6.1	6.7
Control	2.20	2.50	0.366	0.489	6.0	5.1
LSD (0.05)	1.09	0.65	0.068	0.077	-	-

I = Initial   F = final soon after clearing
(Modified from Lal and Cummings, 1979.)

Table 3. Land use effects on soil organic carbon profile of an Alfisol (Iwo) in western Nigeria

Depth (cm)	Bush fallow		10 years of *Cynodon dactylon*	10 years cultivation
	15 - 20 years	20 - 25 years		
		kg/m^2		
0-7	1.83	1.89	2.29	0.70
7-14	0.94	1.07	1.61	0.59
14-21	0.65	0.70	0.82	0.52
21-28	0.42	0.44	0.53	0.38
Total	3.84	4.10	5.25	2.19

- Cultivated plots involved rotation of maize-maize for 2 years, yam-cassava for 2 years, maize-sorghum-cassava for 2 years, maize-maize for 2 yers, and yam-maize-maize for 3 years with low level of fertilizer input.
- Soil organic carbon was measured by the wet combutsion method, and average soil bulk density by the core method.
(Recalculated from Aina, 1979.)

25-year forest regrowth was about 2.6% compared with only 0.6% in a soil cultivated for 10 years (Table 3). Lal (1985) observed that SOC content of 0 to 10 cm soil layer decreased by 20% in 6 years of cultivation. For a coarse-textured soil of low inherent fertility, SOC content initially increased with deforestation and use of fertilizers and then decreased with onset of soil degradation due to intensive cultivation (Figure 1). In an experiment conducted on an Ultisol in southern Nigeria, Ghuman and Lal (1991) observed that SOC content in the non-windrow zone declined during 3 years of cultivation compared with that of the forested control. They also observed that SOC content of 0 to 5 cm layer differed among methods of land clearing, as was also observed in a similar experiment conducted in the Amazon region of Peru (Seubert, 1975; Seubert et al., 1977). Alegre et al. (1988) observed that SOC content declined in mechanically cleared treatments due to increased oxidation rates resulting from tillage and higher soil temperatures.

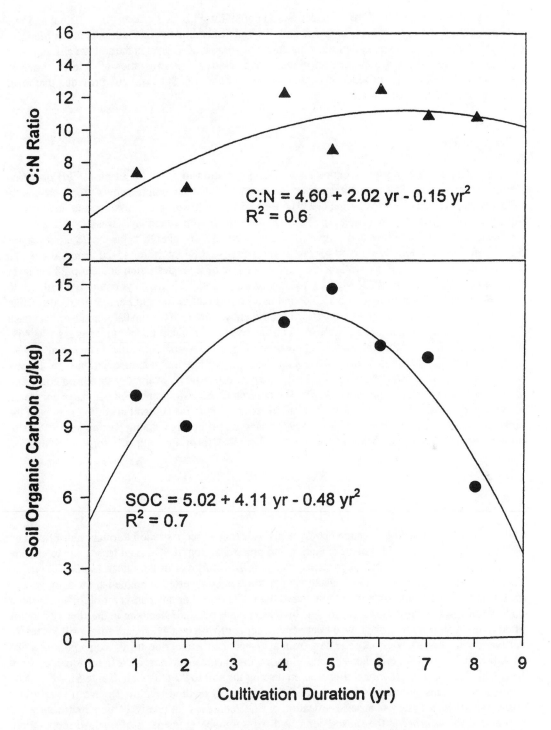

Figure 1. Temporal changes in mean soil organic carbon (SOC), total soil nitrogen (TSN), and C:N ratio. (Adapted from Lal, 1997b.)

B. Cropping Systems

Crop sequences and combinations influence SOC content through their effects on: (i) total and rate of biomass production, (ii) shoot:root ratio, and (iii) ratio of carbon to other nutrients in the biomass, e.g. C:N, C:P, C:S. Cropping systems that maintain a favorable SOC level in the soil are characterized by high total and below ground or root biomass and relatively narrow ratios of C to N, P and S. Cereals with high biomass production capacity can maintain high SOC content, provided that other essential nutrients are not limiting.

1. Tropical Grasses

Growing deep-rooted grasses is important to C sequestration in soil. Lal et al. (1979) observed significant increases in SOC content under grasses compared with leguminous cover crops, especially with *Brachiaria* and *Cynodon*. Aina (1979) observed that cultivation of grasses maintained higher SOC levels than bush fallow (Table 3). The SOC level in the top 10 cm layer was 2.5%, 2.2%, and 0.6% under grass, bush fallow, and cultivated soil, respectively. Total SOC in the top 28 cm layer was 5.25 kg/m^2 for *Cynodon*, 4.10 kg/m^2 for 20-25 years fallow, 3.84 kg/m^2 for 15-20 years fallow, and 2.19 kg/m^2 for 10 years of cultivation. The excessive rate of C sequestration in the top 28 cm layer of soil under *Cynodon* was 0.14 Mg/ha/yr, 0.12 Mg/ha/yr, and 0.31 Mg/ha/yr compared with 15-20 year fallow, 20-25 year fallow, and arable land use, respectively. Juo and Lal (1979) reported that guinea grass (*Panicum maximum*) sequestered significantly more C in the top 0-15 cm layer than bush fallow, *Leucaena* or pigeon pea (Table 4). The relative SOC content under guinea grass was 86% more than that under continuous soybean cultivation. In the Llanos of Colombia, Fisher et al. (1994; 1995) observed that pastures of *Brachiaria humidicola* alone and grown in association with *Arachis pintoi* sequestered 4.1 Mg C/ha/yr and 11.7 Mg C/ha/yr, respectively, over a 6-year period compared with the native savannas. Therefore, *A. pintoi* increased C sequestration by 7.6 Mg/ha/yr compared with the pure grass. Grass-based pastures sequestered more C in the sub-soil in comparison with the native savannas. Thomas and Asakawa (1993) observed differences among pasture species due to differences in C:N ratio which was 88 for B. *decumbens*, 130 for A. *gayanus*, 126 for B. *dictyoneura*, and 117 for B. *humidicola*.

2. Tillage

Soil disturbance and residue management systems influence C sequestration through their effects on biomass returned to the soil, rate of oxidation and mineralization of SOC, and losses due to erosion and leaching. Plow-based tillage systems involving soil turnover usually decrease SOC content. Therefore, when soils under native vegetation or pastures are converted to arable land use, the rate of decline in SOC is usually more with plow-based than with no-till or any conservation tillage system. Lal (1976a) observed that the rate of decline in SOC content was 0.03%/month in the first 12 months of cultivation after fallow for plowed treatments compared with 0.004%/month for no-till for maize cultivation in southern Nigeria. The data from a 17-year study under no-till vs. conventional till of an Alfisol in Western Nigeria showed that soil under no-till contained more SOC in the top 0 to 10 cm layer than that under conventional till based on discing the soil to 20-cm depth (Table 5). The SOC content was consistently more for the no-till than plow-till method (Figure 2). No-till and other conservation tillage systems cause stratification of SOC content with relatively higher concentration in the surface but lower in the subsoil compared with plow-based methods of seedbed preparation. Juo and Lal (1979) observed that ratio of SOC content for no-till:plow-till system was 2.5 for 0-5 cm

Table 4. Land use effects on SOC content of the top 0-15 cm layer 3 years after deforestation and cultivation

Treatment	SOC content (kg/m²)	Relative SOC content
A. Fallow		
Bush regrowth	3.0	136
Guinea grass	4.1	186
Leucaena	2.9	132
Pigeon pea	3.5	159
Mean	3.4	154
B. Cropping		
Maize (with residue mulch)	3.5	159
Maize (residue removed)	2.6	118
Maize-cassava	2.7	123
Soybean (with residue mulch)	2.2	100
Mean	2.8	116

SOC content on weight basis was multiplied with the bulk density for the 0-15 cm layer. (Recalculated from Juo and Lal, 1979.)

Table 5. Tillage effects on temporal changes in SOC content of 0-5 cm depths of an Alfisol in western Nigeria

	Sampling data			
Tillage method	January 1981	September 1983	July 1986	January 1987
	—% on weight basis—			
A. 0-5 cm depth				
No-till	2.07 ± 0.26	1.62 ± 0.22	1.86 ± 0.14	1.07 ± 0.26
Plow till	1.32 ± 0.15	1.33 ± 0.12	0.93 ± 0.03	0.54 ± 0.06
B. 5-10 cm depth				
No-till	1.31 ± 0.20	1.49 ± 0.14	1.08 ± 0.18	0.60 ± 0.07
Plow-till	$1.17 \pm o.23$	1.37 ± 0.13	0.90 ± 0.05	0.53 ± 0.05

(Unpublished data of R. Lal.)

depth, 1.5 for 5-10 cm depth and 1.1 for 10-15 cm depth (Table 6). The SOC content was generally more in the subsoil of plow-till than no-till treatments.

3. Mulch Farming

Residue management is an important aspect of seedbed preparation. Leaving crop residues on the soil surface as mulch is a beneficial practice in improving edaphological properties due to its favorable effects on soil and water conservation, soil temperature regime, activity of soil fauna, and increased SOC content. Lal et al. (1980) reported that the rate of decline of SOC content during the 18-month period after land clearing was 0.103, 0.100, 0.092, 0.083, and 0.078%/month for mulch rates of 0, 2, 4, 6, and 12 Mg/ha/season, respectively. The data in Table 7 and Figure 3 show that the rate of SOC decline decreased linearly with increase in mulch rate. One year after clearing, the rate of SOC decline

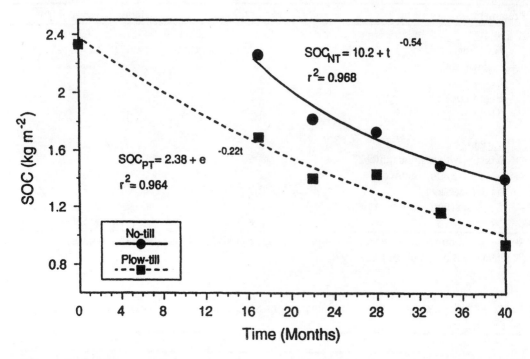

Figure 2. Tillage methods effects on SOC content of an Alfisol in western Nigeria (t = time in months; NT = no-till; PT = plow-till).

Table 6. Tillage effects on SOC profile of an Alfisol after 5 years of continuous cultivation

Depth (cm)	No-till	Plow-till	NT:PT
		—% by weight—	
0-5	2.87	1.17	2.5
5-10	1.77	1.19	1.5
10-15	1.23	1.10	1.1
15-20	1.11	1.09	1.0
20-25	0.71	0.92	0.8
25-30	0.96	0.82	1.2
30-35	0.57	0.82	0.7
35-40	0.47	0.62	0.8
40-45	0.42	0.47	0.9
45-50	0.34	0.37	0.9

(Adapted from Juo and Lal, 1979.)

Table 7. Mulch rate effects on soil organic carbon content of 0-5 cm layer at different times after land clearing

Mulch rate (Mg/ha/season)	Months after land clearing			Rate of SOC decline	
	0	12	18	12 months	18 months
	—kg/m²—			—Mg/ha/yr—	
0	1.71	1.09	0.92	6.2	5.3
2	1.68	1.20	0.89	4.8	5.3
4	1.65	1.19	0.89	4.6	5.1
5	1.68	1.26	1.02	4.2	4.4
12	1.68	1.29	1.02	3.9	4.4
Mean	1.68 ± 0.02	1.21 ± 0.08	0.95 ± 0.07	4.7 ± 0.9	4.9 ± 0.5

(Recalculated from Lal et al., 1980.)

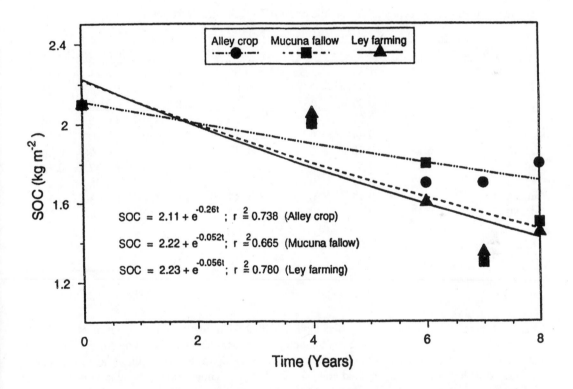

Figure 3. Land use effects on temporal changes in soil organic carbon content of Alfisols in western Nigeria (recalculated from Lal, 1996a, b; t = time in years).

Table 8. Effects of crop residue management on SOC content of an Alfisol in western Nigeria

Tillage method	Sampling data			
	January 1981	September 1983	July 1986	January 1987
	% on weight basis			
A. 0-5 cm depth				
Residue removed	1.68 ± 0.33	1.35 ± 0.14	1.21 ± 0.38	0.67 ± 0.21
Residue returned	1.70 ± 0.59	1.60 ± 0.23	1.58 ± 0.73	0.94 ± 0.41
B. 5-10 cm depth				
Residue removed	1.16 ± 0.26	1.44 ± 0.20	0.95 ± 0.12	0.53 ± 0.05
Residue returned	1.32 ± 0.15	1.39 ± 0.10	1.03 ± 0.20	0.60 ± 0.08

Each figure is an average of 12 separate measurements.
(Unpublished data of R. Lal.)

Table 9. Mulch rate effects on SOC profile of an Alfisol in western Nigeria

Soil depth (cm)	Mulch rate (Mg/ha)					Mean	LSD (0.05)
	0	2	4	6	8		
0-5	1.54	1.83	1.55	1.59	2.01	1.70 ± 0.21	0.57
5-10	1.21	0.94	1.25	1.07	1.14	1.14 ± 0.12	0.23
10-15	0.95	0.89	1.05	0.74	1.02	0.93 ± 0.12	0.36
15-20	0.69	0.57	0.89	0.56	0.87	0.72 ± 0.16	0.30
20-25	0.66	0.44	0.59	0.55	0.68	0.58 ± 0.10	0.19
25-30	0.70	0.45	0.54	0.46	0.55	0.54 ± 0.10	0.20
30-35	0.57	0.46	0.49	0.52	0.37	0.48 ± 0.07	0.28
35-40	0.52	0.47	0.44	0.54	0.40	0.47 ± 0.06	0.24
40-45	0.44	0.47	0.52	0.47	0.47	0.47 ± 0.03	0.12
45-50	0.38	0.42	0.42	0.47	0.47	0.43 ± 0.04	0.11
Mean	0.77 ± 0.37	0.69 ± 0.44	0.77 ± 0.39	0.70 ± 0.36	0.80 ± 0.50		

(Unpublished data of R. Lal.)

was 6.2 Mg/ha/yr for unmulched control compared with 3.9 Mg/ha/yr for mulch rate of 24 Mg/ha/yr. The data in Table 8 on SOC content from another experiment also conducted on Alfisols show that SOC content was consistently more for treatments in which crop residue was returned as surface mulch compared with that of the unmulched control. Surface application of crop residue mulch can also influence the SOC content of the subsoil (Table 9). There was somewhat higher SOC content in subsoil of plots receiving high compared with those receiving low-mulch rates. The translocation of SOC in the subsoil may be facilitated by activity of soil fauna, e.g., earthworms, termites.

4. Nutrient Management

The data in Tables 6 to 9 show that the high rate of decomposition of SOC could not be compensated by addition of crop residues even at the high rates of 8 to 12 Mg/ha/season. The decomposition rate of crop residues can be as high as 25 to 30% in one month for cereals and 40 to 50% for legumes. The

Table 10. Tillage system and alley cropping effect on soil organic dynamics for 0-10 cm depth of an Alfisol in western Nigeria

Treatment	Cultivation duration (years)					SOC decline in 4 years
	0	1	2	3	4	
	kg/m²					Mg/ha/yr
Plow-till	2.1	1.5	1.1	1.3	0.5	4.0
No-till	1.9	2.1	1.1	1.6	1.4	1.25
Leucaena - 4m	2.1	2.2	1.7	2.0	1.4	1.75
Leucaena - 2m	2.1	1.9	0.9	1.5	1.0	2.75
Gliricidia - 4m	2.2	2.1	1.5	1.5	1.0	3.0
Gliricidia - 2m	2.1	1.4	1.3	1.6	1.0	2.75
Mean	2.1 ± 0.1	1.9 ± 0.3	1.3 ± 0.3	1.6 ± 0.2	1.1 ± 0.3	2.5 ± 1.0

(Recalculated from Lal, 1989a,b.)

retention of organic carbon in the soil as humus is also determined by the availability of nutrients, especially those of N, P, and S (Himes, 1997). If the SOC content is to be increased by 10,000 kg/ha, in addition to biomass or straw (67,340 kg @ 45% C in the biomass and only 35% converted to humus), other nutrients required are 833 kg of N (C:N of 12:1), 200 kg of P (C:P of 50:1), and 143 kg of S (C:S of 70:1). Addition of crop residues alone in a nutrient-deficient soils is not enough. For the C in residues to be converted to humus, it is necessary to have adequate amounts of N, P, S, and other essential elements.

5. Agroforestry and Perennial Crops

Growing trees and woody perennials in association with crops and pastures can be a beneficial practice in conserving soil and water, strengthening nutrient cycling mechanisms, and minimizing risks of SOC decline (Lal, 1987; 1989a, b). The data in Table 10 and Figure 4 from a coarse-textured soil in western Nigeria show that the rate of SOC decline was less with *Leucaena* and *Gliricidia* based systems than with the plow-till method of seedbed preparation. The *Leucaena* hedgerows planted at a 4-m interval maintained SOC content equivalent to that of the no-till plot even though soil was disced in the *Leucaena* treatment. Similar trends are apparent from a large watershed experiment conducted with mechanized systems of farm operation (Lal, 1996a, b; Table 11). The rate of SOC decline over an 8-year period was 0.038 kg/m²/yr with alley cropping compared with 0.05 to 0.063 kg/m²/yr with other farming systems including grazed pastures. The rate of SOC decline can be drastically reduced by growing perennials which do not require soil disturbance and cultivation. The data in Table 12 show that the rate of SOC decline was reduced from 2 to 10 times by growing perennials compared with cassava cultivation. In contrast, the SOC content increased at the rate of 50 kg/ha/yr in the top 10 cm depth of an improved forestry plantation of *Cassia siamia* (Ghuman and Lal, 1990). The rate of litter fall under *Cassia* was 5 to 7 Mg/ha/yr which helped maintain high SOC content.

C. Soil Erosion and Carbon

Accelerated soil erosion is a principal factor responsible for depletion of SOC in cultivated soils (Lal, 1995); however, the magnitude of its impact on SOC content is not known. Erosion leads to

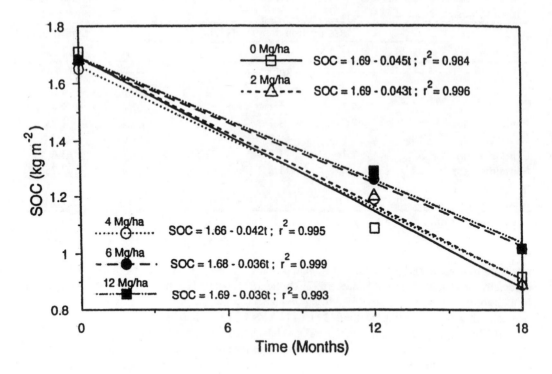

Figure 4. Mulching effects on soil organic carbon content of an Alfisol at different times after deforestation (recalculated from Lal et al., 1980; t = time in months).

Table 11. Deforestation and land use effects on SOC content of 0-10 cm layer of Alfisols under intensive mechanized cultivation

Treatment	under forest	Cultivation duration (years)				Cultivation mean	Rate of SOC decrease
		4	6	7	8		
			kg/m²				kg/m²/yr
Alley cropping	2.1	2.0	1.7	1.7	1.7	1.8	0.038
Mucuna fallow -A	2.1	2.1	1.8	1.2	1.8	1.7	0.05
Mucuna fallow - B	2.1	1.9	1.8	1.4	1.3	1.6	0.063
Ley farming - A	2.1	2.1	1.7	1.1	1.6	1.6	0.063
Ley farming - B	2.1	2.0	1.5	1.6	1.3	1.6	0.063
Mean	2.1	2.0	1.7	1.4	1.5	1.6	

(Recalculated from Lal, 1996a,b.)

Table 12. Effects of land use systems on SOC content of 0 to 10 cm depth Ultisol in southern Nigeria

Land use system	SOC content at different times after clearing		
	2 years	4 years	ΔSOC
		kg/m^2	
Cassava-based	1.75	1.29	-0.46
Oil palm-based	1.47	1.43	-0.04
Alley cropping	1.44	1.36	-0.08
Plantain	1.59	1.37	-0.22
Pasture	1.68	1.50	-0.18
Improved forestry	1.43	1.44	+0.01
Mean	1.56 ± 0.13	1.40 ± 0.07	

Each value was computed by multiplying the SOC content (% on weight basis) with soil bulk density (Mg/m^3) and soil depth (0.1m).
(Calculated from Ghuman and Lal, 1991a,b.)

preferential removal of clay and humus fraction. Several experiments have shown that the enrichment ratio for SOC is about 3 to 5 (Lal, 1976b, c), and there is usually a linear decline in SOC content with accumulative soil erosion (Eq. 1; Lal, 1980).

$$\text{SOC content (\%)} = 1.79 - 0.002E, r = -0.71** \quad \text{...............(Eq. 1)}$$

where E is the annual accumulative soil erosion (Mg/ha). While the on-site effects of soil erosion on SOC are adverse, the overall impact on landscape or watershed level are debatable. The principal discrepancy lies in the unknown fate of SOC on depositional sites. The pathways of SOC deposited in depressional sites, reservoirs, waterways, and oceans are not known. It is likely that C deposited in terrestrial ecosystems is sequestered (Lal, 1995). There is a need to develop a database regarding the fate of carbon translocated by erosion and deposited over the landscape and aquatic ecosystems.

D. Soil Restoration

Restoration of degraded soils involves improvements in total and biomass organic carbon and associated soil properties and processes, e.g., aggregation. Restoration may be achieved by natural fallow, planted fallow or cover crops, use of organic wastes, and soil amendments. Growing cover crops and woody perennials is an important strategy for soil restoration. The data in Table 13 from an Alfisol show that the rate of SOC sequestration in the top 10 cm layer in comparison with uncultivated control was 2.6 Mg/ha/yr for *Cynodon* and 2.4 Mg/ha/yr for *Brachiaria* (Lal et al., 1978). *Stylosanthes* and *Psophocarpus* were also effective in C sequestration (Lal et al., 1979). The rate of SOC sequestration depends on the degree of soil degradation. The greater the depletion of SOC content, the higher the rate of SOC sequestration. The data in Table 14 is from an experiment conducted to evaluate restoration effectiveness of various crop covers. Once again, it is apparent that the rate of SOC sequestration was extremely high under grasses *Glycine* and *Melinis* ranging from 7 to 12 Mg/ha/yr in the top 20 cm depth.

Table 13. Effects of growing 2 years of cover crops on SOC content of top 0-10 cm depth of an eroded Alfisol in western Nigeria

Cover crop	Soil bulk density (Mg/m^3)	SOC content (%)		Carbon sequestration rate (Mg/ha/yr)
		Before	After	
Brachiaria	1.34	1.21	1.57	2.41
Paspalum	1.35	1.23	1.45	1.49
Cynodon	1.30	1.30	1.70	2.60
Pueraria	1.32	1.27	1.50	1.52
Stylosanthes	1.33	1.30	1.63	2.19
Stizolobium	1.33	1.30	1.57	1.80
Psophocarpus	1.14	1.20	1.57	2.11
Centrosema	1.33	1.30	1.53	1.53
Control	1.42	1.33	1.37	0.28
LSD (0.05)	0.04	0.50	0.23	

(Recalculated from Lal et al., 1979.)

Table 14. Effects of 2 years of grass and leguminous cover crops on soil organic carbon content of 0 to 20 cm depth for the Apomu soil series in western Nigeria

Cover crop	SOC content (kg/m^2)	Carbon sequestration rate (Mg/ha/yr)
Panicum	2.44	1.4
Setaria	1.89	-1.3
Brachiaria	2.52	1.8
Melinis	3.57	7.1
Centrosema	2.17	0.1
Pueraria	2.59	2.1
Blycine	4.62	12.3
Stylosanthes	2.24	0.4
Control	2.16	-

Soil bulk density was 1.40 Mg/m^3 for cover crops and 1.60 for the control.
(Recalculated from Lal et al., 1978.)

V. Carbon Dynamics and Strategies for its Sequestration in Soil

The SOC is a dynamic entity and is affected by numerous interacting factors. The SOC accretion and humification are influenced by: (i) rate of biomass return through litter fall and crop residues, (ii) root biomass and its distribution with depth, (iii) biomass organic carbon especially activity of soil macrofauna, e.g., earthworms, termites, and (iv) soil fertility management including use of inorganic fertilizers and organic amendments. In contrast, processes leading to a decline in the SOC content are: (i) mineralization and oxidation of organic substances, (ii) soil erosion, and (iii) leaching. The equilibrium level of SOC depends on the balance between factors and processes that increase and decrease SOC content. The rate of change of SOC in soil depends on the amount present and the management (Eq. 2) (Nye and Greenland, 1960, 1964; Jenkinson and Raynor, 1977; Stevenson, 1982):

$$\frac{dC}{dt} = -kC + A \quad \text{(Eq. 2)}$$

Table 15. Tillage and residue management effects on decomposition constant for a West Nigerian Alfisol

Treatment	Decomposition constant (k)	Accretion constant (A)
No-till	-0.25	0.47
Plow-till	-0.17	0.34
Residue returned	-0.22	0.44
Residue removed	-0.20	0.37

(Unpublished data of R. Lal.)

where k is the decomposition constant, C is the SOC content at a time t, and A is the accretion constant. The magnitude of A depends on land use and management. It is the difference between kC and A that determines the rate of change of SOC. Soil degradative processes decrease SOC and increase the magnitude of the decomposition constant k. In contrast, soil restorative processes increase SOC and decrease the magnitude of k. Agricultural and land use practices that increase k include: mechanized deforestation, plowing, continuous cultivation, residue removal, low-input subsistence agriculture, excessive grazing, etc. In contrast, agricultural practices that decrease k include: fallowing, cover crops, afforestation and perennial plantations, improved pastures with controlled grazing, no-till and conservation tillage, mulch farming, agroforestry techniques, soil fertility management, etc. The decomposition constant k and accretion constant A for different mulch rates on an uncropped soil are shown in Table 15. Both the decomposition constant and the accretion constant increased with increase in mulch rate. These values are in accord with the observation that the rate of decline of SOC decreases with increase in rate of mulch application. It is apparent that k is low for no-till, grass cover crops, and alley cropping systems, and high for plow based and continuous cropping.

If soil is continuously cultivated, the SOC content declines until an equilibrium level (C_e) is achieved. The magnitude of C_e depends on the climate, land use, and cropping system. The difference (ΔC) in the initial (C_o) and the C_e comprises: gaseous emission into the atmosphere, and losses due to soil erosion and leaching as dissolved and particulate C. Considering these factors, the SOC content in the soil at time t is given by Eq. 3:

$$C = C_e + (C_o - C_e)e^{-rt} \quad \text{(Eq. 3)}$$

where C is SOC content at time t, r is a fraction of C decomposed per year, and t is time in years.

All other factors remaining the same, the rate of SOC decomposition is much higher in the tropics than in temperate climate. The Van't Hoff's law states that the rate of any chemical reaction approximately doubles for every 10 °C rise in temperature. Jenkinson and Ayanaba (1977) reported that the rate of decomposition of organic matter at Ibadan, Nigeria was 4 times that at Rothamsted, England. Consequently, the SOC content is inversely related to mean annual temperature (Eq. 4):

$$C = ne^{-kT} \quad \text{(Eq. 4)}$$

where T is mean annual temperature and n is an empirical constant.

Considering important ecological characteristics of the TRF (e.g., high mean annual temperature, high soil erosion risks, soils with predominantly low-activity clays), appropriate land use and management strategies that enhance SOC content are: (i) tree crops and plantations, (ii) deep-rooted grasses and improved pastures, (iii) cover crops, (iv) mulch farming, (v) conservation tillage, and (vi) judicious inputs of inorganic fertilziers and organic amendments. These practices decrease the decomposition constant k, improve soil structure and aggregation, decrease soil degradative processes (e.g., erosion and leaching), and increase nutrient cycling and other ecosystem restorative mechanisms.

References

Aina, P.O. 1979. Soil changes resulting from long-term management practices. *Soil Sci. Soc. Am. J.* 43:173-177.

Alegre, J.C., D.K. Cassel, and D.E. Bandy. 1988. Effect of land clearing method on chemical properties of an Ultisol in the Amazon. *Soil Sci. Soc. Am. J.* 52: 1283-1288.

Cunningham, R.K. 1963. The effect of clearing a tropical forest soil. *J. Soil Sci.* 14: 334-345.

Dale, V.H., R.A. Houghton, A. Grainger, A.E. Lugo, and S. Brown. 1993. Emission of greenhouse gases from tropical deforestation and subsequent uses of the land. p. 215-260. In: *Humid Tropics*. National Research Council, National Academy of Sciences, Washington, D.C.

Eswaran, H., E. Van Den Berg, and P. Reich. 1993. Organic C in soils of the world. *Soil Sci. Soc. Am. J.* 57:192-194.

Fisher, M.J., I.M. Rao, M.A. Ayarza, C.E. Lascano, J.I. Sanz, R.J. Thomas, and R.R. Vera. 1994. Carbon storage by introduced deep-rooted grasses in the South American savannas. *Nature* 371:236-238.

Fisher, M.J., I.M. Rao, C.E. Lascano, J.I. Sanz, R.J. Thomas, R.R. Vera, and M.A. Ayarza. 1995. Pasture soils as carbon sink. *Nature* 376:472-473.

Ghuman, B.S. and R. Lal. 1990. Nutrient addition into soil by leaves of *Cassia siamea* and *Gliricidia sepium* grown on an Ultisol in southern Nigeria. *Agroforestry Systems* 10:131-133.

Ghuman, B.S. and R. Lal. 1991. Land clearing and use in the humid Nigerian tropics. II. Soil chemical properties. *Soil Sci. Soc. Am. J.* 55:184-188.

Grace, J., J. Lloyd, J. McIntyre, A.C. Miranda, P. Meir, and H.S. Miranda. 1996. Carbon dioxide flux over Amazonian rainforest in Rondonia. p. 307-318. In: J.H.C. Gash, C.A. Nobre, J.M. Roberts, and R.L. Victoria (eds.), *Amazonian Deforestation and Climate*, John Wiley & Sons, Chichester, U.K.

Hernani, L.C., E. Sakai, I. Ishimura and I.F. Lepsch. 1987. Effect of clearing a secondary forest on a yellow latosol of the Ribeira Valley, Sao Paulo, Brazil. I. Dynamics of physical and chemical properties and maize production. *Revista Brasileira de Ciencia do Solo* 11:205-213.

Himes, F.L. 1997. Nitrogen, sulfur and phosphorus and the sequestering of carbon. In: R. Lal, J. Kimble, R. Follett, and B.A. Stewart (eds.), *Management of Carbon Sequestration*. CRC, Boca Raton, FL (In press).

Houghton, R.A. 1990. The global effects of tropical deforestation. *Env. Sci. Technol.* 24:414-422.

Hougthon, R.A. 1995. In: R. Lal, J. Kimble, E. Levine, and B.A. Stewart (eds.), *Soils and Global Change*. CRC/Lewis Publishers, Boca Raton, FL.

Jenkinson, D.S. and A. Ayanaba 1977. Decomposition of C^{14}-labelled plant material under tropical conditions. *Soil Sci. Soc. Am. J.* 41:912-915.

Jenkinson, D.S. and J.H. Raynor. 1977. The turnover of soil organic matter in some of the Rothamsted classical experiments. *Soil Sci.* 123:298-305.

Jordon, C.F. (ed) 1989. An Amazon rainforest: The structure and function of a nutrient-stressed ecosystem and the impact of slash-and-burn agriculture. Man in the Biosphere series vol. 2, UNESCO, Paris, and the Parthenon Publ. Group Ltd., U.K.

Juo, A.S.R. and R. Lal. 1979. Nutrient profile in a tropical Alfisol under conventional and no-till systems. *Soil Sci.* 127:168-173.

Kang, B.T. and A.S.R. Juo. 1986. Effect of forest clearing on soil chemical properties and crop performance. p. 383-394. In: R. Lal, P.A. Sanchez, and R.W. Cummings, Jr. (eds.), *Land Clearing and Development in the Tropics*. A.A. Balkema Publishers, Rotterdam, Holland.

Kimble, J.M., T. Cook, and H. Eswaran. 1990. Organic matter in soils of the tropics. p. 250-258. In: *Proc. Symp. on Characterization and Role of Organic Matter in Different Soils.* 14th Int'l Cong. Soil Sci., Kyoto, Japan,

Kira, T. and T. Shidei. 1967. Primary production and turnover of organic matter in different forest ecosystems of Western Pacific. *Japan J. Ecol.* 17:70-97.

Kruijt, B., J. Lloyd, J. Grace, J.A. McIntyre, G.D. Farquhar, A.C. Miranda, and P. McCracken. 1996. Sources and sinks of CO_2 in Rondonia tropical rainforest. p. 331-351. In: J.H.C. Gash, C.A. Nobre, J.M. Roberts, and R.L. Victoria (eds.), *Amazonian Deforestation and Climate*, John Wiley & Sons, Chichester, U.K.

Lal, R. 1976a. No-tillage effects on soil properties under different crops in western Nigeria. *Soil Sci. Soc. Amer. Proc.* 40:762-768.

Lal, R. 1976b. Soil erosion on Alfisols in western Nigeria. I. Effects of slope, crop rotation and residue management. *Geoderma* 16: 363-375.

Lal, R. 1976c. *Soil Erosion Problems on Alfisols in Western Nigeria and Their Control.* IITA Monograph I, Ibadan, Nigeria, 208 pp.

Lal, R. 1985. Mechanized tillage systems effects on properties of a tropical Alfisol in watershed cropped to maize. *Soil and Tillage Res.* 6:149-162.

Lal, R. 1987a. Conversion of tropical rainforest: agronomic potential and ecological consequences. *Adv. Agron.* 39:173-264.

Lal, R. 1987. *Tropical Ecology and Physical Edaphology.* John Wiley & Sons, Chichester, U.K., 707 pp.

Lal, R. 1989a. Agroforestry systems and soil surface management of a tropical Alfisol. III. Changes in soil chemical properties. *Agroforestry Systems* 8:113-132.

Lal, R. 1989b. Agroforestry systems and soil surface management of a tropical Alfisol. IV. Effects on soil physical and mechanical properties. *Agroforestry Systems* 8:197-215.

Lal, R. 1995. Global soil erosion by water and C dynamics. p. 131-142. In: R. Lal, J. Kimble, E. Levine, and B.A. Stewart (eds.), *Soils and Global Change.* CRC/Lewis Publishers, Boca Raton, FL.

Lal, R. 1996a. Deforestation and land use effects on soil degradation and rehabilitation in western Nigeria. I. Soil physical and hydrological properties. *Land Degradation and Development* 7:19-45.

Lal, R. 1996b. Deforestation and land use effects on soil degradation and rehabilitation in western Nigeria. II. Soil chemical properties. *Land Degradation and Development* 7:87-98.

Lal, R., G.F. Wilson, and B.N. Okigbo. 1978. No-till farming after various grasses and leguminous cover crops in tropical Alfisol. I. Crop performance. *Field Crops Res.* 1:71-84.

Lal, R. and D.J. Cummings. 1979. Clearing a tropical forest. I. Effects on soil and microclimate. *Field Crops Res.* 2:91-107.

Lal, R., G.F. Wilson, and B.N. Okigbo. 1979. Changes in properties of an Alfisol produced by various crop covers. *Soil Sci.* 127: 377-382.

Lal, R., D. De Vleeschauwer, and R.M. Nganje. 1980. Changes in properties of a newly cleared Alfisol as affected by mulching. *Soil Sci. Soc. Am. J.* 44:827-833.

Lal, R. and T.J. Logan. 1995. Agricultural activities and greenhouse gas emissions from soils of the tropics. In R. Lal, J. Kimble, E. Levine, and B.A. Stewart (eds.), *Soil Management and Greenhouse* Effect. CRC/Lewis Publishers, Boca Raton, FL: 293-307.

Lugo, A. and S. Brown. 1991. Tropical forests as sinks of atmospheric carbon. *Forest Ecol. Management.*

Medina, E. and E. Cuevas. 1989. Patterns of nutrient accumulation and release in Amazonian forests of the upper Rio Negro basin. p. 217-240. In: J. Proctor (ed.), *Mineral Nutrients in Tropical Forest and Savannas Ecosystems.* Special Publication Series, British Ecological Society No. 9, Blackwell Scientific Publ. Oxford.

Medina, E. and E. Cuevas. 1996. Biomass production and accumulation in nutrient-limited rainforests: implications for responses to global change. p. 221-239. In: J.H.C. Gash, C.A. Nobre, J.M. Roberts, and R.L. Victoria (eds.), *Amazonian Deforestation and Climate*. John Wiley & Sons, Chichester, U.K.

Meir, P., J. Grace, A. Miranda and J. Lloyd. 1996. Soil respiration in Amazonia and in cerrado in central Brazil. p. 319-329. In: J.H.C. Gash, C.A. Nobre, J.M. Roberts, and R.L. Victoria (eds.), *Amazonian Deforestation and Climate*. John Wiley & Sons, Chichester, U.K.

Miranda, A.C., H.S. Miranda, J. Lloyd, J. Grace, J.A. McIntyre, P. Meir, P. Riggan, R. Lockwood, and J. Brass. 1996. Carbon dioxide flux over a cerrado sensu stricto in central Brazil. p. 353-363. In: J.H.C. Gash, C.A. Nobre, J.M. Roberts, and R.L. Victoria (eds.), *Amazonian Deforestation and Climate*. John Wiley & Sons, Chichester, U.K.

NRC 1993. Humid Tropics. National Research Council, National Academy of Sciences, Washington, D.C.

Nye, P.H. and D.J. Greenland. 1960. The Soil Under Shifting Cultivation. Commonwealth Bureau of Soils, Hapenden, England Tech. Comm. No. 51.

Nye, P.H. and D.J. Greenland. 1964. Changes in soil after clearing a tropical forest. *Plant Soil* 21:101-112.

Salati, E. and P.B. Vose. 1994. Amazon Basin: a system in equlibrium. *Science* 225:129-138.

Sanchez, P.A. and I.G. Salinas. 1981. Low-input technology for managing Oxisols and Ultisols in tropical America. *Adv. Agron.* 34:279-406.

Sanchez, P.A., D.E. Bandy, J.H. Villachica, and I.I. Nicholaides. 1982. Amazon Basin soils: management of continuous crop production. *Science* 216:821-827.

Sanchez, P., M.P. Gichuru, and L.B. Katz. 1982. *Organic Matter in Major Soils of the Tropical and Temperate Regions.* Trans. XIIth Int'l Cong. Soil Sci., New Delhi, India: 99-114.

Sanchez, P.A., C.A. Palm, L.T. Scott, E. Cuevas, and R. Lal. 1989. Organic input management in tropical agroecosystems. p. 125-152. In: D.C. Coleman, J.M. Oades, and G. Uehara (eds.), *Dynamics of Soil Organic Matter in Tropical Ecosystems.* Univ. of Hawaii Press, Honolulu.

Seubert, C.A. 1975. Effects of Land Clearing Methods on Crops Performance and Changes in Soil Properties in an Ultisol of the Amazon Jungle in Peru. M.Sc. Thesis, North Carolina State Univ., Raleigh, N.C., 152 pp.

Seubert, C.E., P.A. Sanchez and C. Valvarde. 1977. Effects of land clearing methods on soil properties of an Ultisol and crop performance in the Amazon jungle of Peru. *Trop. Agric.* 54:307-321.

Stevenson, F.J. 1982. *Humus Chemistry: Genesis, Composition and Reactions.* John Wiley & Sons, NY, 443 pp.

Thomas, R.J. and N.M. Asakawa. 1993. Decomposition of leaf litter from tropical forage grasses and legumes. *Soil Biol. Biochem.* 25:1351-1361.

Tiessen, H., P. Chacon, and E. Cuevas. 1994a. Phosphorus and nitrogen status in soils and vegetation along a toposequence of dystrophic rainforests on the upper Rio Negro. *Oecologia (Berl.)* 99:145-150.

Tiessen, H., E. Cuevas, and P. Chacon. 1994b. The role of soil organic matter in sustaining soil fertility. *Nature* 371:783-785.

CHAPTER 10

Arctic Paleoecology and Soil Processes: Developing New Perspectives for Understanding Global Change

Wendy R. Eisner

I. Introduction

The arctic tundra is a major ecosystem which plays a significant role as a major repository of the world's carbon. Whether northern peat is still accumulating soil carbon or may actually be losing carbon is a question with important consequences for global change scenarios. The fundamental relationship between past climate vegetation and soil in the Arctic has important implications for understanding the modern tundra. Many of the soils of the Arctic preserve pollen and other fossil material which record the process of plant succession and carbon dynamics since at least the late Pleistocene, over 13,000 years ago.

Pollen, as well as spores, fungi, algae, and larger plant remains can be analyzed as proxy information of past vegetation, hydrology, nutrient, and temperature variation, offering ecologists and soil scientists a very long-term monitoring device for understanding landscape dynamics in this complex environment. The objective of this chapter is to demonstrate the advantages of studying pollen, spores, and other fossil material in tandem with soil properties, and to present some methodologies which soil scientists and paleoecologists can adopt to further understand soil processes, vegetation, and climatic influences in the Arctic.

The relationship between vegetation and soil has important implications for understanding the modern arctic tundra. Stratification of silt deposits at various degrees of reduction, peat deposition in depressions, and cryogenic structure are all the combined results of past climate, vegetation, loess transport, and other landscape changes. The environmental processes related to carbon accumulation and soil development are tightly linked in the Arctic, and it is logical to study these changes on the time scale that the soils actually developed, the 100 to 1000 year time scale. Pollen analysis, or to use the term referring to the broader discipline, paleoecology, holds the key to understanding long term climatic controls over landscape processes.

The problem–to understand the controls on carbon storage in the Arctic–requires research on how these controls operated in the past. Investigation of carbon accumulation rates is required to recognize rates of change in arctic soil processes on a large geographic scale, but it is equally necessary to investigate the processes which are creating these sinks and potential sources of carbon (Oechel, et al., 1993). Successional processes are very slow in the tundra region, resulting in a mosaic of seral communities associated with various ages of surfaces (Webber, 1978; Billings and Peterson, 1980; Bliss and Peterson, 1992). Loess deposition, important during the late Pleistocene and Holocene, helped stabilize vegetation during the early development of the modern tundra, as it now acts to maintain vegetation at early succession stages (Walker and Everett, 1991).

How can pollen analysis help soil scientists to develop new insights into global change and arctic landscape processes? This chapter demonstrates the advantages of studying pollen and other fossil material in tandem with soil properties, to develop a better understanding of past and present arctic systems, and to use this information to make predictions of future changes.

II. Arctic Paleoecology

Pollen analysis, or palynology, has a variety of different practical applications. For the purposes of this chapter the focus is on the fossil record from various types of soil deposits, which may include peats or organic soils, mineral soils, buried soils, and the complex deposits of outcrop sections, such as riverine bluff formations. Pollen analysis usually refers not only to the study of fossil pollen, but fossil spores as well. Although pollen analysis is the most widely used paleoecological technique, when combined with the study of other fossil material preserved in the soil matrix, such as fungi, algae, insect and plant parts, a much wider range of information on past soil development becomes available. Paleoecological investigations of soils can give information on past vegetation, nutrient status, and soil development, and, in some cases, on landscape and regional scale climate changes. From the perspective of the paleoecologist, pedological analysis can supply critical information about the deposition environment of their research area, which can then be related to soil development.

North American palynologists have tended to place more emphasis on data collected from lacustrine (lake) deposits. Fossil pollen preserved in stratified sediments, especially from lakes, is a valuable source of palaeoclimatic information. Until recently, pollen stratigraphies were largely qualitative indications of climate change, but recent advances in our understanding of modern pollen-climate relationships, coupled with more accurate dating techniques, have enabled arctic palynologists to work toward a more accurate quantitative reconstruction of past environments (Anderson et al., 1991).

Pollen analysis is a powerful tool for understanding vegetation change and landscape processes. However, the particular conditions of the Arctic, with its dynamic hydrological variation, cryoturbation, and lack of sensitive indicator species, make the paleoecological analysis of most environments and sediments of the arctic tundra, including lakes, peats, and mineral soils, particularly challenging.

A key issue in the interpretation of pollen data is distinguishing between climate-induced change and non-climatic change. This problem should be familiar to anyone researching the effects of climate change on soil development, since soil processes, like vegetation, are effected by a myriad of influences which may be directly related to long-term changes in temperature and precipitation, indirectly related (for instance, by hydrologic changes) or independent of climate (human disturbance, vegetation succession).

Typical problems in the interpretation of pollen records include the following:
1. Inability to identify pollen source: regional or locally occurring?
2. Possible focusing or concentrating of pollen in one part of sediment.
3. Inability to identify pollen to species level.
4. Lag time in response of regional vegetation to abrupt environmental change.

The challenge for the paleoecologist working in collaboration with the soil scientist is to develop a feasible research program which will overcome these drawbacks, or at least alleviate them. The researchers must be very precise on what specific range of problems they are studying, and what question they want answered. Although the pollen analysis of arctic lakes is generally a robust signal for regional scale climate change, the paleoecological record from soils, peats, and complex depositional environments will tend to be affected by smaller scale, local processes.

III. Depositional Environments

The character of the depositional environment is probably the single most important consideration in determining the feasibility of conducting a paleoecological analysis of soils. In general, the ideal situation in which to find well preserved, undisturbed pollen and microfossil is where sediments have been accumulating gradually over time, where there is relatively little disturbance to this accumulation, and where there is low biotic activity. Thus, using these constraints, lakes and peat bogs are the most ideal study sites, but stable mineral soils, especially those in permafrost effected regions, where freezing preserves the pollen in place, can also be excellent repositories. Conversely, the character of much of the arctic landscape, with large scale eolian transport, erosion, cryoturbation, and recurring disturbances to stable sediment accumulation, makes such pollen repositories scarce, and often palynologists must work with less than ideal depositional environments. However, the combination of paleoecological pedological analysis in these situations offers insights into the actual development of many of these environments. In this section I will offer some guidelines of how pollen and other fossil material is deposited and preserved in a variety of arctic soils and examples of what to expect from their analysis.

A. Peats and Organic Soils

The same circumstances which encourage the accumulation and preservation of peat deposits favor the accumulation and preservation of pollen, spores, and other indicators of past vegetation changes. Paleoecology as a science had its historical beginnings as the study of past changes in peat stratigraphy, rather than as the study of pollen and microfossils, which was invented by Lennart von Post in 1916. The correlation of periods of peat growth initiation and accumulation, and of humification are still viable areas for the investigation of climatic change, although the direct invocation of climate as the forcing agent for peat growth should be avoided, unless corroborating, independent evidence is also available.

In the Arctic, where peat accumulation is slow and subject to a myriad of disturbances, even more caution must be practiced when using changes in peat growth as indications of climate change. Here again, paleoecological methods can help to sort out local successional events and disturbances. The combination of peat and soil analysis and palynology has rarely been attempted in arctic studies and offers some new insights into the relationship of landscape response to recent climate change.

Vegetation changes at the end of the late Pleistocene are indicated throughout the Arctic by the expansion of birch approximately 14,000 years ago. The synchronous timing of this event indicates a wide-spread response to a climatic event. After about 5000 yr. B.P., there is no strong pollen evidence from northern Alaska to indicate that climate was a major forcing agent on vegetation changes. Vegetation changes are linked, however, to the expansion of peatlands both within the treeline further south and in the tundra regions (Ovenden, 1990; Ritchie, 1987). A study of 108 peatlands in western Canada found that the majority developed under frost-free conditions, with only those furthest to the northwest developing on permafrost (Zoltai and Tarnocai, 1975). The transition to permafrost was dated at about 9,500 B.P. in the Northern Yukon through macrofossil record (Ovenden, 1982).

The first detailed study of long term soil development coupled with pollen stratigraphy on the North Slope (Figure 1) raised a significant question concerning the causes of carbon accumulation in northern environments (Eisner, 1991). The Imnavait core stratigraphy shows a complex record of mineral deposition and organic accumulation and decomposition over the last 11,000 yr. (Figure 2). Peat growth abruptly halts at about 45 cm and is overlain by sapric or highly decomposed soil. This change is not accompanied by any change in pollen percentages, which is the proxy indicator for

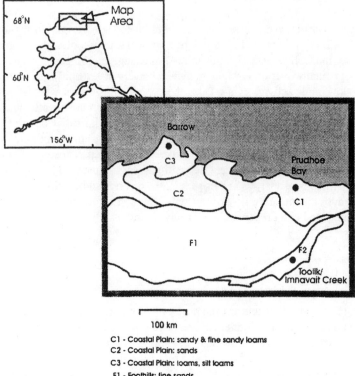

Figure 1. Map of North Slope, with inset of Alaska, showing general soil types and Imnavait Creek site.

regional climate change. This sharply defined humified layer is overlain by a regrowth of peat dated at 2960 ±70 yr. B.P.

The Imnavait Creek study raises some important questions concerning landscape development in the Arctic: 1) What were the controls on peat production in the past? 2) Is local hydrological change, internal peatland processes, or a wider regional disturbance responsible for the peat deterioration 3000 years ago? 3) Is the vegetation represented by the pollen simply insensitive to subtle climate change? 4) What specific parameters can be studied in the peat stratigraphy which can be correlated to peat accumulation?

B. Mineral Soils

European pollen analysts (or palynologists) have long recognized the potential, as well as the various pitfalls, of pollen analysis from mineral soils, and many of their methods can be applied to the Arctic, although the particular characteristics of Arctic soils and cryogenetic processes make collaboration with permafrost experts particularly crucial here.

A recent study in the Chersky region of Yakutia, Northeastern Siberia, in the former Soviet Union, illustrates a simple application of pollen and microfossil analysis which may help to understand the soil development of the region. Samples were taken from soil sections and pit samples at several soil

Arctic Paleoecology and Soil Processes

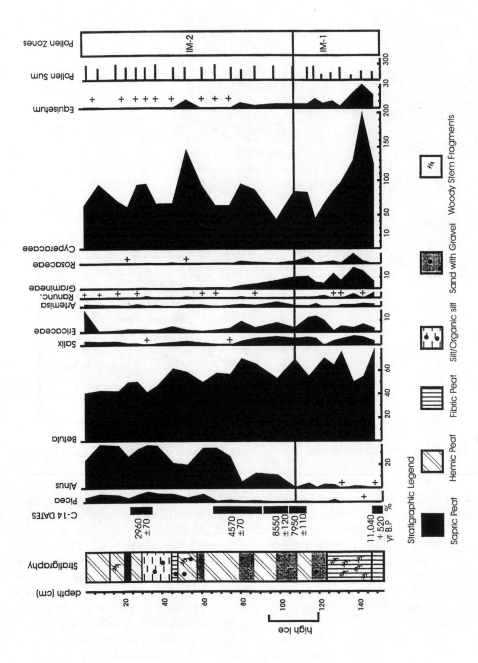

Figure 2. Summary pollen percentage diagram of Imnavait Creek, Alaska (Eisner, 1991). Pollen sum excludes Cyperaceae and Equisetum, but includes all other identified and unknown grains.

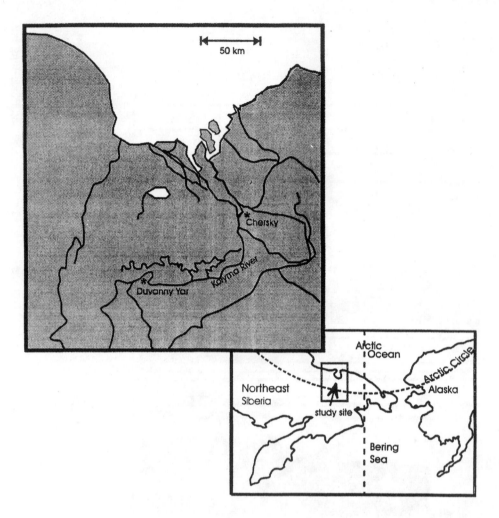

Figure 3. Map of Northern Yacutia, Russia, showing sampling site of Chersky, with inset of Beringia.

sampling sites at Rodinka Mountain (Lat. 68° 44'N/ Long. 161° 30'E) in Chersky (Figure 3). These sites were part of the USDA-National Soil Survey Center world soils database mapping project and included soil genesis, soil geography, and soil morphological description.

The two soil pits analyzed represent two different landscapes in the Chersky region. The first, 004, is situated on the lower foot slope of the mountain (elevation 140 m), with a modern vegetation dominated by shrub tundra, and the other site, 005, in a larch (*Larix dahurica*) woodland (elevation 100 m). Site 004 was sampled for pollen and microfossils from soil horizons Bw3, at 38 to 59 cm depth, Cfm1 (78 to 94 cm) and Cfm2 (94 to 110 cm). Site 005 was sampled from soil horizon Cfm1 (68 to 84 cm), and Cfm2 (84 to 95 cm). A more precise depth measurement is not given for these samples, since we are assuming that each soil horizon will have one distinct paleoecological "fingerprint".

The soils of both sites have been classified as coarse-silty, mixed, pergelic Cryochrepts, and the parent material of both is described as moderately weathered loess. One major difference, besides the different vegetation regime, is the active layer thickness: the active layer at site 004 is 92 cm deep, although from 78 to 94 cm the soil was ice rich. The active layer at site 005 ends at 84 cm. The expectation was that the best preservation and little vertical movement of pollen grains would occur below the active layer. Both sets of soils are described as highly acidic, the single most important factor for good pollen preservation.

The results of the paleoecological analysis is shown in Figures 4 and 5(a, b). Pollen concentrations were low, but not unusually so for arctic sediments, and pollen preservation was excellent. The best way to depict the taxa assemblages in this case is by percentages, where each taxa type is presented as a percentage of the pollen sum, which is composed of the pollen from the vascular pollen types, with spores, fungi, and algae outside the pollen sum, but still represented here as percentages of the pollen sum. Using traditional palynological methodology, the pollen horizons are first described, then interpreted in terms of changes within each site, and, finally, the two sites are compared and possible reasons for their differences are discussed.

The footslope site shows some significant changes in pollen assemblage through the three horizons. The lowest horizon, Cfm2 has very high grass (Gramineae) pollen, over 30%, with 25% alder (*Alnus*) pollen and about 15% birch (*Betula*). *Artemisia* (sage) is less than 5% and other herbs generally low. The non-pollen grains show high mosses, clubmosses and fern types (Pteriodophytes) at 10%, and low fungal percents (<5%). The middle horizon, Cfm1, shows a significant drop in grasses and a rise in alder to over 35%. The shrubs willow (*Salix*) and the heaths (Ericaceae) are low (5%), but even this is significant, since these pollen are usually underrepresented in the pollen assemblage. *Sphagnum* also increases significantly. The top horizon analyzed, Bw3, shows a rise in *Artemisia* and a decrease in grass, as well as in alder. Fungal remains drop to trace amounts and all spores decrease.

Any interpretation of the pollen from soils must first consider downwashing of grains. In the case of site 004, the pollen assemblages of each soil horizon are quite discreet and pollen types vary independently of one another, so it is unlikely any mix occurred between layers. The rise in alder in the middle Cfm1 layer may represent a regional rise in alder or a local increase in the shrub or may just be a function of a decrease in the input of other pollen. The general trend toward the upper layers, however, indicates increasing aridity, especially in terms of the high Artemisia and *Chenopodia* (goosefoot), two typical steppe-tundra indicators. This may indicate that the Bw3 soil developed in a situation of increased drainage but it is difficult to know, with this small data set, whether local or regional forcing agents were responsible.

Site 005 also shows significant changes between the two sampled horizons. Cfm2 has high grass, but also high percentages of *Artemisia* with other herbs also significant. Birch, alder and stone pine (*Pinus pumila*) are also significant contributors to the pollen sum. Notable here are the fungal remains which are high at 10%. Cfm1 shows some major changes, with *Artemisia* decreasing to less than 5%, and all the trees and shrubs (birch, alder, stone pine, and heather) increasing dramatically, and also sedge (Cyperaceae) increasing to over 5%. Ferns and mosses also increase to very high levels, and fungal remains decrease slightly to 7%.

Site 005 shows a much more dramatic change between horizon Cfm2 and Cfm1 than at the other site. The high *Artemisia* levels indicate dry or disturbed conditions and the subsequent changes to dominance of shrubs, supported by the presence of sedges and pteriodophytes, strongly argue for a shift to higher moisture conditions. The pine rise is notable: pine is a prodigious pollen producer, and amounts of less than 5% are not considered significant, since they can have their source hundreds of kilometers away. But the rise to over 5% may indicate that pine was migrating closer to the site. There is also no indication that a larch woodland existed at the time of the development of the Cfm1 soil, but larch pollen is produced in extremely small amounts, with low dispersal, and thus is notoriously difficult to find, even in a larch forest.

A comparison between the two sites shows interesting differences. While the trend at site 004 is toward decreasing moisture, that at site 005 is the opposite. There is a considerable difference in fungal levels between the two sites, which is related to biotic activity. Obviously, landscape aspect is important in understanding the basic differences in the two sites but both sites also responded quite differently to more regional or even climatic changes.

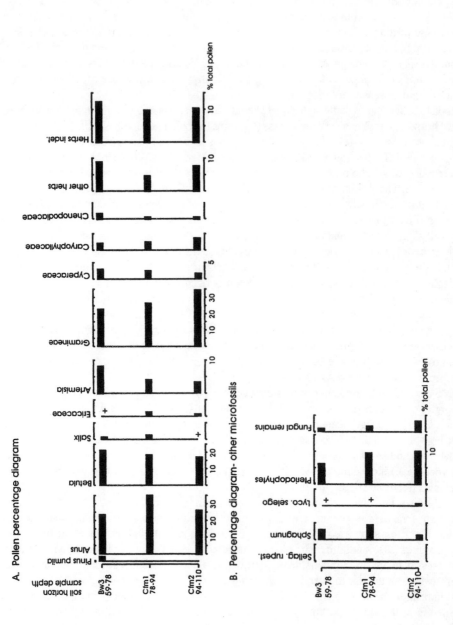

Figure 4. Pollen percentage diagram-pollen taxa. Site 004, Rodinka Mt., footslope, Chersky. Pollen sum includes all identified and unidentified pollen grains. Note changes in scale.

Arctic Paleoecology and Soil Processes

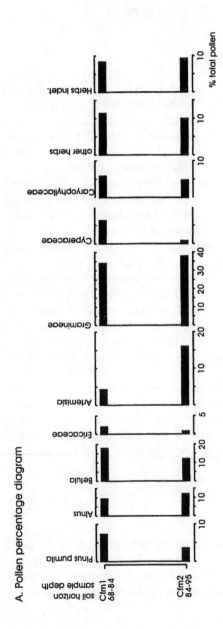

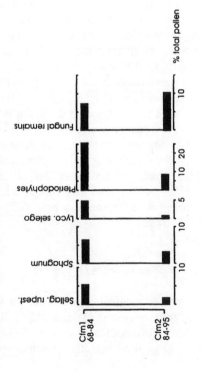

Figure 5. Pollen percentage diagram of Site 005, Larch woodland, Chersky. Pollen sum includes all identified and unidentified pollen grains. Note changes in scale.

Radiocarbon samples were taken for these soils and dated by accelerator mass spectrometry at the Lawrence Livermore National Laboratory, with inconclusive results; it appears that the soil samples have been contaminated. Three out of four samples were found to be ^{14}C enriched at slightly higher levels than occur in nature, and this may be caused by radioactive fallout (T. Brown, personal communication). Unfortunately, samples dated by accelerator at other Northeast Siberian sites have yielded similar results, a worrisome situation which must be investigated further.

C. Fungi as an Estimate of Soil Decomposition

Fungal spores are ubiquitous in standard pollen preparations and provide a remarkable range of ecological information. The role of fungi in decomposition processes is important for this research strategy. We have noted abrupt changes in frequency of fungal hyphae in microfossil preparations in arctic peat sediments, which may be correlated with changes in carbon accumulation. Bunnel et al (1980) studied fungi and algae densities in arctic sediments, and found variations according to depth and soil environments, finding an inverse relation between bulk density and fungal densities. If the variations in microfossils can be correlated with long term variations in peat vegetation and climate, we will have developed a powerful tool for studying the relationship between vegetation, carbon accumulation rates, and climate. Few studies currently exist in the Arctic which compare pollen changes with fungal abundance, but it has great potential and should be further investigated.

Disturbance of the arctic ecosystem, causing relatively small-scale disruptions in hydrology, thaw depth, and vegetation cover can occur without human intervention, from fire damage, flooding, or over-grazing, as examples. Wildfires place important constraints on vegetation growth and can alter thaw depths and nutrient levels and, therefore, can disrupt or encourage peat accumulation in the arctic tundra. Charcoal, burned plant material, and specific fungal remains are all sources of information about past fire which can be derived from the paleoecological record. Fungal ascospores are associated with the occurrence of local bog fires (van Geel, 1978). Fungal spore frequencies may be a more reliable method of recognizing fire disturbance through time than the current, rather unsatisfactory method of counting charcoal remains.

Danish investigators have shown how hyphae fragments may be used to determine stages of soil development and the relation of soil development to biological activity (Anderson, 1986). Anderson found a characteristic difference between the length of hyphae fragments in brown earth soils as opposed to podzolic soils, and from this determined that former soil conditions could be traced by measuring these fungal fragments, a possibility that could be useful for future paleoecological soil analysis in the Arctic.

D. Overview of Microfossil Analysis

A variety of paleoecological data sets can be studied as interrelated parts of a whole (Table 1). Depending on the preservation of these in the soil, the way in which the soil was sampled in the field, and the availability of qualified experts, the following may be studied: (more) microfossils including pollen, spores, fungi, algae, and other palynomorphs, and macroscopic plant remains such as seeds and fruits.

Advantages of studying these parameters in tandem with pollen:
1. Determination of source of indicator species now possible: these are mostly locally occurring.
2. Greater understanding of soil development, vegetation, and climate changes are possible.
3. Less lag time in the response of fungi and algae to abrupt environmental change.

Table 1. Summary of palaeoecological parameters and their paleoecological significance

Method	Ecological significance
Paleoecological analysis	
Pollen	Regional vegetation, regional correlation, climate
Macrofossils	Local vegetation, soil and landscape development
Fungi	Decomposition, soil development, fire
Algae	Nutrients, hydrology
Physical analyses	
Organic content	Decomposition, local vegetation, plant community development
Humification	Decomposition, hydrology
Bulk density	Mineral content, silt influx, disturbance
Radiometric dating	Peat and pollen accumulation rates, timing of climatic events
$\delta 13$ C isotopic analysis	Decomposition and diagenesis of soils in permafrost regions

Other types of depostional environments can be more briefly described, because of their relative rarity, or else because of their more restricted value as reliable paleoecological depositories. Buried soils, or paleosols, are an infrequent depositional environment which can be extremely useful paleoclimate "snapshots" of past vegetation and climate. Soils may be buried because of rapid climatic or local changes such as rapid eolion deposition, volcanic activity, waterlogging and subsequent peat growth, or flooding. The plant remains preserved in the soil are a reliable indication of vegetation, but only at that particular time; no time series is available. The same restrictions in interpretation for mineral soils also apply to buried soils: there may be considerable mixing, the pollen and plants are only locally representative, and added to this, pollen accumulations will probably be quite low, because of the short deposition time. Buried soils are often easy to date, have excellent preservation of organic material, and may be our only available source of paleoecological information (Hoefle et al., 1995).

Loess deposition was one of the primary forces on soil and landscape development during the late Glacial and early Holocene in the Arctic, and still plays a major role in vegetation and landscape evolution. However, the dynamic nature of eolian activity makes it a poor candidate for pollen or microfossil analysis, because deposition is generally too rapid and discontinuous to allow for the steady accumulation of pollen. Moreover, the source of the loess sediments, which is generally from fluvioglacial river systems, is itself reworked material, and much of the material it contains may be from much older deposits of the Tertiary and earlier ages. Some loess deposits, notably the accumulations of the Yedoma in Northeast Siberia, include not only eolian material, but thaw lake deposits, peaty accumulations, and even alluvial and marine deposits. Obviously, these complex depositional environments require extreme care in sampling techniques and very good dating control.

Pollen analysis of mineral soils is not recommended as a proxy for understanding global climate change. Arctic pollen analysis in any kind of depositional environment is difficult to interpret and pollen deposition in any cryoturbated soil is compounding the problems of interpretation. We need to ask about the post-depositional environment of the microfossils within the sediment matrix. What can the soils tell us about how the proxy data were deposited and then redistributed in the soil? How did cryoturbation or soil formation alter the sediments and can we understand these alterations?

E. Other Considerations in the Study of Microfossil in Soils

The interpretation of pollen records from those sections where soil formation has taken place requires the careful assessment of several potential problems: 1) the downwashing of pollen grains, 2) differential decay of pollen grains, and 3) homogeneity of the pollen profile because of soil mixing (Havinga, 1974). To this we may include problems particularly acute in the Arctic: 4) cryoturbation, and 5) problems of dating.

1. Downwashing of Grains

The accurate interpretation of any pollen, microfossil and macrofossil analyses requires an understanding of the processes of pollen deposition and awareness of the potential of redeposition. Although many experts contend that pollen grains and other fossil material are not displaced downward in water-logged peats, as might happen in soils, virtually no work has been done on this in arctic soils (Janssen, 1992; van Geel, 1986). Downslope transportation of macrofossils occurs during melt-off, and pollen may be effected by internal dynamic processes as well (Glaser, 1981). The affect of underlying frozen soil and snowmelt activity on the stratigraphic integrity of these tundra peat sediments must be considered. At Imnavait Creek, the active layer — the soil layer that thaws during the year — is still frozen and impermeable at snowmelt time, so that the meltwater transports virtually no mineral or organic material downslope (Hinzman et al., 1991). There is also little vertical percolation of water from the snow melt in this upper layer.

The hydrologic properties of peats and soils may indicate that some pollen mixing occurs over short vertical distances within the active layer. Site selection is crucial in order to minimize the downwash effect of vertical soil water movement. Hinkel and Outcalt (1994) found considerable variation in water migration through the active layer at two adjacently situated sites (13 m apart) but with different soil conditions. Monitoring of soil water movement and thermal properties could be extremely useful when determining sampling intervals through the active layer, and would aid in interpretation of results. If the vertical movement of microfossils does occur differentially within the active layer and this can be correlated to soil water movement, it may be possible to extrapolate this relationship to older soils in order to understand past soil temperature and water transport.

The comparison of profiles from several locations will enable us to determine the extent of reliability of some types of pollen fluctuation. One method of identifying the extent of downwashing of pollen grains is to determine pollen accumulation rates (PAR) by adding a known spike of exotic pollen to a known dry weight of sediment. High concentrations of total pollen or pollen types can be inspected to identify characteristic trends which indicate vertical pollen movement within the soil (Janssen, 1974). Certain adaptations to conventional pollen analysis need to be made, such as the calculation of pollen accumulation rates in perennially frozen soils (Wang and Geurts, 1993).The process of cryoturbation obviously results in the mixing of pollen grains and, therefore, limits the utility of a representative pollen stratigraphy. However, if a cryoturbated soil material can be identified as a single unit, a sample from that unit might cautiously be representative of the vegetation at the time of soil development.

An understanding of watershed processes of any peatland is requisite to understand peatland development, and in the Arctic, our complete understanding requires the realization that permafrost and snowmelt "dominate the hydrology of the Arctic"(Kane et al., 1992). Snow tends to accumulate in landscape depressions and in water tracks, where the minimum thaw depth occurs. Thus, the active layer depth is closely related to precipitation, topography, and temperature. Changes in the hydrology of an arctic wetland do not necessarily have to be caused by climate variation, however. The accumulation of peat independently alters the drainage pattern and changes in the vegetation cover will also change hydrological levels. The removal of the insulating vegetation through fire, over-grazing, human disturbance, or erosion can thicken the thaw depth, creating a cycle of hydrological changes with implications for peat accumulation. Temperature change on the order of long term climatic warming can, of course, also increase the active layer depth which could lead to wide scale subsurface drainage, effectively drying out the wetlands (Kane et al., 1992).

F. Dating

Dating reversals and other problems related to low organic levels in lake sediments has made accurate C-14 dating of climatic events problematic in the Arctic. Peats and other soils, which contain higher percentages of organic material than lacustrine sediments and can often be sampled in greater quantity, which solves the problem of obtaining material. However, the problem of root contamination from younger vegetation requires a strategy specific to the dating of peats and other soils. While it is true that the root systems of plants can "contaminate" lower deposits with younger material, it is relatively simple to separate root material from datable material (stems, leaves, seeds). The ability to date small sample sizes by accelerator makes this a viable way to acquire high resolution temporal control of our data sets (van Geel and Mook, 1989). Cryoturbated sediments need to be sampled with caution.

G. Carbon Accumulation in Soils and Peats: The Paleoecological Tool

How is carbon accumulation related to climate change? Is the timing of peat initiation a climatic signal? The perspective of the past long-term record of paleoecology, and the correlation of that record with soil carbon accumulation, provides a powerful tool for understanding the relationship between carbon flux and climate.

Carbon accumulation rates for North Slope soils have shown a striking variation at Prudhoe Bay between the middle Holocene (7.9 gC m^{-2} yr^{-1}) and late Holocene (1.4 gC m^{-2} yr^{-1}) (Marion and Oechel, 1993). There also appears to be a latitudinal variation in carbon accumulation rates, with rates averaging from 23-35 gC m^{-2} yr^{-1} to 9-11 gC m^{-2} yr^{-1} from the Boreal zone to the northerly subarctic zone (Ovenden, 1990; Tarnocai, 1988). Not only vegetation change but soil development (time) and climatic change will also influence carbon accumulation rates. Marion and Oechel (1993) point out that it is difficult to directly compare recent carbon accumulation to older sediments, mainly due to surfaces losing carbon while buried soils are at a steady state.

Questions to ask in the formulation of paleoecological/pedological research:
1. How does pollen and biotic data reflect vegetation assemblages and dynamics through time?
2. How does the modern pollen rain vary between basins, and what does this tell us about the interpretation of heterogeneity in the stratigraphic record?
3. Can vegetation changes be correlated with changes in soil horizon?
4. Are regional patterns detectable from the pollen stratigraphies, once the issues of local effects and soil formation have been resolved?
5. Is there an independent record of climate change available in the area? For example: a lacustrine pollen stratigraphy, lake level history, glacial evidence?
6. Is the carbon accumulation rate a function of local or regional signals in the paleoecological record?

IV. Conclusion

The link between modern Arctic ecosystem research and soil science might be seen as the record of long term vegetation change contained in the peat and soil stratigraphies themselves. A reconstruction of past ecosystem changes, providing it is well-dated, also forges a link between landscape system science and records of climate change from lake sediments.

A number of environments can provide important climatic and paleoecological information, and are recommended for further integrative and collaborative study. The Kolyma Lowland in NE Russia provides an ideal location for a combined study of the cryopedospheric soils and fossil pollen stratigraphy. The Kolyma Lowland was not glaciated during the Pleistocene, thus the Yedoma deposits preserved pollens and features related to past climatic change. The syngenetic permafrost protects this evidence from weathering.

Significant peat deposits from the drained lakes of the North Slope of Alaska are a potential major resource of paleoecological information, recording the process of plant succession and carbon sink since the early Holocene, about 10,000 years ago. Pollen stratigraphies from the infilled peats of thaw lake basins of the North Slope may be actually indicate fluctuations in the thaw lake cycle (Anderson, 1982). Thaw lake sediments may be more prevalent in fossil form and more valuable for vegetation reconstruction than formerly thought (Hopkins and Kidd, 1988). The study of these sediments and the comparison of ancient thaw lake sediments with modern thermocarst environments, can be applied not only to the Alaskan material but to other similar environments.

Tundra soils lose or accumulate carbon because of changes in temperature and precipitation, but are also effected by soil processes and vegetation change due to mechanisms independent of climate, such as drainage, disturbance, nutrient availability, or plant succession. The factors controlling soil carbon accumulation and sequestration may or may not be directly linked to climate change. Until we understand the long term processes which effect the carbon balance in northern ecosystems, we will not be able to accurately model future long term ecosystem response to global change.

Acknowledgment

This chapter is Byrd Polar Research Center contribution 1055.

References

Anderson, P. M. 1982. Reconstructing the past: The synthesis of archaeological and palynological data, northern Alaska and northwestern Canada. Ph.D. Dissertation, Brown University.

Anderson, P. M., P.J. Bartlein, L.B. Brubaker, K. Gajewski, and J.C. Ritchie. 1991. Vegetation-pollen-climate relationships for the arcto-boreal region of North America and Greenland. *J. Biogeography* 18:565-585.

Anderson, S. T. 1986. Paleoecological studies of terrestrial soils. p. 165-177. In: B. E. Berglund (ed.), *Handbook of Holocene Palaeoecology and Palaeohydrology*. John Wiley & Sons, Chicago.

Billings, W. D. and K.M. Peterson. 1980. Vegetational change and ice-wedge polygons through the thaw-lake cycle in Arctic Alaska. *Arctic and Alpine Research* 12:413-432.

Bliss, L.C. and K.M. Peterson. 1992. Plant Succession, Competition, and the Physiological Constraints of Species in the Arctic. p. 469. In: F. S. C. III, R. L. Jefferies, J. F. Reynolds, G. R. Shaver, and J. Svoboda (eds.), *Arctic Ecosystems in a Changing Climate: An Ecophysiological Perspective*. Academic Press, San Diego, CA.

Bunnell, F.L., O.K. Miller, P.W. Fanagan, and R.E. Benoit. 1980. The microflora: composition, biomass, and environmental relations. p. 225-290. In: J. Brown, P. C. Miller, L. L. Tieszen, and F. L. Bunnell (eds.), *An Arctic Ecosystem: The Coastal Tundra at Barrow, Alaska*. Dowden, Hutchinson, and Ross, Inc., Stroudsburg, PA.

Eisner, W.R. 1991. Palynological analysis of a peat core from Imnavait Creek, the North Slope, Alaska. *Arctic* 44:279-282.

Glaser, P.H. 1981. Transport and deposition of leaves and seeds on tundra: a late-glacial analog. *Arctic and Alpine Research* 13:173-182.

Havinga, A.J. 1974. Problems in the interpretation of pollen diagrams of mineral soils. *Geologie en Mijnbouw* 53:449-453.

Hinkel, K.M. and S.I. Outcalt. 1994. Identification of heat-transfer processes during soil cooling, freezing, and thaw in Central Alaska. *Permafrost and Periglacial Processes* 5:217-235.

Hinzman, L.D., D.L. Kane, R.E. Gieck, and K.R. Everett. 1991. Hydrological and thermal properties of the active layer in the Alaskan arctic. *Cold Regions Science and Technology* 19:95-110.

Hoefle, C., M.E. Edwards, C.-L. Ping, D.H. Mann, and D.M. Hopkins. 1995. The Late-Pleistocene environment of the Bering Land Bridge: Buried Soils from Seward Peninsula, NW Alaska. Paper presented at the 25th Arctic Workshop, Centre de'Etudes Nordiques, University of Laval, Quebec, Canada, March 16-18.

Hopkins, D.M. and J.G. Kidd. 1988. Thaw lake sediments and sedimentary environments. Paper presented at the 5th International Permafrost Conference, Trondheim.

Janssen, C.R. 1992. The Myrtle Lake Peatland. p. 223-235. In: H.E.J. Wright, B. A. Coffin, and N. E. Aaseng (eds.), *The Patterned Peatlands of Minnesota*. University of Minnesota Press, Minneapolis, MN.

Janssen, C. R. 1974. *Verkenningen in de Palynologie*. Oosthoek, Scheltema, and Holkema, Utrecht.

Kane, D. L., L.D. Hinzman, M.-k Woo, and K.R. Everett. 1992. Arctic hydrology and climate change. p. 35-57. In: F. S. C. III, R. L. Jefferies, J. F. Reynolds, G. R. Shaver, and J. Svoboda (eds.), *Arctic Ecosystems in a Changing Climate: An Ecophysiological Perspective*. Academic Press. San Diego, CA.

Marion, G.M. and W.C. Oechel. 1993. Mid- to late-Holocene carbon balance in Arctic Alaska and its implications for future global warming. *The Holocene* 3:193-200.

Oechel, W. C., S.J. Hastings, G. Vourlitis, M. Jenkins, G. Riechers, and N. Grulke. 1993. Recent change of Arctic tundra ecosystems from a net carbon dioxide sink to a source. *Nature* 361:520-523.

Ovenden, L. 1982. Vegetation history of a polygonal peatland, northern Yukon. *Boreas* 11: 209-224.

Ovenden, L. 1990. Peat accumulation in northern wetlands. *Quaternary Research* 33:377-386.

Ritchie, J. C. 1987. *Postglacial Vegetation of Canada*. Cambridge University Press, Cambridge.

Tarnocai, C. 1988. Wetlands in Canada: distribution and characteristics. p. 21-25. In: H. I. Schiff and L. A. Barrie (eds.), Global change, Canadian wetlands study, workshop report and research plan. Report No. 018801. Canadian Institute for Research in Atmospheric Chemistry, North York.

van Geel, B. 1978. A palaeoecological study of Holocene peat bog sections in Germany and The Netherlands. *Review Palawobotany and Palynology* 25:1-120.

van Geel, B. 1986. Application of fungal and algal remains and other microfossils in palynological analyses. p 497-505. In: B. E. Berglund (ed.), *Handbook of Holocene Paleoecology and Paleohydrology*. New York: John Wiley & Sons, Ltd.

van Geel, B., and W.G. Mook. 1989. High-resolution C-14 dating of organic deposits using natural atmospheric C-14 variations. *Radiocarbon* 31:151-155.

Walker, D. A. and K.R. Everett. 1991. Loess ecosystems of Northern Alaska: regional gradient and toposequence at Prudhoe Bay. *Ecological Monographs* 61:437-464.

Wang, X.-C. and M.-A. Geurts. 1993. Determination of pollen accumulation rates in frozen sediments. *Permafrost and Periglacial Processes* 4:87-93.

Webber, P. J. 1978. Spatial and temporal variation of the vegetation and its production, Barrow, Alaska. Chapter 3. In: L.L. Tieszen (ed.), *Vegetation and Production Ecology of an Alaskan Arctic Tundra*. Springer Verlag, New York.

Zoltai, S.C. and Tarnocai, C. 1975. Perennially frozen peatlands in the western Arctic and Subarctic of Canada. *Canadian Journal Earth Sciences* 12:24-43.

CHAPTER 11

Soil Carbon Distribution in Nonacidic and Acidic Tundra of Arctic Alaska

J.G. Bockheim, D.A. Walker, and L.R. Everett

I. Introduction

Global climate models predict that greenhouse warming will be several times greater in the arctic regions than the projected global mean of 1.5 to 4.5°C (IPCC, 1992). Indeed, during the past several decades, climate warming has been greater in the arctic than in other regions. The warming has been particularly large over Siberia, northwestern Canada, and much of Alaska, where warming rates over the past 30 years have been approximately 1°C per decade (Oechel et al., 1993).

Most of the high-latitudes are underlain by permafrost, in which the summer thaw is limited to a thin (often <50 cm) active layer. The soil active layer (0 to 35 cm) in the circumarctic contains up to 455 Gt C which is approximately 60% of the ~750 Gt C currently in the atmosphere as CO_2 (Post et al., 1982; Billings, 1989; Oechel et al., 1993). However, the amount of carbon in arctic tundra soils may be even greater than previously believed as recent data show that the upper 30 to 40 cm of permafrost contains as much carbon as the active layer (Michaelson et al., 1996). If global warming induces thawing of permafrost, this carbon could be released to the atmosphere as CO_2, CH_4, and other trace gases, thereby amplifying the greenhouse effect. In fact, the warming of the 1970s and 1980s over the Alaskan North Slope appears to have changed the tundra from a net sink to a net source for the atmosphere (Oechel et al., 1993).

The only comprehensive survey of soil carbon stores in arctic Alaska was by Michaelson et al. (1996). Soil carbon values (to 1 m) commonly ranged from 16 to 94 kg C/m^2, about one-half of which was in the upper portion of the permafrost. Other soil carbon data are contained in reports of the Alaskan coastal plain (Everett and Parkinson, 1977; Parkinson, 1978; Walker et al., 1980) and the arctic foothills of Alaska (Walker et al., 1989; Walker and Barry, 1991).

A unique aspect of the physiography of arctic Alaska is a belt of calcareous loess between 5 and 70 km wide that extends across the arctic foothills and coastal plain provinces north of the Brooks Range (Carter, 1988). The loess exerts a strong effect on the distribution of plant communities and on soil development (Walker and Everett, 1991). Nonacidic landcover types may be more abundant than previously thought on the North Slope, occupying 50% or more of the Kuparuk basin (Auerbach et al., 1996). Nonacidic and acidic landcover types are readily distinguished by the normalized differences in vegetation index (NDVI) from SPOT multispectral digital data (Walker et al., 1995) and false color AVHRR images of northern Alaska (Walker and Everett, 1991).

Mollisols have been observed on south-facing slopes of pingos (Walker et al., 1991), on high-center polygons and rims of well developed low-center polygons (Everett and Parkinson, 1977; Parkinson, 1978), and on loess-affected floodplains of the arctic coastal plain (Walker and Everett,

1991). However, Mollisols may be even more abundant than previously believed in the Kuparuk basin, also occurring in the northern Brooks Range and the arctic foothills.

The objectives of this study are (1) to determine the relative abundance of Mollisols and other nonacidic soils in the Kuparuk basin, the location of the Arctic Research Consortium of the U.S. CO_2 and methane flux study, (2) to compare key soil properties in moist nonacidic tundra and moist acidic tundra, and (3) to elucidate the mechanisms accounting for differences in the depth-distribution of carbon between these two landcover types.

II. Methods and Materials

A. Sites

The study was conducted in the 9,200 km^2 Kuparuk watershed (Figure 1). Pedons were described and sampled in three physiographic provinces, the Brooks Range (4 sites), the arctic foothills (33 sites), and the coastal plain (18 sites) along a north-south gradient from ca. 70°17' to 68°30N. The area occurs within the zone of continuous permafrost (Péwé, 1975).

The climate of the area varies with distance from the Arctic Ocean and elevation. The mean annual temperature ranges from -12.8°C at Prudhoe Bay to -5.9°C in the Brooks Range (Haugen, 1982). Temperature extremes are greater in the Brooks Range and arctic foothills provinces than in the coastal plain. Precipitation declines from 300 to 450 mm/yr in the Brooks Range to 180 to 230 mm/yr in the coastal plain.

The major landcover classes in the Kuparuk watershed (including percentage of area) are moist nonacidic and dry tundra (38.9%), moist acidic tundra (30.8%), shrublands (16.8%), and wet tundra (7.0%), with water and shadows (5.1%) and barrens (1.4%) occupying the remaining areas (Auerbach et al., 1996). Moist nonacidic tundra contains predominantly non-tussock sedges (*Carex bigelowii* and *Eriophorum triste*), a few prostrate shrubs (*Dryas integrifolia, Salix reticulata*, and *S. arctica*), and brown mosses (*Tomenthypnum nitens* and *Hylocomium splendens*). In contrast, the moist acidic tundra contains cottongrass tussocks (*Eriophorum vaginatum*), dwarf-birch (*Betula nana*), and other acidophilous dwarf-shrub species, such as *Ledum palustre* spp. *decumbens, Vaccinium vitis-idaea, V. ulignosum, Salix planifolia* spp. *pulchra*, and *Sphagnum* moss (Walker et al., 1994).

Residual surfaces and dissected uplands dominate the Brooks Range and arctic foothills. Glacial deposits are limited to a region extending 65 km north of the Brook Range and vary from Holocene to early Pleistocene from south to north (Kreig and Reger, 1982). The coastal plain contains primarily drained or thaw lakes with isolated pingos of mid-Holocene age. Parent materials are dominantly loess and silty colluvium in the foothills and lacustrine silts and organics in the coastal plain. Alluvium occurs along the major river courses, including the Kuparuk, Toolik, and Sagavanirtok. Elevations range from 1150 m in the southern foothills to sea level at Prudhoe Bay.

B. Sample Collection

There were two sets of sampling localities. Thirty-two pedons were examined during a close-support helicopter reconnaissance to prepare landcover and soil maps of the watershed (designated as R95-1 through R95-32 on Figure 1). These pedons were located in major landcover types selected randomly from aerial photographs and located using a global positioning system. The pedons were excavated to the surface of the permafrost table in early August when the active layer was at its thickest. The upper 10 cm of the permafrost was sampled using a hammer and cold chisel.

An additional set of 23 detailed pedons (designated as A95-1 through A95-23 on Figure 1) were examined at 11 CO_2 and methane flux-tower measuring sites along a north-south gradient from Betty

Soil Carbon Distribution in Nonacidic and Acidic Tundra of Arctic Alaska

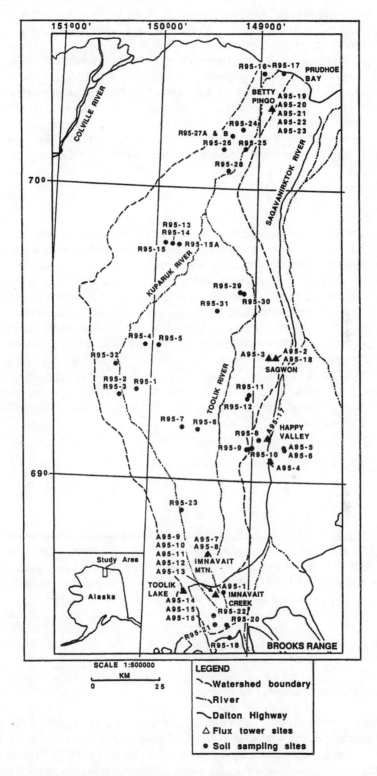

Figure 1. The Kuparuk drainage showing detailed and reconnaissance description and sampling sites.

Pingo near Prudhoe Bay to Imnavait Creek (Figure 1). These pits were dug by hand to the permafrost table (average depth = 51 cm) and additionally excavated to 1 m with a gasoline-powered Pico impact drill. Detailed soil descriptions were taken at all sites, and bulk samples were collected from each horizon and placed in water-tight bags. The soils were classified according to *Soil Taxonomy* (Soil Survey Staff, 1994) and the recently proposed Gelisol order (ICOMPAS, 1996).

Bulk density cores were taken from each horizon of the detailed pedons within the active layer. Bulk density of reconnaissance pedons and permafrost horizons were estimated from the equation, $y = 1.374 * 10^{-0.026x}$, where y = bulk density (g/cm^3) and x = organic carbon (%). The equation was derived from 82 measurements and had an r^2 of 0.823 (p = 0.0001). Soil pH was measured on a saturated paste within 8 hours of sample collection using a portable pH meter.

C. Laboratory Analyses

The samples were returned to the University of Wisconsin where bulk density (reported on samples dried at 105°C) and gravimetric and volumetric field moisture contents were determined. Air-dried samples were ground to pass a 0.5-mm screen and subsamples were sent to the University of Alaska-Fairbanks Agriculture and Forestry Experiment Station at Palmer for carbon analysis. Total carbon and nitrogen were determined by dry combustion on a Leco C and N determinator (LECO Corp., St. Joseph, MI). No adjustments were made for $CaCO_3$ so that the carbon values represent organic and inorganic forms. The carbon and nitrogen contents of the profiles were determined to a depth of 1 m by taking the product of carbon or nitrogen concentration, bulk density, and horizon thickness. The percentage of coarse fragments was very low so that no corrections were necessary for skeletal material. The carbon and nitrogen data are reported for the active layer, the upper part of the permafrost, and to a depth of 1 m.

Comparisons in soil properties between nonacidic and acidic landcover types were done by one-way analysis of variance.

III. Results

A. Soil Classification

Nonacidic soils comprise 54% of the Kuparuk basin, including nonacidic Pergelic and Histic Pergelic Cryaquepts (26.6%), Pergelic Cryoborolls (16.9%), and lesser amounts of Cryaquolls and nonacidic Cryorthents (Table 1). These soils occupy primarily moist nonacidic tundra and, to a lesser extent, shrublands, wet tundra, and barrens. Moist acidic tundra contains almost exclusively Pergelic and Histic Pergelic Cryaquepts. Histosols comprise only 3.9% of the watershed.

In the recently proposed Gelisol order (ICOMPAS, 1996), about 38% of the soils in the Kuparuk drainage are classified as Turbels, primarily Aquaturbels; 58% are Haplels, primarily Aquahaplels and Histohaplels; and about 3.9% are Histels (Table 2).

B. Morphological and Chemical Soil Properties

Although data are shown for soils of all of the landcover types, the results for the moist nonacidic tundra and moist acidic tundra will be emphasized. The active layer thickness was significantly greater for the moist nonacidic tundra than for the moist acidic tundra (Table 3). The organic layer was thicker for the moist acidic tundra than the moist nonacidic tundra, but the differences were not significant. There was a highly significant (p = 0.0004) correlation between active layer thickness and

Table 1. Distribution of soils in the Kuparuk drainage according to Soil Taxonomy (1994)

Soil taxonomy	Landcover type	% of Kuparuk[a]
Perglic, Histic Perglic Cryaquepts	Moist acidic tundra	40.8
Perglic, Histic Perglic Cryaquepts, nonacidic	Moist nonacidic tundra, shrublands	26.6
Perglic Cryoborolls	Barrens, moist nonacidic tundra	16.9
Perglic, Histic Perglic Cryaquolls	Moist nonacidic tundra, wet tundra	8.1
Perglic Cryofibrists	Wet tundra	2.7
Perglic Cryorthents, nonacidic	Shrublands	2.3
Perglic Cryochrepts	Barrens	1.0
Perglic Cryohemists	Wet tundra	0.6
Perglic Cryosaprists	Wet tundra	0.6
Perglic Cryumbrepts	Barrens	0.2

[a]Reported for land area only.

Table 2. Distribution of soils in the Kuparuk drainage according to the recently proposed Gelisol order

Soil taxonomy	Landcover type	% of Kuparuk[a]
Aquaturbels	Moist nonacidic tundra, moist acidic tundra	25.4
Aquahaplels	Moist nonacidic tundra, moist acidic tundra, shrublands	22.0
Histohaplels	Moist nonacidic tundra, moist acidic tundra, shrublands, wet tundra	20.4
Mollihaplels	Barrens, moist nonacidic tundra, shrublands	9.5
Molliturbels	Moist nonacidic tundra	8.2
Orthohaplels	Barrens, moist nonacidic tundra	4.9
Fibristels	Wet tundra	2.6
Haploturbels	Barrens, moist nonacidic tundra	2.2
Umbrihaplels	Barrens	0.2
Histoturbels	Moist nonacidic tundra	2.0
Hemistels	Wet tundra	0.6
Sapristels	Wet tundra	0.6

[a]Reported for land area only.

organic layer thickness for soils in the moist nonacidic tundra and moist acidic tundra. The A horizon averaged 19 cm thick for moist nonacidic tundra soils; an A horizon was not present in soils of the moist acidic tundra. The organic layers in the moist acidic tundra commonly overlaid a mottled, dilatant-prone Bg horizon. There were no significant differences in thickness of the B horizon or the solum between the two landcover types.

The pH values of the surface organic layer and the uppermost B horizon averaged 7.2 and 7.0, respectively, for the moist nonacidic tundra soils and were significantly greater than the average values of 4.5 and 5.3 for the moist acidic tundra soils (Table 3).

There were no significant differences in the amount of soil carbon in the upper 1 m between the two landcover types; however, there were significantly greater amounts and proportion of carbon in the active layer of soils in the moist nonacidic tundra than in soils of the moist acidic tundra (Table 3). Similarly, the amounts of nitrogen in the active layer and in the upper 1 m were significantly greater in soils of the moist nonacidic tundra than in soils of the moist acidic tundra. Whereas the

Table 3. Properties of soils in the Kuparuk drainage, northern Alaska

Profile number	Active layer (cm)	Organic layer (cm)	A hor. (cm)	B hor. (cm)	pH O hor.	pH B hor.	N (g/m²) Active	N (g/m²) Perm.	N (g/m²) 100 cm.	% Active	C (kg/m²) Active	C (kg/m²) Perm.	C (kg/m²) 100 cm	% Active
Barrens														
A95-1	>100	2	4	35		4.3	943	0	943	100	14.2	0.0	14.2	100
A95-7	>85	0	4	16		5.3	913	0	913	100	7.5	0.0	7.5	100
A95-8	>100	0	8	14		4.6	865	0	865	100	12.7	0.0	12.7	100
A95-18	>80	5	9	17	4.7	5.8								
R95-23	>60	1	5	26		5.6								
R95-21	>70	3	15	5	6.3	6.4								
Avg.	>60	2	8	19	5.5	5.3	907	0	907	100	11.5	0.0	11.5	100
SE		1	2	4.2	0.5	0.3	23	0	23	0	2.0	0.0	2.0	0
Moist nonacidic tundra and dry tundra														
A95-2	72	24	0	45	7.4	7.3	2464	1261	3725	66	35.5	18.6	54.1	66
A95-18	71	3	0	32	8.0	7.9	2697	1194	3890	69	41.1	20.1	61.2	67
A95-19	43	0	21	0	7.3	7.4	2136	260	2396	89	32.5	4.1	36.6	89
A95-20	41	25	0	0	7.0		1575	2497	4072	39	27.6	40.8	68.5	40
A95-21	49	17	41	>41	7.7	7.3	3452	2681	6133	56	55.0	42.9	97.8	56
R95-4	83	1	17	16	7.0	6.9	2344	1132	3476	67	35.4	13.4	48.8	73
R95-8	30	23	10	0	6.6		523	3352	3874.1	13	11.5	52.8	64.3	18
R95-11	50	26	8	16	7.1	6.3	1980	1485	3464.9	57	30.5	21.7	52.2	58
R95-12	60	14	6	30	7.3	6.3	1528	1991	3519	43	22.2	29.1	51.3	43
R95-13	64	8	43	13	7.1	7.4	2909	3133	6042.8	48	43.6	32.6	76.2	57
R95-15	46	7	9	16	7.0	6.8	1917	3594	5511.2	35	29.7	42.5	72.2	41
R95-15A	53	2	5	26		7.8								
R95-17	41	5	8	28		6.0								
R95-24	>92	3	20	0	7.0		933	0	933.4	100	14.0	0.0	14.0	100
R95-26	43	13	42	0	7.0		1549	266.8	1815.5	85	24.5	18.3	42.8	57
R95-27A,B	61	1	41	0	7.0		1981	185.6	2166.5	91	34.6	11.0	45.6	76
R95-28	78	1	10	0										
R95-29	52	9	32	27	7.1	7.1	1942	1733	3674.1	53	28.1	19.5	47.6	59
R95-30	>70	0	13	0										
R95-32	70	5	55	0		6.7	2464	1261	3725.1	66	30.7	21.9	52.6	58
Avg.	56	9	19	13	7.2	7.0	2025	1627	3651	61	31.0	24.3	55.4	60

Table 3. continued--

Profile number	Active layer (cm)	Organic layer (cm)	A hor. (cm)	B hor. (cm)	pH O hor.	pH B hor.	N (g/m²) Active	N (g/m²) Perm.	N (g/m²) 100 cm.	% Active	C (kg/m²) Active	C (kg/m²) Perm.	C (kg/m²) 100 cm	% Active
Moist acidic tundra														
A95-3	45	16	0	24	4.0	5.5	1151	1763	2915	40	19.2	34.4	53.6	36
A95-9	35	18	0	12	4.6	4.9	1052	1263	2315	45	21.7	25.5	47.2	46
A95-10	48	23	0	17	4.5	5.2	831	1057	1888	44	20.5	23.6	44.1	46
A95-15	44	11	0	24	4.5	5.2	1058	705	1763	60	21.9	14.6	36.5	60
A95-17	39	10	0	18	4.4	5.2	451	1146	1597	28	10.2	23.9	34.1	30
R95-1	34	32	0	0	4.1	5.3	496	2540	3036	16	15.3	43.4	58.7	26
R95-6	41	9	0	6	4.3	5.4	1062	2402	3464	31	19.4	47.2	66.6	29
R95-7	46	22	0	0	4.7	6.4								
R95-22	61	11	0	28	5.1	5.2								
R95-31	47	4	0	32		5.0								
Avg.	44	16	0	16	4.5	5.3	872	1554	2425	38	18.3	30.4	48.7	39
SE	2.4	2.6	0	3.6	0.11	0.13	109	265	272	5.3	1.6	4.4	4.4	4.6
Prob. level[a]	0.013	0.08	0.002	0.77	0.000	0.000	0.001	0.88	0.044	0.022	0.005	0.35	0.40	0.017
Shrublands (acidic)														
A95-6	40	21	0	19	4.4	4.4	934	0	934	100	17.0	0.0	17.0	100
A95-14	30	14	0	14	5.0	5.0	803	1559	2362	34	15.7	27.9	43.6	36
A95-16	72	18	0	0	6.7		1571	554	2125	74	26.6	12.1	38.8	69
R95-2	>50	0	8	8		5.6								
Avg.	47	13	2	10	5.4	5.0	1103	704	1807	69	19.8	13.3	33.1	68
SE	11	5	2	4	0.7	0.3	237	456	442	19	3.4	8.1	8.2	18
Shrublands (nonacidic)														
A95-5	59	26	0	33	5.5	6.6	1844	631	2475	74	31.2	17.8	49.0	64
A95-4	>192	0	1	0		7.1	573	0	573	100	27.9	0.0	27.9	100
R95-10	82	12	0	0	5.9	6.4								
R95-26	85	0	35	25			1774	33	1807	98	27.1	6.5	33.6	81
Avg.	75	10	9	15	5.7	6.7	1397	221	1618	91	28.7	8.1	36.8	81
SE	7.1	6.2	9	8.5	0.2	0.2	412	205	557	8.2	1.3	5.2	6.3	10

continued next page--

Table 3. continued--

Profile number	Active layer (cm)	Organic layer (cm)	A hor. (cm)	B hor. (cm)	pH O hor.	pH B hor.	N (g/m²) Active	N (g/m²) Perm.	N (g/m²) 100 cm.	% Active	C (kg/m²) Active	C (kg/m²) Perm.	C (kg/m²) 100 cm.	% Active
Wet tundra (mineral soils only)														
A95-23	39	21	18	0	7.5									
R95-9	33	25	>20	0	5.8									
R95-14	49	20	29	0	6.9									
R95-16	46	32	0	0	6.7									
R95-20	29	33	0	0	5.5	6.5								
Avg.	39	26	12	0	6.5	6.5	2287	3131	5418	42	34.8	50.4	85.2	41
SE	3.8	2.7	6.4	0	0.4									
Wet tundra (organic soils)														
R95-3	54	54+	0	0	5.5									
R95-5	36	42+	0	10	5.5	5.3	955	6780	7735	12	19.1	29.0	48.1	40
A95-11	65	65+	0	0	4.6		1621	4956	6577	27	27.0	39.7	66.7	41
A95-12	62	62+	0	0	4.5		666	1825	1159	15	7.9	10.7	18.6	0.5
A95-13	36	40+	0	0	4.1		625	2931	3556	18	12.8	57.0	69.8	18
A95-22	35	35	0	0	7.3		1740	1482	3221	54	33.1	50.4	85.2	39
Avg.	48	35+	0	2	5.3	5.3	2433	2432	4865	50	46.3	29.9	76.2	61
SE	4.7		0	2	0.5		2458	678	3137	78	45.4	11.6	57.0	80
							1684	0	1684	100	26.7	0.0	26.7	100
							1788	1505	3293	60	32.9	29.8	63.0	60
							334	540	508	14	6.2	10.9	10.2	14

SE = standard error of the mean.
[a]Based on single-factor analysis of variance between moist nonacidic tundra and moist acidic tundra.

amount and proportion of field moisture in the active layer were significantly greater in moist nonacidic tundra soils, the amounts of field moisture in the upper permafrost and in the entire profile were significantly greater in moist acidic tundra soils.

There also were comparable differences in morphological and chemical properties in nonacidic and acidic shrublands (Table 3); however, there were an insufficient number of sites to do meaningful statistical comparisons.

IV. Discussion

A. Cryoturbation and Active Layer Dynamics

A greater proportion (60%) of soils in the moist nonacidic tundra were cryoturbated than in the moist acidic tundra (40%). This is also reflected by a greater percentage of frost scars on the surface of the moist nonacidic tundra. For example, at Sagwon, frost scars comprised 36% of the moist nonacidic tundra (pedons A95-2 and A95-18) and <1% (pedon A95-3) of the moist acidic tundra. Cryoturbation may be inhibited in the moist acidic tundra by the thicker organic mat. This mat also insulates the soil and results in higher volumetric moisture contents and a thinner active layer. The organic mat, which is often dominated by *Sphagnum*, produces strongly acidic conditions (Sjors, 1963).

In contrast, cryoturbation causes mixing of the organic matter throughout the active layer and exposes the dark-colored mineral soil, enabling greater thermal diffusivity in the moist nonacidic tundra than in the moist acidic tundra. The incorporation of organic matter causes the development of an A horizon, which often qualifies as mollic in most soils of the moist nonacidic tundra; an A horizon is lacking in the moist acidic tundra soils. The amounts of carbon and nitrogen are significantly greater in the active layer of the moist nonacidic tundra than in the moist acidic tundra (Table 3). Whereas 60% of the carbon and nitrogen present in the upper 1 m exists in the active layer of the moist nonacidic tundra, only 40% is in the active layer of the moist acidic tundra. A preliminary model of the influence of cryoturbation on carbon dynamics and soil development in the moist tundra of northern Alaska is given in Figure 2.

B. Origin of Alkalinity

According to Walker and Everett (1991), the distribution of moist nonacidic tundra closely parallels the zone of calcareous loess deposition. The silt originates from limestone deposits of the Lisburne Group in the Brooks Range and is transported by Sagavanirktok River and its tributaries to the coastal plain. The calcareous silt is then transported as loess by strong, predominantly east-northeasterly winds. Moist acidic tundra exists in areas receiving lesser amounts of snowfall and calcareous loess, or on older surfaces where a *Sphagnum* layer has developed during plant succession (Walker et al., 1989; 1995).

C. Implications

This study suggests that nonacidic soils may comprise 54% of the Kuparuk drainage, which is greater than previous reports. The Kuparuk basin may represent a modern analogue of steppe tundra that existed across Alaska and Siberia during Pleistocene glaciations (Hopkins, 1982; Guthrie, 1990; Walker et al., 1991). Cryoturbation is an important soil-forming process that causes deep mixing of carbon and results in dark-colored, organic-enriched mineral soils. Our estimates suggest that soils

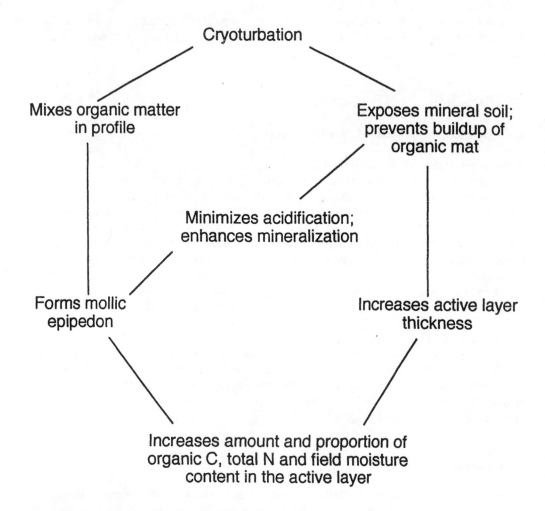

Figure 2. Conceptual model of the influence of cryoturbation on carbon distribution and soil development in moist nonacidic tundra in arctic Alaska.

of the Kuparuk basin may contain 435 Tg of carbon in the upper 1 m (Table 4) which is slightly greater than the 381 Tg estimate of Michaelson et al. (1996).

We propose that in the event of global warming there will be a greater release of CO_2 from soils of the moist nonacidic tundra than from soils of the moist acidic tundra because of the greater carbon and field moisture contents in the active layer. Soils in the moist acidic tundra are protected by a thick organic mat (average thickness = 16 cm) that will buffer against sudden changes in soil moisture and temperature. However, the moist acidic tundra may be more susceptible to thermokarst because of a greater proportion of ice wedges and other forms of massive ice than in moist nonacidic tundra. Additional studies are needed to monitor CO_2 and CH_4 dynamics in moist acidic and moist nonacidic tundra.

Table 4. Distribution of soil carbon to 1 m by landcover type in the Kuparuk basin

Landcover type	Area %	Area ha	Soil C (kg m⁻²) Active	Soil C (kg m⁻²) Permafrost	Soil C (kg m⁻²) Total	Soil C (Tg)
Barrens	1.4	129	11.5	0	11.5	1.5
Moist nonacidic	38.9	3579	31.0	24.3	55.4	198.3
Moist acidic	30.8	2834	18.3	30.4	48.7	138.0
Shrublands (nonacidic)	9.0	828	28.7	8.1	36.8	30.5
Shrublands (acidic)	7.8	718	19.8	13.3	33.1	23.8
Wet tundra (organic)			33.2	33.2	66.4	
	7.0	644				43.0
Wet tundra (mineral)			19.1	29.0	48.1	
Water and shadows	5.1	469				
Clouds and ice	0.1	9				
Total	100.1	9210				435.1

V. Summary and Conclusions

Nonacidic soils comprise 54% of the Kuparuk basin primarily in association with moist nonacidic tundra, shrublands along rivers, and wet tundra landcover types. Based on analysis of variance, the following soil properties were significantly greater in soils of moist nonacidic tundra than in soils of moist acidic tundra: active layer thickness, the thickness of the A horizon, pH of the surface organic and the uppermost B horizon, and the amount and proportion of carbon, nitrogen, and field moisture content in the active layer.

The depth-distribution of carbon in soils of arctic Alaska is controlled largely by the presence or absence of an organic mat and its effect on the degree of cryoturbation (frost churning). Cryoturbation in moist nonacidic tundra exposes the mineral soil, prevents the buildup of the thick organic mat that is characteristic of moist acidic tundra, and slows down the rate of soil acidification. Cryoturbation also mixes organic matter into the soil, contributes to the formation of a mollic epipedon, and results in a deeper active layer in soils of the moist nonacidic tundra than in soils of moist acidic tundra. Additional studies are needed to determine whether soils of moist nonacidic tundra and moist acidic tundra will act as a source or a sink of CO_2 in a climate-warming scenario.

Acknowledgments

Funds for this project were provided by the National Science Foundation Arctic System Science Land-Atmosphere-Ice Interactions Program. G. Michaelson and C.L. Ping kindly analyzed N and C on the samples. J. Munroe assisted in the field work. This chapter is Byrd Polar Research Center contribution 1058.

References

Auerbach, N.A., D.A. Walker, and J.G. Bockheim. 1996. Landcover of the Kuparuk river basin, Alaska. Unpublished Draft Map, Joint Facility for Regional Ecosystem Analysis, University of Colorado, Boulder, CO.

Billings, W.D. 1989. Carbon balance of Alaskan tundra and taiga ecosystems: past, present and future. *Quarterly Science Review* 6:165-177.

Carter, L.D. 1988. Loess and deep thermokarst basins in Arctic Alaska, p. 706-711. In: Proceedings of the Fifth International Conference on Permafrost, Vol. 1. Tapir, Trondheim, Norway.

Everett, K.R. and R.J. Parkinson. 1977. Soil and landform associations, Prudhoe Bay area, Alaska. *Arctic and Alpine Research* 9:1-19.

Guthrie, R.D. 1990. *Frozen Fauna of the Mammoth Steppe*. University of Chicago Press, Chicago, IL. 323 pp.

Haugen, R.K. 1982. Climate of Remote Areas in North-Central Alaska, 1975-1979 Summary. CRREL Report #82-35, U.S. Army Cold Regions Research & Engineering Laboratory, Hanover, NH.

Hopkins, D.M. 1982. Aspects of the paleography of Beringia during the late Pleistocene, p. 3-28. In: Hopkins, D.M., J.V. Matthews, C.E. Schweger, and S.B. Young (eds.), *Paleoecology of Beringia*. Academic Press, New York.

ICOMPAS (International Committee on Permafrost Affected Soils). 1996. Circular Letter No. 5. May 2, 1996, c/o J.G. Bockheim, Department of Soil Science, University of Wisconsin, Madison, WI 53706-1299. 30 pp.

IPCC (Intergovernmental Panel on Climate Change). 1992. Climate Change 1992, The Supplementary Report to the IPCC Scientific Assessment. Cambridge Univ. Press, Cambridge. 200 pp.

Kreig, R.A. and R.D. Reger. 1982. Air-Photo Analysis and Summary of Landform Soil Properties along the Route of the Trans-Alaska Pipeline System. Alaska Division of Geology and Geophysical Surveys. Geological Report 66. 149 pp.

Michaelson, G.J., C.L. Ping, and J.M. Kimble. 1996. Carbon content and distribution in tundra soils in arctic Alaska. *Arctic and Alpine Research*: in press.

Oechel, W.C., S.J. Hastings, G. Vourlitis, M. Jenkins, G. Riechers, and N. Grulke. 1993. Recent change of Arctic tundra ecosystems from a net carbon dioxide sink to a source. *Nature* 361:520-523.

Parkinson, R.J. 1978. Genesis and Classification of Arctic Coastal Plain Soils, Prudhoe Bay, Alaska. Institute of Polar Studies Report Number 68, Ohio State University, Columbus. 147 pp.

Péwé, T.L. 1975. Quaternary Geology of Alaska. U.S. Geological Survey Professional Paper 835. 145 pp.

Post, W.M., W.R. Emanuel, P.J. Zinke, and A.G. Stangenberger. 1982. Soil carbon pools and world life zones. *Nature* 298:156-159.

Sjors, H. 1963. Bogs and fens on Attawapiskat River, northern Ontario. *Bulletin of the National Museum of Canada* 186:45-103.

Soil Survey Staff. 1994. Keys to Soil Taxonomy (6th edit.). U.S. Department of Agriculture, Soil Conservation Service, Washington, D.C.

Walker, D.A., N.A. Auerbach, and M.M. Shippert. 1995. NDVI, biomass, and landscape evolution of glaciated terrain in northern Alaska. *Polar Record* 31:169-178.

Walker, D.A. and N.C. Barry. 1991. Toolik Lake permanent vegetation plots: site factors, soil physical and chemical properties, plant species cover, photographs, and soil descriptions. Dept. of Energy, R4D Program Data Report, Institute of Arctic and Alpine Research, University of Colorado, Boulder.

Walker, D.A. and K.R. Everett. 1991. Loess ecosystems of northern Alaska: regional gradient and toposequence at Prudhoe Bay. *Ecological Monographs* 61:437-464.

Walker, D.A., K.R. Everett, P.J. Webber, and J. Brown. 1980. Geobotanical Atlas of the Prudhoe Bay Region, Alaska. CRREL Report 80-14, United States Army Cold Regions Research and Engineering Laboratory, Hanover, NH.

Walker, M.D., D.A. Walker, and N.A. Auerbach. 1994. Plant communities of a tussock tundra landscape in the Brooks Range foothills, Alaska. *Journal of Vegetation Science* 5:843-866.

Walker, M.D., D.A. Walker, and K.R. Everett. 1989. Wetland Soils and Vegetation, Arctic Foothills, Alaska. Biological Report 89(7), June 1989, U.S. Fish & Wildlife Service, Washington, D.C.

Walker, M.D., D.A. Walker, K.R. Everett, and S.K. Short. 1991. Steppe vegetation on south-facing slopes of pingos, central arctic coastal plain, Alaska, U.S.A. *Arctic and Alpine Research* 23:170-188.

CHAPTER 12

Characteristics of Soil Organic Matter in Arctic Ecosystems of Alaska

C.L. Ping, G.J. Michaelson, W.M. Loya, R.J. Chandler, and R.L. Malcolm

I. Introduction

Northern ecosystems contain an estimated 350 to 455 pg of C in the active layers of tundra soils (Miller et al., 1983; Post, 1990; Oechel and Billings, 1992) which is equivalent to 25 to 33% of the total world C pool (Billings, 1992). Arctic tundra ecosystems alone contain about 192 pg of soil C, or approximately 14% of the global soil C pool (Billings, 1987). However, a recent study (Michaelson et al., 1996) indicated that the soil C stores in Arctic ecosystems could be significantly underestimated when the C stores cryoturbated into the upper permafrost layers are ignored. If these stores are considered, then the Arctic ecosystems alone could contain nearly 30% of the global soil C pool.

With an observed increase in atmospheric CO_2 concentration of 20% in recent years, concern has arisen over the possible climatic effects (Smith and Shugart, 1993). With a warming climate, the Arctic tundra ecosystems are expected to show the most pronounced effects (IPCC, 1992). The high latitude ecosystems are particularly vulnerable to climate change because of the predominance of permafrost and their large C storage. Historically, northern ecosystems have been net sinks for atmospheric C and there has been a net soil organic matter (SOM) accumulation due to the combined effects of cold climate and wet soils (Oechel et al., 1993). Climate change resulting in higher temperature and less precipitation could increase the depth of the active layer and, hence, zone of biological activities. With moisture and temperature being the major controllers of SOM decomposition, higher decomposition rates would likely result from warming conditions. If soil carbon decomposition increases more rapidly than net primary production, the systems cease to be carbon sinks and become carbon sources (Billings et al., 1982; Oechel et al., 1993; Webb and Overpeck, 1993). The increased CO_2 emission of the tundra systems could further increase atmospheric concentrations of the trace gases and serve to accelerate climate change. It has been realized that relatively little is known of the response or even the current state of controls for the arctic ecosystems. This has made it difficult to model and predict potential effects of global climate change on the arctic region or even know its present function in the global carbon cycle.

Most research in soil carbon of arctic ecosystems has been limited to the storage or total C content, and very few studies have included the characterization of soil organic matter. Much of the available soil organic matter characterization data is from temperate regions (Stevenson, 1986, 1994) and there are very limited data from arctic soils (Kononova, 1961; Beyer et al., 1995). This study is part of the integrated Flux Study founded through the National Science Foundation (NSF) Arctic System Science (ARCSS) program. The primary objectives of the Flux Study are to determine the rates and controls over trace gas fluxes in arctic Alaska and to predict how these fluxes will change in response to climatic warming (Weller, et al., 1995). This research is critical with the Arctic being the region of greatest expected warming, containing significantly large soil carbon stores and extensive wetlands. These conditions give the arctic systems the potential to act as a large positive feedback to global warming. The purposes of this study are to evaluate the quality of the SOM and to determine the distribution of different organic fractions and access the bioreactive fractions in some arctic tundra ecosystems.

II. Study Sites and Soil Description

Three of the sites selected for this study are associated with plot level CO_2 flux measurements in arctic Alaska. They represent the three major landform/vegetation complexes in the arctic Alaska. Geographical settings and soil classifications for the sites are listed in Table 1 and the soil horizon sequences are listed in Table 2. The Betty Pingo site is located at Prudhoe Bay, east of the Kuparuk River, in a moist nonacidic tundra. This site represents the coastal plain of the arctic slope, characterized by oriented thaw lakes with flat and low-centered polygons. The dominant vegetation includes Carex aquatilis, Eriophorum angustifolium, Drepanocladus spp., and scorpidium. This poorly drained soil is formed in thick organic matter over discontinuous fluvial deposits with massive ice wedges occurring at 60 cm.

The West Dock Site is located on a coastal plain drained lake characterized by low hummocks. The microrelief ranges from 5 to 20 cm. The wet tundra vegetation is dominated by Carex aquatilis, Eriophorum angustifolium, Eriophorum scheuchzeri, Dryas integrifolia, and Salix spp. This poorly drained soil formed in recent lake sediment and the horizonation is weak. The thickness of the organic horizons ranges from 2 to 12 cm. The A horizon is weakly cryoturbated.

The Happy Valley site is located on low hillslopes south of the coastal plain on moist acidic tundra. The landscape is dominated by cotton grass tussocks with scattered nonsorted circles (mud boils). The microrelief ranges from 15 to 30 cm. The dominant vegetation includes Eriophorum vaginatum, Carex bigelowii, Betula nana, Salix pulchra, Vaccinium uliginosum, Vaccinium vitis-idaea, Ledum palustre, Sphagnum spp., Aulacomium turgidum, and Hylocomium splendens. This somewhat poorly drained soil is formed in loess over glacial till. It is highly cryoturbated. Organic horizons are absent on the fresh mud-boils and the soil appears well drained (Bwj horizon). The upper boundary of the permafrost table contains cyoturbated humus-rich zones interspersed. Ice wedges were found in the Cf horizon of this pedon.

The Toolik Lake site is on the rolling foothills of the northern Brooks Range. It is located on the slope just east of Toolik Lake with a low hummock microrelief. The dominant vegetation includes Carex bigelowii, Dryas integrefolia, Salix reticulata, Tomentypnum nitens, and Hylocomium splendens. The soil is formed in glacial till. The depth to permafrost is about 35 cm with some cryoturbated humus occurring on the top of the Bcjf horizon.

The landform position and vegetation of the Imnaviat Creek Site are similar to those of the Toolik Lake site. It is located on the west facing slope above Imnaviat Creek with a low hummock microrelief.

III. Methods of Investigation

A. Soil Sampling and Analysis

A pit approximately 1m² was opened to a 1m depth at each site using a gasoline powered chisel for the permafrost layers. Soil samples were taken and profiles described from the pits according to the

Table 1. Study site characteristics and locations

Site ID	Land cover class[a] Soil classification[b]	Microrelief Landform	Active layer (cm)	Latitude Longitude	Elev. (m)
Betty Pingo	Moist nonacidic tundra Loamy, mixed nonacid Histic Pergelic Cryochrept	Flat-Polygon Center Coastal Plain, moist	35	70° 16' 58.87"N, 148° 53' 20.69"W	21
West Dock	Moist nonacidic tundra Coarse-loamy, mixed Pergelic Ruptic-Histic Cryaquept	Flat-Polygon Center Drained, thaw-lake basin	45	70° 16' 01"N, 149° 38' 09"W	1
Happy Valley	Moist acidic tundra Fine-loamy, mixed Pergelic Cryaquept	Tussock tundra Hilltop	40	69° 08' 47"N, 148° 51' 14"W	300
Toolik Lake	Moist acidic tundra Loamy, mixed Ruptic Histic Pergelic Cryaquept	Mid-slope, some nonsroted circles Moraine mid-slope	35	68° 37' 27.54"N, 149° 37' 0.22"W	741
Imnaviat Creek	Moist Acidic Tundra Loamy-skeletal, mixed Pergelic Cryaquept	Moraine mid-slope Moraine	26	68° 36' 41.70"N, 149° 18' 26.46"W	909

[a]According to Auerbach and Walker (1995); [b]According to Soil Survey Staff (1994).

Table 2. Selected properties of study soils (ExtC-0.1N NaOH extractable-C, TOC-total organic-C, TN-total nitrogen., SOM-soil organic matter, TOC/SOM-carbon content of the soil organic matter as % and B.D.-soil bulk density)

Site ID	Hor.	Depth (cm)	pH 1:1	C-stores kgC/m²	ExtC %	TOC %	TN %	C/N	SOM %	TOC/SOM	Clay %	Sand %	B.D. Mg/m³
Betty Pingo	Oa1	0-18	7.1	25	25	21.2	1.7	12	43.7	49	--	--	--
	Oa2	18-39	6.6	43	27	12.7	1.0	13	25.5	50	10	36	0.61
	Cf	39-100	7.2	1	3	5.0	0.3	17	10.6	47	1	89	1.81
West Dock	A/Bw	5-57	8.2	28	43	3.3	0.3	11	8.4	39	9	70	1.53
	Cf	57-100	8.1	15	9	1.1	0.2	6	4.5	24	9	66	1.70
Happy Valley	Oe	0-10	5.1	2	16	31.8	1.4	23	62.7	50	7	54	--
	Bw	10-40	4.8	18	42	3.3	0.2	17	8.1	41	27	9	1.40
	Cf	70-100	4.9	38	42	9.7	0.5	19	18.6	52	16	20	1.42
Toolik Lake	O	7-13	5.9	14	17	25.6	2.0	13	78.0	33	--	--	0.73
	A	13-20	5.3	9	26	12.9	0.9	14	39.2	33	38	24	0.90
	Bw	20-30	5.0	6	59	2.7	0.3	9	6.5	42	24	36	1.41
	Cf	30-100	5.1	8	24	7.0	0.4	7	15.5	45	26	33	
Imnaviat Creek	Oe	0-12	5.3	3	24	36.3	2.1	17	71.9	51	--	--	0.44
	Oa/A	12-17	5.5	9	51	22.8	1.4	16	44.4	51	--	--	0.12
	Bw	17-24	5.0	4	60	3.4	0.2	17	7.6	45	22	45	0.26 0.75

Soil Survey Manual (Soil Survey Division Staff, 1994). Samples from each genetic horizon were sent to the USDA National Soil Survey Laboratory in Lincoln, NE for characterization analyses according to Soil Survey Laboratory Staff (1992). Bulk samples were shipped in coolers to the University of Alaska Fairbanks, Plant and Soil Analysis Laboratory of the Palmer Research Center. Soil organic matter (SOM) was estimated by loss on ignition at 450°C. Total C (TOC) and N analyses were performed using a LECO 1000 CHN analyzer with soils after removal of inorganic C. Carbon storage (kgC/m^2) was calculated from the horizon bulk density, %TOC, and thickness by the method of Kimble et al. (1993).

B. Fractionation of Soil Organic Carbon

Soil organic fractions (Table 3) were determined on field moist samples extracted sequentially, first with 0.1N NaOH (Ping et al., 1995 and Michaelson and Ping, 1997) and then on the residual soil, by the fiber analysis sequence of Van Soest and Robertson (1980). The SOM dissolved in the NaOH (extractable-C) was fractionated using the tandem XAD-8/XAD-4 resin technique as described by Malcolm and MacCarthy (1992). This technique allowed for the determination of extractable-C as humic acid (HA), fulvic acid (FA), hydrophobic neutrals (HON), low molecular weight acids or XAD-4 acids (LMA), and hydrophilic neutrals (HIN). The fiber analysis sequence allowed the determination of hemicelluloses (detergent soluble SOM), cellulose (72% H_2SO_4 soluble SOM), and insoluble SOM (considered humin) in the residual soil extracted by NaOH.

IV. Results and Discussion

A. Soil Properties and Carbon Distribution

Selected soil properties are presented in Table 2. Soil reactions are nearly neutral to slightly alkaline (Table 2) due to the high base status of the loess deposits and the lack of leaching in the arctic environment (Parkinson, 1977). Soil carbon percentage decreased drastically with depth as mineral contents increased, with an average for the five pedons of 28, 13, 3, and 5% in the O, A, B, and Cf horizons, respectively. However, when total C stores are calculated on a volume basis to a 1m depth for each horizon, the decreasing C with depth trend is lessened or reversed. This C sequestration at depth resulted from cryoturbation in the relatively thick and dense upper permafrost layers (Ping et al., 1996; Michaelson et al., 1996).

By convention, the Van Bemmelen factor of 1.724 is used to estimate SOM content from soil C content assuming that organic matter contains 58% organic carbon. However, it has been found, based on data from temperate regions, that SOM can range from 48 to 58% C depending on soil type and depth in the profile (Nelson and Sommers, 1982). The tundra soils of this study showed a much wider range of 24 to 52% C in their soil organic matter (Table 2). Lower values obtained for some of these arctic soils indicate less humified SOM than for soils of the temperate zones. Beyer et al., (1995) reported the total organic carbon content of the SOM in Histosols of coastal continental Antarctica ranging from 36-50%. They found that the SOM became increasingly C-enriched in the deeper horizons. The lack of such a relationship, except at the Toolik Lake site, is likely due to the role of cryoturbation in carbon sequestration.

B. C:N Ratio of Soil Organic Matter

The data for C distribution among the fractions are presented in Table 3. The C:N ratio narrows following the humification process from more than 20 in fresh organic matter to 8-20 in humus (Tan, 1993). Studying Aquepts (Gleysols) from Germany, Beyer (1995) measured C:N ratios of 5.9 to 7.7 and attributed such low values to the relatively high amounts of microbial biomass in both the whole soils and the TOC. The C:N ratio of the tundra soil horizons of this study averaged 14 with a range of 6 to 19, indicating a relatively low overall degree of humification relative to the temperate cultivated soils (Beyer, 1995; Gotoh, 1983; and Scheldrick, 1986). The lowest values found in the permafrost layer (Cf) of the West Dock site and the Bw and Cf horizons of the Toolik Lake sites were in or close to the range of Beyer (1995). In most of the O and A horizons, the C:N ratios are higher indicating the lower proportion of microbial biomass and, thus, the lower degree of decomposition or humification; however, soil microbial biomass (SMB) was not measured in this study. In addition to C:N ratio, Follett (1996) pointed out that SMB is one of the indexes that can be used to assess soil organic carbon (SOC) dynamics as SOC itself. Since seasonal patterns of increased plant residue C during periods of warm temperatures and adequate soil moisture result in increased SMB, measuring of SMB in different soil horizons in future studies may shed light on the past environment changes in these soils.

C. Extractability of Soil Organic Matter

The low degree of humification in the O and A horizons is further supported by the generally lower extractability. The NaOH extractable fractions include both humic and nonhumic substances which are the product of biochemical degradation of plant materials and microbial biomass. This percentage extractable C can thus be used as an indicator of the degree of decomposition or humification of SOM within a soil profile. The SOM of the B horizons averaged 52% of TOC extractable, nearly double the extractability of the O, A, and Cf horizons at 27, 27, and 20%, respectively. These relatively high values for the B horizons are, however, much lower than those found for the spodic horizons (Bhs, Bs) of soils from cryic temperature regimes which ranged from 68 to 86% of TOC extractable (Ping et al., 1995). In addition to climate factors acting to preserve C, the formation processes are different in these two groups of soils. In Spodosols, the carbon in the spodic horizons are largely translocated as organo-metal complexes (DeConinck, 1980) and thus more easily solubilized. Whereas, indicated by this data, the carbon in the tundra O and A horizons is largely accumulated *in situ* and contain large amounts of fine roots and organic remains from vegetation. In the B horizons of this study, there is likely some dissolved organic carbon moving in with percolating water in late summer, the wet season in the study area. The low carbon extractability of the upper permafrost layers (Cf) may be influenced by the source of organic matter. Most of the SOM in this portion of the soil profile is cryoturbated or frost-churned from the surface horizons and remains frozen, thus slowing the humification process.

D. Fractions of Soil Organic Matter

The results of the fractionation of SOM are presented in Table 3. The hemicellulose fraction contained the highest amount of SOM carbon in all soils. The total amount of hemicellulose-C was inversely proportional to the C-extractability. The unextracted fractions were mostly hemicelluloses which were also the dominant organic fractions in all of the soils. On the average, hemicellulose, cellulose, and humin fractions account for 53, 5, and 10% of the TOC, respectively. As compared with the organic fractions of the temperate region soils, the arctic tundra soils have high proportions of hemicellulose indicating the relatively low degree of decomposition in the arctic tundra environment

Table 3. Total organic carbon (TOC) and C-fractions data (HA-humic acids, FA-fulvic acids, LMA-low molecular weight acids, HON-hydrophobic neutrals, HIN-hydrophilic neutrals, Hc-hemicellulose, Cel-cellulose, and Hm-humin) for soils studied

Site ID	Hor.	Depth (cm)	Soil TOC %	HF:FA[a]	HA[b]	FA[b]	LMA[b]	HON[b]	HIN[b]	Hc[c]	Cel[c]	Hm[d]
							% of TOC					
Betty	Oa1	0-18	21.2	1.4	11.3	5.3	3.0	1.0	4.5	50.3	7.5	17.3
Pingo	Oa2	18-39	12.7	0.9	10.0	7.3	4.3	1.1	4.3	43.1	8.0	21.9
	Cf	39-100	5.0	0.9	1.1	0.9	0.4	0.0	0.7	76.6	6.8	13.6
West	A/Bw	5-57	3.3	1.3	19.4	12.5	2.2	0.4	12.5	53.6	1.1	2.3
Dock	Cf	57-100	1.1	0.9	3.3	2.3	1.4	0.3	1.6	83.7	2.7	5.5
Happy	Oe	0-10	31.8	1.3	6.9	3.8	1.6	0.2	3.8	55.4	3.4	26.0
Valley	Bw	10-40	3.3	1.9	23.1	9.2	2.9	0.4	6.3	48.7	3.5	5.2
	Cf	70-100	9.7	1.3	19.7	11.3	3.8	0.4	7.1	48.7	4.1	5.2
Toolik	O	7-13	25.6	1.4	7.5	3.4	2.0	0.5	3.6	57.3	12.5	13.3
Lake	A	13-20	12.9	1.2	10.7	6.0	2.9	0.5	6.2	59.9	3.7	10.4
	Bw	20-30	2.7	1.2	23.0	12.4	7.1	3.0	13.6	34.9	2.5	3.7
	Cf	30-100	2.7	0.9	9.6	7.0	3.4	0.2	4.1	63.1	4.6	8.4
Imnaviat	Oe	0-12	36.3	0.5	5.5	7.0	4.1	0.2	7.7	54.0	7.1	14.9
Creek	Oa/A	12-17	22.8	2.5	21.4	2.6	5.6	8.7	12.8	37.7	3.4	7.8
	Bw	17-24	3.4	0.9	23.4	16.2	9.0	1.2	10.8	35.2	2.0	2.8

[a]FA includes fulvic acid fractions (FA+LMA); [b]Carbon extractable in 0.1N NaOH, fractions determined according to Malcolm, 1992; [c]Carbon not extracted in 0.1 N NaOH, fractions determined in residual soil by Van Soest and Robertson (1980); [d]Considered lignin in procedure for plant samples by Van Soest and Robertson (1980).

due to wetness and cold climate. Data from Michaelson and Ping (1997) supports this contrast between arctic and more temperate region SOM, with the B horizons of arctic soils having an average C extractability of 47% compared with 93% for temperate B horizons. Beyer et al., (1995) also concluded that in the Antarctica Histosols the SOM transformations were retarded due to extreme climate conditions and high water contents. Hemicellulose is relatively easily decomposed and it is likely the substrate for methane production. During anaerobic decomposition, soluble simple compounds such as acetic, formic, lactic, and butyric acids are produced (Stevenson, 1986). This explains the correlation of general trend of decreasing hemicellulose and increasing hydrophilic acids plus XAD-4 acids (lower molecular weight fulvic acids) for all horizons ($r = -0.86$). This indicates that hemicellulose is likely the bioactive fraction in these arctic tundra soils in addition to the readily decomposed simple acids and sugars in the HIN fraction. In most horizons, hydrophobic neutrals are present in trace amount except the Oa/A horizon of the Imnaviat Creek site. Since this fraction includes mostly paraffins, lipids, and tannins, the extra-high HON in this horizon could be the result of vegetation succession with this site dominated by shrubs.

The humin fraction may include lignins not differentiated in this study. Humin fractions of the O horizons range from 13 to 26% TOC, with the higher values in the horizons formed under vegetation dominated by sedges. The humin fractions in the mineral A, B, and Cf horizons range from 3 to 8% TOC. This carbon distribution pattern is quite different from that found for temperate region soils. In recent studies of soils from northern Germany, Beyer (1995) and Beyer et al., (1992) found that hemicellulose and cellulose combined was only a minor component of SOM (ranging 6-24%). Soils studied included Aquepts, Ochrepts, and Histosols where the dominant organic carbon fraction was humic acid, followed in decreasing order by humins, lignins, fulvic acids, and polysaccharides. Both Hatcher et al. (1985) and Beyer et al. (1992) indicated that lignin decreased with depth and humin increased with depth due to increased degree of humification. In this study the humin fractions are presumed composed mostly of lignin and decreased with depth.

Soils with less development have been found to have a lower humic-to-fulvic acid ratio (HA:FA) and this has been used to partition different groups of soils (Kononova, 1966). The tundra soil in Kononova's study had a HA:FA ratio of 0.3 which was the lowest of all soils studied. Although Stevenson (1985) pointed out that in the calculation of this ratio, there has seldom been an allowance made for nonhumic substances present as impurities, particularly in the fulvic acid fractions. Therefore, this ratio may not reflect the true humic substances present and the nonhumics may vary considerably among soils or within a profile. In this study, on the average, the humic fraction accounts for 64% of the extracted organic carbon with a HA:FA ratio averaging (Table 3).

Humic acids remain the dominant fraction of extractable C followed by fulvic acids or hydrophilic neutrals/acids, low-molecular-weight acids, and hydrophobic neutrals. The difference between these fractions within the soil profile and among soil profiles are likely due to the differences in climate as well as vegetation succession in the tundra ecosystems. The HA:FA ratios decreased with depth reflecting the reduced mobility of the HA fraction. The ratios found in these soils (Table 2) are considerably higher than those of Kononova (1966), in part because the fulvic fractions were further fractionated into hydrophilic acids and neutrals not included in the ratio calculation here. Since the FA fractions are more mobile, their proportions often increased with depth in the active layer which decreases the HA:FA ratio. The exception to this trend was the Imnaviat Creek site where the ratio was highest in the 12-14cm Oa/A horizon. This condition was a result of the depressed FA content of this horizon which was likely due to the slope position of this site. Movement of water in this horizon could account for the removal/low levels of FA.

V. Summary and Conclusion

The low extractability of the organic horizons by alkali solution is attributed to the presence of fresh or partially decomposed organic matter including roots and litter. Conventionally, the alkali-insoluble fraction was considered as humin. However, after acid digestion, more than 78% of this humin fraction or 53% of the TOC are hemicelluloses which are more easily decomposed than humin. Hemicelluloses are 2-3 times higher than has been reported for temperate region soils. The high percentage of hemicelluloses and the relatively low extractability reflect the arctic environment characterized by cold climate and high water content which retard SOM transformation. Soil TOC generally decreased with depth on a weight basis. However, there is a noticeable increase in the Cf horizons of the Happy Valley and Toolik Lake sites when the TOC was expressed on a weight-per-unit-area basis due to carbon sequestration caused by cryoturbation or frost-churning. Based on the results of this study, the relatively fresh and partially decomposed materials comprise a high percentage of the arctic soils. These materials are likely to have a high potential for bioactivity with warming soil temperatures and increased nutrient inputs. Experiments designed to evaluate the bioactivity of SOC of different soil horizons and different fractions are under way.

Acknowledgments

This research was funded through grants from the National Science Foundation Arctic System Science Program (NSF-DPP-9318534), the USDA Hatch Project, and the USDA Global Change Initiative Program. The Laboratory support by Corinne Reintjes and Pete Robinson is gratefully acknowledged.

References

Auerbach, N.A. and D.A. Walker. 1995. Preliminary vegetation map, Kaparuk Basin, Alaska: a landsat-derived classification. Institute of Arctic and Alpine Research, Univ. of CO., Boulder, CO.

Beyer, L. 1995. Soil microbial biomass and organic matter composition in soils under cultivation. *Biol. Fertil. Soils.* 19:197-202.

Beyer, L., H.R. Schulten, and R. Frund. 1992. Properties and composition of soil organic matter in forest and arable soils of Schleswig-Holstein: 1. Comparison of morphology and results of wet chemistry, CPMAS-13C-NMR spectroscopy and pyrolysis-field ionization mass spectrometry. *Z. Pflanzenernahr. Bodenk.* 155:345-354.

Beyer, L., C. Sorge, H.P. Blume, and H.R. Schulten. 1995. Soil organic matter composition and transformation in Gelic Histosols of coastal continental Antarctica. *Soil Biol. Biochem.* 27:1279-1288.

Billings, W.D. 1987. Carbon balance of Alaskan tundra and taiga ecosystems: past, present, and future. *Quat. Sci. Rev.* 6:165-177.

Billings, W.D., J.O. Luken, D.A. Mortensen, and K.M. Peterson. 1982. Arctic tundra: a source or sink for atmospheric carbon dioxide in a changing environment. *Oecologia* 53:7-11.

DeConinck, F. 1980. Major mechanisms in formation of spodic horizons. *Geoderma* 24:101-128.

Follett, R.F. 1997. CRP and microbial biomass dynamics in temperate climates. In: R. Lal, J.Kimble, R.F. Follett, and B.A. Stewart (eds.) *Advances in Soil Science - Carbon Sequestration in Soil.* Lewis Publishers. (in press).

Gotoh, S. and H. Koga. 1983. Soil organic matter and nitrogen contents of Saga polder rice fields: changes with depth and time. *Soil Sci. Plant Nutr.* 29:47-61.

Hatcher, P.G., I.A. Breger, G.E. Maciel, and N.M. Szeverenyi. 1985. Geochemistry of humin. p. 275-302. In: G.R. Aiken, D.M. McKnight, R.L. Wershaw, and P. MacCarthy (eds.), *Humic Substances in Soil, Sediment, and Water: Geochemistry, Isolation, and Characterization.* John Wiley & Sons, Inc.

IPCC (Intergovernmental Panel on Climate Change). 1992. Climate change 1992. The supplementary report to the IPCC Scientific Assessment. Cambridge University Press, Cambridge. U.K.

Kimble, J.M., C. Tarnocai, C.L. Ping, R. Ahrens, C.A.S. Smith, J. Moore, and W. Lynn. 1993. Determination of the amount of carbon in highly cryoturbated soils. p. 277-291. In: D.A. Gilichinsky (ed.), Proceedings of the 1st International Conference on Cryopedology, Joint Russian-American Seminar on Cryopedology and Global Change, Nov. 15-16,1992 Pushchino, Russia. Russian Academy of Sciences, Pushchino, Russia.

Kononova, M.M. 1961. *Soil Organic Matter, its Nature, its Role in Soil Formation and in Soil Fertility.* Translated from Russian by T.Z. Nowakowski, and G.A. Greenwood. Pergamon Press. 450 pp.

Malcolm, R.L. 1990. Evaluation of humic substance from Spodosols. In: J. Kimble and R.D. Yeck (eds.), Proceedings of the fifth International Soil Correlation Meeting (ISCOM V). Characterization, classification, and utilization of Spodosols. USDA, Soil Conservation Service, Lincoln, NE.

Malcolm, R.L. and P. MacCarthy, 1992. Quantitative evaluation of XAD-8 and XAD-4 resins used in tandem for removing organic solutes from water. *Environ. Internat.* 18:597-607.

Mathur, S.P. and R.S. Farnham. 1985. Geochemistry of humic substances in natural and cultivated peatlands. p. 53-86. In: G.R. Aiken, D.M. McKnight, R.L. Wershaw, and P. MacCarthy (eds.) *Humic Substances in Soil, Sediment, and Water: Geochemistry, Isolation, and Characterization.* John Wiley & Sons, Inc.

Michaelson, G.J. and C.L. Ping. 1997. Comparison of 0.1 M sodium hydroxide with 0.1 M sodium-pyrophosphate in the extraction of soil organic matter from various soil horizons. *Communication in Plant and Soil Analysis* (in press).

Michaelson, G.J., C.L. Ping, and J.M. Kimble. 1996. Carbon storage and distribution in tundra soils associated with C-Flux Study in Arctic Alaska. *Arctic and Alpine Research* 28:414-424.

Miller, P.C., R. Kendall, and W.C. Oechel. 1983. Simulating carbon accumulation in northern ecosystems. *Simulation* 40:119-131.

Nelson, D.W. and L.E. Sommers. 1982. Total carbon, organic carbon, and organic matter. p. 539-580. In: A.L. Page, R.H. Miller, and D.R. Keeney (eds.), *Methods of Soil Analysis, Part 2; Chemical and Microbiological Properties.* 2nd ed. Agronomy No. 9. ASA, SSSA, Madison, WI.

Oechel, W.C. and W.D. Billings. 1992. Effects of global change on the carbon balance of arctic plants and ecosystems p. 139-168. In: F.S. Chapin III, R.L. Jefferies, J.F. Reynold, G.R. Shaver, and J. Svoboda (eds.), *Arctic Ecosystems in a Changing Climate.* Academic Press Inc., San Diego, CA.

Oechel, W.C., S.J. Hastings, G. Vourlitis, M. Jenkins, G. Riechers, and N. Gruelke. 1993. Recent change of arctic tundra ecosystems from a net carbon dioxide sink to a source. *Nature* 361:520-523.

Parkinson, R.J. 1977. Genesis and classification of Arctic Coastal Plain soils, Prudhoe Bay, Alaska. M.S. diss. Ohio State Univ., Columbus, OH.

Ping, C.L, G.J. Michaelson, and J.M. Kimble. 1996. Carbon storage along a latitudinal transect in Alaska. Special issue Nutrient Cycling in Agro-Ecosystems. In: D.C. Reicosky (ed.), *Fertilizer Research.* (in press).

Ping, C.L., G.J. Michaelson, and R.L. Malcolm. 1995. Fractionation and carbon balance of soil organic matter in selected cryic soils in Alaska. p. 307-314. In: R. Lal, J. Kimble, E. Levine, and B.A. Stewart (eds.), *Advances in Soil Science - Soils and Global Change.* Lewis Publishers, Boca Raton.

Scheldrick, B.H. 1986. Test of the Leco CHN-600 determinator for soil carbon and nitrogen analysis. *Can. J. Soil Sci.* 66:543-545.

Schnitzer, M. 1982. Organic matter characterization. p. 581-594. In: A.L. Page, R.H. Miller, and D.R. Keeney (eds.), *Methods of Soil Analysis, Part 2 ; Chemical and Microbiological Properties*. 2nd ed. Agronomy No. 9. ASA, SSSA, Madison, WI.

Smith, T.M. and H.H. Shugart. 1993. The transient response of terrestrial carbon storage to a perturbed climate. *Nature* 361:523-526.

Soil Survey Division Staff. 1994. Soil survey manual. USDA Soil Conservation Service. U.S. Gov. Printing Office, Washington D.C.

Soil Survey Laboratory Staff. 1992. Soil survey laboratory methods manual. Soil Survey Investigation Report No. 42. USDA Soil Conservation Service National Soil Survey Center, Lincoln, NE.

Soil Survey Staff. 1994. *Keys to Soil Taxonomy*. 6th ed. USDA Soil Conservation Service. Washington, D.C.

Stevenson, F.J. 1985. Geochemistry in soil humic substances. p. 13-52. In: G.R. Aiken, D.M. McKnight, R.L. Wershaw, and P. MacCarthy (eds.), *Humic Substances in Soil, Sediment, and Water: Geochemistry, Isolation, and Characterization*. John Wiley & Sons, Inc. 692 pp.

Stevenson, F.J. 1986. *Cycles of Soil: Carbon, Nitrogen, Phosphorus, Sulfur, Micronutrients*. John Wiley & Sons. 380 pp.

Stevenson, F.J. 1994. *Humus Chemistry: Genesis, Composition, Reactions*. 2nd ed. John Wiley & Sons, Inc. 496 pp.

Tan, K.H. 1993. *Principles of Soil Chemistry*. 2nd ed. Marcel Dekker. 362 pp.

Thurman, E.M. and R.L. Malcolm. 1981. Preparative isolation of aquatic humic substances. *Environ. Sci. Technol.* 15:463-466.

Van Soest, P.J. and J.B. Robertson. 1980. System of analysis for evaluating fibrous feeds. p. 49-60. In: W.J. Pigden, C.C. Balch, and M. Graham (eds.), Standardization of Analytical Methodology for Feeds. Intern. Dev. Res. Center, Ottawa, Canada. Publ. IDRC-134e.

Webb, R.S. and J.T. Overpeck. 1993. Carbon reserves released? *Nature*, 361:497-498.

Weller, G., F.S. Chapin, K.R. Everett, J.E. Hobbie, W.C. Oechel, C.L. Ping, W.S. Reeburgh, D. Walker, and J. Walsh. 1995. The Arctic Flux Study: a regional view of trace gas release. *J. Biogeography* 22:365-374.

CHAPTER 13

Soil Structure and Organic Carbon: A Review

B.D. Kay

I. Introduction

Soil structure has a major influence on the ability of soil to support root development, to receive, store, and transmit water, to cycle carbon and nutrients, and to resist soil erosion and the dispersal of chemicals of anthropogenic origin. Particular attention must be paid to soil structure in managed ecosystems because of its sensitivity to land use practices. Management practices can alter soil structure directly by processes such as tillage and traffic. Many of these changes are relatively short term and reversible. Of greater importance, however, are the management-induced changes in the quantity and characteristics of organic carbon in the soil, since such changes lead to changes in soil structure that are much more persistent. Land use practices which are sustainable must maintain the structure of soil, over the long term, in a state that is optimum for a range of processes related to crop production and environmental quality. A key consideration in designing such practices must, therefore, be management of the organic carbon of soils.

This chapter will explore the relations between soil structure, organic carbon, and land use practices. Emphasis will be placed on soils containing less than 40 to 50% clay, since, it is on these soils that the influence of organic carbon on structure is the greatest. The focus will be restricted to soils of the temperate regions of the world where the relations between soil structure, organic carbon, and land use have been studied most extensively. Although the nature of these relations change with depth in the profile, attention will be directed to the A horizon since the organic carbon content is greatest at this depth and the effects of management most pronounced.

II. Soil Structure

The critical role that soil structure plays in a broad range of processes was recognized by researchers more than 150 years ago and the large volume of research on the nature of soil structure that has resulted in the intervening period has been summarized in a number of comprehensive reviews (e.g., Harris et al., 1966; Oades, 1984; Dexter, 1988; Kay, 1990; Horn et al., 1994). Consequently, the fundamental characteristics of soil structure and the influence of different factors on these characteristics will be examined only in enough detail in this report to provide the reader with a basis for the subsequent discussion on organic carbon and soil structure.

A. Fundamental Aspects

The structure of soils can be considered from four different and fundamental aspects: form, stability, resiliency, and vulnerability. Each aspect of soil structure can be considered across a range of scales.

The term structural form applies to the group of characteristics that describe the heterogeneous arrangement of void and solid space that exists in soil at any given time. Examples of characteristics of structural form that are related to the void space include total porosity, pore size distribution, and continuity of the pore system. Examples of characteristics of structural form related to the solid space include the arrangement of primary particles into secondary structures (aggregates or peds) that are distinguished from adjacent structures on the basis of failure zones of different strength. The relation between the characteristics of the void and the solid space are closely related to the characteristics of the failure zones. For instance, fragmentation of the soil matrix by processes such as tillage occurs via the failure zones. The spatial distribution of failure zones of different strengths leads to the creation of aggregates with characteristic size distributions and characteristic distribution of pore sizes between aggregates. The application of increasing energy causes fragmentation to occur along zones of increasing strength and closer proximity thereby resulting in smaller aggregates. Failure zones are made up of elemental volumes in which the bonding between the elementary particles is small and, therefore, must include pores as well as elemental volumes containing less cementing materials than the surrounding matrix. The cementing materials may include amorphous and crystalline inorganic materials as well as different organic materials.

The term structural stability is used to describe the ability of soil to retain its arrangement of solid and void space when exposed to different stresses. Stability characteristics are specific for a structural form and the type of stress applied. Stresses may arise from processes as diverse as tillage, traffic, and rain drop impact. Extensive research has been directed to understanding the stability of aggregates when they are exposed to stresses arising from wetting under different conditions and then exposed to mechanical stresses arising from sieving or shaking in water or exposed to ultrasonic energy. Consequently, the term structural stability is often considered to be synonymous with aggregate stability. There are, however, important characteristics that reflect the ability of a soil to retain its arrangement of solid and void space when other stresses are applied and conventional measurements of aggregate stability may not provide meaningful measures of these properties. For instance, the compressibility index is a measure of the ability of the total porosity to withstand compressive stresses. This parameter is particularly important where changes in agricultural practices are leading to greater traffic. There is no obvious reason that the compressibility of a soil is strongly correlated with its aggregate stability.

The term structural resiliency describes the ability of soil to recover its structural form through natural processes when the stresses are reduced or removed. Recovery relates, in the first instance, to changes in pore characteristics and the distribution of strengths of failure zones. Resiliency arises from a number of processes such as freezing/thawing, wetting/drying, and biological activity (e.g., root development and activity of soil fauna). The maximum change or recovery that is possible and the rate at which this change can occur are particularly important characteristics of soils (Kay and Rasiah, 1994).

The term structural vulnerability reflects the combined characteristics of stability and resiliency (Figure 1). Soils which are least stable under a given stress and which do not recover when the stress is reduced or removed (or which recover very slowly) are most vulnerable to stress, whether the origin of the stress is natural or anthropogenic. Conversely, soils which have a great resistance to stress and are very resilient are the least vulnerable.

Characteristics of soil structure can be considered at scales ranging from angstroms to meters. There is a degree of similarity of behavior at different scales, as evident from the success of applying fractal theory to different aspects of soil structure (see review by Perfect and Kay, 1995). Structure should, nevertheless, be considered at a range of scales that is most appropriate to the processes under

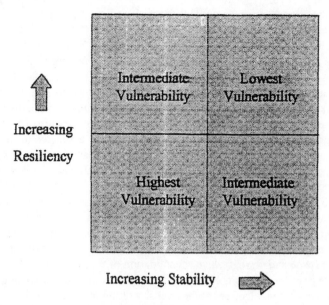

Figure 1. Variation in vulnerability with stability and resiliency.

consideration. For instance, the processes that are most important to plant growth include the penetration of the soil matrix by plant organs, the storage of water, and the movement of oxygen and water. The scales of soil structure that are most relevant to these processes extend from microns to centimeters. Hydrological processes such as infiltration, drainage, storage, or evapotranspiration are influenced by structure at scales ranging from microns to meters.

B. Factors Influencing Soil Structure

The dominant factors influencing structure are soil characteristics such as texture, clay mineralogy, composition of exchangeable ions, and organic carbon content. Other factors influencing soil structure include climate, biological processes, management, and depth in the profile. Few, if any, of these factors can be considered in isolation. An assessment of the influence of organic carbon on soil structure must, therefore, be considered in the context of the other factors.

Soil texture has a major influence on the form, stability, and resiliency of soil structure as well as the response of soil structure to weather, biological factors, and management. The nature of the influence of texture on structural characteristics is related to the nature of the soil matrix. For instance, sands may be viewed, in the simplest case, as being made up of single grains and all characteristics of structural form are determined by the distribution of grain sizes, and modifications introduced by tillage or traffic. Clay or silt-sized materials that may be present, exist as coatings on the sand grains or fill the interstices between the sand grains. The structure does not shrink or swell and is not responsive to freezing. Organic materials, together with fine clays and other amorphous and crystalline inorganic materials, provide what little cementation exists between the grains. The structure is relatively stable in that the arrangement of solid and void space is not very sensitive to the application of stress. The structural form, however, possesses limited resiliency and has, therefore, intermediate vulnerability (Figure 1). As the clay content increases, the characteristics of the soil matrix (including both structural form and stability) are increasingly dominated by the characteristics of the clay (including mineralogy and exchangeable ions), the nature and quantities of cementing materials and the composition of the pore fluid. Structural resiliency, related to weather, biological processes, and management play an increasingly important role and the magnitude of the impacts of these processes are determined by the characteristics of the matrix. For instance, the formation of pores and other failure zones by biological processes may be more important in medium textured soils that undergo limited shrinking and swelling than is the case for soils that are often finer textured and are particularly responsive to wetting and drying (Oades, 1993). In the latter soils, biological processes may complement abiotic processes which exert the dominant control over soil structure.

Climate controls the temporal variation in the water content and temperature of soils and, therefore, has a direct effect on a range of physical and biological processes that are linked to soil structure. The impact of temporal variation in water content on the structural characteristics of soils is strongly influenced by the rate at which the water content changes and the swelling characteristics of the soils. Slow changes in water content arising from low intensity rainfall events, drainage, and evapo-transpiration can cause changes in the pore characteristics in both swelling and nonswelling soils. In the case of swelling soils the changes in pore characteristics are a result of swelling or shrinkage and crack formation. In the case of nonswelling soils, the change in pore characteristics are restricted to the seedbed where decreasing water content and a concomitant increase in effective stress causes a loss of interaggregate pore space and a progressive consolidation of the seedbed.

Fast rates of change in water content normally occur in the field only under wetting conditions and the rate of change in water content increases as the intensity of the rainfall increases and/or the initial soil water content decreases. As the rate of wetting increases, internal stresses arising from differential swelling and air entrapment can become larger and these stresses may be released through the creation of an increasingly extensive network of failure zones. The seasonal variation in structural characteristics that are strongly influenced by failure zones (e.g., tensile strength and aggregate stability) is often much larger than changes caused by different cropping practices and this variation can be related to wetting events preceding the time of sampling (Kay et al., 1994). The extent of development of failure zones is strongly dependent on soil characteristics; soils that exhibit this trait to the greatest extent are referred to as "self mulching" soils (Grant and Blackmore, 1991). The magnitude of the variation in water content and its rate of change decreases rapidly with depth in the profile. It is, therefore, not surprising that major differences in the strength of failure zones can be found with differences in depth within the A horizon as small as 5 cm (Kay et al., 1994).

Changes in the liquid water content of surface soils, as a consequence of ground freezing, is common in many soils. The pore water may freeze *in situ* and, where the soil is saturated, the increase in the volume of the water undergoing phase change can create stresses that may be expected to result in failure zones. A much more common feature, particularly on medium and fine textured soils, is the migration of water in response to gradients in water potential at the freezing front and the subsequent accumulation and freezing of water in the form of lamella or ice lenses just behind the freezing front (photographic evidence of ice lenses in soils of different texture is provided in Kay et al., 1985). The pores that are created under these conditions do not appear to be stable and most are lost as the soil consolidates during the thaw period (Kay et al., 1985). The sites of ice lenses must, however, represent zones of very low strength and undoubtedly contribute to the loss in stability (Willis, 1955; Sillanpaa and Webber, 1961; Bullock et al., 1988) and strength (Utomo and Dexter, 1981a; Voorhees, 1983; Douglas, 1986) of large aggregates that are often observed as a consequence of freezing. Zones between ice lenses are desiccated and the rearrangement and flocculation of clay-sized particles (Rowell and Dillon, 1972; Richardson, 1976) in these areas may account for the increased stability of small aggregates that is often observed after freezing (e.g., Perfect et al., 1990a,b).

Biological processes that influence soil structure include root development, the activity of soil fauna ranging in size from earthworms to amoeba, and the activity of soil flora. These processes are, in turn, also strongly regulated by soil structure. Root development is strongly influenced by soil strength. Although the pores created by drying or by freezing may not be stable, the failure zones that are created by these processes or by rapid wetting may be important in reducing the resistance to penetration by growing roots in soils with high strength, and may, therefore, provide sites for root development. Once the roots penetrate the failure zones, the deposition of carbon will subsequently lead to stabilization of the root channels and the adjacent matrix. Pores may also be created and stabilized by soil fauna. In the case of earthworms, ingestion of carbon and the selective ingestion of fine textured mineral material results in intimate mixing of organic and mineral material and the excretion of "casts" which have structural characteristics that are much different from the bulk soil. Soil flora are primarily responsible for the processing of carbon from plant and animal sources (including soil fauna). The amount and the characteristics of carbon in soil at any given time reflect the rate of addition of carbon and the rate of processing of this carbon by the soil organisms. The use of land for agriculture involves a suite of management practices. Management of soil, water, and crops influence soil structure by controlling the form and amount of carbon entering the soil and the spatial distribution of this carbon in the soil. These practices also influence the populations of soil organisms (macro- as well as micro-) and the rate of mineralization of soil carbon. The nature of soil cover provided by plants or crop residue influences soil structure through its influence on water content, the rate of wetting, and the depth of freezing. The nature of the root system of the crops selected will influence the depth of water extraction and, therefore, the depth to which shrinkage may occur. In addition, large pores (macropores) can be created by tillage and destroyed by traffic.

Soil properties, climate, biological processes, and management practices vary in both space and time. It is, therefore, not surprising that soil structural characteristics are also variable in a given soil. This variation occurs at different scales in both space and time. Organic carbon influences the response of soil structure to all of these factors. Consequently, the structural characteristics that exist at any given time in the field reflect the cumulative interaction of organic carbon with other soil properties, climate, biological processes, and management.

III. Influence of Organic Carbon on Soil Structure

A. Forms of Organic Carbon

Organic carbon in soil represents materials of plant, animal, and microbial origin that are in various stages of decomposition and are associated with the mineral fraction with different degrees of intimacy. This gives rise to different forms of carbon that can be distinguished on the basis of their physical, chemical, or biological properties. The properties of organic materials that are of greatest relevance to soil structure are: (a) the pore characteristics of these materials and their capacity for absorbing water, (b) their capability to strengthen the failure zones between primary soil particles or secondary structures (microaggregates), and (c) their persistence (recalcitrance or protection against attack by microorganisms).

Plant and animal materials are attacked by soil flora and fauna at the outset of the decomposition process. At this stage, these materials have little association with the mineral phase and can be separated from the rest of the soil on the basis of density. Organic carbon that is "free", or has least association with the mineral fraction, is the lightest fraction (the density of this fraction may be less than 1.6 g cm^{-3}). Golchin et al. (1995) distinguished between free particulate organic carbon and occluded organic carbon in the light fraction (occluded organic carbon required ultrasonic dispersion of the aggregates to allow separation). They found the two fractions to be present in roughly similar proportions in the light fraction of five cultivated soils and to make up as much as 38% of the total organic carbon. In agricultural ecosystems, the proportion of the total organic carbon in the light fraction is highest under perennial forages and declines with time under arable crops (Christensen, 1992, and references cited therein) or summer fallow (Janzen et al., 1992). The light fraction has been found to be a very sensitive indicator of management-induced effects on soil organic carbon (Bremer et al., 1994). The proportion of soil carbon in the light fraction has been observed to increase with decreasing clay content (Greenland and Ford, 1964; Richter et al. 1975) and increasing moisture deficit (Janzen et al., 1992). Increases in free particulate carbon would be expected to coincide with increases in the volume fraction of space filled with low density material. This space would have dimensions corresponding to the dimensions of the particulate material. The free particulate organic carbon may have internal porosity and the contribution that these pores make to the water release characteristics of soil will depend on their continuity with soil pores. The extent of continuity would be expected to vary with the source of the carbon. Although the water release characteristics of free particulate materials have not been measured within the soil matrix, the internal pores in root and fungal materials would be expected to exhibit greater continuity with pores in the soil than would crop residue incorporated into the soil by tillage. Occluded organic carbon in the light fraction may contribute more to the water held by soil because of stronger association with the mineral material and, therefore, greater continuity of internal pores with soil pores. The occluded organic carbon in the light fraction may, however, initially have a greater effect on structural stability than on structural form. Fine roots and fungal hyphae will be included in this fraction and these materials are particularly important in strengthening failure zones between microaggregates and thereby stabilizing macroaggregates >250 µm (Tisdall and Oades, 1979; Miller and Jastrow, 1990). The light fraction would be expected to be very labile (the respiration rate and microbial N content of soils has been found by Janzen et al., 1992, to be strongly correlated with the light fraction content), and Tisdall and Oades (1982) have referred to these materials as temporary stabilizing materials. The chemical characteristics of the light fraction would be expected to be similar to those of the original plant material.

The presence of readily mineralizable carbon in particulate material will result in a growth in the biomass of microbes and their predators and the production of extracellular materials including polysaccharides. Electron micrographs show that the microorganisms and extracellular material can become intimately associated with the mineral material at this stage (Foster, 1988). Part of the microbial biomass may grow into pores adjacent to the plant residue or may be squeezed into this space along with the extracellular material if the water absorbed by the growing population of organisms and increasing quantities of extracellular polysaccharides exceeds the pore space where the particulate carbon is located. Dense mucilage may impregnate the soil fabric as much as 50 µm from the plant residue (Foster, 1988). The matrix of decomposing plant residues, microorganisms, extracellular materials, and mineral material has been proposed by Golchin et al. (1994) to coincide with material in the 1.8 to 2.0 g cm^{-3} density fraction and organic material that is intimately associated with the mineral phase would appear in the >2.0 g cm^{-3} fraction. The latter fraction accounted for 25 to 50% of the organic carbon in the soils considered by Golchin et al. (1994). Analyses of this fraction using ^{13}C CP/MAS NMR spectroscopy indicate that about half of the carbon in this fraction is O-alkyl or carbohydrate carbon. There is no microscopic evidence of particulate material with a cellular structure and spectroscopic analyses indicate that the composition is relatively uniform irrespective of the composition of the original carbon source (Golchin et al.,1994). These analyses suggest that the material is largely of microbial origin. This material would be expected to exhibit water release characteristics that are much different from those of soil materials. For instance, analyses of data on the water release characteristics of pure extracellular polysaccharides adsorbed on sand and pure clay that were collected by Chenu (1993), show that addition of scleroglucan increased the available water by about 26g water per g adsorbed organic carbon irrespective of whether the addition was to sand or Ca-montmorillonite.

Polysaccharides are strongly adsorbed on mineral materials and, therefore, are particularly effective, at a molecular scale, in strengthening failure zones. This characteristic accounts for the high correlation that has been observed between aggregate stability and the content of polysaccharides or readily extractable carbohydrates reported in a large number of studies (e.g., Chaney and Swift, 1984; Haynes and Swift, 1990; Haynes et al., 1991; Angers et al., 1993a,b).

Polysaccharides are readily mineralized and, therefore, their effects on water absorption and the strength of failure zones will be transient (Tisdall and Oades, 1982) unless the residue source is continually renewed or the polysaccharides are physically protected from attack by microorganisms and extracellular enzymes. Steric considerations related to the size of bacteria (e.g., Bakken and Olsen, 1987; Foster, 1988) and the relation between the size of the bacteria and the size of the pore neck that they can pass through (e.g., Kilbertus, 1980) suggest that few, if any, bacteria would penetrate pores with necks smaller than about 0.2 µm. The rate of mineralization of polysaccharides located in pores smaller than 0.2 µm would then be controlled by the diffusive flux of exocellular enzymes into these pores. The concentration gradient and, therefore, the diffusive flux would decline with increasing distance between the bacteria and the polysaccharides and therefore polysaccharides located in these pores would be provided a measure of protection against mineralization. The volume fraction of pores < 0.2 µm is strongly positively-correlated with clay content and, therefore, the proportion of carbon existing as polysaccharides closely associated with the mineral phase should be expected to increase as clay content increases.

A matrix of plant residues, microbial biomass, and extracellular material that becomes encrusted by mineral material dominated by small pores may become the center of water stable aggregates. Electron micrographs (Figure 2) of aggregates show convincing evidence of roots and plant cells encrusted with mineral material and show cellular remnants in the center of aggregates surrounded by layers of inorganics (Waters and Oades, 1991). The extracellular materials that are produced during the initial stages of the decomposition process would increase the strength of the bonds between the mineral grains and would increase the strength of the failure zone in which the organic particle was originally located. The increase in strength would lead to increased aggregate stability. The decomposition process would slow down if the pore characteristics of the mineral layer limited supplies of oxygen to the zone of active aerobic decomposition or if the mineral layer physically protected the microorganisms from predation. Chemical analyses of the carbon associated with fractions of different density after ultrasonic dispersion led Golchin et al. (1994) to propose that during decomposition of the organic cores of these aggregates, the more labile portions of the cores (proteins and carbohydrates) are consumed and the more resistant plant structural materials (which have low O-alkyl, high alkyl, and aromatic content) are concentrated as occluded particles within the aggregates. These materials would be increasingly hydrophobic and exhibit less strong bonding with the mineral material. The electron micrographs of aggregates obtained by Waters and Oades (1991) show an abundance of void spaces within the aggregates that would fall in the range between 0.2 and 30 µm diameter (corresponding to pores draining at potentials between -1.5 and -0.01 Mpa, respectively). Some of these pores are associated with cellular debris while others represent voids within aggregates. These pores may be completely closed, have one entrance (bottle-shaped pores), or may be tubes with two or more entrances. The contribution that these pores make to the potentially available water content of soil would be expected to depend on their continuity with other soil pores and the degree to which any hydrophobic organic cores reduce the ability of these pores to absorb and retain water.

The most general characteristic of organic carbon in soils is the total organic carbon content, and the impact of organic carbon on soil structure is often described using this parameter. Implicit in such descriptions are the assumptions that the proportions and characteristics of different forms of carbon remain constant irrespective of: (a) the total carbon content (the different forms of carbon may have different effects on structural characteristics but the impact of these forms will vary linearly with total carbon content if the relative proportions remain constant) and (b) the cause of the change in total carbon content. These assumptions are acceptable as a "first approximation" in relating structural characteristics to organic carbon when a large part of the carbon is intimately associated with the mineral fraction after being processed by microorganisms (the composition of this material is relatively uniform irrespective of soil properties or the nature of the carbon entering the soil) and the structural characteristics are controlled primarily by these materials. Neither assumption, however, is strictly valid and the limitations in these assumptions are greatest when: (a) there are significant differences

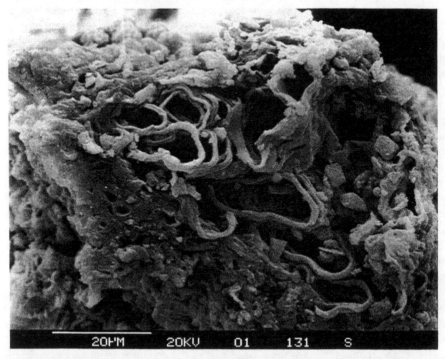

Figure 2. Selected electron micrographs of microaggregates: (a) microaggregates 20-90 μm; Oxosol. Large void showing remnants of plant anatomy completely coated with inorganics; and (b) microaggregates 90-150 μm; Alfisol. Partly decayed vascular bundle surrounded by inorganics. (From Waters and Oades, 1991.)

in the proportions of carbon in the different forms (particularly in the light fraction and the fraction most intimately associated with the mineral material), and (b) structural characteristics are considered that are strongly influenced by a form of carbon that is labile or temporary and represents a very small proportion of the total carbon. Notwithstanding these limitations, meaningful relations between structural characteristics and total carbon can be obtained if the relations are interpreted in the context of these assumptions.

B. The Role of Organic Carbon on Different Aspects of Soil Structure

The interdependence of factors influencing soil structure (soil characteristics, climate, biological processes, management, and depth of sampling) has already been noted. This reality has particular relevance as research on the role of organic carbon on soil structure is reviewed.

Two approaches can be used to study the influence of organic carbon on soil structure. The most common approach involves studies in a geographic region in which climate is constant, soil characteristics (e.g., texture, mineralogy, composition of exchangeable ions) are either constant or vary over a relatively narrow range and samples are taken from the A horizon. The organic carbon contents, therefore, vary as a consequence of differences in management. Studies in which there are large differences in carbon content involve differences in management that have existed for considerable lengths of time. An alternative way to look at the effect of organic carbon on soil structure involves the statistical analysis of data sets made up of a large number of soils and the development of pedotransfer functions (Bouma and van Lanen, 1987) which quantitatively relate soil structure to organic carbon content and other soil properties. Limitations exist in both approaches. In the former approach, the studies are often of limited scope and preclude an assessment of whether the relations that are observed are sensitive to the variation in soil properties such as texture (an additional limitation is inadequate replication in the case of some long term studies set up decades ago). Pedotransfer functions have been underutilized in assessing the influence of organic carbon on soil structure. Limitations in the use of pedotransfer functions primarily arise from data bases that include too many undefined variables. To be of greatest value, the functions should, at least initially, be based on samples from the A horizons of soils derived from similar parent materials under the same climatic conditions. Variations in carbon content then arise from differences in management, drainage, or position in the landscape. Care must be taken to ensure that sensitivity analyses using pedotransfer functions do not exceed the range in variation in the variables within the data base. Use of pedotransfer functions complement an analysis of data from site-specific studies and both approaches are used in this review to assess the influence of organic carbon on soil structure.

Pedotransfer functions are often sufficiently complex that the sensitivity of a dependent variable to changes in an independent variable is not immediately obvious. In order to illustrate the sensitivity of soil structural characteristics to changes in organic carbon contents that are predicted by pedotransfer functions, five "test" soils with different textures and organic carbon contents (see Table 1 for details) have been used. The characteristics of the test soils coincide with the characteristics of the A horizon of five soils from the data set collected by da Silva and Kay (1996a). The calculation procedure has involved calculating the structural characteristic for each of the five soils using the pedotransfer function, increasing the carbon content of each soil by 1 g carbon per 100 g soil, recalculating the structural characteristic, and then calculating the change in the structural characteristic. Changes in the structural characteristic are then graphically related to the clay content of each soil.

Table 1. Properties of test soils used in pedotransfer functions to predict the impact of structural characteristics of increasing the organic carbon content

Soil texture	Loamy sand	Fine sandy loam	Loam	Silty clay loam	Silty clay loam
Clay (%)	7	14	21	29	35
Silt (%)	13	32	47	55	51
Sand (%)	80	54	32	16	14
Organic carbon (%)	1.57	1.62	1.68	1.71	1.74

1. Structural Form

a. Total Porosity

The total porosity (or bulk density) that is measured at any given time in nonswelling soils is strongly influenced by soil characteristics such as texture and organic carbon content and by management. When other factors are constant, site-specific studies show that the total porosity generally increases with increasing organic carbon content (e.g., Anderson et al., 1990; Lal et al., 1994; Schjonning et al., 1994). Similar trends have been illustrated using pedotransfer functions. One of the first attempts to relate bulk density to texture and organic carbon was based on data for 2721 soils from the United States (Rawls, 1983) and took the form:

$$\rho_b = 100/\{(\% \text{ organic matter/organic matter bulk density}) + ([100 - \% \text{ organic matter}]/\text{mineral bulk density})\} \quad (1)$$

where ρ_b is the bulk density of the soil (g cm^{-3}), the average organic matter bulk density was taken to be 0.224 g cm^{-3} and the mineral bulk density varied with textural class. The analysis has been extended using multiple regression analyses to 12,000 soil pedons grouped according to their Soil Taxonomy classification by Manrique and Jones (1991). The latter investigators found that the influence of organic carbon varied with the soil order. More recently, multiple regression analyses have been applied by da Silva et al. (1997) to data from 36 pairs of sites in a toposequence involving soils of a single soil order, in order to assess the combined influences on bulk density of tillage, position relative to the corn (*Zea mays L*) row, and soil properties (clay contents ranged from 6 to 37%, organic carbon content varied from 0.9 to 3.9%). The tillage treatments had existed for eleven years prior to initiation of the study and involved zero tillage and conventional tillage (primary tillage in the fall followed by secondary tillage in the spring). Additional information on the sites is provided in da Silva and Kay (1996a). The following function was derived from the analyses:

$$\rho_b = 1.573 - 0.0640T + 0.367P + 0.0686T*P - 0.125C - 0.031Cl + 0.0021C*Cl \quad (2)$$
$$F=109.7, \text{ Prob} >F=0.0001, R^2=0.83, N=144$$

where T is tillage (a class variable assigned values of 0 and 1 for zero and conventional tillage, respectively), P is position (a class variable assigned values of 0 and 1 for the row and interrow positions, respectively), C is the organic carbon content (%), and Cl is the clay content (%). This function illustrates the role of management on bulk density but can also be used to assess the effect of organic carbon. The changes in bulk density that are predicted for an increase in the organic carbon content of 1 g organic carbon per 100 g soil in the five test soils using Rawls' model and Eqn. 2 (for the zero till, interrow position) are shown in Figure 3. Although the two pedotransfer functions were based on very much different data sets and developed using different analytical procedures, the sensitivity of bulk density to changes in organic carbon content and the variation in this sensitivity with clay

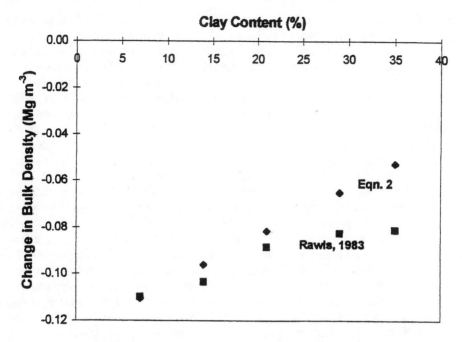

Figure 3. Change in the bulk density of the test soils arising from increasing the organic carbon content by 1% predicted by the Rawls function (Rawls, 1983) and by Eqn. 2 (zero till and interrow position).

content that are predicted by the two functions are qualitatively similar. In both cases, the magnitude of the decrease in bulk density caused by the increase in organic carbon is predicted to diminish as the clay content increases. Corresponding predictions based on the pedotransfer functions developed by Manrique and Jones (1991) have not been included since a clay x organic carbon interaction was not considered in their pedotransfer functions.

The decrease in bulk density and concomitant increase in porosity with increasing organic carbon may be considered to be due to "dilution" of the soil matrix with less dense or more porous material. The organic material may exist in different stages of decomposition and degrees of association with the mineral matrix, but the dilution concept implies that the organic material does not substantially alter the packing of mineral grains or their spatial relation to one another. This "mixture" model appears to have been the basis for Rawls' model. Larger proportions of free particulate material in coarser textured soils may have been expected to result in the changes in bulk density being larger in these soils than in finer textured soils, but the mixture model indicates that this is not a prerequisite for larger decreases in coarser textured soils. Cases can be envisaged where the mixture model may be inappropriate in describing the decrease in porosity with increasing carbon content. For instance, if the additional carbon also coincides with a large increase in soil fauna, the mixture model would not account for the increase in porosity due to burrows. Analyses of data from Denmark (Schjonning et al., 1994) and from Missouri (Anderson et al., 1990) where the increase in carbon was largely due to long term manure applications and where earthworms were believed to have been very active, give values of the change in porosity with carbon content of 9.7 and 9.8 cm^3 of pores per g of organic carbon, respectively. The inverse of these numbers, 0.103 and 0.102 $g\ cm^{-3}$, respectively, correspond to Rawls' bulk density of the organic carbon but are significantly smaller than the value of 0.130 derived by Rawls (after converting from organic matter to organic carbon).

The effect of organic carbon on the total porosity of swelling soils is less well documented. The total porosity of swelling soils is related to water content and shrinkage characteristics. The latter characteristics are controlled by soil properties (including mineralogy, exchangeable ions, and composition of the pore fluid) and the magnitude of previous stresses that the soils have experienced. The relations between total porosity and water content can be also be time-dependent and this dependence may be influenced by organic carbon contents. For instance, addition of extracellular polysaccharides to pure clay minerals has been shown to reduce the rates of both drying and rehydration (Chenu, 1993).

The total porosity is made up of pores of different size. For the purpose of this discussion, macropores will be defined as pores larger than 30 µm in diameter, mesopores as 0.2 to 30 µm, and micropores as < 0.2 µm diameter. The volume fraction of pores in different size classes can be calculated from the water release curve of nonswelling soils, and a number of pedotransfer functions have been developed to describe the water release curve of a wide range of soils. An assessment of 12 of these functions by Tiedje and Tapkenhinrichs (1993) indicated that the model of Vereecken et al. (1989) was the best, even for soils with high organic matter contents where errors in the predictions by other models were often high. The Vereecken model was based on a modification of the van Genuchten (1980) function for the water release curve with the constants in the function defined by regression equations. that included bulk density, along with clay, sand, and organic carbon contents as variables. The regression equations were derived from data on 182 soil profiles from Belgium (no information was provided on management practices on the sites). Characteristics of the five test soils will be used to illustrate the effects of organic carbon on the different size classes of pores by using Rawls' function to predict bulk density or total porosity and then using the Vereecken pedotransfer function to predict the water release curve. The difference between total porosity and the volumetric water content at -0.01 MPa is taken as the macroporosity, the difference in water contents between -0.01 and -1.5 MPa as the mesoporosity, and the water content at -1.5 MPa as the microporosity.

b. Macroporosity

Macropores can represent as much as a third of the total porosity of soils. Pores in this size class span a large range in sizes and include biopores, shrinkage cracks and other interaggregate pores, and the largest pores within aggregates and peds. This porosity is included in the class referred to as structural porosity (Stengel, 1979; Derdour et al., 1993) or aeration capacity (Thomasson, 1978). These pores have a major influence on a range of soil mechanical characteristics (e.g., Carter, 1990), aeration (Thomasson, 1978), water and solute flow (Blackwell et al., 1990; White,1985; Ghodrati and Jury, 1992), and on root development (Jakobsen and Dexter, 1988).

Macropores are the most dynamic of any size class of pores. The strongest correlations between macroporosity and organic carbon would be expected to be found where there is greatest uniformity in soil properties, climate, and management conditions. The increase in macroporosity arising from an increase in the carbon content of 1% in the five test soils, as predicted by the Vereecken pedotransfer function, is shown in Figure 4. The magnitude of the increase is smaller than for the other two size classes and declines with increasing clay content.

c. Mesoporosity

Mesopores are less strongly influenced by management (particularly tillage and traffic) and exhibit less temporal variation than macropores. These pores are included in the class referred to as textural pores (Stengel, 1979) and the water retained by these pores has been referred to as plant available water (Veihmeyer and Hendrickson, 1927). The role of organic carbon on this class of pores has

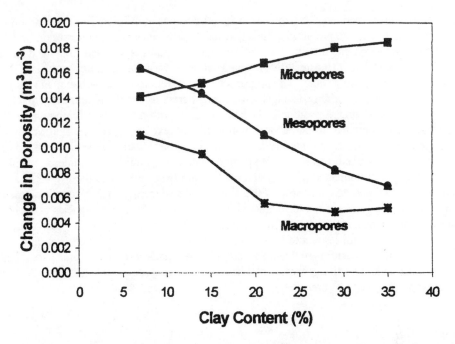

Figure 4. Changes in the volume fractions of different classes of pores in the test soils predicted by the Vereecken model (Vereecken et al., 1989) with the bulk density predicted using the Rawls model (Rawls, 1983) when the organic carbon content is increased by 1%.

frequently been overlooked or underestimated. Emerson (1995) considered several site-specific studies from various parts of the world and showed that an increase in the organic carbon content resulted in increases in the available water content (AWC) ranging from 1 to 10 g of water per g organic carbon. Other site-specific studies not included in Emerson's review provide complementary values, e.g., calculations based on data obtained by Schjonning (1994) and Karlen et al. (1994) show increases of 6.9 and 1.0 g water per g organic carbon, respectively. However, Anderson et al. (1990) found that, although increases in organic carbon due to the annual additions of manure to a montmorillonitic soil for 100 years increased the volume fraction of pores larger than 25 µm, the volume fraction of pores between 5 and 25 µm decreased, suggesting that the mesoporosity of soils which are dominated by swelling clay minerals may be less responsive to increases in organic carbon. Hudson (1994) reevaluated data on the AWC of three textural groups of surface soils from the U.S.A. (each group containing about 20 soils) and developed pedotransfer functions which showed that increasing organic carbon contents resulted in increasing water contents at field capacity (-0.01 MPa) as well as at the permanent wilting point (-1.5MPa) with the largest increase occurring at field capacity. These results are compatible with the site-specific data reviewed by Emerson. Substitution of the properties for the five test soils into the Vereecken model (Figure 4) confirm the conclusions drawn by Emerson and Hudson, i.e., that increasing organic carbon leads to an increase in the AWC.

There are inconsistencies in the conclusions drawn by Emerson (1995), Hudson (1994), and the analyses using the function of Vereecken model, however, regarding the influence of texture on the change in AWC. The Vereecken model predicts that the increase in AWC with carbon becomes smaller as clay content increases (Figure 4) whereas Emerson concluded that the influence of organic carbon on AWC becomes greater as the texture becomes finer. Hudson concluded that the impact of organic carbon was maximum in medium textured soils and declined as the texture got either finer or coarser. The causes of these inconsistencies are not immediately obvious but may relate to differences in the range of textures considered, differences in the proportions of different forms of carbon present in the soils, or differences in parent materials that led to differences in the interaction between the organic carbon and the mineral matrix. The widest range in textures is represented in the data set used for the Vereecken model, and Hudson (1994) argued that wide variation in texture may mask the effect of organic carbon on available water. The proportions of different forms of carbon may change when the variation in the organic carbon content is caused by different means (residue management, tillage, crop or manure inputs, drainage, position in the landscape). If different forms of carbon have different effects on the AWC, comparisons across soils with different textures in which management has also been variable may complicate the interpretation of the data, particularly if the size of the data set is relatively small. Soil characteristics related to parent material (e.g., mineralogy, amorphous inorganic materials exchangeable ions, and composition of the pore fluid) influence the water release curve (Williams et al., 1992) and it is not known if these factors influence the interaction of organic carbon and texture on the AWC. However, having noted possible reasons for inconsistencies in the literature, it is noteworthy that recent studies by da Silva and Kay (1996a) on 256 cores from the A horizons of the 72 sites referred to with respect to Eqn. 2 (on which the clay content varied from 6 to 37%, the organic carbon content varied from 0.9 to 3.9%, the soils were derived from the same parent materials, and the sites were under two tillage treatments) have resulted in pedotransfer functions that generate trends similar to those of the Vereecken model.

d. Microporosity

Micropores are least influenced by management of any of the pore classes. Although these pores remain water-filled for a much larger proportion of time than the other pore classes, little of this water is available to plants and the rate of water flow through them is very slow. These pores remain largely biologically inactive because they are not penetrated by roots or by most microorganisms and are, therefore, only accessible to molecular sized byproducts of biological activity through diffusion. The relative inaccessibility of these pores to microorganisms may also give rise to one of their most important functions, i.e., the physical protection of organic carbon. However, increasing the volume fraction of these pores can also result in increased strength of failure zones and decreased available water when this increase occurs at the expense of mesopores. The microporosity of soils, as for the other pore classes, tends to increase with increasing organic carbon. Emerson (1995) found that the increase in the water content at -1.5 MPa ranged from 1 to 3 g water per g organic carbon. Corresponding values calculated from Schjonning (1994) and Karlen et al. (1994) were 2.3 and zero g water per g organic carbon. Hudson (1994) found that the water content at -1.5 MPa increased about 1g water per g organic carbon in sandy soils, but was unable to establish a significant relation for medium textured soils. Substitution of the data for the five test soils into the Vereecken pedotransfer function shows that an increase in organic carbon results in an increase in the amount of water held at -1.5 MPa and that the increase gets larger as the clay content increases (Figure 4).

e. Origin of Changes in Porosity with Organic Carbon

Macropores can be created by the formation of ice lenses, shrinkage, root growth, the activity of soil animals, or tillage. Organic carbon may have an indirect influence on both the formation and the stabilization of these pores. Among the processes that form macropores, organic carbon content may have the strongest influence on the nature of pores created by shrinkage, soil fauna, and tillage. For instance, it may be speculated that increases in organic carbon content in soils where shrinkage cracks are an important component of macroporosity, will result in decreased shrinkage and, therefore, a decrease in the extent of macropore formation. Increases in organic carbon content that arise from increased input of organic residues may stimulate soil faunal activity (the extent of which will depend on the nature of the carbon source) and increased faunal activity will result in an increase in the volume fraction of biopores. Increasing carbon content will also enhance the friability of soils (see later discussion) with a concomitant change in the aggregate size distribution created by tillage. Changes in aggregate size distribution will result in changes in the pore size distribution and most of the change arising from tillage is in the macropore size range (e.g., Derdour et al., 1993).

Once macropores are created, their persistence is determined largely by the structural stability of the pore walls and surrounding soil and the magnitude of the stresses experienced. Quirk and Williams (1974) showed that the mechanical strength of aggregates from a red brown earth was more than 50% higher when polyvinyl alcohol penetrated pores with effective diameters of 30 to 100 µm than when it was restricted to pores with radii of 15 to 30 µm or 100 to 300 µm. Where the process of creating the pore does not substantially increase the organic carbon content of the pore wall, the stability of the pore wall would be expected to be equivalent to the stability of the bulk soil, and this stability is often inadequate to allow the pore to withstand the effective stress created by drying, rewetting, or swelling pressures or stresses from overburden or traffic. For instance, Kay et al. (1985) found that macropores created through the formation of ice lenses over winter in a nonswelling soil disappeared as the soil dried and reconsolidated after thawing. Many of the macropores created by tillage are equally unstable (e.g., Derdour, 1993). Gusli et al. (1994) showed that the structural collapse of beds of aggregates on wetting and draining was caused by the development of failure zones on wetting followed by consolidation on draining as a result of the development of effective stresses and that the extent of collapse was greater for soils having lower organic carbon contents. Macropores created by shrinkage or by earthworms may also be unstable in some situations. For instance, Mitchell et al. (1995) found that these pores failed to provide preferential flow paths within 10 minutes of the onset of flood irrigation of a swelling soil with low organic carbon content (0.9 %). Pores created by the tap roots of alfalfa (*Medicago sativa* L.) were, however, much more stable under these circumstances, presumably because of increased carbon content of the pore wall and the consequent enhanced stability (the influence of organic carbon on stability is discussed in greater detail later in this report).

The increase in the volume fraction of pores with effective diameters between 0.2 and 30 µm and <0.2 µm may be attributed to several mechanisms, but none have been rigorously assessed. The mechanisms must involve organic carbon stabilizing pores in these size classes after they have been created by other mechanisms and/or the organic material itself contributing pores in these size classes.

Processes that create pores in the 0.2-30 µm diameter class may include the creation of microcracks by shrinkage or by freezing, the collapse of macropores, the plugging of macropores by organic debris or mobile inorganic particles, and, perhaps, the creation of pores through the development of root hairs and fungal hyphae. Processes that create pores <0.2 µm may include shrinkage of the soil matrix and the collapse of mesopores in this matrix. Increasing the organic carbon content may diminish the reversibility of the shrinkage process and the importance of shrinkage would increase with clay content. The intimate association that occurs between clay and organic material in the gut of soil fauna will also result in the formation of micropores when the excreted material, which often has a high water content, shrinks on drying. The degree to which this process results in a net increase in the volume fraction of micropores and a loss of mesopores cannot be ascertained at present.

The possibility that organic material itself contributes pores in the 0.2-30 and <0.2 µm size classes must be considered in relation to the forms of organic carbon. Free particulate organic carbon may, as noted above, contain internal porosity, and the extent to which this form of carbon contributes to the water release characteristics of soils will depend on the continuity of these pores with pores in the mineral matrix. Management practices that result in an increased proportion of the soil carbon in free particulate material may not result in increases in available water, if the increase in free particulate material is largely due to crop residue incorporated by tillage, since the continuity of pores in the organic material and the bulk soil will be low. The intimate association between the organic residue and the mineral matrix that exists with the occluded organic carbon will result in greater pore continuity and the development of voids during the decomposition of the occluded carbon may contribute to an increase in the volume fraction of pores in the two size classes. An increase in the amounts of extracellular gels would also lead to an increase in the amounts of water held between -0.01 and -1.5 MPa. Emerson (1995) has hypothesized that the increase in available water that is observed in soils that have been under grass for a long period is due entirely to these gels. Using the value of 26 g increased available water per g adsorbed extracellular polysaccharide (referred to earlier and calculated from data reported in Chenu, 1993), and considering that 10 to 25% of the carbon in soil is polysaccharide material closely associated with the mineral phase (Golchin et al., 1994), then an increase in the organic carbon content of 1 g per 100 g soil would lead to an increase in available water of 2.6 to 6.5 g water per g organic carbon. These values span the range reported above and are compatible with Emerson's hypothesis.

f. Pore Tortuosity and Continuity

Other important characteristics of pores include pore tortuosity and continuity. These characteristics are particularly relevant to macropores and, therefore, to air and water flow (Groenevelt et al., 1984; Blackwell et al., 1990; Ball and Robertson, 1994). Very little information is available, however, on the influence of organic carbon on these characteristics.

g. Failure Zones, Aggregation, and Strength

Failure zones represent the second major characteristic of structural form. Failure zones can be characterized in terms of their average strength (or strength at a given percentile of failure in a frequency distribution) and the distribution of the magnitude of these strengths. Zones with a given strength will fracture when the force that is applied exceeds this value. The spatial distribution of the zones determines the size distribution of aggregates when a stress is applied. Failure can occur through the application of tensile, shear, or compressive stresses. The increase in the tensile strength of aggregates with decreasing size of aggregates has been used to quantify the friability of soils (Utomo and Dexter, 1981a). Perfect et al. (1995) found that the friability of air-dried soils with a range of textures (7 to 42% clay) and organic carbon contents (1 to 4.4 %) was positively correlated with the strength of aggregates and this relation became more pronounced with increasing aggregate size. Aggregate strength was strongly influenced by organic carbon and sand contents, and an interaction term involving these two variables. In finer textured soils (sand contents less than about 42 %) the strength increased with organic carbon contents, but, in coarser textured soils the reverse occurred. While the nature of this relation may vary with parent materials, it may explain why some researchers have found no relation between organic carbon and tensile strength (Dexter et al., 1984; Guerif, 1988; Ley et al., 1993) while others have found a positive relation (Rogowski et al., 1968; Bartoli et al., 1992). This analysis also suggests that the effect of organic carbon on friability will vary with sand content; in fine textured soils increasing organic carbon will cause an increase in friability, but in coarser textured soils the reverse occurs.

The trends observed by Perfect et al. (1995) need to be considered in the context of possible roles of organic carbon on the dominant factors influencing the tensile strength of dry aggregates. The characteristics of failure zones are controlled by pore characteristics and the nature of cementing materials between the mineral particles. Tensile failure occurs, at least partly, via air-filled pores (Dexter et al., 1984; Snyder and Miller, 1985; Grant et al. 1990). The preceding discussion noted that increasing organic carbon by a given amount causes a bigger change in the porosity of coarse textured soils than fine textured soils. This could contribute to the loss of strength in coarse textured soils with increasing carbon. Increasing carbon content would also be expected to strengthen bonds between mineral materials and this effect would be expected to be important where the matrix is dominated by clay. The strength of some soils, however, is dominated by dispersible clay and silt material which function as cementing material as soils dry. Under these conditions, increasing organic carbon content (and, in particular, the fraction closely associated with the mineral phase) may reduce the dispersibility of these fractions resulting in reduced strength. This mechanism appears to be particularly important in hard setting soils (Young and Mullins, 1991; Chan and Mullins, 1994; Chan, 1995). While one might guess that this mechanism would be most pronounced on fine textured soils, it is of interest to note that the sand content of the soils studied by Chan was 70%, which would be compatible with the trend observed by Perfect et al. (1995).

The distribution of aggregate sizes in the seedbed is related to the type of tillage and the number of passes. When energy input is kept constant, the dry aggregate size distribution is influenced by soil properties and the water content at the time of tillage, and reflects the influence of these factors on the characteristics of the failure zones in the soils. While the effects of tillage or cropping practices on aggregate size distribution can disappear quickly after tillage due to crop growth and climatic factors, these effects can have a significant impact on the germination and early growth of seedlings (see review by Braunack and Dexter, 1989). Cloddiness of the seedbed, as reflected in an increased proportion of large clods or aggregates, increases with soil strength (Rogowski et al., 1968). Unpublished analyses of the tensile strength data of Perfect et al. (1995) and dry aggregate size distribution data of Perfect et al. (1993) showed a highly significant negative relation between the fractile dimension, D, of the seedbed and the logarithm of the tensile strength (smaller values of D coincide with aggregate distributions dominated by large clods or aggregates). The dry aggregate size distribution is most strongly influenced by texture with the cloddiness becoming more common on finer textured soils (Chepil, 1953; Perfect et al., 1993). Increasing organic carbon content results in a greater spread of aggregate sizes with a smaller proportion of large clods or aggregates and more small aggregates (Chepil, 1955; Perfect et al., 1993). Changes in dry aggregate size distribution can also occur due to short term changes in cropping practices. For instance, introduction of forages decreases the relative proportion of large clods and this can occur without measurable changes in organic carbon (e.g., Perfect et al., 1993). These changes reflect a loss in the strength of failure zones that may be caused by different processes. For instance, Caron et al. (1992a) showed that the introduction of bromegrass (*Bromus Inermis* L.) caused initial fragmentation of aggregates that was attributed to root penetration and/or weakening of aggregates by enhanced wetting and drying. Subsequent loss of strength may be due to changes in fractions of organic carbon that are present in small amounts (e.g., extracellular polysaccharides associated with the decay of the fine roots of the grass) and these changes cause a reduction in the level of dispersible mineral material.

Organic carbon can also influence the shape of aggregates (Chapman, 1927). Dexter (1985) found a positive linear relationship between the roundness of aggregates from Australian and Dutch soils and the organic carbon content. Aggregates with low organic carbon contents were much more angular in shape.

Measurements of compressive or shear strength or resistance to penetration also reflect the strength of failure zones. These measurements are normally made on soils containing water and, therefore, are influenced by water potential as well as by the nature of failure zones. In the range of water contents normally encountered in the field, decreasing water potential increases the effective stress (Bishop and Blight, 1963; Groenevelt and Kay, 1981) thereby increasing strength. Care must, therefore, be taken in interpreting relations between organic carbon and the strength of soils in the field. An increase in carbon content can cause an increase in porosity with a concomitant loss of strength. An increase in organic carbon in soil with the same porosity and water content can also cause a decrease in water potential (because of the effect on the water release curve) thereby increasing the strength. This increase in strength may be further enhanced if the increase in organic carbon leads to increasing the level of organic cementing material in the failure zones. Measurements of the shear strength of disturbed soils (clay content of 44% and organic carbon contents ranging from 1.5 to 2.3%) packed to bulk densities of 1.25 g cm^3 have indicated (Davies, 1985) that shear strength increased with organic carbon content when measurements were made at similar potentials. A similar trend has been inferred from measurements of penetration resistance by Emerson (1995). Emerson noted that high levels of polysaccharide gels could lead to compacted parts of the mineral matrix becoming so strongly bonded together that, at a potential of 0.01 MPa, the penetration resistance was sufficiently high to limit root penetration. The different effects of organic carbon can be most readily assessed when strength is described as a function of porosity and potential. Da Silva and Kay (1996a) have described the soil resistance-to-penetration curve (soil resistance as a function of water content) and the water release curve as a function of clay and organic carbon contents and bulk density. These functions were derived from measurements on intact cores from soils with a range of clay contents (6 to 37%) and organic carbon contents (0.9 to 3.9%) that were under two different tillage treatments. Substitution of the properties of the five test soils show that: (a) an increase in the organic carbon content of 1 g per 100 g soil causes a decrease in soil resistance when the potential is constant, and (b) the magnitude of the decrease increases with clay content and with declining potential (Figure 5). The decrease in strength with increasing carbon of these soils could be due to increased porosity and/or reduced strength of failure zones. If Davies' results on disturbed samples are applicable to the soils studied by da Silva and Kay, then the reduced strength must be due to increased porosity.

h. Integration of Porosity and Strength Characteristics: The Least Limiting Water Range

The least (or non-) limiting water range is a characteristic of structural form that has been proposed as a measure of the quality of soil structure for plant growth (Letey, 1985; da Silva et al., 1994). The least limiting water range (LLWR) is defined by water contents at which aeration, water potential, and resistance to penetration reach values that are critical or limiting to plant growth. The upper limit of this range is defined by the water content at field capacity or the water content at which aeration becomes limiting, whichever is smaller. The lower limit is defined by the water content at the permanent wilting point or the water content at which soil resistance becomes limiting, whichever is higher. The LLWR, therefore, incorporates characteristics related to pores and failure zones into a single variable. Using an air-filled porosity (vol. air/total vol.) of 10% as the aeration limitation, and a resistance to penetration of 2 MPa as the strength limitation, da Silva and Kay (1996b) showed that the growth of corn plants decreased as the soil water content fell outside of the LLWR with increasing frequency. The frequency that the water content fell outside of the LLWR increased as the LLWR got smaller. Emerson et al. (1994), using the same limiting values for aeration and resistance to penetration as used by da Silva and Kay, found that small values of the LLWR coincided with a paucity of roots of peach trees in orchard soils. The LLWR can be calculated from the water release curve and the water content-resistance to penetration curve. Both of these relations can be described using pedotransfer functions. Da Silva and Kay (1996a) found that the clay content, organic carbon content, and bulk density were the soil characteristics of greatest importance in nonswelling calcareous soils. Increasing carbon contents were predicted to cause the bulk density to decrease, the water contents at 10% air-filled porosity to increase, and altered both the water release curve and the soil resistance curve. The net effect of these changes is reflected in Figure 6 for the five test soils. Use of the pedotransfer functions suggests that an increase in the organic carbon content of 1% results in an increase in the LLWR of about 0.005 to 0.054 cm^3 cm^{-3} and the increase is greatest on coarse textured soils.

Soil Structure and Organic Carbon: A Review

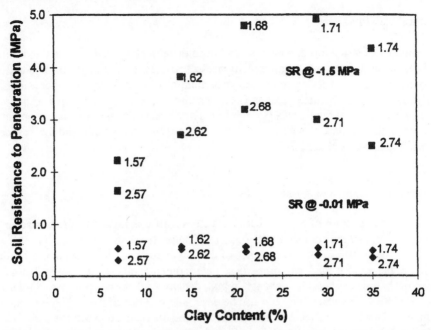

Figure 5. Changes in soil resistance to penetration at different soil water potentials predicted by the pedotransfer functions from da Silva and Kay (1996a) and using Eqn. 1 to predict changes in bulk density (zero till, interrow position) when the carbon content is increased by 1% in the test soils (carbon contents given adjacent to data points).

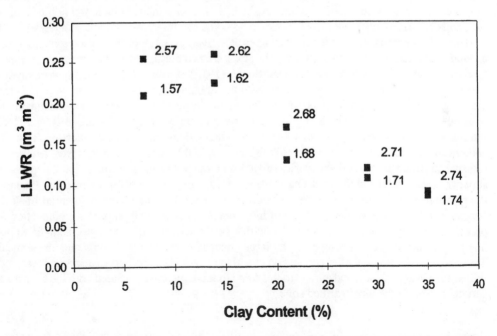

Figure 6. Changes in the least limiting water range (LLWR) predicted by the pedotransfer functions from da Silva and Kay (1996a) and using Eqn. 1 to predict the bulk density (zero till, interrow position) when the organic carbon content is increased by 1% in the test soils (carbon contents are given adjacent to data points).

2. Structural Stability

A large number of structural parameters have been developed to characterize the ability of soil to retain its structural form when exposed to stresses arising from tillage, traffic, and a range of interrelated processes involving rainfall or irrigation, erosion, and drying. Increasing the organic carbon content of soils normally results in an increase in stability irrespective of the origin of the stress. The magnitude of the increase in stability is often related to the soil water content (discussed later). Data on the response of soils to two different types of stress (traffic and mechanical abrasion) will be used to illustrate this generalization.

a. Resistance to Stresses Arising from Traffic

The extent of compaction of topsoil that arises from traffic is related to the contact pressure, the number of passes, and the compactibility of soils. Compaction can be reduced by reducing the load applied to soil or by increasing the resistance of the soil to compaction. The compactibility of soil is a function of the applied load, soil properties, water content, and previous stress history and can be characterized in several ways. The change in bulk density with applied pressure is referred to as the compression index. Compactibility can also be described in terms of the maximum density achieved with a given compaction treatment or in terms of the pressure required to achieve a given bulk density. Irrespective of the methods employed, Soane (1990), in a comprehensive review of the role of organic matter on soil compactibility, has noted that organic carbon is likely to have a greater influence on the compactibility of soil at low stress (e.g., 100 kPa) than at high stress. As compaction occurs, the larger pores are lost first and aeration capacity declines. This trend can create results which may initially appear anomalous when interpreting the effect of organic carbon on the compression index. For instance, if an increase in organic carbon results in an increase in the volume fraction of large pores, such soils can have a higher compression index as some of these pores are lost (Angers et al., 1987). Under such situations, either the bulk densities at a given pressure or the pressure required to reach a given bulk density is more meaningful. The stress created during the Proctor test can be considered a relatively low stress and using this test, Soane (1975) showed that the maximum density of 58 Scottish soils was more closely correlated with the organic carbon content than with any other soil property. An increase in the organic carbon content of 1% resulted in a decrease in the maximum bulk densities of 0.094 g cm^{-3}. Similar trends were observed by Angers and Simard (1986) in studies on the increase in bulk density due to wheel traffic under field conditions; an increase in the carbon content of 1% reduced the change in bulk density by 0.102 g cm^{-3}. Kuipers (1959) determined the pressure to reduce the air-filled porosity to 10% on a range of soils and found that this pressure was strongly positively related to organic carbon content. An increase in organic carbon of 1% increased the required pressure by 64 kPa (when samples were equilibrated at water potential of -7.7 kPa). However, a loss in the volume fraction of large pores is not the only impact of compaction on soil pores. Ball and Robertson (1994) have found that even relatively small changes in air-filled porosity due to compaction were accompanied by large changes in relative diffusivity and air permeability. These authors showed that after application of a given stress, these properties decreased less in direct-drilled than in plowed soil and attributed this trend to the structure associated with the higher organic carbon content in the direct-drilled soil.

a. Resistance to Stresses Arising from Wetting and Abrasion

The ability of aggregates to resist stresses from wetting followed by mechanically shaking in water is referred to as wet aggregate stability and has been studied the most extensively of any stability parameter. Although this parameter was originally used to characterize the erodibility of soil (e.g.,

Yoder, 1936), it has been used increasingly to study the cohesion of aggregates and the dynamics of changes in the nature of bonding between particles (e.g., Tisdall and Oades,1982; Kemper et al., 1987; Bullock et al., 1988; Golchin et al., 1994). The stability of aggregates is strongly dependent on the rate of wetting. Aggregate stability declines as the rate of wetting increases and the decline has been attributed to increased stresses related to air entrapment and differential swelling (Panabokke and Quirk, 1957; Emerson, 1977). In a comprehensive study of (air dried) soils with organic carbon contents ranging from less than 0.6% to about 10%, Kemper and Koch (1966) found that the stability of vacuum saturated soils increased with log (organic carbon content). Different relations were obtained for soil from subsurface layers, surface layers under sod and surface layers under cultivation. The stability changed most rapidly with organic carbon at carbon contents lower than 1.2 to 1.5%. An increase in organic carbon content from 1.5 to 2.5% was found to result in the stable aggregation of surface cultivated soils increasing nearly 9 g stable aggregates per 100 g soil. Subsequent studies have shown that organic matter can influence both the rate of wetting and the resistance to stresses generated during wetting. When wetting occurs from a water source at constant potential, increasing organic carbon contents of nonswelling soils have been found to reduce the rate of wetting at clay contents less than about 30%, whereas at higher clay contents increasing organic carbon is associated with an increase in the rate of wetting (Rasiah and Kay, 1995). The effect of organic carbon content on the rate of wetting declines as the potential of the water source decreases (Quirk and Murray, 1991). At similar wetting rates, increasing organic carbon is correlated with increasing stability (Chan and Mullins, 1994; Rasiah and Kay, 1995). Measurements of the loss of stability due to rapid rates of wetting are most relevant to aggregates at the soil surface in environments where intense rains are experienced when the soil surface is very dry, or where dry soils are wet up quickly due to irrigation. In environments where wetting occurs more slowly (surface aggregates that have high water contents or experience less intense rainfall events, or aggregates beneath the surface), measurements of stability in which air entrapment and differential swelling are minimized are more relevant. Under these conditions, the influence of organic carbon content on aggregate stability is often less pronounced than when rapid wetting occurs (Rasiah and Kay, 1995). Soil samples are often not dried if measurements involve slow wetting and, in these cases, stability is very strongly influenced by the water content of the sample prior to saturation (this relation will be explored in greater detail later in this report).

As is the case for other properties that are strongly dependent on the characteristics of failure zones, the water stability of aggregates from a soil can exhibit large changes due to cropping treatments before changes in the total organic carbon content are observed (e.g., Baldock et al., 1987). These changes have been attributed to changes in the amounts of fine roots and fungal hyphae which strengthen failure zones through physical entanglement, and which also act as sources of carbon for bacteria, thereby contributing to increased production of microbial cementing materials (Tisdall and Oades, 1982; Miller and Jastrow, 1990). These stabilizing materials are very labile and represent only a small part of the total carbon content. Their amounts in the soil at any given time is determined by the rates of input of plant carbon and the mineralization of this carbon and the microbial byproducts. Rasiah and Kay (1994a) have measured the increase in aggregate stability after forages have been introduced on soils of different textures that had previously been used for the production of row crops, and found, as had been reported by Low (1955), that the stability increased exponentially with time. Stepwise multiple regression analyses showed that the stability increased at a faster rate on finer textured soils. Since there were no reasons to expect that the rate of carbon production would vary with texture on the sites, it may be speculated that the increased rates of stabilization on the finer textured soils was due to slower rates of mineralization of the cementing materials. This possibility is compatible with a number of incubation studies that have shown that rates of mineralization of labile carbon sources are slower in fine textured soils (e.g., Amato and Ladd, 1992; Hassink et al., 1993; Rutherford and Juma, 1992).

3. Structural Resiliency

Soils differ in their ability to recover structural form when stresses are reduced or removed. Soils that are most resilient are known as self-mulching soils. The structural form of such soils are not particularly sensitive to mechanical processes. For instance, if much of the macroporosity of the upper few cm are destroyed through tillage when the soil is wet, a desirable structure can be recreated by wetting and drying. Blackmore (1981) has characterized self-mulching soils as those which are soft when wet, do not disperse but have a strong tendency for shrinking and swelling, and have a crumb density > 2g cm^{-3} when dry. Wenke and Grant (1994) have noted that self-mulching is associated with clay contents >35% and defined self-mulching as the ability of a clay soil to re-form aggregates <5 mm diameter during a small number of cycles of drying and gentle wetting after remolding. The role, if any, that organic matter plays in self-mulching is not well understood, although it is probable that it alters the rate of wetting and drying. Other soils, while less resilient than self-mulching soils, can still exhibit an improvement in structural form through the formation of microcracks or biopores. Changes in the structural form of such soils that arise from weather or biological factors can often be as large or larger than the changes created by different cropping or tillage practices. For instance, analyses of tensile strengths of air-dry aggregates measured for five different years on the long term rotation plots at the Waite Institute, Australia, showed that measurements at the 0 to 5 cm depth could vary from 35 to 65 kPa between years within a treatment, whereas the difference between treatments may be as small as 5 to 15 kPa (Kay et al., 1994). The variation between years was attributed to variation in wetting and drying events. The creation of failure zones by wetting events has been referred to as tilth mellowing by Utomo and Dexter (1981b). Large changes in the strength of failure zones can also be caused by freezing and thawing events (e.g., Perfect et al., 1990b). Although the strength of failure zones appears to be particularly sensitive to wetting/drying or freezing/thawing events, there is less evidence that pore characteristics are as responsive and very little is known about the role of organic matter on changes in either failure zones or pore characteristics.

4. Structural Vulnerability

There is a dearth of information on the effects of organic carbon on soil structural vulnerability. Although it may be speculated that the vulnerability of soils decreases with increasing carbon content (since increasing organic carbon generally leads to increased stability and there is no reason to expect that increasing organic carbon leads to decreasing resiliency), research is needed to confirm this speculation. The degree of vulnerability of the structure of soils has an important bearing on the long term impact of land use practices on processes controlling crop productivity and environmental quality and, therefore, the role of organic carbon on structural vulnerability needs to be much better understood.

C. The Role of Organic Carbon on the Interaction Between Water Content and Soil Structure

The water content of soils varies continuously and this variation has direct effects on physical and biological processes that influence a range of soil structural characteristics. Additional affects of variation in water content arise from the concomitant changes in effective stress. These affects contribute to the temporal variability in soil structure. There is, however, an additional cause of variation in structure that arises from variation in water content that has only recently received attention and which may be of critical importance: the strength of the bonds between organic materials and mineral particles may decrease with increasing water content.

Evidence that the strength of bonds involving organic materials varies with water content has been collected primarily in studies on wet aggregate stability and dispersibility of clay, although studies on other characteristics that are strongly influenced by the strength of failure zones would be expected to yield similar observations. Measurements of aggregate stability, made under conditions in which air entrapment and differential swelling are minimized, have shown that the stability of a soil declines with increasing water content (Reid and Goss, 1982; Perfect et al., 1990c; Gollany et al., 1991; Caron and Kay, 1992). The loss in stability applies at the scale of macroaggregates (>250 µm) as well as at the scale of material < 2 µm (Perfect et al., 1990c; Kay and Dexter, 1990; Caron and Kay, 1992). A change in water content initiates changes in stability that may extend over periods as long as 48 to 86 days under laboratory conditions (Caron and Kay, 1992). Spectral analyses of stability measurements made under field conditions (where the water content changes continuously) indicated that stability could be influenced by water contents that existed two days prior to the measurement (Caron et al., 1992b).

When soils with a range of characteristics are considered, the sensitivity of stability to water content (as reflected in the slope of the linear relation between stability and water content) increases with increasing clay content and increasing organic carbon content (Rasiah et al., 1992). The sensitivity of stability to water content also responds to changes in cropping treatment. The increase in stability with time under forages, that has been observed by many investigators, can be described mathematically in terms of how the time under forages influences the intercept and the slope of the relation between stability and water content (Rasiah and Kay, 1994a,b). Increasing time under forages results in an increased intercept (greater stability) and an increased slope (increased sensitivity of that stability to changes in water content).

Studies on a clay loam soil with nonswelling clay minerals have indicated (Caron and Kay, 1992), that the change in stability with time after a change in water content is not due to microbial activity but involves material that is sensitive to Na-borate at a pH of 9.4 (Caron et al., 1992c). Cementing materials that are organic polymers with an abundance of anionic functional groups were postulated to be involved (Caron et al., 1992c). Extracellular polysaccharides most frequently carry a net negative charge due to the presence of glucuronic, galacturonic, or mannuronic acids or to nonsugar units such as pyruvic or succinic acid groups (Chenu, 1995) and such materials may, therefore, be implicated in the sensitivity of stability to water content.

IV. Conclusions

Conclusions that can be drawn from this review include the following:

- Characteristics of the different fractions of organic carbon that have the greatest influence on soil structure include the pore characteristics of these materials and their ability to absorb water, their capability to strengthen failure zones between primary soil particles or secondary structures (microaggregates), and their persistence.

- Changes in the organic carbon content of soils have been shown to correlate with changes in the structural form and stability of soils, and the magnitude of the change in structural characteristics is often strongly dependent on soil texture.

- Pedotransfer functions have been underutilized in studying the relations between soil structural characteristics, organic carbon contents, and other soil properties. These functions will, however, contribute most to our understanding of the influence of changing organic carbon contents in the future if they are based on data from soils of similar parent materials and climate and under similar or well defined management practices.

- The effect of the different fractions of organic carbon on soil structure have been explored most thoroughly in relation to the ability of aggregates to withstand the stresses arising from wetting and mechanical abrasion (wet aggregate stability). Roots, fungal hyphae, and polysaccharides that are intimately associated with the mineral fraction appear to fill a particularly important role in stabilizing aggregates. Less attention has been paid to the influence of different fractions on other stability characteristics.

There have been few studies on the role of different fractions of organic carbon on structural form. Failure to examine the influence of these fractions when changes in carbon content arise from differences in management practices on different soils, may account for inconsistent observations of, for instance, the change in available water content with organic carbon on soils of different texture.

The role of organic carbon or its component fractions in relation to structural resiliency or structural vulnerability remain largely unexplored, yet an understanding of the role of organic carbon on these characteristics is fundamental to understanding the impact that changes in stresses of natural or anthropogenic origin will have on the ability of soils to sustain biological productivity and regulate the terrestrial components of the hydrologic cycle.

Acknowledgments

This chapter was prepared while the author was on leave in the Dept. of Soil Science, Waite campus, University of Adelaide, Australia. Financial support provided by the Cooperative Research Centre for Soil and Land Management, Adelaide, and the Natural Sciences Engineering and Research Council of Canada is gratefully acknowledged. Appreciation is expressed to members of the Cooperative Research Centre who contributed to this chapter through discussions and reviews of earlier versions. Foremost among this group are W. W. Emerson, C. D. Grant, J. A. Baldock, A, Golchin, P. N. Nelson, and J. M. Oades.

References

Amato, M. and J.N. Ladd. 1992. Decomposition of ^{14}C-labelled glucose and legume material in soils: Properties influencing the accumulation of organic residue C and the microbial biomass C. *Soil Biol. Biochem.* 24:455-464.

Anderson, S.H., C.J. Gantzer, and J.R. Brown. 1990. Soil physical properties after 100 years of continuous cultivation. *J. Soil Water Cons.* 45:117-121.

Angers, D.A. and R.R. Simard. 1986. Relations entre la teneur en matiere organique et la masse volumique apparente du sol. *Can. J. Soil Sci.* 66:743-746.

Angers, D.A., B.D. Kay, and P.H. Groenevelt. 1987. Compaction characteristics of a soil cropped to corn and bromegrass. *Soil Sci. Soc. Amer. J.* 51:779-783.

Angers, D.A., N. Bissonnette, A. Legere, and N. Samson 1993a. Microbial and biochemical changes induced by rotation and tillage in a soil under barley production. *Can. J. Soil Sci.* 73:39-50.

Angers, D.A., N. Samson, and A. Legere. 1993b. Early changes in water-stable aggregation induced by rotation and tillage in a soil under barley production. *Can. J. Soil Sci.* 73:51-59.

Bakken, L.R. and R.A. Olsen. 1987. The relationship between cell size and viability of soil bacteria. *Microb. Ecol.* 13:103-114.

Baldock. J.A., B.D. Kay, and M. Schnitzer. 1987. Influence of cropping treatments on the monosaccharide content of the hydrolysates of a soil and its aggregate fractions. 1987. *Can. J. Soil Sci.* 67:489-499.

Ball, B.C. and E.A.G. Robertson. 1994. Effects of uniaxial compaction on aeration and structure of ploughed or direct drilled soils. *Soil Tillage Res.* 31:135-149.

Bartoli, F., G. Burtin, and J. Guerif. 1992. Influence of organic matter on aggregation in Oxisols rich gibbsite or in goethite. ll. Clay dispersion, aggregate strength and water-stability. *Geoderma* 54:259-274.

Bishop, A. and G.E. Blight. 1963. Some aspects of effective stress in saturated and partly saturated soils. *Geotechnique* 13:177-197.

Blackmore, A.V. 1981. Self-mulching soils. p. 4. In: *Soils News.* Aust. Soc. Soil Sci. (July, 1981).

Blackwell, P. S., A. J. Ringrose-Voase, N. S. Jayawardane, K. A. Olsson, D. C. McKenzie, and W. K. Mason. 1990. The use of air-filled porosity and intrinsic permeability to air to characterize structure of macropore space and saturated hydraulic conductivity of clay soils. *J. Soil Sci.* 41:215-228.

Bouma, J. and H.A.J. van Lanen. 1987. Transfer functions and threshold values from soil characteristics to land quality. In: Quantified Land Evaluation. Proceedings of a workshop, ISSS/SSSA. Washington, D.C. ITC Publ., Enschede, The Netherlands.

Braunack, M. V. and A. R. Dexter. 1989. Soil aggregation in the seedbed: a review. 2. Effect of aggregate sizes on plant growth. *Soil and Tillage Res.* 14:281-298.

Bremer, E., H.H. Janzen, and A.M. Johnston. 1994. Sensitivity of total, light fraction and mineralizable organic matter to management practices in a Lethbridge soil. *Can. J. Soil Sci.* 74:131-138.

Bullock, M.S., W.D. Kemper, and S.D. Nelson. 1988. Soil cohesion as effected by freezing, water content, time and tillage. *Soil Sci. Soc. Amer. J.* 52:770-776.

Caron, J. and B.D. Kay. 1992. Rate of response of structural stability to change in water content: Influence of cropping history. *Soil Tillage Res.* 25:167-185.

Caron, J., B.D. Kay, and E. Perfect. 1992a. Short-term decrease in soil structural stability following bromegrass establishment on a clay loam soil. *Plant Soil* 145:121-130.

Caron, J., B.D. Kay, J.A. Stone, and R.G. Kachanoski. 1992b. Modeling temporal changes in structural stability of a clay loam soil. *Soil Sci. Soc. Amer. J.* 56:1597-1604.

Caron, J., B.D. Kay, and J.A. Stone. 1992c. Improvement in structural stability of a clay loam with drying. *Soil Sci. Soc. Amer. J.* 56:1583-1590.

Carter, M. R. 1990. Relationship of strength properties to bulk density and macroporosity in cultivated loamy sand to loam soils. *Soil Tillage Res.* 15:257-268.

Chan, K.Y. 1995. Strength characteristics of a potentially hardsetting soil under pasture and conventional tillage in the semi-arid region of Australia. *Soil Tillage Res.* 34:105-113.

Chan, K.Y. and C.E. Mullins. 1994. Slaking characteristics of some Australian and British soils. *Europ. J. Soil Sci.* 45:273-283.

Chaney, K. and R.S. Swift. 1984. The influence of organic matter on aggregate stability in some British soils. *J. Soil Sci.* 35:223-230.

Chapman, J.E. 1927. The effects of organic matter on the tillage of a clay soil. *Proc. 1st Intl. Cong. Soil Sci.* 1:443-445.

Chenu, C. 1993. Clay- or sand-polysaccharide associations as models for the interface between microorganisms and soil: water related properties and microstructure. *Geoderma* 56:143-156.

Chenu, C. 1995. Extracellular polysaccharides: An interface between microorganisms and soil constituents. p. 217-233. In: P.M. Huang, J. Berthelin, J.-M. Bollag, W.B. McGill, and A.L. Page (Eds.), *Environmental Impact of Soil Component Interactions. Natural and Anthropogenic Organics.* CRC Press, Boca Raton, FL.

Chepil, W.S. 1953. Factors that influence clod structure and erodibility of soil by wind: 1. Soil texture. *Soil Sci.* 75:473-483.

Chepil, W.S. 1955. Factors that influence clod structure and erodibility of soil by wind: V. Organic matter at various stages of decomposition. *Soil Sci.* 80:413-421.

Christensen, B.T. 1992. Physical fractionation of soil organic matter in primary particle size and density separates. *Adv. Soil Sci.* 20:1-90.

da Silva, A.P., B.D. Kay, and E. Perfect. 1994. Characterization of the least limiting water range of soils. *Soil Sci. Soc. Amer. J.* 58:1775-1781.

da Silva, A. and B. D. Kay. 1996a. The influence of tillage and soil properties on the least limiting water range of soils. *Soil Sci. Soc. Amer. J.* (accepted).

da Silva, A. and B. D. Kay. 1996b. The sensitivity of shoot growth of corn to the least limiting water range of soils. *Plant Soil* (accepted).

da Silva, A., B.D. Kay and E. Perfect. 1997. Management versus inherent soil properties effects on bulk density and relative compaction. *Soil Tillage Res.* (manuscript submitted).

Davies, P. 1985. Influence of organic matter content, moisture status and time after reworking on soil shear strength. *J. Soil Sci.* 36:299-306.

Derdour, H., D.A. Angers, and M.R. Laverdiere. 1993. Caracterisation de l'espace poral d'un sol argileaux: effets de ses constituants et du travail du sol. *Can. J. Soil Sci.* 73:299-307.

Dexter, A.R. 1985. Shapes of aggregates from tilled layers of some Dutch and Australian soils. *Geoderma* 35:91-107.

Dexter, A.R. 1988. Advances in the characterization of soil structure. *Soil Tillage Res.* 11:199-238.

Dexter, A.R., B. Kroesbergen, and H. Kuipers. 1984. Some mechanical properties of aggregates of top soils from the Ijsselmeer polders. 1 Undisturbed soil aggregates. *Neth. J. Agric. Sci.* 32:205-214.

Douglas, J.T. 1986. Effects of season and management on the vane shear strength of a clay topsoil. *J. Soil Sci.* 37:669-679.

Emerson, W.W. 1977. Physical properties and structure. p. 78-104. In: J.S. Russel and E.L. Greacen (eds.), *Soil Factors in Crop Production in a Semi-Arid Environment.* Queensland Univ. Press, Brisbane.

Emerson, E. E., R. C. Foster, J. M. Tisdall, and D. Weissmann. 1994. Carbon content and bulk density of an irrigated Natrixeralf in relation to tree root growth and orchard management. *Aust. J. Soil Res.* 32:939-951.

Emerson, W.W. 1995. Water retention, organic C and soil texture. *Aust. J. Soil Res.* 33:241-251.

Foster, R.C. 1988. Microenvironments of soil microorganisms. *Biol. Fert. Soils* 6:189-203.

Ghodrati, M. and W.A. Jury. 1992. A field study of the effects of soil structure and irrigation method on preferential flow of pesticides in unsaturated soil. *J. Contam. Hydrol.* 11:101-125.

Golchin, A., J.M. Oades, J.O. Skjemstad, and P. Clarke. 1994. Soil structure and carbon cycling. *Aust. J. Soil Res.* 32:1043-1068.

Golchin, A., P. Clarke, J.M. Oades, and J.O. Skjemstad. 1995. The effects of cultivation on the composition of organic matter and structural stability of soils. *Aust. J. Soil Res.* 33:975-993.

Gollany, H.T., T.E. Schumacher, P.D. Evanson, M.J. Lindstrom, and G.D. Lemme. 1991. Aggregate stability of an eroded and desurfaced Typic Haplustoll. *Soil Sci. Soc. Amer. J.* 55:811-816.

Grant, C.D. and A.V. Blackmore. 1991. Self-mulching behavior in clay soils: Its definition and measurement. *Aust. J. Soil Res.* 29:155-173.

Grant, C.D., A.R. Dexter, and C. Huang. 1990. Roughness of soil fracture surfaces as a measure of soil microstructure. *J. Soil Sci.* 41:95-110.

Greenland, D.J. and G.W. Ford. 1964. Separation of partially humified organic materials from soils by ultrasonic dispersion. *Trans. 8th Intl. Cong. Soil Sci. Bucharest* 3:137-148.

Groenevelt P.H. and B.D. Kay. 1981. On pressure distribution and effective stress in unsaturated soils. *Can. J. Soil Sci.* 61:431-443.

Groenevelt, P.H., B.D.Kay, and C.D. Grant. 1984. Physical assessment of a soil with respect to rooting potential. *Geoderma* 34:101-114.

Guerif, J. 1988. Tensile strength of soil aggregates: Influence of clay and organic matter content. *Agron. (Paris)* 8:379-386.

Gusli, S., A. Cass, D.A. MacLeod, and P.S. Blackwell. 1994. Structural collapse and strength of some Australian soils in relation to hardsetting: 1. Structural collapse on wetting and draining. *European J. Soil Sci.* 45:15-21.

Harris, R.F., G. Chesters, and O. N. Allen. 1966. Dynamics of soil aggregation. *Adv. Agron.* 18:107-169.

Hassink. J., L.A. Bouwman, K.B. Zwart, J. Bloem, and L. Brussard. 1993. Relationships between soil texture, physical protection of organic matter, soil biota, and C and N mineralization in grassland soils. *Geoderma* 57:105-128.

Haynes, R.J. and R.S. Swift. 1990. Stability of soil aggregates in relation to organic constituents and soil water content. *J. Soil Sci.* 41:73-83.

Haynes, R.J., R.S. Swift, and R.C. Stephen. 1991. Influence of mixed cropping rotations (pasture-arable) on organic matter content, water-stable aggregation and clod porosity in a group of soils. *Soil Tillage Res.* 19:77-87.

Horn, R., H. Taubner, M. Wuttke, and T. Baumgartl. 1994. Soil physical properties related to soil structure. *Soil Tillage Res.* 31:135-148.

Hudson, B.D. 1994 Soil organic matter and available water capacity. *J. Soil Water Cons.* 49:189-194.

Jakobsen, B. and A. R. Dexter. 1988. Influence of biopores on root growth, water uptake and grain yield of wheat. Predictions from a computer model. *Boil. Fert. Soils* 6:315-321.

Janzen, H.H., C.A. Campbell, S.A. Brandt, G.P. Lafond, and L. Townley-Smith. 1992. Light-fraction organic matter in soils from long-term rotations. *Soil Sci. Soc. Amer. J.* 56:1799-1806.

Karlen, D.L., N.C. Wollenhaupt, D.C. Erbach, E.C. Berry, J.B. Swan, N.S. Eash, and J.L. Jordahl. 1994. Crop residue effects on soil quality following 10-years of no-till corn. *Soil Tillage Res.* 31:149-167.

Kay, B.D. 1990. Rates of change of soil structure under different cropping systems. *Adv. Soil Sci.* 12:1-52.

Kay, B.D. and A.R. Dexter. 1990. Influence of aggregate diameter, surface area and antecedent water content on the dispersibility of clay. *Can. J. Soil Sci.* 70:655-671.

Kay, B.D., A.R. Dexter, V. Rasiah, and C.D. Grant. 1994. Weather, cropping practices and sampling depth effects on tensile strength and aggregate stability. *Soil Tillage Res.* 32:135-148.

Kay, B.D., C.D. Grant, and P.H. Groenevelt. 1985, Significance of ground freezing on soil bulk density under zero tillage. *Soil Sci. Soc. Amer. J.* 49:973-978.

Kay, B.D. and V. Rasiah. 1994. Structural aspects of soil resiliency. p. 449-468. In: D.J. Greenland and I. Szabolcs (eds.), *Soil Resilience and Sustainable Land Use*. CAB International, London.

Kemper, W.D. and E.J. Koch. 1966. Aggregate stability of soils from western United States and Canada. Tech. Bull. No. 1355, Agricultural Research Service, USDA, Washington, D.C.

Kemper, W.D., R.C. Rosenau, and A.R. Dexter. 1987. Cohesion development in disrupted soils as affected by clay and organic matter content and temperature. *Soil Sci. Soc. Amer. J.* 51:860-867.

Kilbertus, G. 1980. Etude des microhabitats contenus dans les agregats du sol. Leur relation avec la biomasse bacterienne et la taille des procaryotes presents. *Rev. Ecol. Biol. Sol.* 17:543-557.

Kuipers, H. 1959. Confined compression tests on soil aggregate samples. *Meded. Landbouwhogesch. Opzoekingsstn. Staat Gent* 24:349-357.

Lal, R., A.A. Mahboubi, and N.R. Fausey. 1994. Long-term tillage and rotation effects on properties of a central Ohio soil. *Soil Sci. Soc. Amer. J.* 57:517-522.

Letey, J. 1985. Relationship between soil physical properties and crop production. *Adv. Soil Sci.* 1:277-294.

Ley, G.J., C.E. Mullins, and R. Lal. 1993. Effects of soil properties on the strength of weakly structured tropical soils. *Soil Tillage Res.* 28:1-13.

Low, A.J. 1955. Improvements in the structural state of soils under leys. *J. Soil Sci.* 6:179-199.

Manrique, L.A. and C.A. Jones. 1991. Bulk density of soils in relation to soil physical and chemical properties. *Soil Sci. Soc. Amer. J.* 55:476-481.

Miller, R.M. and J.D. Jastrow. 1990. Hierarchy of root and mycorrhizal fungal interactions with soil aggregation. *Soil Biol. Biochem.* 22:579-584.

Mitchell, A.R., T.R. Ellsworth, and B.D. Meek. 1995. Effect of root systems on preferential flow in swelling soil. *Commun. Soil Sci. Plant Anal.* 26:2655-2666.

Oades, J.M., 1993. The role of biology in the formation, stabilization and degradation of soil structure. *Geoderma* 56:377-400.

Oades, J.M., 1984. Soil organic matter and structural stability: mechanisms and implications for management. *Plant Soil* 76:319-337.

Panabokke, C.R. and J.P. Quirk. 1957. Effect of initial water content on the stability of soil aggregates in water. *Soil Sci.* 83:185-189.

Perfect, E. and B.D.Kay. 1995. Applications of fractals in soil and tillage research: a review. *Soil Tillage Res.* 36:1-20.

Perfect, E., B.D. Kay, J.A. Ferguson, A.P. da Silva, and K.A. Denholm. 1993 Comparison of functions for characterizing the dry aggregate size distribution of tilled soil. *Soil Tillage Res.* 28:123-139.

Perfect, E., B.D. Kay, and A.P. da Silva. 1995. Influence of soil properties on the statistical characterization of dry aggregate strength. *Soil Sci. Soc. Amer. J.* 59:532-537.

Perfect, E., W.K.P. van Loon, B.D. Kay, and P.H. Groenevelt. 1990a. Influence of ice segregation and solutes on soil structural stability. *Can. J. Soil Sci.* 70:571-581.

Perfect, E., B.D. Kay, W.K.P. van Loon, R.W. Sheard, and T. Pojasok. 1990b. Rates of change in soil structural stability under forages and corn. *Soil Sci. Soc. Amer. J.* 54:179-186.

Perfect, E., B.D. Kay, W.K.P. van Loon, R.W. Sheard, and T. Pojasok. 1990c. Factors influencing soil structural stability within a growing season. *Soil Sci. Soc. Amer. J.* 54:173-179.

Quirk, J.P. and B.G. Williams. 1974. The disposition of organic materials in relation to stable aggregation. p. 1:165-173. In: *Trans. 10th Cong., Int. Soc. Soil Sci.* Moscow.

Quirk, J.P. and R.S. Murray. 1991. Towards a model for soil structural behavior. *Aust. J. Soil Res.* 29:828-867.

Rasiah, V. and B.D. Kay. 1994a. Characterizing changes in aggregate stability subsequent to introduction of forages. *Soil Sci. Soc. Amer. J.* 58:935-942.

Rasiah, V. and B.D. Kay. 1994b. Quantifying the changes in clay stabilization after the introduction of forages. *Soil Sci.* 157:318-327.

Rasiah, V. and B.D. Kay. 1995. Characterizing rate of wetting: Impact on structural destabilization. *Soil Sci.* 160:176-182.

Rawls, W.J. 1983. Estimating soil bulk density from particle size analysis and organic matter content. *Soil Sci.* 135:123-124.

Reid, J.B. and M.G. Goss. 1982. Interactions between soil drying due to plant water use and decreases in aggregate stability caused by maize roots. *J. Soil Sci.* 33:47-53.

Richardson, S.J. 1976. Effect of artificial weathering cycles on the structural stability of a dispersed silt soil. *J. Soil Sci.* 27:287-294.

Richter, M., I. Mizuno, S. Aranguez, and S. Uriarte. 1975. Densiometric fractionation of soil organo-mineral complexes. *J. Soil Sci.* 26:112-123.

Rogowski, A.S., W.C. Moldenhauer, and D. Kirkham. 1968. Rupture Parameters of soil aggregates. *Soil Sci. Soc. Amer. J.* 32:720-724.

Rowell, D.L. and P.J. Dillon. 1972. Migration and aggregation of Na and Ca clays by the freezing of dispersed and flocculated suspensions. *J. Soil Sci.* 23:442-447.

Rutherford, P.M. and N.G. Juma. 1992. Influence of soil texture on protozoa-induced mineralization of bacterial carbon and nitrogen. *Can. J. Soil Sci.* 72:183-200.

Schjonning, P., B.T. Christensen, and B. Carstensen. 1994. Physical and chemical properties of a sandy loam receiving animal manure, mineral fertilizer or no fertilizer for 90 years. *European J Soil Sci.* 45:257-268.

Sillanpaa, M. and L.R. Webber. 1961. The effect of freeze-thawing and wetting-drying cycles on soil aggregation. *Can. J. Soil Sci.* 41:182-18.

Snyder, V.A. and R.D. Miller. 1985. Tensile strength of unsaturated soils. *Soil Sci. Soc. Amer. J.* 49:58-65.

Soane, B.D. 1975. Studies on some physical properties in relation to cultivation and traffic. p. 160-183. In: Soil physical conditions and crop production. Min. Agric. Food Fish., Tech. Bull. 29.

Soane, B.D. 1990. The role of organic matter in soil compactibility: A review of some practical aspects. *Soil Tillage Res.* 16:179-201.

Stengel P. 1979. Utilization de l'analyse des systemes de porosite pour la caracterisation de l'etat physique du sol in situ. *Ann. Agron.* 30:27-51.

Thomasson, A.J. 1978. Towards an objective classification of soil structure. *J. Soil Sci.* 29:38-46.

Tiedje, O. and M. Tapkenhinrichs. 1993. Evaluation of pedo-transfer functions. *Soil Sci. Soc. Amer. J.* 57:1088-1095.

Tisdall, J.M. and J.M. Oades. 1979. Stabilization of soil aggregates by the root systems of rye grass. *Aust. J. Soil Res.* 17:429-441.

Tisdall, J.M. and J.M. Oades. 1982. Organic matter and water-stable aggregates in soils. *J. Soil Sci.* 33:141-163.

Utomo, W.H. and A.R. Dexter. 1981a. Soil friability. *J. Soil Sci.* 32:203-213.

Utomo, W. H. and A. R. Dexter. 1981b. Tilth mellowing. *J. Soil Sci.* 32:187-201.

van Genuchten, M.Th. 1980. A closed form equation for predicting the hydraulic conductivity of unsaturated soils. *Soil Sci. Soc. Amer. J.* 44:892-898.

Veihmeyer, F. J. and A. H. Hendrickson. 1927. Soil moisture conditions in relation to plant growth. Pl. *Physiol.* 2:71-82.

Vereecken, H., J. Maes, J. Feyen, and P. Darius. 1989. Estimating the soil moisture characteristic from texture, bulk density and carbon content. *Soil Sci.* 148:389-403.

Voorhees, W.B. 1983. Relative effectiveness of tillage and natural forces in alleviating wheel-induced soil compaction. *Soil Sci. Soc. Amer. J.* 47:129-133.

Waters, A.G. and J.M. Oades. 1991. Organic matter in water stable aggregates. p. 163-174. In: W.S. Wilson (ed.), *Advances in Soil Organic Matter Research: The Impact on Agriculture and the Environment.* R. Soc. Chem., Cambridge.

Wenke, J.F. and C.D. Grant. 1994. The indexing of self-mulching behavior in soils. *Aust. J. Soil Res.* 32:201-211.

White, R. E. 1985. The influence of macropores on the transport of dissolved and suspended matter through soil. *Adv. Soil Sci.* 3:94-120.

Williams, J., R.E. Prebble, W.T. Williams, and C.T. Higgnett. 1992. The influence of texture, structure and clay mineralogy on the soil moisture characteristic. *Aust. J. Soil Res.* 21:15-32.

Willis, W. 1955. Freezing and thawing, and wetting and drying in soils treated with organic chemicals. *Soil Sci. Soc. Amer. Proc.* 19:263-267.

Yoder, R.E. 1936. A direct method of aggregate analysis of soils and a study of the physical nature of erosion losses. *J. Amer. Soc. Agron.* 28:337-351.

Young, I. and C.E. Mullins. 1991. Water-suspensible solids and structural stability. *Soil Tillage Res.* 19:89-94.

CHAPTER **14**

Dynamics of Soil Aggregation and C Sequestration

Denis A. Angers and Claire Chenu

I. Introduction

Upon decomposition, about one third of the C derived from plant residues is still present in the soil after one year and decomposes slowly. Chemical transformations of the C and its close association with the mineral phase, which provides physical protection, have been proposed as general mechanisms to explain this sequestration of C in the soil (Oades, 1988). The actual distribution of soil organic matter (SOM) relative to soil architecture is highly heterogenous and can be studied at different scales varying from whole soil profile to the microscopic level (Kooistra and van Noordwijk, 1995). Using the results of recent studies as examples, the objectives of this chapter are 1) to examine the location of SOM relative to the soil aggregate structure, 2) to illustrate some mechanisms by which SOM, and in particular fresh plant residues, contribute to the formation and stabilization of soil aggregates, and 3) to discuss the possible role of aggregation on the decomposition and fate of SOM.

II. Location of SOM in Soil Structure

In many soils, the solid particles are bound together to form aggregates of various sizes and stability which are, in turn, separated by voids or failure zones, also of various shapes and sizes. Micromorphological and physical fractionation methods can be used to study SOM distribution relative to soil structure at centimetric down to micrometric scales. The former are based on the visual observation and recognition of organic compounds in association with mineral particles at different scales using light, scanning (SEM) or transmission (TEM) electronic microscopy (e.g., Foster and Martin, 1981; Chenu and Tessier, 1995), and are usually not quantitative. On the other hand, fractionation studies most often involve the separation of soil into aggregates or particles of various sizes, using treatments that will disrupt the soil structure at different degrees and thereby select units of different physical stabilities (e.g., Christensen, 1995). Soil organic matter may also be physically separated on the basis of its degree of association with minerals using their respective differences in density: free SOM being recovered in the light fraction whereas mineral-associated SOM is found in the heavy one (e.g., Monnier et al., 1962; Turchenek and Oades, 1979; Puget et al., 1996). The fractionation procedure used will obviously largely determine the nature, size, and stability of the structural units separated.

Most physical separation studies have shown that a large proportion of the total SOM is actually closely associated with soil minerals (e.g., Monnier et al., 1962; Turchenek and Oades, 1979). Nearly 90% of SOM was found to be located within soil aggregates according to Jastrow et al. (1996). In cultivated soils, macroscopic or particulate organic matter (POM) was found to be located mostly

Table 1. Distribution of particulate organic matter (POM) within (occluded) or out (free) of water-stable aggregates. Silty loam cultivated soil, 0-30 cm sampling

Aggregate fractions	POM		C in the POM	
	Occluded	Free	Occluded	Free
(mm)	(mg POM g^{-1} soil)		(mg POM-C g^{-1} soil)	
> 1	0.90	1.30	0.12	0.36
0.2 - 1	9.50	0.70	0.93	0.10
0.05 -0.2	2.80	0.00	0.27	0.00
Total	13.20	2.00	1.32	0.46
% of total POM	87%	13%	74%	26%

(After Puget et al., 1996.)

within the soil aggregates that were separated by wet-sieving with slaking (Table 1). External or free POM which represented 13% of the total POM and 26% of the total POM-associated C would then be located in the interaggregate pore space of the soil.

The distribution of SOM among the aggregate size fractions can be very heterogenous and is subject to variations depending on the techniques used to obtain the aggregates. Sieving soil in its natural state, i.e., with dry sieving, most often leads to an even distribution of C and N among size fractions (Puget et al., 1995; Angers and Giroux, 1996). When higher stress is applied such as with slaking (direct immersion in water of dry aggregates), primary particles, microaggregates, and interaggregate SOM are freed from unstable aggregates and the remaining highly stable macroaggregates are usually enriched in C and N (Elliott, 1986; Puget et al., 1995; Angers and Giroux, 1996). Generally, upon slaking, the aggregates do not break down into primary particles but rather into highly stable smaller units indicating that the aggregates are arranged in a hierarchical fashion (Oades and Waters, 1991). This concept of hierarchy seems to apply to most soils except those for which SOM is not the principal aggregate binding agent, e.g., Oxisols (Oades and Waters, 1991). Not only the distribution of total SOM is heterogeneous in soils, but also that of young vs. old organic matter. Puget et al. (1995) and Angers and Giroux (1996) have shown, using ^{13}C natural abundance, that young C (i.e., C of less than 6, 15, or 22 years) tends to accumulate in larger aggregates where it would act as a binding agent. This young macroaggregate-associated C is composed of both macro-organic matter (decomposing plant residues) and mineral-associated C in approximately equal proportions (D.A. Angers and M. Giroux, unpublished data). As shown in Table 1, the distribution of POM, which constitutes a significant fraction of the labile SOM pool, is also heterogeneous in aggregate fractions.

The observed SOM contents in soil aggregate fractions are the result of (i) the actual quantity of SOM incorporated into the fractions, (ii) its specific turnover rate, which depends on the chemical nature and physical location, and (iii) the turnover rate of the fraction itself (i.e., the rate of aggregate disruption and formation). Hence, the significance of the relative abundance of SOM in physical fractions at a given point in time is difficult to interpret. Further, the mechanisms by which SOM gets incorporated into soil aggregates are still poorly understood. In general terms, root growth and rhizodeposition, faunal activity, tillage, wetting and drying, and consequent formation and destruction of aggregates would be processes by which organic material gets associated with the mineral phase. These processes may be considered as being relatively passive. By contrast, active organic matter incorporation occurs when fresh organic particles, roots, and microorganisms form aggregates around them, and end in an inner position.

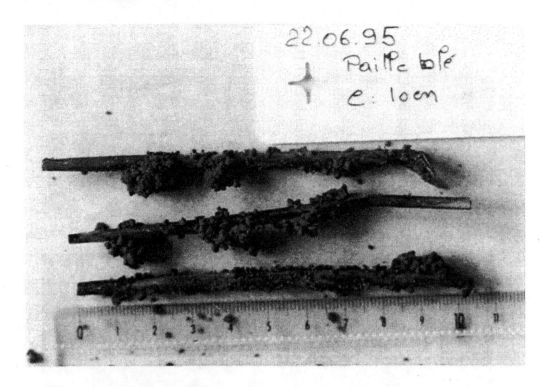

Figure 1. Aggregate formation around wheat straw after 3 months of decomposition in a silty soil. (Source: D.A. Angers and S. Recous, unpublished.)

III. Organic Matter as an Aggregate Binding Agent

The extensively demonstrated role of SOM in the formation and stabilization of soil aggregates will not be reviewed here and we will focus on the effects of decomposing plant residues on aggregation. It has long been shown that when organic residues are incorporated to the soil, the abundance and stability of soil aggregates increase (Harris et al., 1966; Monnier, 1965). Microorganisms decomposing the organic material are largely responsible for this effect (Lynch and Bragg, 1985).

Recently incorporated particulate organic matter was shown to initiate aggregation (Figure 1) by acting as a substrate for the fungi and bacteria which aggregate soil particles through their associated mucilages or physical enmeshment (Figure 2a, b). Such aggregates are also found around older POM (Figure 2c, d). Oades and Waters (1991) have also observed that many aggregates of about 100-200 µm in diameter have cores of plant debris, and encrustation of plant fragments by mineral particles was proposed as a mechanism of aggregate formation and stabilization. Golchin et al. (1994) proposed a model of microaggregate (20-250 µm) formation around plant residues. They suggested that particulate SOM entering the soil is rapidly colonized by a microbial population. The microflora and its by-products have strong adhesive properties, and mineral particles adhere to them as shown in Figure 2. The plant fragments are thereby rapidly encrusted by mineral particles and become the center of water-stable aggregates. The dynamics of association of crop residues with mineral particles can be studied using isotopically-labelled residues in conjunction with aggregate fractionations (Buyanovsky et al., 1994). During decomposition of ^{13}C-labelled wheat straw in the field, the proportion of the soil residual ^{13}C (whole-soil ^{13}C) recovered in the various slaking-resistant aggregate fractions varied during an 18-month period (Angers et al., 1996a). The proportion present in the 50

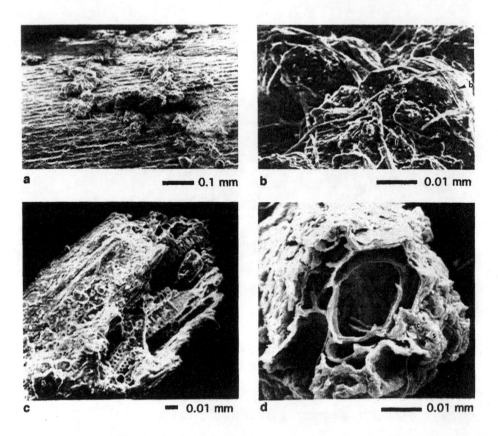

Figure 2. Aggregate stabilization by fungi developing on decomposing wheat straw residues. Straw was incubated for 3 months in a silty soil. Fungi enmesh the soil aggregates within mycelial strands and the newly stabilized aggregates are also colonized by bacteria. Low-temperature SEM. Black bar is 100 μm (a) and 10 μm (b). b = bacteria. f = fungi. (Source: D.A. Angers, C. Chenu, and S. Recous, unpublished.) (c) Particulate organic matter encrusted with mineral soil particles. (d) Remnants of conducting vessels are coated by clay and silt particles. Conventional SEM. Black bar is 10 μm. c = clay particles. s = silt particles. (Source: P. Puget and C. Chenu, unpublished.)

to 250 μm fraction increased steadily, confirming quantitatively the progressive association of plant residues and their derivatives with microaggregates (Golchin et al., 1994), and suggesting an involvement of the plant residue-derived products on aggregate stabilization.

However, organic residues vary in their efficiency to promote the aggregation of mineral particles. Incubation experiments have shown clearly that less decomposable residues such as mature straw have a much smaller effect on stable macroaggregation than a fresh residue like alfalfa which highly promotes microbial activity and aggregation (J. Lafond and D.A. Angers, unpublished results). In plant residue-mediated aggregation, the microorganisms are the active binding agents through both physical enmeshment (see Figure 2a,b), and adhesion of mineral particles to microorganisms surfaces and their associated by-products. Fresh plant residues may also exhibit some adhesive and binding properties per se, e.g., in the case of roots and their associated mucilage (Morel et al., 1991).

As decomposition of plant residues proceeds, less C is quantitatively and qualitatively available to the decomposers and microbial activity decreases. As pointed out by Golchin et al. (1994), the aggregating activity of POM-associated microorganisms must then decrease and the POM is freed from its mineral crust which disrupts. The dynamics of aggregation at this scale are then linked to the dynamics of SOM decomposition. Plant residues and POM can be ascribed different aggregating potential depending on their intrinsic decomposability and stage of decomposition, i.e., on their ability to support microbial growth.

IV. Decomposition of SOM and Soil Aggregation

During decomposition in soil, organic residues and their decomposition products become closely associated with the mineral phase and, simultaneously, their chemical composition changes. Both processes are believed to provide recalcitrancy to further decomposition; however, their respective importance is difficult to determine. The most direct evidence of the role of soil structure in protecting SOM from decomposition probably comes from the observation that when soil aggregates are disrupted, an increase or flush in C mineralization is observed relative to undisrupted aggregates (Rovira and Greacen, 1957; Powlson, 1980; Elliott, 1986; Gupta and Germida, 1988). Further, Beare et al. (1994) have shown that the level of physical protection varies with soil management practices with apparently more aggregate protection in no-till soils than in cultivated ones.

Further support that the location of SOM in the soil matrix influences its decomposition is provided by isotopic tracer studies. In a cultivation sequence, Besnard et al. (1996) found that, upon cultivation, much more C was lost from the free POM fraction than from the occluded POM (located within aggregates). Physical transfers of POM from outside to inside aggregates or vice-versa, with aggregate formation or disruption could not completely account for this observation. Further, after 35 years of cultivation, the 50- to 200-µm microaggregates were relatively enriched in POM-C derived from the initial forest vegetation, as compared to macroaggregates or to nonaggregated soil. Gregorich et al. (1996) separated the light (<1.6 g cm^{-3}) particulate organic matter occluded within soil aggregates from the free light fraction (LF) and determined its relative age using ^{13}C natural abundance. Although LF is most often perceived as a labile SOM fraction, they found that most of the occluded LF found within the aggregates originated from the grass (> 32 yr in age) whereas the free LF was mostly of recent corn origin. These differences in relative accumulation were due to either differences in the chemical composition of SOM in different locations or to physical protection provided by the microaggregates.

The fate of SOM located and protected within aggregates will depend upon its intrinsic decomposability and on the persistence of the aggregates. The protective capacity of soil aggregates should, therefore, be related to their stability to water and to other mechanical stresses, although there is yet no direct evidence of such relationships. On the other hand, SOM contributes to the stability of aggregates and to increasing their life expectancy. Soil organic matter, therefore, indirectly contributes to its "self-protection" against biodegradation.

Management systems involving high C inputs and reduced tillage should, therefore, favor C storage directly by reducing agregate breakdown and indirectly by enhancing the SOM-mediated stable aggregation. There are several examples of such complex interrelationships. Soil aggregation seems to play an important role in the accretion of SOM under aggrading cropping systems. For example, shortly after perennial grass or legume establishment on poorly-structured soils, soil water-stable macroaggregation rapidly increases. This short-term change was related to increases in macro-organic matter and labile polysaccharides, both of which were likely involved in promoting aggregation (Angers and Mehuys, 1989; 1990) through processes already described in this chapter. In return, improved aggregation under perennial crops likely favored the protection of these labile C and a consequent increase in C storage in the longer term (Angers, 1992). Similar observations were made in the case of improvement of soil aggregation and C content with repeated applications of farmyard manure (N'Dayegamiye and Angers, 1990; Aoyama and Taninai, 1992).

Tillage practices have a strong influence on soil aggregation and C storage. Several studies have shown that surface soils (0-10 cm) under no-till show better stable-aggregation, and greater labile and total C storage than their tilled counterparts (Carter, 1992; Beare et al., 1994); however, studies on the effects of tillage on SOM storage should consider the whole soil profile. Tillage not only directly affects soil aggregation but also induces changes in organic matter distribution in the profile, in soil physical and soil climatic conditions, and in microbial populations. All these factors are likely to influence SOM decomposition processes and storage in soils when till vs. no-till systems are compared. If it is true that moldboard plowing results in a reduction in C content of the very surface soil, it may also result in an increase in deeper layers, particularly in soils where compaction and climatic conditions may reduce decomposition at depth (Angers et al., 1996b). Further, Balesdent et al. (1990) have shown, using ^{13}C natural abundance, that plowing increased the decomposition of native organic matter but, on the other hand, also favored the incorporation of crop residue-derived organic matter. The importance of tillage and the mechanisms by which it influences SOM dynamics under various conditions requires more research.

V. Conclusion

Upon decomposition in soil, plant residues and their decomposition products undergo chemical and physical transformations and become closely associated with mineral particles to form stable aggregates. This close association with mineral particles in the form of stable aggregates and the chemical stabilization provide organic matter with "self-protection" from further decomposition. However, the relative contribution of these two mechanisms to the process of C sequestration in soil is still unclear. In order to better understand the role of aggregation on SOM dynamics, more research is necessary to quantify the actual turnover time of natural soil aggregates and the relationship with its stability, as well as the mechanisms by which SOM gets incorporated into aggregates.

References

Angers, D.A. 1992. Changes in soil aggregation and organic carbon under corn and alfalfa. *Soil Sci. Soc. Am. J.* 56:1244-1249.
Angers, D.A. and G.R. Mehuys. 1989. Effects of cropping on carbohydrate content and water-stable aggregation of a clay soil. *Can. J. Soil Sci.* 69:373-380.
Angers, D.A. and G.R. Mehuys. 1990. Barley and alfalfa cropping effects on carbohydrate contents of a clay soil and its particle size fractions. *Soil Biol. Biochem.* 22:285-288.
Angers, D.A. and M. Giroux. 1996. Recently-deposited organic matter in soil water-stable aggregates. *Soil Sci. Soc. Am. J.* 60:1547-1551.
Angers, D.A., S. Recous, and C. Aita. 1996a. Fate of carbon and nitrogen in water-stable aggregates during decomposition of wheat straw *in situ*. *Europ. J. Soil Sci.* (in press)
Angers, D.A., M. A. Bolinder, M.R. Carter, E.G. Gregorich, R.P. Voroney, C.F. Drury, B.C. Liang, R.R. Simard, R.G. Donald, R. Beayert, and J. Martel. 1996b. Impact of tillage practices on organic carbon and nitrogen storage in cool, humid soils of eastern Canada. *Soil Tillage Res.* (in press).
Aoyama, M. and Y. Taninai. 1992. Organic matter and its mineralization in soil particle and aggregate size fractions of soils with four years of farmyard manure applications. *Jpn. J. Soil Sci. Plant Nutr.* 63:571-580.
Balesdent, J., A. Mariotti, and D. Boisgontier. 1990. Effect of tillage on soil organic carbon mineralization estimated from ^{13}C natural abundance in maize fields. *J. Soil Sci.* 41:587-596.

Beare, M.H., P.F. Hendrix, and D.C. Coleman. 1994. Water-stable aggregates and organic matter fractions in conventional- and no-tillage soils. *Soil Sci. Soc. Am. J.* 58:777-786.

Besnard, E., C. Chenu, J. Balesdent, P. Puget, and D. Arrouays. 1996. Fate of particulate organic matter in soil aggregates during cultivation. *Europ. J. Soil Sci.* 47:495-503.

Buyanovsky, G.A., M. Aslam, and G.H. Wagner. 1994. Carbon turnover in soil physical fractions. *Soil Sci. Soc. Am. J.* 58:1167-1173.

Carter, M.R. 1992. Influence of reduced tillage systems on organic matter, microbial biomass, macro-aggregate distribution and structural stability of the surface soil in a humid climate. *Soil Tillage Res.* 23:361-372.

Chenu, C. and D. Tessier. 1995. Low temperature scanning electron microscopy of clay and organic constituents and their relevance to soil microstructures. *Scanning Microscopy* 9:989-1010.

Christensen, B.T. 1995. Carbon in primary and secondary organomineral complexes. p. 97-165. In: M.R. Carter and B.A. Stewart (Eds.) *Structure and Organic Matter Storage in Agricultural Soils*. *Adv. Soil Sci.* CRC Press, Boca Raton, FL.

Elliott, E.T. 1986. Aggregate structure and carbon, nitrogen and phosphorus in native and cultivated soils. *Soil Sci. Soc. Am. J.* 50:627-633.

Foster, R. C. and J.K. Martin. 1981. *In situ* analysis of soil components of biological origin. p. 75-111. In: *Soil Biochemistry* Vol. 5. E. A. Paul and J. N. Ladd (eds.), Marcel Decker, N.Y.

Golchin, A., J.M. Oades, J.O. Skjemstad, and P. Clarke. 1994. Soil structure and carbon cycling. *Austr. J. Soil Res.* 32:1043-1068.

Gregorich, E.G., C.F. Drury, B.H. Ellert, and B.C. Liang. 1996. Fertilization effects on physically-protected light fraction organic matter. *Soil Sci. Soc. Am. J.* 60:472-476.

Gupta, V.V.S.R. and J.J. Germida. 1988. Distribution of microbial biomass and its activity in different soil aggregate size classes as affected by cultivation. *Soil Biol. Biochem.* 20:777-786.

Harris, R.F., G. Chesters, and O.N. Allen. 1966. Dynamics of soil aggregation. *Adv. Agron.* 18:107-169.

Jastrow, J.D., T.W. Boutton, and R.M. Miller. 1996. Carbon dynamics of aggregate-associated organic matter estimated by ^{13}C- natural abundance. *Soil Sci. Soc. Am. J.* 60:801-807.

Kooistra and van Noordwijk. 1995. Soil architecture and distribution of organic matter. p. 15-56. In: M.R. Carter and B.A. Stewart (eds.), *Structure and organic matter storage in agricultural soils*. *Adv. Soil Sci.* CRC Press, Boca Raton, FL.

Lynch, J.M. and E. Bragg. 1985. Microorganisms and soil aggregate stability. *Adv. Soil Sci.* 2:133-171.

Monnier, G., L. Turc, and C. Jeanson-Lusinang. 1962. Une méthode de fractionnement densimétrique par centrifugation des matières organicuqes des sols. *Ann. Agron.* 13:55-63.

Monnier, G. 1965. Action des matières organiques sur la stabilité structurale des sols. *Annales Agronomiques* 16: 327-400.

Morel, J. L., L. Habib, S. Plantureux, and A. Guckert A. 1991. Influence of maize root mucilage on soil aggregate stability. *Plant Soil* 136:111-119.

N'Dayegamiye, A. and D.A. Angers. 1990. Effets de l'apport prolongé de fumier de bovins sur quelques propriétés physiques et biologiques d'un loam limoneux Neubois sous culture de maïs. *Can. J. Soil Sci.* 70:259-262.

Oades, J.M. 1988. The retention of organic matter in soils. *Biogeochemistry*. 5:35-70.

Oades, J.M. and A.G. Waters. 1991. Aggregate hierarchy in soils. *Aust. J. Soil Res.* 29:815-828.

Powlson, D.S. 1980. The effects of grinding on microbial and non-microbial organic matter in soil. *J. Soil Sci.* 31:77-85.

Puget, P., C. Chenu, and J. Balesdent. 1995. Total and young organic matter distributions in aggregates of silty cultivated soils. *Europ. J. Soil Sci.* 46:449-459.

Puget, P., E. Besnard, and C. Chenu. 1996. Une méthode de fractionnement des matières organiques particulaires des sols en fonction de leur localisation dans les agrégats. *C.R. Acad. Sci. (Paris) Série II* 322:965-972.

Rovira, A.D. and E.L. Greacen. 1957. The effect of aggregate disruption on enzyme activity of microorganisms in the soil. *Aust. J. Agric. Res.* 8:659-673.

Turchenek, L.W. and J. M. Oades. 1979. Fractionation of organo-mineral complexes by sedimentation and density techniques. *Geoderma* 21:311-343.

CHAPTER 15

Soil Aggregate Stabilization and Carbon Sequestration: Feedbacks through Organomineral Associations

J.D. Jastrow and R.M. Miller

I. Introduction

Primary production (specifically, the rate and quality of C transfer belowground) and soil microbial activity (specifically, the rates of C transformation and decay) are recognized as the overall biological processes governing soil organic C (SOC) dynamics. These two processes and, hence, SOC cycling and storage are controlled by complex underlying biotic and abiotic interactions and feedbacks, most of which can be tied in one way or another to the influences of the five state factors related to soil formation (Jenny, 1941), many of which are sensitive to management practices. Overall, C input rates and quality are largely dependent on climate (especially temperature and precipitation), vegetation type and landscape, soil type, and management practices. Decomposition processes and turnover rates, however, are greatly influenced by climate, the type and quality of organic matter (e.g., N content and the ratios of C:N and lignin:N), chemical or physicochemical associations of organic matter (OM) with soil mineral components, and the location of OM within the soil.

The mechanisms responsible for stabilizing SOC (Figure 1) may be categorized as (1) biochemical recalcitrance, (2) chemical stabilization, and (3) physical protection (Christensen, 1996). Biochemical recalcitrance may be due to the chemical characteristics of the substrate itself — e.g., lignin derivatives (Stott et al., 1983) or melanins produced by fungi and other soil organisms (Martin and Haider, 1986) — or may result from transformations during decomposition, including incorporation into the excrement of soil meso- and microfauna (Kooistra and van Noordwijk, 1996). Chemical stabilization occurs because of chemical or physicochemical associations between what would otherwise be decomposable compounds and soil mineral components. For example, organic compounds sorbed to clay surfaces, often by polyvalent cation bridges, or those intercalated between expanding layers of clays are quite resistant to degradation (Martin and Haider, 1986; Christensen, 1996; Tisdall, 1996). In addition, the drying of organics may cause them to be denatured or polymerized, thereby protecting them chemically from decomposition (Dormaar and Foster, 1991). Soil structure, however, plays a dominant role in the physical protection of soil organic matter (SOM) by controlling microbial access to substrates, microbial turnover processes, and food web interactions (Elliott and Coleman, 1988; van Veen and Kuikman, 1990). Relatively labile material may become physically protected from decomposition by incorporation into soil aggregates (Oades, 1984; Gregorich et al., 1989; Golchin et al., 1994a,b) or by deposition in micropores inaccessible even to bacteria (Foster, 1985).

ISBN 0-849307441-3
©1997 by CRC Press LLC

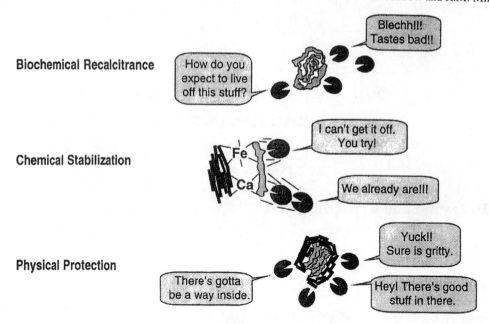

Figure 1. Mechanisms of soil organic matter stabilization.

Because of the physical protection afforded by soil structure, significant interactions exist between SOC dynamics and the formation, stabilization, and degradation of soil aggregates. In soils where OM is the major aggregate binding agent, plant growth and the decomposition of organic inputs lead to the development of a hierarchical aggregate structure (Tisdall and Oades, 1982; Oades and Waters, 1991). The exact nature and stability of this structure in a given soil depend on the relative amounts and strengths of various types of organomineral associations that function as aggregate binding and stabilizing agents at each hierarchical level of organization. At the same time, the nature of these organomineral associations and their spatial locations within the aggregate hierarchy determine the degree to which SOC is physically protected from decomposition and, consequently, result in organic pools with various input and turnover rates.

Thus, feedbacks between SOC cycling and aggregate cycling can occur. In aggrading systems, organic inputs lead to the formation and stabilization of aggregates, which in turn can protect SOM from decomposition, leading to further aggregate stabilization. In degrading systems, the disruption of aggregates exposes previously protected but relatively labile OM to decomposers, resulting in a loss of SOM and further destabilization of aggregates. In soils at or near equilibrium in terms of the amount of SOC in the whole soil, neither OM nor aggregates are static, and the turnovers of aggregates and various OM pools are still interrelated. In any case, the feedbacks between SOC cycling and aggregate cycling appear to be controlled by the formation or destruction of organomineral associations functioning as aggregate binding agents. Furthermore, in soils exhibiting a hierarchical aggregate structure, the feedbacks likely cross spatial scales affecting binding agents at different hierarchical levels of organization.

In this chapter, we will examine the relationship between the function of various organomineral associations as aggregate binding agents and C sequestration in soils. Our discussions will focus on soils that exhibit an aggregate hierarchy in which aggregate stability is controlled primarily by organic materials. After briefly examining the nature and function of organomineral associations in macro- and microaggregates, we will present evidence from our own studies of restored prairies on Illinois

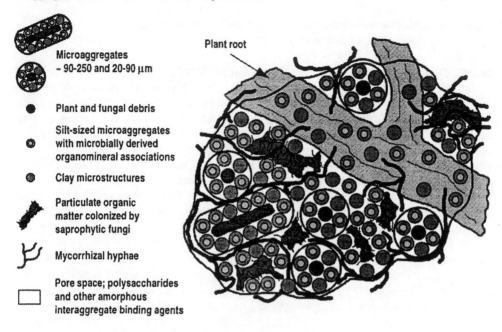

Figure 2. Conceptual diagram of soil aggregate hierarchy.

mollisols as a case study to demonstrate the occurrence of feedbacks between aggregation and C accrual in an aggrading system.

II. Aggregate Hierarchy

Attempts to understand the formation and stabilization of aggregates in soils where OM is the major binding agent have been influenced considerably by the hierarchical view of the aggregation process proposed by Tisdall and Oades (1982) and elaborated by Dexter (1988), Kay (1990), and Oades and Waters (1991), among others. In this conceptual model (Figure 2), the mechanisms of aggregate formation and stabilization and their relative importance change with spatial scale. Primary particles and clay microstructures are bound together with bacterial and fungal debris into extremely stable silt-sized microaggregates (2-20 μm in diameter), which may be bound together with fungal and plant debris and fragments into larger microaggregates (20-250 μm in diameter). The organic binding agents involved in stabilizing microaggregates are believed to be relatively persistent and to consist of humic materials or polysaccharide polymers strongly sorbed to clays, with the most persistent clay-organic associations being strengthened by bridges of polyvalent metal cations (Tisdall and Oades, 1982; Tisdall, 1996). Microaggregates, in turn, are bound into macroaggregates (>250 μm in diameter) by (1) transient binding agents (i.e., readily decomposable organic materials, the most important being microbial- and plant-derived polysaccharides) and (2) temporary binding agents (i.e., fine roots, fungal hyphae, bacterial cells, and algae). As macroaggregates increase in diameter, transient binding agents appear to be less important, whereas temporary binding agents generally increase in importance.

All pools of organic binding agents are subject to the simultaneous effects of loss through decomposition and inputs from the creation of new materials. However, the long-term existence of persistent agents (perhaps tens to hundreds of years) may be due more to protection from decomposition by their physicochemical associations with inorganic soil components than to any inherent biochemical inertness (Kay, 1990).

The distinction between micro- and macroaggregates is based both on size and on susceptibility to slaking, i.e., rapid wetting (Edwards and Bremner, 1967; Tisdall and Oades, 1982). In soils exhibiting aggregate hierarchy, macroaggregates subjected to slaking break down into microaggregates, rather than primary particles. Stabilized microaggregates are not disrupted by rapid wetting and mechanical disturbance, including cultivation. In contrast, the stability of macroaggregates is generally controlled by management practices and other disturbances or by factors affecting root and hyphal growth and rhizosphere organisms (Tisdall and Oades, 1980, 1982).

A consequence of aggregate hierarchy is the porosity exclusion principle outlined by Dexter (1988). If aggregate hierarchy exists, then smaller aggregates should have smaller pores, greater contact between particles, and higher bulk densities than larger aggregates because the latter also contain larger pores between the smaller aggregates that comprise them (Oades and Waters, 1991; Oades, 1993). As such, the effectiveness of various binding mechanisms will depend on their physical dimensions relative to those of the pores (i.e., planes of weakness) being bridged (Kay, 1990).

Although it is helpful to think of aggregate hierarchy in the sense of the binding together of increasingly larger aggregated units, in most cases, aggregates are probably not formed sequentially. In fact, evidence exists to suggest that although plant roots and fungal hyphae provide the mechanical framework for the formation of macroaggregates, it is the decomposition process that leads to the development of microaggregates and an aggregate hierarchy (Elliott and Coleman, 1988; Tiessen and Stewart, 1988; Oades and Waters, 1991; Beare et al., 1994b). Hence, microaggregates may form as a result of biological activity within or on the edges of relatively stable macroaggregates or may develop when macroaggregates turn over or fragment as the roots or hyphae binding them together are decomposed. Of course, in cultivated or other disturbed systems, a significant proportion of macroaggregates may be reformed from relatively intact microaggregates as a result of root and hyphal growth. All levels of aggregate development, however, may occur simultaneously as each aggregate size class forms and turns over at its own rate, depending on management practices and on the degree of protection from decomposition afforded the organic binding agents of each size class. Oades (1993) speculated that aggregate hierarchy occurs in soils as a legacy of long-term exploration by roots, particularly those of grasses, and that it is not likely to occur in very young soils (e.g., polders) or to apply in soils where inorganic cements predominate (e.g., oxisols).

III. Organomineral Associations

Organomineral associations can occur at a variety of spatial scales with varying degrees of stability against physical, chemical, or biological disruption or degradation. In his extensive review of the subject, Christensen (1996) divided organomineral complexes into primary and secondary associations. He defined primary organomineral complexes as those related to the primary structure of soils and associated with primary particles isolated after complete dispersion of the soil. Secondary organomineral complexes were defined as consisting of aggregates of primary organomineral complexes that form the secondary structure of soils. We will consider here secondary organomineral associations within macroaggregates (but external to microaggregates) and those within microaggregates 2-250 μm in diameter and also how these organomineral associations promote the stabilization of both soil aggregates and SOM.

A. Macroaggregates

Because of the porosity exclusion principle, the stabilization of hierarchically formed macroaggregates requires organomineral associations large enough to bridge the gaps and pores between the microaggregates and primary particles composing macroaggregates. Larger macroaggregates can also be composed of smaller macroaggregates, making the size of the gaps to be stabilized even larger. Organomineral associations at this scale are largely related, either directly or indirectly, to the growth and decomposition of plant roots and the hyphae of mycorrhizal fungi (Tisdall, 1996; Jastrow et al., 1997).

The lengths of roots and mycorrhizal hyphae are often directly related to the percentage of soil in water-stable macroaggregates, particularly in aggrading systems (Tisdall and Oades, 1980; Miller and Jastrow, 1992a). The direct effects of living roots and hyphae (Figure 2) may be conceptualized by viewing the three-dimensional network of roots and hyphae as a "sticky string bag" that physically entangles or enmeshes smaller aggregates and particles, creating rather stable macroaggregates (Oades and Waters, 1991). Not only do roots and hyphae form a network that can serve as a framework for macroaggregate formation, but extracellular mucilage coatings on root and hyphal surfaces can strongly sorb to inorganic materials, helping to stabilize aggregates (Tisdall and Oades, 1979; Gupta and Germida, 1988; Tisdall, 1991; Dorioz et al., 1993). Furthermore, encrustation of roots and hyphae with inorganics is believed to physically slow decomposition, thereby preserving the enmeshing framework for a time even after the roots and hyphae senesce (Oades and Waters, 1991). The pressures exerted by growing roots and by localized drying caused by plant water uptake are physical forces that promote both the formation and the degradation of aggregates (Kay, 1990). Hence, the types of roots produced by different plants and their densities and architectures can influence macroaggregate size distributions (Miller and Jastrow, 1990).

The evidence for the role of extracellular mucilage coatings (the "sticky" part of the string bag) is based largely on microscopy and selective staining of ultrathin sections (e.g., Tisdall and Oades, 1979; Gupta and Germida, 1988; Oades and Waters, 1991; Foster, 1981, 1994). These mucilages are generally believed to consist mainly of polysaccharides but may also contain polyuronic acids and amino compounds (Foster, 1981; Tisdall, 1996). However, recent studies suggest that arbuscular mycorrhizal fungi produce a previously undescribed class of aggregate binding agents (R.M. Miller, S.F. Wright, J.D. Jastrow, and A. Upadhyaya, unpublished data). As identified by using immunofluorescent monoclonal antibody techniques, arbuscular mycorrhizal hyphae produce a hydrophobic glycoproteinaceous substance, named glomalin, on the surfaces of both internal and external hyphae (Wright et al., 1996). Glomalin is also deposited on soil surfaces, where its resistance to extraction suggests it might be relatively persistent (Wright and Upadhyaya, 1996). The amounts of glomalin extracted from soils of a prairie restoration chronosequence were directly related to both external hyphal lengths and the mean weight diameter of slaked, water-stable aggregates (R.M. Miller, S.F. Wright, J.D. Jastrow, and A. Upadhyaya, unpublished data). Glomalin may polymerize and develop its hydrophobicity in response to drying and air exposure (Wessels, 1996), thereby protecting the hyphae from desiccation. However, this polymerization process may extend to and incorporate contacts with other soil components, essentially adhering fungal walls to soil and particulate organic matter (POM) surfaces. In addition to the glue-like properties of glomalin, its hydrophobicity may contribute indirectly to aggregate stability by dampening the disruptive forces of rapid water movement into the pores between and within aggregates.

Most of the indirect effects of roots and hyphae are related to decomposition processes. The exudates and rhizodeposits of growing roots are rapidly turned over by active microbial populations that are dominated by gram-negative bacteria (Foster, 1983), resulting in the deposition of extracellular microbial polysaccharides in the rhizosphere. Furthermore, as roots and hyphae die, organic substrate is deposited throughout the macroaggregate structure of the soil, where the activities of soil fauna, saprophytic fungi, and bacteria reduce this material to POM, resulting in further deposition of microbial polysaccharides. In addition, colonization of POM or fecal pellets by saprophytic fungi and the consequent proliferation of hyphae (Figure 2) contribute to stabilizing the pores between microaggregates and other particles (Haynes and Beare, 1996). The coating of aggregate surfaces and intramacroaggregate pores with microbially derived polysaccharides and other transformation products, such as aliphatic compounds, helps to stabilize macroaggregates (Haynes and Swift, 1990; Martens and Frankenberger, 1992; Haynes and Francis, 1993). However, because of the lability of most of these materials as a substrate for further microbial attack, the persistence of this mechanism depends on the balance between production and degradation and the extent of chemical or physical stabilization (Haynes and Beare, 1996).

The OM binding microaggregates and other particles together into macroaggregates is believed to be physically protected from decomposition. Evidence for the protection of relatively labile, intramacroaggregate OM and its function as an intermicroaggregate binding agent comes from observations of the loss of both aggregate stability and OM following the cultivation of undisturbed soils (e.g., Tisdall and Oades, 1980; Carter, 1992; Cambardella and Elliott, 1993) and from laboratory mineralization studies of intact and crushed macroaggregates (Elliott, 1986; Gupta and Germida, 1988; Beare et al., 1994a).

Cambardella and Elliott (1992, 1993, 1994) suggested that much of the SOM lost when long-term grasslands are cultivated is POM or a relatively labile, silt-sized organomineral fraction (2-20 μm) with a density of 2.07-2.22 g cm^{-3}, which together may account for about 40% of the total SOC. Because large shares of both of these organic pools were found within small macroaggregates (250-2000 μm), these authors proposed that both pools serve as intermicroaggregate binding agents. Furthermore, they suggested that together the two SOM fractions account for a significant portion of the conceptually defined intermediate (or slow) pool of current SOM models (Parton et al., 1996).

Theoretically, aggregate hierarchy predicts that the organic C (OC) content of macroaggregates should be greater than that of microaggregates because the former are composed of the latter plus organic intermicroaggregate binding agents. In summarizing a range of studies, Angers and Carter (1996) concluded, however, that the associations between aggregate size and C contents are not clear or consistent. Direct, inverse, and nonexistent relationships have been documented. The results apparently depend both on operational differences between studies (e.g., aggregate separation procedures, whether POM is removed, or whether concentrations are corrected for sand content) and on actual differences between systems (e.g., vegetation and soil types, management practices, and climatic conditions).

Hypothetically, the OM in macroaggregates also should be younger and more labile. In most studies, the relationships between the C:N ratio of aggregate OM and aggregate size support this contention, as do most of the laboratory mineralization studies (Angers and Carter, 1996). Recent studies using the natural abundance of stable C isotopes following a switch in the photosynthetic pathway of the vegetation as a tracer of recent organic inputs demonstrate that macroaggregates have a higher proportion of C derived from the most recent vegetation than do microaggregates (Skjemstad et al., 1990; Puget et al., 1995; Angers and Carter, 1996). Similarly, Jastrow et al. (1996) observed that the proportion of recent C in large macroaggregates (>1 mm) was greater than in microaggregates, but unlike the other studies, they also found the proportion of recent C in microaggregates (53-212 μm) was essentially equal to the proportions in small macroaggregates (up to 1 mm). The differing results may be due to management differences; Jastrow et al. (1996) studied an aggrading temperate pasture, whereas Puget et al. (1995) and Angers and Carter (1996) worked in temperate cultivated systems. Skjemstad et al. (1990) also studied a pasture (although under wet, subtropical conditions), but they separated and analyzed all macroaggregates as one size class.

B. Microaggregates

As noted earlier, the decomposition process is probably the driving force that leads to the development of microaggregates and an aggregate hierarchy. The stability of microaggregates depends mostly on the strength by which clays and other inorganic components of the soil are sorbed to POM, microbial debris, and a variety of other organic colloids and compounds primarily of microbial origin. Sorption occurs via a variety of organomineral associations, such as polyvalent cation bridges, H-bonding, van der Waals forces, and interactions with hydrous oxides and aluminosilicates (Edwards and Bremner, 1967; Oades, 1984; Haynes and Beare, 1996; Tisdall, 1996). Clays are generally the mobile component, with much of their movement and reorientation around organics resulting from localized drying and the mechanical actions of plant and fungal growth. Drying also appears to play an important role in increasing the stability of organomineral associations by enhancing bonding forces and the polymerization and denaturing of the organics (Oades, 1984; Chenu, 1989; Dormaar and Foster, 1991; Dorioz et al., 1993).

A hierarchy of microaggregates, based on the scale of the organic nucleating agents, develops in many soils (Figure 2). The C:N ratios of both the light (<2 g cm^{-3}) and mineral-associated fractions of microaggregates from a mollisol generally decreased with aggregate size, indicating increasing levels of decomposition (Baldock et al., 1992). Particulate OM (mostly plant and fungal debris) is often found at the core of microaggregates 90-250 μm in diameter, where it is protected from rapid decomposition by encrustation with inorganic material (Oades and Waters, 1991; Waters and Oades, 1991). This POM is eventually decomposed, often leaving cavities surrounded by smaller microaggregates believed to be stabilized by the metabolic products and bodies of microbes that used the POM as a substrate. Plant fragments are more difficult to recognize in microaggregates less than 90 μm in diameter. The cores of microaggregates 20-90 μm in diameter may include the resistant remnants of plant debris, such as lignin particles, or may be empty. Microaggregates less than 20 μm in diameter appear to consist of clay microstructures stabilized by microbially derived polymers, hyphal fragments, and bacterial cells or colonies encrusted with clay particles. Baldock et al. (1992) found that both the light and mineral-associated fractions of aggregates 2-20 μm in diameter had C:N ratios (11.5 and 8.8, respectively) within the range measured for soil microbial biomass (8-12).

Golchin et al. (1994b) proposed a conceptual model of the changes OM undergoes during decomposition, from its entry into the soil through its incorporation into microaggregates to its eventual rendering into microbial metabolites and association with clay minerals (Figure 3). These authors proposed that when plant (surface and root) debris enters the soil, it is initially colonized by soil microbes and begins to adsorb mineral particles. As plant fragments become encrusted with mineral particles, they become the organic cores of stable microaggregates and are protected from rapid decomposition. While these organic cores are still rich in carbohydrates and chemically attractive to microorganisms, microbially produced mucilages and metabolites permeate the encrusting mineral particles, resulting in very stable aggregates. However, once the more labile portions of the organic cores are consumed, decomposition of more resistant plant structural materials proceeds more slowly. Eventually, decomposition of the organic cores renders the aggregates unstable, and relatively recalcitrant POM is released from intimate association with mineral particles, because the production of stabilizing microbial products declines as the organic cores become less palatable. Mineral particles and organomineral associations that coated the cores along with adsorbed microbial products may then become associated with more labile POM.

This model was developed from studies employing ultrasonic dispersion and sequential density fractionation techniques to isolate (1) free POM located between aggregates, (2) occluded POM from within aggregates, and (3) colloidal or clay-associated (amorphous) OM (Golchin et al., 1994a,b). The chemical composition of the isolated fractions was characterized by solid-state ^{13}C nuclear magnetic resonance spectroscopy, and the degree of mineral association of each fraction was determined on the basis of density, C and N contents, and scanning electron microscopy. Additional support for the

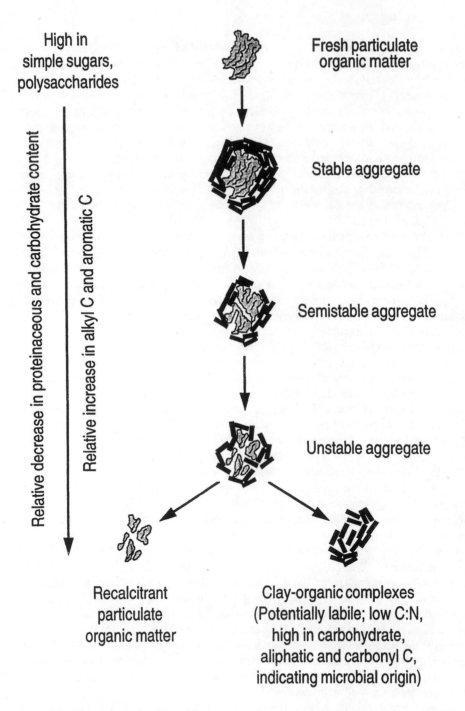

Figure 3. Conceptual model of microaggregate turnover as proposed by Golchin et al., 1994b.

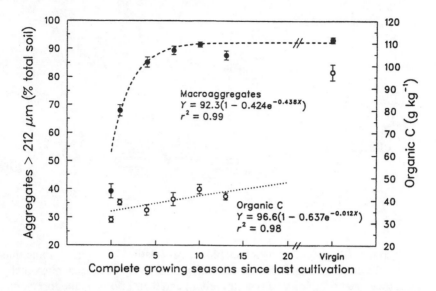

Figure 4. Changes in percentage of macroaggregates and accumulation of total soil organic carbon with time since last cultivation and planting to prairie. Error bars indicate standard errors ($n = 10$). (Reprinted with permission from Jastrow, 1996.)

model was obtained from subsequent studies using the natural abundance of stable C isotopes as a tracer of OM cycling through fractions isolated from soils sampled 35 and 83 years after a switch in the photosynthetic pathway of the vegetation (Golchin et al., 1995).

IV. Interrelationships between Aggregation and Organic Carbon Sequestration: A Case Study

Our studies of a chronosequence of prairie restorations on mollisols that were cultivated for over 100 years can be used as a case study to examine the interrelationships of aggregation and OC sequestration in an aggrading system. Because the restoration chronosequence provides gradients of water-stable macroaggregate formation and total SOC (Figure 4), along with gradients of root and hyphal proliferation and of various other SOC pools (Cook et al., 1988; Miller and Jastrow, 1992b; Jastrow, 1996; Jastrow et al., 1997), the site allows us not only to investigate how different organic binding mechanisms interact to promote the development of soil aggregate hierarchy, but also to determine how and where OC is sequestered within this hierarchy. Similarly, we believe that the site provides evidence for the existence of feedbacks between the formation and stabilization of soil aggregates and SOC sequestration.

A. Organomineral Associations and Development of Soil Aggregate Hierarchy

The relative contributions of roots, mycorrhizal hyphae, SOM, microbial biomass, and microbially derived polysaccharides to the direct and indirect mechanisms of water-stable macroaggregate formation were investigated by using path analysis techniques (Jastrow et al., 1997). Path analysis enables the heuristic examination of causal processes underlying observed relationships. After constructing a conceptual model of the interrelationships among multiple independent and dependent

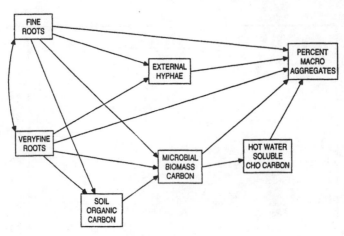

Figure 5. Conceptual path model of hypothesized causal relationships among roots, the external hyphae of mycorrhizal fungi, soil organic carbon, microbial biomass, and microbially derived polysaccharides (hot-water soluble carbohydrate carbon) and their effects on the formation of water-stable macroaggregates. Causal relationships are indicated by single-headed arrows, and existing but unanalyzed correlations are indicated by double-headed arrows. (Adapted from Jastrow et al., 1997.)

variables (Figure 5), the total correlations between independent and dependent variables can be partitioned into causal (direct and indirect effects) and noncausal components (Asher, 1983).

In the restoration chronosequence, external mycorrhizal hyphae had the strongest direct effect on the percentage of water-stable macroaggregates (Table 1). Roots exhibited the next strongest direct effects on aggregation, with essentially equal contributions for each of two measured diameter size classes. However, the indirect effects of the two size classes differed substantially, with fine roots (0.2-1 mm in diameter) having the overall largest total effect of all binding agents evaluated. The indirect effects of fine roots on macroaggregates was due primarily to strong positive associations of fine roots with mycorrhizal hyphae and microbial biomass. Interestingly, the strongest indirect contribution of veryfine roots (<0.2 mm in diameter) was via SOC, probably as the major source of accruing POM. Hence, the path analysis approach strongly supported the importance of the various contributions of roots and mycorrhizal hyphae to the stabilization of macroaggregates in aggrading systems. The roles of microbially derived extracellular polysaccharides (as measured by hot-water soluble carbohydrate C) and of POM (best represented by the indirect effects of roots through SOC) in stabilizing macroaggregates were overwhelmed by the contributions of roots and hyphae via the "sticky string bag" mechanism in this system.

The results of further separate path analyses for each of three macroaggregate size fractions (Jastrow et al., 1997) support the role of the porosity exclusion principle as a factor controlling the types of organic binding agents that function at different spatial scales within aggregates (Tisdall and Oades, 1982; Oades and Waters, 1991; Dorioz et al., 1993). In general, the relative importance of the binding mechanisms for each aggregate size fraction was related to the physical size of the mechanism and the strength of its associations with other mechanisms. For example, fine roots had their greatest effects on aggregates greater than 2 mm in diameter, whereas veryfine roots exerted their strongest effects on smaller aggregates. Also as expected, the strongest direct effect in the smallest macroaggregate size fraction was due to microbial biomass. However, in contrast to what their physical size would suggest, mycorrhizal hyphae were of greatest relative importance in the two largest aggregate size fractions, because of their strong associations with fine roots.

Table 1. Observed correlations between the percentage of water-stable macroaggregates (>212 μm) and measurements of selected organic binding agents and their partitioning into direct, indirect, and total causal effects on the basis of path analysis ($n = 49$); the difference between the correlation and the total effect is due to conceptually noncausal relationships between variables (see text)

Measured parameter	Correlation (r)	Direct effect	Indirect effect	Total effect
Fine root length	0.91	0.25	0.47	0.72
Veryfine root length	0.85	0.26	-0.04	0.22
External hyphal length	0.89	0.38	0	0.38
Soil organic C	0.43	0	0.09	0.09
Microbial biomass C	0.65	0.14	0.03	0.17
Hot-water soluble carbohydrate C	0.55	0.05	0	0.05

(Data from Jastrow et al., 1997.)

B. Aggregate-Associated Carbon Sequestration

Because of the high densities of roots and hyphae occurring in restored prairie (Cook et al., 1988; Miller and Jastrow, 1992b), the formation of macroaggregates stable enough to withstand slaking and wet sieving is rapid. At the same time, OC is accumulating across the chronosequence at a much slower rate (Figure 4). In fact, the rate constant (k) of the exponential model describing changes in aggregation (0.438 yr^{-1}) is more than 35 times the k for total SOC (0.012 yr^{-1}). Consequently, the calculated average turnover time ($1/k$) for macroaggregates is 2.3 yr compared with 83 yr for total SOC. Given that the longevity of a majority of prairie grass roots is about three growing seasons (Weaver and Zink, 1946), the rapid turnover time for macroaggregates is not surprising.

The relatively rapid turnover time for macroaggregates suggests that physical protection of intermicroaggregate binding agents (such as POM) inside macroaggregates may not be a major mechanism influencing C sequestration in this system. However, the lack of major seasonal changes in aggregation for older restorations (J.D. Jastrow and R.M. Miller, unpublished data) indicates that new macroaggregates are probably being formed just as rapidly as old macroaggregates are being degraded. Hence, the majority of intermicroaggregate binding agents are not likely to be unprotected for lengthy periods of time.

An increase in the C:N ratios of macroaggregates with time since planting to prairie also suggested that the accumulating OM is not highly processed and could include relatively large quantities of POM trapped inside macroaggregates (Jastrow, 1996). However, the C content of the light-fraction POM (<1.85 g cm^{-3}) released by relatively low ultrasonic energy from within macroaggregates was a relatively constant proportion of the total OC in whole soils across the chronosequence. Rather, C accrual with time since planting to prairie occurred in the heavy fraction (>1.85 g cm^{-3}), suggesting that C was accumulating as organic cores of undispersed microaggregates located within macroaggregates and/or as microbially produced mineral-associated complexes (Jastrow, 1996). These findings are consistent with the conceptual model of OM cycling through microaggregates proposed by Golchin et al. (1994b).

Other studies from the same site also support this conceptual model. The inputs and turnover of OC in different size classes of water-stable aggregates were estimated by using the natural abundance of stable C isotopes following a switch in vegetation from long-term corn (C4) on soils formed under tallgrass prairie (dominated by C4 grasses) to C3 pasture grasses (Jastrow et al., 1996). The average turnover time for old (C4-derived) C was 412 yr for microaggregates, compared with an average turnover time for macroaggregates of 140 yr, indicating that old C associated with microaggregates may be more biochemically recalcitrant and better physically protected than that associated with macroaggregates. Net inputs of new (C3-derived) C increased with aggregate size (0.73 to

1.13 g kg^{-1} yr^{-1}). Together, these data support the concept of an aggregate hierarchy. The faster turnover times for old C and greater inputs of new C in macroaggregates suggest that they are composed of a mixture of microaggregates (with much longer turnover times) and of other C pools (such as intermicroaggregate binding agents) that are both turning over and accumulating more rapidly than the C associated with microaggregates. However, net inputs to microaggregates were equal to those for small macroaggregates (up to 1 mm in diameter), suggesting (1) that the formation and degradation of microaggregates may be more dynamic than predicted by their resistance to mechanical dispersion or by the turnover times for old C and (2) that at least two different OC pools characterized by differing turnover rates exist within microaggregates. Golchin et al. (1995) also hypothesized, on the basis of changes in the natural abundance of stable C isotopes, that microaggregates contain two C pools with differing turnover rates.

For soils *in situ*, microaggregates whose stability is weakened by the lack of microbial activity may be more susceptible to degradation caused by chemical dispersion or by physical stresses. Thus, as microbial activity slows, clays associated with microaggregate surfaces may be released and dispersed, enabling them to become reoriented around and sorbed to the mucilages and metabolites on the surfaces of new (in our case, C3-derived) chemically attractive POM supporting an active microbial population. At the same time, the slow turnovers observed for old (C4-derived) C occur because (1) additional time is needed for all of the C4-derived POM contained within microaggregates to degrade and (2) significant quantities of C4-derived microbial metabolites likely remain strongly associated with mineral particles encrusting new (C3-derived) microaggregate cores.

Alternatively, a significant proportion of the C3-derived C associated with microaggregates may be located on aggregate surfaces. In soils with very stable macroaggregates, such as those in this study, the few microaggregates released by slaking were probably bound into macroaggregates *in situ* but were not stabilized sufficiently to survive the slaking process. Hence, they could be coated with intermicroaggregate binding agents of relatively new origin. This scenario is consistent with the observed deposition of glomalin by mycorrhizal hyphae on aggregate surfaces (Wright and Upadhyaya, 1996; R.M. Miller, S.F. Wright, J.D. Jastrow, and A. Upadhyaya, unpublished data), as discussed in Section IIIA. The scenario may also be related to the relatively labile, silt-sized organomineral fraction (density = 2.07-2.22 g cm^{-3}) isolated by Cambardella and Elliott (1994) from within macroaggregates, which they hypothesized is of microbial origin and functions as an intermicroaggregate binding agent.

C. Demonstration of Feedbacks

Thus, we have demonstrated for our system that the rapid formation of water-stable macroaggregates is followed by the sequestration and accrual of SOC, and we have examined how and where the C is accumulating. However, because of the rapid incorporation of over 90% of the soil into macroaggregates able to survive slaking and wet sieving (Figure 4), the demonstration that C sequestration feeds back to enhance aggregate stability requires a method capable of assessing greater levels of stability. Hence, we combined wet sieving with turbidimetric techniques in an approach to aggregate stability assessment that differs somewhat from the combined method developed by Pojasok and Kay (1990). For aggregates that have survived slaking and wet sieving, their stabilities should be related to the amounts of clays dispersed by the higher energy and abrasive forces associated with end-over-end rotation (compared with wet sieving).

Size fractions of water-stable macro- and microaggregates were collected by slaking of air-dry soil followed by wet sieving as described by Jastrow et al. (1996). After sieving, all size fractions were air-dried. For each size fraction, 0.25-g subsamples of intact aggregates were weighed into spectrophotometer tubes, 7 ml of deionized water were added, and the tubes were covered with Parafilm. After 5 min in water, the tubes were spun end-over-end about their centers (30 rpm) for

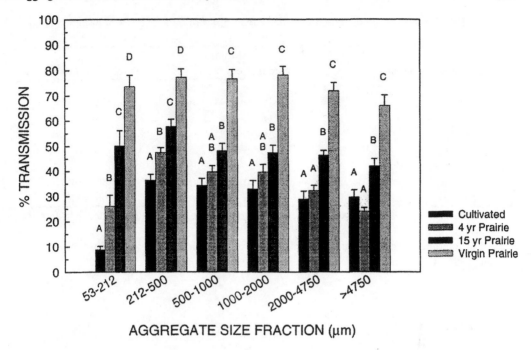

Figure 6. Results of the combination of wet sieving and turbidimetric techniques as an approach to assessing the relative stability of different size classes of aggregates in a chronosequence of restored tallgrass prairie on long-term cultivated soils ($n = 5$). Within each size fraction, bars indicated by the same letter are not significantly different on the basis of Fisher's protected least significant difference ($P \leq 0.05$).

2 min and allowed to settle undisturbed for 30 min. The percent light transmission was measured at a wavelength of 630 nm on a Spectronic 20 spectrophotometer. Two subsamples were measured and averaged for each size fraction from each of five randomly collected cores for each sampled plot.

For each size fraction, the increases in percent transmission observed across the chronosequence indicate relatively lower amounts of dispersed clays and greater aggregate stability with time since disturbance (Figure 6). Improvements in aggregate stability were also associated with increasing SOC ($r \geq 0.83$, $P \leq 0.0001$ for each size fraction). Interestingly, macroaggregates from the cultivated field were significantly more stable than microaggregates, but substantial increases in the stability of microaggregates and small macroaggregates (212-500 μm) occurred only 4 yr after planting to prairie. For larger aggregates (>500 μm), significant improvement in stability was not observed until 15 yr of prairie.

These findings, coupled with those presented in Figure 4, suggest that relatively stable macroaggregates can be formed rapidly in response to the proliferation of roots and hyphae associated with grassland vegetation but that improvements in stability resulting from concomitant increases in SOC appear to occur initially at smaller spatial scales. Presumably this occurs through the deposition of microbial byproducts and the sorption of clays to OM during decomposition.

Table 2. Estimated ranges in the amounts and turnover times of various types of organic matter stored in agricultural soils

Type of organic matter	Proportion of total organic matter (%)	Turnover time (yr)
Litter	—	1-3
Microbial biomass	2-5	0.1-0.4
Particulate	18-40	5-20
Light fraction	10-30	1-15
Intermicroaggregate [a]	20-35	5-50
Intramicroaggregate [b]		
Physically sequestered	20-40	50-1000
Chemically sequestered	20-40	1000-3000

[a] Within macroaggregates but external to microaggregates, including particulate, light fraction, and microbial C.
[b] Within microaggregates, including sequestered light fraction and microbially derived C. (From Carter, 1996, with permission.)

V. Conclusions

Soil aggregation and OM storage are intimately associated with each other. Consequently, changes in either of these processes often result in feedbacks on the other. These feedbacks are mediated through organomineral associations, which function as aggregate binding and stabilizing agents. The nature of various organomineral associations and their spatial locations within soil aggregate structure determine the extent to which SOC is physically protected and chemically stabilized, resulting in organic pools with varying input and turnover rates (Table 2). Similarly, when OM is the major stabilizing agent and aggregate structure is hierarchical, different types of aggregates are being formed and turned over at different rates related to the turnover rates of their organomineral binding agents.

Thus, better information on the nature and dynamics of organomineral associations will lead to a greater understanding of soil structural dynamics and of C cycling and sequestration in soils. Consequently, such information will also contribute to improved approaches to soil management. Similarly, a better understanding of organomineral associations may provide a key to better defining or quantifying the conceptual pools used by SOM simulation models and could serve as the basis for development of a new generation of such models.

Acknowledgments

This work was supported by the U.S. Department of Energy, Office of Energy Research, Office of Health and Environmental Research, Environmental Sciences Division, Global Change Research, under contract W-31-109-Eng-38.

References

Angers, D.A. and M.R. Carter. 1996. Aggregation and organic matter storage in cool, humid agricultural soils. p. 193-211. In: M.R. Carter and B.A. Stewart (eds.), *Structure and Organic Matter Storage in Agricultural Soils*. CRC Press, Inc., Boca Raton, FL.

Asher H.B. 1983. Causal Modeling. Sage University Paper Series on Quantitative Applications in the Social Sciences, 07-003. Sage Publications, Beverly Hills, CA, 96 pp.

Baldock, J.A., J.M. Oades, A.G. Waters, X. Peng, A.M. Vassallo, and M.A. Wilson. 1992. Aspects of the chemical structure of soil organic materials as revealed by solid-state ^{13}C NMR spectroscopy. *Biogeochem.* 15:1-42.

Beare, M.H., M.L. Cabrera, P.F. Hendrix, and D.C. Coleman. 1994a. Aggregate-protected and unprotected organic matter pools in conventional and no-tillage soils. *Soil Sci. Soc. Am. J.* 58:787-795.

Beare, M.H., P.F. Hendrix, and D.C. Coleman. 1994b. Water-stable aggregates and organic matter fractions in conventional- and no-tillage soils. *Soil Sci. Soc. Am. J.* 58:777-786.

Cambardella, C.A. and E.T. Elliott. 1992. Particulate soil organic-matter changes across a grassland cultivation sequence. *Soil Sci. Soc. Am. J.* 56:777-783.

Cambardella, C.A. and E.T. Elliott. 1993. Carbon and nitrogen distribution in aggregates from cultivated and native grassland soils. *Soil Sci. Soc. Am. J.* 57:1071-1076.

Cambardella, C.A. and E.T. Elliott. 1994. Carbon and nitrogen dynamics of soil organic matter fractions from cultivated grassland soils. *Soil Sci. Soc. Am. J.* 58:123-130.

Carter, M.R. 1992. Influence of reduced tillage systems on organic matter, microbial biomass, macro-aggregate distribution and structural stability of the surface soil in a humid climate. *Soil Tillage Res.* 23:361-372.

Carter, M.R. 1996. Analysis of soil organic matter storage in agroecosystems. p. 3-11. In: M.R. Carter and B.A. Stewart (eds.), *Structure and Organic Matter Storage in Agricultural Soils*. CRC Press, Inc., Boca Raton, FL.

Chenu, C. 1989. Influence of a fungal polysaccharide, scleroglucan, on clay microstructures. *Soil Biol. Biochem.* 21:299-305.

Christensen, B.T. 1996. Carbon in primary and secondary organomineral complexes. p. 97-165. In: M.R. Carter and B.A. Stewart (eds.), *Structure and Organic Matter Storage in Agricultural Soils*. CRC Press, Inc., Boca Raton, FL.

Cook, B.D., J.D. Jastrow, and R.M. Miller. 1988. Root and mycorrhizal endophyte development in a chronosequence of restored tallgrass prairie. *New Phytol.* 110:355-362.

Dorioz, J.M., M. Robert, and C. Chenu. 1993. The role of roots, fungi and bacteria on clay particle organization. An experimental approach. *Geoderma* 56:179-194.

Dormaar, J.F. and R.C. Foster. 1991. Nascent aggregates in the rhizosphere of perennial ryegrass (*Lolium perenne* L.). *Can. J. Soil Sci.* 71:465-474.

Dexter, A.R. 1988. Advances in characterization of soil structure. *Soil Tillage Res.* 11:199-238.

Edwards, A.P. and J.M. Bremner. 1967. Microaggregates in soils. *J. Soil Sci.* 18:64-73.

Elliott, E.T. 1986. Aggregate structure and carbon, nitrogen, and phosphorus in native and cultivated soils. *Soil Sci. Soc. Am. J.* 50:627-633.

Elliott, E.T. and D.C. Coleman. 1988. Let the soil work for us. *Ecol. Bullet.* 39:23-32.

Foster, R.C. 1981. Polysaccharides in soil fabrics. *Science* 214:665-667.

Foster, R.C. 1983. The plant root environment. p. 673-684. In: *Soils: An Australian Viewpoint*. CSIRO, Melbourne/Academic Press, London.

Foster, R.C. 1985. *In situ* localization of organic matter in soils. *Quaest. Ent.* 21:609-633.

Foster, R.C. 1994. Microorganisms and soil aggregates. p. 144-155. In: C.E. Pankhurst, B.M. Doube, V.V.S.R. Gupta, and P.R. Grace (eds.), *Soil Biota: Management in Sustainable Farming Systems*. CSIRO Information Services, East Melbourne, Australia.

Golchin, A., J.M. Oades, J.O. Skjemstad, and P. Clark. 1994a. Study of free and occluded particulate organic matter in soils by solid state ^{13}C CP/MAS NMR spectroscopy and scanning electron microscope. *Aust. J. Soil Res.* 32:285-309.

Golchin, A., J.M. Oades, J.O. Skjemstad, and P. Clark. 1994b. Soil structure and carbon cycling. *Aust. J. Soil Res.* 32:1043-1068.

Golchin, A., J.M. Oades, J.O. Skjemstad, and P. Clark. 1995. Structural and dynamic properties of soil organic matter as reflected by ^{13}C natural abundance, pyrolysis mass spectrometry and solid-state ^{13}C NMR spectroscopy in density fractions of an oxisol under forest and pasture. *Aust. J. Soil Res.* 33:59-76.

Gregorich, E.G., R.G. Kachanoski, and R.P. Voroney. 1989. Carbon mineralization in soil size fractions after various amounts of aggregate disruption. *J. Soil Sci.* 40:649-659.

Gupta, V.V.S.R. and J.J. Germida. 1988. Distribution of microbial biomass and its activity in different soil aggregate size classes as affected by cultivation. *Soil Biol. Biochem.* 20:777-786.

Haynes, R.J. and M.H. Beare. 1996. Aggregation and organic matter storage in meso-thermal, humid soils. p. 213-262. In: M.R. Carter and B.A. Stewart (eds.), *Structure and Organic Matter Storage in Agricultural Soils*. CRC Press, Inc., Boca Raton, FL.

Haynes, R.J. and G.S. Francis. 1993. Changes in microbial biomass C, soil carbohydrate composition and aggregate stability induced by growth of selected crop and forage species under field conditions. *J. Soil Sci.* 44:665-675.

Haynes, R.J. and R.S. Swift. 1990. Stability of soil aggregates in relation to organic constituents and soil water content. *J. Soil Sci.* 41:73-83.

Jastrow, J.D. 1996. Soil aggregate formation and the accrual of particulate and mineral-associated organic matter. *Soil Biol. Biochem.* 28:665-676.

Jastrow, J.D., T.W. Boutton, and R.M. Miller. 1996. Carbon dynamics of aggregate-associated organic matter estimated by carbon-13 natural abundance. *Soil Sci. Soc. Am. J.* 60:801-807.

Jastrow, J.D., R.M. Miller, and J. Lussenhop. 1997. Interactions of biological mechanisms contributing to soil aggregate stabilization in restored prairie. *Soil Biol. Biochem.* (accepted).

Jenny, H. 1941. *Factors of Soil Formation. A System of Quantitative Pedology*. McGraw-Hill Book Co., Inc., New York, 281 pp.

Kay, B.D. 1990. Rates of change of soil structure under different cropping systems. *Adv. Soil Sci.* 12:1-52.

Kooistra, M.J. and M. van Noordwijk. 1996. Soil architecture and distribution of organic matter. p. 15-56. In: M.R. Carter and B.A. Stewart (eds.), *Structure and Organic Matter Storage in Agricultural Soils*. CRC Press, Inc., Boca Raton, FL.

Martens, D.A. and W.T. Frankenberger, Jr. 1992. Decomposition of bacterial polymers in soil and their influence on soil structure. *Biol. Fertil. Soils* 13:65-73.

Martin, J.P. and K. Haider. 1986. Influence of mineral colloids on turnover rates of soil organic carbon. p. 283-304. In P.M. Huang and M. Schnitzer (eds.), *Interactions of Soil Minerals with Natural Organics and Microbes*. Spec. Publ. 17, Soil Sci. Soc. Am., Madison, WI.

Miller, R.M. and J.D. Jastrow. 1990. Hierarchy of root and mycorrhizal fungal interactions with soil aggregation. *Soil Biol. Biochem.* 22:579-584.

Miller, R.M. and J.D. Jastrow. 1992a. The role of mycorrhizal fungi in soil conservation. p. 29-44. In: G.J. Bethlenfalvay and R.G. Linderman (eds.), *Mycorrhizae in Sustainable Agriculture*. Spec. Publ. 54. Am. Soc. Agron., Madison, WI.

Miller, R.M. and J.D. Jastrow. 1992b. Extraradical hyphal development of vesicular-arbuscular mycorrhizal fungi in a chronosequence of prairie restorations. p. 171-176. In: D.J. Read, D.H. Lewis, A.H. Fitter, and I.J. Alexander (eds.), *Mycorrhizas in Ecosystems*. C.A.B. International, Wallingford, Oxon, United Kingdom.

Oades, J.M. 1984. Soil organic matter and structural stability, mechanisms and implications for management. *Plant Soil* 76:319-337.

Oades, J.M. 1993. The role of biology in the formation, stabilization and degradation of soil structure. *Geoderma* 56:377-400.

Oades, J.M. and A.G. Waters. 1991. Aggregate hierarchy in soils. *Aust. J. Soil Res.* 29:815-828.

Parton, W.J., D.S. Ojima, and D.S. Schimel. 1996. Models to evaluate soil organic matter storage and dynamics. p. 421-448. In: M.R. Carter and B.A. Stewart (eds.), *Structure and Organic Matter Storage in Agricultural Soils.* CRC Press, Inc., Boca Raton, FL.

Pojasok, T. and B.D. Kay. 1990. Assessment of a combination of wet sieving and turbidimetry to characterize the structural stability of moist aggregates. *Can. J. Soil Sci.* 70:33-42.

Puget, P., C. Chenu, and J. Balesdent. 1995. Total and young organic matter distributions in aggregates of silty cultivated soils. *Eur. J. Soil Sci.* 46:449-459.

Skjemstad, J.O., R.P. Le Feuvre, and R.E. Prebble. 1990. Turnover of soil organic matter under pasture as determined by ^{13}C natural abundance. *Aust. J. Soil Res.* 28:267-276.

Stott, E., G. Kassin, W.M. Jarrell, J.P. Martin, and K. Haider. 1983. Stabilization and incorporation into biomass of specific plant carbons during biodegradation in soil. *Plant Soil* 70:15-26.

Tiessen, H. and J.W.B. Stewart. 1988. Light and electron microscopy of stained microaggregates: The role of organic matter and microbes in soil aggregation. *Biogeochem.* 5:312-322.

Tisdall, J.M. 1991. Fungal hyphae and structural stability of soil. *Aust. J. Soil Res.* 29:729-743.

Tisdall, J.M. 1996. Formation of soil aggregates and accumulation of soil organic matter. p. 57-96. In: M.R. Carter and B.A. Stewart (eds.), *Structure and Organic Matter Storage in Agricultural Soils.* CRC Press, Inc., Boca Raton, FL.

Tisdall, J.M. and J.M. Oades. 1979. Stabilization of soil aggregates by the root system of ryegrass. *Aust. J. Soil Res.* 17:429-441.

Tisdall, J.M. and J.M. Oades. 1980. The effect of crop rotation on aggregation in a red-brown earth. *Aust. J. Soil Res.* 18:423-433.

Tisdall, J.M. and J.M. Oades. 1982. Organic matter and water stable aggregates in soils. *J. Soil Sci.* 33:141-163.

van Veen, J.A. and P.J. Kuikman. 1990. Soil structural aspects of decomposition of organic matter by micro-organisms. *Biogeochem.* 11:213-233.

Waters, A.G. and J.M. Oades. 1991. Organic matter in water-stable aggregates. p. 163-174. In: W.S. Wilson (ed.), *Advances in Soil Organic Matter Research: The Impact on Agriculture and the Environment.* Spec. Publ. 90, Roy. Soc. Chem., Cambridge, United Kingdom.

Weaver, J.E. and E. Zink. 1946. Length of life of roots of ten species of perennial range and pasture grasses. *Plant Physiol.* 21:201-217.

Wessels, J.G.H. 1996. Fungal hydrophobins: Proteins that function at an interface. *Trends Plant Sci.* 1:9-15.

Wright, S.F., M. Franke-Snyder, J.B. Morton, and A. Upadhyaya. 1996. Time-course study and partial characterization of a protein on hyphae of arbuscular mycorrhizal fungi during active colonization of roots. *Plant Soil* 181:193-203.

Wright, S.F. and A. Upadhyaya. 1996. Extraction of an abundant and unusual protein from soil and comparison with hyphal protein of arbuscular mycorrhizal fungi. *Soil Sci.* 161:575-586.

CHAPTER 16

Impact of Variations in Granular Structures on Carbon Sequestration in Two Alberta Mollisols

Z. Chen, S. Pawluk, and N.G. Juma

I. Introduction

Soils are open, dynamic systems which occur as three-dimensional natural bodies on the landscape. The organic component of soil consists of living organisms, undecomposed and partially decomposed organic matter, and humified organic substances. Decomposition of organic matter is controlled by organisms, physical environment, and resource quality and nutrient availability (Anderson and Flanagan, 1989). These factors are incorporated in a number of conceptual and simulation models describing the dynamics of soil organic matter, however, one of the major limitations of these models is the lack of attention to the scale and the sites at which processes occur in soil (Brussaard and Juma, 1996). In order to correctly understand the complex issue of C sequestration in soil, it is necessary to comprehend the spatial scales of particles, aggregates, pores, and biota (Waters and Oades, 1991) and their relationship to the processes occurring in soil. Therefore, it is important to integrate micromorphological observations with process-oriented research.

Soil structural development has been extensively reviewed in the literature (Emerson, 1959; Tisdall and Oades, 1982; Pawluk and Bal, 1985; Oades, 1993). Factors affecting soil structure can be grouped as abiotic (clay minerals, sesquioxides, exchangeable cations), biological (soil organic matter, activities of plant roots, soil fauna, and soil microorganisms), and environmental (soil temperature and moisture). Specifically, polyvalent cations are required to bridge the negatively charged soil organic substances and clay minerals (Edwards and Bremner, 1967a; Duchaufour, 1982); differences in the origin and microbial-resistance of soil organic matter are at least partially responsible for the hierarchy in soil aggregation (Tisdall and Oades, 1982; Miller and Jastrow, 1990), and freeze/thaw processes could reorganize soil materials into granular or platy aggregates (Van Vliet-Lanoe et al., 1984; Coutard and Mücher, 1985).

Granular structures refer to soil aggregates that have a spherical arrangement. Development of stable granular aggregates usually requires intimate mixing between the mineral matrix and organic substances formed after decomposition and transformation of soil organic matter and results in C sequestration in soils. For soils in the temperate and cold regions, soil organic matter (Duchaufour, 1982) and soil fauna (Pawluk, 1985) have a dominant influence in soil aggregation processes. Among mineral soils, Mollisols are particularly effective in sequestration and stabilization of C through humification and aggregation processes.

A unique granular structure (shot structure), present as nearly perfect spherical discrete units in lower Ah and upper AB horizons, was reported for the Raven series, a Typic Cryaquoll developed under grass vegetation (Peters and Bowser, 1960; Chen, 1995). Similar but less morphologically developed structures were observed in Ah horizons of the Malmo series, an Aquic Cryoboroll in the grass-forest transition zone of central Alberta (Sanborn and Pawluk, 1983; Chen 1995). The objectives of this chapter were: (1) to describe the distribution of humified organic C, mineralogical properties, and distribution and stability of granular structure; and (2) to relate soil aggregation and micromorphological properties to C sequestration in the two Alberta Mollisols.

II. Materials and Methods

A. Site Description and Soil Characterization

Soil samples were collected from the Raven series, a Typic Cryaquoll (Soil Survey Staff, 1994) developed in a level-to-depressional area near Banalto (52°23' N, 114°42' W), and from a Malmo soil, an Aquic Cryoboroll at the Ellerslie Research Station of the University of Alberta (53°25' N, 113°33' W). The Raven and Malmo soils are also classified as Orthic Humic Gleysol and Gleyed Black Chernozem, respectively, according to the Canadian System of Soil Classification (Agriculture Canada Expert Committee on Soil Survey, 1987). The former soil (Table 1) is developed from lacustrine sediments under a grassland plant community and has a thick Ah horizon with a distinctive 'shot structure' in lower Ah and AB horizons. The Malmo soil is also developed from lacustrine sediments under open aspen woodland with a dense grass and forb understorey. It has a well-developed organic layer and a thick Ah horizon dominated by granular structure (Sanborn and Pawluk, 1983).

Routine soil analyses were conducted on air-dried samples after passing through a 2.0 mm sieve. Particle size distribution was determined by the pipette method (Sheldrick and Wang, 1993). Soil pH was measured in distilled water at 1:2 mass to volume ratio. The exchangeable cations were displaced with buffered 1N NH_4OAc (pH 7) and determined with a Perkin-Elmer 2380 atomic absorption spectrophotometer. Total CEC of soils was calculated from the measured amount of exchangeable NH_4^+ adsorbed as determined with an autoanalyzer. Soil subsamples were further ground (< 0.25 mm) for determination of soil organic C (Nelson and Sommers, 1982), and for various forms of Fe and Al extracted with sodium hydrosulfite (dithionite)-citrate-bicarbonate (NaDCB) (Mehra and Jackson, 1960), acid ammonium oxalate (0.11 M, pH 3.0), and sodium pyrophosphate (0.10 M, pH 10.6) (Ross and Wang, 1993).

B. Humic Acid Carbon

Humic acids (HAs) from sola of the two soils were extracted under N_2 using 0.10 M sodium pyrophosphate (pH 10.6). The soil samples were repeatedly extracted until the extracts reached a light brown color. The extracts were combined and acidified with 6N HCl to pH 1 to separate HAs and fulvic acids through high speed centrifugation. The HAs were treated with 0.5% (v/v) HF-HCl to remove silicate impurities, then dialyzed with 'Spectra/Por-3 molecularporous dialysis membrane' (3,500 molecular weight cut-off, Fisher Scientific) against deionized water to a chloride-free state. The samples were freeze-dried under vacuum, and ground into fine powder for HA weight and C content determination. The HA yield was calculated on an ash- and moisture-free basis. Detailed extraction, purification, and analytical procedures were reported elsewhere (Chen and Pawluk, 1995).

C. Clay Mineralogy

Clay (< 2 μm) was separated from soil samples of major horizons of the two soils by wet sieving (53 μm sieve) and gravity sedimentation (Jackson, 1969) after soil samples were ultrasonically dispersed

Table 1. Macromorphology of soils

Horizon	Depth (cm)	Color of soil matrix (moist)	Texture	Structure
Raven soil				
Sod	0-10	nd[a]	SiC[b]	Strong fine granular
Ah1	10-17	10 YR 2/2	SiC	Strong fine granular
Ah2	17-24	10 YR 2/2	SiC	Strong fine granular (shot structure)
AB	24-36	10 YR 3/2-3/3	SiC	Strong fine granular (shot structure)
Bg	36-52	10 YR 4/3-5/3	SiC	Moderate fine subangular to granular
Ckg	52+	2.5 Y 5/2	SiC	Massive
Malmo soil				
LF	18-13	7.5 YR 3/2	nd	Matted litter
FH	13-5	5 YR 3/3	nd	Fibrous-amorphous
H	5-0	5 YR 2/2	nd	Amorphous
Ah1	0-10	10 YR 2/1	SiC	Strong medium to fine granular
Ah2	10-29	10 YR 3/1-4/1	SiC	Strong medium to fine granular
Ah3	29-35	2.5Y 3/2	SiC	Stong fine granular, fine subangular blocky
ABgj	35-42	2.5 Y 5/2	SiC	Weak fine platy, fine granular
Bmgj	42-57	2.5 Y 4/2	SiC	Moderate fine subangular
Ckgj	103+	2.5 Y 4/2	HC[c]	Massive

[a]nd = not determined; [b]SiC = silty clay; [c]HC = heavy clay.

in distilled water with a Braun-sonic 1510 (B. Braun, Melsungen, U.S.A.) vibrator probe for 6 min (2 times for 3 min) at 380 watts (Edwards and Bremner, 1967b). The clay mineral identification was conducted by conventional X-ray diffraction (XRD) with Co-Kα radiation generated at 50 kV and 25 mA and focused with a curved LiF monochromator in a Philips PW1730 X-ray diffractometer. The scanning step was 0.053° 2θ every 2 seconds. The clay samples were separately saturated with Ca^{++} and K^+ and slides were prepared according to the paste method (Theisen and Harward, 1962). The K-saturated slides were scanned after heating to 105°C and run at 0% and 54% relative humidity (r.h.). They were further heated to 300°C and 550°C and run at 0% r.h. Ca-saturated slides were scanned at room temperature at 54% r.h. and then at 0% r.h. after ethylene glycol or glycerol solvation (Xing and Dudas, 1994). The scanned ranges were 3-36° 2θ for K-saturated samples heated to 105°C and for Ca-saturated samples that were equilibrated at 54% r.h., and 3-19° 2θ for other treatments.

D. Aggregate Analysis

Wet sieving of soil aggregates was performed in triplicates with field-moist and air-dried samples after passing through an 8 mm sieve. The aggregate samples were sieved in water with a set of five-sieve nests (4.0, 2.0, 1.0, 0.5, 0.25 mm) at an oscillating aptitude of 2.5 cm. The aggregates in the highest sieve remained submerged at the upper strike position. The speed was 60 rotations per min. The aggregates left on the sieve after 30 min were washed quantitatively with distilled water into pre-weighed drying tins. Percentage of dry weight of each aggregate group based on the total dry weight of soil was calculated. No correction for sands was made because microscopic observations showed a low content of sand particles for each aggregate group and the particle size analysis indicated low sand contents (<11%) throughout the two pedons. The mean weight diameter (MWD) for each horizon was determined according to Angers and Mehuys (1993). Statistical analyses for the aggregate data were conducted using SAS software (SAS Institute Inc., 1987).

E. Micromorphology

Undisturbed soil monoliths were collected with Kubiena boxes (Brewer, 1976), air-dried in the laboratory, and then impregnated with 3M Scotchcast electrical resins. Duplicates of 30 μm thick thin-sections were prepared with the aid of a Logitech PM2 polisher (Logitech Ltd., Dunbartonshire, Scotland). Thin-section descriptions followed the terminology defined by Bullock et al. (1985) for microstructure, Brewer (1976) for void, pedofeature and plasma fabric, and Brewer (1976) and Pawluk (1983) for related distribution patterns. General terminology for frequency, size, shape, and smoothness of micromorphological components also followed Bullock et al. (1985). Descriptions were made with the aid of a Zeiss microscope (Carl Zeiss Ltd., Germany). The energy dispersive X-ray analysis (EDX) for selected micromorphological features was conducted with a Cambridge Stereoscan 250 scanning electron microscope (SEM) (Cambridge Instrument Ltd., Cambridge, England) equipped with a Tracor Northern X-ray analyzer.

III. Results

A. Basic Soil Properties

The two soils have high clay and low sand contents. The upper sections are enriched with organic matter that decreases gradually with depth in the Raven soil but abruptly in the Malmo soil. Soil reaction is slightly acidic to neutral throughout the sola for both soils. Exchangeable Ca^{++} dominates the exchange complexes, followed by Mg^{++}, and minor amounts of K^+ and Na^+ (Table 2). The sodium pyrophosphate extractable Fe and Al compounds are higher in the solum, especially in the Ah2 and AB horizons, of the Raven soil than Malmo soil. These organic-bonded Fe and Al have lower concentrations in B and C horizons of both soils (Table 3). The acid-ammonium oxalate extractable Fe and Al are higher in the Sod to Ah2 horizons of the Raven soil and decease with depth. In the Malmo soil, this type of Fe compounds has lower concentrations from Ah2 to Bmgj horizons, and concentration of this type of Al compounds decreases with depth. Concentration of sodium hydrosulfite (dithionite)-citrate-bicarbonate (NaDCB) extractable Fe shows little variation in the Raven soil; its content is lower in the upper solum of the Malmo soil. The NaDCB extractable Al decreases with depth in the Raven soil, but no apparent variation is evident in the Malmo soil.

B. Distribution of Humic Acid Carbon

The HA yield decreases with depth in the Raven soil. In comparison, the HA yield increases with depth in the organic layers, then drops more abruptly with depth in the mineral horizons of the Malmo soil (Table 4). For the Raven soil, the humic acid C (HA-C)-to-soil organic C ratio is highest in the Ah horizons. About 25 to 30% of soil organic C was extracted in Ah horizons. For the Malmo soil, more organic C was extracted as HA-C with soil depth as a result of progressive humification from LF to H layers. Similar to the Raven soil, the highest ratio of HA-C to soil organic C is found in the Ah horizons and ranges from 40 to 50%. It decreases sharply below the Ah horizons. Thus the organic matter in both soils below Ah horizons becomes less extractable.

Table 2. Chemical and physical properties of soils

Horizon	Depth (cm)	Sand (%)	Clay (%)	Org. C (g/100g)	pH	CEC cmol/kg	Exchangeable cations (cmol/kg)			
							Ca^{++}	Mg^{++}	K^+	Na^+
Raven soil										
Sod	0-10	10	45	6.47	6.9	43.5	30.4	7.06	0.72	0.07
Ah1	10-17	8	43	4.57	6.5	41.7	25.1	6.30	0.57	0.09
Ah2	17-24	10	42	3.09	6.3	37.8	22.7	6.07	0.65	0.12
AB	24-36	9	45	1.39	6.5	32.7	21.5	6.61	0.63	0.11
Bg	36-52	10	44	0.56	7.7	27.8	25.2	7.06	0.58	0.10
Ckg	52+	1	53	0.44	8.2	27.3	29.9	7.31	0.48	0.11
Malmo soil[a]										
LF	18-13	nd[b]	nd	48.5	7.2	138.6	70.4	22.5	3.20	0.40
FH	13-5	nd	nd	39.4	6.9	140.7	74.6	21.7	2.60	0.40
H	5-0	nd	nd	41.1	6.6	159.9	76.2	20.0	1.20	0.10
Ah1	0-10	9	48	5.9	6.1	54.2	68.6	28.8	2.40	0.10
Ah2	10-29	10	46	2.1	6.6	40.6	61.3	36.2	1.70	0.20
Ah3	29-35	9	49	0.5	6.7	35.9	58.8	39.0	1.90	0.30
ABgj	35-42	11	42	0.3	6.9	24.1	60.6	34.9	1.70	0.40
Bmgj	42-57	7	53	0.5	7.3	39.1	60.1	32.5	1.50	0.30
Ckgj	103+	3	63	0.6	7.8	51.1	73.4	22.1	1.20	1.20

[a] Data adapted from Sanborn and Pawluk (1983) or Sanborn (1981) for Malmo soil; [b] nd = not determined.

Table 3. Extractable Fe and Al content (g/100g soil) in two soils

Horizon	Fe_p[a]	Fe_o[b]	Fe_d[c]	Al_p	Al_o	Al_d
Raven soil						
Sod	0.22	0.22	0.90	0.22	0.52	0.19
Ah1	0.29	0.24	0.89	0.32	0.52	0.17
Ah2	0.36	0.22	0.82	0.53	0.50	0.24
AB	0.38	0.18	0.90	0.60	0.46	0.08
Bg	0.12	0.14	0.84	0.12	0.25	0.10
Ckg	0.04	0.08	0.91	0.05	0.16	0.12
Malmo soil[d]						
Ah1	0.37	0.31	0.44	0.34	0.20	0.09
Ah2	0.10	0.21	0.46	0.11	0.17	0.09
Ah3	0.11	0.23	0.87	0.13	0.13	0.10
Abgj	0.13	0.26	0.90	0.14	0.12	0.09
Bmgj	0.09	0.27	1.14	0.10	0.13	0.10
Bckgj	0.06	0.32	1.07	0.08	0.10	0.06
Ckgj	0.07	0.32	1.12	0.10	0.09	0.06

[a] p = sodium pyrophosphate extraction; [b] o = acid ammonium oxalate extraction; [c] d = sodium hydrosulfite dithionite-citrate-bicarbonate extraction; and [d] data of Malmo soil were adopted from Sanborn (1981).

Table 4. Distribution of humic acid (HA) and humic acid carbon (HA-C) to organic carbon ratio in soils

Horizons	HA yield (g/100g soil)	C content of HA (%)	HA-C / Soil Org. C ratio
Raven soil			
Sod	2.61	54.5	0.22
Ah1	2.16	55.0	0.26
Ah2	1.75	55.3	0.31
AB	0.54	55.4	0.22
Bg	0.10	56.2	0.10
Malmo soil			
LF	3.61	53.7	0.04
FH	6.84	52.2	0.09
H	15.50	50.3	0.19
Ah1	4.22	54.3	0.39
Ah2	1.63	56.8	0.44
Ah3	0.40	56.5	0.45
Abgj	0.08	56.5	0.15
Bmgj	0.06	59.9	0.06

C. Mineralogy

The mineral composition of these two soils is quite similar in the major horizons. Sand and silt fractions are dominated by quartz that accounts for more than 90% for the orthogranic units according to petrographic observations using optical criteria (Kerr, 1977) and frequency approximation (Bullock et al., 1985). Clay mineral composition is dominantly smectite, with lesser amounts of mica and kaolinite (Figure 1). Smectite clay shows a 001 reflection of 1.0 nm for K-saturated samples run at 0% r.h. after oven-drying, 300°C and 550°C treatments. The spacing expands to 1.2 to 1.4 nm after equilibrating at 54% r.h. An expansion to about 1.7 to 1.8 nm for the Ca-saturated sample after glycerol solvation compared to expansion to 1.6-1.7 nm after ethylene glycol solvation, indicates the smectite mineral is largely montmorillonite. Sharp 1.0 nm 001 reflection for all treatments suggests the presence of mica. A well resolved secondary peak at 0.50 nm indicates that the mica mineral is most likely muscovite. Presence of hydrous mica and/or soil vermiculite is evident since the 1.0 nm peak for the Ca-saturated clay is smaller than for the K-saturated clay. Presence of kaolinite is evident since the 0.72 nm peak is stable to all the treatments except for K-550°C where its crystal lattice collapses. Lack of symmetry of the 0.72 nm peak is likely due to a minor co-reflection of a secondary peak of chlorite at about 0.71 nm. A minor amount of chlorite is also evident from the weak peak at 1.42 nm for all treatments. Moderate enhancement of the 1.42 nm reflection by the Ca-glycerol treatment, as compared to other treatments, reveals possible contribution from soil vermiculite and/or beidellite. The small peak at 0.33 nm arises from clay-sized quartz. These clay mineral assemblages are typical of Alberta soils (Warren and Dudas, 1992).

The smectite peaks (1.7-1.8 nm) for the Ca-glycerol treatment for the Bg and Ckg horizons of the Raven soil are well resolved, whereas for the Ah2 horizon only a poorly defined broad peak is observed (Figure 1). Removal of organic matter by heating with 30% H_2O_2 slightly improves the resolution of this peak (Figure 2). Removal of amorphous Fe and Al oxides by acid ammonium oxalate extraction shows similar results. Repeated NaDCB extraction further improves the resolution. The best resolution results from a combined H_2O_2-NaDCB treatment. Repeated pyrophosphate

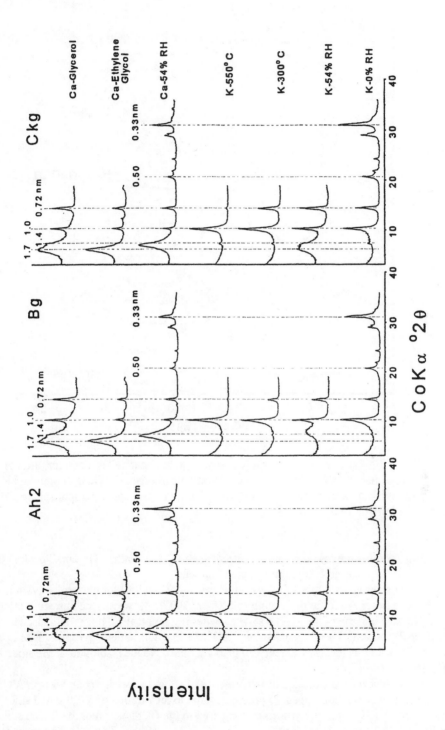

Figure 1. X-ray diffractograms of K- and Ca-saturated clay in the Raven soil.

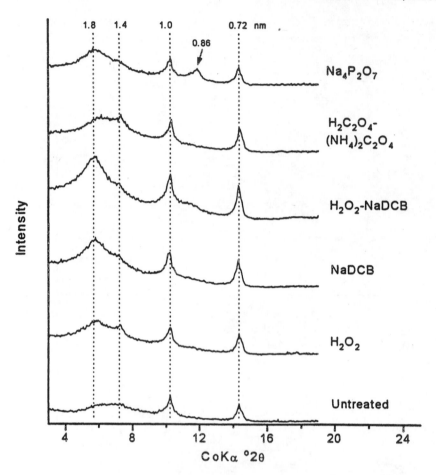

Figure 2. X-ray diffractograms of the Ca-glycerol treated organo-clay complex after hydrogen peroxide (H_2O_2), sodium hydrosulfite (dithionite)-citrate-bicarbonate (NaDCB), combined H_2O_2 and NaDCB, acid ammonium oxalate ($H_2C_2O_4$-$(NH_4)_2C_2O_4$), and sodium pyrophosphate ($Na_4P_2O_7$) treatments, Ah2 horizon of the Raven soil.

extraction also improves the resolution of the smectite peak (1.7-1.8 nm). The small peak at 0.86 nm is probably an artifact caused by pyrophosphate treatment.

Similar differences in resolution of the smectite peaks (1.7-1.8 nm) for glycerol solvation are also observed but to a lesser extent for the Malmo soil (Chen, 1995). However, the untreated organo-clay complex in the Ah2 horizon of the Malmo soil has better resolution at 1.8 nm as compared to the Ah2 horizon of the Raven soil. The H_2O_2 oxidation, NaDCB extraction, sodium pyrophosphate extraction, and the combined H_2O_2-NaDCB treatment also markedly improve the resolution of the smectite peak at 1.8 nm after glycerol solvation.

These mineralogical observations indicate presence of organo-clay edge complexes in Ah horizons of both soils. Such complexes appear to prevent complete rehydration of K-saturated smectite clay to 1.2-1.4 nm for the Ah2 and Bg horizons as compared to the Ckg horizon of the Raven soil (Figure 1). Similar results were reported for the Malmo soil (Chen, 1995). The H_2O_2 treatment increases rehydration of the K-saturated samples in the Ah2 horizons of both soils (Figure 3). Thus, organic-clay complexation may enhance water stability of the granular aggregates. Furthermore, presence of

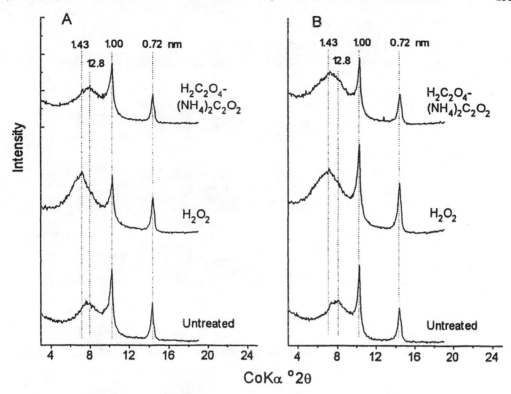

Figure 3. X-ray diffractograms of the K-54% r.h. treated organo-clay complex, before and after removal of organic matter by hydrogen peroxide (H_2O_2) and removal of amorphous sesquioxide by acid ammonium oxalate ($H_2C_2O_4$-$(NH_4)_2C_2O_4$), Ah2 horizons of the Raven (A) and Malmo (B) soils.

stable aggregates reduces biological oxidation of soil organic matter and enhances its storage and stabilization in soils (Elliot, 1986; Jastrow et al., 1996).

D. Aggregate Distribution and Stability

Field-moist aggregates in the two Mollisols are generally water-stable. Aggregate stability, defined as the sum of percent dry weight of the aggregates retained on the sieve (8-0.25 mm) over the dry weight of soil for wet-sieving, ranges from 84 to 89% for the Raven soil and 88 to 95% for the Malmo soil (Chen, 1995).

Distribution of air-dried aggregates shows marked variations with soil depth in the Raven soil (Figure 4). Large aggregates (8-2 mm) decrease progressively with soil depth and are nearly absent in the Bg and Ckg horizons. The content of 2.0-0.5 mm aggregates is also low in the lower section (Bg and Ckg horizons) of the soil. Aggregates of 1.0-0.50 mm size are dominant in the Ah1 to AB horizons. As a result, the aggregate stability declines with soil depth: Sod > Ah1 > Ah2 = AB > Bg > Ckg horizons ($p<0.05$). The MWD follows the same trend: Sod > Ah1 > Ah2 = AB > Bg > Ckg ($p<0.05$) (Table 5).

Distribution of air-dried aggregates in the Malmo soil is similar to the Raven soil (Figure 5). The 8.0-2.0 mm aggregates that are dominant in the Ah1 horizon show the greatest decrease with soil depth. Aggregates of 1.0-0.50 mm size are dominant from the Ah2 to the Ckgj horizon. More finer

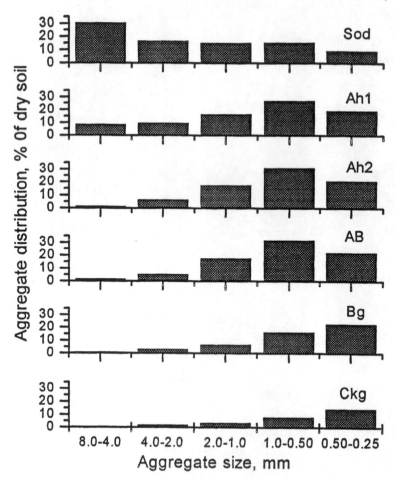

Figure 4. Aggregate distribution of the air dried soil, the Raven soil.

Table 5. Air dried aggregate stability and mean weight diameter (MWD) of two soils

Parameter	Soil horizon					
Raven soil	Sod	Ah1	Ah2	AB	Bg	Ckg
Stability (%)	87a	80b	77c	78c	50d	28e
MWD	271a	131b	86c	85c	47d	24e
Malmo soil	Ah1	Ah2	Abgj	Bmgj	Ckgj	
Stability (%)	90a	81b	75c	88a	68d	
MWD	313a	150b	111c	148b	64d	

Means in the same row followed by different letters are significantly different at 0.05 level.

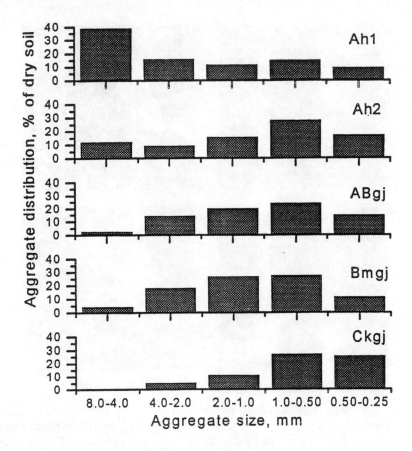

Figure 5. Aggregate distribution of the air dried soil, the Malmo soil.

(< 0.50 mm) aggregates are present in the Ckgj horizon than in the upper horizons. Aggregate stability (Table 5) in descending order is as follows: Ah1 = Bmgj > Ah2 > ABgj > CKgj ($p<0.05$). The high aggregate stability of Bmgj horizon is likely related to a higher clay content (Table 2). The MWD declines as follows: Ah1 > Ah2 = Bmgj > ABgj > Ckgj ($p<0.05$). The granular structures in the Ah2 horizon of the Malmo soil have similar or slightly higher water stability (81%) as compared with the 'shot structure' (77-78%) in the Ah2 and AB horizons of the Raven soil (Table 5). The former have MWD values (150-313) which are two to four times greater than the latter (85-86).

Comparison of results for field-moist samples with air-dried samples shows a strong influence of soil moisture. Air-dried samples generally have fewer large aggregates (8-1.0 mm) but more small aggregates (1-0.25 mm) as compared to the field-moist samples. This trend is more pronounced with increase in soil depth ($p<0.01$), an exception is the upper most horizons (Sod horizon for the Raven and Ah1 for the Malmo soil), where aggregate distribution and stability are not affected by moisture (Chen, 1995). Differences in aggregate stability between the field-moist and air-dried samples are negatively correlated with organic C content ($r = -0.630$, $p<0.05$) for the two soils.

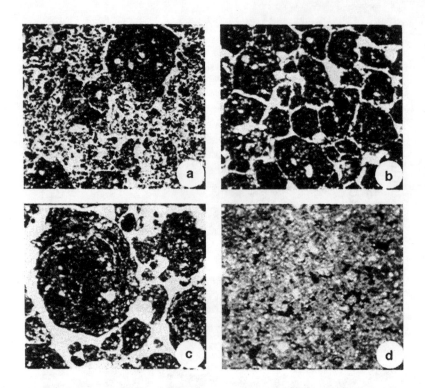

Figure 6. Micromorphological features of the Raven soil: (a) Bimodal distribution of mullgranic units (1120-1750 μm) and humigranic units (42-80 μm). Sod layer. Plane light (PL). Frame length (FL) 4.9 mm. (b) Matriorbiculic fabric of subangular blocky aggregates, Ah2 horizon. PL, FL = 4.9 mm. (c) Typical matriconglomeric fabric of the shot structure; note the edge of the original granular unit that is weakly separated from the mull outer portion. The original granular unit has smooth lower boundary and poorly homoginized matrix. Lower Ah2 to AB horizon. PL. FL = 2.4 mm. (d) Typical vughy microstructure showing initiation of aggregates. Ckg horizon, Crossed polarizers (XPL). FL = 4.9 mm.

E. Micromorphology

The surface of the Raven soil is covered with grass debris mixed with faecal pellets of various soil mesofauna. Large (0.5-1.0 mm) crumbs and granules of mull (dark-colored organo-clay complex) composition that are likely earthworm casts, and smaller (35-80 μm), discrete, organic, faecal pellets, dominantly of collembola and/or enchytraeids, form a complex microstructure with a bimodal distribution (Figure 6a). Several large aggregates with a tubular external form (aggrotubules), likely of similar origin to the large crumbs and granules, were also observed (Chen, 1995). Soil substances in these aggrotubules and granules are well homogenized and have silasepic to moderate skeli-mosepic plasma fabric that is strongly masked by dark colloidal humic substances. Mammilated vughs are their main intra-aggregate voids. The Ah1 horizon has a mixed mullgranoidic and metafragmoidic fabric that is associated with a dominantly subangular blocky microstructure. The Ah2 horizon has a mixed mullgranic and metafragmic fabric comprised of partially accommodated subangular blocky and minor amounts of unaccommodated granular structural units. A concentric distribution pattern of these metafragmic units delineates a 'matriorbiculic' fabric (Fox, 1979; Fox and Protz, 1981) that dominates the Ah2 to AB horizons (Figure 6b). Typical structural units in the Ah2 and AB horizons are dense and their plasma are poorly homogenized. Some aggregates contain diffuse Fe nodules in the center that are encompassed by a mull edge forming a 'matriconglomeric' fabric (Figure 6c) as defined by Fox (1979) and Fox and Protz (1981). A few vesicles and metavughs dominate their intra-aggregate voids and these structural units have smooth lower boundaries. Semiquantitative EDX analyses on a few representative areas (50 μm by 50 μm each) of the thin section show that Fe content is about two to three times higher in the central zones than the edge portion of these typical metafragmic and matriconglomeric units. These metafragmic and matriconglomeric units are the representative micromorphological forms of the unique 'shot structures' in the Raven soil. A sharp decrease in faunal features is most evident in the Ah2 and AB horizons where 'shot structures' are dominant.

The upper Bg horizon has a spongy microstructure and the lower Bg grades to a spongy-vughy microstructure with an increase in the abundance of interconnected and isolated vughs and vesicles. The Ckg horizon has a dominant vughy microstructure (Figure 6d) and crystic plasma fabric.

The major micromorphological change in the organic layers of the Malmo soil is characterized by a decrease in size and abundance of plant litter as a result of an increase in abundance of mesofaunal excrements from the LF to the H layers. The soil fauna involved in the comminution include oribatid mites, beetle larvae, millipedes, dipterous larvae, collembola, and/or enchytraeids (Chen, 1995). The H layer is dominated by fine humigranic-granoidic complex fabric, comprised almost entirely of mesofaunal faecal pellets or casts (Figure 7a) that are infected by fungi to various extents.

The upper Ah horizon of Malmo soil shows considerable heterogeneity in composition and in its microfabric, which is generally a separated modergranic-fragmic complex resulting from poor mixing of soil materials. Better mixing of plasma in the lower section of Ah1 to Ah2 horizons is reflected in the prevailing mullgranoidic or banded mullgranoidic fabrics. The major microstructure of the middle to lower Ah2 is a crumb/granular complex formed by large mullgranic and small humigranic and mullgranic units (Figure 7b) that is similar to the bimodal distribution in the Raven soil. These aggregates have a weak silasepic to inundulic plasma fabric. The Ah2 horizon is also abundant in chambers that are usually filled with some faecal pellets of collembola and/or enchytraeids (Chen, 1995). The Ah3 horizon has a separated complex fabric of mullgranoidic and banded mullgranoidic fabric zones. Platy aggregates in the banded mullgranoidic fabric zones have a weak silasepic to inundulic plasma fabric (Figure 7c) that is identical to those in the mullgranic zones of Ah3 and Ah2, due to an intimate association of colloidal humus to clay minerals that masks the faint interference color usually associated with clay domains. Change in intensity of interference color of clay domains may reflect the degree of organo-clay complexation and humus types in soils. Some zones in the Ah3 horizon show an obliquely banded mullgranoidic-porphyric fabric (Figure 7d). The detailed micromorphological observations for both soils are reported by Chen (1995).

IV. Discussion

A. Interrelationships between Humified Organic C, Mineralogical Properties, and Aggregation

The distribution of HA yields in both soils reflects influence of vegetation. The less abrupt reduction of HA yields with soil depth in the Raven soils is characteristic of the grass communities under which they developed. The major mode of organic matter addition in the Raven soil is through the grass root system that usually humifies *in situ* in the rhizosphere (Stevenson, 1982) and in root channels which are usually filled with mesofaunal excrements (Chen, 1995). Fine grass roots usually form a close association with mineral soil in the rhizosphere resulting in a lower yield of HAs. Mesofaunal excrements also have a low yield of HAs. The HA yield distribution of Malmo soils is typical of soils found in forest-grassland transition zones, where most of the HAs were extracted from the organic layers and upper mineral horizons. Organic matter is added mainly to forest soils by surface addition,

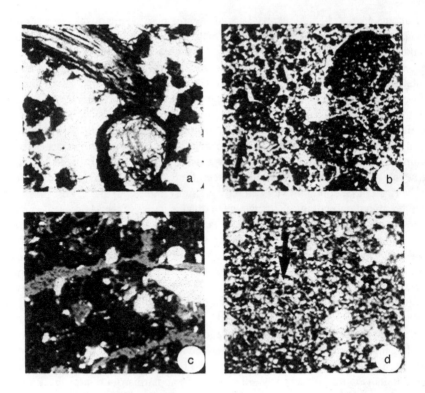

Figure 7. Micromorphological features of the Malmo soil: (a) Mesofauna invasion of phytofragments excreting them as mesofauna faecal pellets (40-120 μm, organic). The latter serves as a substrate for fungi as evident from the hyphae invasion of the faecal pellets (likely of collembola/enchytraeids). The black object (320 μm) in lower middle position resembles a mycorrhiza mantle. H horizon. PL. FL = 0.62 mm. (b) Granular aggregates of variable size (100-2000 μm) in the Ah2 horizon of the Malmo soil. Colloidal humic substances strongly masked the aggregates and form mull. The dark aggregate in tubular form (930 x 1400 μm, marked with arrow) resembles earthworm cast and the other granular aggregates are largely welded or disintegrated earthworm casts. Partially crossed polarizers (PPL), FL = 4.9 mm. (c) Weak silasepic to inundulic plasma fabric for the banded mullgranoidic zone. Ah3 horizons. PPL. FL = 0.62 mm. (d) Obliquely banded mullgranoidic-porphyric intergrade fabric (marked with arrow). Ah3 horizon. PPL. FL = 4.9 mm.

thus an abrupt decrease in HA yield is not surprising (Stevenson, 1982). An increase in HA-C to soil organic C ratio from organic layers to Ah horizons is a result of increased humification. This is in agreement with findings by Kögel-Knabner et al. (1990). Below Ah horizons, the small values of HA-C to soil organic C ratio indicates only a small fraction of soil organic matter is HAs. Thus, the build-up of HA-C in Ah horizons is likely a result of C sequestration from the plant community during pedogenesis rather than inheritance from organic C within the parent geological materials. Differences in physical and chemical structures and in origin of these HAs were reported elsewhere (Chen and Pawluk, 1995).

Clay mineral diffractograms indicated complexation of organics on the edges of smectite minerals in Ah2 horizons of both soils and revealed a mechanism of C sequestration in soil. There does not appear to be interlayer complexation because a reflection greater than 1.8 nm was not observed in this study. The organo-clay edge complexation likely prevented complete penetration of glycerol into the

Ca-saturated clay and allowed the complex to swell partially resulting in a poorly resolved peak at 1.6 to 1.8 nm for the organo-smectite in the Ah2 horizon of the Raven soil. The limited improvement for the 1.8 nm peak by 30% H_2O_2 treatment suggests some of the organic substances are spatially protected against oxidation by inorganic substances. Partial improvement in peak intensity by acid-ammonium oxalate extraction implies limited interaction with amorphous Fe and Al oxides or hydroxides. Further improvement with NaDCB extraction suggests two possibilities: some Fe and Al in micro-crystalline forms also participate in the complexation, and/or more tightly bonded organics are also released by partial dissolution of Fe and Al from clays as suggested by Pawluk (1972). The most improved resolution occurred after combined H_2O_2-NaDCB treatment and further suggested the involvement of microcrystalline Fe and Al in complexation. Improvement of the 1.8 nm peak by repeated pyrophosphate extraction indicates that some of the organic matter in the organo-clay complex is HAs and/or fulvic acids. The remaining organic fraction in the complex is humin that is partially removed by the combined H_2O_2-NaDCB treatment.

Dudas and Pawluk (1969) observed the presence of organo-clay edge complexation but no interlayer complexation in Ah horizons of Chernozems in Alberta. In the current study, mineralogical analyses revealed the Fe and Al in microcrystalline forms likely participated in organo-clay complexation in the Ah2 horizon of Raven soil. The Raven soil has more organic-complexed Fe and Al in the solum than the Malmo soil (Table 3). Theng et al. (1986, 1992) reported the occurrence of interlayer organo-clay complexation for two New Zealand Spodic soils. They proposed three conditions for interlayer organo-clay complexation: (1) presence of smectite minerals; (2) acidic soil reaction; and (3) accumulation of organic matter associated with low microbial activity. Acidic conditions rarely occur in Mollisols because these soils are characterized by a high base saturation dominance of Ca^{++} on the exchange complex and near neutral pH. Therefore, organic-clay edge complexation is likely more common in Mollisols.

The results of aggregate analyses in our study support the observations of Haynes and Swift (1990) that moisture content of the soils is the most important factor affecting wet-sieving, and the effect of moisture is related to soil organic matter content. In this study, correlation analyses further revealed that difference in aggregate stability between the field-moist and air-dried samples is negatively related to neutral saccharides yield ($r = -0.652$, $p<0.05$), to the (galactose + mannose) over (arabinose + xylose) ratios ($r = -0.603$, $p<0.05$) and to the pyrophosphate-extractable Fe content ($r = -0.609$, $p<0.05$). The distribution of neutral saccharides and the (galactose + mannose) over (arabinose + xylose) ratios were reported elsewhere for the two soils (Chen, 1995). Similarly, differences in MWD between the field-moist and air-dried samples were also negatively correlated with the previously mentioned soil properties (r values ranged from -0.646 to -0.714, $p<0.05$). The HA-C to soil organic C ratios are highest in Ah horizons of both soils. According to Jouany (1991) and Giovannini et al. (1983), HAs are capable of reducing wetability of the smectite-HA complex and soil aggregates. The presence of high amounts of HA-C in the Ah horizons likely promotes aggregate stability in both soils used in this study.

B. Impact of Micromorphological Attributes to C Sequestration in Soil

The micromorphological observations enable us to visualize the processes of organic matter transformation and interaction of organic substances with mineral components *in situ*. The microfabric analysis reveals the arrangement and the internal composition of soil substances in thin sections.

In the Raven soil, change in dominant fabric sequence is as follows: matriporphyric (Ckg), matriporphyric to metafragmoidic (Bg), metafragmic (AB to Ah2), metafragmic to metafragmoidic (Ah1), mullgranic, humigranic, and phytogranic (Sod layer). This corresponds to a shift from soil physical processes to biological processes proceeding from the parent materials to the surface soil. The dominance of the metafragmic fabric in the Ah2 and AB horizons of the Raven soil reflects freeze/thaw effects. The presence of vesicles and metavughs, the circular arrangement of the fine sand- to silt-size mineral grains (Chen, 1995), and the associated fabrics (orbiculic and conglomeric) support this conclusion, because these fabric features are typical of soils subjected to frost influence (Fox, 1979; Pawluk, 1988). The dense appearance, poorly homogenized plasma composition, and smooth lower boundary of typical aggregates in the Ah2 and AB horizons closely resemble aggregates generated by freeze/thaw effects (Van Vliet-Lanoe et al., 1984; Pawluk, 1988; FitzPatrick, 1993), though mesofaunal impact cannot be completely excluded as reflected by minor mull-granic/granoidic components.

Aggregates generated through freeze/thaw are usually not water-stable in soils low in organic matter content (Baver et al., 1972). Presence of diffuse Fe nodules in the center of some matriconglomeric units is likely related to the cementation effects of Fe colloids during the initial stage of aggregation due to psuedo-gley processes. These Fe cemented aggregates are likely further stabilized through attachment of the mull materials along the edge of the matriconglomeric units that dominate in the AB and Ah2 horizons of the Raven soil. Thus, a compound effect of Fe colloids and biologically transformed mull substances likely promoted the stability of the 'shot structure'. Enhanced faunal activity in the Sod layer is reflected by the abundance of their faecal pellets. The mammilated vughs associated with these faecal pellets are characteristic of biological origin (Brewer, 1976; Mermut, 1985) and their dense, well-homogenized appearance is characteristic of earthworm faecal pellets.

The microfabrics in the Malmo soil reflect a series of processes in organic matter transformation and a decisive role of soil mesofauna in comminution of plant litter. The H horizon is comprised almost entirely of their organic excrements. Invasion of mesofauna excrements by soil fungi and decrease in abundance of mesofaunal excrements in mineral horizons reflect a transient nature of these organic substances in organic C transformations. Formation of mull humus type in the Ah horizons results largely from earthworm activity. This supports the general view that macrofauna are the major biological driving force in mull formation (Rusek, 1985; Brussaard and Juma, 1996). Formation of persistent organo-mineral complexes and the physical occlusion of organic C in faeces retards further decomposition of organic C in soils (Duchaufour, 1982; Brussaard and Juma, 1996); therefore, it promotes C sequestration in soils.

The presence of platy microstructures in the banded mullgranoidic zones in the Ah horizon of the Malmo soil is likely related to the influence of freeze/thaw as well. Pawluk (1988) showed that extended freeze/thaw processes could produce banded mullfragmoidic and mullgranic fabrics similar to those found in Chernozemic Ah horizons. An 'obliquely banded mullgranoidic-porphyric fabric' in the lower Ah horizon of Malmo soil is probably related to a localized freeze heave developed by vertical frost pressure.

The organic matter in most dark colored mollic epipedons is dominantly humin and HAs (Kononova, 1966; Duchaufour, 1982). Duchaufour (1982) reported that the good structure of mollic epipedons is largely attributable to clay-humus complexation. The strong masking effect by colloidal humus on aggregates, as reported in our study, provided micromorphological evidence that sequestered C promotes soil aggregation.

V. Implications

Carbon sequestration involves physical, chemical, and biological processes which are directly linked to soil aggregation. Soil organic matter resides in soils as undecomposed or partially decomposed phytofragments, excrements of soil mesofauna, and organic-mineral complexes in the two Mollisols studied. Organo-clay complexation, largely with smectite, in the Ah horizons contributed to stabilization of granular aggregates in Mollisols. The compound effect of freeze/thaw processes acting on inorganic substances generated by pseudo-gley and on biologically transformed soil substances is largely responsible for the genesis of the 'shot-structure' in the Raven soils. The granular structure in surface horizons of the Raven soil and those in Ah horizons of the Malmo soil are largely influenced by soil biota, especially soil meso- and macro-fauna. Both granular structures are considerably water stable and consequently enhance sequestration and stabilization of C in soils. The micromorphological study and processes oriented research are required to systematically explore the mechanisms of soil aggregation and C sequestration.

Acknowledgments

The financial support from the Natural Sciences and Engineering Research Council of Canada is greatly appreciated. The authors give special thanks to Dr. Sanborn for permitting us to use some published background data of the Malmo soil. They authors are grateful to P. Yee and M. Abley for laboratory assistance.

References

Anderson, J.M. and P.W. Flanagan. 1989. Biological processes regulating organic matter dynamics in tropical soils. p. 97-123. In: D.C. Coleman, J.M. Oades, and G. Uehara (eds.), *Dynamics of Soil Organic Matter in Tropical Ecosystems*. Univ. of Hawaii Press, Honolulu.

Agriculture Canada Expert Committee on Soil Survey. 1987. *The Canadian System of Soil Classification*. Agri. Can. Publ. 1646, 2nd ed., Ottawa. 164 pp.

Angers, D.A. and G.R. Mehuys. 1993. Aggregate stability to water. p. 651-657. In: M.R. Carter (ed.), *Soil Sampling and Methods of Analysis*. Lewis Publishers, Boca Raton, Florida.

Baver, L.D., W.H. Gardner, and W.R. Gardner. 1972. *Soil Physics*. 4th edition. John Wiley & Sons. New York. 498 pp.

Brewer, R. 1976. *Fabric and Mineral Analysis of Soils*. Robert E. Krieger Publishing Company, Huntington, New York. 482 pp.

Brussaard, L. and N.G. Juma. 1996. Organisms and humus in soils. p 329-359. In: A. Piccolo (ed.), *Humic Substances in Terrestrial Ecosystems*. Elsevier. Amsterdam.

Bullock, P., N. Fedoroff, A. Jongerius, G. Stoops, T. Tursina, and U. Babel. 1985. *Handbook for Soil Thin Section Description*. Waine Research Publications. Wolverhampton, England. 152 pp.

Chen. Z. 1995. Soil organic matter and granular structure in selected Alberta Mollisols. Ph. D. Thesis. University of Alberta. Edmonton, Alberta.

Chen, Z. and S. Pawluk. 1995. Structural variations of humic acids in two sola of Alberta Mollisols. *Geoderma* 65:173-193.

Coutard, J.P. and H.J. Mücher. 1985. Deformation of laminated silt loam due to repeated freezing and thawing cycles. *Earth Surface Processes and Landforms* 10:309-319.

Duchaufour, P. 1982. *Pedology: Pedogenesis and Classification*. English version, George Allen & Unwin, London. 448 pp.

Dudas, M.J. and S. Pawluk. 1969. Naturally occurring organo-clay complexes of orthic Black Chernozems. *Geoderma* 3:5-17.

Edwards, A.P. and J.M. Bremner. 1967a. Microaggregates in soils. *J. Soil Sci.* 18:67-73.

Edwards, A.P. and J.M. Bremner. 1967b. Dispersion of soil particles by sonic vibration. *J. Soil Sci.* 18:47-63.

Elliott, E.T. 1986. Aggregate structure and carbon, nitrogen and phosphorus in native and cultivated soils. *Soil Sci. Soc. Am. J.* 50:627-633.

Emerson, W.W. 1959. The structure of soil crumb. *J. Soil Sci.* 10:235-244.

FitzPatrick, E.A. 1993. *Soil Microscopy and Micromorphology*. John Wiley & Sons, New York. 304 pp.

Fox, C.A. 1979. The soil micromorphology and genesis of the Turbic Cryosols from the MacKenzie River valley and Yukon Coastal plain. Ph. D. thesis. University of Guelph. Guelph, Ontario.

Fox, C.A. and R. Protz. 1981. Definition of fabric distributions to characterize the rearrangement of soil particles in the Turbic Cryosols. *Can. J. Soil Sci.* 61:29-34.

Giovannini, G., S. Luchesi, and S. Cervelli. 1983. Water repellent substances and aggregate stability in hydrophobic soils. *Soil Sci.* 41:73-83.

Haynes, R.J. and R.S. Swift. 1990. Stability of soil aggregates in relation to organic constituents and soil water content. *J. Soil Sci.* 41:73-83.

Jackson, M.L. 1969. *Soil Chemical Analysis — Advanced Course*. 2nd ed. Madison, WI. 895 pp.

Jastrow, J.D., T.W. Boutton, and R.M. Miller. 1996. Carbon dynamics of aggregate-associated organic matter estimated by carbon-13 natural abundance. *Soil Sci. Soc. Am. J.* 60:801-807.

Jouany, C. 1991. Surface free energy components of clay-synthetic humic acid complexes from contact angle measurements. *Clays Clay Miner.* 39:43-49.

Kerr, P.F. 1977. *Optical Mineralogy*. McGraw-Hill Inc. 4th ed., New York. 492 pp.

Kögel-Knabner, I., P.G. Hatcher, and W. Zech. 1990. Decomposition and humification processes in forest soils: implications from structural characterization of forest soil organic matter. p. 218-223. In: Trans. 14th Int. Congr. Soil Sci. Vol. 5, Kyoto.

Kononova, M.M., 1966. *Soil Organic Matter: Its Nature, Its Role in Soil Formation and in Soil Fertility*. Pergamom Press, 2nd English ed., Oxford. 544 pp.

Mehra, O.P. and M.L. Jackson. 1960. Iron oxide removal from soils and clays by a dithionite-citrate system buffered with sodium bicarbonate. *Clays Clay Miner.* 7:317-327.

Mermut, A.R. 1985. Faunal influence on soil microfabrics and other soil properties. *Quaestiones Entomologicae* 21:595-608.

Miller, R.M. and J.D. Jastrow. 1990. Hierarchy of root and mycrorrhizal fungi interactions with soil aggregation. *Soil Biol. Biochem.* 22:579-584.

Nelson, D.W. and L.E. Sommers. 1982. Total carbon, organic carbon, and organic matter. p. 536-577. In: A.L. Page, R.H. Miller, and D.R. Keeney. (eds.), *Methods of Soil Analysis, Part 2—Chemical and Microbiologica Properties*. American Society of Agronomy, Inc., Soil Science Society of America, Inc. 2nd ed., Madison, WI.

Oades, J.M. 1993. The role of biology in the formation, stabilization and degradation of soil structure. *Geoderma* 56:377-400.

Pawluk, S. 1972. Measurements of crystalline and amorphous iron removal in soils. *Can. J. Soil Sci.* 52:119-123.

Pawluk, S. 1983. Fabric sequences as related to genetic processes in two Alberta soils. *Geoderma* 30:233-242.

Pawluk, S. 1985. Soil micromorphology and soil fauna: problems and importance. *Quaestiones Entomologicae* 21:473-496.

Pawluk, S. 1988. Freeze-thaw effects on granular structure reorganization of soil materials of varying texture and moisture content. *Can. J. Soil Sci.* 68:485-494.

Pawluk, S. and L. Bal. 1985. Micromorphology of selected mollic epipedons. p. 63-83. In: L.A. Douglas and M.L. Thompson (ed.), *Soil Micromorphology and Soil Classification*. SSSA. Spec. Publ. 15, Madison, WI.

Peters, T.W. and W.E. Bowser. 1960. Soil survey of the Rocky Mountain House Sheet. University of Alberta Bull. No. 55-1. Alberta soil survey report No. 19. University of Alberta. Edmonton, Alberta.

Ross, G.J. and C. Wang. 1993. Extractable Al, Fe, Mn and Si. p. 239-246. In: M.R. Carter (ed.), *Soil Sampling and Methods of Analysis*. Lewis Publishers. Boca Raton, FL.

Rusek, J. 1985. Soil microstructures—contributions on specific soil organisms. *Quaestiones Entomologicae.* 21:497-514.

Sanborn, P. 1981. Master Thesis. Department of Soil Science, University of Alberta. Edmonton, Alberta.

Sanborn, P. and S. Pawluk. 1983. Process studies of a Chernozemic pedon, Alberta (Canada). *Geoderma* 31: 205-237.

SAS Institute Inc. 1987. *SAS/STAT Guide for Personal Computers.* Version 6 ed., SAS Institute Inc., Cary, NC, U.S.A. 1028 pp.

Sheldrick, B.H. and C. Wang. 1993. Particle size distribution. p. 499-511. In: M.R. Carter (ed.), *Soil Sampling and Methods of Analysis.* Lewis Publishers. Boca Raton, FL.

Soil Survey Staff. 1994. *Keys to Soil Taxonomy.* U.S. Department of Agriculture, Pocahontas Press, Inc., 6th ed., Blacksburg, Virginia. 524 pp.

Stevenson, F.J. 1982. *Humus Chemistry, Genesis, Composition, Reactions.* John Wiley & Sons, New York. 443 pp.

Theisen, A.A. and M.E. Harward. 1962. A paste method for preparation of slides for clay mineral identification by x-ray diffraction. *Soil Sci. Soc. Am. Proc.* 26:90-91.

Theng, B.K.G., G.J. Churchman, and R.H. Newman. 1986. The occurrence of interlayer clay-organic complexes in two New Zealand soils. *Soil Sci.* 145 (5):262-266.

Theng, B.K.G., K.R. Tate, and P. Becher-Heidmann. 1992. Towards establishing the age, location, and identity of the inert soil organic matter of a spodosol. *Z. Pflanznernahr. Bodenk.* 155:181-184.

Tisdall, J.M. and J.M. Oades. 1982. Organic matter and water stable aggregates in soils. *J. Soil Sci.* 33:141-163.

Van Vliet-Lanoe, B., J.P. Coutard, and A. Pissart. 1984. Structures caused by repeated freezing and thawing in various loamy sediments: A comparison of active, fossil and experimental data. *Earth Surface Processes and Landforms* 9:553-565.

Warren, C.J. and M.J. Dudas. 1992. Acidification adjacent to an elemental sulfer stockpile: I. Mineral weathering. *Can. J. Soil Sci.* 72:113-126.

Waters, A.G. and J.M. Oades. 1991. Organic matter in water-stable aggregates. p. 163-174. In: W.S. Wilson (ed.), *Advances in Soil Organic Matter Research: the Impact on Agriculture and Environment.* Royal Society of Chemistry, Cambridge.

Xing, B. and M.J. Dudas. 1994. Characterization of clay minerals in White Clay Soils, People's Republic of China. *Soil Sci. Soc. Am. J.* 58:1253-125

CHAPTER 17

A Model Linking Organic Matter Decomposition, Chemistry, and Aggregate Dynamics

A. Golchin, J.A. Baldock, and J.M. Oades

I. Introduction

The interaction of organic materials with mineral particles is a fundamental process in the surface horizons of most soils. Organo-mineral interactions not only influence the dynamics of soil organic matter (Oades, 1988; Amato and Ladd, 1992; Golchin et al., 1995a; Oades, 1995; Chenu et al., 1996), but also contribute to the formation and stabilization of soil aggregates (Tisdall and Oades, 1982; Oades, 1993). Interactions between small organic molecules and clay surfaces have been described and reviewed at length (Mortland, 1970; Theng, 1974) and interactions of organic polymers with clays are reasonably well understood (Theng, 1979; 1982). Interaction of particulate organic matter (POM) with mineral particles, however, has received less attention and the turnover, composition, and distribution of POM within the soil matrix are not well known. In classical fractionation schemes, SOM has been extracted from soils using alkaline solutions and the unextractable fraction or humin, which includes POM, has not been studied extensively. In this chapter we will focus on the interaction of POM with mineral particles and consider the role of POM in the formation of aggregates of different sizes. We present a conceptual model describing the involvement of POM and microbial metabolites derived from its decomposition in soil aggregation. The importance of biological processes associated with the decomposition of POM and the associated chemical changes will be identified and discussed with respect to their involvement in the proposed model of aggregation. The conceptual model, which is based on our previous works and selected literature results, may be applied generally to soils where organic matter is an important agent responsible for binding soil mineral particles together creating an aggregate hierarchy (Oades and Waters, 1991; Oades, 1993). In the model, we proposed the existence of three levels of aggregation (< 20 µm, 20-250 µm, and > 250 µm) in which the mechanisms of stabilization differ.

II. Concepts of Soil Structure

Soil structure refers to the 3-dimensional arrangement of individual mineral grains and organic constituents present in a soil. The individual grains or organic materials may remain as discrete structural units or be held together by various aggregating agents in compound particles, referred to as aggregates, separated from one another by planes of weakness. The planes of weakness may exist as either pores of variable diameter or as weaknesses within the soil matrix along which preferential fracturing occurs when stresses are applied. The positioning of planes of weakness within the soil matrix arises in response to the spatial distribution of soil aggregating agents and the distance over which these agents are capable of operating. The effectiveness with which the aggregating agents hold soil particles together forming planes of weaknesses is influenced by soil water content.

Edwards and Bremner (1967) proposed the existence of two size classes of soil aggregates: macroaggregates >250 µm diameter and microaggregates <250 µm diameter. Subsequent evidence presented by Tisdall an Oades (1982), Elliot (1986), and Miller and Jastrow (1990) and others has supported such a division. However, a prerequisite to the formation and stabilization of micro- and macroaggregates in soils containing clay is the aggregation of soil clays into stable packets referred to as quasicrystals, domains, or assemblages (Oades and Waters, 1991). If clay particles are not aggregated into stable packets, they will slowly erode from aggregate surfaces resulting in a gradual deterioration of both microaggregates and macroaggregates. On this basis, three levels of structural organization should be recognized in soils containing appreciable contents of clay and where organic matter is an important binding agent: (1) the binding together of clay particles into stable packets <20 µm, (2) the binding of clay packets into stable microaggregates 20-250 µm, and (3) the binding of microaggregates into stable macroaggregates >250 µm. Such a model of soil structure led to the concept of aggregate hierarchy described by Hadas (1987) and Dexter (1988) and further developed by Oades and Waters (1991) and Oades (1993). Aggregates contained in soils exhibiting aggregate hierarchy should break down in a stepwise manner as the magnitude of an applied disruptive force increases. Such a stepwise disintegration of soil aggregates from macroaggregates through microaggregates to clay packets was observed by Oades and Waters (1991) for a mollisol and an alfisol where organic matter was the dominant stabilizing agent. In an oxisol, however, where inorganic cements were important stabilizing agents, the destruction of macroaggregates resulted in the release of particles <20 µm with no indication of a stable microaggregate fraction. On the basis of these results, Oades (1993) has speculated that aggregate hierarchy will exist where soils have a long history of being exploited by roots and will not exist in young soils or soils where inorganic cementing agents are primarily responsible for binding soil particles together.

A conceptual approach which can be used to explain aggregate hierarchy relates to the spatial distribution and persistence of aggregating agents within the soil matrix. At the smallest level of soil aggregation, the binding together of individual clay particles into packets, aggregation is dictated principally by soil chemical properties, such as those controlling clay dispersion, and is often a function of pedological processes (e.g., the formation and deposition of iron oxides/hydroxides in oxisols). As a result, the extent of aggregation of clays throughout the soil matrix should be homogeneous, with the possible exception of clays found in specific positions where differences in chemical properties from that of the bulk soil matrix would be expected (e.g., as cutans along walls of macropores). Without large shifts in soil chemical properties induced by anthropogenic inputs, the level of aggregation of clays should be maintained for long periods of time and not be influenced significantly by management practices unless major shifts in soil chemical properties are induced.

At the two larger scales of aggregation, microaggregation, and macroaggregation, various organic materials have been implicated as the dominant agents stabilizing aggregates. Humified and biologically processed organic materials are involved in the binding of clay packets and silt particles <20 µm into aggregates <53 µm and particulate organic materials, POM, form a core around which clay packets and small microaggregates are bound into larger microaggregates 53-250 µm and small macroaggregates 250-2000 µm (Tisdall and Oades, 1982; Oades, 1984; Elliot, 1986; Beare et al., 1994; Golchin, 1994b). In aggregates with POM cores, metabolic products produced by soil microorganisms utilizing POM act as cements binding smaller aggregations of soil particles and silt and sand sized primary particles together. The size of aggregate formed will depend on the size of the POM and the amount of binding materials secreted by microorganisms as decomposition proceeds. Roots, fungal hyphae, and fragments of plant debris account for the formation and stabilization of larger macroaggregates >2000 mm via mechanisms of physical entanglement and the production of mucilaginous materials during their growth and subsequent fragmentation and decomposition (Tisdall and Oades, 1982; Oades, 1984).

The dominant mechanisms accounting for the deposition of organic material in soil include root and hyphal growth and soil faunal activity in uncultivated systems and tillage where cultivation is practiced. Such mechanisms result in a heterogeneous distribution of POM and its decomposition products within the soil. The results of this will be the production of zones of intense microbial activity around the heterogeneously distributed organic materials and the formation of aggregates in specific regions of the soil matrix. Regions between such zones will be structurally weaker and form planes of weakness along which fracturing occurs when the soil is subjected to disruptive forces. As the POM is decomposed, its ability to maintain the structural integrity of macroaggregates decreases, structural weaknesses are formed, and on exposure to disruptive forces, the macroaggregate breaks down to form microaggregates stabilized by humified or biologically processed organic materials formed during POM decomposition. POM deposition and decomposition are ongoing processes in soil. Therefore, the extent of aggregation measured at any time represents the net difference between rates of aggregate formation and destruction, and a continuous supply of POM is required to maintain a given level of soil aggregation.

In a subsequent section of this chapter, we have used the principles of soil structure described above to develop a model which describes the role of POM in soil aggregation.

III. Conceptual Model of the Role of Particulate and Adsorbed Organic Matter in Soil Structure

A. Free POM

The model (Figure 1) assumes that as particulate organic matter enters the soil from root or plant debris it becomes colonized by the rhizosphere microbial population and at the same time adsorbs mineral materials. Adsorption of mineral particles is facilitated by the close contact between plant residues and mucilages and the mineral phase resulting from an interaction between inorganic materials and products of microbial metabolism (e.g., microbial polysaccharide mucilages) produced on the surface of the decomposing POM. At this stage, the level of association between the POM and soil mineral matrix is low and limited to the adsorption of individual primary particles on POM surfaces. Such POM can be floated out of the soil with gentle shaking using a solution of density 1.6 Mg m^{-3} (Figure 2). In terms of our model, we have labeled this fraction of soil POM as free POM.

B. Macroaggregate-POM

With time, the microbial colonization of free POM and subsequent metabolism of POM carbon and nutrient resources increases, resulting in the further production of metabolic binding agents which strengthen the interaction between POM and surrounding mineral colloid. Electron micrographs presented by Tisdall and Oades (1979) and Waters and Oades (1991) clearly show the encrustation of particulate organic matter with clay and silt particles via mucilages on the surface of roots or organic residues (Figure 3). Due to the presence of a readily available source of C (fresh POM), increases in the length of fungal hyphae and contents of water and acid-extractable carbohydrate, microbial biomass in macroaggregates can be expected (Tisdall and Oades, 1979, 1980; Angers and Mehuys, 1989; Capriel et al., 1990; Miller and Jastrow, 1990; Hynes et al., 1991), all of which contribute to macroaggregate stabilization, especially the ability to withstand slaking in water (Tisdall and Oades, 1982). As the encrustation of POM with soil mineral particles increases due to the continued decomposition of POM and the production of various soil binding agents, POM forms

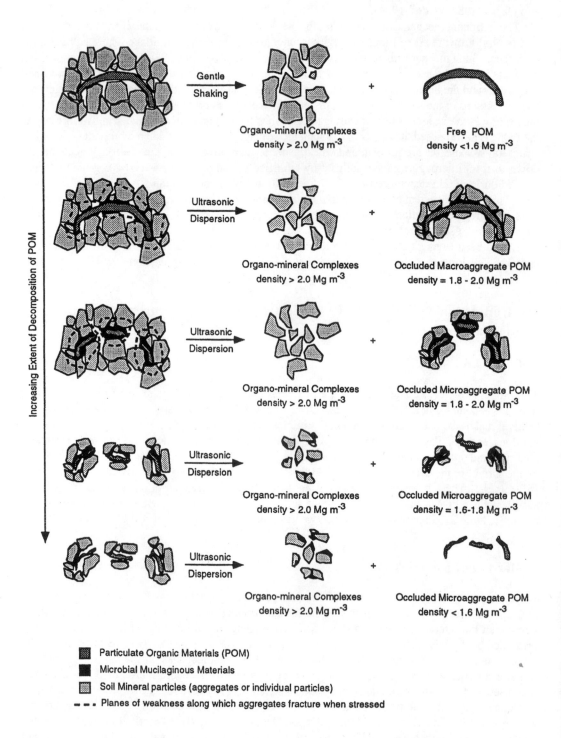

Figure 1. A model linking particulate organic matter decomposition and aggregate dynamics.

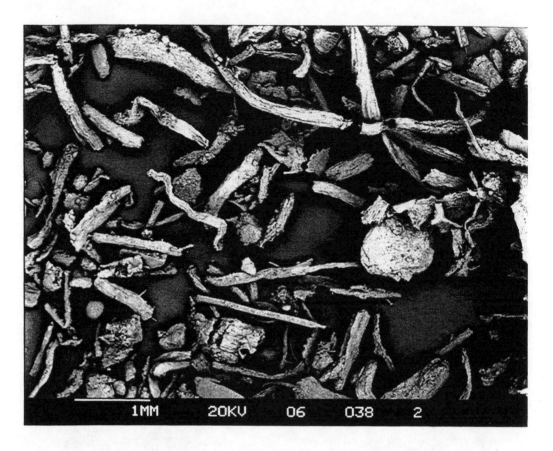

Figure 2. Free POM separated from a red-brown earth using a solution of density 1.6 Mg m^{-3}.

centres of intense biological activity around which aggregations of mineral particles are stabilized. Studies by Angers et al. (1995), Aita et al. (1995) and Puget et al. (1995) which showed that, after addition of labeled (^{13}C and ^{15}N) plant material to soil and incubation under field condition, labeled plant materials were initially incorporated into soil macroaggregates, support this concept. Thus, during the initial stage of POM decomposition which occurs while the POM is still physically intact and acting as a macroaggregate core, macroaggregates remain stable.

The size of macroaggregates stabilized would be a function of the size, geometry and mode of deposition of POM. For example, long pieces of debris derived from roots and fungal hyphae form associations with soil particles during their growth and have the potential to span across groups of microaggregates and pores to form macroaggregates through processes of physical entanglement and the production of metabolic binding agents. Observation of macroaggregates using optical microscopy and scanning electron microscopy have shown that the action of hyphae and root enmeshing soil particles and microaggregates is important in stabilizing macroaggregates (Molope and Page, 1986; Gupta and Germida, 1988; Tisdall and Oades, 1982). In contrast, POM mixed into the soil matrix by processes such as cultivation are probably less homogeneously distributed than root-derived POM and have little prior association with soil particles. As a result, cropping systems which include the production of species with extensive fibrous root systems (e.g., grass pastures) would produce the highest levels of macroaggregation, consistent with the observations of Tisdall and Oades (1982).

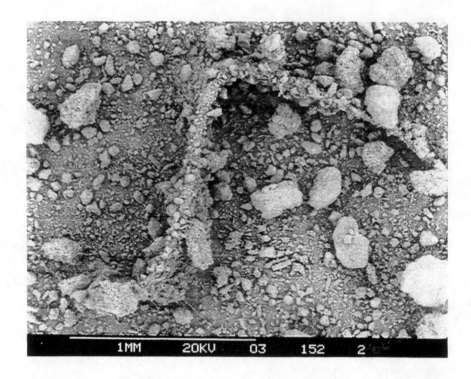

Figure 3. Macroaggregate > 250 μm formed via encrustation of wheat root with inorganic soil particles. (From Waters and Oades, 1991.)

Studies related to the decomposition of plant residues indicate that free POM is readily decomposed in soils (Christensen, 1987). Dalal and Mayer (1986) found that carbon losses from POM contained in light fractions (< 2.0 Mg m^{-3}) were 11 times greater than those associated with heavy fractions (> 2.4 Mg m^{-3}) upon cultivation of virgin soils. In studies where changes in the natural abundance of ^{13}C in response to shifts between C$_3$ and C$_4$ vegetation have been used to monitor decomposition, enhanced decomposition of light fraction POM have been noted relative to organic materials associated with mineral particles (Gregorich et al., 1995; Golchin et al., 1995a; Cadish et al., 1996). Macroaggregate stabilization by POM is, therefore, a transient process, and maintenance of a given level of macroaggregation is dependent upon the continual addition of POM to soil. Although the initial phase of POM decomposition is viewed as having a beneficial effect on macroaggregation through the production of metabolic soil binding agents, as POM decomposition continues, the organic cores holding macroaggregates together are broken down into smaller pieces. Fragmentation of POM is thought to result from the action of soil fauna and the selective microbial oxidation of portions of the POM which are exposed in large pores where oxygen and nutrient movement are not limiting. A point is reached where the POM is no longer of a size sufficient to maintain the integrity of macroaggregates against applied disruptive forces. As a result, the macroaggregate fractures on exposure to applied stresses and produces smaller macroaggregates, microaggregates, or a combination of both. The size distribution of the aggregates released on macroaggregate fracture will be a function of the size of the residual POM and the content of soil binding agents. However, with continued decomposition of the POM, the stability of small macroaggregates will also decrease such that only microaggregates will remain.

In systems where POM is continually being added to the soil, the destruction of macroaggregates by the decomposition of POM is offset by the continual addition of new POM. Thus, as the stability of a portion of macroaggregates is decreasing due to decomposition processes, other macroaggregates are being formed and stabilized, and, in fact, the stabilization of new macroaggregates may occur at the expense of the breakdown of older less stable macroaggregates as depicted in the model in Figure 1. In a given volume of soil, the build up of soil binding agents around a fresh piece of POM may bind some of the soil associated with an older macroaggregate strongly to the fresh POM. The result of exposure to disruptive forces would be the fracturing of the older less stable macroaggregate and the formation of a newer more stable macroaggregate around the fresh POM.

C. Microaggregate-POM

Microaggregates released due to the fracturing of macroaggregates are thought to consist of small fragments of partially degraded plant debris bound in a matrix of mucilages and mineral materials (Figure 4) (Waters and Oades, 1991). Upon treatment with ultrasonic energy, Golchin et al. (1994a, 1994b) demonstrated that solutions with densities of 1.6, 1.8, and 2.0 Mg m^{-3} could be used to separate microaggregate-POM into different fractions based on the degree to which the POM was associated with mineral particles. When compared to microaggregate-POM of a lower density, higher density microaggregate-POM would be more strongly associated with mineral materials and confer greater stability to microaggregate structures (Figure 5). Thus, we have indicated a progressive decrease in the contribution of microaggregate-POM to the stabilization of microaggregates as the density of POM, released by ultrasonic treatment, decreases from 1.8-2.0 Mg m^{-3} to 1.6-1.8 Mg m^{-3} and to then to < 1.6 Mg m^{-3}. This is demonstrated by the decrease in the mineral adsorption onto microaggregate-POM as illustrated in Figure 5 and detailed in Figures 6 and 7. The fraction of ultrasonically dispersed soil contained in the > 2.0 Mg m^{-3} fraction is composed of mineral particles and adsorbed microbial metabolites or other organic molecules which were separated from microaggregate-POM when unstable microaggregates were stressed with ultrasonic energy (Figure 8). These densities are applied to soils where the density of mineral particles is greater than 2.6 Mg m^{-3}.

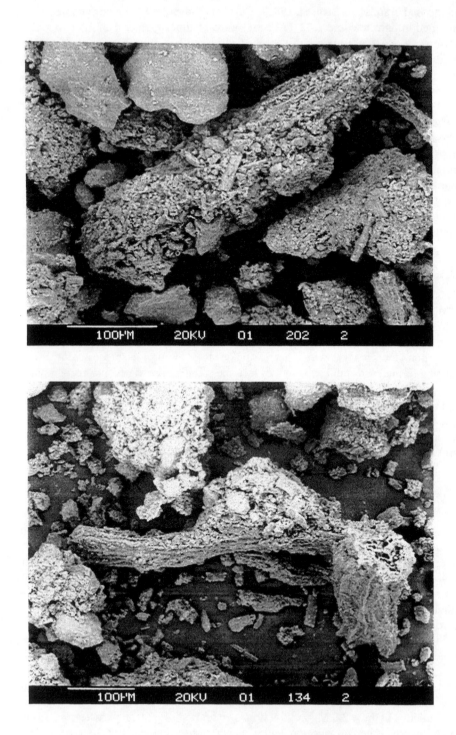

Figure 4. Microaggregate 90-250 µm showing elongate plant fragments coated with inorganic particles.

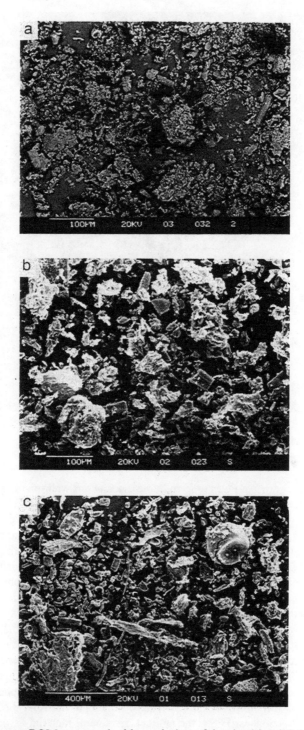

Figure 5. Microaggregate POM separated with a solution of density (a) 1.8-2.0, (b) 1.6-1.8, and (c) < 1.6 Mg m^{-3} showing increase in association of POM and mineral particles with increasing density; from Golchin et al. (1994b).

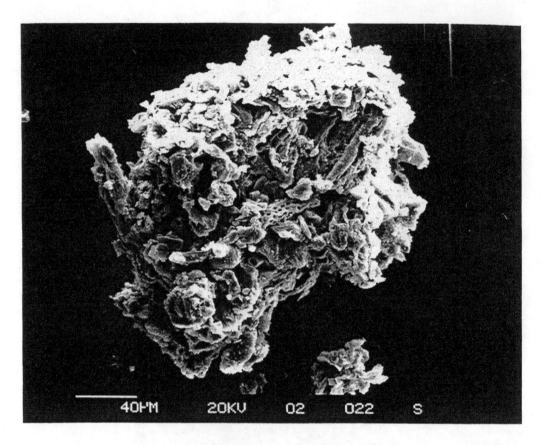

Figure 6. Magnified view of microaggregates contained in fraction 1.8-2.0.

The ability of microaggregate-POM to form stable associations with soil mineral particles is related to the extent of POM decomposition. Microaggregates with a core of carbohydrate rich, relatively undecomposed POM would exhibit a greater stability because the POM is chemically attractive to microorganisms and its decomposition results in the production of mucilages and metabolites capable of binding soil particles. The mucilages and metabolites permeate the coating of mineral particles surrounding the POM and stabilize microaggregates. As POM decomposition proceeds within microaggregates, the more labile portions of POM, such as protein and carbohydrates, are consumed by the decomposers leaving a POM core of organic matter which is biologically more recalcitrant. In microaggregates containing recalcitrant POM, the production of mucilages and metabolites is reduced to the point where it no longer exceeds or equals the rate at which these materials are decomposed and the stability of the microaggregate decreases. SEM has been used to demonstrate the progression from fresh to decomposed materials and the concomitant decrease in the extent of association between the POM and soil mineral particles (Figure 5). Microaggregate-POM found in the 1.8 - 2.0 Mg m^{-3} fraction is seen as intact plant residues showing little morphological alteration and a strong association with soil mineral materials. In progressing to the 1.6-1.8 Mg m^{-3} and then to the <1.6 Mg m^{-3} density fractions of POM, the amount of mineral material associated with the POM decreases and morphological modification indicative of decomposition becomes more apparent. When the POM cores of microaggregates are lost due to complete degradation of plant debris cavities within

A Model Linking Organic Matter Decomposition, Chemistry and Aggregate Dynamics

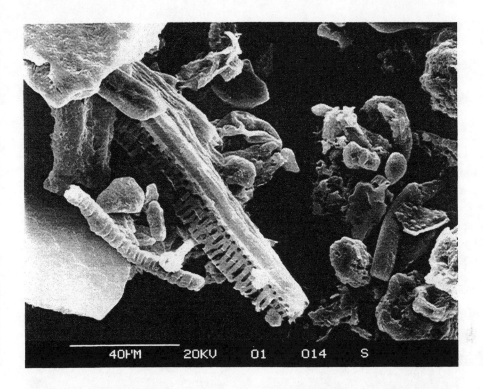

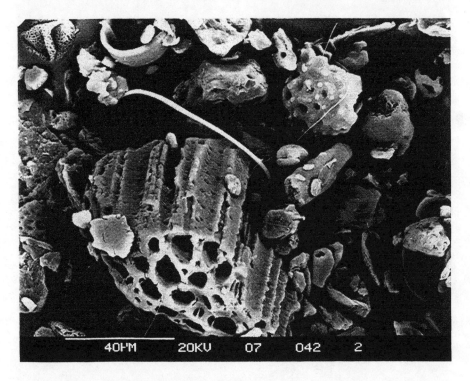

Figure 7. Magnified view of organic particles contained in fraction < 1.6 occluded.

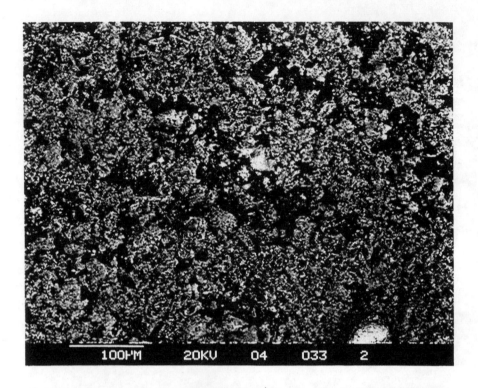

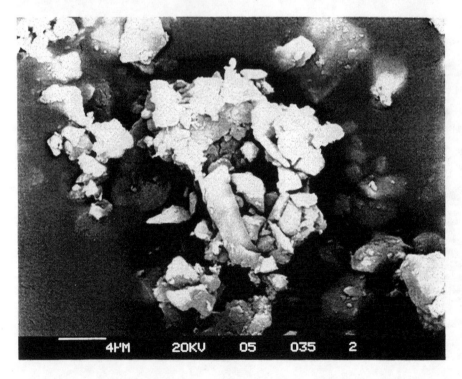

Figure 8. Microaggregates of fraction > 2.0 Mg m^{-3}.

microaggregates may develop, as demonstrated by Waters and Oades (1991) using electron microscopy. Such aggregates would accumulate in the > 2.0 Mg m^{-3} density fractions.

It is important to note that although we have divided microaggregate-POM up into three discrete classes on the basis of density in the proposed conceptual model of soil aggregation, in reality a continuum of POM density would exist and POM particles exhibiting different stages of decomposition could all exist in one density fraction. Our discussion of microaggregate-POM has to be viewed as pertaining to the mean composition of the various density fractions. It is also suggested that the POM found in the 1.8-2.0 Mg m^{-3} density fraction could also be derived from macroaggregates in which decomposition of the POM has progressed to the point that a significant interaction with soil mineral particles has occurred, but the POM has not yet been fragmented.

D. Adsorbed Organic Materials

Adsorption of organic materials on to the reactive surfaces of clay minerals and, to a lesser extent, other mineral particle surfaces in soil is an important mechanism for binding soil particles together. One group of organic compounds which have been implicated in the stabilization of groups of mineral particles are the extracellular polysaccharides produced by soil microorganisms as a metabolic product during the decomposition of various organic substrates. These molecules most frequently carry a net negative charge due to the presence of glucuronic, galacturonic, or mannuronic acids or to nonsugar units (pyruvic or succinic acid groups) (Chenu, 1995), and, therefore, have the potential to adsorb strongly to clay surfaces through polyvalent cation bridging. In addition, uncharged polysaccharides may strongly adsorb to mineral surfaces through hydrogen bonding or van der Waals forces. The strong affinity of such extracellular polysaccharides to the surfaces of clay and other soil minerals, combined with the ability to form intermolecular linkages, make them very effective binding agents (Cheshire et al., 1983; Lynch and Bragg, 1985).

In our model of aggregation, it is envisaged that adsorbed organic molecules, such as microbially derived polysaccharides, play an important role in binding soil particles together over distances <50 μm; that is, the binding together of packets of clays and other small primary particles into small microaggregates with densities >2.0 Mg m^{-3} (Figure 8). It has been shown that some microaggregates are resistant to the dispersive effect of ultrasonic energy and remain in the form of silt and clay sized aggregations of primary soil particles. By transmission electron microscopy, some of these small microaggregates can be identified as bacteria coated with extracellular polysaccharides and clay minerals (Jocteur-Monrozier et al., 1991, Tiessen and Stewart, 1988).

IV. Chemical Changes Associated with Decomposition of POM in Soil Aggregates

Based on the morphology and chemical structure of organic materials contained in the free POM, occluded POM forming the core of aggregates, and colloidal or clay-associated OM, it is proposed that the extent of decomposition of organic materials in soils follows a continuum from relatively fresh plant material in free POM through partially degraded POM residues occluded within aggregates to degraded residues adsorbed on mineral materials. The compositional differences observed between free POM, occluded POM, and clay-associated OM, therefore, represent the changes that OM undergoes during decomposition when it enters the soil, is enveloped in aggregates, and eventually is incorporated into microbial biomass and metabolites and becomes associated with clay minerals.

The free POM is dominated by plant carbohydrates, mainly cellulose, and has a chemical composition similar to plant litter after separation from soil with a liquid of density < 1.6 M g m^{-3}. The C:N ratios of free POM in soils are often between 12 to 30 (Greenland and Ford, 1964); however,

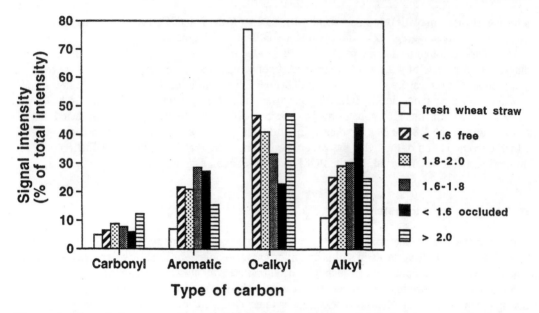

Figure 9. The relative proportions of different types of carbon in fresh wheat straw and density fractions of urbrae red-brown earth under wheat-fallow rotation. Standard deviation associated with the proportion of signal intensity attributed to each type of carbon in replicate analyses of different aliquots of the same sample was < 2% of the total signal intensity.

fresh plant materials added to soil usually have C:N ratios exceeding 30. This indicates that most light fractions or coarse plant debris isolated from soils have already undergone some decomposition and have a 2- or 3-fold concentration of N during the early stages of decomposition before being incorporated into soil aggregates as occluded POM. During the initial stages of the decomposition of plant materials, carbohydrate (cellulose and hemicellulose) and protein structures are degraded. A portion of the degraded carbohydrate and protein carbon is mineralized to carbon dioxide while the remainder is assimilated and converted into microbial tissues and metabolites. The mineralization process accounts for the decrease in the amount of O-alkyl carbon observed in ^{13}C NMR spectra of free particulate organic materials compared with fresh plant materials (Figure 9). As O-alkyl carbon disappears from POM, concentration of recalcitrant organic structures such as alkyl and aromatic carbon increase. The increase observed in the content of alkyl and aromatic carbon during this stage of decomposition probably results from the relatively faster rate of decomposition of carbohydrates. The rapid decomposition of carbohydrate-rich POM is retarded when it is coated by mineral particles. After coating, however, the chemistry of POM controls the rate of decomposition. It has been shown (Golchin et al., 1995a) that the O-alkyl carbon content of occluded organic materials is inversely related to resistance to microbial attack. Thus, as decomposition continues slowly within the soil aggregates, the more labile fractions of POM (mostly carbohydrates) are decomposed by microorganisms, leaving behind the more resistant components such as lignin and alkyl structures (Figure 9). Partially decomposed occluded organic materials separated with a density of < 1.6 Mg m^{-3} represent recalcitrant fractions of plant debris that accumulate in soil aggregates in the early stages of decomposition (Figure 9).

The types and amounts of occluded organic materials are dependent upon the nature of organic matter input to soil and the microenvironment within soil aggregates (Golchin et al., 1995b). A comparison of the relative proportion of different types of carbon in density fractions of two grey clay soils under different vegetative cover indicates the effects of input chemistry on chemical composition of organic materials preserved in soil aggregates (Figure 10). The occluded organic materials from a grey clay under brigalow are high in polymethylene materials and low in aromatic carbon due to the aliphatic nature of organic input to this soil (Figure 10) (Golchin et al., 1994a). In a grey clay under wheat fallow rotation, however, the occluded organic materials have higher aromatic carbon content (Figure 10). The aromatic nature of organic matter in soil under wheat was noted previously (Oades, 1988; Turchenek and Oades, 1979) and was suggested to originate from lignin and its decomposition residues derived from wheat straw and roots (Oades, 1988). It seems the differences in the chemistry of OM inputs exerts a strong influence over the changes in chemical structure observed for the occluded carbon and override the effects of soil type or climate on the chemistry of these materials. Figure 10 also shows the chemistry of occluded organic materials from a red-brown earth under a wheat-fallow rotation. Although this soil has different chemical, physical, and biological properties compared with the cultivated grey clay, the nature of occluded organic materials in both soils is aromatic due to similar organic inputs to these soils (Figure 10).

Due to the recalcitrant nature of aromatic and alkyl carbon structures, occluded organic materials show an accumulation of these structures and have higher alkyl and aromatic carbon content than free organic materials. A comparison of occluded organic materials separated with different densities (Figure 9) indicate that, once the O-alkyl carbon associated with the occluded plant materials has been degraded, the second stage in the decomposition process of occluded POM is initiated. Lignin molecules, previously surrounded by O-alkyl carbon (e.g., cellulose and hemicellulose), are exposed to microbial decomposition and a reduction in the content of aromatic carbon results when they are extensively altered and converted to microbial products (Figure 9). The most persistent organic carbon found in occluded organic materials is that contained in alkyl carbon structures. The chemical shift values associated with these materials (32-35 ppm) indicate that polymethylene type structures are an important component of occluded POM and detailed chemistry has indicated that some of these materials are directly derived from plants. However, as shown by Baldock et al. (1989) and Golchin et al. (1996) microorganisms can also produce significant quantities of alkyl C while utilizing a carbohydrate substrate.

POM occluded within soil aggregates is eventually consumed by microorganisms and microbial products and metabolites are adsorbed to mineral particles which coat the occluded organic particles. In a soil under two different vegetative covers, differences in the chemistry of occluded organic materials become negligible and plant materials from different sources become similar in chemistry when extensively altered by soil organisms and converted to microbial products associated with clay particles. The organic materials adsorbed to clay particles are nitrogen-rich aliphatic materials and include appreciable amounts of polysaccharide (Figure 9). The sharp decreases in aromatic carbon content and increases in O-alkyl carbon in passing from occluded organic materials to clay-associated organic materials indicate that aromatic structures preserved in soil aggregates are eventually degraded by microorganisms and there is a conversion from plant derived O-alkyl carbon to microbially derived O-alkyl carbon (Figure 9).

V. The Effect of Aggregation on Turnover of Organic Matter in Soil

The flux of carbon to and from a SOM pool is a direct function of the bioavailability of the organic materials in that pool. Bioavailability of a SOM pool depends upon the chemistry of the organic materials comprising that pool and external factors operating on that pool as immediate controls. External factors include the environmental (temperature, moisture, oxygen), edaphic (matrix effect, pH, redox potential), and biological effects on bioavailability. Classical approaches for categorizing the bioavailability of SOM pools have been based largely on chemistry by dividing the SOM into humic and nonhumic substances. Humic substances are considered to have low bioavailability because of their complex chemical structure. Although humic molecules are undoubtedly more

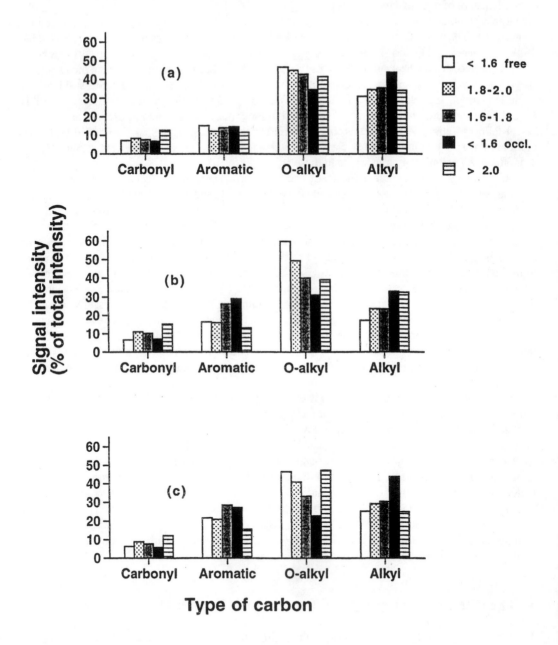

Figure 10. The relative proportions of different types of carbon in density fractions of (a) a grey clay under brigalow, (b) a grey clay under wheat-fallow rotation, and (c) a red-brown earth under wheat-fallow rotation. Standard deviation associated with the proportion of signal intensity attributed to each type of carbon in replicate analyses of different aliquots of the same sample was < 2% of the total signal intensity.

recalcitrant than natural biopolymers, their intrinsic chemical recalcitrance does not account for the observed stabilization of organic matter in soils. Duxbury et al. (1989) showed that old humic acid fractions of soil organic matter, with ages in the range of thousands of years, have a half life on the order of weeks when extracted and added to unextracted soils. Thus, the chemical structures of humic substances do not account for the persistence of organic materials in soils. Recent studies on SOM dynamics, however, indicate that the bioavailability of SOM pools is controlled primarily by external factors (Scholes and Scholes, 1995) and interaction of SOM with the mineral matrix is crucial to SOM behaviour (Greenland, 1965; Huang and Schnitzer, 1986). Interaction of organic substrates with inorganic matrices that result in increased persistence of SOM can occur through occlusion within soil aggregates or adsorption to clay surfaces.

A comparison between the size of bacteria and fungi with that of pore spaces existing in a soil suggests that aggregation should limit microbial access to SOM. Calculations made by Van Veen and Kuikman (1990) indicate that 95% of the pore spaces in a silt loam should be accounted for as pores too small to be accessible to bacteria. Adu and Oades (1987) produced synthetic aggregates that were labeled uniformly with ^{14}C substrates. Aggregates of a sandy loam lost less CO_2 from labeled starch than did unaggregated soil, which they took as evidence for the presence of inaccessible micropores in the aggregates. The inhibitory effect of aggregation on mineralization was not observed for clay soils, or when the substrate was glucose. Similarly, Bartlett and Doner (1988) incorporated lysine and leucine either homogeneously throughout sterilized synthetic aggregates or only on their surfaces. After adding inoculum, more of the amino acid was respired from aggregate surfaces than from within aggregates, indicating a delay in microbial access to substrate within the aggregates. Killham et al. (1993) also observed that ^{14}C-labelled glucose turned over faster when incorporated into larger pores (6-30 μm) than into small pores (< 6 μm).

An approach recently developed by Skjemstad et al. (1993) utilizes high energy ultraviolet photooxidation to remove organic matter external to soil microaggregates (< 50 μm). By applying ultraviolet energy to soil microaggregates, it was shown that in clay-size aggregates, about one fifth of the organic matter was internal and inaccessible to ultraviolet light. In silt-sized aggregates, the content of protected organic matter was higher and accounted for about one third of the total material present. By measuring the age of organic materials using a ^{14}C accelerator mass spectrometer, it was demonstrated that both clay and silt fractions contained modern carbon which was presumably associated with the outsides of aggregates. After 4 h of photooxidation, the internal organic material was shown to be old, with mean residence times near 200 and 300 years BP, respectively. If the accessibility of organic material to ultraviolet light is considered similar to the accessibility to enzymes and/or soil organisms, it must be concluded that incorporation of organic material into soil microaggregates decreases its accessibility to soil organisms as confirmed by the presence of old carbon located within the microaggregates.

The concept of impenetrable microsites has been proposed to account for the increased respiration when soil is disturbed and the inaccessible carbon entrapped within the aggregates is exposed to soil microflora (Rovira and Greacen, 1957). Hassink (1992) measured C and N mineralization in coarse-sieved and fine-sieved samples and concluded that the increased mineralization observed after fine sieving was due to the release of a part of the physically protected organic matter, located in small pores in (or between) aggregates. Fine sieving appeared to expose a physically protected organic fraction which had a lower C-to-N ratio than the total soil organic matter. Additional evidence that aggregation limits the accessibility of organic substrates to soil organisms comes from transmission electron microscopy (TEM) studies. For example, Foster (1988) showed the presence of organic residues of cellular origin within the ultrastructure of aggregates. The cellular materials were physically isolated from decomposer organisms.

Changes in the content and isotopic composition ($\delta^{13}C$) of organic carbon as a consequence of deforestation and pasture establishment were used by Golchin et al. (1995a) to relate the dynamics of SOM to its location within the soil matrix. The free or noncomplexed POM decomposed most

Table 1. Turnover times of C in different density fractions obtained from $\delta^{13}C$ values for forest C in an Oxisol after forest was replaced by grasses. Analytical precision determined as standard deviation obtained on different combustion of the same homogenized sample was better than 0.05‰ $\delta^{13}C$.

Density fractions (Mg m^{-3})	Years under pasture	
	0-35	0-83
< 1.6 free	19	27
1.8-2.0 occluded	24	41
1.6-1.8 occluded	24	53
< 1.6 occluded	40	49
> 2.0 = clay-associated OM	41	73

(Adapted from Golchin et al., 1995a.)

quickly and formed a significant pool for SOM, which turned over rapidly when forest was replaced by pasture (Table 1). Compared with the free POM, the POM occluded within microggregates had a slower turnover time (Table 1). The organic materials associated with clay particles showed the highest stability. The occluded organic materials were in different stages of decomposition and had different chemical compositions. Comparison of the chemistry and isotopic composition of occluded organic materials indicated that O-alkyl carbon content of the occluded organic materials was inversely related to their stabilities whereas their aromatic C content was directly related to their stabilities. At the moment it is not clear to what extent these two modes of stabilization, physical and chemical, are responsible for the stability of POM in soil, but it seems the chemistry or stage of decomposition of POM is an important factor in this respect. In soil, the rapid decomposition of carbohydrate-rich POM is retarded when it is coated by mineral particles. As decomposition proceeds within the microaggregates, POM loses its more labile constituents and becomes more resistant to microorganisms because of its chemistry. Thus, it is difficult to separate the influence of chemical structure and physical protection provided by the soil matrix on bioavailability. This is also true for clay associated organic matter. Clays may impart recalcitrance to a compound by binding the potential substrate and/or by retaining or possibly inactivating enzymes catalyzing the decomposition. Clays also contain polymethylene components with low C:N ratio that are recalcitrant to decomposition (Baldock et al., 1992; Derrene et al., 1993).

VI. Summary and Conclusions

In soils where structural stability is controlled by organic matter, a link exists between organic matter decomposition and the dynamics of soil aggregation. There is also an influence of soil structure on biological activity and thus organic matter decomposition. In this chapter an attempt has been made to identify and integrate the mutual interrelationships of soil structure and organic matter decomposition into a conceptual model. The proposed model assumes that free POM, entering the soil from root or litter, is colonized by microorganisms and adsorbs mineral particles. The free POM is encrusted by mineral particles and becomes the center of water-stable macroaggregates and creates a hierarchy. The size of macroaggregates formed is a function of the size, geometry, and mode of deposition of free POM. Macroaggregate stabilization by POM is a transient process and, due to rapid decomposition and subsequent fragmentation of organic cores of macroaggregates by soil organisms the stability of macroaggregates, declines at about the same rate at which plant material decomposes in soil.

The degradation of macroaggregates creates microaggregates from 20 to 250 μm diameter which are considerably more stable than the macroaggregates. Microaggregates released due to the break down of macroaggregates consist of small fragments of partially degraded plant debris bound in a matrix of mucilages and mineral materials. In contrast to organic cores of macroaggregates, organic cores of microaggregates decompose slowly and are protected from rapid decomposition by encrustation with inorganic materials.

As decomposition continues within microaggregates, the more labile fraction of organic cores such as carbohydrate and protein structures are decomposed by microorganisms, leaving behind the more resistant components such as lignin and alkyl structures. Microbial products, metabolites, and biomass formed as a result of utilization of organic cores are adsorbed to mineral particles which coat the organic cores. Eventually the organic cores are consumed by microorganisms leaving a cavity surrounded by aggregation of particles stabilized by the microorganisms and their metabolic products.

Decomposition of the organic cores renders microaggregates unstable. Microaggregates with relatively undecomposed cores would exhibit a greater stability, because the core is rich in carbohydrate and attractive to microorganisms and its decomposition results in the production of mucilages and metabolites capable of binding soil particles. The produced metabolites and mucilages permeate the coating of mineral particles surrounding the core and stabilized microaggregate. As decomposition of organic cores proceeds, the cores become biologically more recalcitrant and the production of metabolites and mucilages is reduced and the stability of microaggregates decreases.

When unstable microaggregates (< 250 µm) are subjected to disruptive forces, microaggregates of < 20 µm diameter are released. The < 20 µm microaggregates appear to be a random mixture of clay microstructures, biopolymers, and microorganisms.

It is suggested that these three steps in aggregation may apply to soils in which organic matter is an important agent responsible for binding soil particles and creating aggregate hierarchy.

Acknowledgments

The authors would like to thank Ms. Angela Waters for providing the electron micrographs of macroaggregates and Dr. Philip Clarke for NMR analysis.

References

Adu, J.K. and J.M. Oades. 1978. Physical factors affecting decomposition of organic materials in soil aggregates. *Soil Biol. Biochem.* 10:109-115.

Ahmed, M. and J.M. Oades. 1984. Distribution of organic matter and adenosine triphosphate after fractionation of soils by physical procedures. *Soil Biol. Biochem.* 16:465-470.

Aita, C., S. Recous, and D.A. Angers. 1995. Characterization of wheat straw decomposition by the combined use of ^{13}C ^{15}N tracing and soil particle-size fractionation. Conference Abstract, Driven By Nature, on plant litter quality and decomposition. Wye College, University of London, September 17-20.

Amato, M. and J.N. Ladd. 1992. Decomposition of ^{14}C-labelled glucose and legum material in soils: properties influencing the accumulation of organic residue C and microbial biomass C. *Soil Biol. Biochem.* 24:455-464.

Angers, D.A. and G.R. Mehuys. 1989. Effects of cropping on carbohydrate content and water stable aggregation of a clay soil. *Can. J. Soil Sci.* 69:373-380.

Angers, D.A., S. Recous, and C. Aita. 1995. Incorporation of newly-formed organic matter in soil aggregates during labelled wheat straw decomposition. Conference Abstract, Driven By Nature, on plant litter quality and decomposition. Wye College, University of London, September 17-20.

Baldock, J.A., J.M. Oades, A.G. Waters, X. Peng, A.M. Vassallo, and M.A. Wilson. 1992. Aspects of the chemical structure of soil organic materials as revealed by solid-state ^{13}C NMR spectroscopy. *Biogeochem.* 16:1-42.

Baldock, J.A., J.M. Oades, A.M. Vassallo, and M.A. Wilson. 1989. Incorporation of uniformly labelled ^{13}C-glucose carbon into the organic fractions of a soil. Carbon balance and CP/MAS ^{13}C NMR measurements. *Aust. J. Soil Res.* 27:725-746.

Bartlett, J.R. and H.E. Doner. 1988. Decomposition of lysine and leucine in soil aggregates: adsorption and campartmentalization. *Soil Biol. Biochem.* 20:755-759.

Beare, M.H., P.F. Hendrix, and D.C. Coleman. 1994. Water-stable aggregates and organic matter fractions in conventional and no-tillage soils. *Soil Sci. Soc. Am. J.* 58:777-786.

Cadisch, G., H. Imhof, S. Urquiaga, B. Boddey, and K.E. Giller. 1996. Carbon turnover (d^{13}C) and nitrogen mineralization potential of particulate light soil organic matter after rainforest clearing. *Soil Biol. Biochem.* (in press)

Capriel, P., T. Beck, H. Borchert, and P. Harter. 1990. Relationship between aliphatic fraction extracted with supercritical hexane, soil microbial biomass and soil aggregate stability. *Soil Sci. Soc. Am. J.* 54:415-420.

Chenu, C. 1995. Extracellular polysaccharides: An interface between microorganisms and soil constituents. p. 217-233. In: P.M. Huang, J. Berthelin, J.-M. Bollag, W.B. McGill, and A.L. Page (eds.), Environmental Impact of Soil Component Interactions. Natural and Anthropogenic Organics. CRC Press, Boca Raton.

Chenu, C., E. Besnard, J. Balesdent, P. Puget, and D. Arrouays. 1996. Particulate organic matter dynamics and location in the soil structure across a cultivated sequence. *Eur. J. Soil Sci.* (in press).

Cheshire, M.V., G.P. Sparling, and C.M. Mundie. 1983. Effect of periodate treatment of soil and carbohydrate constituents and soil aggregation. *J. Soil Sci.* 34:105-112.

Christensen, B.T. 1987. Decomposability of organic matter in particle size fractions from field soils with straw incorporation. *Soil Biol. Biochem.* 19:429-435.

Dalal, R.C. and R.J. Mayer. 1986. Long-term trends in fertility of soils under continuous cultivation and cereal cropping in southern Queensland. IV. Loss of organic carbon from different density fractions. *Aust. J. Soil Res.* 24: 31-39.

Derrene, S., C. Largeau, and F. Taulelle. 1993. Occurrence of non-hydrolyzable amides in the macromolecular constituent of *Scenedesmus quadricauda* cell wall as revealed by ^{15}N NMR of *n*-alkylnitriles in pyrolysate of ultralaminae containing kerogens. *Geochim. Cosmochim. Acta* 57:851-858.

Dexter, A.R. 1988. Advances in characterization of soil structure. *Soil Tillage Res.* 11:199-238.

Duxbury, J.M., M.S. Smith, and J.W. Doran. 1989. Soil organic matter as a source and a sink of plant nutrients. p. 33-67. In: D.C. Coleman, J.M. Oades, and G. Uehara (eds.), Dynamics of Soil Organic Matter in Tropical Ecosystems. University of Hawaii Press, Honolulu.

Edwards, A.P. and J.M. Bremner. 1967. Domains and quasicrystalline regions in clay systems. *Soil Sci. Soc. Am. Proc.* 35:650-654.

Elliott, E.T. 1986. Aggregate structure and carbon, nitrogen and phosphorus in native and cultivated soils. *Soil Sci. Soc. Am. J.* 50:627-633.

Foster, R.C. 1988. Microenvironments of soil microorganisms. *Biol. Fert. Soils* 6:189-203.

Golchin, A., J.M. Oades, J.O. Skjemstad, and P. Clarke. 1994a. Study of free and occluded particulate organic matter in soil by solid state ^{13}C CP/MAS NMR spectroscopy and scanning electron microscopy. *Aust. J. Soil Res.* 32:285-309.

Golchin, A., J.M. Oades, J.O. Skjemstad, and P. Clarke. 1994b. Soil structure and carbon cycling. *Aust. J. Soil Res.* 32:1043-1068.

Golchin, A., J.M. Oades, J.O. Skjemstad, and P. Clarke. 1995a. Structural and dynamic properties of soil organic matter as reflected by ^{13}C natural abundance, pyrolysis mass spectrometry and solid-state ^{13}C NMR spectroscopy in density fractions of an Oxisol under forest and pasrure. *Aust. J. Soil Res.* 33: 59-76.

Golchin, A., P. Clarke, and J.M. Oades. 1996. The heterogeneous nature of microbial products as shown by solid-state ^{13}C NMR spectroscopy. *Biogeochem.* (in press).

Golchin, A., P. Clarke, J.M. Oades, and J.O. Skjemstad. 1995b. The effects of cultivation on the composition of organic matter and structural stability of soils. *Aust. J. Soil Res.* 33:975-993.

Greenland, D.J. 1965. Interaction between clays and organic compounds in soils. Part 2. Adsorption of soil organic compounds and its effect on soil properties. *Soils Fert.* 28:521-532.

Greenland, D.J. and G.W. Ford. 1964. Separation of partially humified organic materials from soils by ultrasonic dispersion. *Trans. 8th Int. Congr. Soil Sci.* 3:137-148.

Gregorich, E.G., B.H. Ellert, and C.M. Monreal. 1995. The turnover of soil organic matter and storage of corn residue carbon estimated from natural ^{13}C abundance. *Can. J. Soil Sci.* 75:161-167.

Gupta, V.V.S.R. and J.J. Germida. 1988. Distribution of microbial biomass and its activity in different soil aggregate size classes as affected by cultivation. *Soil Biol. Biochem.* 20:777-786.

Hadas, A. 1987. Long-term tillage practice effects on soil aggregation modes and strength. *Soil Sci. Soc. Am. J.* 51:191-197.

Hassink, J. 1992. Effects of soil texture and structure on C and N mineralization in grassland soils. *Biol. Fert. Soils* 14:126-134.

Huang, P.M. and M. Schnitzer (eds.). 1986. Interaction of Soil Minerals with Natural Organic and Microbes. SSSA Special Publication No. 17. Soil Sci. Soc. Am., Madison.

Hynes, R.J., R.S. Swift, and R.C. Stephen. 1991. Influence of mixed cropping rotations (pasture-arable) on organic matter content, water stable aggregation and clod porosity in a group of soils. *Soil Tillage Res.* 19:77-87.

Jocteur Monrozier, L., J.N. Ladd, R.W. Fitzpatrick, R.C. Foster, and M. Raupach. 1991. Components and microbial biomass content of size fractions in soils of contrasting aggregation. *Geoderma* 49:37-62.

Killham, K., M. Amato, and J.N. Ladd. 1993. Effect of substrate location in soil and soil pore-water regime on carbon turnover. *Soil Biol. Biochem.* 25:57-62.

Lynch, J.M. and E. Bragg. 1985. Microorganisms and soil aggregate stability. *Adv. Soil Sci.* 30:133-171.

Miller, R.M. and J.D. Jastrow. 1990. Hierarchy of root and mycorrihizal fungal interactions with soil aggregation. *Soil Biol. Biochem.* 22:579-584.

Molope, M.B. and E.R. Page. 1986. The contribution of fungi, bacteria and physical processes in the development of aggregate stability of a cultivated soil. p. 147-163. In: J. M. Lopez-Reed and R. D. Hodges (eds.), The Role of Microorganisms in a Sustainable Agriculture. Academic Publishers, London.

Mortland, M.M. 1970. Clay-organic complexes and interactions. *Adv. Agron.* 22:75-117.

Oades, J.M. 1984. Soil organic matter and structural stability: mechanisms and implications for management. *Plant Soil* 76:319-337.

Oades, J.M. 1988. The retention of organic matter in soils. *Biogeochem.* 5:35-70.

Oades, J.M. 1993. The role of biology in the formation, stabilization and degradation of soil structure. *Geoderma.* 56:377-400.

Oades, J.M. 1995. Recent advances in organomineral interactions: Implication for carbon cycling and soil structure. p. 217-233. In: P.M. Huang, J. Berthelin, J.-M. Bollag, W.B. McGill, and A.L. Page (eds.), *Environmental Impact of Soil Component Interactions. Natural and Anthropogenic Organics.* CRC Press, Boca Raton.

Oades, J.M. and A.G. Waters. 1991. Aggregate hierarchy in soils. *Aust. J. Soil Res.* 29:815-828.

Puget, P., C. Chenu, and J. Balesdent. 1995. Total and young organic matter distribution in aggregates of silty cultivated soils. *Eur. J. Soil. Sci.* 46:449-459.

Rovira, A.D. and E.L. Greacen. 1957. The effect of aggregate disruption on the activity of microorganisms in the soil. *Aust. J. Agric. Res.* 8:659-679.

Scholes, R.J. and M.C. Scholes. 1995. The effect of land use on nonliving organic matter in soil. pp. 209-225. In: R.G. Zepp and Ch. Sonntag (eds.), The Role of Nonliving Organic Matter in the Earth's Carbon Cycle. John Wiley & Sons, Chichester.

Skjemstad, J.O., L.J. Janik, M.J. Head, and S.G. McClure. 1993. High energy ultraviolet photo-oxidation: a novel technique for studying physically protected organic matter in clay- and silt-size aggregates. *J. Soil Sci.* 44:485-499.

Theng, B.K.G. (ed.). 1974. The Chemistry of Clay-organic Reactions. John Wiley & Sons, New York.

Theng, B.K.G. (ed.). 1979. Formation and Properties of Clay-polymer Complexes. Developments in Soil Science 9. Elsevier Scientific Publishing Company, Amsterdam, 362 pp.

Theng, B.K.G. 1982. Clay-polymer interactions: Summary and perspectives. *Clays Clay Miner.* 7:1-17.

Tiessen, H. and J.W.B. Stewart. 1988. Light and electron microscopy of stained microaggregates: the role of organic matter and microbes in soil aggregation. *Biogeochem.* 5:312-322.

Tisdall, J.M. and J.M. Oades. 1979. Stabilization of soil aggregates by the root systems of ryegrass. *Aust. J. Soil Res.* 17:429-441.

Tisdall, J.M. and J.M. Oades. 1980. The effect of crop rotation on aggregation in a red-brown earth. *Aust. J. Soil Res.* 18:423-433.

Tisdall, J.M. and J.M. Oades. 1982. Organic matter and water stable aggregates in soils. *Soil Sci.* 33:141-163.

Turchenek, L.W. and J.M. Oades. 1979. Fractionation of organo-mineral complexes by sedimentation and density technique. *Geoderma* 21:311-343.

Van Veen, J.A. and P.J. Kuikman. 1990. Soil structural aspects of decomposition of organic matter by microorganisms. *Biogeochem.* 11:213-233.

Waters, A.G. and J.M. Oades. 1991. Organic matter in water stable aggregates. p. 163-175. In: W. S. Wilson (ed.), Advances in Soil Organic Matter Research. The Impact on Agriculture and Environment. Roy. Soc. Chem., Cambridge.

CHAPTER 18

Soil Organic Carbon Dynamics and Land Use in the Colombian Savannas I. Aggregate Size Distribution

W. Trujillo, E. Amezquita, M.J. Fisher, and R. Lal

I. Introduction

Soil organic carbon (SOC) affects plant available nutrients and water holding capacity of soils in natural and managed ecosystems and its dynamics influence the environment quality. Therefore, quantitative assessment of SOC in relation to soil aggregates and particles size fractions is required for understanding its dynamics under different ecosystems. Different conceptual and mathematical models have been developed to describe the processes of SOC formation and turnover (Van Veen et al. 1984; Parton et al., 1987). Most of the models include two to three SOC pools that are kinetically defined with different turnover rates. In general, these pools are conceptualized as one small pool with a rapid turnover rate and one to several large pools but of slower turnover rate. It is widely recognized that SOC fractions of 53-200 nm may provide an accurate estimate of the "slow" pool, while those < 53 nm may provide an accurate estimate of the "passive" pool (Woomer et al., 1996).

Several studies have been conducted to evaluate the extent and rate of decrease of SOC and nitrogen contents of the surface layer caused by the conversion of native ecosystems, such as forests to crop (Martins et. al. 1991) and pasture systems (Cerri et al., 1991; Chrone et al. 1991; Desjardins et al. 1994). Land clearing and cultivation induce a lower equilibrium level of SOC because of reduced biomass inputs. Martin et al. (1990) observed 33-45% decrease in SOC after 5 years of continuous cropping. This decline was attributed to mineralization of about 60% of the coarse fractions of SOC. Pastures and vegetated fallow systems have a great potential to increase SOC because of the high root biomass production. Root biomass tends to be less decomposable than that of shoot because of its higher lignin content. Albrecht et al. (1986) reported an increase by 106% in SOC on a 10-year-old pasture (*Digitaria decumbens*) following sugarcane (*Saccharum* spp.) cultivation on a Vertisol in Martinique. They also observed an improvement in soil aggregation and aggregate stability. Comparing the SOC distribution among particle size fractions of a pasture soil to that of a soil under intensive cropping, Feller (1988) observed that grass cover may significantly increase SOC in both coarse and fine fractions.

The objective of this research was to study SOC dynamics under different land use systems in the acid soil of the Colombian savannas. More specifically, field experiments were conducted to evaluate the relative distribution of SOC in different aggregate size fractions in relation to natural ecosytems and improved pastures.

II. Materials and Methods

The field experiment was located at the Centro Nacional de Investigaciones Carimagua (4° 37' N, 71° 19' W) [Corporacion Colombiana de Investigacion Agropecuaria (CORPOICA)/Centro Internacional de Agricultura Tropical (CIAT)]. The region is characterized by mean annual rainfall and temperature of 2240 mm and 26° C, respectively. Sanchez and Salinas (1981) reported that these soils are Oxisols (kaolinitic isohyperthermic Haplustox), deficient in phosphorus, nitrogen, sulphur, magnesium, and zinc, extremely low in exchangeable base contents, especially calcium and potassium, high in trivalent aluminum (saturation up to 90%), and low soil pH (4.0 to 4.8).

A. Sites

There were five soil sampling sites, described below. In each case soil samples were taken from an immediately-adjacent savanna, which was burned frequently (as often as twice a year) and grazed at normal stocking rate for the region (e.g., 5 to 10 ha animal^{-1}).

1. Alegria

Alegria is characterized by a sandy loam soil. Soil samples were taken from an area sown to *Brachiaria decumbens* and *Stylosanthes capitata* in May 1988. At the establishment of the pasture, fertilizer (100 kg P ha^{-1}, 60 kg K ha^{-1}, 40 kg Mg ha^{-1}, and 60 kg S ha^{-1}) was applied. Grazing has been under a rotational system (15:15 days) at a normal stocking rate (2 animals ha^{-1}) for the region. The introduced pasture has never been burned.

2. Yopare

The Yopare location is characterized by a clay loam soil. The native savanna was replaced with *Melinis minutiflora* in 1974. Since *M. minutiflora* failed to regrow after an accidental fire, the pasture was replanted with *Brachiaria humidicola/Desmodium ovalifolium* in 1981. *Desmodium ovalifolium* failed two years later and has not been resown. Fertilizer (100 kg P ha^{-1}, 60 kg K ha^{-1}, 40 kg Mg ha^{-1}, and 60 kg S ha^{-1}) was applied in 1981. A rotational grazing system has been used with a normal stocking rate (2 animals ha^{-1}). The introduced pasture has not been burned since 1981.

3. Introductions II

In 1984, this site with a clay loam soil was sown to *Brachiaria humidicola* with the legume *Desmodium ovalifolium*, which failed. In 1987, it was resown to *Brachiaria humidicola* alone or with *Arachis pintoi*. Fertilizer (100 kg P ha^{-1}, 60 kg K ha^{-1}, 40 kg Mg ha^{-1}, and 60 kg S ha^{-1}) was applied at the time of pasture establishment in 1984 and every other year thereafter. A rotational grazing system (7:7 days) has been used at a stocking rate of 2 animals ha^{-1} since 1988.

4. Culticore

This location with a clay loam soil has been under a crop rotational system since May 1993. At the establishment of rice (*Oryza sative*), plots were fertilized as follows: 60 kg P ha^{-1}, 100 kg K ha^{-1}, 10 kg S ha^{-1}, 10 kg Zn ha^{-1}, 500 kg Ca ha^{-1}, and 50 kg Mg ha^{-1}. Immediately after harvesting, cowpea *(Vignia unguiculata)* was planted and later (November) incorporated into the soil as green manure. The same cropping scheme was used in each subsequent year. Soil samples were manually taken from areas sown to rice and rice/cowpea in rotation in May 1995.

5. Tabaquera

This site is characterized by a sandy clay loam soil. Soil samples were taken from a native savanna area frequently burned (at least twice a year; last fire in May 1995) and from an area which has not been burned for at least 11 years. The burned area is dominated by native grasses, while the unburned area is dominated by native shrubs and trees.

B. Soil Sampling

Soil samples were taken at six depths (0-10, 10-20, 20-40, 40-60, 60-80, and 80-100 cm) at four places within a 1-ha plot in each sampling site (Table 1) using a hydraulic core probe attached to a tractor. Each soil depth was represented by 16 samples. Soil samples were fractionated into four aggregate sizes, 4-2, 2-1, 1-0.125, and < 0.125 mm, by dry sieving using a rotary sieve (Chepil, 1962; Lyles et al., 1970) and analyzed for total SOC and total soil nitrogen (TSN) contents. Total SOC was determined by the wet combustion method of Walkley-Black (Nelson and Sommers, 1982) and TSN was determined by the Kjeldahl method (Bremner and Mulvaney, 1982). The data were analyzed using the SAS procedures (SAS, 1994) for a randomized complete block design with a factorial arrangement. Means were compared using the least significant difference (LSD) procedure at 0.05 probability level. The data for each depth of each site were treated as independent samples, and the standard error of each mean and standard error for the difference between means were calculated in order to compare the systems between and within sites. Differences between the means were tested for statistical significance with Student's t-test at 0.05 probability level.

III. Results and Discussion

A. Alegria

1. *B. decumbens / S. Capitata*

There was a significant interaction between soil depth and aggregate size fraction for total SOC (Figure 1a) and C:N ratio (Table 2). The highest SOC was observed in the finest aggregate size fraction (<0.125 mm) for all soil depths. The differences among the aggregate size fractions were much greater in the upper 40 cm depth than lower in the soil profile, where no significant differences among the aggregate size fraction were observed. There were no significant differences in soil C:N ratios between the aggregate size fractions at 10-20, 20-40, and 80-100 cm depths (Table 2). At 0-10 cm depth, the C:N ratio (12.0) in the 2-1 mm fraction was up to 31% lower than that in other fractions. At 40-60 cm depth, the C:N ratio decreased with decreasing aggregate size fraction. There was a significantly lower C:N ratio (20.4) in the <0.125 mm. At 60-80 cm, the 1-0.125 mm fraction had a lower soil C:N ratio (17.5) than those in the 4-2 (22.5), 2-1 (23.6), and <0.125 (29.0) mm.

Table 1. Soil texture and age of the different land use systems sampled at Carimagua, Meta Colombia

Site	Soil texture[1]	Pasture/crop	Age (yrs)
Algeria	Sandy loam	B. decumbens / S. capitata	7
		Savanna	
Yopare	Clay loam	B. humidicola	14
		Savanna	
Introductions II	Clay loam	B. humidicola	10
		B. humidicola / A. pintoi	8
		Savanna	
Culticore	Clay loam	Rice	May 1995
		Rice / Cowpea	May 1995
		Savanna	
Tabaquera	Sandy clay loam	Unburned	May 1995
		Burned	11

[1]Kaolinitic isohyperthermic Haplustox

Table 2. C:N ratios for different aggregates size fractions and depths of a sandy loam soil under *B. decumbens/S. capitata* and native savanna

Soil depth (cm)	Aggregate size fraction (mm)									
	4-2	2-1	1-0.125	<0.125	Mean	4-2	2-1	1-0.125	<0.125	Mean
	—*B. decumbens / S. capitata*—					—Native savanna—				
0-10	17.5	12.0	13.5	15.6	14.7	16.9	14.3	18.4	12.6	15.6
10-20	15.2	19.1	20.4	17.2	18.0	22.6	20.4	21.2	18.9	20.8
20-40	22.7	22.9	21.1	20.1	21.7	33.2	23.4	30.0	29.3	24.0
40-60	25.9	25.9	23.1	20.4	23.8	21.2	22.9	26.1	25.4	23.9
60-80	22.5	23.6	17.5	29.0	23.2	30.0	22.3	28.6	29.7	27.7
80-100	23.6	26.0	22.6	26.7	24.7	20.3	25.3	22.9	25.7	23.6
Mean	20.7	21.6	19.7	21.5		24.0	21.4	24.5	23.6	$LSD_{0.05}$= 4.0
	—$LSD_{0.05}$ = 5.0—					—$LSD_{0.05}$ = 4.2—				

2. Native Savanna

There was no significant interaction between soil depth and aggregate size for total SOC and soil C:N ratios. Total SOC varied with aggregate size fractions and soil depths, while C:N ratio significantly changed with soil depth only. Total SOC was significantly higher in the 4-2 mm fraction (0.95%) as compared to those in the 2-1 (0.78%) and 1-0.125 mm (0.82%) fractions (Fig. 1b). As the size fraction decreased from 2-1 mm to < 0.125 mm, total SOC significantly increased by about 4.9%. With increasing depth, SOC increased significantly by 18.4% from the soil surface (0.93%) to 40 cm (1.14%), but decreased drastically by 36.8% at 60 cm (0.72) and remained approximately constant (0.70 to 0.71%) at deeper layers. The same trend was observed with C:N ratios (Table 2), which increased by 33.9% from the soil surface (15.6) to 100 cm depth (23.6).

A significant difference in SOC between *B. decumbens/S. capitata,* and the native savanna was observed. The differences ranged from 0.45% at the soil surface to 0.1% at 20 cm depth. With increasing depth, the differences were much smaller (0.01%) as well as not significant. In general, SOC under the improved pasture was 2.3% (average weighed to 100 cm depth) higher than that under native savanna.

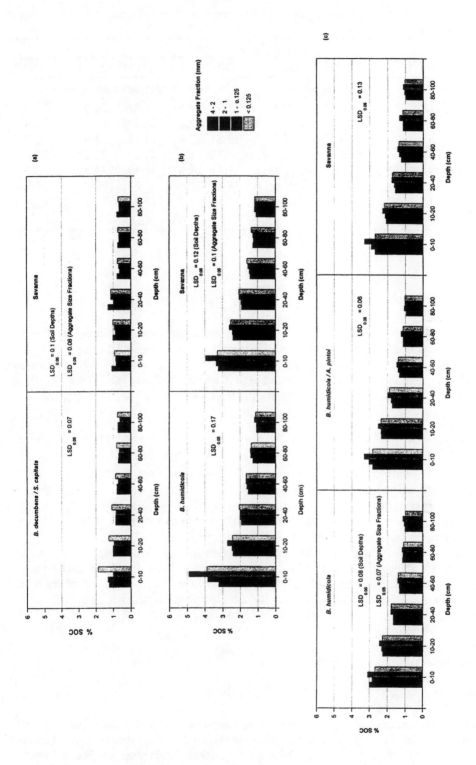

Figure 1. Total soil organic carbon (SOC) in different aggregate size fractions and soil depths in a (a) sandy loam, (b) and (c) clay loam soils under different land use systems

Table 3. C:N ratios for different aggregates size fractions and depths of a clay loam soil under *B. humidicola* and native savanna

Soil depth (cm)	Aggregate size fraction (mm)									
	4-2	2-1	1-0.125	<0.125	Mean	4-2	2-1	1-0.125	<0.125	Mean
	—————B. humidicola—————					—————Native savanna—————				
0-10	20.5	21.1	21.9	19.8	20.8	21.7	20.5	21.0	23.0	21.5
10-20	20.5	23.5	21.5	22.2	21.9	22.2	22.0	22.5	23.7	22.6
20-40	22.7	20.3	21.4	23.4	22.1	23.6	23.6	22.1	23.4	23.2
40-60	23.6	22.4	22.5	25.4	23.4	24.7	21.6	22.0	24.0	23.1
60-80	22.5	24.4	21.9	22.3	22.8	21.8	21.6	20.3	24.4	22.0
80-100	18.5	19.2	23.2	22.9	20.9	28.2	28.6	26.0	32.4	28.8
Mean	21.3	21.8	22.1	22.8	$LSD_{0.05}$=2.3	22.8	23.7	23.0	22.3	$LSD_{0.05}$=2.7
	———$LSD_{0.05}$ = 1.8———					———$LSD_{0.05}$ = 2.2———				

B. Yopare

1. *B. humidicola*

There was a significant interaction between soil depth and aggregate size fraction for total SOC (Figure 1b), while C:N ratio was not affected by either soil depth or aggregate size fraction (Table 3). Total SOC decreased with increasing soil depth for all aggregate size fractions. Differences among the aggregate size fractions were only significant at 0-10 cm depth. Total SOC in the 1-0.125 mm fraction was higher (by 20.8-34.1%) than that in other fractions.

2. Native Savanna

There were significant effects of aggregate size fraction and soil depth on SOC and C:N ratio. The 1-0.125 mm size fraction had 10.5 to 7.6% more SOC than 4-2 mm and 2-1 mm size fractions, repectively. In addition, SOC decreased 66.8% from the surface (3.46%) to 100 cm depth (1.15%), while C:N ratio increased 25.3% from the surface to 100 cm depth with no significant differences among the aggregate size fraction (Table 3).

Brachiaria humidicola made a significant contribution to total SOC compared to the native savanna. Under the improved pasture (1.85% average weighed to 100 cm depth), total SOC was 3.8% higher than that under native savanna (1.78% average weighed to 100 cm depth). C:N ratios under the improved pastures (23.0 average weighed to 100 cm depth) were not significantly different from those under native savanna (23.9 average weighed to 100 cm depth).

C. Introductions II

1. *B. humidicola*

Aggregate size fraction and soil depth had significant effects on total SOC. There was a significant 63.9% decrease in SOC from the soil surface (2.91%) to 100 cm (1.05%) depth (Fig. 1c). Total SOC was up to 7.6% higher in the 1-0.125 mm (1.85%) fraction than that in other fractions. There were no significant differences among 4-2 (1.74%), 2-1 (1.73%), and <0.125 mm (1.71%) fractions. The

C:N ratio increased by 14.5% with increasing depth and decreased by 6.3% with decreasing aggregate size fraction (Table 4).

2. *B. humidicola/A. pintoi*

There was a significant interaction between soil depth and aggregate size for total SOC. Total SOC decreased with increasing soil depth regardless of the aggregate size (Figure 1c). There were significant differences among aggregate size fractions at 0-10 and 10-20 cm depths. At 0-10 cm depth, the 1-0.125 mm size fraction contained (3.95%) 18.0 to 14.9% more SOC than the other fractions, while at 10-20 cm depth, it was higher by only 8.0 to 3.1%. Aggregate size fraction had no significant effect on C:N ratio, while increasing depth significantly increased C:N ratio by 15.5% from the soil surface (18.5) to 60 cm (21.3) (Table 4). At deeper layers, C:N ratios did not significantly change.

3. Native Savanna

There was a significant interaction between soil depth and aggregate size fraction for total SOC. With increasing depth, SOC decreased regardless of the aggregate size fraction. (Figure 1c). Total SOC was significantly higher by 17.4 to 6.8% in the 1-0.125 mm aggregate size fraction as compared to that in other size fractions from the soil surface to 80-cm depth. At 80-100 soil depth, there were no significant differences among aggregate size fractions. C:N ratio was not affected by soil depth and aggregate size fraction (Table 4).

The improved pastures (*B. humidicola* alone and *B.humidicola/A. pintoi*) made a significant contribution to SOC compared to that of native savanna. The contribution of the grass-legume association was higher than that of the grass alone. Total SOC under native savanna (1.57% average weighed to 100 cm depth) were 2.5 and 3.7% lower than that under *B. humidicola* (1.59% average weighed to 100 cm depth) and *B. humidicola/A. pintoi* (1.63% average weighed to 100 cm depth), respectively.

D. Culticore

1. Rice

Soil depth and aggregate size fraction had significant effects on total SOC, while C:N ratio did not significantly change (Table 5). Total SOC decreased by 56.3% with increasing depth (Figure 2a). A higher reduction in SOC was observed from the soil surface (3.0%) up to 60 cm (1.56%) depth. Higher SOC was associated with the 1-0.125 mm (2.0%) aggregate size fraction as compared with that in other fractions. Lower SOC was found in the coarse (4-2 mm) fraction (1.85%).

2. Rice/Cowpea Rotation

Total SOC decreased with increasing soil depth regardless of the aggregate size (Figure 2a). There were significant differences among the aggregate sizes at 0-10 cm depth. Total SOC in 1-0.125 mm aggregate size fraction (2.0%) was 9.5%, 6%, and 2.5% higher than that in 4-2 mm, 2-1 mm, and <0.125 mm size fractions, respectively. There were no significant effects of soil depth and aggregate size fraction on soil C:N ratio (Table 5).

Table 4. C:N ratios for different aggregate size fractions and depth of a clay loam soil under *B. humidicola*, *B. humidicola/A. pintoi*, and native savanna

Soil depth (cm)	Aggregate size fraction (mm)														
	B. humidicola					*B. humidicola / A. pintoi*					Native savanna				
	4-2	2-1	1-0.125	<0.125	Mean	4-2	2-1	1-0.125	<0.125	Mean	4-2	2-1	1-0.125	<0.125	Mean
0-10	21.0	18.4	18.5	17.4	18.9	18.9	18.7	18.0	18.4	18.5	19.1	20.7	22.0	22.4	21.0
10-20	21.5	21.8	20.0	21.3	21.1	21.7	20.4	18.8	20.4	20.3	22.5	22.2	23.6	22.4	22.7
20-40	21.3	21.6	21.2	23.3	21.8	21.0	20.0	19.7	19.8	20.1	21.1	22.3	21.3	21.3	22.5
40-60	21.3	18.5	20.0	20.8	20.1	20.5	22.9	21.0	23.2	21.9	24.7	23.2	20.4	20.7	22.2
60-80	23.7	21.5	23.5	19.6	22.1	21.5	20.8	22.6	21.1	21.5	19.8	20.8	21.8	20.7	20.8
80-100	24.6	20.8	20.4	22.4	22.1	20.5	23.1	19.8	21.6	21.3	20.8	24.9	23.3	22.4	22.9
Mean	22.2	20.4	20.6	20.8		20.8	21.0	20.0	20.7		21.4	22.3	22.1	21.6	
	$LSD_{0.05}$ = 1.6				$LSD_{0.05}$ = 1.9	$LSD_{0.05}$ = 1.3				$LSD_{0.05}$ = 1.6	$LSD_{0.05}$ = 1.7				$LSD_{0.05}$ = 2.1

Table 5. C:N ratios for different aggregate size fractions and depths of a clay loam soil under rice, rice/cowpea, and native savanna

Soil depth (cm)	Aggregate size fraction (mm)														
	Rice					Rice/cowpea					Native savanna				
	4-2	2-1	1-0.125	<0.125	Mean	4-2	2-1	1-0.125	<0.125	Mean	4-2	2-1	1-0.125	<0.125	Mean
0-10	21.7	20.1	20.4	20.4	20.7	21.3	20.6	21.3	20.8	21.0	23.7	22.9	26.9	23.3	24.2
10-20	22.1	23.1	23.0	22.0	22.5	19.8	19.8	19.4	19.3	19.7	28.2	26.4	28.2	27.2	27.5
20-40	25.9	24.3	22.9	24.0	24.3	19.3	18.2	20.9	20.7	19.3	27.1	26.3	30.1	34.0	30.5
40-60	26.2	25.3	27.2	27.3	26.5	16.9	17.6	16.4	16.0	16.8	41.0	37.6	38.0	35.0	37.9
60-80	24.0	24.4	23.9	26.6	24.7	22.0	18.0	19.1	21.1	20.1	37.0	44.1	43.9	51.7	44.2
80-100	27.4	25.6	31.3	26.0	28.6	19.1	19.6	20.5	20.5	19.9	41.0	40.6	48.0	45.5	43.8
Mean	24.6	24.5	24.8	24.4	LSD$_{0.05}$ = 2.6	19.5	19.1	19.5	19.7	LSD$_{0.05}$ = 1.8	32.9	36.6	36.1	36.1	LSD$_{0.05}$ = 6.4
	LSD$_{0.05}$ = 2.1					LSD$_{0.05}$ = 1.5					LSD$_{0.05}$ = 5.2				

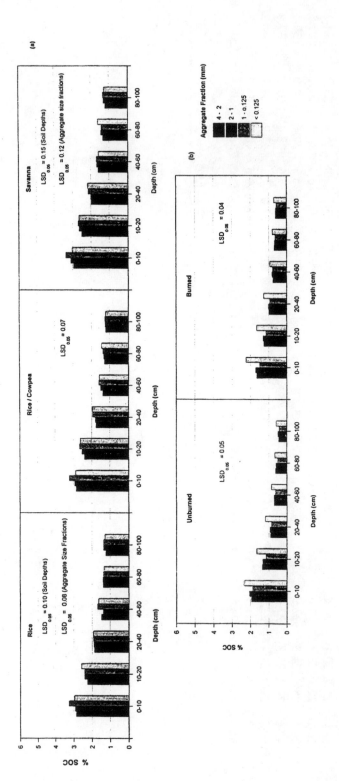

Figure 2. Total soil organic carbon (SOC) in different aggregate size fractions and soil depths in (a) clay loam, and (b) loamy soils under different land use systems.

3. Native Savanna

Total SOC was significantly affected by soil depth and aggregate size fraction, which decreased with increasing depth, while increased with decreasing size fraction (Figure 2a). There was a 47.6% decline in SOC content from the soil surface (3.11%) to 60 cm depth (1.63%). A significantly lower SOC (1.92%) was associated with the coarse fraction (4-2 mm) as compared to that in the 2-1 mm (2.01%), 1-0.125 mm (2.10%), and <0.125 mm (2.05%) fractions (Figure 2a). The C:N ratio increased by 44.7% with increasing soil depth with no significant differences among the aggregate size fractions (Table 5).

There was no significant difference between the crop systems (rice alone and rice and cowpea in rotation) in SOC. The rice/cowpea system showed a lower average weighed SOC (1.74% up to 100 mm) as compared to that under rice alone (1.77% up to 100 cm) as well as under native savanna (1.85% up to 100 cm). Therefore, there was a 5.9 to 4.3% reduction in SOC up to 100 cm depth with cultivation during a period of 3 years. A lower average weighed C:N ratio up to 100 cm depth was found under rice/cowpea (19.3) as compared to rice alone (25.1) and native savanna (36.5).

E. Tabaquera

1. Unburned Native Savanna

The SOC significantly decreased with increasing soil depth regardless of the aggregate size fraction (Figure 2b). A significantly higher (29.2 to 17.0%) SOC was associated with the finer fraction (<0.125 mm) up to 100 cm depth. In contrast, C:N ratios increased by 39.6% and 13.3% with increasing soil depth and decreasing aggregate size fraction, respectively (Table 6).

2. Burned Native Savanna

Total SOC decreased with increasing soil depth regardless of the aggregate size fraction (Figure 2b). A significant higher (by 19.1 to 33.2%) SOC was associated with the finer aggregate size fraction (<0.125 mm) as compared to that in other fractions. Soil C:N ratio increased by 42.6% with increasing soil depth (Table 6), but did not change significantly with decreasing aggregate size fraction.

There were no significant differences in total SOC between unburned (average weighed 0.90% up to 100 cm depth) and burned (average weighed 0.95% up to 100 cm depth) native savannas.

The soil under the 14-year-old pasture at Yopare contained greater SOC (average weighed 1.87% to 100 cm) than the 10-year-old pasture (average weighed 1.59% to 100 cm) at Introductions II. The contribution of the oldest pasture to total SOC was greater in the upper 40 cm than at deeper layers. There was no a significant difference between the two pastures with regard to C:N ratio (22.1 at Yopare vs. 21.2 at Introductions II).

IV. General Discussion and Conclusions

Total SOC diminished with soil depth regardless of the grass species (introduced and native grasses), crop, soil texture, pasture, and burning treatments. Reductions in total SOC with increasing depth are highly correlated with the reduction of root biomass with increasing depth (data not presented) probably coupled with lower microbial activity as a result of low O_2 concentration and biomass inputs.

Table 6. C:N ratios for different aggregates size fractions and depths of a loam soil under unburned and burned savanna

Soil depth (cm)	Aggregate size fraction (mm)									
	4-2	2-1	1-0.125	<0.125	Mean	4-2	2-1	1-0.125	<0.125	Mean
	Unburned savanna					Burned savanna				
0-10	17.1	17.8	17.6	18.3	17.7	20.8	20.8	20.8	20.1	20.6
10-20	21.6	19.9	16.2	22.4	20.0	22.8	24.1	21.3	24.9	23.3
20-40	23.6	23.9	21.8	24.0	23.3	25.1	29.6	26.6	25.6	26.7
40-60	25.2	28.4	27.6	35.9	29.3	33.3	32.9	33.3	30.4	32.5
60-80	37.8	35.7	28.3	33.7	33.9	39.6	38.5	30.4	42.4	37.7
80-100	26.9	29.6	25.6	35.4	29.3	36.0	35.7	32.4	39.4	35.9
Mean	25.4	29.6	25.6	29.3	$LSD_{0.05}=3.6$	29.6	30.3	27.5	30.5	$LSD_{0.05}=2.9$
	$LSD_{0.05}=2.9$					$LSD_{0.05}=2.4$				

The results also showed a decrease in total SOC with cultivation. Land management altered the pattern of residue input and affected SOC in both cultivated and grazed land. Cultivation induced a lower equilibrium level of SOC, partly because reduced biomass input and removal of biomass (Woomer et al., 1994a and b). Martins et al. (1991) also suggested that the declines in SOC were primarily due to rapid mineralization of the coarse fraction of organic matter. Resck et al. (1993) also reported declines of SOC contents under continuous cultivation of soybean *(Glycine max)* as a result of a decrease in total residue inputs and increase in microbial activities resulting from liming and fertilization in the Cerrados of Brazil.

The highest SOC was associated with aggregates size fractions < 1 mm regardless of the soil texture and depth, pasture (grass alone and grass-legume mixture), and crop systems. This finding indicates that the aggregate size fractions < 1 mm appear to have an important role in C sequestration in the Colombian savanna ecosystem under the current land use practices. The high SOC in this aggregate size fraction might have been due to a greater proportion of silt and clay particles in relation to sand particles. Bonde et al. (1992) indicated that the sand size fraction only accounts for 7 to 17% of whole soil C, while the silt size fraction accounts for 23-35%. In contrast, the clay size fraction contained most of the SOC (48-67%). In addition, Dalal and Mayer (1986) and Zhang et al. (1988) reported that the C associated with the sand size fraction decomposed more rapidly than that of other size fractions. It is widely accepted (Elliot, 1986; Cambardella and Elliot, 1988; Woomer et al., 1994b) that the clay size fraction provides protection (physical and/or chemical) for the C against microbial and enzymatic degradation.

This study also confirmed the potential of improved pastures to increase total SOC as suggested by Albrecht et al. (1986) and Fisher et al. (1994). The differences in total SOC contents between improved pastures and native savanna species may be the result of a greater growth rate and net primary productivity of the improved pastures, probably stimulated by the applied fertilizers (50 kg P ha^{-1}) at the establishment and coupled with lower quality and decomposition dynamics of the belowground residues.

The soil C:N ratios under improved pasture, crop, and native savanna (average value 21.5) systems were wider than the commonly accepted values (10-12) reported for other tropical and temperate soils (Jenny, 1980; Parton et al. 1987; Schlesinger, 1995) and suggest that the C and N flows in these soils are radically different. Fisher et al. (1995) reported above and below ground-littler C:N ratios for introduced pastures in the Colombian savannas in the range of 72-194 and 158-224, respectively, and a C:N ratio of 33.2 after 9 years, and concluded that the C:N ratio of the newly SOC accumulated would be very high and, thus, more resistant to microbial decomposition and consequently with longer residence times. In order to narrow the C:N ratio to the accepted figure of 10 to 12, an amount of 1.5 Mg N ha^{-1} yr^{-1} is required, which under the Colombian savanna conditions (low soil N) and management practices (no N fertilization) is difficult to achieve since it seems unlikely that pasture with an associated legume component could have net N inputs in excess of about 160 kg ha^{-1} yr^{-1} (Thomas et al. 1995). In the case of grass alone, the only sources of N are from rainfall (about 8-10 kg N ha^{-1} yr^{-1}) and associative fixation (about 5-35 kg N ha^{-1} yr^{-1}, Boddey and Doberiner, 1988). In order to understand the C sequestration dynamics under the Colombian savannas, further research must be conducted to quantify the amount of N and determine the flow of N to narrow the C:N ratio to 10-12.

References

Albrecht, A., M. Brossard, and C. Feller. 1986. Etude de la matiere organique des sols par fractionnement granulometrique. 2. Augmentation par une prairie *Digitaria decumbens* du stock organique de Vertisols cultives en Martinique. In: Transactions 13th Congress, Hamburg, Germany.

Bonde, T. A, B. T. Christensen, and C. C. Cerri. 1992. Dynamics of soil organic matter as reflected by natural ^{13}C abundance in particle size fractions of forested and cultivated Oxisols. *Soil Biology and Biochemical* 23:275-277.

Boddey, R. M. and J. Dodereiner. 1988. Nitrogen fixation associated with grasses and cereals; recent results and perspectives for future research. *Plant and Soil* 108:53-65.

Bremner, J. M. and C.S. Mulvaney. 1982. Nitrogen - Total. p. 595-624. In: A.L. Page, R. H. Miller, and D. R. Keeney (eds.), *Methods of Soil Analysis. Part 2. Chemical and Microbiological Properties*. American Society of Agronomy, Inc. and Soil Science Society of America, Inc. Madison, WI.

Cambardella, C. A. and E. T. Elliot. 1988. Carbon and nitrogen distribution in aggregates from cultivated and native grassland soil. *Soil Sci. Soc. Amer. J.* 57:1071-1076.

Cerri, C. C., B. Volkoff, and F. Andreux. 1991. Nature behavior of organic matter in soil under natural forest and after deforestation, burning, cultivation near Manaus. *Forest Ecology and Management* 38:247-257.

Chepil, W. S. 1962. A compact rotary sieve and the importance of dry sieving in physical soil analysis. *Soil Sci. Soc. Amer. Proc.* 26:4-6.

Chrone, T.F., F. Andreux, J.C. Correa, B. Volkoff, and C.C. Cerri. 1991. Changes in organic matter in Oxisols from the Central Amazonia forest during eight years as pasture determined by ^{13}C isotopic composition. In: J. Berthelin (ed.), *Diversity of Environmental Biogeochemistry*. Elsevier, Amsterdam, The Netherlands.

Dalal, R.C. and R.J. Mayer. 1986. Long-term trends in fertility of soils under continuous cultivation and cereal cropping in southern Queensland: III. Distribution and kenetics of soil organic carbon in particles size fractions. *Australian J. Soil Res.* 70:395-402.

Desjardins, T., F. Andreux, B. Volkoff, and C.C. Cerri. 1994. Organic carbon and ^{13}C in soil and soil-size fractions, and their changes due to deforestation and pasture installation in Eastern Amazonia. *Geoderma* 61:103-118.

Elliot, E. T. 1986. Aggregate structure and carbon nitrogen and phosphorus in native and cultivated soil. *Soil Sci. Soc. Amer. J.* 50:627-633.

Feller, C. 1988. Effect de different systemes de culture sur les stock organicque de sols argileux tropicaux des Petites Antilles. *Cahoers ORSTO, Series Pedologies* 24:341-343.

Feller, C., A. Albrecht, and A. Tessier. 1996. Aggregation and organic matter storage in Kaolinitic and Smectitic tropical soils. In: M.R. Carter and B.A. Stewart (eds.), *Structure and Organic Matter Storage in Agricultural Soils*. Advances in Soil Science. CRC Press, Boca Raton, FL.

Fisher, M. J., I. M. Rao, M. A. Ayarza, C.E. Lascano, J.I. Sanz, R.J. Thomas, and R.R. Vera. 1994. Carbon storage by introduced deep-rooter grasses in the South America savannas. *Nature* 371:236-238.

Fisher, M.J., I.M. Rao, M.A. Ayarza, C.E. Lascano, J.I. Sanz, R.J. Thomas, and R.R. Vera. 1995. Pasture soils as carbon sink. *Nature* 376:472-473.

Jenny, H. 1980. *The Soil Resource*. Springer, New York, N.Y.

Klute, A. 1986. Methods of Soil Analysis. Part 1. Physical and Mineralogical Methods. American Society of Agronomy, Inc. and Soil Science Society of America, Inc. Madison, WI.

Lyles, L., J.D. Dickerson, and L.A. Disrud. 1970. Modified rotary sieve for improved accuracy. *Soil Sci.* 109:207-210.

Martin, A., A. Mariotti, J. Balesdent, P. Lavelle, and R.Vuattoux. 1990. Estimates of the organic matter turnover rate in a savanna by the ^{13}C natural abundance. *Soil Biology and Biochemical* 22:517-523.

Martins, P.F., C.C. Cerri, B. A. Volkoff, and A. Chauvel. 1991. Consequences of clearing and tillage on the soil of a natural Amazonian ecosystems. *Forest Ecology and Management* 38: 273-82.

Nelson, D. W. and L. E. Sommers. 1982. Total carbon organic carbon, and organic matter. p. 539-579. In: A.L. Page, R. H. Miller, and D. R. Keeney (eds.), Methods of Soil Analysis. Part 2. Chemical and Microbiological Properties. American Society of Agronomy, Inc. and Soil Science Society of America, Inc. Madison, WI.

Parton, W.J., D.S. Schimel, C.V. Cole, and D.S. Ojima. 1987. Analysis of factors controlling soil organic levels in Great Plains grasslands. *Soil Sci. Soc. Amer. J.* 51:1173-1179.

Resck, D.V.S., J. Pereira, and J.E. Silva. 1993. Dinamica da materia organica na Regiao dos Cerrados. Documento 36. Planaltina, Brazil: EMBRAPA/CPAC.

Sanchez, P. A. and J.G. Salinas. 1981. Low-input technology for managing Oxisols and Ultisols in tropical America. *Advances in Agronomy* 34:279-406.

SAS Institute Inc. 1994. SAS User's Guide. Fourth Edition. SAS Institute Inc. Cary, N.C.

Schlesinger, W. H. 1995. An overview of the carbon cycle. p. 9-25. In: R. Lal, J. Kibble, E. Levine, and B. A. Stewart (eds.), Soils and Global Change. Advances in Soil Sciences. CRC Press, Boca Raton, FL.

Thomas, R. J., M. J. Fisher, M. A. Ayarza, and J. I Sanz 1995. The role of forage grasses and legumes in maintaining the productivity of acid soil of Latin America. p. 61-83. In: R. Lal and B.A. Stewart (eds.). Soil Management: Experimental Basics for Sustainability and Environmental Quality. Advances in Soil Science. CRC Press, Boca Raton, FL.

Van Veen, J. A., J. N. Ladd, and M. J. Frissel. 1984. Modeling carbon and nitrogen turn over through microbial biomass in soil. *Plant and Soil* 76:257-274.

Woomer, P.L. and M.J. Swift. 1994a. The Biological Management of Tropical Soil Fertility. John Wiley & Son. New York, N.Y.

Woomer, P.L., A. Martin, A. Albrech, D.V.S. Resck, and H.W. Scharpenseel, 1994b. The importance and management of soil organic matter in the tropics. p. 47-80. In: P.L. Woomer and M.J. Swift (eds.), The Biological Management of Tropical Soil Fertility. John Wiley & Son. New York, N.Y.

Zhang, H., M. L. Thompson, and J. A. Sandor. 1988. Compositional differences in organic matter among cultivated and uncultivated Argiudolls and Hapludalfs derived from loess. *Soil Sci. Society Amer. J.* 52:216-222.

CHAPTER 19

Dissolved Organic Carbon: Sources, Sinks, and Fluxes and Role in the Soil Carbon Cycle

T.R. Moore

I. Introduction

Dissolved organic carbon (DOC) is operationally defined as the organic carbon passing through a filter of 0.45 µm pore size and consists of a wide range of molecules, ranging from simple acids and sugars to complex humic substances with large molecular weights (Thurman, 1985). DOC contributes to the acidity of water, the mobility and toxicity of metals and organic pollutants, and the availability of nutrients in soils and aquatic systems. It is a source of energy for organisms: about 10% of the DOC in stream and lake waters is labile (Sønergaard and Middelboe, 1995). When exposed to ultra-violet radiation, recalcitrant fractions are broken down into simple substrates readily available to bacteria (Wetzel et al., 1995).

At the global scale, DOC plays a role on the C cycle primarily through the very large amount (ca. 700 Pg) stored in the oceans (Figure 1). In contrast, the DOC stored in soil, groundwater, and streams and lakes can be estimated to be 1.1, 8.0, and 1.1 Pg, respectively. In terms of global fluxes, precipitation deposits about 0.1 Pg DOC yr^{-1} to the land surface and the riverine export to the ocean has recently been estimated to be 0.2 Pg DOC yr^{-1}, mainly from tundra/taiga and tropical forest systems, and similar in magnitude to the export of particulate organic carbon from erosion (Ludwig et al., 1996). These fluxes are small compared to those of carbon in photosynthesis and respiration and fossil fuel emissions. At the global scale, however, the importance of DOC fluxes is lost because there can be internal movement of DOC within systems, such as soils, without export. The continental export of DOC is equivalent to an average of 2 g m^{-2} yr^{-1}, yet studies of small catchments reveal exports to be much larger than this, commonly 3 to 10 g DOC m^{-2} yr^{-1} (see Hope et al., 1994) and DOC release from vegetation and soil organic horizons can range from 1 to 50 g DOC m^{-2} yr^{-1}. Thus, much of terrestrially-produced DOC is microbially consumed, photo-degraded, or precipitated in soils and sediments as it passes to the ocean. In using the terms sources and sinks, it should be made clear that a sink for one system may be a source for another and that sinks may be of a temporary, transient nature, with much of the DOC produced ultimately being returned to the atmosphere as CO_2.

Much of the variation in DOC concentrations and export from terrestrial catchments may be explained by differences in soils within catchments. For example, differences in DOC concentrations of three very different regions can be explained, in part, by the proportion of the catchment occupied by peatlands (Figure 2). Therefore, the mechanisms whereby DOC is produced by vegetation and soil, the ability of soils to retain DOC, and the hydrologic pathways that DOC takes from precipitation to the stream channel are critical in understanding DOC dynamics and the role of DOC in the soil C cycle.

ISBN 0-8493-7441-3
©1997 by CRC Press LLC

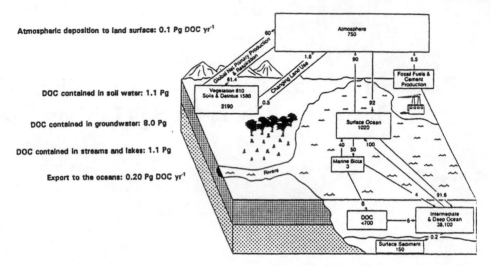

Figure 1. The global carbon cycle, showing the reservoirs and fluxes in Pg and Pg yr^{-1}, from Schimel (1995). The figures on the left side of the diagram are estimates of the deposition of DOC onto the land surface by precipitation, the storage of DOC in soil water, groundwater, and streams and lakes water (assuming DOC concentrations of 15, 2 and 5 mg L^{-1}, respectively), and the riverine export of DOC from the land surface to the oceans. (From Ludwig et al., 1996.)

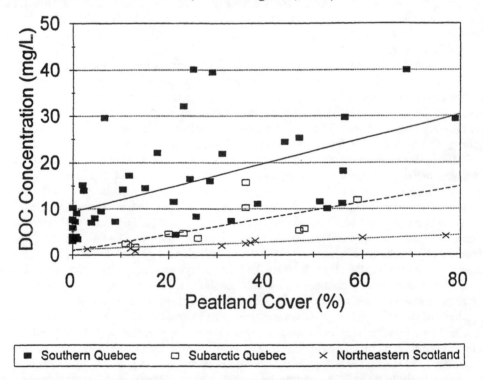

Figure 2. Relationships between DOC concentration in streams and the proportion of the catchment occupied by peatlands in southern Quebec (Eckhardt and Moore, 1990), subarctic Quebec (Koprivnjak and Moore, 1992), and northeast Scotland (Hope et al., 1994). Lines represent the regressions between DOC concentration and peatland cover for the three regions ($r^2 \geq 0.30$).

A series of compartments can be identified within the pedosphere, linked by hydrologic pathways (Figure 3). Figure 4 presents a collation of DOC concentrations in each of these compartments, from published studies, reveals that concentrations increase from precipitation (generally 2 mg DOC L^{-1}), through the vegetation canopy (throughfall and stemflow) to soil organic surface horizons (generally 50 mg DOC L^{-1} in forest soils). As the water passes through the soil profile, DOC concentrations generally decrease, suggesting sorption of DOC by mineral subsoils and parent materials. Streams have extremely variable concentrations of DOC, from < 1 to > 40 mg L^{-1}, reflecting differences in DOC concentrations in source water and catchment soils (Nelson et al., 1993), with the highest concentrations generally in boreal forest and wetland streams (Hope et al., 1994).

An understanding of DOC dynamics in soils thus requires a knowledge of the controls on the production of DOC from vegetation and organic horizons, the characteristics which control DOC retention in subsoils, as well as the hydrologic pathways within catchments. I examine controls on DOC production and retention in soils in the following sections and conclude with a discussion of the ways in which disturbance may affect the soil DOC budget and catchment export.

II. DOC Production

DOC is released from vegetation and soil organic matter. Many field studies show considerable spatial and temporal variation in the concentrations of DOC in soil water and identifying the controls on these variations is difficult, though temperature, substrate characteristics, and water content appear to be important. DOC concentrations appear to be high during warm periods, where water content is low, and after the input of fresh plant material, such as leaves (e.g., Dalva and Moore, 1991; Vance and David, 1991).

In an attempt to identify some of the important controls on the rate of DOC production, an experiment was performed in which 7 types of organic matter, ranging from leaves to soil, were incubated under laboratory conditions, with varying temperature (4 and 22° C) and either aerobic or anaerobic conditions. Samples were placed in water and incubated for 18 days; then the water was replaced and the cycle repeated twice, to determine whether flushing affects DOC release rates, which was measured from changes in DOC concentration in the sample porewater.

The amount of DOC released by the seven samples over the 60 day incubation period ranged from 0.4 to 260 mg DOC g^{-1} (Table 1). Fresh maple leaves produced the largest amount of DOC and the O/Ah sample of the Inceptisol beneath the maple trees produced the smallest amount of DOC. In general, the DOC release rate decreased as the degree of decomposition of organic matter increased.

The influence of the flushing, aerobic/anaerobic, and temperature treatments is presented in Figure 5 for individual treatments. The effect of flushing varied a great deal between samples, most samples producing the same amount of DOC through each of the three phases. However, the fresh plant tissues, maple leaves, and *Sphagnum* declined in their DOC production rate through the three phases, and the Inceptisol soil sample increased DOC production rate through the 3 phases.

There was no significant difference between DOC production rates under aerobic and anaerobic conditions, which was surprising and suggests that the anaerobic conditions of peat have little direct effect of the high DOC concentrations found in peats and the streams draining them. Finally, the influence of temperature on DOC release varied among the treatments but, in general, the Q_{10} values over the 18°C temperature range used lie between 1.5 and 2.5, lower than observed for many microbial processes, and similar to that observed by Christ and David (1997) in a red spruce Spodosol, suggesting that part of the process of DOC production may be physical leaching, rather than being entirely dependent on microbial activities.

This experiment gives some indication of the potential of different plant and soils to release DOC and the environmental conditions which promote this release, which results in DOC concentrations of between 20 and 50 mg L^{-1} in the surface horizons of many soils (Figure 4).

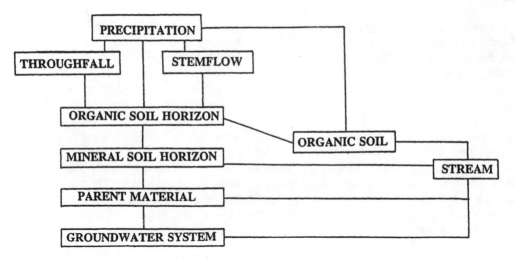

Figure 3. Compartments of the DOC cycle in the pedosphere and hydrologic linkages.

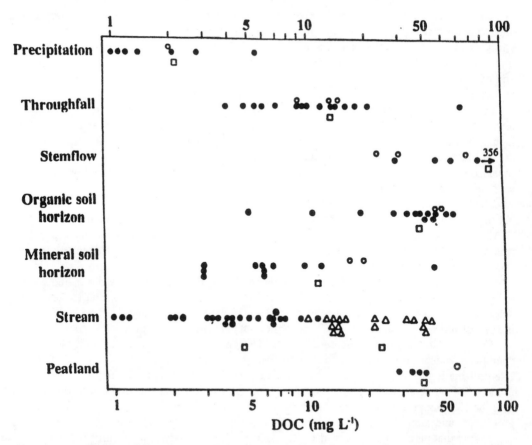

Figure 4. A collation of literature on DOC concentrations in pedosphere compartments, from precipitation to streams; open squares are the mean of each category; triangles represent catchments with >5% covered by wetland. (From Dalva and Moore, 1991.)

III. DOC Retention in Soils

The field DOC concentrations suggest that many soils have the ability to adsorb, or sequester, DOC as water passes down through the soil profile (Figure 4). DOC, as organic anions, will be subject to sorption, in competition with other anions in the soil solution, such as sulfate and phosphate (e.g., Courschene and Landry, 1994; Evans and Anderson, 1990; Inskeep, 1989; Sibanda and Young, 1986). DOC has a similar affinity for soils as sulfate, is more strongly adsorbed than nitrate or chloride and less strongly than phosphate (Moore et al., 1992; Nodvin et al., 1986). The surface characteristics of the soil will also control the sorption of DOC. Mechanisms for this sorption include anion and ligand exchange, cation bonding, and physical adsorption (Jardine et al., 1989; Schulthess and Huang, 1991). A further property controlling DOC sorption will be the chemistry of the DOC.

Although there is field evidence for DOC sorption by mineral subsoils (McDowell and Wood, 1984; Moore, 1989; Ugolini et al., 1977), controls on DOC sorption are best determined in laboratory experiments where conditions can be carefully controlled and a wide range of soil samples and DOC solutions can be tested, such as developed by Nodvin et al. (1986).

To assess the relative sorption potential of soils and identify the major properties responsible for this sorption, an experiment was conducted in which soil samples were treated with solutions containing 0 to 50 mg DOC L^{-1} (Moore et al., 1992). After 24 hr, the suspension was filtered and the change in DOC concentration in the filtrate calculated as retention or loss from the soil. The sorption isotherm, derived from regression of changes in DOC against the initial concentration of DOC in the solution, includes a slope and a null point, the latter being the concentration of DOC at which there is neither gain nor loss of DOC by the soil.

Using an extract of peat, the null point ranged from 5 to 85 mg DOC L^{-1} for 48 soils collected from southern Quebec (Figure 6). When organized by horizon notation and relative position in the profile, the null point decreases from an average of 75 mg L^{-1} in the upper subsoil to 20 to 25 mg L^{-1} in the lower subsoil and parent material. Thus, as the DOC-rich water percolates down through the soil profile, it becomes increasingly strongly adsorbed, through exposure to horizons with lower null point DOC concentrations. Analysis of the soils revealed that a multiple regression including organic carbon content, oxalate-extractable Al, and dithionite-extractable Fe could account for most of the variation in null point values. Soils containing large amounts of Fe and Al and little organic carbon are able to adsorb large quantities of DOC, whereas soils rich in organic carbon and low in Fe and Al (as occurs in many gleysol and peaty soils) will allow DOC to pass through the subsoil with little DOC sorption. DOC sorption is also likely to be strong in soils containing calcium carbonate (e.g., Heilman and Gass, 1974), possibly through the release of carbon dioxide from the soil solution (Otsuki and Wetzel, 1973).

A second factor controlling DOC retention by soils is the character of the DOC. To illustrate this, we determined DOC sorption isotherms for 4 soils (2 sands, 2 clays) using several types of DOC. The results show that, for extracts of an aspen soil, a lichen-pine soil, fresh maple leaves, and a peat soil, there is a threefold variation in both null point and slope (Table 2). Thus, the origin and chemistry of the DOC will affect DOC dynamics, though we have not been able to identify a DOC fraction which causes this variation in DOC sorption. Although DOC is produced in large quantities from fresh maple leaves, it is also relatively strongly adsorbed by soils.

IV. DOC Budget for Soils

Few attempts have been made to produce a DOC budget for soils, though the literature is replete with DOC budgets for streams draining catchments containing varied soils. As an example of the role of soils in producing, transporting, and retaining DOC, tentative DOC budgets are presented for two contrasting, DOC-rich soils, one an upland forest soil and the other a peat (Table 3).

Table 1. The release of DOC from plant tissues and soils incubated for 60 days under aerobic laboratory conditions at 22°C

Sample	DOC release (mg g^{-1})
Fresh maple leaves	260
Old maple leaves	21
Inceptisol O/ah horizon	0.4
Sphagnum moss	5.6
Fibric peat	2.9
Mesic peat	2.2
Humic peat	1.8

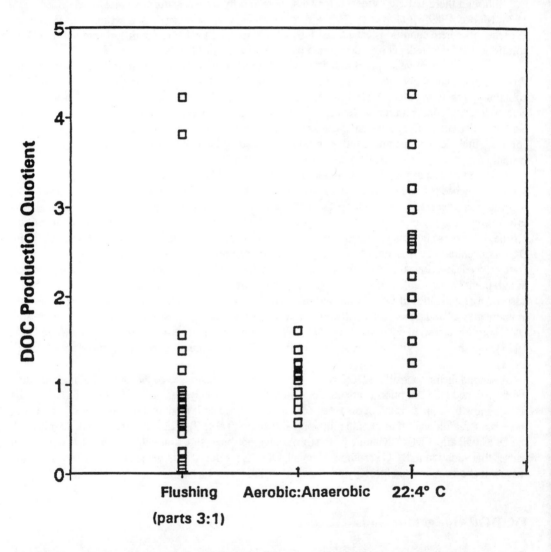

Figure 5. The effect of leaching, aerobic/anaerobic, and temperature (22:4°C) treatments on the release of DOC by decomposing plant tissues and soils, expressed as the quotient between the treatments. Each point represents an individual treatment.

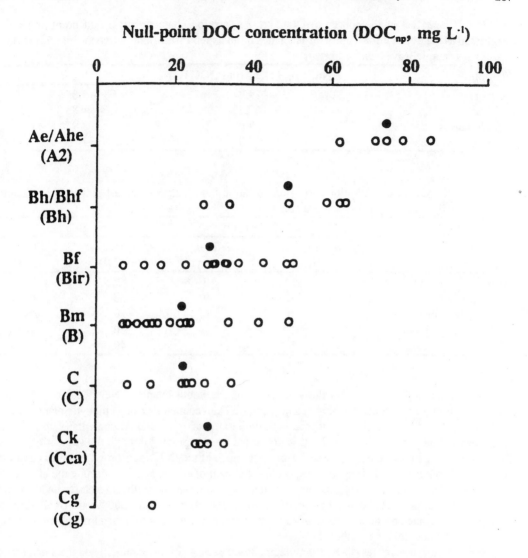

Figure 6. Null-point DOC concentrations in soil samples, organized by horizon notation and depth in the soil profile (from Moore et al., 1992). Solid circles represent the mean for each category. For the 48 samples analyzed, the null point could be represented by the regression:

$DOC_{np} = 32 + 16\log(\text{org. C}) - 13 (Al_o) - 9 \log(Fe_d)$

where: DOC_{np} = null point DOC concentration (mg L^{-1})
org. C = organic carbon content of the soil (%)
Al_o = oxalate-extractable Al content of the soil (%)
Fe_d = dithionite-extractable Fe content of the soil (%)
$r^2 = 0.70$, n = 48, standard error of estimate = 11.0 mg DOC L^{-1}

Table 2. The effect of the source of DOC on the sorption characteristics: null point and slope, expressed as the average for 4 soils for leachates of an aspen forest floor, a lichen-pine forest floor, fresh sugar maple leaves, and a fibrisol

Sample	Null point DOC (mg DOC L^{-1})	Slope
Aspen	19.8	5.83
Lichen	67.2	1.83
Maple leaves	27.9	3.97
Fibrisol	54.0	2.08

Table 3. The DOC concentrations and flux for an upland temperate forest soil in New Zealand (Moore, 1989) and a bog in central Canada

	Forest soil		Bog	
Component	Conc. (mg L^{-1})	Flux (g m^{-2}yr^{-1})	Conc. (mg L^{-1})	Flux (g m^{-2}yr^{-1})
Precipitation	1	3	2	2
Vegetation	16	40	nd	nd
Surface horizon	56	80	50	25
Subsurface horizon	12	18	60	nd
Stream	5	7	40	25

The upland soil is located on the west coast of the South Island of New Zealand, with high precipitation and a rich beech forest with a high litterfall and containing many epiphytic plants (Moore, 1989). The soil is moderately weathered with high extractable Fe and Al contents (1 to 2%). The results show that a large amount of DOC is generated by rain passing through the vegetation canopy and from the organic horizons of the soil. However, much of the DOC is adsorbed as water percolates through the soil profile, resulting in an export in the stream of only 7 g m^{-2} yr^{-1}, despite the high runoff. Thus, up to 70 g DOC m^{-2} yr^{-1} appears to be retained by the solum, a significant sink for DOC. Yet the solum contains only 15 kg m^{-2} of organic carbon, representing only 200 yr of DOC sorption, suggesting that much of the adsorbed DOC must be mineralized back into the atmosphere, or lost as CO_2 carried by percolating water.

The peat soil receives less DOC from precipitation, because of lower rainfall and the small canopy of plants on the peat probably provides little DOC through rainfall leaching of the canopy. The DOC concentrations in the peat porewater are high (40 - 70 mg L^{-1}), and in the absence of any mechanism to retain DOC in the profile, large amounts of DOC are exported by the stream. In the case of the peat, the export of DOC of 20 - 30 g m^{-2} yr^{-1} is similar to the long-term accumulation of organic carbon in the profile, resulting in carbon storage (Gorham, 1991). Although peats are regarded as sinks for organic carbon, it should be recognized that they also export large quantities of DOC to streams, affecting their water chemistry, the DOC ultimately being microbially utilized, sedimented on the floor of aquatic systems, or transported to the ocean.

V. Effect of Disturbance on DOC Dynamics and Export

What effect does anthropogenic disturbance have on the DOC budget of soils and ecosystems? A summary of results is presented in Table 4.

Table 4. Effect of disturbance on DOC dynamics and export

Disturbance	Effect on soil DOC and export
Acid precipitation	Reduced DOC production in organic horizons and reduced sorption in subsoil through competition with SO_4^{2-}.
Agriculture	Possibly increased DOC production in systems where plant material remains on the soil surface; increased DOC mobilization by liming; strong sorption by subsoils; effect unknown.
Forest clearance	Trash left on soil surface and near streams will increase DOC production; general increase in stream DOC export associated with higher DOC concentrations and larger runoff.
Peatland drainage	Small and variable effect on DOC concentrations; general increase in DOC export because of higher runoff.
Reservoir creation	Death and decomposition of vegetation increases DOC production; increased export of DOC.
Climatic change	Increased plant production and higher temperatures will lead to increased DOC production; decreased runoff will result in smaller DOC exports.

Acid precipitation has been shown to have a profound effect on the nutrient cycling of forested and agricultural soils, but the effect on DOC is relatively minor. Acidification of organic horizons produces a small reduction in the release of DOC (e.g., David et al., 1989), though the microbial stimulation by large amounts of N in acid rain may increase rates of DOC release (Guggenberger, 1994). The subsoil retention of DOC by Fe and Al oxides and hydroxides may be reduced through the competition between DOC and SO_4^{2-} for anion adsorption sites (e.g., Vance and David, 1989; 1992). There is strong evidence that acid rain reduces the DOC concentrations in lakes (e.g., Schindler et al., 1996).

The effect of agricultural activities may change DOC dynamics, but there are no studies which have examined this. Increased biological activity and changes in tissue quality may increase rates of DOC production in systems where most of the plant material remains on the soil surface, and liming should mobilize DOC. However, most agricultural subsoils are clay-rich and likely to adsorb DOC. Major changes in DOC are likely a function of changes in hydrologic pathways, but this has rarely been examined (Hope et al., 1994).

Several studies have assessed the effect of forest clearance on DOC dynamics, usually through changes in export in streams draining cut-over catchments. The creation of trash on the soil surface will increase rates of DOC production and import into the soil. In some cases, the subsoils have the ability to retain the released DOC, but in most cases, changes in hydrologic pathways, such as the creation of throughflow and surface runoff, result in elevated concentrations of DOC in streams (e.g., Hobbie and Likens, 1973; Meyer and Tate, 1983; Moore, 1989; Tate and Meyer, 1983). In part, this may be due to trash left in or close to the stream channel (Moore, 1989). Higher DOC concentrations combined with greater runoff generally lead to increased DOC export from cutover catchments, compared to their controls.

Peatlands have been drained for the production of fuel, horticultural moss, and the growth of crops and trees; the change from a partially anaerobic to aerobic soil has created a significant change in the soil C budget (Armentano and Menges, 1986). Studies have revealed small and variable differences in DOC concentration in natural and drained peatlands (e.g., Laine et al., 1995; Moore, 1987). Increased runoff resulting from the drainage will increase DOC export from drained peatlands (e.g., Moore and Jackson, 1989), accelerating the change from sink to source in the peatland C budget.

The creation of reservoirs, particularly in northern environments, results in the death and decomposition of plant material and changes in hydrologic pathways, which circumvent potential sorption in mineral horizons. This results in increases of DOC in waters after flooding, as well as increases in DOC export (e.g., Moore and Matos, 1997).

Finally, climate change will have an effect on DOC dynamics, but the overall effect is uncertain. For example, increased vegetation productivity should increase DOC release, as will elevated temperatures: Melillo et al. (1993) predict an increase in net primary production of global ecosystems of 24% arising from doubled atmospheric CO_2 concentrations and changes in temperature and precipitation from GCM scenarios. This will increase the inputs of DOC into the soil profile. The export of DOC will depend on the capacity of the soil to adsorb DOC, as well as changes in hydrology: increased runoff will increase DOC export. Some empirical evidence for this change has been obtained by Clair and Ehrman (in press), who compared export of DOC in 15 Atlantic rivers over a 10-year period. Neural network analysis revealed that an increase in mean annual temperature of 3°C would result in a decrease in DOC export of 26%, whereas a temperature increase of 3°C and a precipitation increase of 10% would result in a decrease in DOC export of only 0.5%. Changes in evapotranspiration and precipitation, and the resulting runoff, are likely to dominate changes in DOC export in response to climatic change.

VI. Conclusion

At the global scale, fluxes of DOC from the land surface are small compared to the movement of CO_2 or the DOC stored in oceans: DOC is unlikely to be a major part of the "missing carbon sink". At the scale of the pedosphere, however, DOC does play an important role in the C budget of many ecosystems, with production and fluxes ranging from 1 to 50 g C m^{-2} yr^{-1}. These rates are similar to the variations in carbon sequestration in soils brought about by changes in land use. Major controls on DOC production that include temperature, plant production, tissue quality, and DOC retention in subsoils is controlled by Fe, Al and organic carbon contents. Hydrologic pathways play a major role in determining whether DOC produced in soils is retained by the solum or exported. Human-derived disturbances such as forest clearance, drainage, and flooding tend to increase rates of DOC production, transport, and export. The effect of climate change on DOC is uncertain: DOC production will likely be increased because of higher net primary production and temperatures, but changes in export from soils and catchments will be primarily dependent on variations in runoff, reflecting changes in precipitation and evapotranspiration.

References

Armentano, T.V. and E.S. Menges. 1986. Patterns of change in the carbon balance of organic-wetland soils of the temperate zone. *J. Ecol.* 74:755-774.

Christ, M. and M.B. David. 1997. Temperature and moisture effects on the production of dissolved organic carbon in a Spodosol. (Submitted).

Clair, T.A. and J.M. Ehrman. In press. Variation in discharge, dissolved organic carbon and nitrogen export from terrestrial basins with changes in climate: a neural network approach. *Limnol. Oceanogr.*

Courschesne, F. and R. Landry. 1994. Sulfate retention by Spodosols in the presence of organic ligands. *Soil Sci.* 158:329-336.

Dalva, M. and T.R. Moore. 1991. Sources and sinks of dissolved organic carbon in a forested swamp catchment. *Biogeochem.* 15:1-19.

David, M.B., G.F. Vance, J.M. Rissing, and F.J. Stevenson. 1989. Organic carbon fractions in extracts of O and B horizons from a New England spodosol: effect of acid treatment. *J. Environ. Qual.* 18:212-217.

Eckhardt, B.W. and T.R. Moore. 1990. Controls on dissolved organic carbon concentrations in streams, southern Québec. *Can. J. Fish. Aquat. Sci.* 47:1537-1544.

Evans, A., Jr. and T.J. Anderson. 1990. Aliphatic acids: Influence on sulfate mobility in a forested Cecil soil. *Soil Sci. Soc. Amer. J.* 50:1576-1578.

Gorham, E. 1991. Northern peatlands: role in the carbon cycle and probable responses to climatic warming. *Ecol. Appl.* 1:182-195.

Guggenberger, G. 1994. Acidification effects on dissolved organic matter mobility in spruce forest ecosystems. *Environ. Intern.* 20:31-41.

Heilman, P.E. and C.R. Gass. 1974. Parent materials and chemical properties of mineral soils in southeast Alaska. *Soil Sci.* 117:21-27.

Hobbie, J.E. and G.E. Likens. 1973. Output of phosphorus, dissolved organic carbon and fine particulate carbon from Hubbard Brook watersheds. *Limnol. Oceanogr.* 18: 734-742.

Hope, D., M.F. Billett, and M.S. Cresser. 1994. A review of the export of carbon in river water: fluxes and processes. *Environ. Poll.* 84:301-324.

Inskeep, W.P., 1989. Adsorption of sulfate by kaolinite and amorphous iron oxide in the presence of organic ligands. *J. Environ. Qual.* 18:37-385.

Jardine, P.M., N.L. Weber, and J.F. McCarthy. 1989. Mechanisms of dissolved organic carbon adsorption on soil. *Soil Sci. Amer. J.* 53:1378-1385.

Koprivnjak, J-F. and T.R. Moore. 1992. Sources, sinks, and fluxes of dissolved organic carbon in subarctic fen catchments. *Arct. Alp. Res.* 24:204-210.

Laine, J., H. Vasander, and T. Sallantaus. 1995. Ecological effects of peatland drainage for forestry. *Environ. Rev.* 3:286-303.

Ludwig, W., J-L. Probst, and S. Kempe. 1996. Predicting the oceanic input of organic carbon by continental erosion. *Global Biogeochem. Cycles* 10:23-41.

McDowell, W.H. and T. Wood. 1984. Podzolization: Soil processes control organic carbon concentrations in streams. *Soil Sci.* 137: 23-32.

Melillo, J.M., A.D. McGuire, D.W. Kicklighter, B. Moore, III, C.J. Vorosmarty, and A.L. Schloss. Global climate change and terrestrial net primary production. *Nature* 363:234-240.

Meyer, J.L. and C.M. Tate. 1983. The effects of watershed disturbance on dissolved organic carbon dynamics of a stream. *Ecology* 64:33-44.

Moore, T.R. 1987. A preliminary study of the effects of drainage and harvesting on water quality in ombrotrophic bogs near Sept-Iles, Quebec. *Water Resour. Bull.* 23: 785-791.

Moore, T.R. 1989. Dynamics of dissolved organic carbon in forested and disturbed catchments, Westland, New Zealand. I. Maimai. *Water Resourc. Res.* 25:1321-1330.

Moore, T.R. and R.J. Jackson. 1989. Dynamics of dissolved organic carbon in forested and disturbed catchments, Westland, New Zealand. II. Larry River. *Water Resourc. Res.* 25: 1331-1340.

Moore, T.R. and L. Matos. 1997. The effect of shallow flooding on the dynamics and chemistry of dissolved organic carbon in a boreal wetland catchment. *Can. J. Fish. Aquat. Sci.* (Submitted)

Moore, T.R., W. de Souza, and J-F. Koprivnjak. 1992. Controls of dissolved organic carbon sorption by soils. *Soil Sci.* 154:120-129.

Nelson, P.N., J.A. Baldock, and J.M. Oades. 1993. Concentrations and composition of dissolved organic carbon in streams in relation to catchment soil properties. *Biogeochem.* 19:27-50.

Nodvin, S.C., C.T. Driscoll, and G.E. Likens. 1986. Simple partitioning of anions and dissolved organic carbon in a forest soil. *Soil Sci.* 142:27-35.

Otsuki, A. and R.G. Wetzel. 1973. Interaction of yellow organic acids with calcium carbonate in freshwater. *Limnol. Oceanogr.* 18: 190-493.

Schimel, D.S. 1995. Terrestrial ecosystems and the carbon cycle. *Global Change Biol.* 1: 77-91.

Schindler, D.W., P.J. Curtis, B.R. Parker, and M.P. Stainton. 1996. Consequences of climate warming and lake acidification for UV-B penetration in North American boreal lakes. *Nature* 379:705-708.

Schulthess, C.P. and C.P. Huang. 1991. Humic and fulvic acid adsorption by silicon and aluminum oxide surfaces on clay minerals. *Soil Sci. Soc. Amer. J.* 55:34-42.

Sibanda, H.M. and S.D. Young. 1986. Competitive adsorption of humus acids and phosphate on goethite, gibbsite and two tropical soils. *J. Soil Sci.* 37:197-204.

Sønergaard, M. and M. Middelboe. 1995. A cross-system analysis of labile dissolved organic carbon. *Mar. Ecol. Progr. Ser.* 118:283-294.

Tate, C.M. and J.L. Meyer. 1983. The influence of hydrologic conditions and successional state on dissolved organic carbon export from forested watersheds. *Ecology* 64:25-32.

Thurman, E.N. 1985. *Organic Geochemistry of Natural Waters*. Nijhoff/Junk, Boston.

Ugolini, F.C., R. Minden, H. Dawson, and J. Zachara. 1977. An example of soil processes in the *Abies amabilis* zone of Central Cascades, Washington. *Soil Sci.* 124:291-302.

Vance, G.F. and M.B. David. 1989. Effect of acid treatment on dissolved organic carbon retention by a spodic horizon. *Soil Sci. Amer. J.* 53: 1242-1247.

Vance, G.F. and M.B. David. 1992. Dissolved organic carbon and sulfate sorption by spodosol mineral horizons. *Soil Sci.* 154: 136-144.

Wetzel, R.G., P.G. Hatcher, and T.S. Bianchi. 1995. Natural photolysis by ultraviolet irradiance of recalcitrant dissolved organic matter to simple substrates for rapid bacterial metabolism. *Limnol. Oceanogr.* 40: 1369-1380.

CHAPTER 20

Geochemical History of Carbon on the Planet: Implications for Soil Carbon Studies

Dimitrii Ye. Konyushkov

I. Introduction

A century ago, an outstanding Swedish chemist, S. Arrhenius, was the first to surmise that the presence of carbon dioxide in the atmosphere has a certain effect on air temperature and that the increase in concentration of CO_2 caused by burning of fossil fuels can lead to a progressive warming of global climate (Arrhenius, 1896). Also, he supposed that glacial epochs on the planet could be connected with fluctuations in concentration of CO_2 in the atmosphere. Since the 1940s, the problem of possible global climatic changes induced by anthropogenic activities has been attracting the attention of scientists; in the last decade, it has become a problem of public concern and a topical issue of political discussions. Much attention was paid to it during the 1992 Conference on Sustainable Development in Rio-de-Janeiro.

Investigations performed in the 20th century have confirmed the hypothesis of Arrhenius. Indeed, data on air temperature and concentration of CO_2 within the last 160 thousand years derived from the studies of ice cores in different regions of the globe (Vostok station, Antarctica, Barrow cape, Alaska, Greenland) (Barnola et al., 1987, 1991; Kotlyakov, 1994) evidence that a good correlation exists between these indices.

At present, the problems of global climatic changes and the corresponding changes in the whole natural environment and in the types of land management are acknowledged as the problems of the "threat of global warming." A large number of studies are devoted to them. Different scenarios of climate changes are discussed, different measures to mitigate the adverse effects of such changes and to slow the global warming (changes in the industrial cycles, reutilization of energy wastes, use of new sources of energy, substitution of biomass burning for fossil fuels, afforestation, increase in the productivity of forests, conversion of cultivated lands into pastures and forests, the measures aimed at rising humus content in soils, choice of the most efficient and sustainable cropping systems, etc.) are proposed and partly implemented into practice (Trexler, 1991; Elliot, 1994; Izac and Whitman, 1994; Zybach, 1993). A lot of data on the main pools of carbon and their sequestration potential have been accumulated, both in regional and in the global aspects (Tans et al., 1990; Schlesinger, 1984; Zinke et al., 1984; Post et al.,1992; Kurz and Apps, 1994; Vinson and Kolchugina, 1993; Nabuurs and Mohren, 1993; Tarnocai et al., 1993; Tarnocai and Ballard, 1994; Rozhkov et al., 1996). A considerable amount of effort is paid to the development of modeling studies (Kimmins et al., 1990; Kurz et al., 1992; Price and Apps, 1993; Rastetter et al., 1991). The scientists of different specialties—climatologists, biologists and physiologists of plants, soil scientists, ecologists, physicists, mathematicians, technologists, and many others—are actively involved in this work.

However, in spite of all these efforts, we should admit that the scientific community is not yet ready to give exact and prompt answers to the questions about the role of different pools of carbon in carbon cycling, about the proportion and interrelations between mineral- and organic-bound carbon, about the exact patterns of climate changes in the future, and their assessment from the ecological and economical standpoints. Suffice it to say, the mystery of "missing carbon" (the imbalance between the net release and net accumulation of carbon in different pools estimated at 1.8 ± 1.4 Gt. C yr^{-1} (Lal et al, 1995)) remains unsolved for more than two decades. As shown below, many other problems related to the fate of carbon in the biosphere are open to argument and call for further studies.

Probably the reason for the relatively slow progress in our knowledge about the mechanisms governing carbon cycling and the response of environment to the changes in these mechanisms lies not so much in technical difficulties and the lack of adequate data on carbon reserves (it is interesting to note that most of the figures characterizing different sources and sinks of carbon in the biosphere have not changed much since the 1950s) as in a certain isolation of scientists of different specialties and in the lack of really comprehensive, integral, and versatile studies into the problem. Insufficient attention is paid to historical aspects of the problem of global warming and carbon cycling, especially to the problem of geochemical history of carbon on the planet. An interesting work of Scharpenseel and Becker-Heidmann (1990) and the monograph of Budyko, Ronov, and Yanshin (1985) are the rare exclusions. There is a kind of a paradox in this situation, since the principle of "searching the keys to the future in the past" is well known and is firmly established in many natural sciences.

It is interesting to note that the bulk of publications devoted to the problem of global climate change and carbon sequestration are written by multiskilled ecologists involved in the studies of surface spheres of the planet. Meanwhile, the problem of the fate of carbon is a classical problem of geochemistry and geological sciences studying the interiors of the earth. Indeed, the problem of carbon is closely connected with the problems of the origin of fossil fuels and calcareous rocks, volcanic emissions, gaseous emanations from the earth's crust, etc. It should be considered as the problem of not only biological, but also geological cycling of substances on the planet. Its solution is possible only if the scientists of different specialties closely cooperate, work together, and actively exchange ideas. Unfortunately, such joint and integral works are scarce in number. At this stage of investigations, it is necessary to elaborate some common platform for the discussion, to realize the main gaps in our knowledge, and to formulate the problems for further studies.

Certainly, soil scientists would greatly benefit from such joint studies. In tackling the problem of the role of soils in the global cycle of carbon and their sequestration potential, one should bear in mind that the soil is just one of the links in the long chain of carbon transformations. The knowledge of the whole cycle, of all sources and sinks of carbon, and their history and modern functioning is necessary for correct assessments and elaboration of land management strategies.

The aim of this work is to elucidate the geochemical history of carbon on the planet and to reveal the role of soil cover in the global carbon cycling. Also, some methodical issues of soil carbon studies are considered.

II. The Amount and Forms of Carbon on the Earth

Carbon, together with oxygen and nitrogen shares the third to fifth place (after hydrogen and helium) in abundance in the Universe (*Spravochnik po geokhimii*, 1990), and ranks among the second ten of the elements composing our planet. Carbon compounds (CH, CN, C_2) are found in cosmic space and in composition of meteorites and comets; the atmosphere of such planets as Venus and Mars is almost completely composed of CO_2; methane is considered to be the dominant substance in composition of gigantic planets (Jupiter, Saturn, Neptune, and Uran). The content of carbon in the Earth makes up approximately $n \cdot 10^{18}$ tons (Uspenskii, 1956, Kovda, 1985), that is just a split percent in comparison with the total mass of the Earth ($6 \cdot 10^{21}$ tons). However, the geochemical role of carbon on the planet

Table 1. Average content of carbon compounds in the earth's crust

Forms of carbon	Concentration, %	Mass, n•10^{16} t
C_{org}	0.07	2.12
CO_2	1.44	40.8

Table 2. The content of carbonate- and organic-bound carbon in different types of sediments, n•10^{16} t

Groups of sediments	C_{org}	C_{carb}	$C_{org} : C_{carb}$
Continental	0.809	4.72	1:5.8
Shelf	0.268	1.76	1:6.6
Ocean bottom	0.018	0.76	1:42
Sedimentary layer (total)	1.095	7.24	1:6.6

can hardly be overestimated. Carbon is the most important of biophilous elements. Owing to its ability to form various types of chemical bonds and to compose long molecular chains and aromatic structures, carbon builds up the framework of complex organic molecules, including the amino acids and, thus, serves the basis of life on the planet. Though the reserves of carbon in the living matter of the planet do not exceed n•10^{12} t, its role in the development of life on our planet can hardly be overestimated. According to Perel'man (1979), the average concentration of carbon in the living matter is about 18%, i.e., it is by 780 times higher than the clarke value. According to Perel'man, the ratio between the concentration of an element in the living matter and its concentration in the lithosphere (clarke) characterizes the biophilic capacity of this element. This value for carbon (780) is much higher than that for nitrogen (160), hydrogen (70), oxygen (1.5), chlorine (1.1), and all other elements (< 1). Considering the annual production of carbon by the living matter (n•10^{11} t) and the long history of life on our planet (n•10^9 yr.), we can say that all the mass of carbon on the planet could have passed through the biological cycling. As it was vividly stated by Vernadsky (1994), "geochemistry of carbon cannot be perceived beyond the phenomena of life."

However, the main reserves of carbon on the planet are concentrated in the form of carbonates and reduced organic compounds. According to data on chemical composition of the earth's crust (Ronov and Yaroshevskii, 1976), the main pool of carbon is represented by the carbonate-bound form (Table 1).

Carbon of the lithosphere is mainly concentrated in the rocks of sedimentary shell, where the content of organic carbon averages 0.5% and the content of carbonate-bound carbon is about 3–4%. The latest assessments (Ronov, 1993) give the values for the mass of carbon in sedimentary layer (Table 2).

Evident traces of life (stromatoliths) are known for the sediments that formed about 3.5•10^9 yr. BP. Thus, the sedimentary shell of the earth, or the stratisphere (Ronov, 1993) represents a unique object for studying the history of life and, correspondingly, the history of carbon cycling on the planet. The stratisphere was considered by Vernadsky as a reflection of former biospheres. Studying the changes in composition of the stratisphere can help to reveal some important features of carbon geochemistry. Though the reserves of carbon in the stratisphere and in the earth's interiors are considered as a stable inert pool, by far exceeding in size the mobile pool of carbon involved in biogeochemical cycling (i.e., the carbon of the atmosphere (750•10^9 t), hydrosphere (40,000•10^9 t), soils (2,000•10^9 t), and terrestrial biota (590•10^9 t), (see more detailed considerations below), the interactions between these two pools cannot be ignored if we want to understand the mechanisms of sustainable development of the biosphere.

III. Geochemical History of Carbon on the Planet: the Main Milestones

A. The Cosmic Stage

To understand the features of modern carbon cycling and to reveal the general trend in the geochemical history of carbon, we should know the conditions of the zero-moment, i.e., what forms of carbon existed on the planet before the appearance of the biosphere, what origin did they have, and what type of transformations they were subjected to.

It is believed that during the pre-biogenic period of evolution of the Earth, accumulation of simple carbon-containing molecules (CO_2, CO, CH_4) in the Earth's crust, atmosphere, and hydrosphere was conditioned by the process of differentiation of mantle and degasification of magma. Immense amounts of carbon were delivered to the surface in the course of volcanic eruptions, especially during the periods of activation of ancient tectonomagmatic cycles. The total amount of carbon, accumulated in the upper geospheres of the planet, is assessed at $n \cdot 10^{16-17}$ tons (Kovda, 1976, 1985; Ronov, 1993). Emanation of carbon from the deep layers of the planet takes place at present as well. Its intensity was assessed by Goldshmidt (1954) at $1.5-3.0 \cdot 10^7$ tons per year (for CO_2), that is approximately $n \cdot 10^6$ tons C annually. Already, during the first stages of the evolution of the planet, these processes resulted in the formation of the primordial reducing atmosphere composed of methane and ammonia. The release of oxygen during the photodissociation of water and, later, during the reactions of photosynthesis gradually changed the composition of the initial reducing atmosphere producing an N_2 and CO_2 atmosphere (Scharpenseel and Becker-Heineman, 1990).

Quite a different hypothesis about the origin of the initial carbon on the planet was suggested by Galimov (1988), who studied the isotopic composition of carbon in different natural objects. According to his data, there are two groups of carbon different in their isotopic composition. The carbon contained in the rocks of the Earth's mantle is enriched in "light" isotope (^{12}C), whereas the carbon contained in the rocks of the Earth's crust is enriched in "heavy" (^{13}C) isotope and, according to the ratio between isotopes, is similar to the carbon contained in carboniferous stony meteorites (chondrites). The author explains this by the heterogeneity of the initial material that composed the planet, attributing its origin to different stages of cosmic nucleogenesis. He suggests that carbon of mantle rocks could originate from the initial protoplanet substances (cloud) that had experienced a stage of high-temperature accretion ($\sim 5 \cdot 10^9$ yr. BP) with the release and loss of most of its volatile components. The reason for the enrichment of the rocks of the Earth's crust with "heavy" isotopes of carbon, according to Galimov, could be the meteorite "bombing" of the planet after the end of the main stage of accretion $(4.2-3.9) \cdot 10^9$ yr. BP). This bombing could be synchronous with the stage of analogous bombing of the Moon, which has already been established by scientists. As a result of this process, the Earth could gain most of its volatile components, including water vapor, and the initial hydrosphere was formed. Certainly, this hypothesis does not exclude the role of volcanic processes in the accumulation of carbon at the surface of the planet. It is interesting to note that according to the hypothesis of Galimov, the end of the active cosmic stage in the evolution of the planet was almost concurrent with the beginning of the biotic stage.

B. The Pre-Biotic Stage

The first evident features of life on the planet (stromatolithes of *Cyanophyceae* in Western Australia) date back to $3.5 \cdot 10^9$ yr. BP. So, at that time, carbon dioxide dissolved in the ocean water already participated in photosynthetic reactions with the release of oxygen. This oxygen could be spent for the oxidation of a big reserve of reduced forms of iron, sulfur, hydrogen, and carbon in the upper layers of the Earth's crust, and, thus, could only gradually accumulate in the atmosphere. More active accumulation of oxygen in the atmosphere could take place only after complete exhaustion of the initial reserve of reduced forms of elements. Thus, it is believed that the general trend in the evolution of the atmosphere was a gradual decrease in concentration of CO_2, parallel to the increase in concentration of oxygen (Scharpenseel and Becker-Heidmann, 1990). However, this hypothesis is open to argument. Thus, Yaroshevskii (1988) points to the fact that the release of oxygen during the photosynthesis obligatory implies the concurrent accumulation of reduced forms of organic matter. Therefore, the total degree of oxidation of the lithosphere–atmosphere system cannot change. Judging the relatively constant ratio between the organic- and carbonate-bound sediments within the whole geologic history of the planet, he believes that the dynamics of oxygen accumulation in the ancient atmosphere were principally the same as nowadays. From his point of view, geochemical conditions in the atmosphere and hydrosphere have been principally the same during the whole period of evolution of the biosphere; the changes in composition of sediments could be connected not with the general trend in geochemical conditions, but with alternations in the facial conditions of sedimentation. Thus, the principle of geochemical stability of the biosphere since the very first stages of its development is declared. In this context, a sacramental question about the origin of life on our planet gains especial importance. According to the hypothesis of Galimov (1988), life on the Earth occurred after the appearance of free oxygen. The first traces of the presence of free oxygen date back to the formation of oxidized iron ores (iron quartzites), i.e., to $3.8 \cdot 10^9$ yr. BP (their maximum development took place $2.0–2.6 \cdot 10^9$ yr. BP). This oxygen could be the result of protophotosynthetic reactions of the destruction of water molecules performed on the base of porhyrin nuclei bound with lipid membranes, i.e., on the base of chemical compounds that had been delivered to the planet together with other carboniferous substances of chondrites. Further complication of the reacting system could give rise to the origin of chlorophyll molecules, and, thus, to the beginning of the proper biotic stage of carbon cycling. This interesting hypothesis requires further examination; however, it is important that it entails the problem of the origin and evolution of life on our planet and the problem of geochemical history of carbon with the more general problems of the evolution of the Universe.

C. The Biotic Stage

With the appearance of life on the Earth (3.2–3.8 billion years ago) and the beginning of the development of the biosphere, the fate of carbon on the planet was conditioned by its involvement into biogeochemical cycling of substances. The main features of this cycling are well known and consist of (a) the uptake of carbon (in the form of CO_2) from the mobile pools (atmosphere and hydrosphere) and its sequestration in the living matter with a simultaneous release of free oxygen, (b) partial decomposition and oxidation of organic remains accompanied by the return of CO_2 in the mobile pools and consumption of oxygen, and (c) fossilization (withdrawal) of carbon from the mobile pools and its burying in the sediments in the form of carbonates and organic substances. The latter process is accompanied by the accumulation of free oxygen that can accumulate in the atmosphere or participate in various oxidation reactions. Along with these biochemical processes, the concentration of carbon in the mobile pools is also controlled by the processes of dissolution and precipitation of carbonates, i.e., by the reactions of carbonate equilibrium:

$$Ca(HCO_3)_2 \rightleftharpoons CaCO_3 \downarrow + H_2O + CO_2 \uparrow$$

These reactions, because of the immense mass of carbonates in the ocean and in the upper part of terrestrial lithosphere (including the soil cover), are of especial importance for sustaining the relatively stable concentration of carbon dioxide in the mobile pools. Finally, the additional supply of carbon to the biosphere from the interiors of the planet (volcanism) and, from burning of fossil fuels (in the modern epoch) should be mentioned. Carbon cycling is closely associated with the cycling of oxygen and, thus, dictates geochemical conditions of sedimentation. Therefore, studying the sediments we can draw some important conclusions about the geochemical history of carbon. Below, a brief characterization of the main geologic events in the history of the biosphere is given with respect to the geochemical history of carbon.

The Early Proterozoic era $(2.0–2.6) \cdot 10^9$ yr. BP. Widespread sedimentation of iron quartzites owing to the release of oxygen in shallow marine environment in the course of photosynthetic reactions; accumulation of carbon in living organisms with a corresponding decrease in the partial pressure of carbon dioxide in the ocean, which, in turn, led to intensive chemogenic and biochemogenic sedimentation of carbonates.

The Late Proterozoic era $(1.7–0.7) \cdot 10^9$ yr. BP. The development of photosynthetic organisms gradually involved the accumulation of free oxygen in the atmosphere. The establishing of oxidizing conditions caused the origin of "red-earth" oxidized continental formations. Further sequestration of carbon in chemogenic and biochemogenic (stromatolithic) calcareous sediments took place. Probably this strong uptake of carbon dioxide from the atmosphere was one of the reasons of the first (Early Proterozoic and Riphean) glaciations of the planet.

The Early Paleozoic era, the Cambrian period $(0.6–0.5) \cdot 10^9$ yr. BP. This period was marked by the widespread development of calcareous sediments. According to the hypothesis of A.P. Vinogradov (1967), the continuing decrease in concentration of CO_2 in sea waters during the Vendian period (due to sequestration of carbon in the living matter) resulted in relative stability of calcium carbonates in the upper layers of the ocean. This ensured active development of marine organisms with calciferous shells during the Cambrian period. During the following evolution of life within the whole Phanerozoic time, carbonate sedimentation in oceans was predominantly biogenic; with time, an increase in the proportion of calcite minerals in carbonate sediments was observed.

The Late Paleozoic era, the Devonian and Carboniferous periods $(0.35–0.28) \cdot 10^9$ yr. BP. Because of the rapid development of terrestrial vegetation and intensive fossilization of organic matter, huge masses of carbon were accumulated in buried organic residues that served as the sources of coal deposits, especially in the Carboniferous period. Coal deposits of this period enclose about 17% of the total reserves of coal on the planet. A strong uptake and sedimentation of organic-bound carbon induced the accumulation of free oxygen in the atmosphere. The following Permian period was marked by a widespread development of red-earth continental deposits.

The Mesozoic era $(180–60) \cdot 10^6$ yr. BP was characterized by the increased concentration of oxygen in the atmosphere and, hence, accelerated process of oxidation, the intensity of formation of coal deposits was lowered, whereas that of marl deposits substantially increased. This stage is denoted by the appearance of new forms of life in the ocean and on the land surface.

The beginning of the Cenozoic era was marked by an intensive inflow of CO_2 into the atmosphere connected with the Alpine tectonogenesis. Due to intensification of volcanic eruptions, the concentration of CO_2 in the atmosphere amounted up to 0.8%. As a consequence, the rapid development of photosynthetic plants, the propagation of the new forms of plants (*Angiospermae*), widespread of forest landscapes, accumulation of lignite and lime formations took place in the Paleogene period.

However, as volcanic activity weakened and CO_2 got accumulated in living matter and the products of its vital activity, the concentration of carbon dioxide in the atmosphere steadily decreased. As a result, in the Quaternary period it experienced periodic fluctuations within the limits, close to its current level. As evidenced by the data derived from the studies of ice cores from "Vostok" station in the Antarctica (Barnola et al., 1994), within the last 160 thousand years the concentration of CO_2 in the entrapped bubbles of atmospheric air ranged between 190 and 270-290 ppm; a high degree of correlation is observed between the increased or decreased levels of CO_2 concentration and the periods of interglaciations or glaciations, respectively.

After the last glaciation, the concentration of CO_2 in the atmosphere began to rise. The maximal rates of this process have been observed during the past two centuries and, especially, during the last fifty years. This rapid growth of CO_2 concentration is associated with the effects of economic activity of mankind, primarily, with the use of the reserves of organic carbon accumulated during the past geological periods (burning of fossil fuels) and with drastic decrease in the area of forest biomes on the planet, because of the needs of extensive agriculture.

At the same time, according to some estimates (Adams et al., 1990), the total amount of carbon in terrestrial ecosystems of the Earth within the last 18 thousand years has increased by 1300 billion tons. Taking into account that the concentration of carbon in the atmosphere has increased as well, the authors come to conclusion that the only source of the "additional" carbon is the ocean, the reserves of carbon in which (in the form of dissolved CO_2) exceed the amount of carbon in the atmosphere and in terrestrial ecosystems by many times. It should be noted, however, that this estimation does not take into account the possible flux of carbon to the surface from the deep layers of the lithosphere and the mantle of the Earth in the course of volcanic activity.

Judging the amount of CO_2 buried in the sediments of different geologic periods and considering that the rate of accumulation of carbonate-bound sediments was directly proportional to the concentration of CO_2 in the atmosphere, and, also, taking into account the content of CO_2 in the Late Quaternary sediments and the data on the concentration of CO_2 in the atmosphere for this period, M.I. Budyko (1984) calculated the concentration of CO_2 in the atmosphere for the whole Phanerozoic Eon. Assuming that the doubling CO_2 concentration (for the modern epoch) would cause the rise of the global temperature by 2.5°C and that the dependence of air temperature on the concentration of CO_2 in the atmosphere has a logarithmic pattern ($\Delta T = 3 \cdot \ln\alpha / \ln 2$, where ΔT is the shift in the global average temperature of the low atmosphere, °C (as compared to the present-day value), and α is the ratio of CO_2 concentration in hypothetical atmosphere to that in the modern one), and taking into account the relationship between the shifts in temperature, the albedo of the planet, and the intensity of solar radiation, he also calculated the global temperature curve for this period (Table 3, Figure 1).

Thus, the concentration of CO_2 in the atmosphere during the last 600 million years ranged within relatively narrow limits, being generally higher than in the Quaternary time and reaching its maximum in the Carboniferous period.

This short review of the geochemical history of carbon on the Earth testifies that the main factor of redistribution of carbon between different components of the geosphere was the living matter. Biotic component of carbon cycle is "responsible" not only for the appearance of oxygen in the atmosphere and the formation of the immense reserves of carbon in the lithosphere (in the form of buried organic substances and carbonates), but also for the geochemical conditions of formation of the other types of sedimentary rocks. Thus, the living matter sustains the complex system of regulation of geochemical conditions in the biosphere, providing their relative stability and the possibility for existence and evolution of life on the planet. However, as mentioned above, this general picture of geochemical history of carbon is far from being complete. First, it has been derived from the studies of only continental deposits; the sediments of ocean bottoms and continental slopes were not taken into account. Second, the available stratigraphic record of geologic events is incomplete. At present, the pre-Jurassic stage of sedimentation in the oceans is absolutely unknown. Assessment of the role of ocean bottom sediments in the sequestration of carbon can be different in the models accepting the hypothesis of subduction of the ocean floor under the continents (and deeper into the upper mantle) and rejecting it. Little is known about the earliest (Archean) stages of sedimentation and "the starting conditions" of carbon cycling. Probably, some of the values cited above may change in the future.

In general, this brief review testifies that the problem of carbon cycling in the biosphere represents a real challenge for the specialists of many natural sciences, including geology, geochemistry, climatology, meteorology, oceanology, biology, soil science, and ecology. Obviously, this problem calls for joint efforts. Precise assessment of the reserves of carbon in different components of the environment is possible if the data from different sources are correlated with each other. The brilliant example of such correlation is presented in the monograph of Uspenskii (1956). Extensive data on carbon geochemistry in various geospheres of the planet can be found in the works (Budyko et al., 1985; Galimov, 1988; Gorshkov, 1980; Grigor'eva, 1980; Kobak, 1988; Kovda, 1976, 1985; Lisitsyn, 1978; Post et al., 1990; Ronov, 1976, 1982, 1993; Uspenskii, 1956; Vernadskii, 1994; Vinogradov, 1967; etc.).

Table 3. Changes in the average global air temperature (as compared with the modern epoch) during the Phanerozoic Eon

Geologic period	Abs. age, n•10^6 yr.	CO_2, %	ΔT, °C
Early Cambrian	545–570	0.27	7.5
Middle Cambrian	515–545	0.19	6.2
Upper Cambrian	490–515	0.19	6.2
Ordovician	435–490	0.16	6.4
Silurian	400–435	0.13	5.4
Early Devonian	376–400	0.14	6.1
Middle Devonian	360–376	0.40	9.9
Late Devonian	345–360	0.41	10.1
Early Carboniferous	320–345	0.42	10.6
Middle and Late Carboniferous	280–320	0.18	6.5
Early Permian	255–280	0.37	9.2
Late Permian	235–255	0.14	7.1
Early Triassic	220–235	0.11	6.2
Middle Triassic	210–220	0.20	8.2
Late Triassic	185–210	0.17	7.7
Early Jurassic	168–185	0.19	8.3
Middle Jurassic	153–168	0.23	9.3
Late Jurassic	132–153	0.24	9.5
Early Cretaceous	100–132	0.19	8.8
Late Cretaceous	66–100	0.26	10.2
Paleocene	58–66	0.12	7.6
Eocene	37–58	0.19	9.1
Oligocene	25–37	0.055	4.5
Miocene	9–25	0.11	7.0
Pliocene	2–9	0.055	3.6

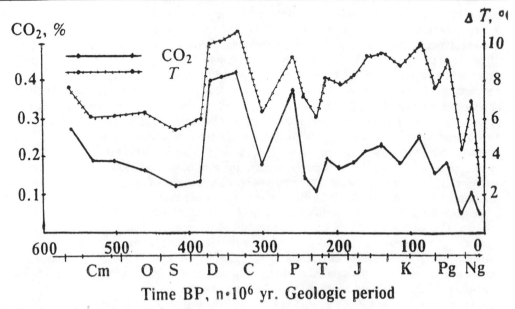

Figure 1. Changes in the average air temperature and CO_2 concentration. (Adapted from Budyko, 1984.)

IV. Sustainable Development of the Biosphere and Carbon Cycling

Probably, the most important and fascinating conclusion from the discussion above lies in the fact of very durable (3.8 billion years) sustainable development of the biosphere on our planet. Budyko (1984) considers this sustainability of the biosphere really surprising, taking into account the extremely high susceptibility of most of the living organisms to changes in the environment (suffice it to remind of the fast and "easy" extinction of many species in the anthropogenic epoch), the possibility of short-term but drastic changes in the global climate under the impacts of active volcanic eruptions or falls of large meteorites, the nearly critical propagation of glaciations in the Quaternary period (when the further increase in their area could cause their irreversible self-development due to the global changes in the albedo of the planet), and, finally, the very low average density of the biomass (for the whole planet) which does not exceed $1 g/cm^2$. From his point of view, the sustainable development of the biosphere is the result of lucky fortune and coincidence of favorable conditions.

At the same time, one should not forget about the powerful ability of living organisms to regulate geochemical conditions in the environment. J. Lovelock (1979) considers the biosphere as an integral organism with such inherent dynamic processes that ensure its stability and well-balanced state (the Gaia hypothesis). It is believed that it is possible to create the closed (for exchange of substances) and opened (for exchange of energy) small artificial ecosystems that would be of indispensable use for space travels, and that the whole biosphere of the Earth represents such a system at the global scale. However, the careful examination of the history of carbon on the planet suggests that this belief can be wrong.

Considering the problem of carbon cycling in the biosphere, A.B. Ronov (1976, 1993) pointed to the importance of geological processes of carbon exchange between the biosphere and the interior of the planet (lithosphere and the Earth's mantle). On the basis of data on the composition and volume of sediments (for the continents) in different geologic periods, he calculated separately the masses of volcanic rocks, organic, and mineral (carbonate-bound) carbon (Table 4, Figure 2).

As is evident from this figure, a good correlation is observed between the rates of accumulation of volcanic rocks, carbonates (including carbonate impurities in non-calcareous types of rocks), and organic substances in the sediments. Maximums in volcanic activity (O_1, C_1, K_2) occurred in the middle phases of tectonic cycles (Caledonian, Hercynian, and Alpine) and were accompanied by intensive accumulation of carbonates and organic substances. Geologically, these processes were separated; active volcanism was mainly characteristic for geosynclinal belts, whereas sedimentation of carbonates and organic substances took place in shallow epicontinental seas. Ronov suggested two reasons explaining this correlation. First, volcanic eruptions were accompanied by active efflux of endogenous CO_2 into the atmosphere and ocean water; thus more favorable conditions for photosynthetic reactions and the growth of the biomass were created. Second, the middle phases of tectonic cycles were characterized by leveling of the relief and widespread transgressions of the ocean; the absence of orogenic barriers and general humidification of climate favored the development of terrestrial vegetation.

A similar picture was obtained by Ronov (1993) for the whole stratosphere of the planet (including ocean sediments) for the Late Jurassic–Pliocene periods (Table 5, Figure 3).

Thus, a kind of feedback relations exists between the supply of CO_2 in mobile pools (atmosphere and hydrosphere) and the development of biota that consumes this mobile CO_2 for the biomass production and finally converts it into the immobile pool represented by carbonates and fossilized organic remains. It is important to note that the main source of CO_2 in the mobile pool (volcanism) is external with respect to the biosphere. Thus, our biosphere is not a closed system. Active exchange of substances between the biosphere, lithosphere, and the mantle of the planet exists. According to the hypothesis of Ronov (1976) this exchange is a necessary prerequisite of sustainable development of life. His calculations evidence that the total reserve of "labile" carbon accumulated in the atmosphere, hydrosphere, living and extinct organisms of the biosphere ($43.5 \cdot 10^{12}$ tons) constitutes just a 0.00054

Table 4. The distribution of masses of volcanic rocks, carbonate- and organic-bound carbon buried in continental sediments* by geological periods

Geologic period	Time interval, $n \cdot 10^6$ yr. BP	Mass, $n \cdot 10^{15}$ t		
		Volcanic rocks	C_{carb}	C_{org}
Pliocene (Ng_2)	5.3–1.6	2.21	0.14	0.10
Miocene (Ng_1)	23.7–5.3	5.54	0.81	0.31
Oligocene (Pg_3)	36.6–23.7	2.04	0.26	0.20
Eocene (Pg_2)	57.8–36.6	7.78	1.85	0.40
Paleocene (Pg_1)	66.4–57.8	2.46	0.46	0.10
Late Cretaceous (K_2)	97.5–66.4	22.06	4.06	0.62
Early Cretaceous (K_1)	144–97.5	21.48	3.17	0.74
Late Jurassic (J_3)	169–144	7.53	2.04	0.49
Middle Jurassic (J_2)	187–169	8.20	1.20	0.44
Early Jurassic (J_1)	208–187	5.46	0.96	0.26
Late Triassic (T_3)	230–208	8.26	1.89	0.18
Middle Triassic (T_2)	240–230	4.54	1.12	0.05
Early Triassic (T_1)	245–240	6.22	0.92	0.07
Late Permian (P_1)	258–245	6.38	1.22	0.07
Early Permian (P_2)	286–258	9.41	1.38	0.19
Late–Middle Carboniferous (C_{2-3})	320–286	8.09	2.72	0.35
Early Carboniferous (C_1)	360–320	11.31	1.72	0.53
Late Devonian (D_3)	374–360	5.66	1.86	0.37
Middle Devonian (D_2)	387–374	8.54	1.44	0.23
Early Devonian (D_1)	408–387	9.35	1.09	0.07
Late Silurian (S_2)	421–408	5.07	0.74	0.08
Early Silurian (S_1)	438–421	7.67	1.19	0.10
Late Ordovician (O_3)	458–438	7.78	1.15	0.17
Middle Ordovician (O_2)	478–458	18.34	1.81	0.35
Early Ordovician (O_1)	505–478	8.68	1.67	0.21
Late Cambrian (Cm_3)	523–505	5.04	1.34	0.06
Middle Cambrian (Cm_2)	540–523	10.25	1.44	0.12
Early Cambrian (Cm_1)	570–540	9.72	1.08	0.17
Phanerozoic eon (total)	570–1.6	235.07	40.73	7.03

*Except for Antarctica
(Adapted from Ronov, 1993)

fraction of the amount of carbon accumulated in the Phanerozoic sediment rocks. Calculation of the rate of carbon accumulation in sediments ($1.32 \cdot 10^8$ tons per year) testifies that the labile carbon of the biosphere can be expended on the processes of sedimentation of carbonates less than in one million years, provided that there is no additional inflow of carbon from the Earth's interior. After this, the reserves of carbon available for the synthesis of organic matter would be exhausted, the oxygen of the atmosphere (without its renewal in the course of photosynthesis) would be expanded on the processes of oxidation, and the life on the planet would cease. This scenario is possible in case that the reserves of radioactive elements in the Earth would be exhausted and the tectonic activity of the planet would terminate. These considerations enabled Ronov to formulate the general geochemical principle of sustainable development and preservation of life: "The life on the Earth, as well as on the other planets is possible only until these planets remain geologically active, i.e., the energy and matter exchange

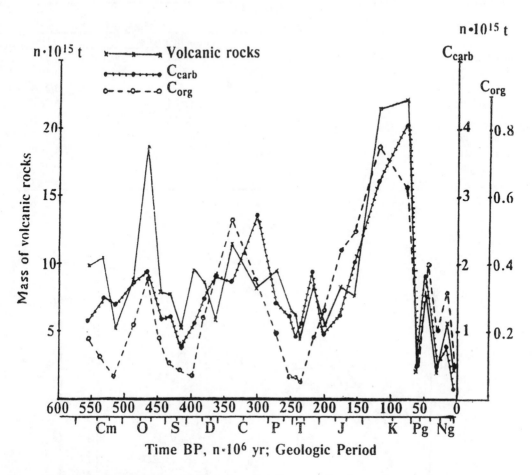

Figure 2. Distribution of masses of volcanic rocks, C_{org}, and C_{carb} in continued sediments. (Adapted from Ronov, 1993.)

exists between the interiors and the surfaces of the planets. The exhaustion of energy resources in the interiors of a planet would inevitably lead to the extinction of life." An interesting question arises: "What is the minimum rate of volcanic activity and what is the minimum concentration of CO_2 in the atmosphere sufficient for sustainable development of the biosphere?" And, also, how far is (and was) the biosphere from these limiting values?

Evidently, the development of the biosphere on our planet lends credit to the hypothesis of Ronov. However, the hypothetical possibility of substitution of "closed" (for substances) biotic cycles for the "open" ones still exists.

The threat of termination of volcanic activity and the exhaustion of available pools of CO_2 is just one of two dangers that were luckily (by chance?) escaped (overcome?) by the life on our planet. Another threat is of the opposite character and is connected with the reserves of CO_2 in the hydrosphere by far exceeding the atmospheric pool of this greenhouse gas. According to recent estimates (Tans et al., 1990), oceans represent the net sink for atmospheric carbon (2 Gt. annually). The partial pressure of CO_2 dissolved in the ocean water and, hence, the exchange of carbon dioxide

Table 5. The distribution of masses of volcanic rocks, carbonate- and organic-bound carbon buried in the stratosphere of the Earth by geological periods

Geologic period	Time interval, $n \cdot 10^6$ yr. BP	Mass, $n \cdot 10^{15}$ t		
		Volcanic rocks	C_{carb}	C_{org}
Pliocene (Ng_2)	5.3–1.6	4.71	1.33	0.214
Miocene (Ng_1)	23.7–5.3	15.76	4.67	0.508
Oligocene (Pg_3)	36.6–23.7	7.07	2.50	0.286
Eocene (Pg_2)	57.8–36.6	14.03	5.41	0.545
Paleocene (Pg_1)	66.4–57.8	6.54	1.35	0.126
Late Cretaceous (K_2)	97.5–66.4	46.64	7.23	0.827
Early Cretaceous (K_1)	144–97.5	50.27	5.66	1.789
Late Jurassic (J_3)	169–144	19.61	3.26	0.792
Late Jurassic–Pliocene (total)	169–1.6	164.63	31.41	5.087

(Modified from Ronov, 1993)

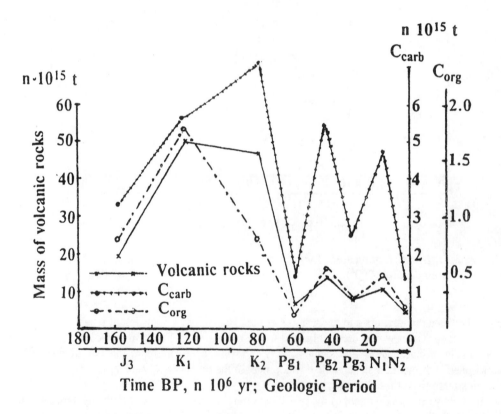

Figure 3. Distribution of masses of volcanic rocks, C_{org}, and C_{carb} in the rocks of the stratosphere. (Adapted from Ronov, 1993.)

between the atmosphere and the ocean is controlled by the reactions of carbonate equilibrium (see above). Also, it is known that the solubility of gases in water strongly depends on temperature conditions. With increasing temperature, solubility considerably decreases. This fact was noted by N.N. Moiseev, who, analyzing the models of carbon cycling suggested in the 1930s by Russian scientist V.A. Kostitsyn (1984), came to the conclusion that a certain increase in temperature (at least in the upper layer of water adjacent to the atmosphere) caused by the greenhouse effect can cause an irreversible change in the direction of the ocean "pump," i.e., instead of being the sink for atmospheric CO_2, the ocean may become the source of new emission of this gas. This would cause the further increase in temperature and, hence, the more active emission of CO_2 from the ocean into the atmosphere and the strong heating of the atmosphere. Fortunately, this mechanism of self-heating of the planet has yet not been realized in the history of the biosphere. The activity of living matter bonding the CO_2 dissolved in the ocean water, the establishment of the system of carbonate equilibrium, and the immense heat capacity of water preserved relatively stable temperature conditions on the planet. However, the potential threat is not eliminated. Human-induced increase in concentration of CO_2 in the atmosphere makes this scenario rather probable. Again, the problem of the threshold upper limits of CO_2 concentration in the atmosphere arises. Certainly, modeling of these processes is a very complex task because of the extreme complexity of feedback relations in the environment. Many of them are still unknown to us or cannot be easily subjected to quantification.

Modern human society has become a powerful geochemical force on the planet. The rate of contemporary human-induced efflux of carbon into the atmosphere (5–8 Gt. C/yr., including the emissions caused by burning of fossil fuels (5–6 Gt. C/yr.) and changes in the system of land management—conversion of forest biomes into plow lands, degradation of forests, peats, etc.—2–2.6 Gt. C/yr.) (Tans et al., 1990; Lal et al., 1995) is at least an order of magnitude higher than the average rate of carbonate sedimentation or volcanic efflux of carbon. This means that mankind bears responsibility for the further fate of the biosphere in its continuous development, which has been always endangered by two opposite "hazards."

This brief review of some general geochemical concepts and hypotheses related to the problem of carbon cycling and sequestration in different pools, including soils, is mainly aimed to attract attention of ecologists and soil scientists to the problem of the role of carbon in preservation of the biosphere and to the need to perform joint cooperative and really multidisciplinary research. This seems rather important, because most of the studies of carbon cycling in the biosphere performed by ecologists ignore the role of geological processes and do not take into account extensive data on carbon geochemistry accumulated in geology. Also, the broadening scope of the research allows us to reevaluate some concepts related to the role of soils in the global carbon cycles.

V. Global Carbon Cycles

A lot of attempts to suggest the models of carbon cycling on the planet are known (Bolin et al., 1986; Uspenskii, 1956; Scharpenseel and Becker-Heidmann, 1990; Downing and Cataldo, 1992; Post et al., 1990, Trexler, 1991 (Data of WRI); Gorshkov, 1980, etc.). Probably, the most sophisticated scheme with quantitative parameters was proposed by Gorshkov et al., 1980. However, in spite of all these efforts, the problem of net imbalance of the annual carbon cycle is yet not solved. It is interesting to note that most authors tried to balance the net efflux of carbon into the atmosphere with carbon sequestration in different pools, i.e., to represent the carbon cycling as a closed one. From my point of view, this approach is open to argument. Several considerations are important in the context of the problem of modeling carbon cycles.

(1) Estimations of the sizes of the main pools of carbon (atmosphere, hydrosphere, biota, soils, sediments) have a high degree of uncertainty, by far exceeding the net annual imbalance. The only

exception is the atmosphere, whose relative homogeneity in composition (owing to active mixing by winds and turbulent flows) together with a wide network of observational stations ensure relatively accurate assessments. Also, the atmosphere is the most sensitive natural object with respect to fluctuations in its gaseous composition. Annual, seasonal, and even daily fluctuations in concentration of CO_2 and other greenhouse gases can be easily traced in atmospheric records.

(2) All the pools of carbon are characterized by principally different sizes, rates of carbon exchange with the neighboring pools, rates of carbon accumulation in (or release from), and the residence times of carbon in them. In other words, each of the pools of carbon has its own dynamics. There are two different ways to study these dynamics. First, we can directly study exchange processes at the key sites and judge the dynamics on the basis of data obtained without measuring the changes in the total pool size (a dynamic, process-oriented approach). Second, we can evaluate the changes in the size and structure of a pool and thus make conclusions about the exchange processes (without studying them) and the net release or accumulation of carbon in the pool (a static, resource-oriented approach). Unfortunately, neither of these approaches can ensure the required accuracy. The first one is extremely labor-consuming and demands a wide network of key sites. Because of the natural variability of local conditions, correct interpretation of the results and their interpolation are always subject to argument. The second one can give quite accurate data only for relatively long periods of time, because the annual changes in size of most pools are negligible as compared to the total reserves. Thus, we can obtain only average data for long periods of time. These data do not necessarily correspond to the modern changes, since each pool has its own dynamics. An assumption that the processes of carbon exchange between different pools are relatively stable in time may be the false one. It means that while using these data for balance calculations, we should take into account the time scale, which will be different for different pools. So, it might be wrong to combine in one model the values characterizing annual industrial emissions and changes in the amount of carbon in the atmosphere (which are more or less accurate and "up-to-date") with the values characterizing the efflux of carbon in volcanic processes or the rate of carbonate sedimentation (which are accurate only for long periods of time and do not necessarily reveal the current intensity of the processes).

(3) As follows from the above discussion, the long-term global cycle of carbon is not closed at all, at least at the geological time scale. Therefore, the general trend toward continuous sequestration of carbon in sediments should be somehow revealed in the existing models. Basically, the modern dynamics of carbon in the biosphere is controlled by superimposing cycles of absolutely different duration—from geologic periods and eras ($n \cdot 10^{6-8}$ yr.) to annual, seasonal, daily, and hourly fluctuations caused by the changes in the intensity of photosynthesis, respiration, and exchange of gases between the atmosphere, biota, ocean water, and soils.

Therefore, the more realistic approach to reveal the patterns of the global carbon cycle is to separately distinguish the cycles with different characteristic times, to reveal modern trends in each of these cycles (using both the static and the dynamic approaches), and to superimpose the resulting curves. This is a difficult challenge. A lot of data are unknown; many processes can be judged only qualitatively; many unsolved problems arise. The discussion below points to some of these complicated issues.

A. Lithospheric (Geological) Cycling of Carbon

The more or less accurate assessments of the lithospheric pool of carbon are available only for the sediments of the Phanerozoic eon (Tables 2, 4). These data testify that the major source of carbon in the lithosphere is represented by calcareous sediments, which account for ¼ of the total volume of sedimentary layer (Ronov and Yaroshevskii, 1976). Taking into account the mass of Pre-Cambrian sediments, we can tentatively assess the total mass of carbon in the earth's crust at $n \cdot 10^{17}$ t. This is the largest pool of carbon, by far exceeding in size the pools of hydrosphere, atmosphere, soils, and biota. Evidently, even slight changes in the lithospheric pool of carbon can considerably affect carbon content in the other pools. Carbon content in the lithosphere is conditioned by the processes of carbon exchange between the surface layer of sediments and the neighboring environments and, also, by the matter exchange between the mantle and the lithosphere. Data on carbon content in the sediments of the Phanerozoic eon allow calculation of the absolute rate of carbon accumulation in them. On the average, it amounts to $72 \cdot 10^6$ t/year, ranging from 20 to $133 \cdot 10^6$ t/year for different geological periods. Evidently, this carbon could originate both from dissolution and secondary sedimentation of the carbon, which had been stored in the Pre-Cambrian sediments, and from the new portions of carbon compounds released from the mantle in the course of volcanism. It is interesting to note that the value of modern sedimentation of carbonate sediments in the ocean ($1361 \cdot 10^6$ t CO_2/ year, or $371 \cdot 10^6$ t C/year) is much higher (Lisitsyn, 1978). That means that a significant portion of carbon of the lithosphere is subjected to continuos renewal. Dividing the total carbon pool in the lithosphere ($n \cdot 10^{17}$ t) by the amount of annually renewable carbon ($n \cdot 10^8$ t), we can obtain the value characterizing the mean residence time of carbon in the lithosphere ($n \cdot 10^9$ yr.); for some types of sediments, it can be smaller ($n \cdot 10^{6-9}$ yr.). The processes of renewal can consist of (1) erosion and dissolution of carbon-bound compounds in terrestrial sediments; (2) direct efflux of gaseous carbon-bound compounds from the lithosphere into the hydrosphere and atmosphere (these processes were named as "respiration of the earth's crust"); (3) dissolution of carbonate-bound sediments submerged below the compensation depth of carbonates (approximately 5000 m). All these carbon fluxes are very important in the context of soil carbon studies, since the soils (or the surface layer of bottom sediments) serve as the main zone of transformation of lithospheric (stable) carbon-bound compounds into more mobile forms that participate in the global carbon cycle within the biosphere. Obviously, we should find more accurate assessments of the rate of these processes.

B. Carbon-Exchange Processes in the Hydrosphere

Oceans represent the main pool of mobile carbon, i.e., the carbon that actively participates in the exchange processes. The total amount of carbon dissolved in the ocean water is assessed at $37-40 \cdot 10^{12}$ t; over 91% of this carbon is represented by carbon dioxide; nearly 9% is represented by the carbon of dissolved organic matter (Uspenskii, 1956). The existing estimates (Tans et al., 1990) suggest that the annual uptake of carbon by the ocean equals to 2 Gt. However, the particular items of carbon balance in the ocean are far from being firmly established. There is a need to have more substantiated values of the annual discharge of carbon into the oceans with hydrochemical flows from terrestrial ecosystems, the annual rate of dissolution and precipitation of ocean carbon-containing sediments, and the rate of direct gas exchange between the ocean and the atmosphere. Considering the size of the pool and the area of oceans on the planet (about 71% of the total surface), we can assume that the ocean plays the most significant role in carbon dynamics and deserves more careful studies. This conclusion becomes especially evident if we consider the role of ocean biota. Its distinctive feature is the highest rate of carbon exchange, especially in phytoplancton organisms. On the average, the total biomass in the ocean (phytomass + zoomass) is equal to $7.2 \cdot 10^{12}$, whereas the NPP of ocean biota constitutes about $76 \cdot 10^{12}$ t annually (Ryabchikov et al., 1988). That means that ocean biota serves as a powerful pump for the transformation of carbon-bound compounds, whose role has yet not been properly evaluated. An important characteristics of carbon behavior in the hydrosphere is the mean residence time of carbon; judging the rate of sedimentation of carbonates, this time can be estimated at $n \cdot 10^2$ yr. Carbon cycle in the biosphere and the soils.

The soils represent the largest pool of carbon in terrestrial ecosystems. Most of the existing estimates of this pool take into account only the upper meter of soil profile, whereas for many soil types this layer may be considered as insufficient, since the traces of soil formation and accumulation of organic- and carbonate-bound carbon occur in deeper soil horizons as well. The most "popular" value characterizing the reserves of organic-bound carbon in the soils is $1500 \cdot 10^{12}$ t. If we take into account the amount of carbonate-bound carbon (Schlesinger, 1985; Glazovskaya, 1996) and enlarge the thickness of soil layer at least to 2 m, then the rough assessment of the total reserve of carbon in the soils will be no less than $3000 \cdot 10^{12}$ t. Partly, this carbon, especially the carbonate-bound one, is inherited from the rocks. Glazovskaya (1997) considers the soils as a reservoir of fossilized carbon-bound compounds. The debates on whether the soils serve as a sink or a source of carbon in modern conditions and on the strategies that should be used in order to improve the sequestration potential of soils have a long history. However, one important aspect of this problem escapes the attention of scientists. What is the role of soils in the global carbon cycle for the steady-state ecosystems (and for the biosphere as a whole)? Considering that the major geological trend of carbon cycling consists of a continuous sequestration of carbon in the sediments and that the only source of replenishment of carbon reserves in the mobile biospheric pool is the volcanic activity (not taking into account the human-induced efflux of carbon), and, also, that the actual rate of carbon accumulation in the sediments considerably exceeds the rate of carbon emission in the course of volcanism, we can state that the source of "additional" carbon is the surface of the continents (i.e., the soil cover). It is the soil, where the lithospheric (geological) carbon cycle starts over and over again. To start this cycle, it is necessary that the initially immobile carbon compounds, which are stored in the rocks of the lithosphere, become mobilized to participate in the migration. In contrast to ocean sediments, the soils cannot accumulate carbon infinitely. At a certain level, which is achieved rather quickly ($n \cdot 10^{2-3}$ yr.), the soil ceases to accumulate humus (or this accumulation proceeds at a very slow rate). Does this mean that carbon cycling within such a steady-state ecosystem becomes completely closed? The answer is "yes," if one considers only the input of organic residues into the soil and the efflux of greenhouse gases in the course of soil respiration. However, in this assessment, the hydrochemical discharge of carbon-bound compounds is not taken into account. Certainly, the formation of a composition of hydrochemical flow takes place not only in the soils proper, but in a considerably deeper layer of rocks (the zone of supergenic processes). However, the role of soils as the most favorable environment for the transformation of the immobile lithospheric carbon pool into the mobile biospheric one cannot be overestimated. Also, it is important that not all the carbon consumed by plants is taken from the atmosphere. Partly, this carbon is taken up from soil solutions. Thus, the soil serves as a medium that provides for accumulation (deposition) of carbon, which was initially captured by terrestrial biota and, at the same time, ensures the hydrochemical discharge of carbon compounds into the basins of sedimentation and, also, the direct efflux of carbon-containing greenhouse gases into the atmosphere. The soil occurs at the cross point between the lithospheric (geological) and the biospheric cycles of carbon and serves as a connection link between them. Also, it is the main zone of transformation of carbon-bound compounds. It is important that different carbon-bound compounds are characterized by different times of retention in the soil. Thus, the mean residence time of carbon in the soils cannot be characterized by a certain value without specifying the particular form of soil carbon and its location in the profile. The range of variation is very wide ($n \cdot 10^{-2}$–$n \cdot 10^{4}$ yr.).

Some important implications for soil carbon studies are evident from the discussion above:

(1) Because of the extremely complicated structure and dynamics of the reserves of carbon in the soils, it is necessary to develop the multidisciplinary studies of soil carbon behavior, including not only humus substances, but also soil carbonates (both lithogenic and pedogenic), dissolved organic and mineral forms of carbon-bound compounds, gaseous phase of the soils, and the living matter of soils. The concept of the carbon profile of soils should be elaborated.

(2) More attention should be paid to soil functioning with respect to carbon cycles. One of the important items of carbon budget in the soils—hydrochemical discharge (or accumulation in certain positions) of carbon-bound compounds—is not taken into account in most of the

assessments of the role of soils in carbon cycling. The role of soils in conversion of more mobile (gaseous) carbon-bound compounds into the less mobile (solid and water-dissolved) ones and vice versa should be more carefully examined. It is possible that even the soils with steady-state humus reserves can serve as the sinks for atmospheric carbon dioxide via its conversion into water-dissolved forms (with their subsequent lateral discharge) or into the form of soil carbonates.

(3) It would be interesting to consider the ecological meaning of accumulation of particular forms of carbon-bound compounds in different environments and to reveal the relationships between the productivity of phytocenoses and the reserves of carbonate-bound compounds (both organic and inorganic) in the soils. The general trends (accumulation of organic substances in humid conditions and of carbonates in arid conditions; the higher the rate of biological turnover, the lower the rate of organic matter accumulation) are well known. However, the comparative geographic analysis of these regularities deserves further studies.

(4) The problem of stability and of the residence time of different forms of carbon-bound compounds in the soil profile calls for new methods of study. In particular, it can be very fruitful to combine the radiocarbon studies of soil humic substances with the studies of their distribution by density fractions.

Understanding the regularities of soil carbon profiles and the role of soils in carbon cycles is necessary to make our predictions of soil response to the global climate changes more substantiated and valid.

VI. Conclusions

The problem of climate change and disturbed carbon cycling in the biosphere calls for the development of multidisciplinary investigations. Carbon cycling in the biosphere proceeds against the background of the geological processes of carbon release in the course of volcanic activity, dissolution of calcareous sediments in the upper part of the Earth's crust, and mineralization of the previously accumulated reserves of organic matter in the lithosphere (burning of fossil fuels can be considered as an anthropogenic modification of this geological process), and of carbon sequestration in the newly forming sediments.

(1) Geological evidences attest to a close relationship between the rate of volcanic activity and accumulation of calcareous sediments. Geological cycle of carbon provides the basis for sustainable development of the biosphere.

(2) The problem of the net imbalance of the annual carbon budget can be a result of combining of the averaged data on carbon fluxes for the pools with principally different dynamics within one model. These data for the pools with a great residence time of carbon sequestration can be misleading with respect to modern dynamics of carbon.

(3) The soil cover of the planet can be considered as a connection link between the geological and the biological cycles of carbon. The structure and profile distribution of carbon-bound compounds in the soils are characterized by the greatest diversity. Currently, most of soil carbon studies are concentrated on the problem of soil organic matter. Meanwhile, the fate of soil carbonates and the problem of the discharge of carbon-bound compounds from the soils via the exchange of gases with the atmosphere and via the lateral hydrochemical flows deserve more careful examination. The concept of carbon profile of the soils should be elaborated. Ecological meaning and regularities of sequestration of different forms of carbon-bound compounds in the soils should be revealed.

(4) The problem of soil response to climatic changes can be solved on the basis of more profound knowledge on the stability of different forms of carbon-bound compounds in the soils. Radiocarbon studies should be applied not only to different chemical fractions of soil humus, but, also, to the fractions separated by means of a densimetric analysis.

References

Apps, M.J., W.A. Kurz, and D.T. Price. 1993. Estimating carbon budgets of Canadian forest ecosystems using a national scale model. p 243–252. In: T.S. Vinson and T.P. Kolchugina (eds.), *Carbon Cycling in Boreal Forest and Subarctic Ecosystems: Biospheric Responses and Feedbacks to Global Climate Change*. Department of Civil Engineering, Oregon State University, Corvallis, OR.

Arrhenius, S. 1896. On the influence of carbonic acid in the air upon the temperature of ground. *Philos. Mag.* 41:237–257.

Barnola, J.M, D. Raynaud, and Y.S. Korotkevich. 1991. CO_2-climate relationship as deduced from the Vostok ice core: A re-examination based on new measurements and on a re-evaluation of the air dating. *Tellus* 43(B):83–90.

Barnola, J.M., C. Lorius, and Y.S. Korotkevich. 1994. Historical CO_2 record from Vostok ice core. p. 7–10 In: T.A. Boden, D.P. Kaizer, R.J. Speranski, and F.W. Stoss (eds.), *Trends '93: A Compendium of Data on Global Change*. ORNL/CDIAC-65. Carbon Dioxide Information Analysis Center, Oak Ridge National Laboratory, Oak Ridge, TN.

Barnola, J.M., D. Raynaud, Y.S. Korotkevich, and C. Lorius. 1987. Vostok ice core provides 160,000-year record of atmospheric CO_2. *Nature* 329:408–414.

Billings, W.D., J.O. Luken, D.A. Mortensen, and K.A. Peterson. 1982. Arctic tundra: a source or sink for atmospheric carbon dioxide in a changing environment? *Oecol.* 52:7–11.

Budyko, M.I. 1984. *Evolyutsiya Biosfery (The Evolution of the Biosphere)*. Leningrad: Gidrometeoizdat.

Budyko, M.I., A.B. Ronov, and A.L. Yanshin. 1985. *Istoriya Atmosfery (The History of the Atmosphere)*, Leningrad: Gidrometeoizdat.

Cihlar, J. and M. Apps. 1993. Carbon budget and succession dynamics of Canadian vegetation. p. 215–220 In: T.S. Vinson and T.P. Kolchugina (eds.), *Carbon Cycling in Boreal Forest and Subarctic Ecosystems: Biospheric Responses and Feedbacks to Global Climate Change*. Department of Civil Engineering. Oregon State University. Corvallis, OR.

Downing, J.P. and D.A. Cataldo. 1992. Natural sinks of CO_2: Technical synthesis from the Palmas Del Mar workshop. *Water, Air, and Soil Pollution* 64:439–453.

Elliot, E.T. 1994. Embodying process information in models evaluated with site network information: Nairobi workshop. p. 163–178. In: *Transactions of the 15th World Congress of Soil Science*, Acapulco, Mexico, July, 1994. Vol. 9 (Supplement).

Galimov, E.M. 1988. Problems of carbon geochemistry. *Geokhimiya* 2:246–258.

Glazovskaya, M.A. 1996 The Role and Functions of the Pedosphere in the Geochemical Cycles of Carbon. *Eurasian Soil Sci.* 2:152–164.

Glazovskaya, M.A. 1997. Fossilizing functions of the pedosphere in continental cycles of organic carbon. *Eurasian Soil Sci.* 3 (submitted).

Goldschmidt, V.M. 1954. *Geochemistry*, Oxford Univ. Press.

Gorham, E. 1991. Northern peatlands: role in the carbon cycle and probable responses to climatic warming. *Ecol. Applic.* 1:182–195.

Gorshkov, S.P., A.G. Sushchevskii, and G.N. Shenderuk. 1980. Turnover of Organic Substances. p. 153–181 In: A.M. Ryabcikov (ed.), *Krugovorot Veshchestva V Prirode I Ego Izmenenie Khozyais tevennoi Deyatel'nost'yu Cheloveka (Cycling of Matter in Nature and its Alteration under the Impact of Anthropogenic Activities)*. Moscow: Mosk. Gos. Univ.

Grigor'eva, T.V. 1980. Carbon. p. 56–67 In: A.M. Ryabcikov (Ed.), *Krugovorot Veshchestva V Prirode I Ego Izmenenie Khozyaistevennoi Deyatel'nost'yu Cheloveka (Cycling of Matter in Nature and its Alteration under the Impact of Anthropogenic Activities)*. Moscow: Mosk. Gos. Univ.

Houghton, R.A. 1991. Tropical deforestation and atmospheric carbon dioxide. *Climatic Change* 19:99–118.

Houghton, R.A. and G.M. Woodwell. 1989. Global climate change. *Scientific American* 260 (4): 36–44.

Houghton, R.A. 1993. Is carbon accumulating in the northern temperate zone? *Global Biogeochemical Cycles,* 7(3):611–617.

Izac, A.M.N. and C. Whitman. 1994. Policy options for soil carbon management in tropical regions. p. 179–199. In: *Transactions of the 15th World Congress of Soil Science,* Acapulco, Mexico, July, 1994. Vol. 9 (Supplement).

Jenkinson, D.S. The turnover of organic carbon and nitrogen in soil. *Phil Trans. Roy. Soc.* (London), 1255:361–368.

Kimmins, J.P., K.A. Scoullar, M.J. Apps, and W.A. Kurz. 1990. The FORCYTE experience: a decade of model development. p. 60–67. In: B.J. Boughton and J.K. Samoil (eds.), Forest Modeling Symposium (Proceedings of a symposium held March 13-15, in Saskatoon, Saskatchewan). Forestry Canada. Northwest Region. Northern Forestry Centre. Information report NOR-X-308.

Kobak, K.P. 1988. *Boticheskie Komponenty Uglerodnogo Tsikla (Biotic Components of the Carbon Cycle),* Leningrad: Gidrometeoizdat.

Kolchugina, T.P. and T.S. Vinson. 1993. Carbon sources and sinks in forest biomes of the former Soviet Union. *Global Biogeochemical Cycles* 7:291–304.

Kolchugina, T.P. and T.S. Vinson. 1994. Production of greenhouse gases in the former Soviet Union. *World Resource Review* 6:291-303.

Kostitsyn, V.A. 1984. *Evolyutsiya Atmosfery, Biosfery I Klimata (Evolution of the Atmosphere, Biosphere, and Climate),* Moscow: Nauka.

Kotlyakov, V.M. 1994. *Mir snega i l'da* (The World of Snow and Ice), Moscow: Nauka.

Kovda, V.A. 1976. Biogeochemical Cycles in Nature and their Deterioration by Man. *Biogekhimicheskie tsikly v biosfere* (Biogeochemical Cycles in the Biosphere), Moscow: Nauka, pp. 19–85.

Kovda, V.A. 1985. *Biogeokhimiya pochvennogo pokrova* (Biogeochemistry of Soil Cover), Moscow: Nauka.

Kurz, W.A. and M.J. Apps. 1993. Contribution of northern forests to the global C cycle: Canada as a case study. *Water, Air, and Soil Pollution* 70:163–176.

Kurz, W.A., M.J. Apps, T.M. Webb, and P.J. McNamee. 1992. The carbon budget of the Canadian forest sector: Phase 1. For. Can., Northwest Reg., North. For. Cent., Edmonton, Alberta. Inf. Rep. NOR-X-326.

Kurz. W. and M. Apps. 1994. The carbon budget of canadian forests: a sensitivity analysis of changes in disturbance regimes, growth rates, and decomposition rates. *Environmental Pollution* 83:55–61.

Lal, R., J. Kimble, and B.A. Stewart. 1995. World soils as a source or sink for radiatively-active gases. p. 1–8. In: R. Lal, J. Kimble, E. Levine, and B.A. Stewart (eds.), *Soil Management and Greenhouse Effect. Advances in Soil Science,* CRC Press, Boca Raton, FL.

Lisitsyn, A.P. 1978. *Protsessy Okeanskoi Sedimentatsii. Litologiya I geokhimiya (Processes of Sedimentation in Oceanic Basins. Lithology and Geochemistry).* Moscow: Nauka.

Lovelock, J. 1979. *Gaia: A New Look at Life on Earth.* Oxford Univ. Press.

Lugo, A.R. and J. Wisniewski. 1992. Natural sinks of CO_2 conclusions, key findings and research recommendations from the Palmas Del Mar workshop. *Water, Air, and Soil Pollution* 64:455-459.

Nabuurs, G.J. and G.M.J. Mohren. 1993. Carbon stocks and fluxes in Dutch forest ecosystems. IBN research report 93/1. ISSN: 0928-6896. Wageningen, The Netherlands.

Odum, E.P. 1986. *Ecologia* (Basic Ecology). Russian translation, Moscow: Mir.

Perel'man A.I. 1979. *Geokhimiya* (Geochemistry). Moscow: Vyshaya shkola.

Post, W.M., J. Pastor, A.W. King, and W.R. Emanuel. 1992. Aspects of the interaction between vegetation and soil under global change. *Water, Air, and Soil Pollution* 64:345–363.

Post, W.M., T-H. Peng, W.R. Emanuel, A.W. King, V.H. Dale, and D.L. DeAngelis. 1990. The global carbon cycle. *American Scientist* 78: 310–326.

Post, W.M., W.R. Emanuel, P.I. Zinke, and A.G. Stangenberger. 1982. Soil carbon pools and world life zones. *Nature* 298(5870):156–159.

Price, D.T. and M.J. Apps. 1993. Integration of boreal ecosystem-process models within a prognostic carbon budget model for Canada. *World Resource Review* 5:15–31.

Price, D.T., M.J. Apps, W.A. Kurz, M. Wesbrook, and R.S. Curry. 1994. A "model forest model": steps toward detailed carbon budget assessments of boreal forest ecosystems. *World Resource Review* 6:462–476.

Rastetter, E.B., M.G. Ryan, G.R. Shaver, J.M. Melillo, K.J. Nadelhoffer, J.E. Hobbie, and J.D. Aber. 1991. A general biogeochemical model describing the responses of the C and N cycles in terrestrial ecosystems to changes in CO_2, climate, and N deposition. *Tree Physiology* 9:101–126.

Ronov, A.B. 1976. Volcanism, sedimentation of carbonates, and life (regularities of global geochemistry of carbon). *Geokhimiya*, 8:1252–1277.

Ronov, A.B. 1982. Global carbon balance during the neogeya. *Geokhimiya* 7:920–932.

Ronov, A.B. 1993. *Stratisfera ili osadochnaya Obolochka Zemli (Stratosphere, the Sedimentary Mantle of the Earth)*, Moscow: Nauka.

Ronov, A.B. and A.A. Yaroshevskii. 1976. The new model of the Earth's crust chemical structure. *Geokhimiya* 12:1763–1795.

Rozhkov, V.A., V.B. Wagner, B.M. Kogut, D.Ye. Konyushkov, S. Nilsson, V.B. Sheremet, and A.Z. Shvidenko. 1996. Soil Carbon Estimates and Soil Carbon Map for Russia. IIASA Working Paper 96-60. Austria.

Ryabchikov, A.M. (ed.), 1988. *Fizicheskaya geografiya materikov i okeanov* (Physical Geography of Continents and Oceans). Moscow: Mosk. Gos. Univ.

Scharpenseel, H.W. and P. Becker-Heidmann. 1990. Overview of the greenhouse effect. Global change syndrome, general outlook. p. 1–14 in Scharpenseel, H.W., M. Schomaker, and A. Ayoub (eds.), *Soils on a Warmer Earth*. Elsevier Science Publishers B.V., The Netherlands.

Schlesinger, W.H. 1977. Carbon balance in terrestrial detritus. *Ann. Rev. Ecolog. Syst* 8:51–81.

Schlesinger, W.H. 1984. Soil organic matter: A source of atmospheric CO_2. p. 111–127 In: G.M. Woodwell (ed.), *The Role of Terrestrial Vegetation in the Global Carbon Cycle*. John Wiley & Sons, New York, N.Y.

Schlesinger, W.H. 1985. The formation of caliche in soils of the Mojave Desert, California. *Geochem. and Cosmochim. Acta* 49:57–66.

Tans, P.P., J.Y. Fung, and T. Takahashi. 1990. Observational constraints on the global atmospheric CO_2 budget. *Science* 247:1431–1438.

Tarnocai, C. and M. Ballard. 1994. Organic carbon in Canadian soils. p. 31–45 In: R. Lal, J.M. Kimble, and E. Levine (eds.), *Soil Processes and Greenhouse Effect*. USDA, Soil Conservation Service, National Soil Survey Center, Lincoln, NE.

Tarnocai, C., J.A. Shields, and B. MacDonald (eds.). 1993. Soil Carbon Data for Canadian Soils (Interim Report). Centre for Land and Biological Resources Research, Ottawa, Canada. Contribution No. 92-179.

Uspenskii, V.A. 1956. *Balans Ugleroda V Biosfere V Svyazi S Voprosom O Raspredelenii Ugleroda V Zemnoi Kore (Balance of Carbon in the Biosphere in the Context of the Problem of Carbon Distribution in the Earth's Crust)*, Leningrad: Gostoptekhizdat.

Vernadsky, V.I. 1994. *Trudy po Geokhimii (The Papers on Geochemistry)*, Moscow: Nauka.

Vinogradov, A.P., 1967. *Vvedenie V Geokhimiyu Okeana (The Basics of Ocean Geochemistry)*, Moscow: Nauka.

Vinson, T.S. and T.P. Kolchugina (eds.). 1993. *Carbon Cycling in Boreal Forest and Subarctic Ecosystems: Biospheric Responses and Feedbacks to Global Climate Change*. Department of Civil Engineering. Oregon State University. Corvallis, OR.

Voitkevich, G.V., A.V. Kokin et al. (Spravochnik po geokhimii) (*Reference Book on Geochemistry*), Moscow: Nedra, 1990.

Yaroshevskii, A.A. 1988. About the geochemical evolution of the biosphere. *Priroda* (Moscow) 2:59–67.

Zinke, P. J., A. G. Stangenberger, W. W. Post, W. R. Emanuel, and J. S. Olson. 1984. Worldwide Organic Soil Carbon and Nitrogen Data. ORNL/TM-8857. Oak Ridge National Laboratory, Oak Ridge, TN.

Zybach, B. 1993. Selecting a model to estimated the effects of a climate change and management activities on carbon cycling in temperate forest regions. p. 193–202 In: T.S. Vinson and T.P. Kolchugina (eds.), *Carbon Cycling in Boreal Forest and Subarctic Ecosystems: Biospheric Responses and Feedbacks to Global Climate Change.* Department of Civil Engineering, Oregon State University, Corvallis, OR.

CHAPTER 21

Nitrogen, Sulfur, and Phosphorus and the Sequestering of Carbon

F.L. Himes

1. Introduction

Carbon is only one of several constituents of soil organic matter. The sequestration of carbon in soils involves the availability of the other building blocks. The elemental composition of soil organic matter has been studied and reported by many soil scientists. The carbon-to-nitrogen ratios (usually by weight), the carbon-to-sulfur ratios, and the carbon-to-phosphorus ratios are frequently used to compare the compositions of organic matter in soils. The quantities of nitrogen, sulfur, and phosphorus needed to sequester a given quantity of carbon can be estimated from these ratios. The exact quantities of these elements needed to sequester a given amount of carbon are very difficult to calculate because the efficiencies of the conversion processes vary with sources of carbon, soil properties, and climate.

Through the centuries, farmers developed soil management practices that accelerated the rate of decomposition of soil organic matter. These practices included frequent tillage and drainage to increase aeration and to promote drying. In addition, very acid soils were limed to increase the microbial activities and the availabilities of plant nutrients. After mineralization, these nutrients could be absorbed by the plants. When the harvested crop was removed from the field, the quantities of these nutrients in the soil were decreased, and the concentration of organic matter in the soil also decreased (Jenkinson, 1981, 1988). Increasing the concentration of soil organic matter to improve soil tilth and to sequester carbon involves the sequestering or immobilizing plant nutrients as well as carbon.

The sequestration of carbon in soil organic matter is more like building a village than building a house or store. In a village, some of the buildings are built to last a few years and others are built to last centuries. Soil organic matter is composed of a large variety of compounds; some are decomposed within a few years and others last centuries (Jenkinson, 1988). The building blocks for the more slowly decomposed organic matter (referred to as humus in the remainder of this chapter) are carbon, hydrogen, oxygen, nitrogen, phosphorus, sulfur, and trace amounts of some other elements. In soils, the quantities of hydrogen and oxygen have not been considered limiting for the production of humus. The factors that influence the quantities of humus in soils have been extensively studied. Recent reviews of these studies include Jenkinson (1981, 1988), Stevenson (1982), Glendining and Powlson (1995), Lucas and Vitosh (1978), Bosatta and Berendse (1984), and Carleton et al. (1988). The quantity of humus in soil is influenced by the quantities of carbon compounds added, the availabilities of nitrogen, sulfur, and phosphorus compounds; and the rates of decomposition. This chapter describes the quantities of elements needed to construct humus and not the factors that influence the rates of decomposition. The sequestration of carbon in humus not only requires the availability of nitrogen, sulfur, and phosphorus compounds in soils to combine with carbon to produce humus, but also additional quantities of these nutrients for the growing plants. The transformation processes for sequestering carbon are not 100% efficient, and uptakes of nitrogen, sulfur, and phosphorus by plants are likewise not 100% efficient. Therefore, the quantities of nitrogen, sulfur, and phosphorus needed both to produce excellent yields and to sequester carbon are difficult to calculate. The following calculations are for sequestering carbon into humus, and the quantities of N, S, and P calculated are the quantities needed to build the humus structure.

II. Calculations

The values used in the following calculations are the general values often used in textbooks. The transformation processes are also assumed to be 100 per cent efficient. The slowly decomposed soil organic matter is assumed to be humus for these calculations.

Humus -- approximately 58% C by weight (Stevenson, 1982)

$C/N = 12/1$ (w/w) or 14/1 (atom/atom) (Allison, 1973)

$C/P = 50/1$ (w/w) or 130/1 (atom/atom) (Jenkinson, 1988)

$C/S = 70/1$ (w/w) or 187/1 (atom/atom) (Jenkinson, 1988)

To sequester 10,000 kg of carbon in humus, 833 kg of nitrogen, 200 kg of phosphorus, and 143 kg of sulfur are needed. This amount of humus will increase the concentration of humus in a hectare-furrow slice of soil an additional 0.7%. The reverse of this implies that when the humus concentration decreased, for example, from 2.7% to 2%, 833 kg of N, 200 kg of P, and 143 kg of S were mineralized for plant use.

The challenge of sequestering carbon is to increase the concentration of humus while producing good crop yields. Using the above figures, one must not only add residue that contains 10,000 kg of C, but the additional carbon used by the microbes (for tissue and for respiration) decomposing the residues. Under aerobic conditions, approximately 35% (Jenkinson 1981) of the carbon in the residue becomes incorporated in humus. Therefore, the residues must contain approximately 28,000 kg of carbon for 10,000 kg of C to be sequestered in humus. On an oven-dry weight basis, most residues are approximately 45% carbon. (ibid) The weight of residue required to provide the 28,000 kg C is approximately 62,000 kg. Since residues applied to soils are not oven-dried, the total weight of applied residue is greater. The results of these calculations are summarized in Table 1.

The plant residues will contain some of the nutrients. The factors that influence the rates of decomposition of residues are summarized by Jenkinson (1981). When the rate of decomposition is moderate to fast, immobilization of nitrogen usually occurs when the C/N ratio of the residue is greater than 30/1. This ratio can be decreased by adding fertilizers or by using the residue as bedding and the waste nitrogen compounds of the animals will decrease the C/N ratio of the residue. Jenkinson (1981) summarizes data from the Broadbalk plots at Rothamsted. Continuous wheat plots were compared and the data covers 1843 to 1963. The concentration of organic carbon in the unmanured plot did not vary much in the 120 years. The annual application of N, P, K, Na, and Mg, increased the total soil carbon about 15%. The increase in organic matter is due to the increased yields and more residue was produced. The plot that received 34 tonnes of manure per year nearly tripled in soil organic matter. This plot not only received additional residue but the residue in the manure had already undergone some decomposition and the carbon compounds in the manure were the more stable materials. Therefore, 34 tonnes of "composted" manure is equivalent to more than 34 tonnes of crop residue.

Table 1. Calculations for the sequestering of 10,000 kg of carbon into humus

Assumptions:		
1.	Humus	58% carbon
2.	Humus	C/N ratio is 12/1
3.	Humus	C/P ratio is 50/1
4.	Humus	C/S ratio is 70/1
5.	Residue-C to humus-C	35%
6.	Residue (over-dry)	45% carbon
7.	2,240,000 kg soil in 1 ha to a depth of 15 cm	

Quantities:
1. 10,000 kg C sequestered into humus requires:
    a. 833 kg N
    b. 200 kg P
    c. 143 kg S

2. 17,241 kg of humus produced
3. 28,000 kg of C in residue
4. 62,000 kg of residue (oven-dry)
5. New humus to total soil -- 0.77%

Table 2. Carbon and nitrogen sequestered in the fertilized Broad balk plot at Rothamsted

Continuous wheat since 1843	Organic C (%)	Nitrogen (%)	kg C ha^{-1}	kg N ha^{-1}
Unmanured	0.90	0.098	20,160	2,195
Manured NPK Mg annually	1.08	0.112	24,192	2,590
Element sequestered			4,032	314

(Adapted from Jenkinson, 1981.)

The increased quantities of carbon and nitrogen sequestered in the fertilized Broadbalk plot over the nonfertilized plot are shown in Table 2. The ratio of the carbon to nitrogen sequestered was 12.8/1. To sequester the 4,000 kg of C, 314 kg of nitrogen was required. Glendining (1995) summarizes the changes in the total soil nitrogen content for many long term field experiments in various countries. He selected the experiments that involved the addition of inorganic nitrogen fertilizers. The plots receiving the inorganic nitrogen fertilizers were compared to plots receiving phosphorus and potassium fertilizers or to the plots receiving no fertilizer. Of the 35 experiments that he lists, the total soil nitrogen increased in 32. In some of these experiments, all of the top growth was removed each year. The length of the experiments ranged from 11 to 136 years. A second summary includes 25 different research locations, and the treatments varied from 7 to 40 years. Of the data presented, the addition of inorganic nitrogen increased the total soil nitrogen 68 times. Glendining (1995) presents evidence that the addition of nitrogen fertilizer increased the quantities of mineralizable nitrogen. The fraction of the nitrogen that is a constituent of humus cannot be calculated from the data presented.

In some of the experiments summarized by Glendining (ibid), the addition of PK fertilizers increased the quantity of total nitrogen in the soil as compared to the control. These nutrients, no doubt, increased the dry matter production so that more carbon was added to the soil. In addition, the additional phosporus may have been necessary for a maximum amount of carbon and nitrogen to be converted to humus. One can postulate similar situations for sulfur. Figure 1 summarizes the carbon cycle and the sequestering of carbon in humus.

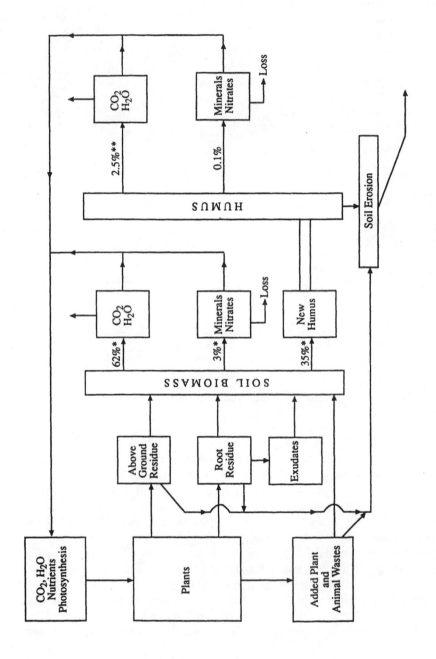

Figure 1. Modified carbon cycle (gas, plant, humus). (Modified chart from Lucas and Vitosh, 1978.)

III. Summary

The sequestering of carbon in soil organic matter involves the production of complex organic structures. Besides carbon, these structures contain hydrogen, oxygen, nitrogen, phosphorus, and sulfur. The sequestering of 10,000 kg of carbon requires a very large quantity of residue because the fresh residue is less than 45% carbon and less than 35% of the carbon will be converted to humus. Also, the quantity of carbon sequestered will be limited if there is insufficient quantities of nitrogen, phosphorus, and sulfur. The sequestering of 1000 carbon atoms requires approximately 71 atoms of nitrogen, approximately 8 atoms of phosphorus, and approximately 5 atoms of sulfur. The production of humus not only sequesters carbon but also nitrogen, phosphorus, and sulfur. These elements are essential for plant growth. Maintaining good crop growth, maintaining low concentrations of plant nutrients in the drainage water, and sequestering carbon in the soil is a challenge to the farmer. Some of the field research experiments can serve as guides to farming practices that increase the sequestering of carbon.

References

Allison, F.E. 1973. *Soil Organic Matter and its Role in Crop Production.* Elsevier Scientific Publishing Co., NY.

Bosatta, F. and F. Berendse. 1984. Energy or nutrient regulation of decomposition: implications for mineralization-immobilization response to perturbations. *Soil Biology & Biochem.* 16:63-67.

Carleton, S.W., D.I. Moore, J.D. Horner, and J.R. Gosz. 1988. Nitrogen mineralization-immobilization response to field N or C perturbations: An evaluation of a theoretical model. *Soil Biology and Biochem.* 20:101-105.

Glendining, M.J. and D.S. Powlson. 1995. The effects of long continued application of inorganic nitrogen fertilizer on soil organic nitrogen – a review. p. 385-446. In: R. Lal and B.A. Stewart (eds.), *Soil Management – Experimental Basis for Sustainability and Environmental Quality. Advances in Soil Science*, CRC Lewis Publishers, Boca Raton.

Jenkinson, D.S. 1981. The fate of plant and animal residues in soil. p. 505-561. In: D.J. Greenland and M.H.B. Hayes (eds.), *The Chemistry of Soil Processes.* John Wiley & Sons, NY.

Jenkinson, D.S. 1988. Soil organic matter and its dynamics. p. 564-607. In: A. Wild (ed.), *Russell's Soil Conditions and Plant Growth.* Longman Scientific & Technical, Essex CM20 2JE, U.K.

Lucas, R.E. and M.L. Vitosh. 1978. Soil organic matter dynamics. Research Report 358. Mich. State Univ. E. Lansing.

Stevenson, F.J. 1982. *Humus Chemistry, Genesis, Composition, Reactions.* John Wiley & Sons, NY.

CHAPTER 22

Management of Soil C by Manipulation of Microbial Metabolism: Daily vs. Pulsed C Additions

D.C. Jans-Hammermeister, W.B. McGill, and R.C. Izaurralde

I. Introduction

The concentration of C in soil is a function of C input and output. Excluding erosion losses, C output is mainly in the form of CO_2, a by-product of heterotrophic microbial activity. Managing microbial metabolism by manipulating C inputs to minimize the rate of C mineralization may have potential for increasing C storage in soil. In the short-term, slower rates of substrate mineralization are correlated with slower rates of microbial biomass turnover (Ladd et al., 1992). Low concentrations of available C in soil result in smaller proportions of substrate mineralization compared to high concentrations (Bremer and Kuikman, 1994; Wu et al., 1993). This response is attributed to higher efficiencies of microbial growth or less turnover of microbes or microbial products in the presence of low C concentrations. These responses have been well documented for pulsed infusions of substrate: we hypothesized that continuous additions of low C concentrations would also result in slower rates of substrate mineralization. It follows that C added slowly, as in the form of root exudates, would have greater potential for storage in soil compared to C added in a large pulse, as in the form of manure or straw amendment.

Deterioration of soil structure is correlated with losses of organic C and N (Elliot, 1986). Tisdall and Oades (1982) suggest three types of cementing agents to be responsible for soil aggregation: (i) transient, mainly in the form of microbial-derived polysaccharides; (ii) temporary, including fungal hyphae; and (iii) persistent, aromatic humic material in association with polyvalent metal cations. Microaggregates (50-250 µm) are bound together to form macroaggregates (>250 µm) by the above mentioned transient and temporary agents. Consequently, microorganisms play an important role in the formation and persistence of aggregates. Conversely, aggregates influence the spatial relationships among substrates, microorganisms and soil environment and, therefore, are an important factor in the potential stabilization of soil C. Surprisingly few laboratory studies on soil C dynamics use soil cores with intact structure and the importance of spatial relationships is often overlooked.

Our hypotheses were (1) that the timing of glucose C addition determines the metabolic state of the microbial biomass, and (2) that the metabolic state of the microbial biomass determines the potential for C storage in soil by controlling the rate of mineralization of carbonaceous substrates and by influencing the spatial distribution of added C within soil size fractions.

II. Materials and Methods

A. Soil

Soil samples were collected September 21, 1995 from experimental plots on a Gray Luvisol (Typic Cryoboralf) located at Cooking Lake (SE 3-51-21-W4) in the Luvisolic soil zone of Alberta. Plots had been seeded to spring wheat (*Triticum aestivum* L.) or barley (*Hordeum vulgare* L.) since 1991 and were not fertilized. Samples were collected from between rows of standing stubble within one month of harvest. Soil was removed from the field as intact cores within ABS (acrylonitrile butadiene styrene) cylinders (11 cm length, 5 cm i.d.). The soil was characterized as Cooking Lake loam with an organic C content of 4.0%, N content of 0.35%, pH of 6.5, clay content of 22%, and bulk density of 1.25 Mg m^{-3}.

Soils were conditioned at 22°C for 19 weeks. Prior to conditioning, crop residues were removed from the soil surface and moisture was adjusted to 45% WHC (as defined by Harding and Ross, 1964) by increasing water content from 22% (field moist) to 29%. Soil cores were stored inside 1.9 l glass jars and were aerated every 10-14 days. Gravimetric moisture content was maintained at 45% WHC throughout the conditioning period.

B. Treatments

To avoid contamination of specialized equipment such as the soluble C analyzer, the experiment was duplicated: one set of samples was treated with nonlabeled glucose and the other set was treated with [U-^{14}C]-glucose (Sigma) at 7.4 Bq ug^{-1} C. Glucose was added to soil cores at two rates: 30 ug C g^{-1} soil day^{-1} for 15 days or 450 ug C g^{-1} soil administered as a single pulse at time zero. The daily rate of C addition was chosen to balance the rate of respiration based on estimates of the microbial metabolic quotient in similar agricultural soils (Wardle and Ghani, 1995). The pulsed C addition was equivalent to the total C added in daily increments over the 15 day incubation. Glucose solutions also contained sufficient KNO_3 and $Na_2PO_4 \cdot 2H_2O$ to maintain the C:N ratio at 10 and the C:P ratio at 80 (Bremer and Kuikman, 1994). A nonamended control treatment was also included.

Glucose was injected into the soil core with a fine-needled syringe. Each soil core received the equivalent of 1.2 ml d^{-1} which maintained the water content below 60% WHC (38% moisture) during the course of the incubation. The placement of the needle was determined by a template indicating 4 locations and 6 depths for a total of 24 sites per core. A pilot-hole was made with a fine wooden probe before inserting the needle to avoid plugging the needle tip. Each injection site received approximately 0.05 ml and amendment was placed in the same location at each application. Glucose solution was injected at 24 hour intervals in the daily treatment. Distilled water was injected at 48 hour intervals in the pulse (following initial dose of glucose) and control treatments.

C. Respiration

Three replicate cores of each treatment (control, [U-^{14}C]-glucose pulse and [U-^{14}C]-glucose daily) were incubated in 1.9 l jars each containing sufficient 0.5M NaOH to trap CO_2. NaOH was replaced 12 and 24 hours after initial amendment and every 24 hours after that. To determine total CO_2-C trapped, a 1 ml subsample of NaOH was titrated to the phenolphthalein end-point with HCl in the presence of excess $BaCl_2$. ^{14}C respired as CO_2 was determined from a 2 ml subsample of NaOH which was mixed with 10 ml OptiPhase "Hisafe" scintillation cocktail and counted on a Packard 2000CA TRI-CARB liquid scintillation counter.

D. Microbial Biomass

Three jars per treatment were destructively sampled after 1, 4, 7, 11, and 15 days. Soil and cylinder were weighed, soil was carefully removed from the cylinders, and passed through an 8 mm sieve to obtain a more uniform sample. A subsample was further sieved with a 2 mm sieve for microbial biomass analysis by liquid chloroform extraction (Gregorich et al., 1990) as follows: Two 20 g samples were placed in 125 ml HDPE bottles. Ethanol-free chloroform (1.0 ml) was pipetted directly onto the soil in one sample; 40 ml of $0.5M$ K_2SO_4 was then added to both samples. Soil suspensions were shaken in a reciprocating shaker for 30 minutes. The samples were filtered through Whatman No. 42 filter paper which had been rinsed with deionized water to remove soluble C contaminants (C. Gomez, pers. comm.). N_2 gas was bubbled through the chloroform-treated extracts for 45 seconds to remove chloroform. Nonlabeled extracts were analyzed for soluble C by an automated u.v.-persulphate oxidation method (Astro 2001, Astro International, Texas, U.S.A.). Biomass C was calculated as (C extracted from chloroform-treated soil C extracted from nontreated soil) / k_{EC}. A value of 0.17 was used for k_{EC} (Gregorich et al., 1990). The use of liquid chloroform in the procedure did not appear to increase the C content of the extracts and, therefore, no correction factor was required.

Aliquots (2 ml) of labeled extracts were mixed with 10 ml OptiPhase 'Hisafe' scintillation cocktail and counted on a Packard 2000CA TRI-CARB liquid scintillation counter. Biomass ^{14}C was calculated as (^{14}C extracted from chloroform-treated soil ^{14}C extracted from nontreated soil) / 0.17.

E. Water-Stable Aggregates

On the final sampling day, a 100 g subsample of 8 mm-sieved soil from each core was used for the collection of water-stable aggregates following the method of Elliott (1986). Nested sieves (250 μm and 53 μm) were submersed in a tub of water. Moist soil was placed in the 250 μm sieve and submersed for 5 minutes before sieving commenced. Floating organic matter was removed during this time. Soils were sieved under water by gently moving the sieves 3 cm vertically 50 times in 2 minutes. Material remaining on the sieves was allowed to drain for 5 minutes. The sieve and soil were weighed. Subsamples were removed for biomass analyis by liquid chloroform extraction as previously described. Due to the small quantity of material captured on the 53 μm sieve, samples equivalent to approximately 10 g dry mass (in comparison to the 20 g collected from other fractions) were extracted with 20 ml of $0.5M$ K_2SO_4. Our preliminary studies had indicated that soil water content was not a significant factor in biomass determination with the liquid chloroform extraction procedure.

Within 5 hours of filtering, $CaSO_4$ precipitates began to form in the 53-250 μm aggregate extracts and in some of the > 250 μm aggregate extracts. These precipitates were resuspended with a vortex mixer. A 5 ml aliquot was quickly removed and mixed with 10 ml Na metaphosphate (5% w/v) which immediately dissolved the precipitate (Wu et al., 1990). These samples were then analyzed for soluble C, as previously described.

Subsamples of nonlabeled whole soil and aggregate fractions were oven-dried at 60°C and analyzed for C and N (Carlo-Erba). Subsamples (approximately 0.5 g) of labeled whole soil and aggregate fractions were quickly air dried and oxidized in a Harvey Biological Oxidizer (Model OX-300). ^{14}C released during oxidation was trapped in Harvey's ^{14}C cocktail and counted on a Packard 2000CA TRI-CARB liquid scintillation counter.

F. Calculations

The first order decay constant (day^{-1}) for nonlabeled biomass was calculated as:

$$k = -\ln(A_t/A_0)/t \tag{1}$$

where A_t is nonlabeled biomass at time t, A_0 is nonlabeled biomass at time 0, and t is time.

Storage efficiency represents the proportion of added substrate ^{14}C which is metabolized and remains in the soil. Storage efficiency (%) was calculated as:

$$(\text{residual } ^{14}C\text{-soluble } ^{14}C)/\text{added } ^{14}C*100\% \tag{2}$$

Residual ^{14}C was calculated as added ^{14}C - respired ^{14}C. Soluble ^{14}C was calculated as the amount of K_2SO_4-extractable ^{14}C in the nonfumigated soil.

The proportion of residual ^{14}C released by liquid chloroform extraction (k_{EC}) was calculated as:

$$(^{14}C_f - ^{14}C_{nf})/(\text{residual } ^{14}C\text{-soluble } ^{14}C) \tag{3}$$

where $^{14}C_f$ is ^{14}C extracted with K_2SO_4 from the liquid chloroform treated soil and $^{14}C_{nf}$ is ^{14}C extracted from the nonfumigated soil.

III. Results

A. Mineralization of Glucose and Native Soil Organic C

A significantly (P<0.01) larger proportion of the pulsed addition of glucose C was evolved as CO_2 compared to the daily additions (Figure 1a). Initially the rate of glucose-derived C mineralization was greater following the pulse addition (Figure 1b). This rate fell quickly and after day 3, mineralization rate was greater for the daily treatment. The rate of mineralization of daily glucose remained fairly constant at about 10 mg C kg^{-1} day^{-1} during the 15 days of glucose addition.

Both daily and pulsed additions of glucose increased the rate of native soil organic C mineralization. As a result, after 29 days of incubation, the soil receiving the daily glucose treatment had mineralized 76 mg C kg^{-1} soil and the pulse treatment had mineralized 18 mg C kg^{-1} more nonlabeled C than the control (Figure 2a). The increase in the rate of organic matter mineralization was short-lived in the pulsed treatment and by day 9, mineralization rates were similar to the control (Figure 2b). Mineralization in the daily treatment remained higher than the control for the duration of the incubation. Considering C additions and losses, the net change in soil C after 29 days was -219 mg C kg^{-1} for the control, -58 mg C kg^{-1} for the daily addition, and -28 mg C kg^{-1} for the pulsed addition. Statistically, however, there was no significant difference (P<0.05) in the change in soil C between the daily and pulsed treatments.

B. Microbial Biomass

Microbial biomass C was significantly greater in the pulsed glucose treatment on sampling days 1 and 4 (Figure 3). The increase in biomass C was not sustained, however, and was similar to that in the control and daily treatment after day 7. There was no significant difference between biomass C in the daily and control treatments on any of the sampling dates. On day 1, the metabolic quotient (defined as CO_2-C respired per unit biomass C) was significantly higher in the pulsed treatment (Figure 4). The metabolic quotient was significantly higher in the daily treatment on sampling days 4 to 15.

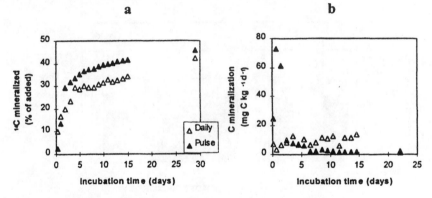

Figure 1. Cumulative mineralization of ^{14}C (a) and mineralization rate of glucose-derived C (b) following glucose amendment.

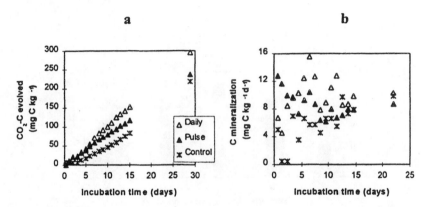

Figure 2. Cumulative mineralization (a) and mineralization rate (b) of nonlabeled C following amendment with glucose.

Microbial ^{14}C as a proportion of added ^{14}C was significantly greater in the daily glucose treatment on days 1 and 15 (Figure 5a). The lower recovery of ^{14}C in the microbial biomass on day 1 in the pulsed treatment coincided with a greater recovery of soluble ^{14}C (Figure 5b). Calculated values of k_{EC} were greater in the daily treatment on day 1 and were significantly different over time ($P<0.05$) (Table 1).

In the daily glucose treatment, labeled biomass increased over time while nonlabeled biomass decreased slowly (Figure 6). In the pulsed treatment, the increase in biomass C on sampling day 1 was mainly nonlabeled: glucose-derived biomass as a percent of total biomass was highest at the day 7 sampling. The pulsed treatment had a greater decay constant ($k = -0.041$) than the daily treatment ($k = -0.024$) indicating a more rapid turnover in the pulsed treatment. Storage efficiency was significantly greater in the daily treatment on the last three sampling dates (Figure 7).

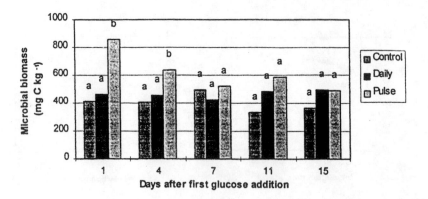

Figure 3. Microbial biomass C estimated by liquid CHCl$_3$-extraction. On each day, treatments with the same letter do not differ significantly (P<0.05).

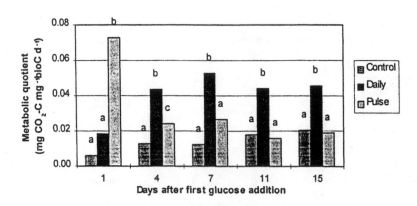

Figure 4. Metabolic quotients of microbial populations. On each day, treatments with the same letter do not differ significantly (P<0.05).

C. Water-Stable Aggregate Fractions

The distribution of soil dry mass among aggregate fractions was similar in all treatments (Table 2). Significantly more (P<0.05) dry mass was recovered in the >250 μm size fraction. The daily glucose treatment had significantly lower concentrations of C in both the aggregate fractions compared to the control and pulse treatments (Table 2). This C depletion resulted in a significantly lower C:N ratio in the large aggregates of the daily treatment (Table 2). There was no significant difference (P<0.05) in C contents or C:N ratios between size fractions. Microbial biomass C was similar among treatments and aggregate size fractions (Table 2).

The >250 μm aggregates contained a significantly greater proportion of ^{14}C added as a pulse compared to that added daily (Table 3). Approximately 75% of ^{14}C added as a pulse was accounted for in the two aggregate fractions. In contrast, only 49% of ^{14}C added daily was accounted for in aggregates.

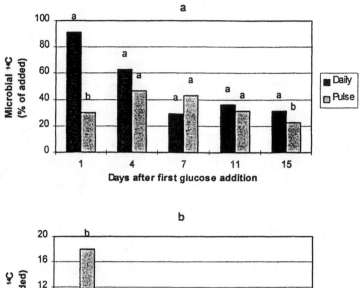

Figure 5. Proportion of glucose ^{14}C as microbial ^{14}C (a) and K_2SO_4-soluble ^{14}C (b) microbial ^{14}C calculated with $K_{EC} = 0.17$. On each day, treatments with the same letter do not differ significantly (P<0.05).

Table 1. Calculated k_{EC} values

	Daily	Pulse
Day 1	0.19	0.07
Day 4	0.15	0.12
Day 7	0.07	0.12
Day 11	0.09	0.09
Day 15	0.08	0.07

the remainder is thought to be associated with material finer than 53 µm. Of the ^{14}C which was recovered in each aggregate fraction, significantly more of the ^{14}C added daily was associated with microbial biomass (Table 3). Significantly (P<0.05) more microbial ^{14}C was recovered in the >250 µm fraction.

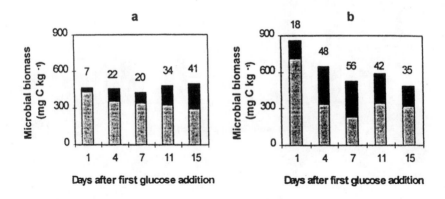

Figure 6. Glucose-derived (dark bars) and native organic matter-derived (light bars) microbial biomass C following daily (a) and pulsed (b) glucose additions. Numbers above bars indicate percent of total biomass derived from glucose.

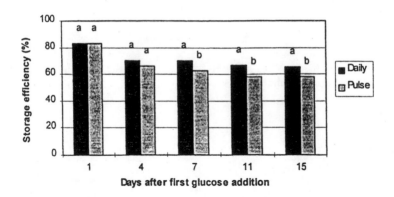

Figure 7. Storage efficiency represents the proportion of metabolized substrate C which remains in the soil. On each day, treatments with the same letter do not differ significantly (P<0.05).

IV. Discussion

The experimental data support the hypothesis of distinct physiological responses to different rates of substrate infusion. First, differences between treatments were evident in kinetic analysis of ^{14}C mineralization. In the pulsed treatment, mineralization was best described by a combination of two classes of kinetics: Michaelis-Menten and first-order. A similar scheme was used in the Phoenix simulation model to describe the decomposition of the labile and resistant portions of introduced residues in soil (McGill et al., 1981). In contrast, ^{14}C mineralization in the daily treatment was best

Table 2. Characteristics of water-stable aggregate fractions 15 days after first glucose addition

	Aggregates >250 μm		
	Control	Daily	Pulse
Mass (% of total dry mass)	56 a	61 a	54 a
C content (%)	4.3 a	3.1 b	4.8 a
C:N ratio	11.5 a	10.4 b	12.5 a
Microbial C (mg C kg^{-1})	369 a	566 a	452 a

	Aggregates 53 - 250 μm		
	Control	Daily	Pulse
Mass (% of total dry mass)	18 a	15 a	21 a
C content (%)	4.8 a	2.8 b	5.5 a
C:N ratio	12.2 a	11.2 a	11.9 a
Microbial C (mg C kg^{-1})	307 a	436 a	350 a

Values in each row followed by the same letter do not differ significantly (P<0.05).
n = 3 for mass and microbial C; n = 2 for C content and C:N ratio.

Table 3. Distribution of ^{14}C in water-stable aggregate fractions 15 days after first glucose addition

	Aggregates >250 μm	
	Daily	Pulse
^{14}C content (% of added)	36 a	58 b
Microbial ^{14}C (% of aggregate ^{14}C)	43 a	9 b
K$_2$SO$_4$-soluble ^{14}C (% of aggregate ^{14}C)	0.3 a	0.2 a

	Aggregates 53 - 250 μm	
	Daily	Pulse
^{14}C content (% of added)	13 a	17 a
Microbial ^{14}C (% of aggregate ^{14}C)	26 b	16 a
K$_2$SO$_4$-soluble ^{14}C (% of aggregate ^{14}C)	0.2 a	0.2 a

Values in each row followed by the same letter do not differ significantly (P<0.05).

described by first-order kinetics. First-order kinetics are generally used to describe processes with more gradual transitions in rate over time.

Changes in metabolic quotient were also different between treatments. The metabolic quotient immediately increased in response to the pulsed glucose addition and then fell to levels similar to the control. In contrast, the metabolic quotient in the daily treatment increased more slowly and reached an intermediate level by day 4. These observations support the hypothesis that a pulsed infusion of labile C shifts the microbial population towards energy-wasteful zymogenous species which quickly decline once the available substrate is exhausted (Paul and Clark, 1989; Bradley and Fyles, 1995). The daily infusions of small quantities of labile C, on the other hand, encourage autochthonous species which have greater C use efficiencies.

Previous studies have used *in situ* labeling of microbial cells to calibrate the chloroform fumigation-extraction method (Sparling and West, 1988; Gregorich et al., 1990). The flush of ^{14}C resulting from chloroform addition represents the proportion of microbial C which is extractable and has been termed the k_{EC} factor. To be meaningful, these calibrations require the complete metabolism and uniform incorporation of substrate ^{14}C into the soil microbial biomass within a specified time. Complete glucose assimilation has been assumed to occur between 24 hours (Voroney and Paul, 1984) and 72 hours (Bremer and Kuikman, 1994) after addition. An important assumption in estimating k_{EC} is that non-soluble ^{14}C remaining in soil is in the form of viable microbial biomass and not stabilized nonmicrobial compounds. Physical stabilization prior to microbial uptake is unlikely: approximately 98% of glucose ^{14}C remained K_2SO_4-extractable after 24 hours in sterile soil (Bremer and van Kessel, 1990). Post-metabolism products, however, are likely to accumulate as extra-cellular polysaccharides and cellular debris associated with clay particles and aggregates. The decrease in calculated k_{EC} values is evidence for stabilization of post-metabolism compounds over time and emphasizes the importance of proper timing when calculating k_{EC} as a calibration factor.

At 24 hours, only one daily amendment had been made. The comparison between treatments was, therefore, simply a high rate of C addition (450 µg g^{-1} soil) vs. a low rate (30 µg C g^{-1} soil). Assuming that all nonsoluble ^{14}C remaining in soil at 24 hours was in the form of microbial biomass, we found significantly more ^{14}C from the low rate of addition to be chloroform-extractable. These results are consistent with those reported by Bremer and Kuikman (1994) who suggest that microorganisms have a threshold requirement for substrate C: at low available concentrations, microorganisms assimilate C but do not metabolize it due to metabolic arrest. As a result of this nonuniform labeling, a greater proportion of cellular ^{14}C is chloroform-extractable. The k_{EC} of 0.19 estimated from the low rate of glucose addition is, however, in close agreement with the k_{EC} of 0.17 suggested by Gregorich et al. (1990) for the liquid chloroform-extraction method. Gregorich et al. (1990) used a relatively high rate of ^{14}C-labeled glucose (1 mg C g^{-1} soil) and an incubation time of 30 hours prior to extraction. Why, then, at the high rate of glucose addition, was their calculated k_{EC} so low? We suggest that a substantial portion of the large dose of glucose ^{14}C was present in soil as post-metabolism nonmicrobial materials. Extra-cellular polysaccharides are immediate by-products of active microbial metabolism and are closely associated with soil aggregates where they act as binding agents (Roberson et al., 1995). The use of soil cores in our experiment ensured that the spatial relationship between microorganisms and aggregates remained intact. The initial rapid rate of substrate metabolism in the pulsed treatment coupled with the microorganisms' proximity to soil aggregates may have enhanced the production and stabilization of extra-cellular polysaccharides. By day 4, there was a large increase in the calculated k_{EC} for the pulsed treatment. This can be explained by secondary metabolism of the stabilized by-products of primary decomposition. Cellular incorporation of the processed ^{14}C by slower growing autochthonous species would increase the chloroform-extractable proportion of residual ^{14}C.

Differences in the physiology of the decomposers influenced the storage of added glucose C. The daily glucose additions maintained a more stable biomass with a lower decay rate compared to the pulsed treatment. Consequently, a greater proportion of metabolized substrate C remained stabilized as microbial biomass. Soil texture influences decomposition in a similar way: the slower rates of substrate decomposition in soils of heavier texture are the result of relatively slower rates of microbial biomass decay (Ladd et al., 1992).

Both daily and pulsed additions of glucose increased the mineralization of nonlabeled soil C. The increase in response to the pulsed addition of glucose was short-lived and coincided with the synthesis of new microbial biomass. Additional mineralization of nonlabeled C in the daily treatment was evident throughout the incubation and coincided with the continuous replacement of nonlabeled biomass with labeled material. The additional mineralization was, therefore, most likely the result of conversions in biomass as suggested by Dalenberg and Jager (1989). Over the 29-day incubation, the sustained higher rate of nonlabeled C mineralization in the daily treatment offset the greater storage of glucose C and, as a result, the net change in soil C was not significantly different between the daily and pulsed treatments.

The distribution of dry mass among the aggregate size fractions was in close agreement with results from a cultivated Dark Brown Chernozemic soil (Gupta and Germida, 1988). Microbial C concentrations were higher in the macroaggregates, although not significantly higher; this is in agreement with previous findings of increasing microbial C concentrations with increasing aggregate size (Gupta and Germida, 1988; Singh and Singh, 1995). The significantly lower C content of both aggregate fractions following daily additions of glucose was unexpected. The C depletion cannot be explained by increased nonlabeled C mineralization in the daily treatment as the difference in C content between the control and the daily treatment was more than 100 times the increase in C mineralization. Whether the C depletion was actually the result of glucose treatment or merely an artifact of the variability inherent in soil cores remains unknown.

The use of intact soil cores made possible a study of the spatial distribution of glucose-derived ^{14}C. Significantly more ^{14}C was recovered in the >250 µm aggregates compared to the 53-250 µm aggregates, especially in the pulsed treatment. Singh and Singh (1995) suggest that macroaggregates may favor greater activity of fungal-based food web organisms and, thus, may allow greater retention of microbial C. Our observations of more chloroform-extractable ^{14}C associated with macroaggregates support this hypothesis. After fifteen days in the soil, ^{14}C from the pulsed treatment would be associated with older and less active biomass compared to the daily treatment. ^{14}C associated with older fungal hyphae and fungal debris may explain the lower proportion of aggregate ^{14}C which was chloroform-extractable in the pulsed treatment.

The majority of the ^{14}C in the daily treatment was unaccounted for in the aggregate fractions and was thought to be associated with the <53 µm fraction. It is postulated that with the low daily input of substrate C, actively metabolizing microorganisms located on the outer surfaces of aggregates and in the pore spaces between them would have had ready access to the incoming ^{14}C; these microbes would have been intimately associated with clay and silt-size particulates and would have separated out into this finer fraction upon sieving. Organic matter associated with silt and clay is generally considered more stable than that present as cementing agents associated with aggregates, especially macroaggregates (Christensen, 1992).

V. Conclusions

1. Experimental data support the hypothesis of distinct physiological responses from decomposer microorganisms to the timing of glucose addition. In the pulsed treatment, two phases of decomposition were observed: the first was short-lived and was distinguished by rapid mineralization of ^{14}C coupled with a large increase in biomass C and a high metabolic quotient. The second phase was distinguished by low and steady ^{14}C mineralization rates coupled with biomass C and metabolic quotients similar to the control treatment. In the daily glucose treatment, ^{14}C mineralization rates remained steady throughout the incubation and biomass C did not increase above that of the control. After day 4, the metabolic quotient stabilized at an intermediate level.

2. Differences in the physiology of the decomposers influenced the storage potential of added glucose C. The daily input of glucose C maintained a more stable biomass thus reducing microbial turnover and increasing the proportion of metabolized glucose which remained in the soil.

3. Both daily and pulsed additions of glucose increased the mineralization of nonlabeled soil C. Increases were most likely the result of conversions in the microbial biomass. The sustained higher rate of nonlabeled C mineralization in the daily treatment offset the greater storage of glucose C and, as a result, the net change in soil C was not significantly different between the daily and pulsed treatments.

4. There was a significantly greater recovery of pulsed ^{14}C in the macroaggregates. The majority of the ^{14}C added as daily treatments was associated with material <53 μm in size. Most of the pulsed ^{14}C in aggregates was present as chloroform-resistant material. In contrast, much of the daily treatment ^{14}C in aggregates was in the form of viable biomass.

Acknowledgements

We thank the Canada-Alberta Environmentally Sustainable Agriculture (CAESA) Agreement and the Natural Sciences and Engineering Research Council of Canada for financial support and C.T. Figueiredo for technical assistance.

References

Bradley, R.L. and J.W. Fyles. 1995. A kinetic parameter describing soil available carbon and its relationship to rate increase in C mineralization. *Soil Biol. Biochem.* 27:167-172.
Bremer, E. and P. Kuikman. 1994. Microbial utilization of $^{14}C[U]$glucose in soil is affected by the amount and timing of glucose additions. *Soil Biol. Biochem.* 26:511-517.
Bremer, E. and C.van Kessel. 1990. Extractability of microbial ^{14}C and ^{15}N following addition of variable rates of labelled glucose and $(NH_4)_2SO_4$ to soil. *Soil Biol. Biochem.* 22:707-713.
Christensen, B.T. 1992. Physical fractionation of soil and organic matter in primary particle size and density separates. *Advances in Soil Science* 20:1-90.
Dalenberg, J.W. and G. Jager. 1989. Priming effect of some organic additions to ^{14}C-labelled soil. *Soil Biol. Biochem.* 21:443-448.
Elliott, E.T. 1986. Aggregate structure and carbon, nitrogen, and phosphorus in native and cultivated soils. *Soil Sci. Soc. Am. J.* 50:627-633.
Gregorich, E.G., G. Wen, R.P. Voroney, and R.G. Kachanoski. 1990. Calibration of a rapid direct chloroform extraction method for measuring soil microbial biomass C. *Soil Biol. Biochem.* 22:1009-1011.
Gupta, V.V.S.R. and J.J. Germida. 1988. Distribution of microbial biomass and its activity in different soil aggregate size calsses as affected by cultivation. *Soil Biol. Biochem.* 20:777-786.
Harding, D.E. and D.J. Ross. 1964. Some factors in low-temperature storage influencing the mineralisable-nitrogen of soils. *J. Sci. Food Agric.* 15:829-834.
Ladd, J.N., L. Jocteur-Monrozier, and M. Amato. 1992. Carbon turnover and nitrogen transformations in an alfisol and vertisol amended with [U-^{14}C] glucose and [^{15}N] ammonium sulphate. *Soil Biol. Biochem.* 24:359-371.
McGill, W.B., H.W. Hunt, R.G. Woodmansee, and J.O. Reuss. 1981. Phoenix-A model of the dynamics of carbon and nitrogen in grassland soils. p. 49-115. In: F.E. Clark and T. Rosswall (eds.), *Terrestrial nitrogen cycles*. Ecol. Bull. 33, Stockholm.
Paul, E.A. and F.E. Clark. 1989. *Soil Microbiology and Biochemistry*. Academic Press, Inc. San Diego, CA. 273 pp.
Roberson, E.B., S. Sarig, C. Shennan, and M.K. Firestone. 1995. Nutritional management of microbial polysaccharide production and aggregation in an agricultural soil. *Soil Sci. Soc. Am. J.* 59:1587-1594.
Singh, S. and J.S. Singh. 1995. Microbial biomass associated with water-stable aggregates in forest, savanna and cropland soils of a seasonally dry tropical region, India. *Soil Biol. Biochem.* 27:1027-1033.

Sparling, G. and A.W. West. 1988. A direct extraction method to estimate soil microbial C: calibration *in situ* using microbial respiration and ^{14}C labelled cells. *Soil Biol. Biochem.* 20:337-343.

Tisdall, J.M. and J.M. Oades. 1982. Organic matter and water-stable aggregates in soils. *J. Soil Sci.* 33:141-163.

Voroney, R.P. and E.A. Paul. 1984. Determination of k_C and k_N *in situ* for calibration of the chloroform fumigation-incubation method. *Soil Biol. Biochem.* 16:9-14.

Wardle, D.A. and A. Ghani. 1995. A critique of the microbial metabolic quotient (qCO_2) as a bioindicator of disturbance and ecosystem development. *Soil Biol. Biochem.* 27:1601-1610.

Wu, J., P.C. Brookes, and D.S. Jenkinson. 1993. Formation and destruction of microbial biomass during the decomposition of glucose and ryegrass in soil. *Soil Biol. Biochem.* 25:1435-1441.

Wu, J., R.G. Joergensen, B. Pommerining, R. Chaussod, and P.C. Brookes. 1990. Measurement of soil microbial biomass C by fumigation-extraction: an automated procedure. *Soil Biol. Biochem.* 22:1167-1169.

CHAPTER 23

Investigations of Carbon and Nitrogen Dynamics in Different Long-Term Experiments by Means of Biological Soil Properties

A. Weigel, E.-M. Klimanek, M. Körschens, and S. Merčik

I. Introduction

Soil organic matter, its quality and structure, influences all matter fluxes both in the soil and between soil, plants, water, and atmosphere. The knowledge about these processes, their connections and interactions is prerequisite to estimate the effect of management practices, climate and land use changes on these fluxes. The characterization and quantification of the mineralizable part of soil organic matter has a key position in this connection, since changes as a consequence of varied management (e.g., fertilization, cultivation) practically effect only this mineralizable fraction. Results from long-term experiments show that after constant treatment over many years, an equilibrium of the C and N content in the soil on a certain site-and-management-dependent level will adjust. The comparison of long-term results of different C and N levels offers the possibility to quantify the C and N dynamics of the soil and their relationship to chemical, physical, and biological soil properties. Soil microbial biomass plays a major role as a catalyst in the decomposition and transformation of SOM (Smith, 1994). There is a growing evidence that soil biological parameters may hold potential as early and sensitive indicators of soil ecological stress or restoration (Dick, 1994). One objective of the presented investigations is the identification of sensitive biological indicators for changes in soil organic matter.

II. Material and Methods

A. Material

As part of an EU project, long-term experiments at different locations were investigated with regard to carbon and nitrogen dynamics. Besides the evaluation of long-term data series, an extensive assay program was accomplished on selected treatments to compare the effectiveness of methods for characterisation and quantification of C-N dynamics.

Presented are the results of the Static Fertilization Experiment in Bad Lauchstädt (Germany) and those of two long-term experiments in Skierniewice (Poland). A short survey of the most important site chartacteristics and the selected test treatments is given in Table 1. The experiments are comparable on account of similar experimental questions (organic fertilization, mineral fertilization,

Table 1. Site characteristics and selected treatment of the long-term experiments in Bad Lauchstädt (Germany) and Skierniewice (Poland)

Experiment (Country)	Skierniewice (Poland)		Bad Lauchstädt (Germany)
Year of initiation	1921	1922	1902
Location: latitude/longitude	51°58'N, 20°10'E		51°24'N, 11°53'E
Altitude/slope	128 m / no slope		110 m / no slope
Annual precipitation	520 mm		484 mm
Temperature	7.9°C		8.7°C
Soil type	loamy sand		loam (chernozem)
USDA classification	Hapludalf		Mollisol
Carbon content	0.8%		2.1%
Clay content ($<2\mu m$)	7-10%		21%
Cropping scheme	Field A; arbitrary rotation (SA*) 70% cereals, 12% brassicacee, 15% potatoes, 3% maize Field E; 5-field crop rotation with legumes (SE*) w.-rye, potato, spring barley, legumes, w.-wheat	Vegetable arbitrary crop rotation (SV*); tomatoes, carrots, onions etc.	Arable crop rotation: sugarbeet, s.-barley, potato, w.-wheat (L*)
Tillage	Conventional tillage	Conventional tillage	Conventional tillage
Residue management	Tilled under	Tilled under	Tilled under
Irrigation	None	None	None
Selected treatments	Only lime - Nil plot (SA1) NPK (SA2) 30 t ha⁻¹ 5 yr⁻¹ FYM (SE3) 30t ha⁻¹ 5yr⁻¹ FYM+NPK (SE4) 20t ha⁻¹ FYM+NPK (SE5) min. N ($\bar{x}$ = 77.5 kg N ha⁻¹)	NPK (SV6) 20 t ha⁻¹yr⁻¹FYM (SV7) 40 t ha⁻¹yr⁻¹ FYM (SV8) 60 t ha⁻¹yr⁻¹ FYM (SV9)	Without fertilization- Nil plot (L1) NPK[1] (L2) 20 t ha⁻¹2yr⁻¹ FYM (L3) 20 t ha⁻¹ 2 yr⁻¹ FYM+NPK[2] (L4) 30 t ha⁻¹ 2 yr⁻¹ FYM (L5) 30 t ha⁻¹ 2 yr⁻¹ FYM+NPK[2] (L6) min. N ([1]$\bar{x}$=112.3 kg N ha⁻¹, [2]$\bar{x}$=82.5 kg N ha⁻¹
Literature	Merčik, 1994	Rumpel, 1993	Körschens and Müller, 1996

*Treatment abreviation, FYM = farmyard manure.

combined organic-mineral fertilization). On the other hand the local conditions differ considerably. The Bad Lauchstädt soil is a loess black earth which counts among the best soils of Germany. Indeed, due to its situation in the dry area of Middle Germany, the high yield potential of this soil is limited by water deficiency. The Skierniewice soil is a loamy sand (Hapludalf), which receives 520 mm annual precipitation in the dry continental conditions of central Poland.

B. Methods

In spring 1994 and 1995, soil samples of comparable treatments were taken and an extensive analysis program of chemical, physical, and biological soil characteristics was carried out to detect their interactions. In the following is presented the C and N content depending on location and fertilization and its relation to selected results of biological investigations.

The organic carbon content of the soil (SOC) was determined by dry combustion after Stroehlein (1957), the total nitrogen content (N_t) after Kjeldahl (1883).

Biomass was determined by substrate induced respiration (SIR) according to Anderson and Domsch (1978) in an automatic facility developed by Heinemeyer et al. (1991). The biological activity was described by means of different enzymes to draw conclusions on the activity of different groups of organisms. Among the enzyme activities, the determination of dimethylsulfoxide reduction (DMSO) after Alef and Kleiner (1989) is a measure for the characterization of the total activity of soil flora and fauna, since almost all microorganisms can reduce DMSO to DMS. The ß-glucosidases are involved in the decomposition of cellulose and were determined by the method of Hoffmann and Dedeken (1965). The phosphatase activity was analysed with the p-Nitrophenyl-phosphate test according to Tabatabai and Bremner (1969).

III. Results and Discussion

A. Carbon and Nitrogen Dynamics

The C and N contents of the variants of the presented long-term experiments have reached, as a consequence of constant management for many years, an equilibrium. That means that in both locations, the C content of the variants without fertilization does not further decline. According to Körschens (1980), the minimum C content possible under cultivation of humus consuming crops or fallow without fertilization can be considered as "inert" in the sense of "not involved in transformation processes". Under this assumption, the mineralizable C and N content of different fertilization treatments in long-term experiments can be calculated as the difference in the C content between Nil plot and the respectively fertilized plot. In Figures 1 and 2, the SOC content and the total N content of the different fertilization treatments of both locations is compared to the mineralizable C (C_m) and N (N_m), respectively, calculated as described before. While the SOC and the N_t content of the experiments clearly reflect the different soil conditions, the mineralizable C and N portions as dependent on fertilization are of comparable magnitude at both locations. The highest C_m and N_m are found at the highest fertilization level.

Under consideration of the applied nutrient amount, differences between the sites occur due to a different mineralization intensity. In Bad Lauchstädt the difference between the Nil plot and the plot with the highest fertilization (30 t ha^{-1} 2yr^{-1} FYM + NPK) amounts to 0.8% C_m and 0.078% N_m, respectively. In Skierniewice in the arable crop rotation the difference between 20 t ha^{-1} yr^{-1} + NPK and the Nil plot was 0.46% C_m and 0.04% N_m, respectively.

The highest C_m and N_m (1.0% and 0.01%) was calculated for the vegetable crop rotation at a FYM input of 60 t ha^{-1}yr^{-1}. This high nutrient input amounts to the fourfold quantity of the FYM treatments

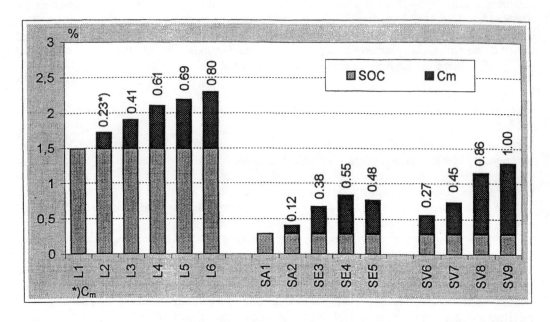

Figure 1. SOC and C_m in % (0-20 cm) of different fertilization treatments in three long-term experiments, 1995.

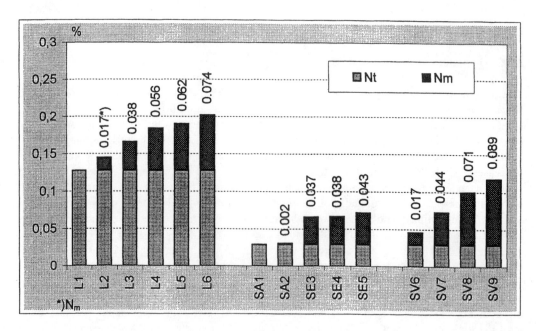

Figure 2. N_t and N_m in % (0-20 cm) of different fertilization treatments in three long-term experiments, 1995.

in Bad Lauchstädt and is not common in agricultural practice. Körschens et al. (1995), who investigated these relationships in several long-term experiments of Europe calculated at a nutrient input of 15 to 20 t ha^{-1} yr^{-1}, a C_m content between 0.2 and 0.4% for sandy soils and 0.6% for clay soils. The total SOC content of the individual locations was, therefore, not of importance and had an effect only over the forming of the physical soil characteristics.

B. Biological Properties

Biomass, DMSO reduction, alcaline phosphatase, and ß-glucosidase dependent on fertilization and location are presented in Figure 3.

The microbial biomass was positively influenced at both locations by mineral and by organic fertilization, with organic fertilization having a larger increase. Similar results were found by Friedel (1993). Despite comparable amounts of C_m, differences in biomass amount existed between the locations. In the Bad Lauchstädt loess soil, observed biomass values were higher on average than in the loamy sand in Skierniewice. Thus, a more narrow correlation with the SOC content ($r = 0.90$) than with the C_m content ($r = 0.55$) of the various fertilization treatments resulted from (Table 2).

Anderson and Gray (1991), who compared the microbial biomass of more than 130 plots with a SOC content between 0.5 and 3.0%, also found a linear correlation of these parameters. Manzke (1995), who correlated the results of three long-term experiments, did not find any significant relationships between biomass and SOC content, but very high correlation coefficients for the decomposable C. However, all three experiments involved in this evaluation were loamy substrates with a clay content of 17 to 21% and lowest SOC levels of 1.1 to 1.7%, so that, at least at the initial state, the physical soil properties were similar. In contrast, the loamy sand in Skierniewice with a clay content of 7-10% differs considerably in its physical soil properties from the loess black earth in Bad Lauchstädt. SOC and clay content are strongly correlated with dry matter density and water capacity (Table 2).

The favorable physical soil properties of the loess black earth result in a high yield capacity even on the Nil plot and allow, in connection with the clay content, the formation of a large, relatively stable biomass pool. The biomass in the Nil plot at Bad Lauchstädt is more than twice the size of that at Skierniewice. The relationships between SOC, clay content, and biomass, as examples, for these multidimensional effects are presented in Figure 4.

The phophatase activity is also strongly correlated with the SOC content ($r = 0.91$). The values are considerably lower on the loamy sand in Skierniewice than in the Static Experiment in Bad Lauchstädt. Comparable values were determined only for the treatments with the highest fertilization (vegetable crop rotation, 40 and 60 t ha^{-1} yr^{-1} FYM). A close relationship between the phophatase activity and the soil organic matter was also found by Beck (1984). Also, its ability to form complexes with humic substances was frequently described in the literature (Nannipieri et al., 1988; Speir and Ross, 1978). This can be confirmed as well by the close correlation of alcaline phosphate and the clay content ($r = 0.79$).

When comparing the different kinds of fertilization, a clearly positive effect of the organic fertilization can be observed, while the mineral fertilization obviously inhibited the phosphatase activity. The reason is an inhibition effect of phosphate, which has been described by Appiah et al. (1985) and others.

The DMSO reduction as a criterion for the total microbial activity was increased both by mineral and by organic fertilization (Figure 3). The ß-glucosidase activity was higher in the treatments with mineral fertilization. Both enzymes did not show any differences between the locations but reached similar values on both soils as function of the decomposable C content. The correlation coefficients (Table 2) of the results of the two sites show closer relationships of these enzyme activities to the C_m

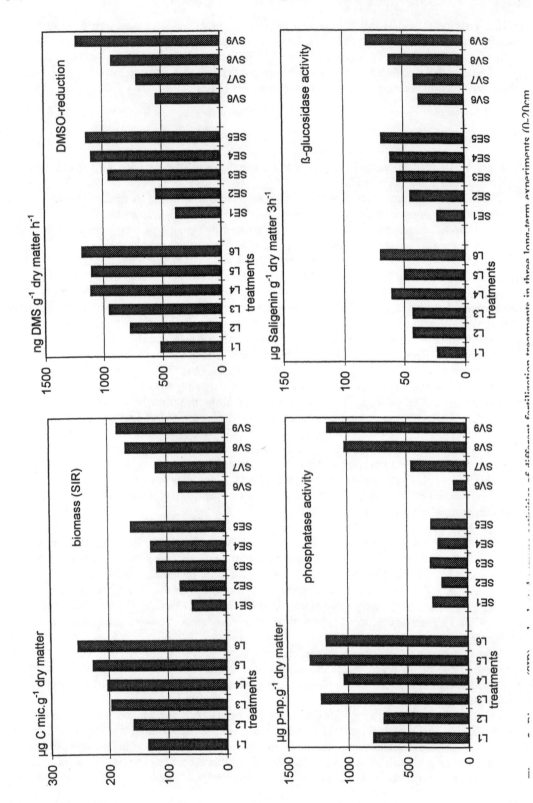

Table 2. Correlation of selected chemical, physical, and biological soil properties of two different sites (n=30)

Variable	SOC	C_m	Clay	DMD	W	SIR	P	DMSO	GLU
SOC	1.00	0.51*	0.91*	-0.95*	0.95*	0.90*	0.91*	0.33	0.51*
C_m		1.00	0.15	-0.33	0.26	0.55*	0.56*	0.75*	0.75*
Clay			1.00	-0.94*	0.96*	0.76*	0.79*	0.23	0.23
DMD				1.00	-0.97*	0.76*	0.79*	-0.24	-0.41
W					1.00	0.85*	0.83*	0.21	0.37
SIR						1.00	0.82*	0.46	0.63*
P							1.00	0.26	0.56*
DMSO								1.00	0.74*
GLU									1.00

DMD=dry matter density; W=water capacity; SIR-biomass (substrate induced respiration); P=phosphate activity; DMSO=dimethylsulfoxide reduction; GLU=β-glucosidase activity.

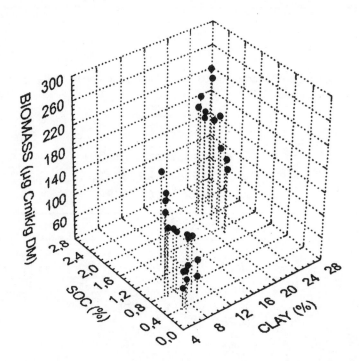

Figure 4. Relation between clay, SOC, and biomass (0-20 cm) in three long-term experiments (1994 and 1995, n=30).

than to the SOC. These were described by means of linear regression and are presented in Figures 5 and 6.

Klimanek (1980), who investigated mineralization capacity of different soils as a function of fertilization by soil respiration, found that the large differences in the C_t content dependent on the soils location had no significant effect on the respiration, while fertilization caused a considerable differ-

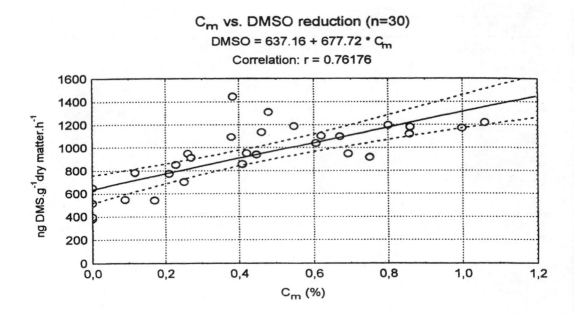

Figure 5. Relation between DMSO reduction (0-20 cm) and decomposable C-content (C_m) in three long-term experiments (1994 and 1995, n=30).

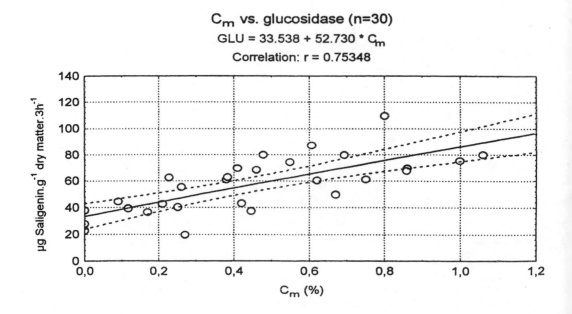

Figure 6. Relation between β-glucosidase activity (0-20 cm) and decomposable C-content (C_m) in three long-term experiments (1994 and 1995, n=30).

entiation. This confirms the hypothesis that the differences in locations predominantly concern the "inert" C content, while the decomposable C amount, as a function of the fertilization, is almost of the same magnitude.

The efficiency of substrate consumption which, according to Anderson and Domsch (1990), is the quotient of specific enzyme activity and the SOC content is, in Skierniewice, almost three times as in Bad Lauchstädt. The comparison of the metabolic quotients (q CO_2), as a measure for the physiological capacity of the biomass, shows differences in locations. The latter is much higher in Skierniewice showing rates between 5.5 and 18.8 compared to Bad Lauchstädt with 3.2 to 4.8. This emphasizes the increased metabolism activity of the biomass on the loamy sand and is proof for the higher mineralization intensity in sandy soils.

IV. Summary

As part of a EU project, long-term experiments on loess black earth in Germany and a loamy sand in Poland were investigated with regard to relationships between SOC and N_t content, their mineralizable portions, and biological soil properties. Dependent on fertilization, the decomposable C and N amounts were almost the same despite large differences in SOC and N_t contents at the two locations. At the same nutrient supply, the decomposable C and N portions are lower in the sandy soil which indicates a higher mineralization intensity. The highest increase in the C and N content (1.0% and 0.01%) was calculated for the vegetable crop rotation at a farmyard manure input of 60 t ha^{-1} yr^{-1} over the last 70 years. This high nutrient input is not common in agricultural practice and causes losses via leaching and gaseous emissions.

DMSO reduction and ß Glucosidase activity among the biological soil characteristics show a close correlation to the decomposable carbon, while biomass differed in different location and, thus, revealed a close connection to the total organic carbon content. The metabolic quotient points to a higher metabolism activity of biomass in the sandy location.

Acknowledgment

The authors would like to thank the European Commission for the funding of this project.

References

Alef, K. and D. Kleiner. 1989. Rapid and sensitive determination of microbial activity in soils and in soil aggregates by dimethylsulfoxid reduction. *Biol. Fertil. Soils* 8:349-355

Anderson, J.P.E. and K.H. Domsch. 1978. A physiological method for the quantitative measurements of microbial biomass in soils. *Soil Biol. Biochem.* 10:215-221

Anderson, T.H. and K.H. Domsch. 1990. Application of eco-physiological quotients (q CO_2 and q D) on microbial biomass from soils of different cropping histories. *Soil Biol. Biochem.* 22:251-255

Anderson, T.-H. and T.R.G. Gray. 1991. The influence of soil organic carbon on microbial growth and survival. p. 253-266. In: W.S. Wilson (ed.), *Advances in Soil Organic Matter Research. The Impact on Agriculture and the Environment.* The Royal Society of Chemistry, Cambridge

Appiah, M.R., B.J. Halm, and Y. Ahenkorah. 1985. Phosphatase activity of soil as effected by cocoa pod ash. *Soil Biol. Biochem.* 17:823-826

Beck, Th. 1984. Mikrobiologische und biochemische Charakterisierung landwirtschaftlich genutzter Böden. *Zeitschrift für Pflanzenernährung und Bodenkd.* 147:456-466

Dick, R.P. 1994. Soil Enzyme Activities as Indicators of Soil Quality. p. ix. In: J.W. Doran, D.C. Coleman, D.F. Bezdicek, and B.A. Stewart (eds.), Defining Soil Quality for Sustainable Environment. SSSA Special Publication Number 35, Soil Science Society of America Inc., American Society of Agronomy Inc., Madison, WI.

Friedel, J. 1993. Einfluß von Bewirtschaftungsmaßnahmen auf mikrobielle Eigenschaften im C- und N-Kreislauf von Ackerböden. Diss. agr. 201 pp. In: U. Babel, W.R. Fischer, K. Roth, K. Stahr (eds.), Hohenheimer Bodenkundliche Hefte. 11.

Heinemeyer, O., E.-A. Kaiser, and H. Insam. 1991. Kalibration eines neuen Meßsystems zur Erfassung der mikrobiellen Biomasse in Bodenproben durch substratinduzierte Respiration (SIR). VDLUFA-Schriftenreihe, Kongreßband: 701-706

Hoffmann, G. and Dedeken, M. 1965. Eine Methode zur colorimetrischen Bestimmung der β-Glucosidaseaktivität im Boden. *Zeitschrift für Pflanzenernährung, Düngung und Bodenkunde.* 108:193-198

Kjeldahl, J. 1883. Neue Methode zur Bestimmung des Stickstoffs in organischen Körpern. *Zeitschrift Anal.Chemie* 22:366-382

Klimanek, E.M. 1980. Mineralisierungsleistung unterschiedlicher Böden in Abhängigkeituon der Düngung. *Arch. Acker- u. Pflanzenbau und Bodenkd.*, Berlin 24:225-232

Körschens, M. 1980. Beziehungen zwischen Feinanteil, C_t- und N_t-Gehalt des Bodens. *Arch. Acker- u. Pflanzenbau u. Bodenkd.*, Berlin 24:585-592

Körschens M. and A. Müller. 1996. The Static Experiment Bad Lauchstädt. In: D.S. Powlson, P. Smith, and J.U. Smith (eds.), Evaluation of Soil Organic Matter Models. NATO ASI Series, Vol. 138, Springer-Verlag Berlin, Heidelberg, New York

Körschens, M., A. Müller, and E.-M. Ritzkowski. 1995. Der Kohlenstoffhaushalt des Bodens in Abhängigkeit von Standort und Nutzungsintensität. *Mitt. Deutsche Bodenkundl.Gesellsch.* 76:847-850

Manzke, F. 1995. Bodenmikrobiologische und bodenchemische Kenngrößen zur Beurteilung des Umsatzes organischer Bodensubstanz in unterschiedlichen Bodennutzungssystemen. Diss. agr. Univ. Göttingen. 175 pp.

Mercik, S. 1994. Long-term continuous experiments in Poland, Bulgaria, Czech Republic and Slovakia. p. 211-219. In: R.A. Leigh and A.E. Johnston (eds.), *Long-Term Experiments in Agricultural and Ecological Sciences.* CAB International Wallingford

Nannipieri, P., R.L. Johnson, and E.A. Paul. 1978. Criteria for measurement of microbial growth and activity in soil. *Soil Biol. Biochem.* 10: 223-229

Rumpel, J., O. Nowosielski, and M. Paul. 1993. The long-term fertilization experiment with vegetable crops in Skierniewice. Effect of organic and mineral fertilization on nutrients and heavy metals content and pH in soil. p. 183 - 200. In: Proceedings of the International Symposium, Long-Term Static Fertilizer Experiments, Warszawa-Krakow, Poland, Part II.

Smith, J. L. 1994. Cycling of Nitrogen through Microbial Activity. p. 91-120. In: J.L. Hatfield and B.A. Stewart (eds.), *Soil Biology: Effects on Soil Quality.* Advances in Soil Science, Lewis Publishers, CRC Press, Inc., Boca Raton, FL.

Speir, T.W. and D.J. Ross. 1978. Soil phosphatase and sulphatase. p. 197-250. In: Burns (ed.), *Soil Enzymes.* Academic press. London

Stroehlein. In: E. Lademann. 1957. Verfahren zur schnellen Bestimmung des Kohlenstoffs. *Z. landw. Versuchs- und Untersuchungswesen* 3:224-235

Tabatabai, M.A. and Bremner, J.M. 1969. Use of p-nitrophenyl phosphate for assay of soil phosphatase activity. *Soil Biol. Biochem.* 1:301-307

CHAPTER 24

Effect of Corn and Soybean Residues on Earthworm Cast Carbon Content and Natural Abundance Isotope Signature

Dennis R. Linden and C. Edward Clapp

I. Introduction

The general role of earthworms in cycling plant residues into soil organic matter and CO_2 is well known. Residues are ingested, mixed with soil and water, and egested at some point other than the point of ingestion. During gut transit, microbial activity proceeds with biochemical breakdown releasing CO_2 and H_2O, while providing energy for themselves and for the earthworm (Hartenstein et al.,1981). Most nutrients undergo some degree of change within the gut or in the egested cast (Aldag and Graff, 1975; Elliott et al., 1990; Elliott et al., 1991; Hendriksen, 1991; Mansell et al., 1981; Parkin and Berry, 1994; Svensson et al., 1986) that is caused by high levels of microbial activity (Scheu, 1987; Tiwari et al., 1993; Tiwari et al., 1989; Bhattacharya and Chakrabarti, 1986; Piearce, 1978). Soil organic matter is affected by this physical ingestion and translocation, and by the mixing that occurs within the gut (Martin, 1982). Freshly deposited fecal pellets of mixed soil and residues are structurally unstable and are easily dispersed by rainfall, resulting in nutrients being released into the water (Sharpley and Syers, 1976; Syers et al., 1979). Aged casts, however, become stable aggregates and may protect these organic residues from rapid breakdown (Dutt, 1948; Marinissen and Dexter, 1990; Shipitalo and Protz, 1988).

Release of CO_2 and nutrients from both the soil and the fresh residue of the ingested mix is primarily affected by microbial respiration within the earthworm gut and fecal deposits (casts). These processes are, however, only known in a qualitative sense since there is little experimental evidence to aid in predicting the quantitative impact on the carbon cycle. Precision in modern analytical techniques, including isotope analysis, is now providing us with the tools necessary to quantify these carbon cycling processes. Natural-abundance-carbon-isotope analysis shows the amount of recent residue as distinct from soil organic matter, because of the unique carbon isotope ratios of fresh residue material due to variations in carbon isotope abundance in photosynthetic products between so called C4 and C3 type plants. The unique isotope signatures of soil and residue allow for determination of the source of carbon in earthworm tissue and in casts. Residue-consuming earthworms, in combination with soils and residues of differing carbon signatures, will allow identification of the carbon encapsulated in casts through the action of earthworms. Interpretation of these results in terms of feeding, migration, and defecation habits of earthworm species will shed light upon their role in the carbon cycle.

II. Methods and Materials

A. Earthworm Collection

Earthworms were collected under various cropping conditions in order to identify the carbon isotope signature of their tissue. Conditions ranging from continuous long-term corn, a C4 type plant, to continuous oats, a C3 type plant, were sampled. The collection sites were near each other and had similar soils (typic hapluboralls). Individual earthworms were collected by hand sorting soil. They were then allowed to void their gut contents for several days and washed thoroughly prior to tissue preparation. Carcasses were freeze dried and finely ground in preparation for analysis. Total carbon and isotope abundance signatures were simultaneously determined using a CARLO ERBA[1] carbon-nitrogen analyzer interfaced with a VG-ISOGAS isotope ratio mass spectrometer (Huggins et al., 1995).

B. Cast Collection

Casts from a common earthworm specie, *Lumbricus terrestris*, were collected from controlled laboratory experiments. *Lumbricus terrestris* is a large surface-feeding and deep-dwelling earthworm, with abundant surface casting. Earthworms were incubated in three soils ranging in carbon content between 1.5 to 3.0 percent. Some characteristics of these soils are shown in Table 1. The earthworms were fed either corn or soybean residues for several weeks. The residues were collected from a field in a dry condition and were placed on the surface of the incubation cores filled with packed soils without any further preparation. Residues were added to the surface periodically so that earthworm consumption would not be limited by the total quantity but rather by their feeding choice. The experimental setup thus resulted in 12 independent treatments that included no-earthworm controls. The cultures were replicated, but the collected casts were composited for analysis. Casts were collected on a periodic basis and immediately frozen for later analysis. Soil samples from zero to 1 cm depth were collected At the termination of the experiment, bulk soil and the remaining surface residues were collected for analysis. These samples were dried, finely ground, and subjected to direct total carbon and isotope analysis (Huggins et al., 1995). Strict control and frequent comparison to standards were followed in the analytical procedures. Pooled standard errors for carbon content and isotope signature for earthworm tissue, soil, and residue were determined from replicated paired analysis with 16, 14, and 14 degrees of freedom, respectively. Statistical significance was evaluated by use of pooled standard error of analysis involving a relatively large number of samples rather than the limited number of replications of collected casts. The carbon signature was expressed as (Huggins et al., 1995):

$$\delta^{13}C(‰) = [(R_{SAM}/R_{STD})-1] \times 10^3$$

where R_{SAM} and R_{STD} are the $^{13}C:^{12}C$ ratio of sample and working standard, respectively.

---

[1] Mention of trade names is for reader's convenience only and does not imply endorsement by the USDA-ARS or the University of Minnesota over similar products of companies not mentioned.

Table 1. Characteristics of the soils and residues used in the earthworm cast carbon experiment

Soil	Percent C	^{13}C signature[a]	Ph	Percent clay	Percent sand
Hubbard sand	1.44	-17.84	5.0	8	81
Webster clay loam	2.31	-18.35	6.5	26	45
Waukegan silt loam	2.77	-19.75	5.6	13	27
Corn residues	40.8	-13.2			
Soybean residues	43.6	-27.28			

[a] ^{13}C signature is defined by $\delta^{13}C(‰) = [(R_{SAM}/R_{STD})-1] \times 10^3$, where R_{SAM} and R_{STD} are the $^{13}C:^{12}C$ ratio of sample and working standard, respectively.

III. Results and Discussion

A. Earthworm Tissue

Analysis of earthworm tissue indicates that, as would be expected from ingestion and rapid assimilation, that carbon is largely derived from recently digested residue sources (Table 2). The carbon signature of earthworms found in corn fields were distinctly less negative than the earthworms found under continuous oats. This difference is consistent with the carbon isotope signature difference between corn, a C4 plant, and oats, a C3 plant. The isotope signature of earthworms under corn were somewhat more negative than fresh corn tissue, which may indicate a mixture in the diet involving soil, previous crop residues, or weed species.

B. Earthworm Casts

The carbon content of *L. terrestris* casts was greater than carbon content of the bulk soil by 0.5 to 2.5% C (Figure 1). The ranking for carbon content of casts produced with the same food source remained the same as the three soils. In other words, soils with higher carbon contents also had higher carbon content casts. Casts produced when earthworms were fed corn residues had higher carbon contents than when they were fed soybean residues which reflects less ingestion of soybeans than of corn in relation to equal quantities in soil. Considerable quantities of carbon are thus incorporated into earthworm casts and, in some cases, may double the carbon content of the cast compared to that of the surrounding soil.

For the same soil, the carbon isotope signature of casts from earthworms fed corn residues increased (became less negative), while those from soybean residue decreased (became more negative) (Figure 2). Thus the change in the carbon signature of the cast is largely reflected by the residues ingested and mixed with the soil. The proportion of the cast carbon coming from residue and soil may be computed according to the procedure of Huggins et al. (1995). Using this procedure, 65 to 79 percent of cast carbon was derived from soil and 21 to 35 percent came from residue (Figure 3). Corn and soybean residues had similar carbon contents (Table 1) and yet soybean casts had proportionately less carbon from residues than the corn casts (Figure 3). This difference probably reflects less ingestion of soybean residues compared to corn. The protein (nitrogen) content of soybean residues is higher than of corn and thus may be a primary determinant of the earthworms need to ingest.

Table 2. Carbon content and ^{13}C signature of representative earthworm tissue

Management	Species	Percent C[a]	^{13}C signature[b]
3 yr Corn	A. Tuberculata	41.82	-11.48
Continuous Corn	A. Tuberculata	38.71	-10.93
	O. Tyrtaeum	36.77	-11.18
Continuous Oats	A. Tuberculata	35.86	-23.99
	A. Caliginosa	40.01	-22.98
Corn/grass/sludge	A. Tuberculata	41.78	-16.67
	L. Rubbelus	37.03	-15.87

[a] The pooled standard error for carbon content and ^{13}C signature of earthworm tissue is 1.92 and 1.50 (with 16 degrees of freedom), respectively.

[b] ^{13}C signature is defined by $\delta^{13}C(‰) = [(R_{SAM}/R_{STD})-1] \times 10^3$, where R_{SAM} and R_{STD} are the $^{13}C:^{12}C$ ratio of sample and working standard, respectively.

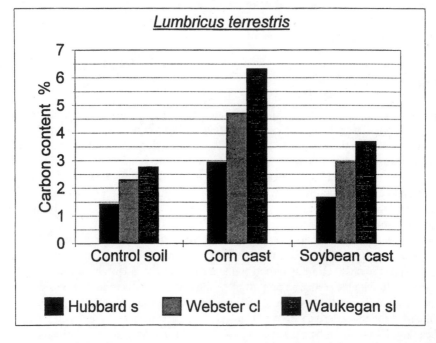

Figure 1. Carbon content of control soils and casts produced by *Lumbricus terrestris* when fed either corn or soybean residue. The pooled standard errors for soil and residue were 0.901 and 0.348 (with 14 degrees of freedom), respectively.

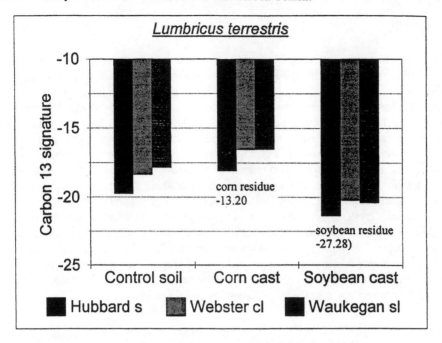

Figure 2. Carbon-13 signature of control soils and casts produced by *Lumbricus terrestris* when fed either corn or soybean residue. The pooled standard errors for soil and residue were 0.131 and 0.0224 (with 14 degrees of freedom), respectively.

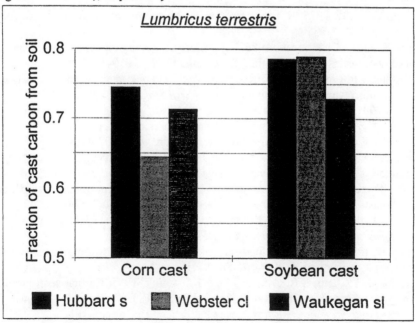

Figure 3. Fraction of carbon derived from soil in casts produced by *Lumbricus terrestris* when fed either corn or soybean residue.

IV. Conclusion

Considerable carbon in sequestered by the ingestion and deposition process of earthworms. The carbon contents of casts were doubled by this inclusion of carbon from fresh corn residues. This immediate encapsulation of fresh crop residues may result in either long term increased or decreased sequestered soil carbon. Increased sequestered carbon may result because of the abundance of casting in soils (Darwin,1881; Cook and Linden, 1996; Sharpley and Syers, 1977), and carbon material isolation and protection within these pellets in a manner similar to soil aggregates shown by Huggins et al., (1995). Decreased sequestered carbon would result if the high microbial respiration rates in the gut and cast, with increased CO_2 evolution rates, exceeded the respiration from soil-residue mixtures without earthworm gut passage. The high respiration rate within the gut and egested casts is usually temporary in nature (Trigo and Lavelle, 1993), but has been suggested to result in net carbon loss from earthworm affected systems compared to non-earthworm controls (Alban and Berry, 1994). The magnitude of carbon encapsulation, the isolation from environmental changes within casts, and the deposition of casts at soil depth by some earthworm species (Cook and Linden, 1995) suggest that the net effect on soil carbon is one of gain rather than loss. Additional research will be required to clarify the balance between these opposing directions for the long term carbon accumulation in soil since this research dealt only with the initial encapsulation of crop residues.

References

Alban, D. H. and E. C. Berry. 1994. Effects of earthworm invasion on morphology, carbon, and nitrogen of a forest soil. *Applied Soil Ecology* 1:243-249.

Aldag, R. and O. Graff. 1975. Nitrogen fractions in earthworm casts and in the surrounding soil. *Pedobiologia* 15:151-53.

Bhattacharya, T. and G. Chakrabarti. 1986. Mycroflora of earthworm cast, adjacent soil, and the gut content of two species of earthworm. *Environ. and Ecol.* 4:619-21.

Cook, S.M.F. and D.R. Linden. 1996. Effect of food type and placement on earthworm (*Apporectodea tuberculata*) burrowing and soil turnover. *Biol Fertil Soils* 21:201-206.

Darwin, C.A. 1881. *The Formation of Vegetable Mould Through the Action of Worms with Observations on Their Habits.* Murray Publishers, London. 326 pp.

Dutt, A. K. 1948. Earthworms and soil aggregation. *Agron. J.* 40:407-410.

Elliott, P.W., D. Knight, and J.M. Anderson. 1990. Denitrification in earthworm casts and soil from pastures under different fertilizer and drainage regimes. *Soil Biol. Biochem.* 22:601-605.

Elliott, P.W., D. Knight, and J.M. Anderson. 1991. Variables controlling denitrification from earthworm casts and soil in permanent pastures. *Biol. Fertil. Soils* 11:24-29.

Hartenstein, F, E. Hartenstein, and R. Hartenstein. 1981. Gut load and gut transit time in the earthworm *Eisenia foetida*. *Pedobiologia* 22:5-20.

Hendriksen, N.B. 1991. Gut load and food-retention time in the earthworms *Lumbricus festivus* and *L. castaneus*: A field study. *Biol. Fertil. Soils* 11:170-173.

Huggins, D.R., C.E. Clapp, R.R. Allmaras, and J.A. Lamb. 1995. Carbon sequestration in corn-soybean agroecosystems. p. 61-68. In: R. Lal, J. Kimble, E. Levine, and B.A. Stewart (eds.), *Soil Management and Greenhouse Effect.* Lewis Publishers, Boca Raton, FL.

Lal, R. and O.O. Akinremi. 1983. Physical properties of earthworm casts and surface soil as influenced by management. *Soil Sci.* 135:114-122.

Mansell, G.P., J.K. Syers, and P.E.H. Gregg. 1981. Plant availability of phosphorus in dead herbage ingested by surface-casting earthworms. *Soil Biol. Biochem.* 13:163-167.

Marinissen, J.C.Y. and A.R. Dexter. 1990. Mechanisms of stabilization of earthworm casts and artificial casts. *Biol. Fertil. Soils* 9:163-167.

Martin, N.A. 1982. The interaction between organic matter in soil and the burrowing activity of three species of earthworms (*Oligochaeta: Lumbricidae*). *Pedobiologia* 24:185-190.

Parkin, T.B. and E.C. Berry. 1994. Nitrogen transformations associated with earthworm casts. *Soil Biol. Biochem.* 26:1233-1238.

Piearce, T.G. 1978. Gut contents of some lumbricid earthworms. *Pedobiologia* 18:153-157.

Scheu, S. 1987. Microbial activity and nutrient dynamics in earthworm casts (*Lumbricidae*). *Biol. Fertil. Soil* 5:230-234.

Sharpley, A.N. and J.K. Syers. 1976. Potential role of earthworm casts for the phosphorus enrichment of run-off waters. *Soil Biol. Biochem.* 8:341-346.

Sharpley, A.N. and J.K. Syers . 1977. Seasonal variation in casting activity and in the amounts and release to solution of phosphorus forms in earthworm casts [Effects in a soil-plant system]. *Soil Biol. Biochem.* 9 :227-231.

Shipitalo, M.J. and R. Protz. 1988. Factors influencing the dispersibility of clay in worm casts. *Soil Sci. Soc. Am. J.* 52:764-769.

Svensson, B.H., U. Bostrom, and L. Klemedston. 1986. Potential for higher rates of denitrification in earthworm casts than in the surrounding soil. *Biol. Fertil. Soil* 2:147-149.

Syers, J.K., A.N. Sharpley, and D.R. Keeney. 1979. Cycling of nitrogen by surface-casting earthworms in a pasture ecosystem. *Soil Biol. Biochem.* 11:181-1

Tiwari, S.C. and R.R. Mishra. 1993. Fungal abundance and diversity in earthworm casts and in uningested soil. *Biol. Fertil. Soils* 16:131-134.

Tiwari, S.C., B.K. Tiwari, and R.R. Mishra. 1989. Microbial populations, enzyme activities and nitrogen-phosphorus-potassium enrichment in earthworm casts and in the surrounding soil of a pineapple plantation. *Biol. Fertil. Soils* 8:178-182.

Trigo, D. and P. Lavelle. 1993. Changes in respiratiom rate and some physicochemical properties of soil during gut transit through Allolobophora molleri (*Lumbricidae, Oligochaeta*). *Biol. Fertil. Soils* 15:185-188.

CHAPTER 25

Soil Organic Carbon Distribution in Aggregates and Primary Particle Fractions as Influenced by Erosion Phases and Landscape Position

R.M. Bajracharya, R. Lal, and J.M. Kimble

1. Introduction

The detrimental consequences of accelerated soil erosion from agricultural land leading to degradation of soil resources, reduced productivity, and adverse environmental impact have received much attention since the 1970s (Brown, 1984; Lal, 1995; Pimentel et al., 1995). Numerous studies have focused on the effects of different management systems on the SOC status of soils with the ultimate objective of improving soil quality and agricultural sustainability (Angers, 1992; Lal et al., 1994; Rasmussen and Parton, 1994; Woomer et al., 1994). In recent years the question has arisen whether deposition of soil organic carbon (SOC), nutrients, and topsoil in depressional areas from adjacent upslope positions may compensate for reduced yields, nutrients lost in the eroded areas, and carbon sequestered in the subsoil (Ebeid et al., 1995; Fahnestock et al., 1995a, b; Lal, 1995). Clearly, landscape position influences the nature and extent of erosion or deposition processes occurring at any given location in a field. The degree of past erosion affects the distribution of SOC in the soil profile and among various aggregate and primary particle size-fractions. Soil SOC, to a large extent, moderates and regulates many processes associated with soil degradation and productivity through its influence on soil structural stability, water holding capacity, nutrient bioavailability, buffering capacity, and soil biodiversity (Tisdall and Oades, 1982; Oades, 1984; Lal et al., 1990; Woomer et al., 1994).

The stable SOC pool in soils, important for carbon sequestration, is normally associated with the colloidal soil fraction and microaggregates and is depleted by soil degradative processes. A commonly observed characteristic of eroded areas is a substantially reduced Ap horizon thickness (Frye et al., 1982; Mermut et al., 1983; William et al., 1989), which in turn affects water-holding capacity, nutrient status, SOC content of the soil and productivity (Frye et al., 1982; Schertz et al., 1989; Chengere and Lal, 1995). Significantly decreased SOC contents in the surface soil layer as a result of erosion (Nizeyimana and Olson, 1988; Mokma and Sietz, 1992), particularly under cultivation, as opposed to grassland (Mermut et al., 1983; William et al., 1989), is well documented in the literature. Decline in SOC from slightly- to severely-eroded phases of 3.03 to 1.86%, 1.89 to 1.51%, and 1.91 to 1.60% for three different soils in Indiana was reported by Schertz et al. (1989). Also, Mermut et al. (1983) observed as much as 41% reduction in of SOC due to erosion on cultivated land compared to grassland in Canada.

ISBN 0-8493-7441-3
©1997 by CRC Press LLC

Land use and vegetation type exert considerable control on SOC quantities as well as the aggregate or particle size-fractions with which they are associated. Management practices offering the least amount of soil disturbance have a positive effect on SOC accumulation, while increasing intensity of cultivation results in lower SOC and a greater proportion associated with the fine size-fractions. Woomer et al. (1994) observed that improved pasture led to greater total SOC and higher proportions of SOC in both the >2.0 mm fraction and <0.1 mm fraction of aggregates at 0.2-0.3 m depth than even natural vegetation. These results suggest enhanced structural stability and ability to sequester carbon under improved tropical pastures. Angers (1992) observed higher SOC contents for alfalfa (*Medicago sativa*) with a larger proportion of it in the macro-aggregates, particularly the 2-6 mm size-range, as compared to corn (*Zea mays*) or fallow treatments. Nelson et al. (1994) reported that 70% of the SOC was associated with the clay fraction at 0.8-1.0 m depth, while at the surface (0-0.2 m) only about 49% was in the clay fraction. The decrease in available SOC with depth was attributed to a decrease in its accessibility due to prevalence in a stable or protected form.

Apart from agricultural and environmental importance of SOC, its impact on the atmospheric carbon pool, and, therefore, global climate change, is a comparatively new yet momentous area of research. World soils may serve as a significant sink for carbon by sequestering SOC in the stable, persistent forms found in micro-aggregates and associated with fine mineral fractions. Conversely, world soils may also be a net source of atmospheric carbon through the release of radiatively-active gases, such as CO_2 and CH_4, due to change in land use and other agricultural activities (Bouwman, 1990; Lal et al., 1994; Lal et al., 1995). Thus, the form and location of SOC in the soil profile and in various particle size-fractions determine its availability to be released to the atmosphere in form of radiatively active gases. An understanding of the processes and mechanisms involving carbon sequestration is vital to identifying management decisions and policy initiatives which capitalize on soil's potential as a sink for atmospheric C. Therefore, the objectives of this study were to identify principal mechanisms of C sequestration and to assess the relative importance of severity of erosion on partitioning of SOC contained in macroaggregates among smaller aggregate and primary particle size-fractions.

II. Material and Methods

A. Site Location and Description

The soils used in this study were located on farmer's fields in Clark County, Ohio, near Springfield. Strawn silty clay loam (fine-loamy, mixed, mesic Typic Hapludalf) occurred at Sites A and B, while Site C consisted of Miamian silty clay loam (fine, mixed, mesic Typic Hapludalf). Five erosion phases: slight, moderate, severe, deposition and noneroded (forest) were identified at each of the sites. The first four phases generally corresponded to summit, shoulder, backslope and footslope/toeslope landscape positions, respectively (Ruhe, 1975; Daniels et al., 1985). A summary of soil and site characteristics, as well as management, is provided in Table 1. Site A was under a corn (*Zea mays* L.)– soybean (*Glycine max* L.)–wheat (*Triticum aestivum* L.) rotation with conventional tillage for corn and no-tillage performed for soybean and wheat. Site B received disking prior to corn planting with subsequent direct drilling of soybean in corn stubble without tillage. Site C was maintained in a no-till corn–wheat, or clover (*Trifolium reppens* L.)–soybean rotation.

A. Soil Sampling

Soil samples were obtained at each of the sites between April and August, 1995, from three or four genetic horizons to a depth of 0.5-0.8 m (depending upon stoniness of subsurface horizons and depth

Table 1. Soil and site information for the Clark County experimental locations

Parameter	Site A	Site B	Site C
Soil series	Strawn silty clay loam	Strawn silty clay loam	Miamian silty clay loam
Classification	Fine-loamy, mixed, mesic Typic Hapludalf	Fine-loamy, mixed, mesic Typic Hapludalf	Fine, mixed mesic Typic Hapludalf
Latitude	39° 57'	39° 58'	39° 51'
Longitude	83° 35'	83° 32'	83° 50'
Slope (%)	1.0-4.0	0.5-4.0	1.0-4.5
Soil properties[a]			
Clay (%)	38-43	37-43	31-40
Silt (%)	30-21	28-18	28-18
Organic C (%)	0.9-3.4	1.1-3.0	0.8-4.0
Bulk density (Mg m^{-3})	1.2-1.5	0.9-1.4	1.0-1.4
Available water capacity (mm/m soil)	170-320	190-320	180-340
pH	6.2-7.6	6.0-6.4	6.3-7.1
Management	Plowed conventional corn, no-till soybean, no-till wheat	Disked corn, no-till soybean	No-till corn, wheat (or clover, soybean)

[a]Soil properties for surface 0-0.15 m.
(From Fahnestock, 1994.)

to C horizon) using a hand bucket auger. The samples were air-dried and clods broken up prior to sieving through an 8 mm sieve. Macroaggregates of size range between 5 and 8 mm diameter were obtained for aggregate fractionation and SOC determination.

B. Fractionation and Organic C Determination

Aggregate fractionation was done using the wet-sieving method of Yoder (1936) with a nest of sieves of 5, 2, 1, 0.5, and 0.1 mm diameter openings. Primary particle fractionation was done on a second sample by washing through the same nest of sieves after dispersion with Na-hexametaphosphate (Na-HMP) and high-speed stirring (as described in Gee and Bauder, 1986). The wet sieving simulated physical disruption which aggregates would be subject to under water erosion, while primary particle separation represented a more rigorous chemical and physical disruption. Soil aggregates and primary particles in the size ranges of : 1-2, 0.5-1, 0.1-0.5, and <0.1 mm were analyzed for SOC using dry combustion (Nelson and Sommers, 1982) at 550°C, at which temperature C from secondary carbonates would be excluded (Dr. W.A. Dick; Dr. T.J. Logan; personal communications). Aggregates larger than 2 mm were excluded as the samples had a large proportion of gravel and organic litter/debris and very little soil. It should be noted that a complete mass balance of soil and organic fractions was not done for the samples. Organic debris/litter which floated in the water during sieving was excluded from SOC analyses. Some suspended colloidal material (mineral and organic) was lost during the separation procedure.

C. Data Analyses

A factorial design in treatment and size-fraction for each horizon was used to statistically analyze the SOC data. There were three replications of each erosion phase, three horizons and four size-fractions. Fisher's least significant differences were calculated for pair-wise comparison of means between erosion phases and size-fractions within horizons (Steel and Torrie, 1980).

III. Results and Discussion

A. Erosion Phase Effects on Whole Soil Organic Carbon

Significant effects of severity of erosion on SOC content of whole soil were most clearly expressed in the surface horizons (Ap or A) as shown in Figures 1 through 3. No statistical differences, as determined by mean comparison using Fisher's $LSD_{0.05}$, existed among slight, moderate, and severe erosion phases at any site (Tables 2-5). The lack of differences were attributed to high degree of variability and low antecedent SOC status of these extensively cultivated soils. Substantial loss of topsoil and lowest SOC contents occurred at site C, which had the most past erosion and exposed subsoil at the surface. By contrast, deposition and noneroded phases had significantly higher SOC than other erosional phases (Tables 2-4), which is in agreement with the findings of other researchers (Burke et al., 1995; Ebeid et al., 1995). Noneroded forest sites had about 30 to 100% more SOC in the surface horizon than deposition and more than twice the amount of SOC in the other phases. Such marked differences between forest and cultivated soils indicated the loss of large quantities of SOC due to tillage and cropping. Relatively high SOC of deposition than other eroded phases reflects significant amounts of SOC lost from upslope areas due to erosion and transport to low lying areas downslope.

B. Soil Depth Effects on Soil Organic Carbon

Changes in SOC from the surface to subsoil horizons followed the expected decreasing trend. The decline in SOC with depth was, however, most pronounced (64 to 70%) for the noneroded phase, which reflected the accumulation and concentration of organic materials near the surface (A horizon) of forest soils. Less rapid decline in SOC with depth under cultivation can be attributed to considerable mixing of the upper 0.2 to 0.3 m of soil and incorporation of crop residue (Blevins and Frye, 1993; Lal et al., 1994; Bajracharya et al., 1996).

Depth in the soil profile had no apparent influence on SOC distribution among aggregate or primary particle fractions. The trend of decreasing SOC with depth generally held across all size-fractions, and, once again, differences were typically less for eroded phases compared to deposition or noneroded phases.

C. Erosion Effects on SOC of Aggregate and Primary Particle Fractions

The effects of severity of erosion on SOC contents of aggregate fractions were not clearly expressed among erosional phases (Tables 2-4), despite statistically significant differences among erosion phases at all three sites (Tables 5-7). The SOC contents of aggregates for deposition and noneroded phases, however, were usually higher than those of slight, moderate and severe erosion phases, which were not different from each other. Similar observations of highest total C and microbial biomass in downslope landscape positions (depositonal areas) were reported by Burke et al. (1995). This trend

Soil Organic Carbon Distribution in Aggregates and Primary Particle Fractions

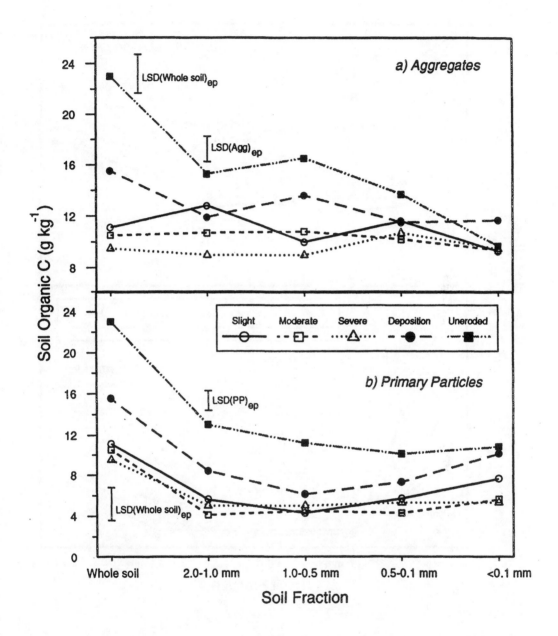

Figure 1. Comparison of soil organic contents on whole soil versus various a) aggregate and b) primary particle size-fractions of surface horizons for five erosion phases at site A. LSD = Fisher's least significant difference; Agg = aggregates; PP = primary particles; ep = erosion phase effects.

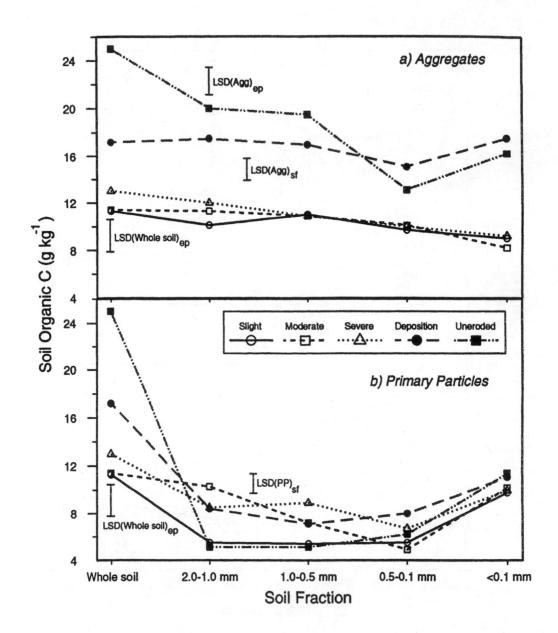

Figure 2. Comparison of soil organic contents on whole soil versus various a) aggregate and b) primary particle size-fractions of surface horizons for five erosion phases at site B. LSD = Fisher's least significant difference; Agg = aggregates; PP = primary particles; ep = erosion phase effects.

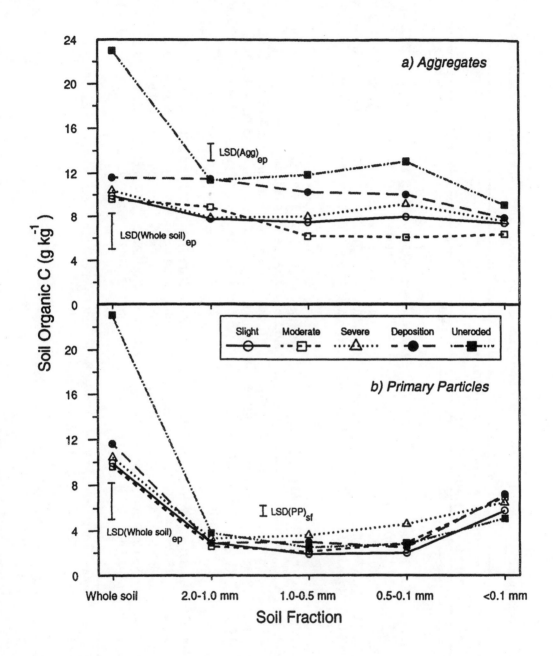

Figure 3. Comparison of soil organic contents on whole soil versus various a) aggregate and b) primary particle size-fractions of surface horizons for five erosion phases at site C. LSD = Fisher's least significant difference; Agg = aggregates; PP = primary particles; ep = erosion phase effects.

Table 2. Soil organic carbon contents of whole soil, aggregates and primary particles from three horizons of five erosion phases at Clark County Site A

Erosion phase	Whole soil SOC	Aggregates size-fractions (mm)				Primary particle fractions (mm)			
		2.0-1.0	1.0-0.5	0.5-0.1	<0.1	2.0-1.0	1.0-0.5	0.5-0.1	<0.1
					(g kg^{-1})				
Ap/A									
Slight	11.1(0.6)a	12.8(1.1)a	10.0(0.4)a	11.6(0.9)a	9.3(1.0)a	5.6(0.7)a	4.3(1.4)a	5.7(0.4)ab	7.6(0.4)a
Moderate	10.5(0.3)a	10.7(1.3)b	10.8(0.9)a	10.2(0.3)a	9.4(0.4)a	4.1(0.6)a	4.5(1.7)a	4.3(0.6)a	5.6(0.7)b
Severe	9.5(1.0)a	9.0(0.8)c	9.0(0.0)a	10.7(0.5)a	9.5(0.4)a	5.0(0.8)a	5.0(0.0)a	5.3(1.6)a	5.3(1.2)b
Deposition	15.5(2.2)b	11.9(3.0)ab	13.6(0.5)b	11.5(4.9)a	11.7(2.3)b	8.4(3.7)b	6.1(1.3)a	7.3(0.9)b	10.1(0.7)c
Uneroded	23.0(0.0)c	15.3(5.0)d	16.5(3.6)c	13.7(0.5)b	9.7(1.6)a	13.0(4.5)c	11.2(2.8)b	10.1(2.0)c	10.8(5.0)c
Bt1/Btk1									
Slight	9.3(1.4)a	9.7(1.2)a	9.8(2.4)ac	9.5(2.4)ab	11.1(2.1)a	4.9(1.5)aAB	3.9(0.6)aA	4.0(0.9)aA	6.0(0.3)aB
Moderate	8.9(0.7)a	8.2(2.3)ab	10.6(0.9)ac	8.0(1.4)ac	8.5(0.5)b	5.0(0.3)aA	4.0(0.5)aA	4.4(0.9)abA	7.2(2.0)abB
Severe	6.9(0.8)a	7.6(2.4)b	7.8(1.1)b	8.1(1.8)ac	10.0(1.5)ab	4.4(1.3)aA	4.2(1.6)aA	5.0(1.2)abA	6.9(1.5)abB
Deposition	13.7(2.0)b	13.0(2.4)c	11.1(3.5)c	10.3(3.5)b	9.3(1.6)ab	7.5(3.0)bAC	6.6(2.5)bA	5.9(3.4)bB	8.4(2.5)bC
Uneroded	8.3(0.0)a	7.9(1.6)b	8.6(1.2)ab	6.9(1.2)c	5.1(0.5)c	5.7(1.2)aA	4.2(1.0)aB	3.7(0.9)aB	5.7(1.3)aA
Bt2/Btk2									
Slight	7.4(0.4)a	10.0(2.5)a	7.2(1.2)a	9.3(2.5)a	8.2(2.0)a	3.4(0.9)aA	4.0(0.7)acA	4.1(0.6)abA	5.5(1.5)aB
Moderate	6.7(0.3)a	7.1(1.2)b	6.3(1.7)a	6.6(0.6)bc	4.9(1.2)b	3.2(0.6)aA	3.6(0.7)acA	3.4(0.3)aA	5.1(1.0)aB
Severe	6.5(1.3)a	5.4(0.3)b	6.5(2.2)a	5.6(0.8)c	5.8(1.9)b	3.9(1.8)abA	3.1(2.0)aA	5.3(1.9)bB	2.8(0.3)bC
Deposition	12.3(3.1)b	12.0(5.6)a	10.2(2.6)b	10.0(3.4)a	10.1(2.7)a	5.0(1.5)bA	7.1(1.9)bB	3.8(1.2)aC	8.6(3.0)cD
Uneroded	11.5(0.0)b	7.3(0.5)a	8.3(0.5)a	8.0(0.8)ab	5.3(0.5)b	5.3(0.5)b	4.7(0.9)cA	4.0(0.8)aA	6.3(0.5)aB

Means followed by standard deviations in parentheses. Values followed by the same lower case in columns for erosion phase effects and upper case letter in rows for size-fraction effects, respectively, are not significantly different.

Table 3. Soil organic carbon contents of whole soil, aggregates and primary particles from three horizons of five erosion phases at Clark County Site B

Erosion phase	Whole soil SOC	Aggregates size-fractions (mm)				Primary particle fractions (mm)			
		2.0-1.0	1.0-0.5	0.5-0.1	<0.1	2.0-1.0	1.0-0.5	0.5-0.1	<0.1
		(g kg^{-1})							
Ap/A									
Slight	11.3(0.6)a	10.1(1.6)aAB	11.0(1.6)aA	9.7(3.8)aAB	9.0(a.a)aB	5.5(2.1)A	5.4(2.3)a	5.5(1.8)A	9.8(3.8)B
Moderate	11.4(0.9)a	11.3(2.2)aA	10.9(1.2)aA	10.1(0.9)aAB	8.2(1.3)aB	10.3(6.0)A	7.2(3.9)B	4.9(1.6)C	10.2(4.2)a
Severe	13.0(0.5)a	12.0(1.7)aA	10.9(0.6)aAB	10.0(1.4)aB	9.2(2.1)aB	8.5(1.3)A	8.9(2.5)A	6.7(2.3)C	10.0(1.0)A
Deposition	17.2(0.6)b	17.5(4.4)bA	17.0(3.1)bA	15.1(3.0)bB	17.5(7.0)bA	8.4(2.1)A	7.1(3.9)A	8.0(5.0)A	11.1(2.5)B
Uneroded	25.0c	20.0(3.6)cA	19.5(3.1)cA	13.1(2.6)bB	16.2(1.4)bC	5.1(1.5)A	5.1(2.6)A	6.2(3.0)A	11.4(5.8)B
Bt1/Btk1									
Slight	9.2(1.3)	6.8(1.1)ab	7.4(2.4)a	7.8(1.2)a	9.0(1.1)a	4.6(1.4)aA	3.6(0.8)aA	3.7(1.8)aA	7.3(3.6)aB
Moderate	9.4(1.1)	8.5(3.5)ad	8.2(2.0)a	8.1(1.5)a	7.7(1.3)ab	8.6(3.0)dA	6.0(0.9)bB	2.9(0.8)aC	6.9(1.2)aB
Severe	8.0(1.1)	6.6(1.9)b	8.6(2.2)a	6.9(2.0)a	6.7(2.0)b	6.9(0.8)bcA	5.9(0.7)bAB	5.3(o.5)bB	8.8(0.6)bC
Deposition	13.3(3.8)	15.5(3.2)c	15.8(3.0)b	14.9(3.6)b	12.9(1.9)c	6.6(0.5)bA	6.0(1.6)bA	3.7(0.5)aB	8.9(0.7)bC
Uneroded	9.0	10.2(1.4)d	7.8(0.5)a	8.7(0.3)a	4.7(0.9)d	7.8(0.9)cdA	5.3(0.2)bB	3.7(0.5)aC	5.3(0.5)cA
Bt2/Btk2									
Slight	6.8(1.3)	6.1(0.7)a	6.3(0.7)a	6.2(1.4)a	5.9(0.9)a	4.4(1.8)AB	5.0(1.4)A	3.5(0.4)B	6.5(3.0)C
Moderate	6.7(1.2)	5.6(0.1)a	5.9(1.6)a	5.0(1.0)a	5.2(1.4)a	5.0(1.6)A	5.1(1.0)A	5.3(1.1)A	4.9(1.1)A
Severe	6.3(0.6)	6.6(1.4)a	6.5(1.2)a	5.3(0.5)a	6.3(1.9)a	6.2(1.2)A	6.6(1.6)A	5.1(1.4)B	6.5(2.0)A
Deposition	11.6(3.2)	10.4(4.8)b	11.4(4.6)b	11.8(3.7)b	9.8(4.9)b	4.9(0.7)A	4.1(1.1)AB	3.4(1.0)B	7.6(0.2)C
Uneroded	6.0	7.4(1.6)c	7.3(1.2)a	6.3(1.8)a	6.7(1.2)a	5.2(0.2)A	4.9(0.8)B	3.0(0.8)B	5.6(0.4)A

Means followed by standard deviations in parentheses. Values followed by the same lower case in columns for erosion phase effects and upper case letter in rows for size-fraction effects, respectively, are not significantly different.

Table 4. Soil organic carbon contents of whole soil, aggregates and primary particles from three horizons of five erosion phases at Clark County Site C

Erosion phase	Whole soil SOC	Aggregates size-fractions (mm)				Primary particle fractions (mm)			
		2.0-1.0	1.0-0.5	0.5-0.1	<0.1	2.0-1.0	1.0-0.5	0.5-0.1	<0.1
					(g kg^{-1})				
Ap/A									
Slight	9.9(0.3)a	7.8(1.4)a	7.5(1.4)ab	8.0(3.6)a	7.4(0.2)a	2.9(0.8)A	1.9(0.3)B	2.0(0.5)B	5.8(0.3)C
Moderate	9.6(0.6)a	8.9(1.2)a	6.2(0.2)a	6.1(1.0)b	6.4(1.5)a	2.6(0.5)A	2.1(0.8)a	2.9(1.0)A	7.0(1.5)B
Severe	10.4(1.7)a	7.9(1.0)a	8.0(2.0)b	9.2(3.0)ac	7.6(1.7)ab	3.3(0.8)A	3.6(2.4)A	4.6(3.4)B	6.5(1.7)C
Deposition	11.4(1.7)a	11.5(1.7)b	10.3(3.7)c	10.1(2.0)c	7.9(1.9)ab	2.9(1.0)A	3.0(1.0)A	2.5(0.5)A	7.2(1.6)B
Uneroded	23.0b	11.4(2.2)b	11.9(1.2)d	13.1(3.4)d	9.1(2.5)b	3.8(1.0)A	2.5(0.5)B	2.8(1.6)B	5.1(2.3)C
Bt1/Btk1									
Slight	6.5(0.1)a	6.2(0.7)aA	6.2(1.6)aA	7.1(2.4)abA	5.5(1.1)abA	2.5(0.6)A	2.6(0.5)A	2.3(0.2)A	5.2(0.7)B
Moderate	7.2(0.7)a	6.4(1.4)aAC	8.5(2.4)bCB	8.0(2.7)aAB	4.9(0.6)aC	3.6(0.8)A	2.8(0.2)B	3.0(0.6)AB	6.5(2.0)C
Severe	7.0(0.5)a	6.4(1.4)aAB	7.8(1.1)abA	6.3(1.5)bAB	5.8(2.5)abB	2.3(0.0)A	3.1(0.9)B	2.8(0.8)AB	4.4(1.4)C
Deposition	11.1(1.9)b	11.6(4.0)bA	10.4(4.0)cA	11.6(1.9)cA	7.4(1.1)bB	2.4(0.4)A	2.8(0.5)A	4.8(1.7)B	5.3(0.5)B
Uneroded	8.0a	7.5(1.1)aA	7.5(2.0)abA	6.8(1.1)abA	4.0(0.9)aB	3.0(0.6)A	1.7(0.2)B	2.8(0.6)A	4.8(1.0)C
Bt2/Btk2									
Slight	5.9(1.0)a	5.0(1.7)a	6.3(0.7)ab	5.7(1.3)ab	6.3(1.8)ac	4.4(0.9)aA	3.3(1.1)aB	2.4(0.4)aC	3.5(0.4)aB
Moderate	5.8(0.5)a	4.5(0.2)a	5.3(0.9)a	5.4(0.8)a	6.0(1.9)ac	2.3(0.4)bA	2.6(0.4)abA	2.7(0.8)aA	6.0(2.2)bB
Severe	5.5(0.6)a	4.3(1.2)a	5.7(1.6)ab	5.8(1.2)ab	4.6(1.1)b	2.0(0.0)bA	2.1(0.8)bA	2.6(0.4)aA	4.4(1.3)cB
Deposition	11.1(1.4)b	8.2(1.3)b	6.6(1.6)bc	6.7(1.1)bc	6.8(1.3)a	4.3(0.9)aA	3.1(0.7)aB	3.8(2.3)bAB	5.1(0.9)cC
Uneroded	6.0a	7.9(2.0)b	7.6(1.0)c	7.4(1.2)c	5.2(0.6)bc	2.5(0.4)bA	2.7(0.5)abA	2.0(0.0)aA	4.1(0.2)aB

Means followed by standard deviations in parentheses. Values followed by the same lower case in columns for erosion phase effects and upper case letter in rows for size-fraction effects, respectively, are not significantly different.

Table 5. Fisher's LSD[1] and F-ratio significance according to erosion phase and size fraction for different soil fractions of three horizons at Clark County Sites A, B, and C

Horizon[2]	Soil fraction	Site A		Site B		Site C	
		Erosion phase	Size-fraction	Erosion phase	Size-fraction	Erosion phase	Size-fraction
Ap/A	Whole soil	3.3***	--	2.4***	--	3.3***	--
	Aggregates	2.0**	NS	2.2**	1.9*	1.6***	NS
	Primary particles	1.9**	NS	NS	1.8***	NS	0.9***
BA/Bt/Btk	Whole soil	3.0**	--	NS	--	2.3**	--
	Aggregates	1.9**	NS	1.9***	NS	2.0**	1.8*
	Primary particles	1.6*	1.4**	1.2*	1.1***	NS	0.8***
Btk/BC	Whole soil	4.9*	--	NS	--	1.8***	--
	Aggregates	2.2***	NS	2.4***	NS	1.2**	NS
	Primary particles	1.3**	1.2*	NS	1.1**	0.9*	0.8***

*, **, *** correspond to significance at $P = 0.05$, 0.01, and 0.001, respectively.
NS = nonsignificant.
[1] Fishers's least significant difference.
[2] Horizons not of equal thickness and profiles not of equal depths.

generally held for all horizons, although sharp drops in SOC occurred below the A horizon for non-eroded sites, with subsoil SOC contents approaching those of subsurface horizons of the eroded phases. The deposition phase, on the other hand, had high SOC contents extending well below the surface horizon. The data indicated that eroded areas had significantly lower (up to 55% less) SOC in aggregates than noneroded soil or depositional areas, but this effect was confined mainly to the surface horizon, i.e., 0.15 to 0.20 m depths.

A significant erosion effect on SOC associated with primary particles was observed mainly at site A (Table 2, Figure 1). The general lack of difference in SOC of primary particles indicated that severity of erosion did not greatly affect the partitioning or redistribution of SOC among primary particle size-fractions upon dispersion and physical disruption. The SOC associated with primary particle fractions appeared to be more a function of the amount of particulate SOC contained in aggregates (Cambardella and Elliott, 1993), and tendency toward redistribution of SOC among fine particles (particularly clay) upon dispersion, leading to a relative enrichment of SOC in these fractions (Bonde et al., 1992).

D. Soil Organic Carbon Distribution in Size-Fractions of Aggregates and Primary Particles

Statistically significant differences due to size-fraction were not observed for SOC content of aggregates with the exception of the surface horizon at site B and the second horizon at site C (Tables 5-7). In general, aggregate SOC values across all size-fractions for eroded and deposition phases were similar to those of corresponding whole soil (Tables 2-4). In contrast, however, noneroded sites had considerably lower (about 20 to 60%) SOC in aggregate fractions compared to whole soil, values being similar, or only slightly higher, than other phases across size-fractions (Figures 1-3). This indicated a high proportion of particulate and less strongly mineral-associated organic materials in macroaggregates of forest soil relative to cultivated soils. Though not statistically different, aggregates in the size-range 0.1 to 2.0 mm appeared to have somewhat higher (0 to 40%) SOC than those of the <0.1 mm fraction, particularly for surface soil layer. No clear trends in SOC distribution among aggregate size-fractions were noted in subsurface horizons.

In the case of primary particles, significant differences in SOC were observed among size-fractions for most horizons (Tables 5-7). Contents of SOC for primary particle fractions were typically substantially lower (by about 9 to 78%) than corresponding aggregate fractions or whole soil, except for the <0.1 mm size-fraction. The SOC contents of the smallest size range of primary particle were, on the contrary, generally higher, by about 7 to 70%, than other size-fractions and were similar to those of the <0.1 mm aggregate fractions. These results support the conclusion that macroaggregates contain larger amounts of SOC, but much of it is less physically stable to disruption than SOC associated with microaggregates and fine soil material. Greater stability of microaggregates compared to macroaggregates (>0.25 mm) have also been noted by other researchers (Cambardella and Elliott, 1993; Tisdall, 1996). Moreover, the data show that upon dispersion and physical disruption, SOC may become redistributed with predominance in the fine particle fractions, especially silt and clay (Bonde et al., 1992; Angers et al., 1993). Similar results were observed by Bajracharya et al. (1996) for two other Ohio soils.

E. Possible Mechanisms of Carbon Sequestration

The results of this study suggest that substantial amounts of SOC are held within stable, slaking-resistant macroaggregates and are likely protected from decomposition through close association with microaggregates and fine mineral fractions (especially clay), as have also been reported by other researchers (Blevins and Frye, 1993; Cambardella and Elliott, 1993; Puget et al, 1995; Tisdall, 1996).

Soil's potential to sequester carbon appears to be linked with formation of durable organo-mineral complexes leading to the stabilization of aggregates and imparting increased resistance to their breakdown by physical and chemical forces. The strength and stability of organo-mineral bonds tend to increase with complexity of organic compounds, possibly through more intimate physical interaction (protection within microaggregates and pores) and chemical protection, such as, increased number of bonds and reduced reactiveness upon drying, or by formation of complexes with polyvalent cations (Tisdall, 1996). Thus, a combination of physical occlusion and chemical recalcitrance (Cambardella and Elliott, 1993) likely results in the stabilization of SOC within the aggregate hierarchy.

IV. Conclusions

The results of this study suggested that while deposition of eroded soil does not necessarily lead to the direct sequestration of carbon, it is likely to increase the overall sequestration of SOC by leading to an accumulation of organic material which has a greater potential to be converted to the stable SOC form. Depositional and noneroded areas increase potential SOC sequestration possibly by providing favorable conditions for aggregation. Carbon sequestration in soil seems to occur within the aggregate structure, particularly in microaggregates within which intimate association (through phyico-chemical bonding) of SOC with colloidal and inorganic soil material lead to high aggregate stability. The microaggregates may occur as subunits of larger aggregates or independent units of the bulk soil. Erosion is likely to lead to a gradual depletion of SOC by exposing the stable carbon pool in microaggregates of not only the surface layer but subsoil as well, to degradative processes by disrupting macroaggregates and removal of successive layers of soil.

References

Angers, D.A. 1992. Changes in soil aggregation and organic carbon under corn and alfalfa. *Soil Sci. Soc. Am. J.* 56:1244-1249.

Angers, D.A., A. N'dayegamiye, and D. Cote. 1993. Tillage-induced differences in organic matter of particle-size fractions and microbial biomass. *Soil Sci. Soc. Am. J.* 57:512-516.

Bajracharya, R.M., R. Lal, and J.M. Kimble. 1996. Long-term tillage effects on soil organic carbon distribution in aggregates and primary particle fractions of two Ohio soils. In: Proceedings of Carbon Sequestration in Soils An International Symposium, July 22-26, 1996, Columbus, OH, USA. The Ohio State Univ., National Soil Survey Center, NRCS and USDA-ARS.

Blevins, R.L. and W.W. Frye. 1993. Conservation tillage: An ecological approach to soil management. p. 33-78. In: D.L. Sparks (ed.), *Advances in Agronomy*, v. 51.

Bonde, T.A., B.T. Christensen, and C.V. Cole. 1992. Dynamics of soil organic matter as reflected by natural ^{13}C abundance in particle size fractions of forested and cultivated oxisols. *Soil Biol. Biochem.* 24:275-277.

Bouwman, A.F. 1990. Exchange of greenhouse gases between terrestrial ecosystems and the atmosphere. In: A.F. Bouman (ed.), *Soils and the Greenhouse Effect*. John Wiley & Sons., Chichester, U.K.

Brown, L.R. 1984. The global loss of topsoil. *J. Soil Water Conser.* 39:162-165.

Burke, I.C., E.T. Elliott, and C.V. Cole. 1995. Influence of macroclimate, landscape position and management on soil organic matter in agroecosystems. *Ecological Applications* 5:124-131.

Cambardella, C.A. and E.T. Elliott. 1993. Carbon and nitrogen distribution in aggregates from cultivated and native grassland soils. *Soil Sci. Soc. Am. J.* 57:1071-1076.

Chengere, A. and R. Lal. 1995. Soil degradation by erosion of a Typic Hapludalf in central Ohio and its rehabilitation. *Land Degradation and Rehabilitation* 6:223-238.

Daniels, R.B., J.W. Gilliam, D.K. Cassel, and L.A. Nelson. 1985. Soil erosion class and landscape position in the North Carolina Pedimont. *Soil Sci. Soc. J. Am.* 49:991-995.

Ebeid, M.M., R. Lal, G.F. Hall, and E. Miller. 1995. Erosion effects on soil properties and soybean yield of a Miamian soil in Western Ohio in a season with below normal rainfall. *Soil Technology* 8:97-108.

Fahnestock, P.B. 1994. Erosion effects on soil physical and chemical properties and crop yield of two central Ohio Alfisols. Unpubl. M.S. Thesis, The Ohio State Univ., Columbus, OH.

Fahnestock, P., R. Lal, and G.F. Hall. 1995a. Land use and erosional effects on two Ohio Alfisols: I. Soil properties. *J. Sustainable Agric.* 7:63-84.

Fahnestock, P., R. Lal, and G.F. Hall. 1995b. Land use and erosional effects on two Ohio Alfisols: II. Crop yields. *J. Sustainable Agric.* 7:85-100.

Frye, W.W., S.A. Ebelhar, L.W. Murdock, and R.L. Blevins. 1982. Soil erosion effects on properties and productivity of two Kentucky soils. *Soil Sci. Soc. J. Am.* 46:1051-1055.

Gee, G.W. and J.W. Bauder. 1986. Particle-size analysis. p. 383-411. In: A. Klute (ed.), Methods of Soil Analysis, Pt. I. Physical and Mineralogical Methods, 2nd Ed. ASA Monograph No. 9, ASA-SSSA, Madison, WI.

Lal, R. 1995. Global soil erosion by water and carbon dynamics. p. 131-142. In: R., Lal, J. Kimble, E. Levine, and B.A. Stewart (eds.), *Soil Management and Greenhouse Effect*. Lewis Publ., Boca Raton, FL.

Lal, R., T.J. Logan, and N.R. Fausey. 1990. Long-term tillage effects on Mollic Ochraqualf in northwestern Ohio. III. Soil nutrient profile. *Soil Tillage Res.* 15:371-382.

Lal, R., A.A. Mahboubi, and N.R. Fausey. 1994. Long-term tillage and rotation effects on properties of a central Ohio soil. *Soil Sci. Soc. Am. J.* 58:517-522.

Lal, R., J. Kimble, and B.A. Stewart. 1995. World soils as a source or sink for radiatively-active gases. p. 1-8. In: R., Lal, J. Kimble, E. Levine, and B.A. Stewart (eds.), *Soil Management and Greenhouse Effect*. Lewis Publ., Boca Raton, FL.

Mermut, A.R., D.F. Action, and W.E. Eilers. 1983. Estimation of soil erosion and deposition by a landscape analysis technique on clay soils in southern Saskatchewan. *Canadian J. Soil Sci.* 63:727-739.

Mokma, D.L. and M.A. Sietz. 1992. Effects of soil erosion on corn yields on Marlette soils in south-central Michigan. *J. Soil Water Conser.* 47:325-327.

Nelson, D.W. and L.E. Sommers. 1982. Total carbon, organic carbon, and organic matter. p. 539-580. In: A.L Page, R.H. Miller, and D.R. Keeney (eds.), Methods of Soil Analysis, Pt. II. Chemical and Microbiological Properties, 2nd Ed. ASA Monograph No. 9, ASA-SSSA, Madison, WI.

Nelson, P.N., M.C. Dictor, and G. Soulas. 1994. Availability of organic carbon in soluble and particle-size fractions from a soil profile. *Soil Biol. Biochem.* 26:1549-1555.

Nizeyimana, E. and K.R. Olson. 1988. Chemical, mineralogical, and physical property differences between moderately and severely eroded Illinois soils. *Soil Sci. Soc. Am. J.* 52:1740-1748.

Oades, J.M. 1984. Soil organic matter and structural stability: Mechanisms and implications for management. *Plant Soil*, 76:319-337.

Pimentel, D., C. Harvey, P. Resosudarmo, K. Sinclair, D. Kurz, M. McNair, S. Crist, L Shpritz, L. Fitton, R. Saffouri, and R. Bair. 1995. Environmental and economic costs of soil erosion and conservation benefits. *Science* 267(5201):1117-1123.

Puget, P., C. Chenu, and J. Balesdent. 1995. Total and young organic matter distribution in aggregates of silty cultivated soils. *European J. Soil Sci.*, 46:449-459.

Rasmussen, P.E. and W.J. Parton. 1994. Long-term effects of residue management in wheat-fallow: I. Inputs, yield, and soil organic matter. *Soil Sci. Soc. Am. J.* 58:523-530.

Ruhe, R.V. 1975. *Geomorphology: Geomorphic Processes and Surficial Geology.* Houghton Mifflin Co., Boston.

Schertz, D.L., W.C. Moldenhauer, S.J. Livingston, G.A. Weesies, and E. A. Hintz. 1989. Effect of past soil erosion on crop productivity in Indiana. *J. Soil Water Conser.* 44:604-608.

Steel, R.G.D. and J.H. Torrie. 1980. *Principles and Procedures of Statistics: A Biometrical Approach.* 2nd ed. McGraw-Hill, New York, NY.

Tisdall, J.M. 1996. Formation of soil aggregates and accumulation of soil organic matter. p. 57-96. In: M.R. Carter and B.A. Stewart (eds.), *Structure and Soil Organic Matter Storage in Agricultural Soils.* Adv. Soil Sci., Lewis Publ., Boca Raton, FL.

Tisdall, J.M. and J.M. Oades. 1982. Organic matter and water-stable aggregates in soil. *J. Soil Sci.* 33:141-163.

William, R.K., K.R. Olson, W.L. Banwart, and D.L. Johnson. 1989. Soil, landscape, and erosion relationships in a northwest Illinois watershed. *Soil Sci. Soc. Am. J.* 53:1763-1771.

Woomer, P.L., A. Martin, A. Albrecht, D.V.S. Resck, and H.W. Scharpenseel. 1994. The importance and management of soil organic matter in the tropics. p. 47-80. In: P.L. Woomer and M.J. Swift (eds.), *The Biological Management of Tropical Soil Fertility.* Wiley-Sayce Publ., Chichester, U.K.

Yoder, R.E. 1936. A direct method of aggregate analysis and a study of the physical nature of erosion losses. *J. Am. Soc. Agron.* 28:337-351.

CHAPTER 26

Carbon Storage in Eroded Soils after Five Years of Reclamation Techniques

R.C. Izaurralde, M. Nyborg, E.D. Solberg, H.H. Janzen,
M.A. Arshad, S.S. Malhi, and M. Molina-Ayala

I. Introduction

Soil erosion is a major form of soil degradation in western Canada (Coote, 1984; Sparrow, 1984). Eroded soils reduce plant yield due to their high bulk density, poor tilth, reduced organic matter content, low nutrient availability, and reduced water-holding capacity (Dormaar et al., 1986; Eck et al., 1965; Frye et al., 1982; Power et al., 1981; and Tanaka and Aase, 1989). Recent studies in Alberta (Larney et al., 1995) using an artificial-erosion approach, demonstrated drastic reductions in crop yield for every increment in erosion level. The study suggested that soil productivity could be restored by replenishing nutrients with commercial fertilizers or manure (Izaurralde et al., 1994). Restoring the productivity of eroded soils brings the potential of not only increasing economic benefits to producers but also storing atmospheric C in soil organic matter (Lal, 1995; Cole et al., 1996). After 23 years of managing an artificially-eroded soil (Eck, 1987), soil organic matter was found to increase with treatments where the topsoil removal had been greater than 20 cm.

Measuring changes in soil organic matter over time and space is challenging because of the inherent variability of this soil property (Post et al., 1985). Soil organic matter concentration is a direct reflection of factors that affect plant growth such as past erosion, topography, texture, as well as soil water amounts and redistribution among others. Determination of changes of soil C mass over time due to a given management (e.g., direct seeding) requires information on the soil C mass present at the beginning of the measurement period (time zero). When this information is not available, then changes are determined by subtracting a 'control' amount of soil C (e.g., treatment without fertilizer) from that present in a management of interest (e.g., treatment with fertilizer). Use of the second approach, however, accounts not only for the change due to the management under consideration but also for changes in soil C elapsed between the times of measurement. To increase the accuracy of comparisons, it has been suggested that these should be made with data of soil C transformed to account for differences in soil bulk density and depth of sampling. This type of proposed comparison is referred to as 'soil-mass equivalency' (Ellert and Bettany, 1995).

Light fraction organic matter has been proposed and used as a sensitive indicator of changes in soil C occurring under specific management (Dalal and Mayer, 1986; Christensen, 1992; Janzen et al., 1992; and Bremer et al., 1995). Light fraction organic matter (LFOM) can be separated from soil by various methods (Christensen, 1992), one of which is by floatation in a heavy liquid such as NaI (Janzen et al., 1992). The amount of LFOM thus obtained is thought to represent organic material of recent incorporation, mostly plant residues and roots. This extracted material is assumed to be in the process of active transformation, have a rapid turnover time, and be associated, therefore, with the most active portions of soil organic matter (Christensen, 1996).

Long-term experiments have been extremely useful to study soil organic matter dynamics (Jenkinson et al., 1992; Buyanovski, et al., 1996; Campbell and Zentner, 1993, Bremer et al., 1994; Nyborg et al., 1995, and Izaurralde et al., 1996). Currently in Canada, there is interest in developing a scientific, credible plan to sequester C in farms of the prairie region using techniques such as direct seeding among others (Ellert and Janzen, pers. comm.; Izaurralde et al., this volume). An assumption in these plans is that changes in soil C should be detectable after relatively short periods of time (e.g., less than a decade). Short-term experiments, therefore, should provide information of this nature.

In this field study, we tested if agronomic practices such as continuous cereal cropping, reduced tillage, and addition of fertilizers and/or organic amendments could change, in a five-year span, the amount and quality of carbon stored in artificially-eroded soils.

II. Materials and Methods

Field studies were established in the fall of 1990 at two sites in the Aspen Parkland ecoregion of the Canadian Prairie, each located approximately 40 km east of Edmonton, Alberta (53° 34' N, 113° 33' W) and about 25 km apart from each other. The soil at the first site (Josephburg) is a Typic Cryoboroll of the Angus Ridge series with an A horizon approximately 30 cm thick. The soil at the second site (Cooking Lake) is a Typic Cryoboralf of the Cooking Lake series with an A horizon only 15 cm thick. The common rotation practiced by the farmer at Josephburg prior to the initiation of the study was canola (*Brassica* spp.), wheat (*Triticum aestivum* L.), and barley (*Hordeum vulgare* L.). In recent years, the producer had applied N in the fall at a rate of 60 kg ha^{-1} and P with the seed at 9 kg ha^{-1}. In contrast, the producer at Cooking Lake had grown oat (*Avena sativa* L.) intercropped with field pea (*Pisum sativum* L.) in rotation with barley. Application of farmyard manure was the norm on this field at approximate rates of 5 Mg ha^{-1}. In addition, the farmer used fertilizer N and P at rates of 60 and 10 kg ha^{-1}. Selected characteristics of both soils at the initiation of the experiment are reproduced in Table 1 (Izaurralde et al., 1993).

The experiment consisted of a factorial combination of five levels of artificial erosion and four restorative amendments arranged in a split plot design with four replications (Izaurralde et al., 1993). Erosion treatments were assigned to the whole plots while amendments were allocated to the subplots. Whole plots were 10 m long by 14.6 m wide to accommodate four subplots of the same length and a width of 3.65 m. The simulated erosion levels (or cuts) were established in fall 1990 by removing topsoil with an excavator with a grading bucket in 5-cm depth increments from 0 to 20 cm. The four amendments treatments were: (i) control (C), (ii) addition of 5 cm of topsoil (T), (iii) addition of fertilizer (F) (N at 100 kg ha^{-1} and P at a rate of 9 kg ha^{-1}), and (iv) addition of farmyard manure (M) (at a rate of 75 Mg ha^{-1} on a dry weight basis).

In 1992, the experiment was modified to further test the residual effects of the amendments applied in 1991. Each plot was split in two halves one of which received fertilizer each year at rates of 100 kg ha^{-1} of N and 9 kg ha^{-1} of P. For the study reported here, we selected three artificial erosion treatments (0, 10, and 20 cm) and four amendments out of the possible eight (control without fertilizer (C-F), fertilizer (F+F), manure without fertilizer (M-F), and manure with fertilizer (M+F).

Tillage was shallow and limited only to either fertilizer incorporation or weed control before seeding. All plots were cropped to hard red spring wheat cv. 'Roblin' from 1991 to 1994 and to spring barley cv. 'Duke' in 1995. Plots were sown in early May each year with a drill equipped with double-disk openers, mounted 22.5 cm apart (Fabro, Swift Current, Saskatchewan). Weeds were controlled with herbicides and by hand. Plots were harvested when wheat was mature with a plot combine (Wintersteiger, Salt Lake City, Utah). In addition, four to six subsamples (1 m^2) per plot were

Table 1. Natural and emergent soil layer properties of a Typic Cryoboroll at Josephburg and a Typic Cryoboralf at Cooking Lake before (0 cm cut) and after imposing two levels of artificial erosion (10 and 20 cm cut)

Cut	Depth	Bulk density[1]	pH	Clay	Cation exchange capacity	Organic N	Total N
cm	cm	Mg m^{-3}		%	cmol(+) kg^{-1}	%	%
			—Josephburg—				
0	0-10	1.17	6.6	30	33.0	4.28	0.360
	10-20	1.21	6.7	30	33.0	3.85	0.296
	20-30	1.31	6.6	30	33.0	1.46	0.125
10	0-10	1.24	6.7	30	33.0	3.71	0.285
	10-20	1.31	6.5	30	29.7	1.69	0.142
	20-30	1.37	6.5	31	22.6	1.20	0.104
20	0-10	1.31	6.8	30	28.7	2.60	0.205
	10-20	1.37	6.6	31	23.3	0.76	0.075
	20-30	1.41	6.5	30	19.2	0.63	0.061
			—Cooking Lake—				
0	0-10	1.25	6.5	22	20.0	3.51	0.300
	10-20	1.30	6.4	19	15.0	1.41	0.139
	20-30	1.36	5.9	22	16.7	0.80	0.100
10	0-10	1.30	6.4	21	16.3	1.90	0.170
	10-20	1.38	6.1	23	27.7	0.90	0.100
	20-30	1.47	5.8	28	30.8	0.80	0.083
20	0-10	1.38	6.2	34	26.0	0.85	0.093
	10-20	1.45	5.6	27	29.0	0.70	0.085
	20-30	1.49	5.5	30	30.0	0.67	0.072

[1]Soil bulk density in the 0 - 10 cm depth was determined with a gamma-ray density probe (Model MC-1, Campbell Scientific Nuclear, WA). Soil bulk density values below 10 cm were estimated from the Soil Inventory Attribute Map - Alberta: Soil Layer Digital Data (Canada Soil Survey Staff, 1989). (Data extracted from Izaurralde et al., 1993.)

harvested by hand for total dry matter determination. Grain and straw yields were calculated and expressed on a dry mass basis. Carbon content of plant samples was taken to be 0.44 kg kg^{-1}.

Izaurralde et al. (1993) reported initial characteristics of soil layers after the erosion treatments had been implemented. Selected results from this initial characterization carried out in May 1991 are reproduced in Table 1. Soil samples were taken again in fall 1995 while the barley crop was still standing. The plots sampled were those resulting from a combination of three erosion levels (0, 10, and 20 cm) and four amendments (C-F, F+F, M-F, and M+F). Sampling depth increments were 0 - 10, 10 - 20, and 20 - 30 cm. The samples were extracted with a soil corer that had an internal diameter of 4.2 cm. Each core extracted, *ca.* 35 cm deep, was laid horizontally and cut into three portions with a sharp knife according to the sampling depth increments. This procedure was repeated six times on each plot and all core segments of a specified depth combined into one sample per plot. Cores that did not meet these specific criteria were discarded and the procedure was repeated. After air-drying the soil samples, their dry weight were determined and these were used to calculate bulk density values. The soil samples were then ground to pass a 2-mm mesh sieve. In the laboratory, the soil samples were analyzed for organic C with a LECO Carbon Determinator (Model CR12, St. Joseph, MI) and total N by colorimetric determinations of Kjeldhal-digested samples on an autoanalyzer (Technicon Industrial Systems, 1977).

Light fraction organic matter (LFOM) was determined on soil samples following a slightly modified density-fractionation procedure described by Janzen et al. (1992). Briefly, a 20-g soil sample was treated with 40 mL of NaI with specific gravity of 1.7 Mg m^{-3}. The solution was shaken vigorously for 30 minutes and allowed to settle for 24 h at room temperature. The suspended material (LFOM) was removed by vacuum and placed in a Millipore filtration unit (Millipore Corp., Bedford, MA) with Whatman no. 2 paper. The LFOM was then rinsed with 50 mL of 0.01 M CaCl$_2$ and 50 mL of deionized water. All LFOM fractions were dried at 45°C for 24 h, scraped from the filter paper, and weighed. The samples were then finely ground in a steel-ball grinder. Simultaneous determinations of C and N on the LFOM samples were carried out by Dumas combustion on a Carlo Erba analyzer (Model NA 1500, Carlo Erba Strumentazione, Italy).

Carbon and nitrogen results were converted to soil-mass equivalent basis as described by Ellert and Bettany (1995) whereby soil bulk densities and depth adjustments were used to calculate soil layers with equal mass across all treatments at each site. Analyses of variance were performed on all basic (bulk density, concentration and mass of C and N) and derived soil parameters (C:N ratios of whole-soil and light fractions). The statistical model used was split-plot with erosion as whole-plot treatment and amendment as subplot treatment.

III. Results and Discussion

A. Crop Productivity (1991 - 1994)

Analysis of variance of grain yields on individual years revealed both erosion and amendment to strongly influence crop productivity (statistical analyses not shown). At both sites, yields decreased with erosion level in all years except in 1994 at Josephburg where the erosion effect was not statistically significant. On that occasion, an infestation of *Septoria graminis* (glum blotch) depressed yields across all erosion levels, but its effect was more severe on noneroded soil, presumably because of higher levels of straw production.

Any of the amendment combinations used (F+F, M-F, and M+F) consistently increased yields over the control (C-F) at both sites and in all years except at Josephburg in 1994 where the M-F treatment (714 kg ha^{-1}) was not significantly different from the C-F treatment (640 kg ha^{-1}). The statistical models showed significant interactions between erosion level and amendment in most years. The statistical interaction was not significant at p<0.05 level of probability in 1991 at Josephburg and in 1994 at Cooking Lake. All significant interactions accentuated the difference in response of crop productivity to amendment as the level of artificial erosion increased from 0 to 20 cm.

At Josephburg, fertilizers N and P, alone (F+F) or in combination with manure (M+F) produced the best average yields (Table 2). The average grain yield increase over the control rose from 1.08 Mg ha^{-1} in the 0 cm cut to 1.81 Mg ha^{-1} in the 20 cm cut. At Cooking Lake, with the addition of manure in 1991 and fertilizers N and P in 1992 thereafter, (M+F) was by far the best treatment across all erosion levels. The average yield increase of this treatment over the control rose from 1700 kg ha^{-1} in the 0 cm cut to 2000 kg ha^{-1} in the 20 cm cut. The yield increase effect caused by the addition of a single dose of manure in 1991 lasted for the entire period of study. For example, in 1991 at Cooking

Table 2. Average (1991 - 1994) grain yield, straw yield, straw:grain ratio, root:shoot ratio and estimated total plant C input to soil at Josephburg and Cooking Lake (Alberta)

Cut	Amendment	Average grain yield	Average straw yield	Straw: grain ratio	Root: shoot ratio	Total plant C input to soil	Total manure C input to soil[2]
cm		Mg ha^{-3}	Mg ha^{-1}			Mg ha^{-1}	Mg ha^{-1}
—Josephburg—							
0	C-F	1.54	2.41	1.57	0.24	6.33	0.0
	F+F	2.68	4.19	1.56	0.12	9.18	0.0
	M-F	2.08	2.98	1.43	0.12	6.58	23.0
	M+F	2.56	3.69	1.44	0.12	8.14	23.0
10	C-F	1.07	1.69	1.59	0.24	4.44	0.0
	F+F	2.56	4.02	1.57	0.12	8.82	0.0
	M-F	1.62	2.62	1.62	0.12	5.74	23.0
	M+F	2.41	3.78	1.57	0.12	8.30	23.0
20	C-F	0.51	0.79	1.55	0.24	2.08	0.0
	F+F	2.32	3.45	1.49	0.12	7.59	0.0
	M-F	1.35	2.08	1.54	0.12	4.57	23.0
	M+F	2.31	3.51	1.52	0.12	7.72	23.0
—Cooking Lake—							
0	C-F	2.05	3.10	1.51	0.24	8.17	0.0
	F+F	3.32	5.06	1.52	0.12	11.13	0.0
	M-F	3.07	4.58	1.49	0.12	10.08	23.0
	M+F	3.75	5.74	1.53	0.12	12.60	23.0
10	C-F	0.81	1.25	1.54	0.24	3.28	0.0
	F+F	2.66	4.03	1.51	0.12	8.87	0.0
	M-F	2.27	3.51	1.54	0.12	7.70	23.0
	M+F	3.20	4.95	1.54	0.12	10.86	23.0
20	C-F	0.32	0.53	1.65	0.24	1.39	0.0
	F+F	1.85	2.83	1.53	0.12	6.21	0.0
	M-F	1.18	1.87	1.59	0.12	4.09	23.0
	M+F	2.32	3.52	1.52	0.12	7.73	23.0

[1]Straw:grain ratio calculated from individual measurements of straw and grain in each plot in all years except in 1992 when a standard straw:grain ratio of 1.5 was used.
[2]Manure C was calculated by multiplying the rate of application (75 Mg ha^{-1} dry matter basis) times the C concentration (0.31 Mg Mg^{-1}).

Lake, the yield increase of the manure treatment (M) over the control (C) in the 20 cm cut was 1349 kg ha^{-1} (1767 - 418 kg ha^{-1}) while in 1994 the difference was 394 kg ha^{-1} (461 - 67 kg ha^{-1}).

Plant C inputs to soil were calculated on the basis of straw yields and estimates of root production using root-to-shoot ratios (Table 2) calculated from standing root and shoot mass measurements in fall, 1991, at both study sites by Izaurralde et al. (1993). Total C inputs from plants, therefore, reflected above-ground productivity except for the increased C input derived from stressed-wheat plants growing on control plots, regardless of erosion level. Under soil-stress conditions, roots demanded proportionally more carbohydrate resources from shoots and altered root-to-shoot ratios accordingly.

B. Whole Soil Organic C

1. Josephburg

In layer 1 (approximate depth range 0 - 10 cm), artificial erosion did not change soil bulk density (1.12 - 1.23 Mg m^{-3}) significantly. On average, treatments with manure had lower soil bulk density than treatments without, regardless of fertilizer addition. From the point of view of C and N concentration, our results indicate that concentration differences were easily detectable; all amendments were different from each other.

As expected, artificial erosion produced drastic changes in whole soil C with the highest value in the 0 cm cut (73.1 Mg ha^{-1}) and the lowest in the 20 cm cut (39.5 Mg ha^{-1}). On average, the largest stock of C in layer 1 was found in the M+F treatment (64.3 Mg ha^{-1}) followed by M-F (61.4 Mg ha^{-1}), F+F (54.2 Mg ha^{-1}), and C-F (49.4 Mg ha^{-1}). The mass of C in the former two treatments was significantly larger than in the latter two. Whole soil N in layer 1 average by erosion followed the trends discussed for whole soil C. The amount of whole soil N found in the various levels of artificial erosion ranged from 3.82 to 6.32 Mg ha^{-1}. Whole-soil N averages by treatment followed closely those discussed for whole-soil C. Artificial erosion lowered the C/N ratio significantly in the 20 cm cut (10.2) with respect to the average of the other two cuts (11.4). Treatment C-F had the narrowest C/N ratio (10.7). The C/N ratio for the other three treatments ranged closely between 11.0 and 11.1.

The equivalent soil masses of C and N for layer 1 in Table 3 follow the results discussed above except for the large difference detected between C-F and C+F in the 20 cm cut. We re-examined the individual values making up the average C concentrations and soil bulk densities, but found not enough evidence to discard any of the measurements as an outlier. Had we used the whole soil N mass (2.88 Mg ha^{-1}) and C/N ratio resulting from the other three treatments for that depth of cut (10.7), then the whole soil C mass would have been estimated at 30.82 Mg ha^{-1}, a more realistic estimate.

In layer 2 (approximate depth range 10 - 20 cm), soil bulk density increased significantly by 21% from 1.26 in the 0 cm cut to 1.53 Mg m^{-3} in the 20 cm cut. Changes in soil bulk density due to amendments were less pronounced than those by cut, with the lowest values measured in the manure treatments. Due to lack of statistical interaction, the individual treatment combinations in Table 3 reflect the differences noted above for the main effects. On average, the mass of whole soil C found in this layer in the 20 cm cut (13.8 Mg ha^{-1}) was only one third of that found in the 0 cm cut (46.1 Mg ha^{-1}). Already in this layer, the kind of amendment used was not able to change whole soil C significantly; the range of soil C adjusted by mass equivalency was 27.1 - 30.0 Mg ha^{-1}. The results for whole soil N in this layer were similar to those of C, but the differences were less pronounced.

Adding together the C masses across soil layers having an approximate depth range of 0 - 30 cm allowed for comparisons of the stock of C five years after the initiation of the experiment. Whole soil C in the 20 cm cut was less than half (61.1 Mg ha^{-1}) that found in the 0 cm cut (130.6 Mg ha^{-1}). When the averages are calculated by amendments, the only significant differences occurred between the C-F (86.9 Mg ha^{-1}) and M+F (104 Mg ha^{-1}) treatments. The method of whole soil C analysis used failed to detect changes in soil C after five years of treatment; not even the increase of 11.9 Mg ha^{-1} detected

Table 3. Partial and total whole-soil organic matter in a Typic Cryoboroll at Josephburg (Alberta) under three levels of artificial erosion and four nutrient amendment combinations

Layer	Cut	Amendment	Soil bulk density	Soil C conc.	Soil C mass	Soil N conc.	Soil N mass	C/N ratio
			Mg m^{-3}	%	Mg ha^{-1}	%	Mg ha^{-1}	
1	0	C-F	1.17	4.67	68.96	0.38	5.68	12.17
1	0	F+F	1.18	4.53	67.63	0.40	5.91	11.45
1	0	M-F	1.08	5.33	76.74	0.47	6.72	11.45
1	0	M+F	1.07	5.51	78.90	0.49	6.96	11.33
1	10	C-F	1.21	4.00	54.23	0.35	4.78	11.35
1	10	F+F	1.15	4.18	56.07	0.37	5.06	11.07
1	10	M-F	1.10	4.89	62.73	0.43	5.59	11.20
1	10	M+F	1.05	5.19	64.57	0.46	5.85	11.05
1	20	C-F	1.37	1.73	24.98	0.19	2.88	8.48
1	20	F+F	1.28	2.87	38.92	0.26	3.59	10.80
1	20	M-F	1.20	3.47	44.60	0.32	4.26	10.44
1	20	M+F	1.09	4.14	49.52	0.37	4.55	10.90
LSD$_{0.05}$			0.17	0.90	14.18	0.08	1.19	1.06
2	0	C-F	1.30	3.88	44.50	0.32	3.80	11.70
2	0	F+F	1.28	4.00	44.91	0.35	4.05	11.09
2	0	M-F	1.21	4.21	46.66	0.36	4.16	11.27
2	0	M+F	1.25	4.29	48.12	0.37	4.44	10.81
2	10	C-F	1.44	1.81	24.72	0.19	2.69	9.09
2	10	F+F	1.43	2.08	27.34	0.21	2.81	9.59
2	10	M-F	1.37	2.11	26.17	0.21	2.75	9.28
2	10	M+F	1.42	2.12	25.70	0.21	2.68	9.46
2	20	C-F	1.59	0.81	12.18	0.14	2.07	5.83
2	20	F+F	1.55	0.91	13.68	0.10	1.57	8.55
2	20	M-F	1.52	0.91	13.07	0.12	1.75	7.36
2	20	M+F	1.45	1.17	16.11	0.12	1.71	8.92
LSD$_{0.05}$			0.15	1.00	14.32	0.08	1.12	2.04
Total	0	C-F			113.45		9.47	11.98
Total	0	F+F			112.54		9.96	11.31
Total	0	M-F			123.40		10.88	11.38
Total	0	M+F			127.02		11.40	11.13
Total	10	C-F			78.95		7.47	10.58
Total	10	F+F			83.40		7.87	10.56
Total	10	M-F			88.90		8.34	10.60
Total	10	M+F			90.26		8.53	10.57
Total	20	C-F			37.15		4.95	7.39
Total	20	F+F			52.60		5.16	10.17
Total	20	M-F			57.67		6.01	9.58
Total	20	M+F			65.63		6.25	10.42
LSD$_{0.05}$					26.25		2.12	1.20

on average in the M-F treatment over the C-F treatment. As with whole soil C, whole soil N differences were detected easily under erosion main effects (range: 5.6 - 10.43 Mg ha^{-1}) but not under amendments (range: 7.30 - 8.73 Mg ha^{-1}). Overall, the C/N ratio decreased by 2 whole units from 11.4 in the 0 cm cut to 9.4 in the 20 cm cut.

2. Cooking Lake

In contrast to the Josephburg site, there was a relatively sharp increase in soil bulk density with erosion level from 1.08 Mg m^{-3} in the 0 cm cut to 1.34 Mg m^{-3} in the 20 cm cut. As at Josephburg, treatments with manure (M+F and M-F) had lower soil bulk density than treatments without (F+F and C-F). Erosion-amendment treatment means in Table 4 reflect the overall trend described.

Whole soil C mass differences created by artificial erosion treatments were large. The 0 cm cut (36.8 Mg ha^{-1}) contained almost three times as much C as treatments with 20 cm erosion (13.7 Mg ha^{-1}). The sampling and analytical methodology employed had no difficulty this time in unraveling the differences in concentration and equivalent soil mass of C when the comparisons were made on an-amendment treatment basis. All four treatments were significantly different from each other. Moreover, there was a significant interaction ($p<0.017$) between erosion level and amendment (Table 4). This interaction resulted from the incremental relative difference observed between manure and nonmanure treatments as the erosion level increased from 0 to 20 cm. On the pair comparisons (C-F vs. F+F and M-F vs. M+F) by erosion level, only the C-F vs. C+F comparison in the 10 cm cut and both comparisons in the 20 cm cut were significant at $p<0.05$ level of probability. This represents, then, the first opportunity to observe an increase in C storage as a result of fertilizer-induced increase in plant productivity. Whole soil N followed the same trends as did whole soil C except that only the difference between C-F and F+F in the 10 cm cut was significant at $p<0.05$ (Table 4). Carbon to nitrogen ratios, when averaged by erosion level, decreased by two and a half units from 11.4 in the 0 cm cut to 8.9 in the 20 cm cut suggesting the presence of fixed ammonium. The M+F treatment had the widest C/N ratio (10.9) and this was significantly higher than the other three.

B. Light Fraction Soil Organic C

1. Josephburg

Concentration of light fraction C in layer 1 did not change with erosion level (range: 23.6% - 24.8%) but did considerably with amendment. The highest average concentration of LFC was measured on M+F treatments (30.2 %) while the lowest was on C-F treatments (18.4 %). Therefore, all LFC erosion-amendment concentration means in Table 5 reflect these differences.

The mass of LFC found on average in 0-cm cut treatments (4.24 Mg ha^{-1}) was not different than that found in 10-cm treatments (3.90 Mg ha^{-1}), but either of these two was greater than that in 20-cm cut treatments (2.95 Mg ha^{-1}). When the mass of LFC was averaged by amendment, it varied fivefold from 1.51 Mg ha^{-1} in C-F treatments to 7.47 Mg ha^{-1} in M+F treatments. Although M+F treatments had consistently more mass of LFC than M-F treatments at all erosion levels, the mass of LFC in F+F treatments was greater than that in C-F treatments only in the 10-cm cut ($p<0.052$). The difference in LFC mass needed to reach this level of statistical significance was 1.52 Mg ha^{-1} (Table 5). Overall, erosion level did not affect the ratio of LFC to whole soil C (LC/TC). In contrast, the LC/TC ratio in M+F treatments (0.135) was, on average, four times greater than that in C-F treatments (0.032). The corresponding measurements of concentration, mass and ratios of LFN in layer 1 closely followed those of LFC.

Table 4. Partial and total whole-soil organic matter in a Typic Cryoboralf at Cooking Lake (Alberta) under three levels of artificial erosion and four nutrient amendment combinations

Layer	Cut	Amendment	Soil bulk density	Soil C conc.	Soil C mass	Soil N conc.	Soil N mass	C/N ratio
	cm		Mg m^{-3}	%	Mg ha^{-1}	%	Mg ha^{-1}	
1	0	C-F	1.15	3.66	33.19	0.32	2.92	11.33
1	0	F+F	1.14	3.75	34.07	0.32	2.93	11.57
1	0	M-F	1.04	4.31	39.16	0.38	3.41	11.46
1	0	M+F	0.99	4.49	40.73	0.39	3.55	11.42
1	10	C-F	1.33	1.69	15.34	0.17	1.51	10.10
1	10	F+F	1.23	2.33	21.16	0.22	1.99	10.54
1	10	M-F	1.12	3.10	28.17	0.29	2.61	10.76
1	10	M+F	1.06	3.37	30.58	0.30	2.72	11.20
1	20	C-F	1.43	1.01	9.15	0.12	1.13	8.09
1	20	F+F	1.38	1.09	9.90	0.13	1.21	8.18
1	20	M-F	1.30	1.80	16.35	0.19	1.74	9.40
1	20	M+F	1.25	2.14	19.43	0.21	1.92	10.12
LSD$_{0.05}$			0.12	0.51	4.67	0.04	0.37	0.74
2	0	C-F	1.46	1.34	22.23	0.15	2.45	9.06
2	0	F+F	1.43	1.69	27.73	0.19	2.87	9.62
2	0	M-F	1.41	1.59	25.05	0.19	2.85	8.80
2	0	M+F	1.40	1.85	29.09	0.22	3.11	9.27
2	10	C-F	1.54	0.73	13.86	0.12	1.80	7.70
2	10	F+F	1.51	0.81	15.59	0.11	1.84	8.48
2	10	M-F	1.49	0.88	16.33	0.12	1.95	8.34
2	10	M+F	1.50	1.03	17.27	0.13	2.06	8.37
2	20	C-F	1.51	0.67	10.94	0.10	1.48	7.39
2	20	F+F	1.51	0.70	11.45	0.10	1.48	7.72
2	20	M-F	1.51	0.69	13.78	0.10	1.72	8.02
2	20	M+F	1.54	0.73	14.71	0.11	1.84	7.99
LSD$_{0.05}$			0.09	0.32	4.44	0.02	0.30	1.00
Total	0	C-F			67.72		7.07	9.55
Total	0	F+F			75.83		7.67	9.86
Total	0	M-F			76.34		7.97	9.56
Total	0	M+F			82.92		8.43	9.81
Total	10	C-F			39.77		4.83	8.23
Total	10	F+F			47.42		5.30	8.94
Total	10	M-F			55.47		6.12	9.04
Total	10	M+F			59.11		6.32	9.33
Total	20	C-F			29.96		3.99	7.51
Total	20	F+F			31.27		4.07	7.68
Total	20	M-F			39.98		4.84	8.26
Total	20	M+F			43.90		5.15	8.52
LSD$_{0.05}$					9.06		0.66	0.74

Table 5. Partial and total Light Fraction (LF) C and N in a Typic Cryoboroll at Josephburg (Alberta) under three levels of artificial erosion and four nutrient amendment combinations

Layer	Cut	Amendment	LFC conc.	LFC mass	LFC/TC ratio	LFN conc.	LFN mass	LFN/TN ratio	LF C/N ratio
	cm		%	Mg ha^{-1}		%	Mg ha^{-1}		
1	0	C-F	18.73	2.32	0.04	1.08	0.13	0.03	17.46
1	0	F+F	23.31	3.06	0.05	1.40	0.18	0.03	16.60
1	0	M-F	23.42	3.87	0.06	1.58	0.26	0.04	14.83
1	0	M+F	29.10	7.70	0.12	2.09	0.55	0.10	13.97
1	10	C-F	16.99	1.30	0.03	0.99	0.08	0.02	17.17
1	10	F+F	24.94	2.82	0.06	1.40	0.16	0.04	17.77
1	10	M-F	25.22	3.39	0.06	1.65	0.22	0.04	15.25
1	10	M+F	30.32	8.10	0.14	2.20	0.59	0.12	13.81
1	20	C-F	19.34	0.91	0.04	1.10	0.05	0.02	17.66
1	20	F+F	24.23	1.94	0.05	1.31	0.10	0.03	18.51
1	20	M-F	24.68	2.36	0.05	1.65	0.16	0.04	14.98
1	20	M+F	31.05	6.60	0.14	2.29	0.49	0.12	13.58
LSD$_{0.05}$			4.09	2.38	0.04	0.31	0.17	0.03	1.80
Total	0	C-F		2.61			0.15		17.29
Total	0	F+F		3.44			0.21		16.60
Total	0	M-F		4.19			0.28		14.95
Total	0	M+F		8.21			0.58		14.07
Total	10	C-F		1.35			0.08		17.15
Total	10	F+F		2.91			0.16		17.75
Total	10	M-F		3.46			0.23		15.23
Total	10	M+F		8.23			0.60		13.85
Total	20	C-F		0.94			0.05		17.64
Total	20	F+F		1.99			0.11		18.51
Total	20	M-F		2.40			0.16		14.96
Total	20	M+F		6.65			0.49		13.59
LSD$_{0.05}$				2.40			0.17		1.72

Results of LFC mass accumulated in the first and a fraction of the second soil layer (Table 5) revealed that F+F treatments (2.78 Mg ha^{-1}) contained on average 1.15 Mg ha^{-1} more LFC than C-F treatments. Individual comparisons by erosion level also revealed the LFC difference to be significant only in the 10 cm cut. Treatments with an application of cattle manure during the first year plus four more years of fertilizer N and P (M+F) had, on average, double the amount of LFC of treatments receiving manure only in the first year (M-F, 3.35 Mg ha^{-1}). This difference was highly significant across all erosion levels. Results of LFN pooled across all soil layers followed those of LFC.

Table 6. Partial and total Light Fraction (LF) C and N in a Typic Cryoboralf at Cooking Lake (Alberta) under three levels of artificial erosion and four nutrient amendment combinations

Layer	Cut	Amendment	LFC conc.	LFC mass	LFC/TC ratio	LFN conc.	LFN mass	LFN/TN ratio	LF C/N ratio
cm			%	Mg ha^{-1}		%	Mg ha^{-1}		
1	0	C-F	21.56	2.94	0.09	1.10	0.15	0.05	19.55
1	0	F+F	22.60	3.47	0.10	1.19	0.18	0.06	18.95
1	0	M-F	23.87	4.62	0.12	1.46	0.28	0.08	16.40
1	0	M+F	25.71	5.75	0.14	1.60	0.36	0.10	16.11
1	10	C-F	19.89	0.91	0.06	0.94	0.04	0.03	21.25
1	10	F+F	22.04	1.86	0.09	1.09	0.09	0.05	20.20
1	10	M-F	24.89	3.00	0.11	1.58	0.19	0.08	15.77
1	10	M+F	25.14	4.38	0.14	1.58	0.27	0.10	15.94
1	20	C-F	24.07	0.34	0.04	0.99	0.01	0.01	24.30
1	20	F+F	20.27	0.77	0.08	0.97	0.04	0.03	21.13
1	20	M-F	26.28	1.42	0.09	1.74	0.10	0.06	15.11
1	20	M+F	27.23	1.56	0.08	1.58	0.09	0.05	17.23
LSD$_{0.05}$			3.20	1.13	0.03	0.20	0.08	0.02	2.02

2. Cooking Lake

Inherent characteristics of the Alfisol at Cooking Lake restricted the extraction of significant amounts of LFOM to the first layer sampled (Table 6). The increases in LFC in treatments with fertilizer (M+F and F+F) over treatments without (M-F and C-F) were consistent with but smaller than those at Josephburg. The largest differences created through fertilizer addition were measured in the 10 cm cut between F+F and C+F (0.95 Mg ha^{-1}) as well as between M+F and M-F (1.38 Mg ha^{-1}). These differences, however, were considerably smaller than those measured at Josephburg (Table 5). The trends observed for LFN were similar to those of LFC. The proportion of C present as LFC was, in general, less than 10% in treatments without fertilizer (C-F), but increased up to 14% in treatments with manure and fertilizer (M+F; 0 and 10 cm cuts) (Table 6). Similar proportions were observed in soil samples from Josephburg (Table 5).

C. Estimating the Change in Soil C and Total Soil Respiration over Five Years

Availability of initial values of soil C at the initiation of the experiment enabled us to estimate the change in whole soil C (ΔC) under conditions of mass equivalency (Table 7). At Cooking Lake, estimated changes followed approximately expected changes with the exception of the C-F treatment in the 10 cm cut where a substantial loss appeared to have occurred (-4.34 Mg ha^{-1}). At Josephburg, estimated C changes suggested unexpected gains of about 4 Mg ha^{-1} in C-F treatments of the 0 and 10 cm cuts. The C loss of 11.2 Mg ha^{-1} is anomalous and suggests a methodology problem. We expected

Table 7. Carbon storage as a function total C added either in plant parts or cattle manure over five years

Cut	Amendment	Josephburg	Cooking Lake
cm		Mg ha^{-1}	Mg ha^{-1}
0	C-F	4.34	-1.43
0	F+F	3.43	3.37
0	M-F	14.29	5.77
0	M+F	17.90	9.47
10	C-F	4.52	-4.34
10	F+F	8.97	3.16
10	M-F	14.47	10.96
10	M+F	15.83	14.26
20	C-F	-11.16	2.05
20	F+F	4.29	3.25
20	M-F	9.35	12.05
20	M+F	17.32	16.05

not to observe major C changes in control treatments, even though the modified soil profiles in the 10 and 20 cm cut had been abruptly brought to nonsteady state conditions.

We further explored our data using a simple model of C turnover (McGill, 1977). In this model, C enters soil either as plant residues or manure (Figure 1). The initial mass of soil C is apportioned into three compartments: a small (f_1 = 0.03 kg kg^{-1}) and active compartment (biomass C), a larger (f_2 = 0.6 kg kg^{-1}) but less active compartment (humad C), and an intermediate (f_3 = 0.37 kg kg^{-1}) but very stable compartment (resistant C). Plant and animal C decomposes at rate k_1 (2.5 y^{-1}) and enters into the biomass C compartment. Sixty percent (f_4 = 0.5 kg kg^{-1}) of biomass C produces CO_2 at rate k_2 (0.8 y^{-1}). The rest of C is transferred to the humad C (f_5 = 0.3 kg kg^{-1}) and resistant C (f_6 = 0.1 kg kg^{-1}) compartments, each step producing CO_2 at rate k_2. Further respiration from humad and resistant C occurs at rates k_3 (0.02 y^{-1}) and k_4 (0.01 y^{-1}), respectively.

Total respired C simulated over a 5-year period was compared against the 'observed' final soil C subtracted from initial soil C plus all C additions (Figure 2). Simulated and observed values correlated well, but the correlation was higher with the Cooking Lake than with the Josephburg data. Consequently, there was also acceptable agreement between simulated and observed ΔC although the estimated C gains at Josephburg were almost half of the 'observed' gains (Figure 2).

IV. General Discussion

In all, there was evidence in this study that soil organic C of artificially-eroded soils increased, over a 5-y span, with management that favored crop productivity. These management techniques were not special and were based on principles of sound agronomic practice such as use of fertilizers, addition of composted manure, and minimal soil disturbance. In spite of similar productivity, the carbon-mass results were less satisfactory for the Typic Cryoboroll at Josephburg than the Typic Cryoboralf at Cooking Lake. This is consistent with results previously reported by Nyborg et al. (1995) on similar

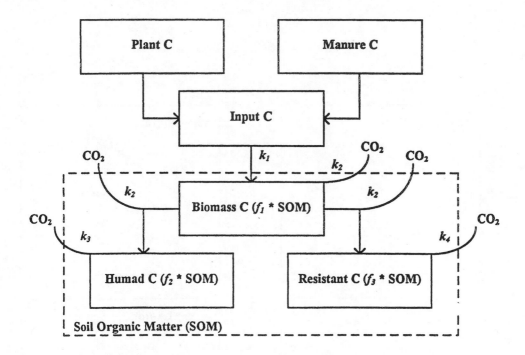

Figure 1. Flow diagram of soil organic matter model. (Adapted from McGill, 1977.)

soils after 11 years of continuous cropping under two tillage regimes. In that study, the high organic matter content of a Typic Cryoboroll (*ca.* 60 g kg^{-1} of organic C) masked much of the influence ascribed to management. Our results also speak of the strong influence that spatial and temporal variability has on C accounting methods.

The LFOM technique was more successful than whole-soil C at detecting increases in C storage. These results agree with those locally reported by Janzen et al. (1992), Bremer et al. (1993), Nyborg et al. (this volume), and Solberg et al. (this volume) but differ from them in the shortness of time in which these differences were detected. This is important if carbon sequestration programs are to be implemented and their results scientifically corroborated (Ellert and Janzen, pers. communication; Izaurralde et al., this volume).

A simple soil organic matter model, applied to the experimental data, predicted annual respiration rates of up to 6.24 Mg ha^{-1} of CO_2-C under productive management systems such as the M+F treatment. Conversely, fertilizer treatments appeared to have emitted considerably less CO_2-C (*ca.* 3.06 Mg ha^{-1}) than manure treatments. Model parameters were not adjusted either to soil properties (e.g., texture) or to environmental variables (e.g., temperature, water content). Despite this, it managed to approximate the patterns of C changes described above. The implication of this is that perhaps models of simple behavior could be suitable for integrating C fluxes across large ecological regions with large combinations of soil and management scenarios (Izaurralde et al., 1996).

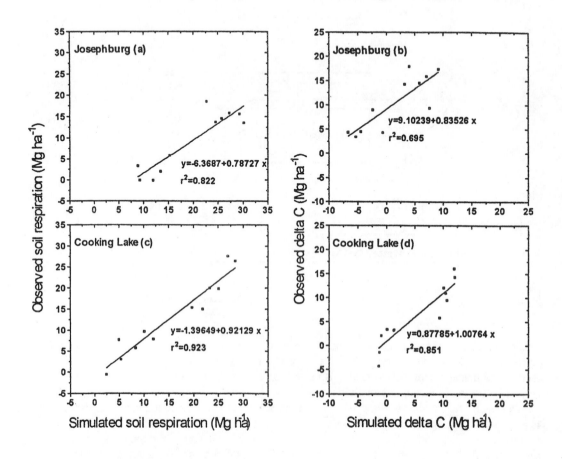

Figure 2. Comparison of observed and simulated soil respiration [(a) and (c)] and C changes [(b) and (d)] after five years of continuous cropping and reduced tillage on two artificially-eroded soils.

V. Summary and Conclusions

This study attempted to document short-term organic-matter changes in two artificially-eroded soils of the Aspen Parkland ecoregion of the Canadian Prairie. Five years of continuous cereal cropping, reduced tillage, and nutrient addition via fertilizers or composted manure favored soil C storage. The accretion results were more evident in the Typic Cryoboralf than in the Typic Cryoboroll tested. Light fraction organic matter was a more sensitive indicator than total soil C at detecting these changes. The increased productivity induced in eroded soils caused a measurable accumulation of light fraction organic matter. A relatively simple model of soil organic matter decomposition accounted for at least

82% in the total variation in soil respiration and 70% of the change in total organic C. We conclude that eroded soils under favorable management could revert their productivity and, at the same time, store atmospheric C in a relatively short period of time. Our results also support the hypothesis that soils most depleted in soil organic matter (e.g., eroded soils) may have the highest potential for C gain. This study also shows that it is possible to measure these changes in soil C in a period of 5 years using sensitive techniques such as light fraction organic matter. The last conclusion is important for developing a scientifically credible trading mechanism of soil C to offset C emissions from other industries.

Acknowledgments

We thank J. Brown, J. DeMulder, B. Hoar, C. Nguyen, J. Thurston, and Z. Zhang for their technical help and Dr. W.B. McGill for valuable discussions. This work was jointly supported by the Agriculture and Agri-Food Canada Greenhouse Gas Research Initiative and the Parkland Agriculture Research Initiative.

References

Bremer, E., H.H. Janzen, and A.M. Johnston. 1994. Sensitivity of total, light fraction and mineralizable organic matter to management practices in a Lethbridge soil. *Can. J. Soil Sci.* 74:131-138.

Bremer, E., B.H. Ellert, and H.H. Janzen. 1995. Total and light-fraction carbon dynamics during four decades after cropping changes. *Soil Sci. Soc. Am. J.* 59:1398-1403.

Buyanovski, G.A., J.R. Brown, and G.H. Wagner. 1996. Soil organic matter dynamics in Sanborn Field (North America). p. 295-300. In: D.S. Powlson, P. Smith, and J.U. Smith (eds.), *Evaluation of Soil Organic Matter Models Using Existing Long-Term Datasets*. NATO ASI Series I, Vol. 38, Springer-Verlag, Heidelberg. 429 pp.

Campbell, C.A. and R.P. Zentner. 1993. Soil organic matter as influenced by crop rotations and fertilization. *Soil Sci. Soc. Am. J.* 57:1034-1040.

Canada Soil Survey Staff. 1989. *Soil Inventory Attribute Map - Alberta: Soil Layer Digital Data. Version 89.09.01.* Canada Soil Survey Staff, Alberta Unit NSDB, CLBRR. Agriculture Canada, Research Branch, Ottawa.

Christensen, B.T. 1992. Physical fractionation of soil and organic matter in primary particle size and density separates. *Adv. Soil Sci.* 20:1-90.

Christensen, B.T. 1996. Matching measurable soil organic matter fractions with conceptual pools in simulation models of carbon turnover: revision of model structure. p. 143-160. In: D.S. Powlson, P. Smith, and J.U. Smith (eds.), *Evaluation of Soil Organic Matter Models Using Existing Long-Term Datasets*. NATO ASI Series I, Vol. 38, Springer-Verlag, Heidelberg. 429 pp.

Cole, V., C. Cerri, K. Minami, A. Mosier, N. Rosenberg, D. Sauerbeck, J. Dumanski, J. Duxbury, J. Freney, R. Gupta, O. Heinemeyer, T. Kolchugina, J. Lee, K. Paustian, D. Powlson, N. Sampson, H. Tiessen, M. van Noordwijk, and Q. Zhao. 1996. Agricultural options for mitigation of greenhouse gas emissions. p. 744-771. In: R.T. Watson, M.C. Zinyowera, and R.H. Moss (eds.), *Climate Change 1995: Impacts, Adaptations, and Mitigation of Climate Change: Scientific-Technical Analyses*. Contribution of Working Group II to the Second Assessment Report of the Intergovernmental Panel on Climate Change. Cambridge University Press, Cambridge and New York. 880 pp.

Coote, D.R. 1984. The extent of soil erosion in western Canada. p. 34-38. In: *Soil Erosion and Land Degradation*. Proc. 2nd Annual Western Provincial Conference. Rationalization of Water and Soil Research and Management. Saskatchewan Inst. of Pedology, Saskatoon, SK.

Dalal, R.C. and R.J. Mayer. 1986. Long-term trends in fertility of soils under continuous cultivation and cereal cropping in southern Queensland. IV. Loss of organic carbon from different density fractions. *Aust. J. Soil Res.* 24:293-300.

Dormaar, J.F., C.W. Lindwall, and G.C. Kozub. 1986. Restoring productivity to an artificially eroded Dark Brown Chernozemic soil under dryland conditions. *Can. J. Soil Sci.* 66:273-285.

Eck, H.V. 1987. Characteristics of exposed subsoil - at exposure and 23 years later. *Agron. J.* 79:1067-1073.

Eck, H.V., V.L. Hauser, and R.H. Ford. 1965. Fertilizer needs for restoring productivity on Pullman silt loam after various degrees of soil removal. *Soil Sci. Soc. Am. Proc.* 29:209-213.

Ellert, B.H. and J.R. Bettany. 1995. Calculation of organic matter and nutrients stored in soils under contrasting management regimes. *Can. J. Soil Sci.* 75:529-538.

Frye, W.W., S.A. Ebelhar, L.W. Murdock, and R.L. Blevins. 1982. Soil erosion effects on properties and productivity of two Kentucky soils. *Soil Sci. Soc. Am. J.* 46:1051-1055.

Izaurralde, R.C., M. Nyborg, E.D. Solberg, S.S. Malhi, and M.C. Quiroga Jakas. 1993. *Relationships of Topsoil Depth to Soil Productivity*. National Soil Conservation Program. Dept. Soil Science, Univ. of Alberta. Edmonton, AB. 47 pp.

Izaurralde, R.C., M. Nyborg, E.D. Solberg, M.C. Quiroga Jakas, S.S. Malhi, R.F. Grant, and D.S. Chanasyk. 1994. Relationships of simulated erosion and soil amendments to soil productivity. p. 15-1,15-5. In: D. F. Acton (ed.), *A Program to Assess and Monitor Soil Quality in Canada*. Centre for Land and Biological Resources Research. Research Branch, Agriculture and Agri-Food Canada. Ottawa.

Izaurralde, R.C., N.G. Juma, J.A. Robertson, W.B. McGill, and J.T. Thurston. 1996. The Breton Classical Plots dataset. p. 351-362. In: D.S. Powlson, P. Smith, and J.U. Smith (eds.), *Evaluation of Soil Organic Matter Models Using Existing Long-Term Datasets*. NATO ASI Series I, Vol. 38, Springer-Verlag, Heidelberg. 429 pp.

Izaurralde, R.C., W.B. McGill, D.C. Jans-Hammermeister, K.L. Haugen-Kozyra, R.F. Grant, and J.C. Hiley. 1996. *Development of a Technique to Calculate Carbon Fluxes in Agricultural Soils at the Ecodistrict Level Using Simulation Models and Various Aggregation Methods*. Agriculture and Agri-Food Canada Greenhouse Gas Initiative. Univ. of Alberta. Edmonton, AB. 67 pp.

Izaurralde, R.C., W.B. McGill, A. Bryden, S. Graham, M. Ward, and P. Dickey. 1997. Scientific challenges in developing a plan to predict and verify carbon storage in Canadian Prairie soils. (*Adv. Soil Sci.*, this volume.)

Janzen, H.H., C.A. Campbell, S.A. Brandt, G.P. Lafond, and L. Townley-Smith. 1992. Light-fraction organic matter in soil from long-term rotations. *Soil Sci. Soc. Am. J.* 56:1799-1806.

Jenkinson, D.S., D.D. Harkness, E.D. Vance, D.E. Adams, and A.F. Harrison. 1992. Calculating net primary production and annual input of organic matter to soil from the amount and radiocarbon content of soil organic matter. *Soil Biol. Biochem.* 24:295-308.

Lal, R. 1995. Global soil erosion by water and carbon dynamics. p. 131-142 In: R. Lal, J. Kimble, E. Lavine, and B.A. Stewart (eds.), *Advances in Soil Science: Soils and Global Change*. Lewis Publishers, CRC Press. Boca Raton, FL.

Larney, F.J., R.C. Izaurralde, H.H. Janzen, B.M. Olson, E.D. Solberg, C.W. Lindwall, and M. Nyborg. 1995. Soil erosion-crop productivity relationships for six Alberta soils. *J. Soil Water Conserv.* 50:87-91.

McGill, W.B. 1977. *Swine Manure Disposal on Land: a Link in the Agricultural Nutrient Cycle*. Supplement to the Proc. of the Alberta Pork Seminar held January 1977 in Banff, Alberta. Edmonton, AB.

Nyborg, M., E.D. Solberg, S.S. Malhi, and R.C. Izaurralde. 1995. Fertilizer N, crop residue, and tillage alter soil C and N contents after a decade. p. 93-100. In: R. Lal, J. Kimble, E. Levine, and B.A. Stewart (eds.), *Advances in Soil Science: Soil Management and Greenhouse Effect*. Lewis Publishers, CRC Press. Boca Raton, FL.

Nyborg, M., M. Molina-Ayala, E.D. Solberg, R.C. Izaurralde, S.S. Malhi, and H.H. Janzen. 1997. Carbon storage in grassland soil and relation to application of fertilizer. (*Adv. Soil Sci.*, this volume.)

Power, J.F., F.M. Sandoval, R.E. Ries, and S.D. Merrill. 1981. Effects of topsoil and subsoil thickness on soil water content and crop production on a disturbed soil. *Soil Sci. Soc. Am. J.* 45:124-129.

Post, W.M., W.R. Emanuel, P.J. Zinke, and A.G. Stangenberger. 1985. Soil carbon pools and world life zones. *Nature* 317:613-616.

Sparrow, H.O. 1984. *Soil at Risk: Canada's Eroding Future*. Report of the Standing Committee on Agriculture, Fisheries, and Forestry. Supply and Services Canada, Ottawa, Ontario. 129 pp.

Solberg, E.D., M. Nyborg, R.C. Izaurralde, S.S. Malhi, H.H. Janzen, and M. Molina-Ayala. 1997. Carbon storage in soils under continuous cereal grain cropping: N fertilizer and straw. (*Adv. Soil Sci.*, this volume.)

Tanaka, D.L. and J.K. Aase. 1989. Influence of topsoil removal and fertilizer application on spring wheat yields. *Soil Sci. Soc. Am. J.* 53:228-232.

Technicon Industrial Systems. 1977. *Individual/Simultaneous Determinations of Nitrogen and/or Phosphorus in Bd Acid Digests*. Industrial method 334-74w/Bt. Technicon Industrial Systems, Tarrytown, NY.

CHAPTER 27

Quantification of Soil Quality

C.A. Seybold, M.J. Mausbach, D.L. Karlen, and H.H. Rogers

I. Introduction

Sustaining soil quality is the most effective method for ensuring sufficient food to support life as we know it. Only after the most basic physiological needs of air, water, and food are satisfied, will humankind begin to consider their safety (Maslow, 1943) and other qualities of life, such as the environment in which they live. Lowdermilk (1953) concluded that the health of a nation's soil directly affects its national security and its people's freedom. He also suggested it is the responsibility of all peoples within a nation to safeguard the integrity of the soil resource.

Daily (1995) estimated that about 43 percent of the earth's vegetated surface has diminished capacity to supply benefits to humankind as a result of improper land use decisions. Willis and Evans (1977) stressed as a national resource, soil is not renewable over the short term, and most definitely not within a generation. They suggested, depending upon the kind of soil, it may take from 30 to more than 1000 years to form one inch of top soil by natural processes. Soil is a resource most people do not think about on a daily basis, especially as society becomes more diversified and fewer people are directly involved in agriculture (Haberern, 1992). Often it takes a catastrophic event, such as the droughts and subsequent dust storms of the 1930s; widespread, highly erosive rainfall events; or famine to highlight the importance of a high-quality soil resource.

The National Research Council (NRC), in their book on soil and water quality (National Research Council, 1993), stated "protecting soil quality, like protecting air and water quality, should be a fundamental goal of national policy." They linked soil quality to water quality and suggested enhancement of soil quality should be the first step toward increasing water quality. The NRC also suggested that methods to measure soil quality should be identified or developed, and indicators of soil quality should be used for monitoring and predicting changes in it. Cox (1995) reinforced those goals by suggesting that management and protection of soil resources should become a cornerstone of national policy with respect to natural resources and the environment.

The objectives of this chapter are to discuss some different approaches to quantifying soil quality and to recommend a framework for measuring and assessing it. The definition of soil quality is fundamental in the development of quantitative measures; therefore, we discuss the evolution of the definition and its implications. A discussion of indicators based on minimum data sets (MDS) is included and interrelationships among indicators and other factors, such as climate and land use patterns are explored. Scale is an important issue in the development of appropriate indicators. We discuss the topic from two points of view: (1) selecting indicators to match the scale of assessment, and (2) expanding from point scale indicators to larger scales using statistical procedures and related spatial information on soils, land use, and management systems. Finally, we provide an overview of soil quality indices and a framework for assessing soil quality.

II. Defining Soil Quality

The term and concept of soil quality evokes various responses depending on our scientific and social backgrounds. For some, soil quality evokes an ethical or emotional tie to the land much like that of Leopold (Flader, 1995), Lowdermilk (1953), and Bennett (1928). To others, soil quality is an integration of soil processes and provides a measure of change in soil condition as related to factors such as land use, climate patterns, cropping sequences, and farming systems (National Research Council, 1993; Doran and Parkin, 1994; Doran et al., 1994; Karlen and Stott, 1994; Larson and Pierce, 1994). To the land manager or farmer, soil quality is often viewed as soil health; some farmers have asked the question, how is my farming system impacting the health of my soil (Romig et al., 1995). In the scientific community, soil quality and soil health are used and defined synonymously (Doran and Parkin, 1994; Larson and Pierce, 1994; Doran et al., 1994; Harris and Bezdicek, 1994; Karlen and Stott, 1994; Acton and Gregorich, 1995; Karlen et al., 1997b). In a national context, soil quality provides a foundation for national policy to protect the environment (National Research Council, 1993; Howard, 1993; de Haan et al., 1993).

Historically, soil quality meant suitability or limitations of a soil for a particular use (Warkentin and Fletcher 1977; Soil Survey Division Staff, 1993). Warkentin and Fletcher (1977) also discussed soil quality from the perspective of soils having value in the biosphere. They concluded that the soil quality concept has both intrinsic and current use components. Warkentin (1995) and others (Izac and Swift, 1994; National Research Council, 1993) have discussed how the concept has changed and been adapted to meet current interests and increased knowledge of soils. They concluded that soil quality is a key element for evaluating the sustainability of agricultural systems. Presently, soil quality has been defined by some scientists as the *"fitness for use"* (Pierce and Larson, 1993; Acton and Gregorich, 1995), and by others as the *"capacity of the soil to function"* (Doran and Parkin, 1994; Karlen et al., 1997b). We use the latter and the definition of soil quality as proposed by Karlen et al. (1997b):

> *"The capacity of a specific kind of soil to function, within natural or managed ecosystem boundaries, to sustain plant and animal productivity, maintain or enhance water and air quality, and support human health and habitation."*

Soil function describes what the soil does. Larson and Pierce (1991) defined three major functions of soil: (1) a medium for plant growth, (2) regulating and partitioning of water flow through the environment, and (3) serving as an effective environmental filter. We will use the five soil functions as described by Karlen et al. (1997b), which are:

1. *Sustaining biological activity, diversity, and productivity;*
2. *Regulating and partitioning water and solute flow;*
3. *Filtering, buffering, degrading, immobilizing, and detoxifying organic and inorganic materials, including industrial and municipal by-products and atmospheric deposition;*
4. *Storing and cycling nutrients and other elements within the earth's biosphere; and*
5. *Providing support of socioeconomic structures and protection for archeological treasures associated with human habitation.*

Collectively, the functions are evaluated with respect to their capacity to sustain plant and animal productivity, maintain or enhance water and air quality, and support human health and habitation in determining soil quality (Doran and Parkin, 1994).

Quality with respect to soil can be viewed in two ways: (1) as inherent properties of a soil; and (2) as the dynamic nature of soils as influenced by climate, and human use and management (Larson and Pierce, 1991; Pierce and Larson, 1993). With respect to inherent properties, a soil is a result of the factors of soil formation — climate, topography, vegetation, parent material, and time (Jenny, 1941). Each soil, therefore, has an innate capacity to function, e.g., some soils will be inherently more productive or will be able to partition water much more effectively than others. This view of the definition is useful for comparing the abilities of one soil against another, and is often used to evaluate the worth or suitability of soils for specific uses. In the National Cooperative Soil Survey Program, soil qualities are defined as inherent characteristics or properties of a soil, such as texture, slope, structure, and soil color (Soil Survey Division Staff, 1993).

The second view of quality relates to the dynamic nature of soils as influenced by human use and management. For example, a farming system that does not protect the surface layer from erosion results in the loss of clay and other finer sized soil particles, organic matter, nutrients, and other beneficial properties. In most cases, this eroded soil would be functioning at less than its original potential for agricultural production, and its condition or health would be considered impaired or of a lesser quality. This view of soil quality requires a reference condition for each kind of soil with which changes in soil condition are compared, and is currently the focal point for the term "soil quality" (National Research Council, 1993; Larson and Pierce, 1994; Doran et al., 1996; Acton and Gregorich, 1995; Karlen et al., 1997b; Howard, 1993; de Haan et al., 1993).

III. Soil Quality Indicators

Soils have chemical, biological, and physical properties that interact in a complex way to give a soil its quality or capacity to function (genesis and classification). Thus, soil quality cannot be measured directly, but must be inferred from measuring changes in its attributes or attributes of the ecosystem, referred to as indicators. Indicators of soil quality should give some measure of the capacity of the soil to function with respect to plant and biological productivity, environmental quality, and human and animal health. They should also be used to assess the change in soil function within land use or ecosystem boundaries. Doran and Parkin (1994) have defined a set of specific criteria that indicators of soil quality must possess; they should (1) encompass ecosystem processes and relate to process oriented modeling, (2) integrate soil physical, chemical, and biological properties and processes, (3) be accessible to many users and applicable to field conditions, (4) be sensitive to variations in management and climate, and (5) where possible, be components of existing soil data bases. Also, indicators should be easily measured and measurements should be reproducible (Gregorich et al., 1994). Arshad and Coen (1992) also suggest that indicators should be sensitive enough to detect changes in soil as a result of anthropogenic degradation.

It would be unrealistic to use all ecosystem or soil attributes as indicators, so a minimum data set (MDS) consisting of a core set of attributes encompassing chemical, physical, and biological soil properties are selected for soil quality assessment (Larson and Pierce, 1991). Several core minimum data sets have been proposed (Larson and Pierce, 1991; Arshad and Coen, 1992; Doran and Parkin, 1994; Gregorich et al., 1994; Larson and Pierce, 1994; Papendick et al., 1995). We propose the minimum data set as described in Table 1. The suite of indicators used for assessing soil quality can vary from location to location depending on the kind of land (e.g., rangeland, wetland, agricultural land) or land use, soil function, and the soil forming factors (Arshad and Coen, 1992; Hellkamp et al., 1995). Different kinds of land uses may require increased capacities of certain soil functions and, thus, require certain indicators over others for assessment. For example, a wetland soil has a different role than a agricultural soil for the soil function, partitioning of water. Therefore, infiltration would not be a useful indicator of wetland soil quality, but would be very useful for assessing the quality of most agricultural soils.

Table 1. Proposed minimum data set of physical, chemical, and biological indicators for screening the quality or health of soils

Indicators	Relationship to soil condition and function: rationale as a priority measurement
Physical	
Texture	Retention and transport of water and chemicals; modeling use, soil erosion and variability estimate
Depth of soil and rooting	Estimate of productivity potential and erosion; normalizes landscape & geographic variability
Infiltration and bulk density	Potential for leaching, productivity, and erosivity; bulk density: SBD needed to adjust analyses to volumetric basis
Water holding capacity	Related to water retention, transport, and erosivity; available H_2O; calculate from SBD, texture, and OM
Chemical	
Soil organic matter (OM)	Defines soil fertility, stability, and erosion extent; matter (OM); use in process models and for site normalization
pH	Defines biological and chemical activity thresholds; essential to process modeling
Electrical conductivity	Defines plant and microbial activity thresholds; presently lacking in most process models
Extractable N, P, and K	Plant available nutrients and potential for N loss; productivity and environmental quality indicators
Biological	
Microbial biomass C and N	Microbial catalytic potential and repository for C and N; modeling: early warning of management effects on OM
Potentially mineralizable N	Soil productivity and N supplying potential; mineralizable N; process modeling (surrogate indicator of biomass)
Soil respiration	Microbial activity measure (in some cases plants); process modeling; estimate of biomass activity

(After Doran et al., 1996 and Larson and Pierce, 1994.)

A. Soil Organic Matter

Soil organic matter (SOM) includes a number of fractions such as the light fraction, microbial biomass, water-soluble organics, and humus (stabilized organic matter) of soils (Stevenson, 1994). It is one of the more useful indicators of soil quality, because it interacts with other numerous soil components; affecting water retention (Hudson, 1994), aggregate formation, bulk density, pH, buffer capacity, cation exchange properties, mineralization, sorption of pesticides and other agrichemicals, color (facilitate warming), infiltration, aeration, and activity of soil organisms (Schnitzer, 1991). It is the interaction of the various components of a soil that produce the net effects and not organic matter acting alone (Stevenson, 1994). In addition to the amount of SOM, its quality is also an important indicator of soil quality. For example, organic matter derived from manure differed in quality from organic matter derived from fertilized plots after 90 years of cropping (Schjonning et al., 1994). The fertilized plots contained organic matter with a greater chemical reactivity, which was attributed to a larger amount in the silt-sized fraction. The biologically active components of SOM (e.g., constituents of soil microbial biomass, and energy sources like organic C and N) have been shown to be some of the more sensitive indicators of initial changes in soils due to management (Kennedy and Papendick, 1995). Because organic matter can have a tremendous effect on the capacity of a soil to function, it has been recommended to be a basic component in every minimum data set for assessing soil quality (Gregorich et al., 1994).

B. Descriptive Indicators

Descriptive indicators, which are inherently qualitative, can be used in assessing soil quality. Arshad and Coen (1992) list several observational and morphological indicators; these include soil crusting/surface sealing, rills, gullies; ripple marks, sand dunes, salt crusting, and standing or ponded water. When farmers are asked to describe the quality of their soil, they use descriptive terms of the senses such as loose, soft, crumbly, loamy, earthy smell, darkly colored, massive, lumpy, dense, and so on (Romig et al., 1995). Farmers tend to rely more heavily on what their senses tell them about soil quality, and how that relates to ease of tillage and crop productivity, more than any one particular measurement or response.

C. Production Indicators

Crop yield (grain or biomass production), plant vigor, rooting patterns, and other aspects of the crop have been used as indicators of soil quality (Parr et al., 1992; Acton and Gregorich, 1995). Crop yield is an important indicator because it gives information about the interacting soil properties of the system as whole as in a bioassay. Kinako (1983) used growth performance of an indicator species (*Caesalpinia pulcherrima*) for the assessment of soil quality. Growth performance of the indicator species was positively correlated with the nutrient status of the soils. This shows promise for using indicator plant species (bioassay) for assessing soil quality. However, it should be noted that productivity should not be the only measure of soil quality, since environmental and animal health aspects of soil also need to be considered.

D. Ecosystem Processes

An ecosystem approach for rapidly assessing changes in soil quality has been suggested by Visser and Parkinson (1992). Their approach focuses on the processes involved in cycling organic matter C and nutrients (mainly N), and on nutrient retention efficiencies. In C cycling, indicators would include organic matter decomposition, soil respiration, and microbial biomass C.

System entropy has been suggested as an indicator of sustainability by Andiscott (1995). In this approach, the soil-plant system is considered to be thermodynamically open with respect to its exchange of energy and matter. In systems where entropy is increasing, processes are degrading the complex, i.e., ordered structure of large molecular weight molecules to small molecules. An example is the oxidation of organic matter to CO_2, NH_3, and H_2O. Conversely, photosynthesis builds small molecules into larger ones and the accumulation of C represents systems that are aggrading and entropy is decreasing. Andiscott (1995) also suggested that minimum production of entropy should be a criterion for sustainability, and an audit of small molecules vs. large molecules could be a way of measuring sustainability. Many of the soil quality indicators in the MDS (Table 1) can indicate the direction of entropy production over time, such as changes in organic matter content, organic N, aggregate stability, microbial biomass, and respiration. A steady state condition of entropy represents a sustainable system (Andiscott, 1995; Elliott et al., 1994).

IV. Scale

The dimensions of scale influence soil quality assessments in both space and time. The area of consideration can be as small as a point on the landscape, a research plot, or as large as a nation or the world. Time frames are important because of the effect of climate, soil moisture conditions, human actions, stage of plant growth, and other factors that give rise to temporal variability in indicator status. Response time for an individual indicator to change (as a result of management), determines the appropriate time interval for measuring indicator changes. Doran and Parkin (1994), Larson and Pierce (1991), and Lal (1994) have discussed considerations of temporal variability with respect to soil quality indicators.

Two approaches dealing with spatial scale have been utilized by researchers; they are to: (1) select indicators to match the geographic scale for which the soil quality assessment is being made; and (2) expand point scale indicator information to larger areas of consideration.

A. Generalization of Indicators

Karlen et al. (1997a) discusses the concept of selecting more general soil quality indicators to match larger scales of assessment. Their approach is similar to that of Hole and Campbell (1985) for scaling maps based on taxonomic generalization. Both authors caution, however, that evaluations at the farm, watershed, and regional scales will become progressively less precise because of fewer actual measurements and a greater dependence on simulation models, remote sensing, and existing databases to estimate measurements.

The loss of precision and detail in soil quality assessment is one of the major disadvantages of this approach. For example, if productivity becomes unstable, assessments of the reasons for instability are difficult, if not impossible, to make without additional data collection. However, the main advantage of this approach is the economy of resources needed to collect the data. Table 2 presents a list of possible indicators for soil quality assessments at the point, field, farm, or watershed scales, and regional or national scales.

B. Expansion of Point Scale Indicator Information

Point scale indicators provide the most detail, but are also the most expensive and time consuming to gather and interpret. Therefore, it becomes necessary to develop methods to generalize or relate information from point scale to the scale under consideration; whether it is a field, watershed, or an entire region. At the field or farm scale, geostatistical methods (Smith et al., 1993, 1994; Burrough, 1989; Bouma, 1988) and landscape models (Pennock et al., 1994) have been used to assess changes in soil quality.

Table 2. Potential indicators of soil quality at different scales

Biological	Chemical	Physical
	Field, farm, or watershed indicators	
Crop yield	Changing quality/quanity of OM	Topsoil thickness
Crop appearance	pH changes	Soil color
Weed pressure	Available P and K changes	Subsoil exposure
Nutrient deficiencies	Change in cation levels	Compaction
Earthworms	Change in N availability	Crusting
Root growth patterns	Change in heavy metals	Infiltration
Biomass production	Change in salinity	Runoff
Canopy cover	Nutrient loss to streams and groundwater	Sediment fans
		Surface cover
	Pesticide loss to streams and groundwater	Rill and gully erosion
		Plant emergence
		Ease of tillage
		Soil structure
	Regional or National Indicators	
Productivity (yield stability)	Organic matter trends	Desertification
Taxonomic diversity at the group level	Acidification	Vegetation cover
	Salinization	Water erosion
Species richness/diversity	Changes in water quality	Wind erosion
Keystone species and ecosystem engineers	Changes in air quality	Siltation of rivers and lakes
		Sediment load in rivers
Biomass, density, and abundance		

(After Karlen et al., 1997a.)

At the regional and national scales, a longitudinal survey of soil, water, and related resources to assess conditions and trends at various time intervals within the U.S. has been employed by the National Resources Inventory (NRI) of the Natural Resources Conservation Service. For a complete discussion of the statistical procedure, readers are referred to Nusser and Goebel (1997). Briefly, the NRI incorporates a flexible two-stage sampling design that accommodates varying intensities for a national survey, as well as more detailed designs for regional or watershed surveys. Extending point scale information to larger geographic areas requires using expansion factors to establish each point within the NRI statistical framework. The statistical design intensity is applicable to multi-county watersheds, regional and national scales. The Environmental Protection Agency used a similar sampling approach in their Environmental Monitoring and Assessment Program (Hellkamp et al., 1995). They chose a probabilistic sampling design over time and space to provide a cost-effective approach for comparing the status and trends in ecological condition among multiple resources at the regional and national levels.

Other geospatial information, such as soil survey information, land use, climate, and other data of compatible scales, can be used to extend the usefulness of point data in the NRI. The use of National Cooperative Soil Survey soil maps to extend point information has been discussed (Hole and Campbell, 1985; Bouma, 1988; Bliss and Reybold, 1990). In this approach, soil at each NRI point is identified, and then information collected at the point is extended using geospatial information from soil survey databases (Reybold and TeSelle, 1989).

We prefer using a probabilistic sampling similar to that of the NRI and coupling the information gathered at the points with other geospatial information. This approach requires more resources for measurement of indicators and analysis of results. However, it also provides more detailed information for quantifying soil quality and interpreting the causes and effects of changes in soil quality. This approach also makes full use of existing geospatial information in the process of expanding point data to an area basis, whether it is at the farming system level or the national level.

V. Soil Quality Indices

One way to integrate information obtained from MDS measurements is to develop a soil quality index. Such an index could be used to monitor and predict the effects of farming systems and management practices on soil quality, or could provide early signs of soil degradation (Parr et al., 1992). Granatstein and Bezdicek (1992) suggest the need for a soil quality index that reflects both the general potential for human use and unique biophysical conditions of a specific location. The first concept of a soil quality index was introduced by Rust et al. (1972) who related soil quality to the environmental impacts of agrochemicals (e.g., soluble N fertilizers). Both quantitative and qualitative soil quality indices have been proposed.

A. Qualitative

Qualitative measures of soil quality tend to be more subjective in their measurement, but can be assessed more easily, and sometimes be more informative to the land manager. Based on the farmer's perception of soil quality, a scorecard was developed for assessing soil quality (Harris and Bezdicek, 1994; Romig et al., 1995; Romig et al., 1997). The scorecard is a farmer-based subjective rating system that places indicators into rating scales of healthy, impaired, and unhealthy. It has 43 soil health indicator properties that integrate observations made throughout the growing season. The indicators are almost exclusively based on sensory observations (e.g., look, feel, smell). Currently, the score card does not recognize the relative importance of indicators, and is only developed for cropping systems in Wisconsin. Modifications of the scorecard to encompass other regions and cropping systems would require structured input from additional farmers (Romig et al., 1997).

To evaluate soil quality as part of the National Cooperative Soil Survey program, an index based on field morphology of the 0 to 30 cm zone was developed (Harms et al., 1995). Properties evaluated include depth, horizon designation, texture (field estimated), water state, structure (including soil crusts), consistence (rupture resistance), and size and quantity of macropores for each horizon described (Soil Survey Division Staff, 1993). Structure, consistence, and macropores are placed in classes with ratings and further evaluated to obtain a single weighted index for each 10 cm depth increment to 30 cm. The indices are then averaged into a single index to represent the quality of the soil. The index is a relative measure of soil quality that can be used to compare soils and to evaluate the same soil over time. This method requires soils to be moderately moist or wetter for evaluation, which might limit its application, but other protocols could be established and added depending on soil use and need (Harms et al., 1995).

B. Quantitative

Larson and Pierce (1991) suggest a concept for quantifying soil quality by expressing soil quality (Q) as a function of measurable soil attributes referred to as soil qualities (q_i): $Q = f(q_{1...n})$, with the magnitude of Q being a function of the collective contribution of all q_i values. They also measured the change in soil quality over time (dQ/dt) and proposed the use of a minimum data set (MDS) of measured soil properties and pedotransfer functions for assessing soil quality. Pierce et al. (1983) used similar concepts in combination with a productivity model developed by Kiniry et al. (1983) to quantify soil productivity and the change in productivity with soil erosion.

A soil tilth index has also been developed and is based on using five physical soil properties or indicators — bulk density, cone index, aggregate size uniformity coefficient, organic matter, and plastic index (Singh et al., 1992). A normalized tilth coefficient is computed for each indicator and is represented by a second order polynomial. The index is a multiplicative combination of the tilth coefficients. This framework for the tilth index could be expanded to include other soil attributes (chemical and biological) and used as part of a soil quality index.

Smith et al. (1993) developed an approach (multiple indicator kriging procedure) that integrates an unlimited number of soil quality indicators, measured spatially, into an overall soil quality index, and then uses that procedure to evaluate soil quality across landscapes. A kriging procedure is used to estimate index values at unsampled locations, which represent the probability that unknown values meet or do not meet the specified criteria for good soil quality. The result is a landscape map showing the probability of having good soil quality according to predetermined criteria. These maps can then be used for land use planning, and compared over time for changes in soil quality (Smith et al., 1994).

Doran and Parkin (1994) presented a framework for the evaluation of soil quality based on the function of soil with respect to (1) sustainable production, (2) environmental quality, and (3) human and animal health. They proposed an index of soil quality as a function of six specific soil quality elements — food and fiber production, erosivity, groundwater quality, surface water quality, air quality, and food quality. The functions of a soil are assessed for each element. They suggested using weighting coefficients for each element and the multiplicative approach that Singh et al. (1992) developed for the soil tilth index. With this in mind, a functional relationship between measurable soil attributes (e.g., organic matter content, infiltration) and the five soil functions must be developed for the index to be operational. They suggest the use of algorithms from existing models as a starting point for further development.

Karlen and Stott (1994) developed a framework for quantifying soil quality using multi-objective analysis principles of systems engineering. They defined critical soil functions and potential chemical and physical indicators of those functions. For each indicator, a scoring function, and realistic baseline and threshold values are established. The scoring function uses similar mathematical principles as assigning membership in a fuzzy set (Burrough, 1989). All indicators affecting a particular soil function are grouped together and assigned a relative weight based on importance. After scoring each indicator, the value is multiplied by the appropriate weight, and an overall soil quality rating is calculated by summing the weighted score for each soil function. The systems engineering approach to quantifying soil quality was tested in two long-term (10 yr) studies that attempt to show the effects of tillage and crop residue effects on soil quality ratings; the results showed that no-till management had an improved soil quality rating over plow and chisel till (Karlen et al., 1994a), and double crop residue treatment had an improved soil quality rating over removal and normal residue treatments (Karlen et al., 1994b). This approach to quantifying soil quality is suggested as a starting point for the development of a soil quality index. We incorporate this approach in our framework for assessing soil quality.

VI. Framework for Measuring Soil Quality

Soil, unlike air and water, is a highly complex and buffered system for which there are no single or "easy" measures of its quality or health. To assess soil quality, indicators (soil properties) are usually linked to soil function (Howard 1993; Doran and Parkin 1994; Larson and Pierce 1994; Acton and Gregorich, 1995; Karlen et al., 1996b; Karlen and Stott, 1994; Doran et al., 1996; Hornung, 1993).

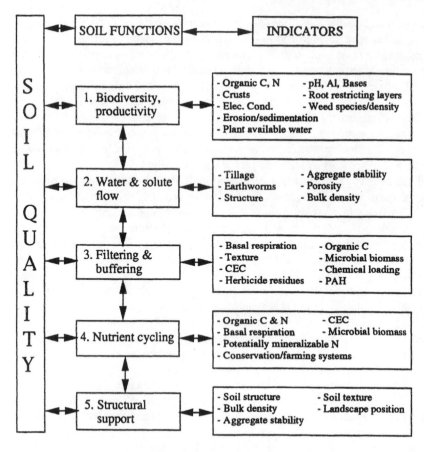

Figure 1. Graphical representation of the concept of soil quality using soil functions and indicators of soil quality.

Izac and Swift (1994) describe a stepwise approach in defining measures of soil quality based on a set of interactive attributes rather than an absolute quality; the framework is shown in Figure 1. Together, the five functions define the capacity of a soil to function, while measurable soil properties indicate the level of soil functioning (Karlen et al., 1994a). Many of the indicators are not unique to a single function, but work in an integrative manner with other indicators or properties of the soil. Once the indicators of soil function are selected, reference or baseline values for these indicators need to be established for comparison, so that a soil's degree of excellence can be ascertained (Larson and Pierce, 1991). We describe two approaches for establishing a reference condition for quantifying and assessing soil quality.

A. Monitoring of Trends

Monitoring soil quality trends requires establishing baseline values for the various indicators, and measuring change in those indicators over time. Changes in the indicators reflect the combined effects of land use and climate. Soil quality, or individual indicators, can be evaluated by trend lines as described by Pierce and Larson (1993) and Larson and Pierce (1994). If the change in a soil quality indicator is positive, and more is of better quality, then the soil can be regarded as improving or

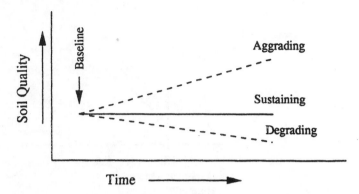

Figure 2. Monitoring trends in soil quality over time can result in aggrading, sustaining, or degrading soil conditions.

aggrading in quality with respect to that indicator (Figure 2). Conversely, if the trend line is negative for that indicator, then quality is degrading. A no-change trend would indicate a sustaining system. Calculating the slope of the trend lines is a way of quantifying change in soil quality or its indicators. A disadvantage of this approach is that it requires measurement of indicators for at least two points in time and therefore, does not provide immediate assessment of soil quality. In addition, this approach is somewhat misleading if a soil is functioning at the highest level attainable and cannot improve, or if it is functioning at its lowest level and cannot go lower; both cases show a static or no-change trend, indicating a sustaining system, but their qualities are completely different. Therefore, it is always desirable to show the state of soil quality on a normalized scale in addition to trends. Pierce and Larson (1993) and Larson and Pierce (1994) describe the use of indicator limits to show if a soil is functioning out of control.

Monitoring trends is useful for the plot, field, and farm scales of evaluation. The land manager or researcher has a relatively small area of consideration and, for the most part, is concerned about the effects of a particular management system or farming system on soil quality over time. They also routinely sample the soil for nutrient requirements and would be able to easily add additional analyses as suggested within the MDS.

B. Reference Values

Indicator reference values are established that represent a soil functioning at full potential, similar to the approaches of Stasch and Stahr (1993), MacDonald et al. (1995), Lal (1994), and Karlen et al. (1994a). Full potential is defined as a soil functioning at its maximum capacity; maximum capacity is defined by land use, soil type, climate, and system inputs (e.g., liming, irrigation). Potential scenarios for a soil functioning at maximum capacity are (1) the native state, (2) the intensively managed state (the capacity to function is usually lower than native state), or (3) the altered state (the capacity to function is usually higher than native state). The scenarios represent three different situations for which a specific kind of soil could have reference values developed. Reference values are developed for each indicator and consist of a maximum value and baseline value, and values in between are target values for good soil quality (Figure 3). The maximum value represents the maximum potential that can be obtained by an indicator under a given land use, and the baseline value represents the minimum acceptable potential (Karlen et al., 1994a). For soils in their native condition, reference values represent the inherent ability of a soil to function as defined by the soil forming factors and processes.

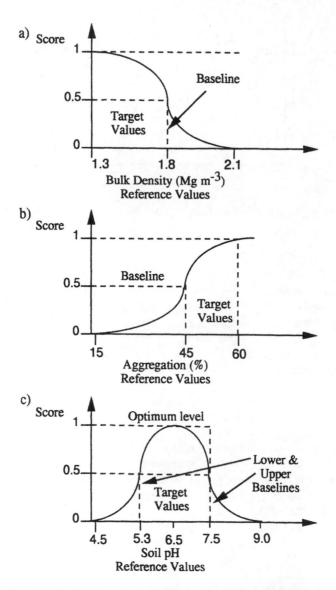

Figure 3. Examples of scoring functions for soil quality indicators developed from reference values. Three situations are shown: a) less is better; b) more is better; c) optimum levels. (After Karlen et al., 1994a.)

The effects of land use change the soil from its native condition, requiring a different set of reference values. For example, human induced changes may result in accelerated erosion resulting in loss of topsoil and rooting depth, lowering its productive potential. On the other hand, reclamation of salt-affected soils, through drainage and irrigation practices, results in leaching of excess salts and renders the soils suitable for intensive agricultural production. In either case, human use has changed the original soil into a new soil. Impact of land use requirements and various management systems have altered soil processes and soil condition (Figure 4). Soils that are intensively managed may be

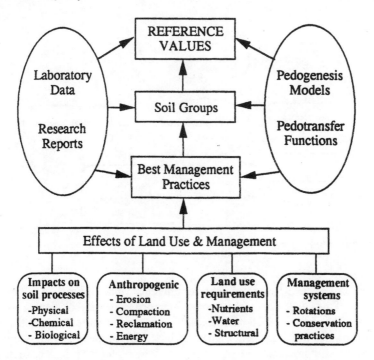

Figure 4. An approach for developing reference values for soil quality indicators is shown.

functioning at their maximum capacity, but are usually functioning at a lower potential than their native condition. Conversely, human reclamation activities can raise the maximum capacity of soils to function from their native state, e.g., reclaimed soils, drained soils, limed soils. Initial reference values are adjusted to reflect human impacts and requirements.

Ideally, reference values could be developed for each soil series and land use. However, development of reference values for more than 18,000 soil series in the U.S. would be monumental, especially with the need to stratify by land use. A more practical approach is to develop reference values for a group of soils that function similarly or are genetically similar as defined by the National Cooperative Soil Survey (Soil Survey Division Staff, 1993). This group of soil may be taxonomically similar, or simply soils of a similar climate and parent material. For example, soils could be grouped at the family level in Soil Taxonomy (Soil Survey Staff, 1975), e.g., fine-silty, mixed, mesic Typic Hapludolls. Understanding the pedogenic processes of a group of similar soils is necessary for development of reference values. A logical point to begin determining reference values for indicator properties is to look at pedogenic models of the soils (Hoosbeck and Bryant, 1992). These models usually address the native conditions under which a soil developed, and are only the starting point from which to address the impacts of land use. Reference values can also be obtained by summarizing data from the literature, research reports, soil survey characterization data, and knowledge about the pedogenesis of a soil as described by Stasch and Stahr (1993) and Karlen et al. (1994a). Laboratory and research data are important in developing these initial reference values. Pedotransfer functions can also be used to generate reference values (Bouma and van Lanen, 1987). Pierce and Larson (1993) discuss the use of pedotransfer functions in establishing limits for soil quality indicators. In Figure 4, we show a framework for developing reference values for soil quality indicators. At the core of Figure 4 is the group of soils that function similarly for a specific land use.

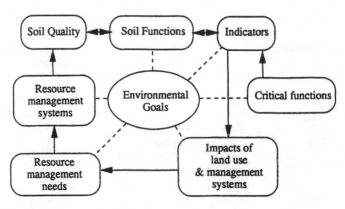

Figure 5. A soil quality management framework.

Scoring functions can be developed from indicator reference values for each group of similar soils. In Figure 3, examples of the endpoints for developing reference values and displaying them as scoring functions are shown. The scoring functions presented do not represent real situations.

In Figure 5, we show a soil management framework based on soil quality, and relationship to soil functions and indicators. Environmental goals are shown as the center for soil management decisions. On the right side, indicators are chosen that represent critical soil functions for various land uses. These indicators are used to monitor the impacts of land use and management systems on the soil resource. Land use impacts are then used as a feedback loop to determine resource management systems that are needed to protect and enhance the quality of the soil resource. This soil quality based framework demonstrates that monitoring of various management decisions is central to evaluation of how a system impacts soil quality.

VII. Summary

Humankind is dependent on a quality soil resource for its survival; therefore, it is a goal of national policy to protect this natural resource. Quantifying soil quality is necessary for defining and managing a quality soil resource. To assess soil quality, indicators that reflect the condition of an effectively functioning soil are established. A core set of indicators are suggested for incorporation into a minimum data set, encompassing biological, chemical and physical attributes of the soil. Components of soil organic matter, productivity indices as well as ecosystem processes and sensory and morphological descriptors have been used as soil quality indicators. In developing indicators, the scale of evaluation needs to be considered. Indicators can be adjusted to match the scale of assessment, or can be expanded to larger scales using various statistical procedures. A few soil quality indices have been generated from manipulation and integration of indicators in some MDSs, and have used both descriptive and analytical indicators of quality employing systems engineering, geostatistics, weighting coefficients, multiplicative approaches, and functional analyses.

Two methods of assessing soil quality from a MDS are presented. In the first, baseline values for indicators are established and monitored for changes over time. In the second, reference values for indicators are established, which represent a soil functioning at full potential within its inherent capacity. Reference values for indicators are suggested along with a framework for their development, which include using models, pedotransfer functions, historical literature values, values in data bases and research publications.

References

Acton, D.F. and L.J. Gregorich. 1995. Understanding soil health. p. 5-10. In: D.F. Acton and L.J. Gregorich (eds.) *The Health of our Soils: Toward Sustainable Agriculture in Canada*. ZCentre for Land and Biological Resources Research, Research Branch, Agriculture and Agri-Food Canada, Ottawa, Ontario.

Andiscott, T.M. 1995. Entropy and sustainablility. *European J. Soil Sci.* 46:161-168.

Arshad, M.A. and G.M. Coen. 1992. Characterization of soil quality: Physical and chemical criteria. *Am. J. Altern. Agric.* 7:25-31.

Bennett, H.H. 1928. Soil erosion a national menace. Circular No. 33. USDA.

Bliss, N.B. and W.U. Reybold. 1989. Small-scale digital soil maps for interpreting natural resources. *J. Soil Water Cons.*. 44:30-34.

Bouma, J. 1988. Land qualities in space and time. p. 3-13. In: J. Bouma and A.K. Bregt (eds.) *Land Qualities in Space and Time*. Proc. of a symposium organized by the International Soc. Soil Sci., Wageningen, The Netherlands. 22-26 August, 1988.

Bouma, J. and H.A.J. van Lanen. 1987. Transfer functions and threshold values: From soil characteristics to land qualities. p. 106-111. In: K.J. Beek, P.A. Burrough, and D.E. McCormack (eds.) *Proc. ISSS/SSSA Workshop on Quantified Land Evaluation Procedures*. Int. Inst. for Aerospace Surv. and Earth Sci., Publ. No. 6. Enschede.

Burrough, P.A. 1989. Fuzzy mathematical methods for soil survey and land evaluation. *J. Soil. Sci.* 40:477-492.

Cox, C. 1995. Soil quality: Goals for national policy. *J. Soil Water Cons.* 50:223.

Daily, G.C. 1995. Restoring value to the world's degraded lands. *Science* 269:350-354.

de Haan, F.A.M., W.H. van Riemsdijk, and S.E.A.T.M. van der Zee. 1993. General concepts of soil quality. p. 155-170. In: H.J.P. Eijsackers and T. Hamers (eds.) *Integrated Soil and Sediment Research: A Basis for Proper Protection*. Proceedings of the First European Conference on Integrated Research for Soil and Sediment Protection and Remediation (EUROSOL), Maastricht, The Netherlands, 1-12 Sept. 1992. Kluwer Academic Publ., Dordrecht.

Doran, J.W. and T.B. Parkin. 1994. Defining and assessing soil quality. p. 3-21. In: J.W. Doran, D.C. Coleman, D.F. Bezdicek, and B.A. Stewart (eds.), *Defining Soil Quality for a Sustainable Environment*. SSSA Spec. Pub. No. 35. ASA, CSSA, and SSSA, Madison, WI.

Doran, J.W., M. Sarrantonio, and R. Janke. 1994. Strategies to promote soil quality and soil health. p. 230-237. In: C.E. Pankhurst, B.M. Doube, V.V.S.R. Gupta, and P.R. Grace (eds.), *Soil Biota: Management in Sustainable Farming Systems*. Proc. OECD Intern. CSIRO, Melbourne, Victoria, Australia.

Doran, J.W., M. Sarrantonio, and M.A. Liebig. 1996. Soil health and sustainability. p. 1-54. In: D.L. Sparks (ed.) *Advances in Agronomy*, Vol. 56. Academic Press, San Diego, CA.

Elliot, E.T., H.H. Janzen, C.A. Campbell, C.V. Cole, and R.J.K. Myers. 1994. p. 35-56. In: R.C. Wood and J. Dumanski (eds.) *Sustainable Land Management for the 21st Century*. Volume 2: Plenary Papers. Proceedings of the International Workshop on Sustainable Land Management for the 21st Century. June, 1993. University of Lethbridge, Lethbridge, Canada.

Flader, S. 1995. Aldo Leopold and the evolution of a land ethic. p. 3-24. In: T. Tanner (ed.) *Aldo Leopold The Man and His Legacy*. Soil and Water Conservation Society, Ankeny, IA.

Granatstein, D. and D.F. Bezdicek. 1992. The need for a soil quality index: local and regional perspectives. *Am. J. Altern. Agric.* 7:12-6.

Gregorich, E.G., M.R. Carter, D.A. Angers, C.M. Monreal, and B.H. Ellert. 1994. Towards a minimum data set to assess soil organic matter quality in agricultural soils. *Can. J. Soil Sci.* 74:367-85.

Haberern, J. 1992. Viewpoint: A soil health index. *J. Soil Water Cons.* 47:6.

Harms, D.S., R.B. Grossman, and G.B. Muckel. 1995. Point soil quality evaluation protocol for the near surface. p. 281. In: *Agronomy Abstracts*. ASA, Madison, WI.

Harris, R.F. and D.F. Bezdicek. 1994. Descriptive aspects of soil quality/health. p. 23-35. In: J. W. Doran, D. C. Coleman, D. F. Bezdicek, and B. A. Stewart (eds.) *Defining Soil Quality for a Sustainable Environment*. SSSA Spec. Pub. No. 35. ASA, CSSA, and SSSA, Madison, WI.

Hellkamp, A.S., J.M. Bay, K.N. Easterling, G.R. Hess, B.F. McQuaid, M.J. Munster, D.A. Neher, G.L. Olson, K. Sidik, L.A. Stefanski, M.B. Tooley, and C.L. Campbell. 1995. *Environmental Monitoring and Assessment Program: Agricultural Lands Pilot Field Program Report-1993*. EPA/620/R-95/004. U.S. Environmental Protection Agency, Washington, D.C. 64 pp.

Hole, F.D. and J.B. Campbell. 1985. *Soil Landscape Analysis*. Rowman & Allanheld Publishers, Totowa, NJ. 196 pp.

Hoosbeek, M.R. and R.B. Bryant. 1992. Towards the quantitative modeling of pedogenesis - a review. *Geoderma* 55:183-210.

Hornung, M. 1993. Defining soil quality for ecosystems and ecosystem functioning. p. 201-211. In: H.J.P. Eijsackers and T. Hamers (eds.) *Integrated Soil and Sediment Research: A Basis for Proper Protection*. Proceedings of the First European Conference on Integrated Research for Soil and Sediment Protection and Remediation (EUROSOL), Maastricht, The Netherlands. 1-12 Sept. 1992. Kluwer Academic Publ., Dordrecht.

Howard, P.J.A. 1993. Soil protection and soil quality assessment in the EC. *Sci. Total Environ.* 129:219-239.

Hudson, B.D. 1994. Soil Organic matter and available water capacity. *J. Soil Water Cons.* 49:189-194.

Izac, A.M.N. and M. J. Swift. 1994. On agricultural sustainability and its measurement in small-scale farming in sub-Saharan Africa. *Ecol. Econ.* 11:105-125.

Jenny, H. 1941. *Factors of Soil Formation*. McGraw-Hill, New York, 281 pp.

Karlen, D.L. and D.E. Stott. 1994. A framework for evaluating physical and chemical indicators of soil quality. p. 53-72. In: J.W. Doran, D.C. Coleman, D.F. Bezdicek, and B.A. Stewart (eds.) *Defining Soil Quality for a Sustainable Environment*. SSSA Spec. Pub. No. 35. ASA, CSSA, and SSSA, Madison, WI.

Karlen, D.L., J.C. Gardner, and M.J. Rosek. 1997a. A soil quality framework for evaluating the impact of CRP. *J. Prod. Agric.* (In press.)

Karlen, D.L., M.J. Mausbach, J.W. Doran, R.G. Cline, R.F. Harris, and G.E. Schuman. 1997b. Soil quality: A concept, definition, and framework for evaluation. *Soil Sci. Soc. Am. J.* 61: (in press).

Karlen, D.L., N.C. Wollenhaupt, D.C. Erbach, E.C. Berry, J.B. Swan, N.S. Eash, and J.L. Jordahl. 1994a. Crop residue effects on soil quality following 10 years of no-till corn. *Soil Tillage Res.* 31:149-167.

Karlen, D.L., N.C. Wollenhaupt, D.C. Erbach, E.C. Berry, J.B. Swan, N.S. Eash, and J.L. Jordahl. 1994b. Long-term tillage effects on soil quality. *Soil Tillage Res.* 32:313-27.

Kennedy, A.C., and R.I. Papendick. 1995. Microbial characteristics of soil quality. *J. Soil Water Cons.* 50:243-248.

Kinako, P.D.S. 1983. Assessment of relative soil quality by ecological bioassay. *Afr. J. Ecol.* 21:291-5.

Kiniry, L.N., C.L. Scrivner, and M.E. Keener. 1983. *A Soil Productivity Index Based upon Predicted Water Depletion and Root Growth*. Res. Bull. 1051. Mo. Agr. Exp. Stn., Columbia, MO. 26 pp.

Lal, R. 1994. *Methods and Guidelines for Assessing Sustainable Use of Soil and Water Resources in the Tropics*. USDA, Soil Conservation Service, Soil Management Support Services, Washington, D.C. SMSS Technical Monograph No. 21. 78 pp.

Larson, W.E. and F.J. Pierce. 1991. Conservation and enhancement of soil quality. p. 175-203. In: *Evaluation for Sustainable Land Management in the Developing World, Vol. 2*: Technical papers. Bangkok, Thailand: International Board for Research and Management, 1991. IBSRAM Proceedings No. 12(2).

Larson, W.E. and F.J. Pierce 1994. The dynamics of soil quality as a measure of sustainable management. p. 37-51. In: J.W. Doran, D.C. Coleman, D.F. Bezdicek, and B.A. Stewart (eds.) *Defining Soil Quality for a Sustainable Environment.* SSSA Spec. Pub. No. 35. ASA, CSSA, and SSSA, Madison, WI.

Lowdermilk, W.C. 1953. *Conquest of the Land Through Seven Thousand Years.* Agriculture Information Bulletin No. 99. USDA, Soil Conservation Service, Washington, D.C. 30 pp.

MacDonald, K.B., W.R. Fraser, F. Wang, and G.W. Lelyk. 1995. A geographical framework for assessing soil quality. p. 19-30. In: Acton, D.F. and L.J. Gregorich, (eds.) *The Health of Our Soils: Toward Sustainable Agriculture in Canada.* Centre for Land and Biological Resources Research, Research Branch, Agriculture and Agri-Food Canada, Ottawa, Ont.

Maslow, A.H. 1943. A theory of human motivation. *Psychological Review* 50:370-396.

National Research Council. 1993. *Soil and Water Quality: An Agenda for Agriculture.* National Academy Press, Washington, D.C., 516 pp.

Nusser, S.M. and J.J. Goebel. 1997. The national resources inventory: A long-term multi-resource monitoring program. *Environ. Ecol. Stat.* (in press).

Papendick, R.I., J.F. Parr, and J. Van Schilfgaarde. 1995. Soil quality: new perspective for a sustainable agriculture. In: *Proceedings for International Soil Conservation Organization.* New Delhi, India, December 4-8, 1994.

Parr, J.F., R.I. Papendick, S.B. Hornick, and R.E. Meyer. 1992. Soil quality: attributes and relationship to alternative and sustainable agriculture. *Am. J. Altern. Agric.* 7:5-11.

Pennock, D.J., D.W. Anderson, and E. de Jong. 1994. Landscape-scale changes in indicators of soil quality due to cultivation in Saskatchewan, Canada. *Geoderma* 64:1-19.

Pierce, F.J. and W.E. Larson. 1993. Developing criteria to evaluate sustainable land management. p. 7-14. In: J. M. Kimble (ed.) *Proceedings of the Eighth International Soil Management Workshop: Utilization of Soil Survey Information for Sustainable Land Use.* May 3, 1993. USDA Soil Conservation Service, National Soil Survey Center, Lincoln, NE.

Pierce, F.J., W.E. Larson, R.H. Dowdy, and W.A.P. Graham. 1983. Productivity of soils: assessing long-term changes due to erosion. *J. Soil Water Cons.* 38:39-44.

Reybold, W.U. and G.W. TeSelle. 1989. Soil geographic data bases. *J. Soil Water Cons.* 44:28-29.

Romig, D.E., M.J. Garlynd, and R.F. Harris. 1997. Farmer-based assessment of soil quality: a soil health scorecard. In: J.W. Doran and A.J. Jones (eds.) *Methods for Assessing Soil Quality.* SSSA Spec. Publ. SSSA, Madison, WI. (In press.)

Romig, D.E., M.J. Garlynd, R.F. Harris, and K. McSweeney. 1995. How farmers assess soil health and quality. *J. Soil Water Cons.* 50:229-236.

Rust, R.H., R.S. Adams, and W.P. Martin. 1972. Developing a soil quality index. *Indic. Environ. Qual.* 1:243-247.

Schjonning, P., B.T. Christensen, and B. Carstensen. 1994. Physical and chemical properties of a sandy loam receiving animal manure, mineral fertilizer or no fertilizer for 90 years. *European J. Soil Sci.* 45:257-68.

Schnitzer, M. 1991. Soil organic matter and soil quality. p. 33-49. In: S.P. Mathur and C. Wang (eds.) *Soil quality in the Canadian Context: 1988 Discussion Papers.* Tech. Bull. 1991-1E. Land Resource Research Centre, Research Branch, Agriculture Canada, Ottawa, Ontario.

Singh, K.K., T.S. Colvin, D.C. Erbach, and A.Q. Mughal. 1992. Tilth index: an approach to quantifying soil tilth. *Trans. ASAE* 35:1777-85.

Smith, J.L, J.J. Halvorson, and R.I. Papendick. 1994. Multiple variable indicator kringing: A procedure for integrating soil quality indicators. p. 149-157. In: J.W. Doran, D.C. Coleman, D.F. Bezdicek, and B.A. Stewart (eds.), *Defining Soil Quality for a Sustainable Environment.* SSSA Spec. Pub. No. 35. ASA, CSSA, and SSSA, Madison, WI.

Smith, J.L, J.J. Halvorson, and R.I. Papendick. 1993. Using multiple-variable indicator kriging for evaluating soil quality. *Soil Sci. Soc. Am. J.* 57:743-749.

Soil Survey Division Staff. 1993. *Soil Survey Manual*. U.S. Dept. Agric. Handbk No. 18, U. S. Govt. Print. Office, Washington, D.C. 437 pp.

Soil Survey Staff, 1975. *Soil Taxonomy: A basic system of soil classification for making and interpreting soil surveys*. USDA-SCS, Agric. Handb. 436. U.S. Govt. Print. Office, Washington, D.C., 754 pp.

Stasch, D. and K. Stahr. 1993. The soil potential concept - A method for the fundamental evaluation of soils and their protection. p. 107-109. In: H.J.P. Eijsackers and T. Hamers (eds.) *Integrated Soil and Sediment Research: A Basis for Proper Protection*. Proceedings of the First European Conference on Integrated Research for Soil and Sediment Protection and Remediation (EUROSOL), Maastricht, the Netherlands, 1-12 Sept. 1992. Kluwer Academic Publishers, Dordrecht.

Stevenson, F.J. 1994. *Humus Chemistry: Genesis, Composition, Reactions*. 2nd ed. John Wiley & Sons, New York, 496 pp.

Visser, S. and D. Parkinson. 1992. Soil biological criteria as indicators of soil quality: soil microorganisms. *Am. J. Altern. Agric.* 7:33-37.

Warkentin, B.P. and H.F. Fletcher. 1977. Soil quality for intensive agriculture. p. 594-598. In: Proceedings of the International Seminar on Soil Environment and Fertilizer Management in Intensive Agriculture. Soc. of Sci. of Soil and Manure. Japan.

Warkentin, B.P. 1995. The changing concept of soil quality. *J. Soil Water Cons.* 50:226-228.

Willis, W.O. and C.E. Evans. 1977. Our soil is valuable. *J. Soil Water Cons.* 32:258-259.

CHAPTER 28

Relationships between Soil Organic Carbon and Soil Quality in Cropped and Rangeland Soils: the Importance of Distribution, Composition, and Soil Biological Activity

Jeffrey E. Herrick and Michelle M. Wander

I. Introduction

The often-cited positive relationship between soil organic carbon (SOC) content and soil quality (Arshad and Coen, 1992; National Research Council, 1993; Doran and Parkin, 1994; Manley et al., 1995; Pikul and Aase, 1995; Karlen and Cambardella, 1996) is consistent with the results of over one hundred years of modern agricultural research (Bauer and Black, 1994) and with thousands of years of on-farm observation and experimentation (Magdoff, 1992). This relationship is based on contributions which SOC makes as a constituent of soil organic matter (SOM) to critical soil properties, processes and functions. The term "SOM" will be used in the remainder of this chapter except where carbon per se is of interest. "SOM" is preferred here because it reflects the reality that impacts of SOC on soil quality are determined by its chemical, physical, and biological configuration within SOM.

The positive correlation between SOM and soil quality holds, in general, for all of the definitions of soil quality listed by Doran and Parkin (1994), as well as for the concepts of soil quality discussed by Warkentin (1995). Soil organic matter content was listed by a group of 28 Wisconsin farmers as the single most important property for characterizing soil health or soil quality based on their personal definitions (Romig et al., 1995). SOM was also cited as "perhaps the single most important indicator of soil quality" in a NRC (National Research Council) report on soil and water quality. The report defined soil quality as "the capacity of the soil to promote the growth of plants; protect watersheds by regulating the infiltration and partitioning of precipitation; and prevent water and air pollution by buffering potential pollutants such as agricultural chemicals, organic wastes, and industrial chemicals" (National Research Council, 1993).

For the purposes of this chapter, soil quality is defined as the capacity of a soil to perform functions which sustain biological productivity and maintain environmental quality. Soil quality depends on current soil functional integrity and on soil's resistance to, and resilience following, perturbation. Resistance is inversely proportional to loss of soil functional integrity following perturbation, while resilience is proportional to the recovery of functional integrity (Figure 1; Pimm, 1984). It is clear that a general relationship exists between SOM content, soil functions, and the capacity of the soil to maintain and/or recover those functions over time. However, a closer examination of the literature

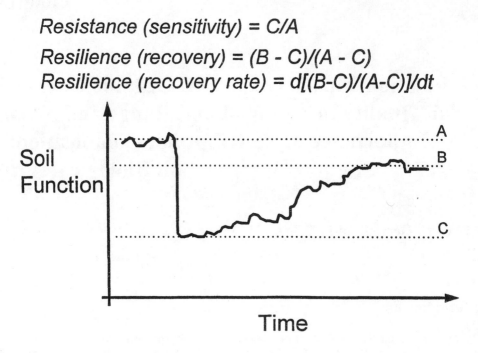

Figure 1. Relationship between soil functional integrity, disturbance, resistance, and resilience. The Y axis represents a generalized soil function, such as infiltration capacity.

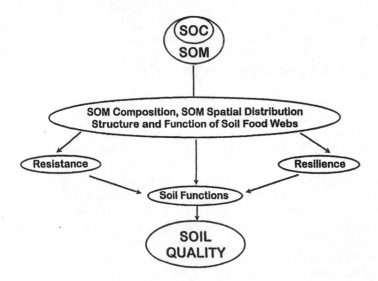

Figure 2. Conceptual model of relationship between soil organic carbon and soil quality.

suggests that the net effect of SOM often depends less on its quantity than on a combination of other factors including (1) its spatial distribution (horizontal, vertical, and location relative to soil aggregates) and (2) its composition, understood in both biochemical and functional terms. These two factors are mediated by a third: the structure and integrity of soil food webs (Figure 2). Organisms have direct and indirect effects on soil physical integrity, fertility, and a variety of soil properties, processes, and functions which are related to environmental quality. Soil food webs are discussed in a number of reviews of the role of soil biodiversity in ecosystem processes (Oades, 1993; Kennedy and Smith, 1995; Whitford, 1996).

The objective of this chapter is to explore the relationships between SOM and soil quality by: (1) defining the basic contributions of SOM to soil functions as related to soil quality (section II), (2) highlighting the role of SOM composition and spatial distribution in determining the effect of SOM on soil quality (section III), and (3) illustrating the dynamic nature of SOM-soil quality relationships through a discussion of resistance and resilience (section IV).

II. Contributions to Soil Function

SOM contributes to soil quality through its effects on specific soil functions. These functions include serving as a medium for the growth of plant roots and root-symbionts, regulating the flow of water, air, and nutrients, partitioning precipitation into plant-available-, ground-, and surface-water, serving as a repository for atmospheric carbon, and mitigating the impacts of pollutants on human and ecosystem health (National Research Council, 1993; Doran and Parkin, 1994; Elliott et al., 1994). SOM affects these functions by causing or mediating changes in soil properties and processes which are related to (1) soil physical integrity, (2) soil fertility and productivity, and (3) environmental quality.

A. Physical Integrity

There is a strong association between soil structure and SOM in the agronomic literature. A change in SOM content has often been cited as the single most important factor contributing to changes in functions related to soil physical integrity. For example, Zwerman (1947) attributed a twelve-fold increase in soil water infiltration rates in kudzu-planted fields to increased organic matter content in those fields relative to fallow soils under broom sedge and weeds. Bruce et al. (1995) identified carbon content as the single soil property which could be managed in degraded Appalachian piedmont soils to limit erosion and improve plant water availability. Hudson (1994) and Emerson (1995) found highly significant positive correlations between SOM content and available water capacity based on both experiments and a reexamination of the literature. Regression-based empirical pedotransfer functions for many soil physical properties include soil organic matter as a key component (Dutartre et al., 1993; Rasiah and Kay, 1994; Bell and van Keulen, 1995). Farmers also frequently associate high levels of SOM with good soil structure as soils with higher SOM tend to be easier to till over a wider range of moisture contents.

These general associations between soil structure and SOM are supported by more specific studies. Micromorphological and process-based studies have demonstrated the mechanistic role of organic matter in the formation and stabilization of soil aggregates (Tisdall and Oades, 1982; Elliott, 1986). Soil aggregate structure and stability, in turn, are related to soil erodibility (Lal and Elliot, 1994), infiltration characteristics (Eldridge, 1993), and soil compactibility (Soane, 1990).

B. Fertility and Productivity

The high correlation between soil organic matter content and soil fertility, the capacity of a soil to supply essential plant nutrients, has long been known (Russel, 1988). SOM contributes to soil fertility by serving as a source of plant nutrients (Broadbent, 1986), by providing exchange surfaces for nutrients from other sources (Mengel and Kirkby, 1987), and through its capacity to act as a pH buffer (Cantrell et al., 1990). In non- or little-fertilized systems, plant N, P, S, and microelement nutrition can depend heavily on mineralization of SOM and residues of plants and animals (Chen and Stevenson, 1986; Broadbent, 1986). Base cations, which are present only in trace amounts in humic substances, are supplied by decay of fresh organic residues. Their availability is influenced by SOM chiefly through its impact on ion exchange (Barber, 1984; Mengel and Kirkby, 1987).

SOM contributes to, and is affected by, biological productivity in natural and managed ecosystems. Zak et al. (1994) demonstrated this using a gradient spanning North America. They showed that net primary productivity of late succession ecosystems was positively correlated with microbial biomass and labile organic matter pools. In managed systems, maintenance of SOM levels benefits crop yields even when mineral nutrition is optimized by fertilizer application (Avinmelch, 1986; Cassman and Pingali, 1995). Laboratory studies have demonstrated that organic substances themselves have direct effects on plant growth. Organic matter can influence plant nutrient uptake through its impact on root physiology, root morphology, and even genome expression (Durante et al., 1994). SOM can also improve productivity by decreasing or preventing disease (Cook and Baker, 1983). The mechanisms involved in disease supression and resistance can be direct or indirect, resulting from biochemical or microbial interactions. In controlled experiments, Chen and Aviad (1990) found that humic substances stimulate plant growth directly. Benefits can arise from hormone-like interactions or can result from increased rooting volumes. SOM can also improve productivity by decreasing or preventing plant disease (Cook and Baker, 1983; Hoytink et al., 1986; Weltzien, 1992).

C. Environmental Quality

Soil quality and environmental quality are conceptually and functionally related. Soils are often viewed as contributing to environmental quality because they influence more traditional indicators such as atmospheric dust and water pollutant concentrations. However, soil quality has not historically been included in environmental quality assessments. Attempts are being made to quantify soil quality as an indicator of environmental quality for terrestrial ecosystems. This trend is illustrated by the Environmental Protection Agency's foray into the arena through its Environmental Monitoring and Assessment Program (EMAP) (Hunsaker and Carpenter, 1990), as well as by the establishment of soil quality standards by the International Standards Organization (ISO) (Hortensius and Welling, 1996).

SOM contributes to environmental quality both through its impact on overall soil quality and through effects on off-site environmental problems that are regulated by soil processes. Two of the most intensively studied SOM-related environmental problems which SOM directly affects are global climate change and water quality (National Research Council, 1993; Murphy and Zachara, 1995). The potential for soils to store additional carbon and thereby reduce the impacts of fossil fuel combustion on global climate change is widely recognized (Post et al., 1990; Schlesinger, 1991; Anderson, 1992), and there is a large body of knowledge on management-induced changes in soil carbon storage (Bolin, 1977; Barnwell et al., 1992; Wallace, 1994). Effects of tillage, drainage, residue management, and range management on SOC have been widely documented (Burke et al., 1989; Schlesinger, 1990; Sims et al., 1994; Manley et al., 1995; Ash et al., 1996). However, the relative magnitudes of the impact of tillage on soil carbon reductions (Burke et al., 1989) and of SOC storage on atmospheric carbon levels (Schlesinger, 1990; Kern and Johnson, 1993; Trumbore et al., 1996) continue to be debated.

SOM affects the quality of both ground and surface water supplies. Soil protects ground water by acting as an active filter in which contaminants are trapped and degraded or transformed into substances which are less toxic (Murphy and Zachara, 1995). SOM provides adsorbing surfaces and supports an active community of microorganisms capable of modifying and degrading contaminants. SOM impacts on adsorption and degradation similarly reduce groundwater pollution. Its effects on soil structure reduce runoff and erosion, thereby reducing transport of contaminants to surface water. Unfortunately, coarse-textured soils, which are among the most susceptible to groundwater pollution, tend to have low SOM contents.

The environmental fate of metals and organic contaminants in soils depends upon chemical lability and soil biology as well as compound adsorption and partitioning characteristics. In general, metal concentrations are positively correlated with clay contents (Holmgren et al, 1993) and, therefore, with SOM contents. The fate of metals is influenced by organic matter complexation, exchange, adsorption, precipitation, dissolution, and acid-base equilibria (Chen and Stevenson, 1986). Similarly, the fate of organic chemicals is also controlled by interacting mechanisms (Rao et al., 1993). For example, Reddy et al. (1995) showed that chlorimuron degradation in conventionally tilled and nontilled soils was regulated by native microbial populations' adaptability to the substrate and not by sorption, desorption, total SOM, or biomass. SOM influences adaptability directly by supporting individual heterotrophic organisms, and indirectly by influencing the physical space within which organisms and substrates interact.

III. Critical Factors

While higher levels of SOM are generally associated with improved physical integrity, higher soil productivity, and enhanced environmental quality, correlations are highly variable. Relationships between SOM and soil quality can be refined by considering two critical factors in the context of soil biological processes: (1) SOM composition, and (2) SOM spatial distribution.

A. Composition (Biochemical and Functional Aspects)

Soil organic matter includes all carbon in soil which is bound in organic forms. Accordingly, SOM refers to a heterogeneous range of living and dead C-based materials, the quantity and composition of which are known to vary significantly within major ecosystem types (Post et al., 1982; Anderson, 1991). The diversity of SOM is reflected by numerous characterization schemes (Hayes et al., 1989; Wilson, 1991). While there is some correlation between SOM characteristics and performance criteria, the extensive information available about the chemical composition of various SOM fractions has not been systematically tied to soil functions. However, existing research does indicate that specific SOM fractions may be superior to total SOM as indices of soil functions and, therefore, soil quality (Gregorich et al., 1994). By linking SOM composition to turnover characteristics, biochemical aspects of SOM can be tied to soil function (Table 1).

The living SOM component embodies only a small percentage of total carbon mass and includes below-ground roots, bacteria, fungi, and soil micro-, meso-, and macro-arthropods. This is the most dynamic portion of all SOM (Coleman and Crossley, 1996). Living SOM carbon is thought to play a regulatory role in soils, acting as a feedback mechanism controlling soil material and energy cycling characteristics (Elliott and Coleman, 1988).

Nonliving organic matter includes a range of materials that vary greatly in age, dynamic characteristics, and functional significance. These materials have been divided into pools or fractions which have different turnover characteristics (Jenkinson and Rayner 1977; Jenkinson and Parry, 1989; Parton et al., 1987; Paustian et al., 1992). The fractions are identified through isotopic tracer studies. The differentiation of organic matter among such dynamically distinct pools is caused not only by the chemical structure of constituent molecules, but also by their physical location in soils and the precise nature of their association with soil minerals (Greenland, 1965a,b; Tisdall and Oades, 1982; Waters and Oades, 1991). The biologically active, physically protected, and recalcitrant or stable fractions are three of the most commonly recognized pools.

Table 1. Management impacts on soil organic matter composition and turnover characteristics. Data are from the Rodale Research Center's Farming Systems Trial after ten years comparing organic-animal, organic-cover cropped, and conventionally-based corn and soybean rotations.

	Total C	Humin	Humic substances	Light fraction
	—————————g C kg^{-1} soil—————————			
Animal	23.4ab	11.7ab	10.1ab	1.7ab
Cover cropped	24.5a	12.4a	10.2a	2.0a
Conventional	21.3b	11.0b	8.9b	1.6b

	Available N†	Mineralizable N†	Respired C†	Biomass
			mmol CO_2 kg soil d^{-1}	nmol PLFA g^{-1} soil
	———mg N kg^{-1} soil———			
Animal	17.3	34.1	14.2a	29.17
Cover cropped	16.5	30.1	12.4b	33.00
Conventional	14.8	26.8	9.4c	26.60

†Values are seasonal averages based on samples collected on five dates in 1990. Available N is NH_4 and NO_3 extracted from fresh soil. Mineralizable N is additional N recovered. Respired C is CO_2 evolved during subsequent aerobic incubation. (Data from Wander et al., 1994, 1995; and Wander and Traina, 1996.)

The most dynamic or rapidly cycled SOM fraction is the biologically active fraction, which is tied to mineralization characteristics and, therefore, soil nutrient supply capacity (Greenland and Ford, 1964). No single method effectively isolates this important nutrient reservoir. The light fraction is a SOM pool defined by density. It is more sensitive to management impacts than total SOM and has been suggested as a measure of the biologically active fraction (Janzen et al., 1992; Wander et al., 1994; Barrios et al., 1996; Gregorich and Janzen, 1996) and, consequently, as an indicator of soil quality. Polysaccharides, which are also part of the active SOM pool, contribute to soil quality by increasing macroaggregation even in the absence of detectable changes in total SOM (Roberson et al., 1991; Breland, 1995; Roberson et al., 1995).

Physically protected SOM (material entrapped in mineral particles and theoretically sequestered away from organisms) is a second pool which is relevant to soil quality because of its importance to soil structure and, therefore, the soil's biological and ecological integrity (Elliott and Coleman, 1988). As with the biologically active fraction, measures of this pool are indirect and imperfect.

Chemically recalcitrant SOM dominates the third pool, which is usually the oldest and largest. This pool is relatively resistant to decay compared to the active and physically protected pools. Both chemical structure (particularly structural randomness), and organo-mineral associations contribute to the stability and longevity of this pool. Recalcitrant SOM affects multiple aspects of soil quality by influencing the fate of ionic and nonionic compounds, contributing to the long term stabilization of microaggregates, increasing soil cation exchange capacity, and influencing soil color.

The specific SOM fraction or fractions selected for study in any system will depend on the function or functions of interest. Hence, one might focus on residue characteristics for erosion, humic substances, and depth distribution for carbon storage potential, and biologically active SOM for nutrient supply. More research is needed to establish clear ties between procedurally defined SOM fractions and soil function as it occurs within individual ecosystems.

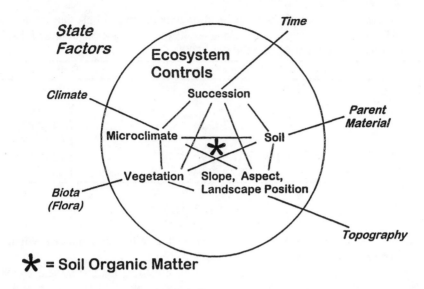

Figure 3. Soil organic matter content is determined by management in the context of local state factors and ecosystem controls. (Adapted from Van Cleve, 1991.)

B. Spatial Distribution

Soil organic matter exhibits strong spatial dependence in both agricultural and rangeland systems at scales ranging from the microaggregate (Kooistra and van Noordwijk, 1996; Tisdall, 1996) to the field (Cambardella et al., 1994) and the landscape (Burke et al., 1995). This spatial heterogeneity affects soil properties, processes, and functions at each of these scales (e.g., Gallardo and Schlesinger, 1992; Fromm et al., 1993; Lafrance and Banton, 1995) and, therefore, has an impact on soil quality. In fact, patterns of SOM distribution can be more important than total SOM content.

Patterns of SOM distribution and the relationship of these patterns to soil properties, processes and functions vary depending on the scale of interest. The scales discussed below (aggregate, root, plant and plant community, and landscape) are defined functionally with respect to the vegetative component of the system to emphasize the dynamic relationship between the plant as the primary source of organic matter and the soil as a medium which supports plant productivity (Figure 3). These scales, however, are by no means distinct. Patterns of variation in SOM content and composition vary continuously and some of the relationships described at the root scale could be equally applied at the plant level.

1. Aggregate Scale

Patterns at the aggregate scale range from submicron associations of clay domain and organic polymers to macroaggregate structures which encompass a cubic centimeter or more of soil volume (Tisdall and Oades, 1982). These macroaggregates may include microaggregates (< 250 um) as well as plant fragments, fine roots, and fungal hyphae. One of the most important functions of the heterogeneous spatial organization of SOM at this scale is to provide structure and stability. Organic materials play a key role in both generating and maintaining soil structure. They function directly as binding agents. They also serve as an energy source for soil biota which create voids and rearrange and redistribute soil primary particles into new soil structures (Table 2). For a more detailed discussion

Table 2. Termite-dominated cattle dung decomposition impacts on soil bulk density in the surface 7 cm in a seasonally-dry Costa Rican pasture. Dung was deposited at the beginning of the 1993 dry season. Means ± S.E. (n = 36 for 0 d and 9 for 12 - 270 d).

Dung patch age	Dung removed	Bulk density
days	%	g cm^{-3}
0	--	1.16 (0.01)
12	12.3 (2.8)	1.14 (0.03)
60	31.8 (2.2)	1.09 (0.03)
140	41.0 (1.8)	1.02 (0.03)
270	83.3 (3.9)	1.05 (0.02)

(Calculated from Herrick and Lal, 1995; 1996.)

of aggregate formation and structure, see related reviews in this volume and in earlier volumes in this series (e.g. Kooistra and van Noordwijk, 1996; Tisdall, 1996).

In addition to its role in aggregate structure and stabilization, the spatial and compositional heterogeneity of aggregate-associated SOM at the aggregate scale regulates the rate and pattern of many soil processes. The retention and release of water at intermediate tensions is regulated by the density and distribution of micropores within aggregates, as well as by characteristics of the organic matter itself. Soil carbon turnover rates are positively correlated with the diameter of the pore in which the carbon is located (Killham et al., 1993; Hassink et al., 1993; Nelson et al., 1994) and with aggregate size (Jastrow et al., 1996). Organic matter associated with macroaggregates is qualitatively more labile and more readily mineralized than SOM associated with microaggregates, and is the primary source of organic matter released when soils are cultivated (Oades et al., 1987; Balesdent et al., 1988; Waters and Oades, 1991). Parkin (1987) found denitrification, associated with microsites enriched in C, coincident with anaerobism. Barriuso and Koskinen (1996) demonstrated that atrazine biodegradation was associated primarily with particulate organic matter and Scow (1993) showed that pesticide decay rates are influenced by aggregate size and SOM content.

2. Root Scale

The root scale is defined as the soil volume explored by a single plant root or a segment of that root. Scale within a soil pedon is traditionally defined in a hierarchical framework of particles, aggregates, and clods. Many processes in soils are structured linearly. Roots generate many of these patterns, establishing zones rich in organic matter and biotic activity. Air- and water-conducting macropores are related linear structures that occur at the same scale. This scale also includes spatial variability associated with clods and organic fragments larger than a macroaggregate, such as corn stover and chunks of manure.

Critical soil properties related to SOM content at the root scale include nutrient availability and water holding capacity. The relationship between SOM content and nutrient availability is generally positive, although organic matter inputs with a high C-to-N ratio can result in a temporary reduction in plant-available nutrients due to microbial immobilization. Soil processes affected by local SOM concentrations include many biochemical transformations (Bonmati et al., 1991) and transport phenomena. While preferential flow through biogenic and physical macropores is frequently cited as a major factor in fate and transport studies, the distribution of SOM in the soil matrix relative to the macropores is also important (Felsot and Shelton, 1993). This illustrates the value of assessing soil quality in three dimensions, even at the intra- and inter-aggregate scales.

Table 3. Shrub-associated patterns of soil organic carbon distribution in the top 5 cm in the semiarid Great Basin and in the Sonoran Desert. The Great Basin data represent an average of eight sampling dates and include a range of interspace cover types from vesicular crust (lowest carbon) to moss-grass (highest carbon). The Sonoran Desert interspace were removed from a distance of one-third times the radius of the shrub from the edge of the shrub canopy.

Dung patch age	Shrub	Interspace
	———% Organic carbon———	
Great Basin		
Sagebrush	2.46	0.65 - 1.98
Sonoran Desert		
Velvet mesquite	0.77	0.30
Palo verde	0.69	0.31
Vine mesquite	0.99	0.56

(Calculated from Blackburn et al., 1992; and Barth and Klemmedson, 1978.)

3. Plant and Plant Community Scale

The plant scale covers an extremely wide range of spatial patterns including higher SOM inputs associated with individual plants (Smith et al., 1994), individual cattle dung patches in a pasture (Herrick and Lal, 1995), and microtopographically-controlled patterns of litter redistribution (Tongway and Ludwig, 1996). A common characteristic of these patterns is that they differentially affect resource availability to individual plants within a plant community. The relative importance of a change in SOM concentration or composition at the plant scale depends both on how different the particular concentration or composition is from that in the surrounding matrix, and on the size of the volume relative to the size of the organism or scale of the process affected by it. These relationships can evolve over time as, for example, with the growth of a seedling to a mature plant which exploits an increasingly large soil volume.

Soil properties, processes, and functions affected by spatial variability at the plant scale include all of those listed above for the root scale; however, the relative importance and persistence of the spatial patterns varies widely among ecosystems. Soil organic matter distribution in perennial rangelands is largely controlled by the spatial pattern of perennial plants. A number of studies have demonstrated "islands of fertility" in arid and semiarid rangelands associated with enhanced litter and SOM accumulation beneath shrubs (Table 3; Barth and Klemmedson, 1978; Santos et al., 1978; Blackburn et al., 1992; Smith et al., 1994; Schlesinger et al., 1996). In addition to adding organic matter directly to the soil through root exudates, root death, and litterfall, plants control the spatial redistribution of inputs through their effects on wind and water flow patterns. Spatial and temporal patterns of accumulation vary with shrub species (Barth and Klemmedson, 1982). Shrub-associated SOM islands are associated with enhanced soil quality, including higher nutrient availability and higher levels of soil biotic activity (Santos et al., 1978; Schlesinger et al., 1996). Soil surface hydrology is also affected. Blackburn et al. (1992) found that vegetation growth form was the most important factor contributing to soil surface properties which control infiltration and erosion in sagebrush dominated communities.

The ultimate impact of plant-scale concentrations of carbon resources on rangeland soil quality depends on local ecosystem dynamics. In water erosion-dominated areas of Australia, increased patchiness associated with banded vegetation systems is viewed positively. Linear patches intercept runoff and retain soil and water resources more effectively than more homogeneous systems in which threshold levels of resource availability required for vigorous vegetative growth are not reached

(Tongway and Ludwig, 1996). Conversely, shrub-associated patchiness on wind erosion-dominated soils in the arid U.S. southwest is cited as a sign of degraded rangeland as these systems are frequently associated with increased rates of soil erosion (Gibbens et al., 1983). Thus, while patch-level spatial distribution of SOM may serve as an excellent indicator of soil quality, the distribution pattern and scale must be carefully interpreted in the context of each ecosystem (Herrick and Whitford, 1995).

In annual cropping systems, the degree of SOM spatial heterogeneity and its impact on soil quality depends on the management system and the average SOM content. In general, reduced tillage systems are associated with a higher level of vertical SOM stratification in the rooting zone. A shift in the distribution of SOM associated with a shift to minimum- or no-tillage practices can dramatically impact other soil processes and ultimately alter soil biotic communities and trophic interactions, soil properties like macroporosity (Lee and Foster, 1991), and soil functions (Fromm et al., 1993; Pikul and Aase, 1995).

4. Landscape Scale

The landscape scale begins at the plant community or field scale. The upper limit depends on both the physiographic characteristics of the region and the properties, processes, and functions of interest. The upper limit can be defined in terms of a watershed. At this scale, soil quality, and the contribution of SOM to it, become more closely linked to social, political, and economic issues. SOM contributes to economic land values. It affects and is affected by the choice of management practice.

The redistribution of SOM by water can contribute to the development of isolated, deep, relatively fertile soils with enhanced physical integrity in regions which otherwise could not support arable agriculture (Forman and Godron, 1986). Where the source of the off-site organic matter inputs can be identified, SOM deposition can serve as a useful indicator of potential soil quality degradation at other landscape positions. In fact, the scale and pattern of variation at the landscape level may be one of the more useful and cost-effective regional indicators of soil quality (Herrick and Whitford, 1995), particularly if it is integrated with information on regional soil-climate-SOM relationships, such as those presented by Burke and others (1995).

5. Scale in Different Systems

At the aggregate and root scales, the impacts of SOM distribution are similar in agricultural and rangeland systems. In both cases, the three-dimensional spatial pattern of different carbon fractions affects soil structure and stability, nutrient availability, and the fate and transport of pollutants. At larger scales, the two systems diverge, with rangelands tending to exhibit more distinct and permanent spatial differences in SOM distribution and, consequently, in the properties, processes, and functions associated with these spatial patterns. In both systems, the spatial patterns of SOM and of carbon inputs at all scales should be explicitly or implicitly considered in the assessment of soil quality. This is particularly true when changes in management to improve soil quality through increased carbon inputs are contemplated. As Kooistra and van Noordwijk (1996) pointed out in their paper on soil architecture and organic matter distribution, "Using larger amounts of organic inputs is not a simple recipe to obtain a better soil structure".

Spatial variability in SOM concentrations at the whole-plant level is often developed and exploited in alley cropping and agroforestry systems to enhance the quality of the soil supporting both an annual and a perennial crop (Nair, 1984). By tapping nutrient stocks which would be inaccessible to shallow-rooted species, deep-rooted perennial crops increase system productivity. In addition to reducing runoff and soil erosion from the annually-cropped areas, the SOM derived from the deep-rooted perennial crop in agroforesty systems provides a supplementary source of nutrients to annual roots which invade the perennial-dominated microsites. The net increase in nutrient availability may be due to increased nutrient content in the surface soil and/or an SOM-associated increase in cation exchange capacity (Rodella et al., 1995).

IV. Resistance and Resilience

Disturbance, or any event which causes a significant change from the normal pattern in an ecosystem (Forman and Godron, 1986), is increasingly recognized as a key component of both natural and managed systems. Agronomists have long recognized the importance of various types of disturbance, such as cultivation (Eghball et al., 1994). The mechanisms of ecosystem adaptation to disturbance have received serious attention only in the last several decades (Sousa, 1984).

The previous discussion of SOM composition and distribution as critical factors largely ignores disturbance. This perspective is temporally static: it is fixed at the particular point in time that samples are taken or measurements are made. When soils are studied at increasingly long time scales, it becomes clear that the current capacity of a system to perform key processes is not necessarily a good indicator of its capacity to maintain functional integrity over time, particularly following perturbation. Organic matter content, like other soil properties, is determined by the characteristics of the ecosystem within which a soil develops and results from the dynamic equilibrium between soil parent material, climate, topography, and vegetation (Figure 3; Jenny, 1980; van Cleve and Powers, 1995; Huggett, 1995). SOM content is unique because it is one of the most rapidly and readily changed soil properties. As a result, SOM frequently plays a key role in both the resistance of the soil to degradation, and to its resilience or capacity to rapidly recover its functional integrity.

A. Relative Importance of Factors

The relative importance of SOM composition and spatial distribution, the structure and integrity of soil food webs, and the resistance and resilience relative to disturbance depend on the function and time scale of interest, and on the anticipated disturbance regime. For example, over relatively short time periods in the absence of major disturbance, soil physical integrity should be closely related to the composition and spatial distribution of SOM. These factors, which are related to aggregate stability and soil strength, also contribute to the resistance of the soil to compactive and erosive disturbances (Soane, 1990; National Research Council, 1993). Over longer periods of time, the resilience, or capacity of the system to recover physical integrity after disturbance, becomes more important as the probability that a major disturbance event will occur increases. Recovery of soil structure, including soil aggregates and macropores, frequently depends on soil biotic integrity and even on the presence of specific components of the soil biota such as earthworms (Blanchart, 1992; Edwards and Bater, 1992) or termites (Elkins et al., 1986; Herrick and Lal, 1995).

Similarly, the pre- and post-disturbance capacity of the soil to supply plant nutrients also depends on SOM composition, spatial distribution, and on the structure and integrity of soil food webs. However, it is SOM content and SOM turnover rates which appear to control the capacity of the soil to continue to supply plant nutrients after repeated harvests (Tiessen et al., 1994). Turnover rates, in turn, are controlled by climate, soil biota, soil structure, the nature of the disturbance, and organic matter composition. Based on organic matter turnover rates, the predicted length of time that existing litter and soil organic matter could supply nutrients to support agriculture without external inputs ranged from three years for a tropical forest soil in the Amazon to six years for a tropical semiarid forest soil and 65 years for a temperate grassland soil (Tiessen et al., 1994). Changes in nutrient supply rates associated with a switch to agriculture may lag several years due to immobilization processes (Burke et al., 1995). Different organic matter sources within a system can also affect turnover rates.

Based on stable carbon-isotope work in Nebraska, Cambardella and Elliott (1992) suggested that wheat-derived particulate organic matter may have a higher turnover rate than organic matter from grass in an undisturbed grassland. These turnover rates are positively correlated with both plant nutrient content and soil nutrient status (Cheshire and Chapman, 1996). Turnover and accumulation rates are particularly important when catastrophic losses of SOM are possible. A single rainfall event can remove SOM accumulated over thousands of years, substantially reducing the inherent nutrient-supplying capacity of one site while enriching another.

The resilience of a soil with respect to its capacity to retain and supply nutrients, therefore, depends on both SOM pool characteristics, including content and turnover rates, and on the nature of the disturbance. Tillage can lead to reductions in SOM, particularly the physically protected fraction, while erosion can cause across-the-board reductions in all fractions.

B. Disturbance Regimes

The response of a system to a particular disturbance, and whether or not a particular disturbance has a catastrophic impact on soil function, depends on the historic disturbance regime under which the system evolved and on the other stresses to which it is currently exposed. Disturbance regimes are defined in terms of disturbance type and the frequency, seasonality, intensity, and predictability or regularity, of each event. While a system may function satisfactorily and have a relatively high level of resistance and resilience under one disturbance regime, it may collapse in response to the introduction of a new disturbance, or in response to an increase in the intensity and/or frequency of existing disturbances (Belnap, 1995). For example, it is believed that Chihuahuan Desert grassland ecosystems became established and persisted under a regime of unpredictable, episodic, large-scale drought disturbances, and more regular small-scale animal-induced soil surface and grazing disturbances. Although bison may have entered the region occasionally in small groups, there is little evidence to suggest that they played a significant role in the evolution of the system.

Based on observations and limited data from remnant grassland areas, soil organic matter, while limited, was relatively homogeneously distributed and was present in sufficient quantities to help stabilize even relatively erodible sandy loams. With the introduction of cattle and technologies for stock-water development, however, the intensity of grazing disturbances increased. This, together with simultaneous invasion of the sandy grassland soils by a shrub (*Prosopis glandulosa*), is hypothesized to have caused the breakdown of the system during severe droughts, including a net loss of soil carbon and nitrogen, redistribution of the remaining SOM to areas under the shrubs, and reductions in soil stability (Schlesinger et al., 1990; Virginia et al., 1992). Thus, soil quality, as defined by function, resistance and resilience, could be regarded as high under the historic disturbance regime; however, resistance turned out to be quite low in the context of the new disturbance regime. A decline in soil quality is evidenced by both reductions in soil stability in the shrub-dominated systems (J.E. Herrick, unpubl. data) and the failure of the grassland to return even where cattle have been excluded for over 50 years (R.P. Gibbens, unpubl. data). Similar examples can be found for many soils which have experienced a change in the disturbance regime, including Amazonian rainforests (Tiessen et al., 1994) and midwestern prairies (Robinson et al., 1996).

Row crop agriculture, which is based on a temporally regular or predictable tillage disturbance, has led to degraded soil function in some systems, but not others. The differences cannot be easily attributed to differences in historical disturbance regimes, as few, if any, North American landscapes evolved under regimes analogous to annual tillage. Climate, including storm characteristics and fundamental soil properties, such as soil texture and mineralogy, more readily explain these differences. Striking examples of system loss of functional integrity are associated with the increased erosion and SOM loss following cultivation of weakly-structured soils in sloping landscapes, such as many soils of the Appalachian piedmont (Bruce et al., 1995). Conversely, SOM levels and soil

Figure 4. Conceptual model of the relationship between soil organic carbon and soil quality, expanded to include additional linkages. (Modified from Figure 2 and Arshad and Coen, 1992.)

structure can be maintained at desirable levels by using cover crops and applications of animal manures and/or composts.

The positive contribution of SOM to the resilience of cropped soils following cessation of tillage has been demonstrated by work on Conservation Reserve Program (CRP) land (Gebhart et al., 1994; Burke et al., 1995). At least one study showed that soil resilience, as indicated by increases in aggregate stability, was highest on CRP land with the highest levels of SOM at the time the land was taken out of annual crop production (Rasiah and Kay, 1994). Breland (1995) showed that improvements in soil structure on CRP land were rapidly erased when CRP land was returned to annual cropping practices. This suggests that while SOM may enhance resilience, the resistance of the improved structure to subsequent tillage disturbances can be quite low.

The proposed relationships between resistance, resilience, SOM, and soil quality is illustrated in Figure 4 in the context of linkages with the land, climate, and humans. The figure emphasizes the complexity of the interactions between SOM and soil quality. To define resistance and resilience, we must consider the time scale of interest, the nature of the disturbance(s), the level of recovery expected, and the type and quantity of inputs which are expected to be added to the system to support recovery. One criterion which is frequently applied to the assessment of rangeland condition is whether or not the system is likely to recover with a reasonable level of inputs. An informal survey of professionals active in developing rangeland assessment tools indicated that there is relatively little consensus on what constitutes a reasonable level. The same question is being asked in agricultural systems where cover crops, long-term rotations, use of manures, and reduced tillage systems are implemented only where they are perceived to be profitable. Even though resistance and resilience are likely to prove even more difficult to define than "a reasonable level of inputs," they add a key dimension to the discussion: system sustainability. Resistance and resilience must be considered, together with current soil functional integrity, in any but the most short-term assessments of soil quality.

As demands and stresses on soil resources grow, it becomes increasingly important that we understand the relationships between SOM composition and distribution and the functional integrity of specific systems. Key elements of soil quality vary among systems (Cole, 1994). The resistance and resilience of individual systems, contributions of SOM to resistance and resilience, and the relationship between SOM composition and distribution and soil function should receive special attention in the future.

V. Summary and Conclusions

A general relationship between SOM and soil quality has been confirmed by over 100 years of formal experimentation and by thousands of years of formal and informal observations and on-farm trials. This relationship is supported by both correlations and process-based mechanistic studies showing that, for most soils, physical integrity, fertility, and environmental quality are positively correlated with total SOM content. The exact nature of each of these relationships, however, is best understood in a site- and ecosystem-specific context. Within each ecosystem, the biochemical composition of SOM and its spatial distribution (microaggregate to landscape) are key factors in determining the ultimate impacts of SOM on current soil functional integrity and, therefore, on soil quality for a particular ecosystem.

The relationship between SOM and soil quality is not limited to current functional integrity, however. SOM plays a unique role in the maintenance and recovery of soil functions following disturbance for at least two reasons. First, it is relatively dynamic and, unlike mineral components, can be quickly regenerated when lost. Second, it supports activity by a wide variety of soil organisms, which themselves contribute to the restoration of soil functions.

While there is a plethora of research supporting the general relationship between SOM and soil quality, many of the mechanisms discussed or alluded to in this chapter remain poorly understood. As competition for organic inputs increases, it will become increasingly important to develop management approaches which optimize the impacts of these inputs on soil quality, and which tend to conserve the most important SOM fractions. These management techniques should ideally be based on an understanding of the mechanistic relationships between SOM and soil functions, and on interactions between organic matter and the soil food webs which are largely responsible for the fate of organic inputs. Unfortunately, many of the relationships and the SOM-food web interactions are poorly understood. Future research on soil quality-soil carbon relationships should, therefore, be increasingly biological and process-based, while maintaining clearly identifiable links to potential management practices for agricultural, range and forest soils.

Acknowledgments

A. de Soyza, K. Havstad, M. Manning, W. Whitford and several anonymous reviewers provided useful comments on earlier versions. The preparation of this chapter was supported by a USDA-NRI grant to J. Herrick and by a Hatch award (#1653990) to M. Wander.

References

Anderson, J.M. 1991. The effect of climate change on decomposition processes in grassland and coniferous forests. *Ecological Applications* 1:326-347.

Anderson, J.M. 1992. Responses of soils to climate change. *Advances Ecological Res.* 22:163-210.

Arshad, M.A. and G.M. Coen. 1992. Characterization of soil quality: physical and chemical criteria. *Am. J. Alternative Agric.* 7:25-32.

Ash, A.J., S.M. Howden, and J.G. McIvor. 1996. Improved rangeland management and its implications for carbon sequestration. p.19-20. In: N.E. West (ed.), *Proceedings of the Fifth International Rangeland Congress.* Salt Lake City, Utah. Society for Range Management, Denver, CO.

Avinmelch, Y. 1986. Organic residues in modern agriculture. p. 1-10. In: Y. Avinmelch and Y. Chen (eds.), *The Role of Organic Matter in Modern Agriculture.* Martinus Nijhoff, Dordrecht, The Netherlands.

Balesdent, J., G.H. Wagner, and A. Mariotti. 1988. Soil organic matter turnover in long-term field experiments as revealed by carbon-13 natural abundance. *Soil Sci. Soc. Am. J.* 52:118-124.

Barber, S.A. 1984. *Soil Nutrient Availability: a Mechanistic Approach.* John Wiley & Sons, NY.

Barnwell, T.O., R.B. Jackson, E.T. Elliott, I.C. Burke, V.C. Cole, K. Paustian, E.A. Paul, A.S. Donigian, A.S. Patwardhan, and A. Roswell. 1992. An approach to assessment of management impacts on soil carbon. *Water, Air and Soil Pollution* 64:423-435.

Barrios, E., R.J. Buresh, and J.I. Sprent. 1996. Organic matter in soil particle size and density fractions from maize and legume cropping systems. *Soil Biology & Biochemistry* 28:185-193.

Barriuso, E. and W.C. Koskinen. 1996. Incorporating nonextractable atrazine residues into soil size fractions as a function of time. *Soil Sci. Soc. Am. J.* 60:150-157.

Barth, R.C. and J.O. Klemmedson. 1982. Amount and distribution of dry matter, nitrogen, and organic carbon in soil-plant systems of mesquite and palo verde. *J. Range Management* 35:412-418.

Barth, R.C. and J.O. Klemmedson. 1978. Shrub-induced patterns of dry matter, nitrogen, and organic carbon. *Soil Sci. Soc. Am. J.* 42:804-809.

Bauer, A. and A.L. Black. 1994. Quantification of the effect of soil organic matter content on soil productivity. *Soil Sci. Soc. Am. J.* 58:185-193.

Bell, M.A. and J. van Keulen. 1995. Soil pedotransfer functions for four Mexican soils. *Soil Sci. Soc. Am. J.* 59:865-871.

Belnap, J. 1995. Surface disturbances: their role in accelerating desertification. *Environmental Monitoring and Assessment* 37:39-57.

Blackburn, W.H., F.B. Pierson, C.L. Hanson, T.L. Thurow, and A.L. Hanson. 1992. The spatial and temporal influence of vegetation on surface soil factors in semiarid rangelands. *Transactions of the ASAE* 35:479-486.

Blanchart, E. 1992. Restoration by earthworms (*Megascolecidae*) of the macroaggregate structure of a destructured savanna soil under field conditions. *Soil Biology and Biochemistry* 24:1587-1594.

Bolin, B. 1977. Changes of land biota and their importance for the carbon cycle. *Science* 196:613-615.

Bonmati, M., B. Ceccanti, and P. Nannipieri. 1991. Spatial variability of phosphatase, urease, protease, organic carbon and total nitrogen in soil. *Soil Biology and Biochemistry* 23:391-396.

Breland, T.A. 1995. Green manuring with clover and ryegrass catch crops undersown in spring wheat: Effects on soil structure. *Soil Use and Management* 11:163-167.

Broadbent, F.E. 1986. Nitrogen and phosphorus supply to plants by organic matter and their transformations. p. 13-23. In: Y. Avinmelch and Y. Chen (eds.), *The Role of Organic Matter in Modern Agriculture.* Martinus Nijhoff, Dordrecht, The Netherlands.

Bruce, R.R., G.W. Langdale, L.T. West, and W.P. Miller. 1995. Surface soil degradation and soil productivity restoration and maintenance. *Soil Sci. Soc. Am. J.* 59:654-660.

Burke, I.C., W.K. Lauenroth, and D.P. Coffin. 1995. Soil organic matter recovery in semi-arid grasslands: implications for the Conservation Reserve Program. *Ecological Applications* 5:793-801.

Burke, I.C., C.M. Yonker, W.J. Parton, C.V. Cole, K. Flash, and D.S. Schimel. 1989. Texture, climate and cultivation effects on soil organic matter content in U.S. grassland soils. *Soil Sci. Soc. Am. J.* 53:800-805.

Cambardella, C.A. and E.T. Elliott. 1992. Particulate soil organic matter across a grassland cultivation sequence. *Soil Sci. Soc. Am. J.* 56:777-783.

Cambardella, C.A., T.B. Moorman, J.M. Novak, T.B. Parkin, D.L. Karlen, R.F. Turco, and A.E. Konopka. 1994. Field-scale variability of soil properties in central Iowa soils. *Soil Sci. Soc. Am. J.* 58:1501-1511.

Cantrell, K.J., S.M. Serkiz, and E.M. Perdue. 1990. Evaluation of acid neutralizing capacity data for solutions containing natural organic acids. *Geochimica Et Cosmochimica Acta* 54:1247-1254.

Cassman, K.G. and P.L. Pingali. 1995. Extrapolating trends from long-term experiments to farmers' fields: the case of irrigated rice systems in Asia. p. 63-84. In: V. Barnett, R. Payne, and R. Steiner (eds.), *Agricultural Sustainability: Economic, Environmental and Statistical Considerations*. John Wiley & Sons, London.

Chen, Y. and T. Aviad. 1990. Effects of humic substances on plant growth. p. 161-86. In: P. McCarthy, C.E. Clapp, R.L. Malcolm, and P.R. Bloom (eds.), *Humic Substances in Soil and Crop Sciences*. ASA-SSSA, Madison, WI.

Chen, Y. and F.J. Stevenson. 1986. Soil organic matter interactions with trace metals. p. 73-116. In: Y. Chen and Y. Avnimelech (eds.), *The Role of Organic Matter in Modern Agriculture*. Martinus Nijhoff, Dordrecht, The Netherlands.

Cheshire, M.V. and S.J. Chapman. 1996. Influence of the N and P status of plant material and of added N and P on the mineralization of C from C-14-labelled ryegrass in soil. *Biology and Fertility of Soils* 21:166-170.

Cole, M.A. 1994. Soil quality as a component of environmental quality. p. 223-37. In: C.R. Cotherin and N.P. Ross (eds.), *Environmental Statistics: Assessment and Forecasting*. Lewis, Boca Raton, FL.

Coleman, D.C. and D.A. Crossley Jr. 1996. *Fundamentals of Soil Ecology*. Academic Press, San Diego, CA.

Cook, R.J. and K.F. Baker. 1983. *The Nature and Practice of Biological Control of Plant Pathogens*. The American Phytopathological Society, St. Paul, MN.

Doran, J.W. and T.B. Parkin. 1994. Defining and assessing soil quality. p. 3-21. In: J.W. Doran, D.C. Coleman, D.F. Bezdicek and B.A. Stewart (eds.), *Defining Soil Quality for a Sustainable Environment*. SSSA Special Publication Number 35. Soil Science Society of America, Madison, WI.

Durante, M., E. Attina and C. Cacco. 1994. Changes induced by humic substances and nitrate in the genomic structure of wheat roots. p. 279-82. In: N. Seresi and T.M. Miano (eds.), *Humic substances in the global environment and implication on human health*. Elsevier, Amsterdam.

Dutartre, P., F. Bartoli, A. Andreux, J.M. Portal, and A. Ange. 1993. Influence of content and nature of organic matter on the structure of some sandy soils from West Africa. *Geoderma* 56:459-478.

Edwards, C.A. and J.E. Bater. 1992. The use of earthworms in environmental management. *Soil Biology and Biochemistry* 24:1683-1689.

Eghball, B., L.N. Mielke, D.L. McCallister, and J.W. Doran. 1994. Distribution of organic carbon and inorganic nitrogen in a soil under various tillage and crop sequences. *J. Soil and Water Conserv.* 49:201-204.

Eldridge, D.J. 1993. Cryptogam cover and surface condition: effects on hydrology on a semiarid woodland soil. *Arid Lands Research and Rehabilitation* 7:203-217.

Elkins, N.Z., G.V. Sabol, T.J. Ward, and W.G. Whitford. 1986. The influence of subterranean termites on the hydrological characteristics of a Chihuahuan Desert ecosystem. *Oecologia* 68:521-528.

Elliott, E.T. 1986. Aggregate structure and carbon, nitrogen, and phosphorus in native and cultivated soils. *Soil Sci. Soc. Am. J.* 50:627-633.

Elliott, E.T., I.C. Burke, C.A. Monz, S.D. Frey, K. Paustian, H.P. Collins, E.A. Paul, C.V. Cole, R.L. Blevins, W.W. Frye, D.L. Lyon, D.A. Halvorson, D.R. Huggins, R.F. Turco, and M.V. Hickman. 1994. Terrestrial carbon pools: preliminary data from corn belt and Great Plains regions. p. 179-191. In: J.W. Doran, D.C. Coleman, D.F. Bezdicek, and B.F. Stewart (eds.) *Defining Soil Quality for a Sustainable Environment.* SSSA Special Publication Number 35, Madison, WI.

Elliott, E.T. and D.C. Coleman. 1988. Let the soil work for us. *Ecological Bulletin* 39:23-32.

Emerson, W.W. 1995. Water-retention, organic-C and soil texture. *Australian J. Soil Res.* 33:241-251.

Felsot, A.S. and D.R. Shelton. 1993. Enhanced biodegradation of soil pesticides: interactions between physicochemical processes and microbial ecology. p. 227-251. In: D.M. Linn, T.H. Carski, M.L. Brusseau, and F.H. Chang (eds.) *Sorption and Degradation of Pesticides and Organic Chemicals in Soils.* SSSA Special Publication Number 32. Soil Science Society of America, Madison, WI.

Forman, R.T.T. and M. Godron. 1986. *Landscape ecology.* John Wiley & Sons, NY., 619 pp.

Fromm, H., K. Winter, J. Filser, R. Hantschel, and F. Beese. 1993. The influence of soil type and cultivation system on the spatial distributions of soil fauna and microoganisms and their interactions. *Geoderma* 60:109-118.

Gallardo, A. and W.H. Schlesinger. 1992. Carbon and nitrogen limitations of soil microbial biomass in desert ecosystems. *Biogeochemistry* 18:1-17.

Gebhart, D.L., H.B. Johnson, H.S. Mayeux, and H.W. Polley. 1994. The CRP increases soil organic carbon. *J. Soil and Water Conserv.* 49:488-492.

Gibbens, R.P., J.M. Tromble, J.T. Hennessy, and M. Cardenas. 1983. Soil movement in mesquite dunelands and former grasslands of Southern New Mexico from 1933-1980. *J. Range Management* 36:145-148.

Greenland, D.J. 1965a. Interaction between clays and organic compounds in soils. Part 1. Mechanisms of interaction between clays and defined organic compounds. *Soils and Fertilizers* 28:412-425.

Greenland, D.J. 1965b. Interaction between clays and organic compounds in soils. Part 2. Adsorption of soil organic compounds and its effect on soil properties. *Soils and Fertilizers* 28:521-532.

Greenland, D.J. and G.W. Ford. 1964. Separation of partially humified organic materials from soils by ultrasonic dispersion. p. 137-148. In: *Transactions of the 8th International Congress of Soil Science.*

Gregorich, E.G., M.R. Carter, D.A. Angers, C.M. Monreal, and B.H. Ellert. 1994. Towards a minimum data set to assess soil organic matter quality in agricultural soils. *Canadian J. Soil Sci.* 74:367-385.

Gregorich, E.G. and H.H. Janzen. 1996. Storage of soil carbon in the light fraction and macroorganic matter. p. 167-190. In: M.R. Carter and B.A. Stewart (eds.) *Structure and Organic Matter Storage in Agricultural Soils.* CRC-Lewis, Boca Raton, FL.

Hassink, J., L.A.B. Bouwman, K.B. Zwart, and L. Brussaard. 1993. Relationships between habitable pore space, soil biota and mineralization rates in grassland soils. *Soil Biology and Biochemistry* 25:47-55.

Hayes, M.H.B., P. McCarthy, R.L. Malcolm, and R.S. Swift (eds.). 1989. *Humic Substances II: In Search of Structure.* John Wiley & Sons, Chichester, England.

Herrick, J.E. and R. Lal. 1995. Evolution of soil physical properties during dung decomposition in a tropical pasture. *Soil Sci. Soc. Am. J.* 59:908-912.

Herrick, J.E. and R. Lal. 1996. Dung decomposition and pedoturbation in a seasonally-dry tropical pasture. *Biology and Fertility of Soils* 23:177-181.

Herrick, J.E. and W.G. Whitford. 1995. Assessing the quality of rangeland soils: challenges and opportunities. *J. Soil and Water Conserv.* 50:237-242.

Holmgren, G.G.S., M.W. Meyer, R.L. Chaney, and R.B. Daniels. 1993. Cadmium, lead, copper and nickel in agricultural soils of the United States of America. *J. Environmental Quality* 22:335-349.

Hortensius, D. and R. Welling. 1996. International standardization of soil quality measurements. *Communications in Soil Sci. and Plant Analysis* 27:387-402.

Hoytink, H.A.J. and P.C. Fahy. 1986. Basis for the control of soilborne plant pathogens with composts. *Annual Review of Phytopathology* 24:93-114.

Hudson, B.D. 1994. Soil organic matter and water holding capacity. *J. Soil and Water Conserv.* 49:189-194.

Huggett, R.J. 1995. *Geoecology: An Evolutionary Approach*. Routledge, London.

Hunsaker, C. T. and D. E. Carpenter. 1990. *Ecological indicators for the Environmental Monitoring and Assessment Program*, EPA 600/3-90/060. US EPA, Office of Research and Development, Research Triangle Park, NC.

Janzen, H.H., C.A. Campbell, S.A. Brandt, G.P. Lanfond, and L. Townley-Smith. 1992. Light fraction organic matter in soils from long term crop rotations. *Soil Sci. Soc. Am. J.* 56:1799-1806.

Jastrow, J.D., T.W. Boutton, and R.M. Miller. 1996. Carbon dynamics of aggregate-associated organic matter estimated by carbon-13 natural abundance. *Soil Sci. Soc. Am. J.* 60:801-807.

Jenkinson, D.S. and L.C. Parry. 1989. The nitrogen cycle in the Boradbalk wheat experiment. *Soil Biology and Biochemistry* 21:523-541.

Jenkinson, D.S. and J.H. Rayner. 1977. The turnover of soil organic matter in some Rothamsted classical experiments. *J. Soil Sci.* 28:298-305.

Jenny, H. 1980. *The Soil Resource. Ecological Studies #37*. Springer-Verlag, NY.

Karlen, D.L. and C.A. Cambardella. 1996. Conservation strategies for improving soil quality and organic matter storage. p. 395-420. In: M.R. Carter and B.A. Stewart (eds.) *Structure and Organic Matter Storage in Agricultural Soils*. CRC-Lewis, Boca Raton, FL.

Kennedy, A.C. and K.L. Smith. 1995. Soil microbial diversity and the sustainability of agricultural soils. p. 75-86. In: H.P. Collins, G.P. Robertson, and M.J. Klug (eds.), *The Significance and Regulation of Biodiversity*. Kluwer Academic, Dordrecht, The Netherlands.

Kern, J.S. and M.G. Johnson. 1993. Conservation tillage impacts on national soil and atmospheric carbon levels. *Soil Sci. Soc. Am. J.* 57:200-210.

Killham, K., M. Amato, and J.N. Ladd. 1993. Effect of substrate location in soil and soil pore-water regime on carbon turnover. *Soil Biology and Biochemistry* 25:57-62.

Kooistra, M.J. and M. van Noorwijk. 1996. Soil architecture and distribution of organic matter. p. 15-53. In: M.R. Carter and B.A. Stewart (eds.), *Structure and Organic Matter Storage in Agricultural Soils*. CRC-Lewis, Boca Raton, FL.

Lafrance, P. and O. Banton. 1995. Implication of Spatial Variability of Organic-Carbon on Predicting Pesticide Mobility in Soil. *Geoderma* 65:331-338.

Lal, R. and W. Elliott. 1994. Erodibility and erosivity. p. 181-208. In: R. Lal (ed.), *Soil Erosion Research Methods*. St. Lucie Press, Delray Beach, FL.

Lee, K.E. and R.C. Foster. 1991. Soil fauna and soil structure. *Australian J. Soil Res.* 29:745-775.

Magdoff, F. 1992. *Building soils for better crops: organic matter management*. University of Nebraska Press, Lincoln, NE, 176 pp.

Manley, J.T., G.E. Schuman, J.D. Reeder, and R.H. Hart. 1995. Rangeland soil carbon and nitrogen response to grazing. *J. Soil and Water Conserv.* 50:294-298.

Mengel, K. and E.A. Kirkby. 1987. *Principles of Plant Nutrition*. International Potash Institute, Bern, Switzerland.

Murphy, E.M. and J.M. Zachara. 1995. The role of sorbed humic substances on the distribution of organic and inorganic contaminants in groundwater. *Geoderma* 67:103-124.

Nair, P.K.R. 1984. *Soil Productivity Aspects of Agroforestry. Science and Practice of Agroforestry 1*. ICRAF, Nairobi, Kenya.

National Research Council. 1993. *Soil and Water Quality: an Agenda for Agriculture.* National Academy Press, Washington, D.C., 516 pp.

Nelson, P.N., M-C. Dictor, and G. Soulas. 1994. Availability of organic carbon in soluble and particle-size fractions from a soil profile. *Soil Biology and Biochemistry* 11:1549-1555.

Oades, J.M. 1993. The role of biology in the formation, stabilization and degradation of soil structure. *Geoderma* 56:377-400.

Oades, J.M., A.M. Vassallo, A. Waters, and M.A. Wilson. 1987. Characterization of organic matter in particle size and density fractions from a red-brown earth by solid state 13C NMR. *Australian J. Soil Res.* 25:71-82.

Parkin, T. 1987. Soil microsites as a source of denitrification variability. *Soil Sci. Soc. Am. J.* 51:1194-1199.

Parton, W.J., D.S. Schimel, C.V. Cole, and D.S. Ojima. 1987. Analysis of factors controlling soil organic matter levels in Great Plains grasslands. *Soil Sci. Soc. Am. J.* 51:1173-1179.

Paustian, K., W.J. Parton, and J. Persson. 1992. Modeling soil organic matter in organic-amended and nitrogen fertilized long term plots. *Soil Sci. Soc. Am. J.* 56:476-488.

Pikul, J.L. and J.K. Aase. 1995. Infiltration and soil properties as affected by annual cropping in the northern Great Plains. *Agronomy J.* 87:656-662.

Pimm, S.L. 1984. The complexity and stability of ecosystems. *Nature* 307:321-326.

Post, W.M., W.R. Emanuel, P.J. Zinke, and A.G. Stangenberger. 1982. Soil carbon pools and world life zones. *Nature* 298:156-159.

Post, W.M., T.H. Peng, W.R. Emanuel, A.W. King, V.H. Dale, and D.L. DeAngelis. 1990. The global carbon cycle. *Am. Scientist* 78:310-326.

Rao, P.S.C. C.A. Bellin, and M.L. Brusseau. 1993. Coupling biodegradation of organic chemicals to sorption and transport in soils and aquifers: paradigms and paradoxes. p. 1-26. In: D.M. Linn, T.H. Carski, M.L. Brusseau, and F.H. Chang (eds.), *Sorption and Degradation of Pesticides and Organic Chemicals in Soils.* Soil Science Society of America Special Publication no. 32, Madison, WI.

Rasiah, V. and B.D. Kay. 1994. Characterizing changes in aggregate stability subsequent to introduction of forages. *Soil Sci. Soc. Am. J.* 58:935-942.

Reddy, K.N., R.M. Zablotowicz, and M.A. Locke. 1995. Chlorimuron adsorption, desorption, and degradation in soil from conventional tillage and no-tillage systems. *J. Environmental Quality* 24:760-767.

Roberson, E.B., S. Sarig, and M.K. Firestone. 1991. Cover crop management of polysaccharide-mediated aggregation in an orchard soil. *Soil Sci. Soc. Am. J.* 55:734-739.

Roberson, E.B., S. Sarig, C. Shennan, and M.K. Firestone. 1995. Nutritional management of microbial polysaccharide production and aggregation in an agricultural soil. *Soil Sci. Soc. Am. J.* 59:1587-1594.

Robinson, C.A., R.M. Cruse, and M. Ghaffarzadeh. 1996. Cropping system and nitrogen effects on mollisol organic carbon. *Soil Sci. Soc. Am. J.* 60:264-269.

Rodella, A.A., K.R. Fischer, and J.C. Alcarde. 1995. Cation exchange capacity of an acid soil as influenced by different sources of organic litter. *Communications in Soil Science and Plant Analysis* 26:2961-2967.

Romig, D.E., M.J. Garlynd, R.F. Harris, and K. McSweeney. 1995. How farmers assess soil health and quality. *J. Soil and Water Conserv.* 50:229-236.

Russel, E.W. 1988. *Soil Conditioners and Plant Growth.* Longman, New York, NY.

Santos, P.F., E. DePree, and W.G. Whitford. 1978. Spatial distribution of litter and microarthropods in a Chihuahuan desert ecosystem. *J. Arid Environments* 1:41-48.

Schlesinger, W.H. 1990. Evidence from chronosequence studies for a low carbon-storage potential of soils. *Nature* 348:232-234.

Schlesinger, W.H. 1991. Climate, environment and ecology. In: J. Jager (ed.), *Proceedings of the Second World Climate Conference.* Geneva, Switzerland.

Schlesinger, W.H., J.F. Reynolds, G.L. Cunningham, L.F. Huenneke, W.M. Jarrell, R.A. Virginia, and W.G. Whitford. 1990. Biological feedbacks in global desertification. *Science* 247:1043-1048.

Schlesinger, W.H., J.A. Raikes, A.E. Hartley, and A.F. Cross. 1996. On the spatial pattern of soil nutrients in desert ecosystems. *Ecology* 77:364-374.

Scow, K.M. 1993. Effect of sorption-desorption and diffusion processes on the kinetics of biodegradation of organic chemicals in soil. p. 73-114. In: D.M. Linn, T.H. Carski, M.L. Brusseau, and F.H. Chang (eds.) *Sorption and Degradation of Pesticides and Organic Chemicals in Soils.* SSSA Special Publication No. 32. Soil Science Society of America, Madison, WI.

Sims, G.K., D.D. Buhler, and R.F. Turco. 1994. Residue management impact on the environment. p. 77-98. In: P.W. Unger (ed.), *Managing Agricultural Residues.* Lewis, Chelsea, MI.

Smith, J.L., J.J. Halvorson, and H. Bolton, Jr. 1994. Spatial relationships of soil microbial biomass and C and N mineralization in a semi-arid shrub-steppe ecosystem. *Soil Biology and Biochemistry* 26:1151-1159.

Soane, B.D. 1990. The role of organic matter in soil compactibility: a review of some practical aspects. *Soil and Tillage Research* 16:179-201.

Sousa, W.P. 1984. The role of disturbance in natural communities. *Annual Review of Ecology and Systematics* 15:353-391.

Tiessen, H., E. Cuevas, and P. Chacon. 1994. The role of soil organic matter in sustaining soil fertility. *Nature* 371:783-785.

Tisdall, J.M. 1996. Formation of soil aggregates and accumulation of soil organic matter. p. 57-96. In: M.R. Carter and B.A. Stewart (eds.) *Structure and Organic Matter Storage in Agricultural Soils.* CRC-Lewis, Boca Raton, FL.

Tisdall, J.M., and J.M. Oades. 1982. Organic matter and water-stable aggregates in soils. *J. Soil Sci.* 33:141-163.

Tongway, D.J. and J.A. Ludwig. 1996. Restoration of landscape patchiness in semi-arid rangeland, Australia. p. 563-564. In: *Proceedings of the Fifth International Rangeland Congress.*

Trumbore, S.E., O.A. Chadwick, and R. Amundson. 1996. Rapid exchange between soil carbon and atmospheric carbon dioxide driven by temperature change. *Science* 272:393-396.

van Cleve, K. and R.F. Powers. 1995. Soil carbon, soil formation, and ecosystem development. p. 155-200. In: W.W. McFee and J.M. Kelly *Carbon Forms and Functions in Forest Soils.* Soil Science Society of America, Madison, WI.

Virginia, R.A., W.M. Jarrel, W.G. Whitford, and D.W. Freckman. 1992. Soil biota and soil properties in the surface rooting zone of mesquite (*Prosopis glandulosa*) in historical and recently desertified Chihuahuan Desert habitats. *Biology and Fertility of Soils* 14:90-98.

Wallace, A. 1994. Strategies to avoid global greenhouse warming: stashing carbon away in soil is one of the best. *Communications Soil Sci. and Plant Analysis* 25:37-44.

Wander, M.M., D.S. Hedrick, D. Kaufman, S.J. Traina, B.R. Stinner, S.R. Kehermeyer, and D.C. White. 1995. The functional significance of the microbial biomass in organic and conventionally managed soils. *Plant and Soil* 170:87-95.

Wander, M.M., S.J. Traina, R.B. Stinner, and S.E. Peters. 1994. The effects of organic and conventional management on biologically-active soil organic matter pools. *Soil Sci. Soc. Am. J.* 58:1130-1139.

Wander, M.M. and S.J. Traina. 1996. Organic matter fractions from organically and conventionally managed soils: I. Carbon and nitrogen distributions. *Soil Sci. Soc. Am. J.* 60:1081-1087.

Warkentin, B.P. 1995. The changing concept of soil quality. *J. Soil and Water Conserv.* 50:226-228.

Waters, A.G. and J.M. Oades. 1991. Organic matter in water stable aggregates. p. 163-74. In: W.S. Wilson (ed.), *Advances in Soil Organic Matter Research: The Impact on Agriculture and the Environment.* The Royal Society of Chemistry, Melksham, England.

Weltzien, H.C. 1992. Biocontrol of foliar fungal diseases with compost extracts. p. 430-450. In: J.H. Andres and S.S. Hirano (eds.), *Soils on a Warmer Earth*. Elsevier, Amsterdam.

Whitford, W.G. 1996. The importance of the biodiversity of soil biota in arid ecosystems. *Biodiversity and Conservation* 5:185-195.

Wilson, W.S. (ed.). 1991. *Advances in Soil Organic Matter Research: The Impact on Agriculture and the Environment*. The Royal Society of Chemistry, Melksham, England.

Zak, D.R., D. Tilman, R.R. Parmenter, C.W. Rice, F.M. Fisher, J. Vose, D. Milchunas, and C.W. Martin. 1994. Plant production and soil microorganisms in late successional ecosystems: a continental-scale study. *Ecology* 75:2333-2347.

Zwerman, P.J. 1947. The value of improved land use as measured by preliminary data on relative infiltration rates. *Journal of the American Society of Agronomy* 39:135-140.

CHAPTER 29

Soil Quality Indices of Piedmont Sites under Different Management Systems

Betty F. McQuaid and Gail L. Olson

I. Introduction

The Agricultural Lands Resource Group of the Environmental Monitoring and Assessment Program (EMAP) examined soil quality of agroecosystems in the Southeast (1992, North Carolina), Midwest (1993, Nebraska), and mid-Atlantic (1994, West Virginia, Maryland, Pennsylvania, Virginia, and Delaware) using field-scale physical and chemical measurements (Campbell et al., 1994; Hellkamp et al., 1994a,b). These studies did not adequately address biotic indicators of soil quality or the relationship between management systems and soil quality. This study was initiated to address: (1) the feasibility of using microbial biomass carbon as a biotic indicator for evaluating soil quality at a field-scale and (2) the sensitivity of microbial biomass carbon and selected soil quality indicators to various management systems.

II. Methods

This study was initiated in September 1994 in the Piedmont physiographic region of Maryland, Pennsylvania, and North Carolina (Figure 1). Twenty-four paired sites were selected by USDA Natural Resources Conservation Service (NRCS) soil scientists and EMAP staff. Each paired site shares the same soil map unit, and consists of a least one conventionally-tilled and one no-till site. Organically managed (minimum input) sites, which may or not be tilled, were compared with pairs in three cases. Sites were selected based on the following criteria:

(1) The sites had to be in close proximity to each other and occupy the same soil map unit.
(2) The sites represented some of the major crop rotations and tillage practices of the Piedmont province. Paired sites contained the same crop or rotation, where possible.
(3) The sites represented the major soil types and crops in the Piedmont province.

Soils on the various field sites include Chester, Glenelg, Manor, Mt. Airy, Penn, Cecil, Appling, Wedowee, Helena, Vance, and Hiwassee (Table 1). Crops on the various field sites include tobacco (*Nicotiana tobacum* L.), corn (*Zea mays* L.), wheat (*Triticum aestivum* L.), soybean (*Glycine max* L. Merr.), and fescue (*Festrica elatior*, L. Linn.) or corn/soybean, corn/wheat, corn/alfalfa (*Medicago sativa* L.), corn/rye (*Secale cereale* L.) rotation, except on the organically managed sites where berries (*Rubus* L. and *Vaccinium corymbosum* L.), cut flowers (*Zinnia elegans* L.), or vegetables

Figure 1. Location of study sites.

(*Lycopersicum esculentum* L. and *Capsicum* L.) were grown. Farm fields were preferred, but due to time constraints and the difficulty of finding pairs, eight plots managed by North Carolina Agricultural Station personnel were included in the study. Field sites were generally between 2.5 and 25 hectares and the plots averaged 0.6 by 15 m.

Soil was composited from 20 subsamples along a random transect in each field. A starting point for the field transect was located by a random number of rows along and paces into the field, starting at the first corner of entry into the field. (Samplers were provided with random row and pace counts before sampling.) The transect ran at a 45 degree angle to the rows, 25 meters to the northeast and 25 meters to the southwest. Plots were transected diagonally, covering the entire length of the plot. Twenty (10 cm) Qakfield probe soil samples were taken along each transect and composited and 1000 g of the composited sample was sent to the USDA-NRCS National Soil Survey Center Laboratory in Lincoln, Nebraska for chemical and physical analysis.

Chemical indicator analyses included organic carbon (6A3), total nitrogen (6B2), total carbon (6A2d), pH (in water, 8c1a), cation exchange capacity (CEC at pH 7, 5A6), sodium absorption ratio (5E), and electrical conductivity (8A1a) (USDA-NRCS, 1994; see method code). Physical indicator analyses included particle-size analysis by pipette (3A1) and wet aggregate stability (4G1) (USDA-NRCS, 1994; see method code). Another 50 g of soil was sent to USDA Agricultural Research Service (ARS) laboratory in Pullman, Washington for microbial biomass analysis (biotic indicator). Samples were analyzed under the direction of Dr. Jeff Smith, USDA-ARS soil microbiologist, using the chloroform fumigation method described in Jenkinson and Powlson (1976). Statistical analyses were computed using PROC GLM in SAS (SAS Institute INC., 1989). The Kruskal-Wallis Test was used to test management treatment differences. A soil quality assessment framework suggested in Karlen and Stott (1994) and the threshold values for the soil quality indicators presented in Karlen et al. (1994) were used to derive a soil quality index for each field site and then to compare soil quality across management system.

Table 1. Taxonomic classification and site characteristics of study soils

Maryland and Pennsylvania Piedmont sites	
Average elevation- 100 to 300 m	
Average annual rainfall- 900 to 1150 mm	
Average annual temperature- 10 to 14 °C	

Soil series	Taxonomic classification[a]
Chester	Fine-loamy, mixed, mesic Typic Hapludult
Glenelg	
Manor	Coarse-loamy, micaceous, mesic Typic Dystrochrept
Mt. Airy	Loamy-skeletal, micaceous, mesic Typic Dystrochrept
Penn	Fine-loamy, mixed, mesic Ultic Hapludult

North Carolina Piedmont sites	
Average elevation- 100 to 400 m	
Average annual rainfall- 1150 to 1400 mm	
Average annual temperature- 14 to 18 °C	

Soil series	Taxonomic classification
Appling	Clayey, kaolinitic, thermic Typic Kanhapludult
Cecil	
Helena	Clayey, mixed, thermic Aquic Hapludult
Hiwassee	Clayey, kaolinitic, thermic Rhodic Kanhapludult
Vance	Clayey, mixed, thermic Typic Hapludult
Wedowee	Clayey, kaolinitic, thermic Typic Hapludult

[a] Soils classified according to Soil Survey Staff, 1996.

III. Results and Discussion

A. Indicators

In this study, the EMAP group researched the possibility of using chemical, physical, and biotic indicators for evaluating soil quality for plant growth. These indicators were used to evaluate the effects of management system on soil quality. Ten indicators in three different categories were measured: physical (two indicators, related to water retention), chemical (seven indicators, related to water retention, nutrient cycling, and toxicity), and biotic (one indicator). The biotic indicator selected, microbial biomass, represents the total amount of viable biomass active in nutrient cycling and organic matter breakdown. Microbial biomass is a good candidate (for a biotic indicator) in this study because the method is affordable and responsive to field-scale stress such as tillage (Karlen et al., 1994 and Doran, 1987). The other chemical and physical indicators were selected from summaries found in NRC (1993). Table 2 shows the median of each indicator by management system.

Overall, conventional sites have significantly lower pH, total carbon, CEC, organic carbon, total nitrogen, aggregate stability, and microbial biomass carbon than no-till sites ($p<0.01$). Untilled and organic sites with lower values of these indicators (except pH and aggregate stability), were not significantly different ($p<0.01$), although sample size is too small to draw definite conclusions. These findings agree with Doran (1987) who found higher microbial biomass in no-till soils than plowed soils. Similar results are seen when the indicator data are analyzed by site pairs (same soil type but different management system). CEC, pH, organic carbon, total carbon, total nitrogen, and aggregate stability are significantly higher for no-till systems ($p<0.01$) than conventional systems within site

Table 2. Median indicator levels by management system[a]

Management system	Organic carbon/clay	Microbial biomass (mg C kg^{-1})	CEC (cmol kg^{-3})	Clay (%)	pH	Total N (mg cm^{-3})	Total C (mg cm^{-3})	Aggregate stability (%)	SAR	EC (ds m^{-1})
Conventional (n = 26)[b]	0.07	453 (n=53)	6.6	14.9	5.6	1.70	12.7	3.5 (n=10)	0.05	1.0
No-till (<3 years) (n=9)	0.09	514 (n=10)	11.0	17.7	5.9	2.34	22.0	9.5 (n=6)	0.11	0.7
No-till (>3 years) (n=18)	0.08	487 (n=21)	12.0	20.3	6.6	2.34	20.9	11.0 (n=14)	0.05	0.6
Untilled (n=7)	0.08	392 (n=18)	4.9	9.3	5.7	1.04	11.3	28.0	0.7	0
Organic (n=7)	0.12	425 (n=30)	3.4	7.2	6.4	1.17	10.4	57.0	0	1.1
All (n=67)	0.08	453 (n=132)	9.5	16.5	5.9	1.82	16.5	8.5 (n=30)	0.06	1.0

[a] All indicators are significantly different at the $p < 0.01$ level for the conventional and no-till treatments.
[b] Population numbers by tillage are the same for all indicators except where noted.

pairs. However, microbial biomass carbon was not found to be significantly different (P<0.01) within paired sites. Perhaps the significant difference between no-till and conventional sites with respect to microbial biomass carbon is an artifact of crop type and rotation rather than tillage. No-till sites in this study supported tobacco, corn, and corn/soybean, corn/wheat, corn/alfalfa, and corn/rye rotations; there were no corn/alfalfa or corn/rye rotations among the conventional sites. Conventional sites in this study were predominantly in monoculture. In concurrence with this is a study by Insam et al. (1989) in which higher microbial biomass carbon was found in rotations than in monoculture. Also, in support of this hypothesis, is work done by the EMAP group in the mid-Atlantic states, which reported higher microbial biomass carbon among hay crops than field corn or soybeans (Hellkamp et al., 1994). Results of this study suggest that a closer investigation (and adjustment) of microbial biomass carbon by soil type, state, and crop rotation is warranted; however, this data set is not large enough to accomplish this level of scrutiny.

B. Soil Quality Index (SQI)

For each site, each indicator was rated as low (1), moderate (2), or high (3) quality for supporting plant growth (Table 3). The factors were averaged to give an overall soil quality index for the site. Mean soil quality indices are presented for all sites and by management system (Table 4). For all sites, the mean soil quality index (SQI) was 2.14, the low end of the moderate range. Tillage seems to a have a negative effect on soil quality. That is, soil quality indices were significantly higher in no-till sites than conventional. In agreement with this are the data presented in Karlen et al. (1994). Mean SQI for untilled sites (pasture and hay fields) was between the mean SQI of the no-till and conventional sites, but not significantly different (p<0.01).

Total carbon and microbial biomass carbon tended to lower the overall SQI for each site but did not affect the relative ranking of the soil quality indices by management (Table 5). Conventional sites had the lowest SQI whether or not the carbon indicators were included in the SQI determination, whereas the no-till sites had the highest SQI.

IV. Summary and Conclusions

Soil quality for plant growth was evaluated at paired management sites using individual physical, chemical, and biotic indicators, and a soil quality index calculated from several indicators. Overall, tillage had a negative effect on soil quality. Conventional sites had significantly lower pH, total carbon, CEC, organic carbon, total nitrogen, aggregate stability, and microbial biomass carbon than no-till sites (p<0.01). Investigation of the differences within site pairs (same soil and crop type) by tillage also showed lower CEC, pH, organic carbon, total carbon, total nitrogen, and aggregate stability for no-till systems (p<0.01) than conventional. In contrast, microbial biomass carbon was not found to be significantly different (P<0.01) within site pairs. These results suggest that the significant difference between no-till and conventional sites (across sites pairs) with respect to microbial biomass carbon appears to be an artifact of crop type and rotation rather than tillage.

Soil quality indices were also lower for conventional management systems than for no-till systems (p<0.01). The addition of microbial biomass carbon and total carbon levels to the soil quality index calculation lowered the soil quality index for all sites but did not affect relative ranking of soil quality index by management. Results of this study warrant a more intensive field-scale assessment of the soil quality indicator, microbial biomass carbon, by soil type, state, and crop rotation.

Table 3. Ratings for each of the indicators used in the soil quality index

Indicator	1 = Low	2 = Moderate	3 = High
Clay	<18% >35%		18 - 35%
Cation exchange capacity	--	<15 cmol kg^{-1}	>15 cmol kg^{-1}
Organic[a] carbon / clay	<0.05	0.05 - 0.09	>0.09
Soil pH	<4 <9	>7.5 <6.1	6.1 - 7.5
Microbial biomass	<35 mg C kg^{-1}	350 - 700 mg C kg^{-1}	>700 mg C kg^{-1}
Total nitrogen	<2.0 mg cm^{-3}	2 - 3 mg cm^{-3}	≥ 3.0 mg cm^{-3}
Total carbon	<20 mg cm^{-3}	<10%	≥ 30 mg cm^{-3}
Aggregate[a] stability	--	--	≥ 10%
Electrical conductivity	>2 ds m^{-1}	--	≤2 ds m^{-1}
Sodium adsorption ratio	>4		--

[a] Thresholds from Karlen et al. (1994) except for aggregate stability and organic carbon/clay ratio which were tailored to account for regional soil differences (personal communication, Dr. Robert Grosman, Soil Scientist, National Soil Survey Center, Lincoln, NE).

Table 4. Mean, median, 25th, and 75th quantiles of the soil quality index by management system

Management system	SQI mean	SQI 25th quantile	SQI median	SQI 75th quantile
Conventional (n=26)[a]	1.97 ± 0.02	1.82	2.00	2.09
No-till (0-3 years) (n=9)[a]	2.26 ± 0.07	2.00	2.31	2.45
No-till (> years) (n=18)[a]	2.33 ± 0.08	2.09	2.36	2.54
Untilled (n=7)	2.14 ± 0.05	1.91	2.00	2.00
Organic (n=7)	2.26 ± 0.03	2.00	2.09	2.36
All (n=6)	2.15 ± 0.07	1.91	2.09	2.30

[a] SQI means are significantly different at the $p \leq 0.01$ level for the conventional and no-till treatments only.

Table 5. Mean soil quality index (SQI) by management system

Management system	SQI all indicator components	SQI without microbial biomass carbon	SQI without total carbon
Conventional (n=26)[a]	1.97 ± 0.02	2.01 ± 0.02	2.06 ± 0.03
No-till (0-3 years) (n=9)[a]	2.26 ± 0.07	2.30 ± 0.07	2.33 ± 0.02
No-till (> years) (n=18)[a]	2.33 ± 0.08	2.37 ± 0.09	2.40 ± 0.06
Untilled (n=7)	2.06 ± 0.06	2.14 ± 0.04	2.16 ± 0.07
Organic (n=7)	2.19 ± 0.05	2.26 ± 0.03	2.26 ± 0.02
All (n=6)	2.14 ± 0.07	2.19 ± 0.07	2.22 ± 0.06

[a] SQI means are significantly different at the $p \leq 0.01$ level for the conventional and no-till treatments only.

References

Campbell, C.L., J. M. Bay, A.S. Hellkamp, G.R. Hess, K.E. Nauman, D.A. Neher, G.L. Olson, S.L. Peck, B.A. Schumacher, K. Sidik, M.B. Tooley, and D.M. Turner. 1994. Environmental Monitoring and Assessment Program-Agroecosystem Pilot Field Program Report-1992. EPA/630/R-94/014. U.S. Environmental Protection Agency, Washington, D.C.

Doran, J.W. 1987. Microbial biomass and mineralizable nitrogen distributions in no-tillage and plowed soils. *Biol Fertil. Soils* 5:68-75.

Hellkamp, A.S., J. M. Bay, K.N. Easterling, G.R. Hess, B.F. McQuaid, M.J. Munster, D.A. Neher, G.L. Olson, K. Sidik, L.A. Stefanski, M.B. Tooley, and C. Lee Campbell. 1994a. Environmental Monitoring and Assessment Program-Agricultural Lands Pilot Field Program Report-1993. EPA/620/R-95/004. U.S. Environmental Protection Agency, Washington, D.C.

Hellkamp, A.S., J. M. Bay, G.R. Hess, B.F. McQuaid, M.J. Munster, G.L. Olson, S.L. Peck, K. Sidik, M.B. Tooley, S. Shafer, and C. Lee Campbell. 1994b. Environmental Monitoring and Assessment Program-Agricultural Lands Mid-Atlantic Interim Study Report 1994. U.S. Environmental Protection Agency, Washington, D.C.

Insam, H., D. Parkinson, and K.H. Domsch. 1989. Influence of macroclimate on soil microbial biomass. *Soil Biol. Biochem.* 2:211-221.

Karlen, D.L. and D.E. Stott. 1994. A framework for evaluating physical and chemical indicators of soil quality. p. 53-72. In: J.W. Doran, D.C. Coleman, D.F. Bezdicek, and B.A. Stewart (eds.), *Defining Soil Quality for a Sustainable Environment.* SSSA Special Publ. 34, Soil Sci. Soc. Am., Madison, WI. 244 pp.

Karlen, D.L., N.C. Wollenhaupt, D.C. Erbach, E.C. Berry, J.B. Swan, N.S. Nash, and J.L.Jordahl. 1994. Crop residue effects on soil quality following 10-years of no-till corn. *Soil and Tillage Research* 31:149-167.

Jenkinson, D.S. and Powlson, D.S. 1976. The effects of biocidal treatments on metabolism in soil- I. Fumigation with chloroform. *Soil Biol. Biochem.*, 8:167-177.

SAS Institute Inc. 1989. SAS User's Guide: Statistics, Ver. 6 Edition, Cary, NC, 856 pp.

Soil Survey Laboratory Staff. 1992. Soil Survey Laboratory Methods. Soil Survey Investigation No. 42. Version 2.0. USDA-Natural Resources Conservation Service. National Soil Survey Center. U.S. Government Printing Office. 1993/758-492/no program. 400 pp.

Soil Survey Staff. 1996. Keys to Soil Taxonomy. USDA Soil Conservation Service. Sixth edition. U.S.Government Printing Office. ISBN 0-16-048848-6. 644 pp.

CHAPTER 30

Impact of Carbon Sequestration on Functional Indicators of Soil Quality as Influenced by Management in Sustainable Agriculture

C.M. Monreal, H. Dinel, M. Schnitzer, D.S. Gamble, and V.O. Biederbeck

I. Introduction

The extensive use of mechanical tillage for weed control and seedbed preparation has induced a deterioration in the quality of Canadian soils (Sparrow, 1984). Cultivation decreased soil organic matter (SOM) in western and eastern Canada (McGill et al., 1981; Martel and MacKenzie, 1980). SOM losses reduced crop nutrient supply (Janzen, 1987) and the size and stability of soil aggregates leaving the soil prone to wind and water erosion (Monreal et al., 1995a,b). Erosion reduced topsoil thickness and the grain yield of wheat (Monreal et al., 1995b). Excess cultivation increased bulk density, decreased porosity, and compacted soils (de Jong, 1981; Low, 1972). The latter affects the water and air regimes of soil that may restrict root growth (Eaves, 1972). Deterioration of chemical, physical, and biological soil properties results in decreased soil quality.

Soil organic matter together with physical properties have been proposed as indicators of soil quality (Larson and Pierce, 1991; Doran and Parkin, 1994). Living and nonliving SOM components influence air and water infiltration and water storage. Other properties influenced by SOM are the tilth, fertility, and structure stability of soils. Gregorich et al. (1994) proposed a minimum data set including various SOM pools to evaluate soil quality. Healthy plant growth also needs adequate physical environment. Soil structure refers to the size, shape, and arrangement of aggregates and pore space. Soil structure influences the function and growth of roots, the movement and storage of air and water, nutrient cycling, and habitat diversity for soil organisms. Measurements of soil aggregate stability and the nonlimiting water range indicate soil structure condition (Gregorich et al., 1994; Topp et al., 1995). Malik et al. (1965) showed that aggregate stability does not always correlate with the total content of SOM, but often relates to specific organic constituents (Mehta et al., 1960) like long-chain aliphatics and carbohydrates (Capriel et al., 1990; Haynes and Swift, 1990), and soil minerals (Nwadialo and Mbagwu, 1991). During aggregate formation, clusters of soil particles may be bound through the physical action of roots and fungal hyphae (Oades and Waters, 1991), and properties of water repellency and adhesiveness of soil organic constituents (Wallis and Horne, 1992; Martens and Frankenberger, 1993).

Recent efforts made to determine the health of Canadian soils used indicators to evaluate changes in SOM, soil structure, erosion, salinity, and contamination by agrochemicals at a national level (Acton and Gregorich, 1995). Prevention of soil deterioration at the national level, however, requires sensitive indicators assessing the impact of management practices at the farm. The latter can be achieved only by establishing indicators capable of evaluating vital soil functions sustaining plant growth. For assessing agricultural sustainability, quantifiable soil quality indicators then need to be integrated in time and space and linked to economic and sociological factors (Park and Seaton, 1996). Our chapter presents a conceptual multidisciplinary soil environmental model to support the use of indicators in relation to agroecological functions. The basis for the conceptual model is supported with published information and data obtained from long-term crop rotation plots and soil quality benchmark monitoring sites in western and eastern Canada, respectively.

II. Materials and Methods

A. Crop Rotations and Soils

The long-term crop rotation plots in western Canada used in this study were established on a Brown Chernozem in 1967 at Swift Current, Saskatchewan. From a total of twelve crop rotations, a subset of two rotations with plots (40.0 m x 10.5 m) replicated three times, was selected for this study. Namely, wheat-fallow (WF) and continuous wheat (CW), each fertilized according to soil tests. Details on experimental design, rates of fertilizer addition, and plot layout were described by Campbell et al. (1983) and Monreal et al. (1995b). In eastern Canada, samples were taken from the 0-8 cm depth of the Ap horizon of a Gray Brown Luvisolic soil at Clinton, Ontario in the fall of 1994. The site is oriented in a northeast-to-southwest direction and has been managed under a corn-soybean-wheat crop rotation since 1976 with conventional tillage (CT) and zero-tillage (ZT). Two classes of soil texture (SiCL and FSL) predominate at the site, and plot size was 561 m x 9 m for each tillage treatment. Samples were also taken from a Gleysol at a soil quality benchmark monitoring site located at St-Marc, Québec. The Ap horizon of plots (size 88 m x 576 m) was sampled in the spring of 1989, 1992, and 1994. Pedological information and details on the experimental design and cropping history since 1960 at St-Marc were reported by Nolin et al. (1995).

Soil sampling involved taking six soil cores 7.5 cm (diam) within a radius of 2 m from the center of each plot at Swift Current. Four cores were taken at a distance of 3 m within each textural class at Clinton, and fifteen cores were taken at 30 m intervals at St-Marc. Cores were pooled and mixed thoroughly to obtain representative samples per plot or treatment. At Swift Current, soil cores were divided according to their genetic horizons and into 7.5 and 15 cm segments to the bottom of the B horizon. Soil samples were air dried, ground to pass a 2 mm sieve, and stored before chemical analysis and measurement of ^{137}Cs activity. At Swift Current, a second set of fresh moist samples were taken from each plot and the 0-7.5 cm depth with a shovel. Moist soil samples were placed in plastic bags and stored in the cold at 3°C for 10 days before wet sieving.

B. Organic Carbon, Microbial Populations, Erosion, and Aggregate Stability

The air-dried Brown Chernozemic samples were ground to pass a 100 mesh (150 µm) sieve before elemental analyses. Separate soil samples were used to measure total and inorganic C. Total C (TC) was measured by dry combustion in a Leco furnace model CHN-600 (Nelson and Sommers, 1982). Inorganic C (IC) was determined by measuring carbon dioxide evolved after digestion with 6M HCl (Tiessen et al., 1983). OC was determined by difference between TC and IC. Aerobic prokaryotes (bacteria + actinomycetes) were enumerated by dilution plate counting on soil extract agar. Aerobic eukaryotes (filamentous fungi + yeasts) were determined by soil dilution plating on rose bengal-streptomycin agar (Biederbeck et al., 1996). Chemoautotrophic nitrifiers were determined by most probable number technique with selective broth medium (Sarathchandra, 1979).

Soil erosion was determined in air-dried samples using the ^{137}Cs technique described by Pennok and de Jong (1987). The ^{137}Cs activity in soil samples was measured using gamma spectroscopy (de Jong et al., 1982). The proportion of water stable macroaggregates (>250 m) and two microaggregate (50-250 m and <50 m) fractions collected from the Brown Chernozemic soil was determined using wet sieving. Further details on the measurement of ^{137}Cs and wet sieving in the Brown Chernozemic soil were described by Monreal et al. (1995a,b). Water stable aggregates at St-Marc were determined in air-dried samples sieved through 2 mm and in aggregates > 150 μm (Dinel et al., 1992).

C. Organic Carbon Dynamics

The Century model version 4.0 of Parton et al. (1994) was used to simulate the dynamics of OC and ON in the 0-30 cm depth of soils from each treatment. Initial SOM pool values were distributed as follows: active (3%), slow (45%), and passive (52%). The model uses a monthly time step and the driving variables include: monthly average maximum and minimum air temperature, monthly precipitation, and soil texture. Climatic data for simulation runs was obtained from databases maintained by Agriculture and Agri-Food Canada and Environment Canada. In our study, the plant production subroutine was parameterized to simulate mean grain yields within 10% of that measured *in situ*. The parameter PRDX was altered to simulate grain production from old and new crop varieties, and to provide an acceptable estimate of the amount of plant residue returned to the soil after harvest. Initial OC content in the 0-30 cm depth at Swift Current was estimated from ON concentration reported by Biederbeck et al. (1984) and assuming a C-to-N ratio of 9.8.

D. Chemical Characterization of Soil Organic Matter

Soil organic matter in the Brown Chernozemic soil was chemically characterized using pyrolysis-field ionization mass spectrometry (Py-FIMS). For Py-FIMS, 5-7 mg of soil sample were transferred to a quartz micro-oven and heated linearly in the direct inlet system of the mass spectrometer from 50 to 750°C at a rate of 1°C s^{-1}. A double-focusing Finnigan MAT 731 mass spectrometer (Finnigan MAT, Bremen, Germany) was used (Schulten, 1987). The ion source was kept at a pressure below 1 mPa and at 250 C. The volatile thermal degradation products of the samples were ionized in high electrostatic fields and electrically recorded by repetitive magnetic scans. To avoid condensation of the volatilized products during the recording of the FI mass spectra, the emitter was flash-heated to 1500°C between magnetic scans. About 40 magnetic scans per sample were recorded over the mass range m/z 50 to m/z 750. Three analytical replicates per sample were run. The FIMS signals of all spectra were integrated and plotted using a Finnigan SS2000 data system to produce summed spectra. Individual ion intensities were assigned to seven classes of compounds according to their mass numbers. The assignment of the pyrolysis products was based on determinations of thermal properties (Schulten, 1987; Leinweber et al., 1992), accurate mass measurements (Hempfling et al., 1988; Hempfling and Schulten, 1990), Curie-point gas chromatography-mass spectrometry (Hempfling and Schulten, 1991; Schnitzer and Schulten, 1992), extensive National Institute of Science and Technology, Wiley library searches, and pyrolysis-mass spectrometry investigations of model polymers. The coefficient of variation for mass signals above 0.2% relative abundance of the summed spectra was 6%. Quantification of classes of SOM compounds was based on the total ion intensities (TII) detected in the mass spectrometer after pyrolysis. The ion intensity of generated molecular ions is proportional to the concentration of molecular ions, and equal to the sum of all individual molecular ion intensities. In this chapter, quantity of compounds is expressed as a number of molecular ions per mg sample detected by the mass spectrometer. Soil lipids in the Gray Brown Luvisolic and Gleysolic soils were extracted and determined using the method of Dinel et al. (1996). Kinetic analysis to describe the dynamics of metal ions followed the method of Lam et al. (1996).

III. Results and Discussion

A. Soil Quality Indicators Model

The premise of the model examined is that a set of measurable indicators must respond to anthropogenic disturbances and their ecological functions associate with crop growth and environmental quality. The multidisciplinary approach uses the physical properties of soil and SOM that influence ecological functions in sustainable agriculture.

The model of Figure 1 includes some physical and SOM components influencing plant growth and C sequestration. Management and organic components influence soil structure, nutrient cycling, the capacity to filter contaminants, and the community structure and viability of soil organisms. These ecological functions are essential for sustained crop production and a safe environment. For soil quality assessments, living and non-living SOM components need to be related to specific ecological functions in terrestrial systems. For example, the supply of available plant N depends on the mineralization of soil organic N compounds (ON) together with trophic interactions between soil microorganisms and fauna (Rutherford and Juma, 1992). The incidence of plant disease is related to the nutrient status of the growing environment of pathogens (Chen et al., 1988). Soil structure stability is associated with SOM molecules and the activity of soil fauna (Monreal et al., 1995a; Juma, 1993). Retention of heavy metals by SOM moieties and organo-mineral complexes (Gamble, 1983; Schnitzer, 1995) enable soils to act as environmental filters. The present conceptual model includes indicators that may be used to establish linkages with economic and sociological facets of sustainable agriculture (Figure 1). Defining functional soil quality indicators helps identify horizontal links between the three facets of sustainable agriculture, and vertical links with policy makers and decision making authorities (Park and Seaton, 1996).

B. Physical Indicators of Soil Quality

Effective soil rooting depth and adequate structural support for plants are functions providing favorable environments for root growth, plant development, and, thus, C sequestration. Rooting growth patterns are mostly a function of the genetic pool of plant species (Taylor and Terrel, 1982). The depth and volume of soil needed for optimum plant growth is controlled by soil forming processes, erosion, and cropping practices. Solum depth and carbonate content were tested as indicators of soil quality and provided information on effective rooting depth in the Brown Chernozemic soil. Figure 2 indicates that the depth of solum was 58 cm under CW, or 16% greater than under WF. A calcareous C horizon with 16% (w/w) calcite equivalent (i.e., 0.07% inorganic-C, w/w) underlies the experimental site (Ayres et al., 1985). The greater solum depth under CW which was paralleled by greater topsoil thickness (Monreal et al., 1995b), provided an improved growing environment for wheat, facilitated root penetration, and increased the capacity of soils to store available water and plant nutrients. Soil erosion reduced the solum depth under WF, and appears to be slowly degrading the chemical environment for root growth. The rate of soil erosion under WF was 22 t ha^{-1} y^{-1} or ten times higher than under CW (Monreal et al., 1995b). In 1990, the concentration of inorganic C in the B horizon of the WF plots was 23% (w/w) calcite equivalent, and 10 times higher than under CW (Figure 2). Carbonates of Mg and Na, which are common to Canadian Prairie soils, induce phytotoxicity and affect crop yield (Szabolcz, 1979). Effective rooting depth may be easily determined by measuring solum depth in the field and conducting chemical analysis of samples in the laboratory.

Impact of Carbon Sequestration on Functional Indicators of Soil Quality as Influenced by Management   439

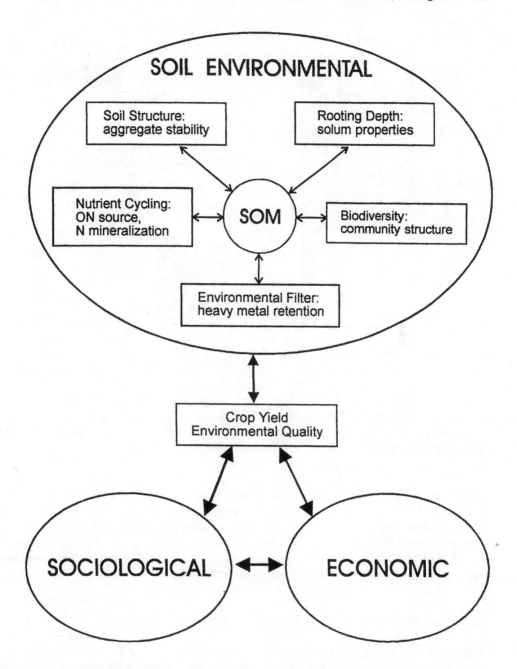

Figure 1. A conceptual model for soil quality indicators with ecological functions in sustainable agriculture.

Aggregate stability is a physical indicator of soil quality because it affects crop establishment, growth, and yield (Rynasiewicz, 1945). The use of this indicator is supported with data on water stable aggregates measured in the Brown Chernozemic soil. Relative to CW, four tillage operations under WF disrupted aggregates and decreased the proportion of water stable macroaggregates > 250 m

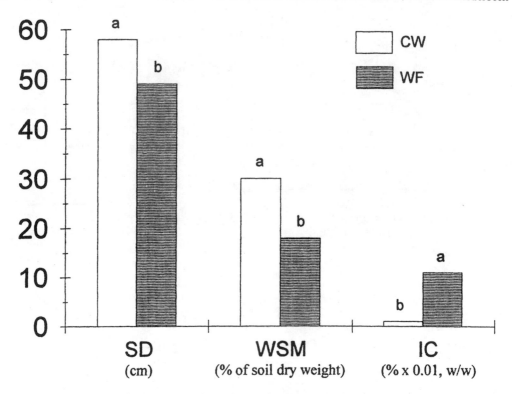

Figure 2. Physical and chemical indicators: solum depth (SD), proportion of water stable macroaggregates (WSM) in the 0-7.5 cm depth, and the content of inorganic carbon (IC) in the B horizon of a Brown Chernozemic soil under wheat-fallow (WF) and continuous wheat (CW). Columns with different letters are significantly different at p=0.05.

(Figure 2). The decline in the proportion of macroaggregates > 250 m under WF was paralleled by higher soil and OC losses. Cultivation for 90 y also reduced the proportion of macroaggregates > 250 m in a Gleysol of Canada (Monreal et al., 1997b). Reducing aggregate stability with cultivation increases compaction and bulk density, and decreases porosity, thereby affecting the air and water regimes of Canadian Prairie soils (Cameron et al., 1981). Our results showed that solum depth and aggregate stability are physical indicators to assess the effects of management on soil quality.

C. Organic Carbon Dynamics an Indicator of Trends

SOM plays a key role in the model of Figure 1. OC is the backbone of SOM and once it enters the soil is subject to numerous biological, chemical, and physical reactions that influence SOM reactivity and thus soil quality. Therefore, examining the dynamic behavior of OC is a first step in assessing the potential impact of climate and management on soil quality. This may be accomplished using SOM simulation models. The Century model has been tested with field data from long-term crop rotations and found to closely simulate OC data (Parton et al., 1994). OC dynamics represented an indicator to assess the impact of land use scenarios on C sequestration in Canada (Dumanski et al., 1997).

We evaluated the influence of crop rotation on OC dynamics in the top 30 cm of the Brown Chernozemic soil. Figure 3 shows that after 24 y of cropping, OC storage increased under CW and decreased under WF. Management significantly affected C sequestration. In 1990, the amount of OC

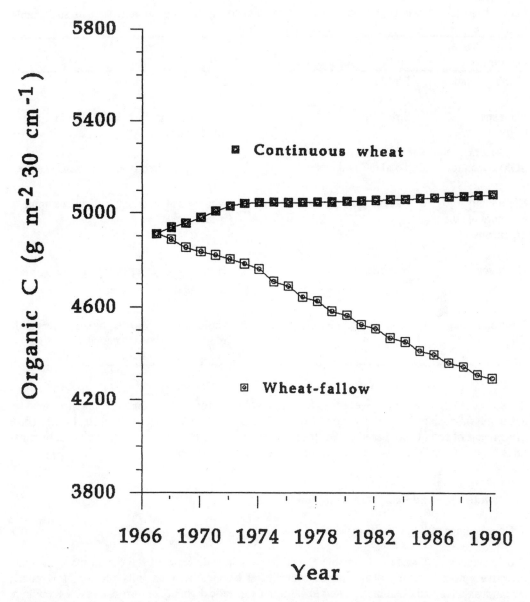

Figure 3. Century simulations describing the dynamics of organic carbon in a Brown Chernozemic soil cropped to continuous wheat (CW) and wheat-<u>fallow</u> (W<u>F</u>).

in the solum (A + B horizons) of the CW plot was 8,604 g m^{-2}, or 31% greater than under W<u>F</u>. Changes in OC level between crop rotations were associated with differences in the rates of soil erosion and plant residue addition. On average, the equivalent annual additions of plant residue (tops + roots) was 1.3 t C ha^{-1} under CW and 0.9 t C ha^{-1} under W<u>F</u>.

Our data shows that OC dynamics can be used as an indicator of trends. Measurements of OC alone, however, do not always provide conclusive information on ecological relationships with soil components and processes important for plant growth. Correlations between aggregate stability and

Table 1. A set of soil quality indicators in relation to agroecological functions in sustainable agriculture

Agroecological function	Soil component/process	Soil quality indicator
Rooting depth	Profile	Solum depth, carbonates
Plant structural support	Structure	Aggregate stability, lipids, lignin dimers, alkylaromatics
SOM dynamics	Total OC and ON	OC and ON trend: +, -, or zero change
Community structure of soil organisms	Bacteria, fungi, insects	Muramic acid, ergosterol, glucosamine, n-fatty acids
N cycling	N mineralization	Heterocyclic and peptide N, rate of mineralization, nitrifiers
Environmental filter	Heavy metal retention	Lipids, carbohydrates, lignin dimers and monomers, alkylaromatics, N compounds

SOM content were negative or weak (Malik et al., 1965; Toogood, 1978); and N mineralization was not always associated with OC content (Monreal et al., 1981; Greer and Schoenau, 1992). Thus, assessments of soil quality based on OC measurements need to be complemented with characterizations of specific SOM pools performing known ecological functions.

D. Bioindicators and Ecosystem Function

1. Communities of Soil Organisms

Crop production systems in sustainable agriculture require a diversity of soil organisms to maintain effective nutrient cycling, plant disease control, gas balance between soils and the atmosphere, decomposition of toxic xenobiotics, and to sustain the genetic diversity of terrestrial systems. Figure 1 and Table 1 show classes of chemical compounds and individual molecules of living and nonliving SOM components as functional indicators used to assess soil quality.

In the model of Figure 1, diversity of soil organisms is represented at the community level by biomarkers of SOM. Both bacteria and fungi synthesize macromolecules which are not normally found in other organisms. Muramic acid has been used for quantification of bacteria, and ergosterol for fungi in soil ecological studies (Rönkkö et al., 1994; Scheu and Parkinson, 1994). Since both muramic acid and ergosterol are end-products of cell wall synthesis, their measurement can assess microbial activity, growth, and viability. N-acetyl glucosamine is found in peptidoglycans of bacteria, and in chitin of fungi and insects (Parsons, 1981). Long chain fatty acids with 26 to 38 carbon chain length are typically synthesized by insects and plants (Salisbury and Ross, 1978), and those with < 18 carbon chain length are associated with bacterial metabolism (Harwood and Russel, 1984). The content and type of phospholipids have also been used to characterize microbial biomass and community structure (Vestal and White, 1989).

Soil biomarkers were chemically characterized to determine the impact of C sequestration on viable biomass and community structure (Table 1). Figure 4 shows the Py-FIMS spectra for soil samples of the WF and CW crop rotations in the Brown Chernozemic soil. In general, there was no qualitative difference in the composition of major classes of compounds between samples of the two crop rotations (Table 2). The total ion intensity (T 11) of pyrolyzable SOM, however, reflected quantitative differences of the effect of crop rotation (Table 2). The main difference was the presence of a wide range of n-fatty acids (up to C_{29}) in samples of the CW rotation relative to those found under WF (up to C_{26}). Detectable biomarkers were m/z 137 (N-acetyl muramic acid), 167 (N-acetyl glucosamine), and 396 (ergosterol), and their relative abundances were higher under CW than under WF. The chemical quality of SOM in whole samples was similar to that reported for aggregate size fractions obtained from the same soil and crop rotation plots (Monreal et al., 1995a).

Differences in the relative abundance of biomarkers indicate that changes in the community structure of soil organisms were induced by the type of crop rotation. The CW rotation with aggraded levels of OC increased the diversity of soil organisms and the number of viable bacteria and fungi. Fatty acids > 29 C chain length were almost undetected by Py-FIMS, suggesting a low number or possible absence of soil insects, especially under WF. The impact of C sequestration on microbial communities was also supported with data from a study using more conventional methods of soil microbiology. Table 2 shows that the number of prokaryots (bacteria + actinomycetes) and eukaryots (fungi + yeasts) was higher under CW than under WF. We hypothesized that the higher addition of plant residue under CW favored a greater viable biomass by increasing the number of growth sites and the supply of carbon and energy sources. In addition, greater frequency of tillage under WF decreased macroaggregate stability (Figure 2), thereby reducing living habitats for soil organisms (Monreal and Kodama, 1997b). The enhanced effect of C sequestration on community structure found under CW in the semiarid Brown Chernozemic soil was similar to that reported for effects caused by improved management on groups of microorganisms, fauna, and similar biomarkers in humid cultivated soils of the world (Andrén et al., 1990; Hendrix et al., 1986). The sensitivity of the tested biomarkers to management show their usefulness as indicators for the viability and community structure of soil organisms in sustainable crop production systems.

2. Nutrient Cycling: N Mineralization

Biogeochemical cycling of the earth's elements provides nutrients, carbon, and energy for soil organisms and nutrients for crops. Nitrogen is an essential macronutrient for plant growth. An important proportion of the soil N taken up by nonlegume crops is derived through the mineralization of SOM (Jansson, 1963). Depolymerization and oxidation of SOM involves metabolic activity of soil organisms and the synthesis, activity, and persistance of extracellular soil enzymes. Figure 5 shows that transformation of ON sources (heterocyclic and aliphatic-N) are catalyzed by ammonifiers and nitrifiers together with enzyme complexes acting sequentially on substrates with specific chemical structures. Enzymes involved in N transformations are subject to various control mechanisms at the molecular level (Baumberg, 1973), and soils (Burton and McGill, 1992). Figure 5 also shows common classes of enzymes involved in the mineralization of aliphatic-N compounds like proteins are proteases (E1) and deaminases (E2). Alternatively, nucleases (E4, E5), deaminases and urease (E6) mineralize soil heterocyclic-N-like nucleic acids.

Based on the latter, we examined three soil components involved in N mineralization to evaluate their potential use as soil quality indicators. The three components were type of ON source, rate of N mineralization, and number of nitrifiers. Figure 6 shows that TII for peptide and heterocyclic-N compounds was higher under CW than under WF. Higher rate of N fertilization and lower erosion

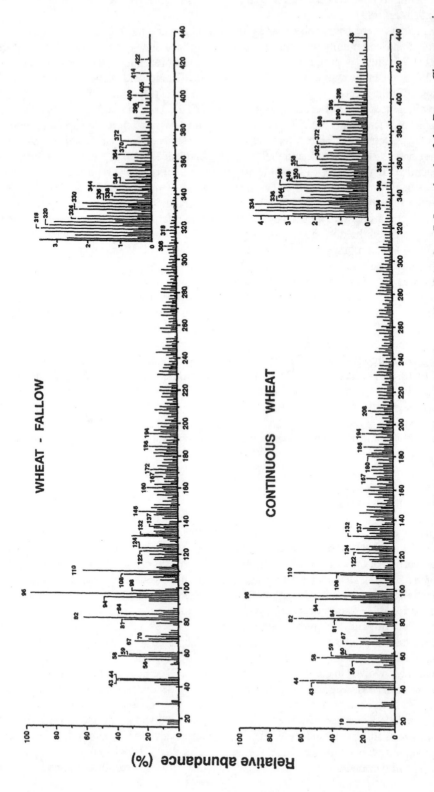

Figure 4. Pyrolysis-field ionization mass spectra of soil organic matter in whole samples of the 0 to 7.5 cm depth of the Brown Chernozemic soil.

Table 2. Microbial and chemical indicators of SOM quality under continuous wheat (CW) and wheat-<u>fallow</u> (WF) crop rotations sequestering different amounts of C

Crop rotation	Community of soil organisms		Biomarkers[3]	Carbo-hydrates	Phenols[4]	Lignin dimers	Lipids[5]	Alkylaro-matics	N com-pounds[6]	Sterols	Peptide N
	Prokaryots[1]	Eukaryots[2]									
				Total ion intensity (counts x 10^3 mg^{-1} sample)							
CW	48.9a	29.3a	++	322	294	84	97	218	217	8	113
WF	37.7b	19.8b	+	300	258	61	74	184	195	2	109

[1] Prokaryots = number of bacteria and actinomycetes x 10^6 g^{-1} soil.
[2] Eukaryots = number of fungi and actinomycetes x 10^4 g^{-1} soil.
[3] Relative abundance for biomarkers characterized by Py-FIMS including m/z 137, 167 and 396: ++ = 10-20%; (+) = <10%.
[4] Class of compounds includes lignin monomers.
[5] Class of compounds includes fatty acids, alkanes, alkenes, n-alkyl esters.
[6] Class of compounds contains mostly heterocyclic N compounds.

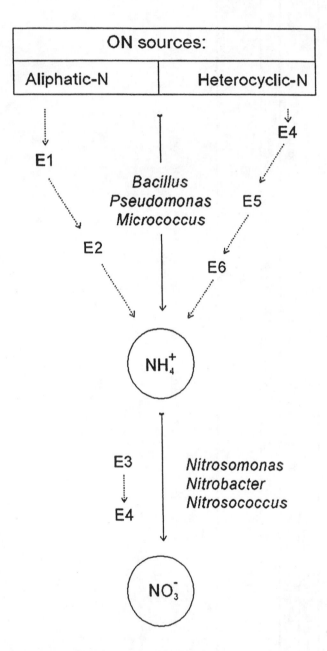

Figure 5. A simplified model with microorganisms and enzyme complexes (E1...E6) controlling the mineralization of two major classes of soil organic N compounds.

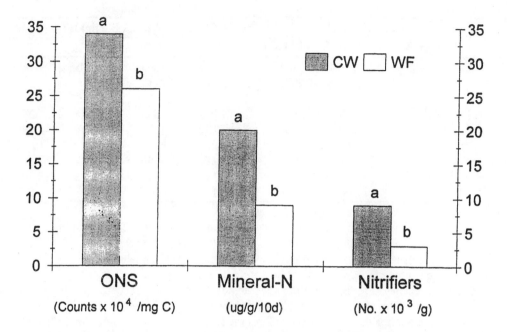

Figure 6. Three biotic indicators assessing N mineralization in the Brown Chernozemic soil: aliphatic and heterocyclic N compounds (ONS), mineralization rate, and number of nitrifiers in the 0-7.5 cm depth of the Brown Chernozemic soil. Columns with different letters are significantly different at p=0.05.

increased the storage of ON in the soil profile under CW (data not shown). The increase in O-N sources under CW was paralleled by significant increases in the rate of N mineralization, and the number of nitrifiers under wheat-fallow (Figure 6) but not under the wheat-fallow phase (data not shown). These data suggest that increased storage of pyrolyzable N due to enhanced C sequestration under CW favoured microbial components and biochemical processes controlling N mineralization. Table 2 also indicates that about 50% of the total N occurs in peptides, so that the remaining N in SOM is nonprotein. Sowden et al. (1977) showed that between 5-7% of the total N in soils is present as amino sugars, which leaves about 40% of the total N to be identified. Recent work by Schnitzer and Schulten (1995) identified more than 50 N-containing compounds in several soils, suggesting these compounds make up a large proportion of the so far unidentified O-N. These compounds included pyrroles, imidazoles, pyrazoles, pyridines, pyrimidines, pyrazines, indoles, quinolines, N-derivatives of benzene, alkyl nitriles, and aliphatic amines. Our data showed that O-N sources, rate of N mineralization and the number of nitrifiers were components of the N cycle that responded to management. These three indicators can be used to evaluate N mineralization in sustainable agriculture.

3. Stability of Soil Structure

Aggregate stability is influenced by organic binding agents like lignin dimers, alkylaromatics, sterols, and other lipids (Monreal et al., 1995a). Aggregate stability may be disrupted by the physical forces

Table 3. Aggregate stability, organic C, and lipid content in the 0-8 cm depth in Gray Brown Luvisolic and Gleysolic soils with different texture, tillage, and crop rotations

Soil	Texture[1]	Tillage[2] and year	WSA[3] (g kg^{-1})	OC (g kg^{-1})	Extractable lipids (mg kg^{-1})[4]		
					DEE	CHCl$_3$	TEL
Luvisolic[5]	Coarse	CT, 1994	591b[7]	35.0a[7]	106a	81b	187b
		NT, 1994	777a	34.0a	111a	129a	240a
Luvisolic	Fine	CT, 1994	353a	17.4a	155a	107a	262a
		NT, 1994	389a	17.3a	155a	104a	259a
Gleysolic	Fine	CT, 1989	643b	29.0a	319b	198b	517b
		CT, 1992	703a	30.0a	447a	257a	704a
		CT, 1994	632b	28.0a	380ab	212ab	592b

[1] Coarse = fine sandy loam; fine = silty clay loam.
[2] Tillage system and year of soil sampling. CT = conventional tillage, NT = zero-tillage.
[3] WSA = water stable aggregates.
[4] DEE = diethyl ether, CHCl$_3$ = chloroform, and TEL = total extractable lipids.
[5] The Luvisolic soil was in a corn-soybean-wheat rotation since 1976.
[6] The Gray Brown Gleysolic soil was cropped to an alfalfa-grass mixture between 1983 and 1990; corn in 1991; sorghum in 1992; and alfalfa-grass hay in 1993 and 1994.
[7] Within columns and for the Luvisolic and Gleysolic fine textured soils, values not followed by the same letter are significantly different at $p < 0.05$. For the Gray Brown Luvisolic coarse textured soil, significant differences in WSA and extractable lipids were found at p 0.055.

of water and tillage implements, and by decomposition of organic binding agents. Disruption of stable soil aggregates reduces the infiltration of water and solutes, increases the risk of erosion, compaction, crusting, and air regime. Therefore, the resistance of soil aggregates to breakdown is an important attribute for the assessment of soil quality. In this chapter, we used aggregate stability as opposed to nonlimiting water range (NLWR) as an indicator of structural stability because it is easily measured and is strongly associated with molecular components of SOM. To the authors' knowledge, relationships between NLWR and SOM or soil mineral components still need to be established.

We examined chemical components of SOM involved in the stabilization of aggregates to assess their response to management in Brown Chernozemic and Gleysolic soils. Table 2 and Figure 2 showed that the higher proportion of water stable aggregates under CW was paralleled by an increase in the TII of the organic binding agents. The higher content of binding agents present under CW than under WF is attributed to higher inputs of crop residues, lower frequency of tillage, and lower rate of erosion (Monreal et al., 1995). Alternatively, increased microbial synthesis and preservation of organic binders in aggregates of the CW may also contribute to these results. We hypothesize that the supply of metabolizable substrates to soil heterotrophs may be lower under WF than under CW, thus inducing greater catabolism of large size molecules under WF. Depletion of labile amino acids under WF with low OC content was induced by rapid turnover of microbial proteinaceous C than in undisturbed soils with high levels of OC storage (Monreal and McGill, 1987).

Table 3 shows that the proportion of water stable aggregates was associated with total soil extractable lipid (TEL) fractions and not with OC in the Gray Brown Luvisolic soil of Ontario and the Gleysolic soil of Québec. Similar results were reported earlier for the role of lipids on soil structure stability in soils of Canada and Germany (Dinel and Gregorich, 1995; Dinel et al., 1992; Capriel, 1990). Noteworthy, unbound lipids favor aggregate stability after increasing the strength of intra-aggregate bonds against the dispersive and dissolution action of water by 72%, and increasing the resistance of aggregates to slaking by 10% (Dinel et al., 1992). In our study, TEL content was influenced by texture, tillage system, and crop rotation, and was more sensitive to management than OC (Table 3). In the Gray Brown Luvisolic soil, areas of fine texture accumulated more TEL than areas with coarse texture. Relative to conventional tillage, sixteen y of cropping with zero-tillage increased TEL content in the coarse but not in the fine textured soil.

Table 3 also shows that unlike OC, TEL content in the Gleysolic soil changed over short periods (2 to 3 y) in response to crop sequence. The content of TEL was not determined in samples taken in 1983; therefore, it is not possible to completely define TEL dynamics between 1983 and 1994. Information for 1989 and 1994 indicates, however, that TEL and water stable aggregate (WSA) fractions increased under the alfalfa-grass mixture, and decreased rapidly following corn and sorghum (Table 3). Changes in TEL content mimicked well those of aggregate stability. The half-life for structural improvement is 4.5 y under forage, and 0.2 y for structural decay under corn production with conventional tillage following a forage crop (Kay, 1990). In general, biosynthesis and storage of soil lipids were favored by cropping to forage. Microbial catabolism of lipids soluble in $CHCl_3$ appeared to be favored in soils cropped to corn and sorghum under conventional tillage. The response of specific macromolecules of SOM to management, especially lipids, indicate their usefulness to assess changes in structure stability in agricultural soils.

4. Heavy Metal Retention

In the model of Figure 1, 'environmental filter' refers to the capacity of SOM to retain chemical pollutants, influence the efficacy and fate of pesticides, and the ability to accept and decompose waste of human origin applied to agricultural land. This section confines its discussion to biophysical properties of SOM that affect the retention of heavy metals, provides examples of kinetics defining the stabilization of heavy metal species, and proposes molecular components of SOM as indicators of environmental quality.

Two key factors influencing the environmental filter function of soils are: the total retention capacity and the rate of adsorption-desorption of metal ions. The adsorption capacity of soils for heavy metals is influenced by SOM content (Schnitzer, 1995), crystalline silicates, and oxides of AL, Fe, and Mn (Tiller et al., 1984; Anderson and Christensen, 1988). The amounts and forms of adsorption and stabilization of metal ions in soils over time may be defined using "*in vivo*" kinetics.

Figure 7a shows data for a kinetic experiment conducted in our laboratories. The dynamics of Zn(II) and Cd(II) were described through soil compartments including the soil solution, adsorbed to surfaces, and stabilized in the interior of soil particles (residue bound). Processes transferring metal ions were the solution phase complex formation and dissociation, labile surface sorption and desorption, and intraparticle diffusion (Gamble et al., 1980). Some of the chemical reactions and mass transfer processes were very fast with large rate constants. Figure 7b shows the impact that binding strengths and process kinetics can have on the chemical speciation of metal ions. This kind of chemical speciation has important consequences for the bio-availability of a metal ion. Kinetic experiments such as depicted on Figure 7a, may also help develop environmental filter indicators for the metal of choice. The complexation of heavy metals with soluble soil organic compounds increases their mobility and plant uptake (Neal and Sposito, 1986). In comparison, organic ligands significantly reduced the thermodynamic activity of metal ions in solution (O'Connor et al., 1984), which suggests that reaction mechanisms between heavy metals and stabilized soil macromolecules may reduce environmental toxicities.

The contributions of SOM components to heavy metal retention can be best understood in terms of the types and numbers of organic chemical functional groups and the metal binding sites derived from them. Two broad categories of binding can exist: weak physical binding or sorption such as outer sphere complexing, ion exchange, hydrogen bonding, and van der Waals interactions; usually,

Table 4. Key for functional groups of SOM in Table 5

Key	Type
1	Carboxyl
2	Alcoholic OH
3	Phenolic OH
4	Amino
5	Alkylamino
6	-SH
7	Aldehyde
8	Ketonic
9	Phosphate
10	Alkylphosphate
11	Porphyrin ring

Table 5. Functional groups in SOM components that bind metal ions

Classes of compounds	1	2	3	4	5	6	7	8	9	10	11	References
Carbohydrates	*	*					*	*				Fruton and Simmonds (1953); Wershaw (1985)
Phenols			*									Mathur and Farnham (1958)
Lignin dimers	*	*	*									Brewster (1948); Wershaw (1985)
Lipids[1]		*							*	*		Fruton and Simmonds (1953)
Alkylaromatics	*	*	*									Mathur and Farnham (1958); Wershaw (1985)
Aromatic-N				*	*							Fruton and Simmonds (1953); Mathur and Farnham (1958)
Proteins	*			*		*						Zeppezauer and Maret (1986); Andreae (1986)
Nucleoproteins				*				*				Fruton and Simmonds (1953)
Sterols	*	*					*					Fruton and Simmonds (1953)
Cytochromes											*	Fruton and Simmonds (1953)

[1] Lipids include phospholipids.

strong binding will be caused by the formation of covalent bonds (Gamble et al., 1980, 1994). The effectiveness of metal ion binding by SOM molecular components is also influenced by the degree of ionization that, in turn, is governed by the solution pH, and the number and type of functional groups in a given macromolecule. Tables 4 and 5 identify some of the functional groups found in various classes of compounds in SOM. The frequency distribution of functional groups among all classes of compounds are: carboxyl > hydroxyl = amino > ketonic = phosphate > aldehyde = porphyrin ring. In terms of numbers of them per gram of OM, the carboxyls and hydroxyls are among the most important. In terms of molecular abundance, lignin dimers and monomers, carbohydrates, alkylaromatics, lipids, proteins, sterols, and heterocyclic N compounds are the most important classes of compounds that bind metal ions.

The chemical structure of SOM is affected by long-term cultivation (Schulten et al., 1995). Changes to the chemical structure of organic adsorbents through management may modify binding sites and thus the mobility of heavy metals. Molecules binding heavy metals may be effective filters

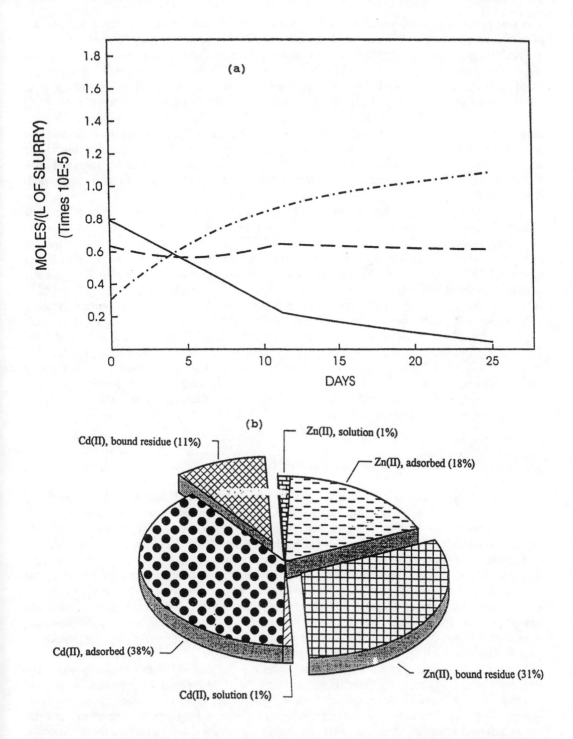

Figure 7. The dynamics of Zn(II) cycling through soil slurry components: solution, – adsorbed, and stable residue (a). Comparison of Zn(II) and (Cd) chemical species in the slurry of a Raisin River soil incubated for 25 d at 25 C (b).

over long periods (many centuries) if their biochemical degradation in soils is minimized. Examples of resistant molecules in SOM are lipids soluble in $CHCl_3$ (Dinel et al., 1992). The potential use of molecular components of SOM as indicators for environmental filter is supported with information reported for long-term crop rotations and soil-plant systems. The abundance of specific molecules of SOM as characterized by Py-FIMS was significantly associated with heavy metal adsorption in "Eternal Rye Cultivation" plots of Germany (Leinweber et al., 1995). Furanone, pentenone, coniferyl aldehyde, coniferyl alcohol, and palmitic acid (nC_{16}) in SOM correlated with Cd adsorption. The latter molecules are better preserved in soils with higher OC storage and tilled less frequently (Monreal et al., 1995a,c). Support for the complexation of heavy metals by amino groups of organic molecules is found in a soil-plant system. Histidine in the root of *(Alyssum lesbiacum)* complexed nickel and drew Ni out of the soil after nitrogen atoms of the amino group donated electrons to Ni to form strong bonds (Krämer et al., 1996).

Based on the latter, we strongly suggest that the mobility and plant uptake of heavy metals may be reduced in crop production systems that favor the synthesis, storage, and preservation of large size molecular components of SOM. The Py-FIMS spectra in Figure 4 and data in Table 2 show that the molecular abundance of the proposed indicators was higher in aggraded than in degraded soils. For example, SOM under CW in the Brown Chernozemic soil was rich in carbohydrates with pentose and hexose subunits (m/z 60, 72, 82, 84, 96, 98, 110, and 132) and in phenols (m/z 94, 110, 122, and 124). In addition, the spectrum shows the presence of a number of monolignins: m/z 180 (coniferyl alcohol), 194 (ferulic acid), 208 (sinapyl aldehyde), 210 (sinapyl alcohol), and 212 (syringyl acetic acid). Present in lower abundance were n-fatty acids at m/z 242 ($n-C_{15}$), 256 ($n-C_{16}$), 270 ($n-C_{17}$), 284, ($n-C_{18}$), and 298 ($n-C_{26}$). Other minor components were n-alkylbenzenes at m/z 330 ($C_6H_5 \cdot C_{18}H_{37}$), 344 ($C_6H_5 \cdot C_{19}H_{39}$), 358 ($C_6H_5 \cdot C_{20}H_{41}$), 372 ($C_6H_5 \cdot C_{21}H_{43}$), 400 ($C_6H_5 \cdot C_{23}H_{47}$) and 414 ($C_6H_5 \cdot C_{24}H_{29}$); and m/z 194 ($n-C_{14}$ alkane), 212 ($n-C_{15}$ alkane), and 422 ($n-C_{30}$ alkane). Preservation of large size molecules needs to be considered when designing management strategies to stabilize heavy metals in soils and to prevent environmental degradation.

IV. Conclusions and Future Research Priorities

A conceptual model assessed soil quality on the basis of: physical, chemical, and biological soil components, and processes of ecological significance in sustainable agriculture. Soil ecological functions were effective rooting depth, structural support for plants, biodiversity, nutrient cycling, and environmental filter. Functional soil quality indicators were tested with data from long-term field trials, and found to be sensitive to farming practices. Adoption of continuous cropping with or without incorporation of forage coupled with zero-tillage and low rates of erosion always increased the content of specific living and nonliving SOM components indicating positive change in soil quality. Some of the latter management practices favored C sequestration only in the Brown Chernozemic and not in the Gray Brown Luvisolic and Gleysolic soils. The ecological functions of SOM molecules depend on their biophysical properties. Our conceptual indicator model sets the basis for developing new technologies for diagnosing soil quality at the farm level.

Solum depth and aggregate stability were sensitive physical indicators of soil quality. The dynamics of OC was an indicator of trends, and specific SOM molecular components were more sensitive than OC content. Basic indicators of soil microbial biodiversity were represented at the community level by biomarkers of soil fungi, bacteria, and insects. Future research needs to relate soil organisms with ecosystem functions vital to sustain crop growth. Modern techniques of molecular biology based on DNA and RNA cellular components need to be used for developing complementary diagnostic tools.

Soil peptide and heterocyclic-N together with mineralization rate and number of nitrifiers were sensitive indicators to management disturbances and assessed the state of N mineralization. Future research needs to identify key soil enzymes responsible for N mineralization and immobilization. There is a need to develop a soil quality index (with key biotic components) for evaluating the N cycle. In this regard, it is important to establish links between biotic and abiotic controls of the N cycle that are expressed at the molecular, microsite, pedon, and landscape level.

Specific molecular components of SOM represented by lipids, alkyl aromatics, lignin monomers and dimers, carbohydrates, and N-containing molecules were sensitive to management and represent indicators of soil structure stability and environmental filter. These SOM molecules together with the kinetics of metal ions may be used to measure and indicate the capacity of soils to filter heavy metals. There is a need to define the components and processes controlling the synthesis, preservation, and decomposition of macromolecules associated with aggregate stability. Also, threshold values at which molecular indicators induce significant changes in the stability of structure and the filtering capacity of soils need to be determined. The latter is facilitated by defining the type and number of heavy metal retention by macromolecules.

While the measurement and use of soil physical indicators appear to use simple and straight forward methodologies, the measurement of chemical and biochemical indicators of soil organic matter quality require conventional and modern technologies of analytical chemistry, microbiology, and biochemistry. The use of effective soil quality indicators requires defining spatial dependencies, and the quantitative relationships existing between bio-indicators and process control (cause - effect) mechanisms. Support for this type of research helps develop technologies useful to farmers, extensionists, policy makers, and all of society.

Acknowledgments

We gratefully acknowledge the contribution of Dr. H.-R. Schulten for the Py-FIMS analysis, and D. Lobb, J. Miller, M. Nolin, and R.P. Zentner for providing the soil samples, and A. Ismaily for his technical support.

References

Acton, D.F. and L.J. Gregorich. 1995. *The Health of Our Soils-Toward Sustainable Agriculture in Canada.* Centre for Land and Biological Resources Research, Research Branch, Agriculture and Agri-Food Canada, Ottawa, Ontario. xiv+138 pp.
Anderson, P.R. and T.H. Christensen. 1988. Distribution coefficients of Cd, Co, Ni, and Zn in soils. *J. Soil Sci.* 39:15-22.
Andrén, O., T. Lindberg, U. Boström, M. Clarholm, A.-C. Hansson, G. Johansson, J. Lagerlöf, K. Paustian, J. Persson, R. Petterson, J. Schnürer, B. Sohlenius, and M. Wivstad. 1990. Organic carbon and nitrogen flows. In: O. Andrén, T. Lindberg, K. Paustian, and T. Rosswall (eds.), Ecology of arable land - organisms, carbon and nitrogen cycling. *Ecological Bulletins* 40:85-126.
Ayres, K.W., D.F. Acton, and J.G. Ellis. 1985. *The Soils of the Swift Current Map Area. 72 J Saskatchewan.* Saskatchewan Institute of Pedology, Publication S6. Extension Division, University of Saskatchewan, Saskatoon, Saskatchewan, Canada.
Baumberg, S. 1973. Co-ordination of metabolism. p 423-493 In: J. Mandelstam and K. McQuillen (eds.), *Biochemistry of Bacterial Growth.* John Wiley & Sons, New York.
Biederbeck, V.O., C.A. Campbell, and R.P. Zentner. 1984. Effect of crop rotation and fertilization on some biological properties of a loam in southwestern Saskatchewan. *Can. J. Soil Sci.* 64:355-362.
Biederbeck, V.O., C.A. Campbell, H. Ukrainetz, D. Durtin, and O.T. Bouman. 1996. Soil microbial and biochemical properties after 10 years of fertilization with urea and anhydryous ammonia. *Can. J. Soil Sci.* 76:7-14.

Brewster, R.Q. 1948. *Organic Chemistry*. Prentice-Hall, Inc. NY.

Burton, D.L. and W.B. McGill. 1992. Spatial and temporal fluctuations in biomass, nitrogen mineralizing reactions and mineral nitrogen in a soil cropped to barley. *Can. J. Soil Sci.* 72:31-42.

Campbell, C.A., D.W.L. Read, R.P. Zentner, A.J. Leyshon, and W.S. Ferguson. 1983. First 12 years of a long-term crop rotation study in southwestern Saskatchewan. Yields and quality of grain. *Can. J. Plant Sci.* 63:91-108.

Cameron, D.R., C. Shaykewich, E. de Jong, D. Chanasyk, M. Green, and D.W.L. Read. 1981. Physical aspects of soil degradation. p 186-255 In: *Agricultural Land. Our Disappearing Heritage.* Proceedings 18th Annual Alberta Soil Science Workshop. Alberta Soil and Feed Testing Laboratory, Edmonton, Alberta.

Capriel, P., T. Beck, H. Borchert, and P. Härter. 1990. Relationship between soil aliphatic fraction extracted with supercritical hexane, soil microbial biomass, and soil aggregate stability. *Soil Sci. Soc. Am. J.* 54:415-420.

Chen, W., H.A.J. Hoitink, A.F. Schmitthenner, and O.H. Tuovinen. 1988. The role of microbial activity on the suppression of damping-off caused by *Pythium ultimum*. *Phytopathology* 78:314-322.

de Jong, E. 1981. Soil aeration as affected by slope position and vegetative cover. *Soil Sci.* 131:34-43.

de Jong, E., Villar H., and J.R. Bettany. 1982. Preliminary investigations on the use of ^{137}Cs to estimate erosion in Saskatchewan. *Can. J. Soil Sci.* 62:673-683.

Dinel, H.M., Lévesque, P. Jambu, and D. Righi. 1992. Microbial activity and long-chain aliphatics in the formation of stable soil aggregates. *Soil Sci. Soc. Am. J.* 56:1455-1466.

Dinel, H. and E.G. Gregorich. 1995. Structural stability status as affected by long-term continuous maize and bluegrass sod treatments. *Biol. Agric. Hort.* 12:237-252.

Dinel, H., M. Schnitzer, and S. Dumontet. 1996. Compost maturity: chemical characteristics of extractable lipids. *Compost Sci. Util.* 4:16-25.

Doran, J.W. and T.B. Parkin. 1994. Defining and assessing soil quality. p. 3-21. In: J.W. Doran, D.C. Coleman, D.F. Bezdicek, and B.A. Stewart (eds.), *Defining Soil Quality for a Sustainable Environment*. Special pub. 35, Soil Sci. Soc. Am. Inc., Madison, WI.

Dumanski, J., R.L. Desjardins, C. Tarnocai, C. Monreal, E.G. Gregorich, C.A. Campbell, and V. Kirkwood. 1997. Possibilities for future carbon sequestration in Canadian agriculture in relation to land use changes. *Global Change* (in press).

Eaves, B.W. 1972. Soil physical conditions affecting seedling root growth. I. Mechanical impedance, aeration and moisture availability as influenced by bulk density and moisture levels in a sandy loam soil. *Plant and Soil* 36:613-622.

Fruton, J.S. and S. Simmonds. 1953. *General Biochemistry*. John Wiley & Sons Inc., London.

Gamble, D.S., A.W. Underdown, and C.H. Langford. 1980. Copper (II) titration of fulvic ligand sites with theoretical, potentiometric, and spectrophotometric analysis. *Anal. Chem.* 52:1901-1908.

Gamble, D.S., M. Schnitzer, and H. Kerndorff. 1983. Multiple metal ion exchange equilibria with humic acid. *Geochim. Cosmochim. Acta* 47:1311-1323.

Gamble, D.S., C.H. Langford, and G.R.B. Webster. 1994. Interactions of pesticides and metal ions with soils: unifying concepts. *Rev. Env. Contam. and Toxicol.* 135:63-91.

Greer, K.J. and J.J. Schoenau. 1992. Soil organic matter content and nutrient turnover in Thin Black Oxbow soils after intensive conservation management. p. 167-173 In: *Management of Agriculture Science*. Soils and Crops Workshop, University of Saskatchewan, Saskatoon, Saskatchewan.

Gregorich, E.G., M.R. Carter, D.A. Angers, C.M. Monreal, and B.H. Ellert. 1994. Towards a minimum data set to assess soil organic matter quality in agricultural soils. *Can. J. Soil Sci.* 74:367-385.

Harwood, J. L. and N. J. Russell. 1984. *Lipids in Plants and Microbes*. George Aleen and Unwin, London.

Haynes, R. J. and R.S. Swift. 1990. Stability of soil aggregates in relation to organic constituents and soil water content. *J. Soil Sci.* 41:73-83.

Hempfling, R., W. Zech, and H.-R. Schulten. 1988. Chemical composition of the organic matter in forest soils: 2. Moder profile. *Soil Sci.* 146:262-276.

Hempfling, R. and H.-R. Schulten. 1990. Chemical characterization of the organic matter in forest soils by Curie-point pyrolysis-GC/MS and pyrolysis-field ionization mass spectrometry. *Org. Geochem.* 15:131-145.

Hempfling, R. and H.-R. Schulten. 1991. Pyrolysis-gas chromatography mass spectrometry of agricultural soils and their humic fraction. *Zeitschrift Pflanzen. Bodenk.* 154:425-430.

Hendrix, P.F., R.W. Parmelee, D.A. Crossley Jr., D.C. Coleman, E.P. Odum, and P.M. Groffman. 1986. Detritus food webs in conventional and no-tillage agro-ecosystems. *BioScience* 36:374-380.

Jansson, S.L. 1963. Balance sheet and residual effects of fertilizer nitrogen in a 6-year study with N^{15}. *Soil Sci.* 95:31-37.

Janzen, H.H. 1987. Effect of fertilizer on soil productivity in long-term spring wheat rotations. *Can. J. Soil Sci.* 67:165-174.

Juma, N.G. 1993. Interrelationships between soil structure/ texture, soil biota/soil organic matter and crop production. *Geoderma* 57:3-30.

Kay, B.D. 1990. Rates of change of soil structure under different cropping systems. p. 1-52 In: B.A. Stewart (ed.), *Advances in Soil Science,* Vol. 12. Springer-Verlag, New York, NY.

Krämer, U., J.D. Cotter-Howells, J. M. Charnock, A.J. Baker, and J.A.C. Smith. 1996. Free histidine as a metal chelator in plants that accumulate nickel. *Nature* 379:635-638.

Lam, M.T., D.S. Gamble, C.L. Chakrabanti, and A. Ismaily. 1996. On-line HPLC micro extraction of Zn(II) and Cd(II) in soil slurries: a chemical speciation method. *Int. J. Envir. Anal. Chem.* 64:217-231.

Larson, W.E. and F.J. Pierce. 1991. Conservation and enhancement of soil quality. p. 175-203 In: *Evaluation for Sustainable Land Management in the Developing World.* Int. Board Soil Res. and Management (IBSRAM). Proc 12 (Vol. 2). Bangkok, Thailand.

Leinweber, P., H.-R. Schulten, and C. Horte. 1992. Differential thermal analysis, thermogravimetry and pyrolysis-field ionization mass spectrometry of organic matter in particle-size fractions and bulk soil samples. *Thermochim. Acta* 194:175-187.

Leinweber, P., C. Paetsch, and H.-R. Schulten. 1995. Heavy metal retention by organo-mineral particle-size fractions from soils in long-term agricultural experiments. *Arch. Acker-Pfl. Boden.* 39:271-285.

Low, A.J. 1972. The effect of cultivation on the structure and other physical characteristics of grasslands and arable soils (1945-70). *J. Soil Sci.* 23:363-380.

Malik, M. N., D.S. Stevenson, and G.C. Russell. 1965. Water-stable aggregation in relation to various cropping rotations and soil constituents. *Can. J. Soil Sci.* 45:189-197.

Martel, Y.A. and A.F. MacKenzie. 1980. Long-term effects of cultivation and land use on soil quality in Québec. *Can. J. Soil Sci.* 60:411-420.

Martens, D. A. and W. T. Frankenberger. 1993. Soil saccharide extraction and detection. *Plant and Soil* 149:145-147.

Mathur, S.P. and R.S. Farnham. 1958. Geochemistry of humic substances in natural and cultivated peatlands. In: G. Aiken, et al. (eds.), *Humic Substances in Soils, Sediment, and Water. Geochemistry, Isolation and Characterization.* John Wiley & Sons, NY.

McGill, W.B., C.A. Campbell, J.F. Dormaar, E.A. Paul, and D.W. Anderson. 1981. Soil organic matter losses. p. 72-133. In: *Agricultural Land. Our Disappearing Heritage.* Proceedings 18th Annual Alberta Soil Science Workshop. Alberta Soil and Feed Testing Laboratory, Edmonton, Alberta.

Mehta, N. C., H. Streuli, M. Muller, and H. Duel. 1960. Role of polysaccharides in soil aggregation. *J. Sci. Food and Agr.* 11:40-47.

Monreal, C.M., W.B. McGill, and J.D. Etchevers. 1981. Internal N cycling compared in surface samples of an Andept and a Mollisol. *Soil Biol. Biochem.* 13:451-454.

Monreal, C.M. and W.B. McGill. 1987. The dynamics of free cystine cycling at steady-state through the solutions of selected cultivated and uncultivated Chernozemic and Luvisolic soils. *Soil Biol. Biochem.* 21:689-694.

Monreal, C.M., M. Schnitzer, H.-R. Schulten, C.A. Campbell, and D.W. Anderson. 1995a. Soil organic structures in macro and microaggregates of a cultivated Brown Chernozem. *Soil Biol. Biochem.* 27:845-853.

Monreal, C.M., R.P. Zentner, and J.A. Robertson. 1995b. The influence of management on soil loss and yield of wheat in Chernozemic and Luvisolic soils. *Can. J. Soil Sci.* 75:567-574.

Monreal, C.M., E.G. Gregorich, M. Schnitzer, and D.W. Anderson. 1995c. The quality of soil organic matter as characterized by solid CP/MAS ^{13}C NMR and Py-FIMS. p. 207-215. In: P.M. Huang, J. Berthelin, J.-M. Bollag, W.B. McGill, and A.L. Page (eds.), *Environmental Impacts of Soil Component Interactions. Land Quality, Natural and Anthropogenic Organics.* Lewis Publishers, Chelsea, MI.

Monreal, C.M., H.-R. Schulten, and H. Kodama. 1997a. Age, turnover and molecular diversity of soil organic matter in aggregates of a Gleysol. *Can. J. Soil Sci.* (submitted).

Monreal, C.M. and H. Kodama. 1997b. Influence of aggregate architecture and minerals on living habitats and soil organic matter. *Can. J. Soil Sci.* (submitted).

Neal, R.H. and G. Sposito. 1986. Effects of soluble organic matter and sewage sludge amendments on cadmium sorption by soils at low cadmium concentrations. *Soil Sci.* 142:164-172.

Nelson, D.W. and L.E Sommers. 1982. Total carbon, organic carbon, and organic matter. p. 539-577. In: A.L. Page, R.H. Miller, and D.R. Kenney (eds.), *Methods of Soil Analysis. Part 2: Chemical and Microbiological Properties.* 2nd ed. American Society of Agronomy – Soil Science Society of America, Madison, WI.

Nolin, M.C., C. Wang, M.J. Deschênes, and C. Lévesque. 1995. *Description des sites repères: 17-QU and 18-QU.* Centre for Land and Biological Resources Research, Research Branch, Agriculture and Agri-Food Canada, Ottawa, Canada. Contribution CLBRR No. 95-105.

Nwadialo, B. E. and J.S.C. Mbagwu. 1991. Analysis of soil components active in microaggregate stability. *Soil Technol.* 4:343-350.

O'Connor, G.A., M.E. Essington, M. Elrashidi and G. Cline. 1984. *Trace Metal Sorption in Sludge-Amended Soils.* p. 225-231. In: CEP Consultants. Environmental Contamination (UNEP). Edinburgh,

Oades, J. M. and A.G. Waters. 1991. Aggregate hierarchy in soils. *Aust. J. of Soil Res.* 29: 815-828.

Park, J. and R.A.F. Seaton. 1996. Integrative research and sustainable agriculture. *Agric. Systems* 50:81-100.

Parsons, J.W. 1981. Chemistry and distribution of amino sugars in soils and soil organisms. *Soil Biochem.* 5:197-227.

Parton, W.J., D.S. Ojima, C.V. Cole, and D.S. Schimel. 1994. A general model for soil organic matter dynamics: sensitivity to litter chemistry, texture and management. p. 147-166. In: *Quantitative Modeling of Soil Forming Processes.* Soil Sci. Soc. Am., special publication 39. Madison, WI.

Pennock, D.J. and E. de Jong. 1987. The influence of slope curvature on soil erosion and deposition in hummock terrain. *Soil Sci.* 144:209-216.

Rönkkö, R., T. Pennanen, A. Smolander, V. Kitunen, H. Kortemaa, and K. Haahtela. 1994. Quantification of *Frankia Strains* and other root-associated bacteria in pure cultures and in the rhizosphere of axenic seedlings by high-performance liquid chromatography based muramic acid assay. *Appl. Env. Microb.* 60:3672-3678.

Rutherford, P.M. and N.G. Juma. 1992. Simulation of protozoa-induced mineralization of bacterial carbon and nitrogen. *Can. J. Soil Sci.* 72:201-216.

Rynasiewicz, J. 1945. Soil aggregation and onion yields. *Soil Sci.* 60:387-395.

Salisbury, F. B. and C. W. Ross. 1978. *Plant Physiology*. Wadsworth Publishing Company, Belmont, CA.

Salton, M.R. and P. Owen. 1976. Bacterial membrane structure. *Ann. Rev. Microb.* 30: 451-482.

Sarathchandra, S.U. 1979. A simplified method for estimating ammonium oxidizing bacteria. *Plant Soil* 52:305-309.

Scheu, S. and D. Parkinson. 1994. Changes in bacterial and fungal biomass C, bacterial and fungal biovolume and ergosterol content after drying, remoistening and incubation of different layers of cool temperate forest soils. *Soil Biol. Biochem.* 26:1515-1525.

Schnitzer, M. and H.-R. Schulten. 1992. The analysis of soil organic matter by pyrolysis-field ionization mass spectrometry. *Soil Sci. Soc. Am. J.* 56:1811-1817.

Schnitzer, M. 1995. Organic-inorganic interactions in soils and their effects. p. 3-20 In: P.M. Huang, J. Berthelin, J.-M. Bollag, W.B. McGill, and A.L. Page (eds.), *Environmental Impacts of Soil Component Interactions. Land Quality, Natural and Anthropogenic Organics*. Lewis Publishers, Chelsea, MI.

Schnitzer, M. and H.-R. Schulten. 1995. Analysis of organic matter in soil extracts and whole soils by pyrolysis-mass spectrometry. *Adv. Agron.* 55:167-217.

Schulten, H.-R., C.M. Monreal, and M. Schnitzer. 1995. Effect of long-term cultivation on the chemical structure of soil organic matter. *Naturwissenschaften* 82:42-44.

Schulten, H.-R. 1987. Pyrolysis and soft ionization mass spectrometry of acquatic/terrestrial humic substances and soils. *J. Anal. Applied Pyrolysis* 12:149-186.

Sowden, F.J., Y. Chen, and M. Schnitzer. 1977. The nitrogen distribution in soils formed under widely differing climatic conditions. *Geochim. Cosmochim. Acta* 41:1524-1526.

Sparrow, H.O. 1984. Soil at risk. Canada's eroding future. Standing Senate Committee on Agriculture, Fisheries and Forestry. Ottawa, Ontario, Canada

Szabolcs, I. 1979. Effects of salts on soil and soil properties. p 31-40 In: *Review of Research on Salt-Affected Soils*. Natural Resources Research, Series 15. Unesco, Paris.

Taylor, H.M. and E.E. Terrell. 1982. *CRC Handbook of Soil Productivity*. Vol 1. Rechigl, J., Jr. (ed.), CRC Press, Boca Raton, FL.

Tiessen, H., T.L. Roberts, and J.W. Stewart. 1983. Carbonate analysis in soils and minerals by acid digestion and two-end point titration. *Commun. Soil Sci. Plant Anal.* 14:161-166.

Tiller, K.G., J. Gerth, and G. Brümmer. 1984. The relative affinities of Cd, Ni and Zn for different soil clay fractions and Goethite. *Geoderma* 34:17-35.

Toogood, J.A. 1978. Relationship of aggregate stability to properties of Alberta soils. p. 211-215 In: W.W. Emerson (ed.) *Modification of Soil Structure*. John Wiley & Sons, NY.

Topp, G.C., K.C. Wires, D.A. Angers, M.R. Carter, J.L.B. Culley, D.A. Holmstrom, B.D. Kay, G.P. Lafond, D.R. Langille, R.A. McBride, G.T. Patterson, E. Perfect, V. Rasiah, A.V. Rodd, and K.T. Webb. 1995. Changes in soil structure. p. 51-60. In: D.F. Acton and L.J. Gregorich (eds.), *The Health of Our Soils-Toward Sustainable Agriculture in Canada*. Centre for Land and Biological Resources Research, Research Branch, Agriculture and Agri-Food Canada, Ottawa, Ontario.

Vestal, J.R. and D.C. White. 1989. Lipid analysis in microbial ecology. Quantitative approaches to the study of microbial communities. *BioScience* 39:535-541.

Wallis, M.G. and D.J. Horne. 1992. Soil water repellency. p. 91-138. In: B.A. Stewart (ed.), *Advances in Soil Science*, vol. 20. Springer-Verlag, New York, NY.

Weshaw, R.L. 1985. Application of nuclear magnetic resonance spectroscopy for determining functionality in humic substances. In: G. Aiken et al. (eds.) *Humic Substances in Soils, Sediment, and Water. Geochemistry, Isolation and Characterization*. John Wiley & Sons, NY.

Zeppezauer, M. and W. Maret. 1986. Does the coordination environment determine the reactivity of metals in enzymes? In: M. Bernhard, F.E. Brinkman, and P.J. Sadler (eds.), *The Importance of Chemical Speciation in Environmental Processes*. Report of the Dahlem Workshop on the importance of chemical speciation in environmental processes. Springer-Verlag, Berlin.

CHAPTER 31

Modeling Soil Carbon in Relation to Management and Climate Change in Some Agroecosytems in Central North America

Keith Paustian, Edward T. Elliott, and Kendrick Killian

I. Introduction

What is the potential for agricultural soils to sequester carbon for mitigating increased anthropogenic CO_2 emissions? If projected changes in climate and atmospheric CO_2 occur, how will they impact the carbon balance of agricultural soils? Will it be possible to significantly increase soil C storage while ensuring that agricultural productivity increases to meet future demands? These are among the main questions being investigated concerning agricultural soils and global change. The answers to these questions will determine the extent to which agriculture can play a role in mitigating CO_2 increase, a role with far reaching economic, social, and political implications.

Globally, agricultural soils contain a significant portion of the terrestrial organic carbon inventory and represent a potentially large carbon sink. The global C stock (to 1 m depth) for the world's cultivated area is about 170 Pg (Cole et al., 1996). Topsoil C losses of 20-50% of precultivation levels are typical for mineral soils converted to agriculture (Mann, 1986; Davidson and Ackerman, 1993) and high and continuing losses of carbon generally occur in organic soils under cultivation (Armentano and Menges, 1986). Thus, as much as 55 Pg C of the original C stocks in global agricultural soils may have been lost (Cole et al., 1996). If some or all of this amount of "lost" C could be restored to agricultural soils, it would constitute significant mitigation.

Agricultural systems are unique among terrestrial ecosystems in the degree to which they are manipulated by man. Virtually every facet of agricultural management has an impact on the soil C balance either by affecting the input of organic materials to soil or the decomposition rate of soil organic matter (SOM), or both (Paustian et al., 1997). Thus, to some extent, changes in management may be able to buffer negative impacts of climate change on soil C balances or even enhance positive effects of climate/CO_2 change. Given the complexity and multifaceted interactions between management, climate, and CO_2 enrichment, simulation models are useful for exploring possible responses of agricultural systems to global change.

In this chapter, we use an agroecosystem model to analyze these interactions for a set of well characterized agricultural systems in the central U.S. and southern Canada. These long-term field experiments span a range of climate and soil conditions and different management systems, which have well-documented management histories and measurements of SOM responses to different management systems through time. Validation of the model using data on historical changes in soil C at these sites allows us to evaluate the model under a range of conditions. This work is part of a longer-term effort in which we will eventually link ecosystem responses to economic and policy mechanisms that influence the adaptation of management systems to global change.

II. Site Description

The six sites used in the analysis are all derived from native grassland vegetation, ranging from the semiarid shortgrass steppe in the western Great Plains to former tallgrass prairie in eastern Kansas and southern Wisconsin (Table 1). The sites represent a range of climatic conditions. The four western wheat-based sites have similar annual precipitation but are arrayed along a north-south temperature gradient, with mean annual temperature (MAT) varying from < 4°C at the northernmost site (Swift Current, Saskatchewan) to 12°C at Walsh in SE Colorado. Both eastern sites receive roughly twice as much precipitation (ca. 800 mm) as the western sites, but the MAT at Arlington, Wisconsin is about 5°C cooler than Manhattan, Kansas.

Soils at all the sites are classified as Mollisols and range in texture from loam to silt loams and clay loams. However, current soil C levels vary from < 20 Mg ha^{-1} (0-20 cm) at Walsh (Peterson and Westfall, 1997) to over 60 Mg ha^{-1} at Arlington (Vanotti et al., 1997). Typical cropping practices can also be differentiated regionally. In the region represented by the four western sites, the dominant systems include small grains, mainly wheat (*Triticum aestivum* L.), grown in summer fallow rotations. The two eastern sites represent areas dominated by warm season row crops, such as corn (*Zea maize* L.), soybeans (*Glycine max* (L.) Merr.), sorghum (*Sorghum bicolor* L.), grown in continuous cropping sequences. More detailed information on specific management histories and experimental treatments at these sites is given in Campbell and Zentner, 1997; Havlin and Kissel, 1997; Lyon et al., 1997; Peterson and Westfall, 1997; and Vanotti et al., 1997.

III. Model Analyses

A. Historical Responses of Soil C

For our analysis we used the Century agroecosystem model, version 4.0 (Metherell et al., 1993). The model incorporates soil organic matter and nutrient dynamics from previous versions of Century (Parton et al., 1987, 1988), with revised routines for simulating cropping systems, including provision for multicrop rotations and a variety of management options for tillage, fertilization, irrigation, and organic amendments. The model also includes the effects of enriched CO_2 on several plant processes which can be parameterized for specific crop species.

The model was initialized using information on climate, soils, and previous landuse compiled for the sites. To establish initial amounts of C and N in the three soil organic matter pools defined in the model (i.e., active, slow, and passive), the model was run to equilibrium under native grassland vegetation, using the long-term climate averages. From an equilibrium state, the model was then run for the period between initial cultivation of the grassland soil until the initiation of field experiments at each site and then forward to 1995 using the documented management practices of the experimental treatments. Initial cultivation occurred about 1880 at Arlington and Manhattan, about 1905 at Swift Current, and about 1920 at the two Colorado sites. The experiment at Sidney was initiated in 1970 with plowing of the native sod. Management histories prior to the establishment of the field experiments were based on historical references and the advice of the site collaborators. Model results from representative 'baseline' treatments at the sites were used for the starting point of the future scenario simulations (beginning in 1995; see below). These baseline treatments were wheat-summer fallow rotations for the four western sites, continuous corn at Arlington, and continuous sorghum at Manhattan.

Table 1. Characteristics of the sites used in the analyses

Site	Native vegetation	Soil type	Soil texture	MAT (°C)	MAP (mm)
Swift Current, Saskat.	Shortgrass	Aridic Haloboroll	Loam	3.7	330
Sidney, NE	Shortgrass	Pachic Haplustoll	Loam	8.2	380
Stratton, CO	Shortgrass	Aridic Agiustoll	Silty clay loam	10.7	410
Walsh, CO	Shortgrass	Torrertic Paleustoll	Clay loam	11.9	400
Manhattan, KS,	Tallgrass	Pachic Haplustoll	Silt loam	12.8	835
Arlington, WI	Tallgrass	Typic Argiudoll	Silt loam	7.6	793

(From Paul et al., 1997.)

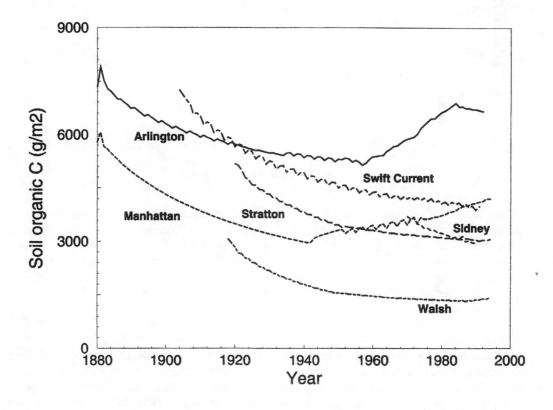

Figure 1. Simulated changes in soil organic C (g m^{-2} for 0-20 cm) from the onset of cultivation of the native grassland to the present. Note that the Sidney site was first cultivated in 1970, when the field experiment was established out of native sod.

Cultivation resulted in the decline of organic matter at all sites, on the order of 30-50% (Figure 1), similar to the observed magnitude of cultivation losses observed for grassland soils of the region (Haas et al., 1957). The model predicts a subsequent recovery of some of the losses in organic matter in the two eastern sites (Arlington, Manhattan), beginning in the 1950-60s, coinciding with increasing productivity and greater residue C inputs due to more intensive management (e.g., increasing

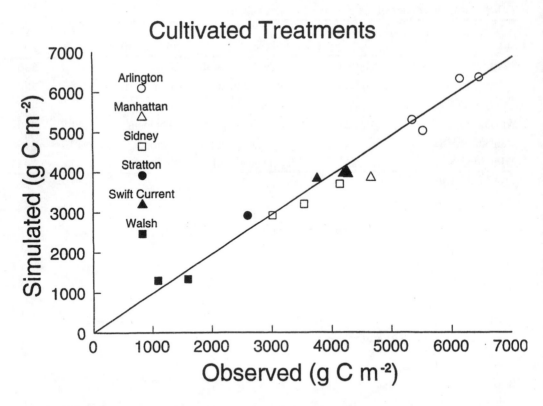

Figure 2. Comparison of simulated vs. observed soil organic C (0-20 cm) for the "baseline" treatments from each site, i.e., continuous corn with 84 kg N ha^{-1} (Arlington), continuous sorghum (Manhattan), wheat-fallow with plow tillage (Sidney), winter wheat-fallow (Stratton and Walsh), and spring-wheat fallow (Swift Current). Multiple symbols for a site denote measurements taken in different years during the experimental period. Details of the experimental treatments are reported in Campbell and Zentner, 1997; Havlin and Kissel, 1997; Lyon et al., 1997; Peterson and Westfall, 1997; and Vanotti et al., 1997.

fertilization rates, improved crop varieties, denser plant spacing). At Arlington, where the site had been poorly managed, with low fertilization rates and residue burning for many years prior to its acquisition as an experimental site (M. Vanotti, pers. comm.), the simulated increase of organic C agrees well with measured increases over the experimental period (Figure 2). At Manhattan, there are no data prior to the first soil sampling in 1986 to confirm a trend at this site. For the four western sites under wheat-fallow management, a slow continuing decline in soil C to the present day is consistent with the data from these and similar sites (Peterson and Westfall, 1997; Campbell and Zentner, 1997). A comparison of simulated vs. observed soil organic C levels (Figure 2) shows that the wide range in soil C levels across the sites, which developed as a function of different climate, soil, and management factors, were reasonably represented by the model. Thus, the model fulfills an important criterion for projecting changes in soil C as a result of changes in climate and other production factors, such as atmospheric CO_2 levels and management systems.

Table 2. Cropping practices simulated for high vs. low cropping intensity treatments in the global change/management scenarios. Spring wheat is used for Swift Current rotations, otherwise, wheat denotes winter wheat. Both maize and sorghum are for grain production (i.e., no silage).

Site	Cropping intensity/C inputs	
	High	Low
Arlington	Continuous maize with winter grass cover crop	Maize-soybean with winter fallow
Manhattan	Continuous sorghum with winter grass cover crop	Sorghum-soybean with winter fallow
Swift Current	Wheat-wheat-fallow	Wheat-fallow
Sidney/Stratton	Wheat-maize-fallow	Wheat-fallow
Walsh	Wheat-sorghum-fallow	Wheat-fallow

B. Global Change/Management Scenarios

1. Model Initialization

Model scenarios to explore the possible effects of changes in climate, CO_2, and management on future organic matter levels were performed using a factorial design including two management variables (cropping intensity and tillage) and variations in climate and CO_2. Simulations were run for 60 years (1995-2055) since an effective doubling of greenhouse gases is projected to occur by the middle of the next century (IPCC, 1990). The simulated treatments for cropping intensity were selected to represent realistic management options at the different sites, with high intensity cropping being characterized by high crop residue production and a reduced fallow period and, conversely, low intensity cropping being characterized by lower residue production and more fallow (Table 2). The tillage contrasts were between plow-based tillage and no-till. Climate treatments consisted of current long-term (1960-1990) climate averages vs. climate conditions under increased CO_2 as simulated by three different general circulation models (GCM). Finally, CO_2 was either assumed constant at current levels (350 µl l^{-1}) or was "doubled" (640 µl l^{-1}) over the course of the simulation (1995-2055). The latter corresponds to the concentration of CO_2 which is projected for the atmosphere in which the total radiative forcing for all greenhouse gases will be twice that at present (Adams et al., 1990). Thus, it represents the CO_2 concentration assumed in the 2XCO_2 GCM simulations. We ran all combinations of the management, climate and CO_2 treatments, allowing us to assess the significance of each individual factor as well as interactions among factors.

To normalize other conditions that can affect the model outcome but which were not factors in the analysis, we re-ran the model for the historical period (pre-1995) assuming a common soil texture (silt loam) and using "generic" management histories (e.g., time of initial cultivation, cropping practices prior to 1995) for the four western sites and the two eastern sites, respectively. For both increasing CO_2 levels and climate change scenarios, the appropriate driving variables (i.e., volumetric CO_2 concentrations, monthly mean maximum and minimum temperatures, and monthly precipitation) were ramped linearly over the course of the simulation. For instance, CO_2 value was increased linearly from 350 µl l^{-1} in 1995 to 640 µl l^{-1} in 2055.

For the climate change simulations, we used scenarios projected by different GCMs. The climates for a doubled CO_2 environment were obtained from the VEMAP (Vegetation/Ecosystem Modeling and Analysis Project) database (Kittel et al., 1995). Included in this database are several GCM outputs which have been interpolated onto a 0.5° latitude/longitude grid for the coterminous U.S. Of the eight model outputs in VEMAP, we chose the CCC (Canadian Climate Centre, high resolution GCM experiment), the GFDL R30 (Goddard Fluid Dynamic Lab R30 2.22 x 3.75 degree grid) and the

Table 3. Relative changes in temperature and precipitation for each site determined from equilibrium simulations of three general circulation models[a]

Site	CCC			GFDL R30			UKMO		
	Summer (°C)	Winter (°C)	Precip. (cm)	Summer (°C)	Winter (°C)	Precip. (cm)	Summer (°C)	Winter (°C)	Precip. (cm)
Arlington	3.5	3.4	15.4	5.3	5.4	19.5	7.6	9.1	9.7
Manhattan	3.7	3.3	13.8	3.6	4.8	-2.7	6.4	7.3	7.9
Swift Current	3.9	4.4	2.6	4.5	3.9	6.1	6.2	6.7	6.6
Sidney	4.4	4.7	5.8	4.5	4.9	0.3	6.0	6.7	5.9
Stratton	4.5	4.9	2.4	4.2	5.0	0.3	5.8	6.5	6.6
Walsh	4.3	4.2	0	3.8	5.0	3.4	5.4	6.3	6.9

[a] CCC (Canadian Climate Centre), GFDL R30 (Goddard Fluid Dynamic Laboratory), and UKMO (United Kingdom Meteorological Office), for doubled CO_2 conditions, as given in the VEMAP (Kittell et al., 1995) database. GCM results were taken for the nearest 0.5° grid cell to the site location. Temperature (°C) changes represent deviations from current mean climate (1960-1990), averaged for half-year periods, i.e., summer (April-September). Precipitation (cm) values are deviations from current average annual precipitation.

UKMO (United Kingdom Meteorological Office) models (Table 3). The CCC and GFDL R30 model results are among the high resolution GCM experiments reported by the IPCC (1990). These three models were chosen to bracket the range of climate change projections, since the UKMO model gave some of the largest projected changes while the CCC predictions gave some of the smallest changes for the experimental sites. We took the deviations from current climate as predicted by the GCMs for the grid cells containing each of our sites and then applied those to the climate averages (1960-1990) at the sites. Since the Swift Current, Saskatchewan site was outside the range of the VEMAP data, we took the deviations from the nearest grid cell in the U.S. (ca. 150 km south) and applied that to the current average climate for Swift Current. As for CO_2, we linearly ramped the changes in temperature and precipitation, as projected by the GCMs for the doubled CO_2 condition (in 2055), over the simulation period.

Finally, to accommodate changes in crop production that could result from CO_2 fertilization and/or climate change, we assumed an increase in N fertilization rates of 50% over current levels by the end of the simulation period. The current levels of N fertilizer were set by the actual applications rates used in the baseline treatments for each of the sites. Fertilization rates were incremented every 10 years to approximate a gradual increase in application rates. Within a site, fertilizer rates were the same for all variations of tillage, climate, and CO_2.

2. Simulated Responses of Soil C to Management, Climate, and CO_2

Baseline changes in soil organic matter (i.e., without changing CO_2 and climate conditions) were predicted to occur over the sixty year period. This reflected a continuation of soil C trajectories which were not yet at steady-state in 1995, the effects of increasing rates of N fertilization on productivity/residue inputs and the fact that some of the future management scenarios (particularly the high crop intensity and no-tillage treatments) represented a significant change in management from the baseline treatments used prior to 1995. The net changes in simulated soil organic C occurring over the period are shown in Table 4. For the low intensity cropping, there was relatively little change in soil C, except for the Manhattan site which was predicted to accumulate C at rates of 15-20 g m^{-2} yr^{-1} even with the low intensity cropping regime. In general, simulated soil C was 100-300 g m^{-2} higher under

Table 4. Simulated changes in SOM C (g m^{-2}) under the ambient (i.e., current climate and CO$_2$) scenario for high and low cropping intensity (see Table 2) and conventional (CT) and no-till (NT) tillage; values represent the change in C occurring from the start (1995) to the end (2055) of the simulation period

Site	Cropping intensity/tillage			
	High		Low	
	CT	NT	CT	NT
Arlington	1300	1400	-30	260
Manhattan	2000	2300	980	1100
Swift Current	290	450	-90	100
Sidney	1170	1320	140	230
Stratton	810	970	-80	30
Walsh	830	980	-20	80

no-till compared to conventional tillage, for both cropping intensities. All systems were predicted to accrue C under the high intensity cropping regimes, with the largest gains occurring in the eastern sites (Arlington and Manhattan) and the least at the Swift Current site.

Despite some differences in projected temperature and precipitation among the three GCM models, simulated soil organic C was not particularly sensitive to which model was used in the climate change scenarios (Figure 3). Although differences in total soil C were slight, there was a consistent pattern between models. For most sites, the UKMO model tended to result in the lowest soil C, the GFDL model predicted the highest soil C, and the CCC model was intermediate. For purposes of displaying the effects of climate and CO$_2$ on the different management systems at each of the sites, we chose to use the results of the CCC model.

The implications of a change in climate of the magnitude projected under the 2XCO$_2$ scenario (for the CCC model) can be visualized by comparing the new climate conditions to their closest analog under present day conditions. These are shown on the map in Figure 4 as a "change" in the geographic location of site. For the four western sites, the increase in temperature and, to a lesser degree precipitation, is analogous to a displacement predominantly to the south. In the case of the Manhattan and Arlington sites, where there is a substantial shift in both temperature and precipitation, the geographic analogs tended to converge, with conditions at Arlington resembling those in present day southern Iowa/northern Missouri and the climate at Manhattan becoming more similar to current conditions in central Missouri. Interestingly, despite climate change, all the sites tended to remain within the same general cropping region as before, i.e., the western sites remain within the wheat belt of the western Great Plains and the two eastern sites still reside within the row-crop dominated corn belt.

The effects of climate and CO$_2$ change on soil organic matter levels are the result of several interactions affecting the amount of residue returned to the soil and the decomposition rate of residues and soil organic matter. In interpreting the model results, we examined a number of outputs including annual (residue and root) C inputs, water balance components, and changes in specific decomposition rates as a function of soil temperature and moisture. The general pattern for most of the site/management combinations was a decrease in soil C with climate change alone, an increase in soil C with enriched CO$_2$ alone, and an intermediate response to the combination of increased CO$_2$ with climate change (Figure 5). There were, however, some interesting exceptions to this general pattern, which largely relate to how CO$_2$ and/or climate affected the productivity of different crops and their C input rates.

Climate change alone reduced soil C levels for all management systems at Manhattan, Swift Current, Stratton, and Walsh (Figure 5). At Manhattan, crop production and C inputs with climate

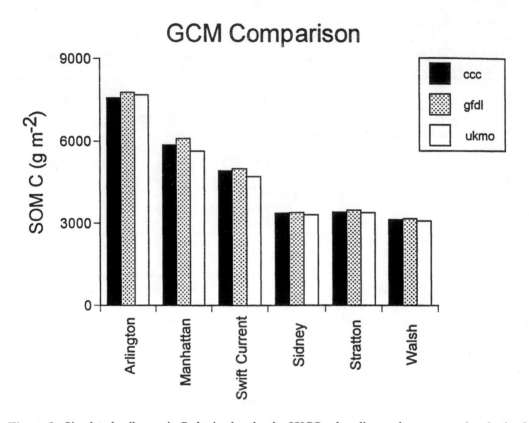

Figure 3. Simulated soil organic C obtained under the 2XCO$_2$ plus climate change scenario, obtained with different GCM models, i.e., the UKMO (United Kingdom Meterological Office), the CCC (Canadian Climate Centre, high resolution GCM experiment), and the GFDL R30 (Goddard Fluid Dynamic Lab R30 2.22 x 3.75 degree grid) models. Values shown are for the high cropping intensity systems under no-till.

change were roughly equal to that under the ambient scenario (i.e., current climate and CO$_2$), while at the other three sites, the change in climate resulted in reduced production and C inputs. This response, together with increased SOM decomposition due to warming, resulted in lower soil C storage. In contrast, the higher temperature and precipitation under a changed climate at Arlington increased crop production and residue return, particularly from corn. In the intensive systems with corn and a winter cover crop, C inputs were increased by about 20% over the ambient scenario. While decomposition rates also increased, the increase in C inputs was large enough to produce an overall increase in soil C levels. However, where there was only one year of corn in the rotation (i.e., corn-soybean), the overall effect of climate change was a small decline in soil C. Similarly, at Sidney there was about a 15% increase in wheat production for the winter wheat-fallow system under the climate change scenario. The net result was a small increase in soil C relative to the ambient scenario. In the more intensive wheat-corn-fallow system, neither wheat nor corn production were enhanced with climate change, due to increased moisture stress (associated with higher temperatures and potential ET) in the systems with a reduced fallow period. Therefore, climate change had an overall negative effect on soil C levels in the intensive systems at Sidney.

Increasing CO$_2$ levels without climate change increased crop (and residue) production in all cases, which tended to result in higher soil C compared to the ambient scenarios (Figure 5). Simulated

Figure 4. Map of central North America showing the present location of the field sites (closed circles) and the geographic analog of the sites with a changed climate (open circles) as projected by the CCC (Canadian Climate Centre) GCM model.

production of wheat and soybean (C_3 crops) increased by 10-30% due to increased photosynthesis and water use efficiency. Production increases for C_4 crops (corn and sorghum) were less since the model assumed no increase in potential photosynthesis for these crops; therefore, production increases were due solely to increased water use efficiency. The magnitude of the simulated CO_2 effects on production and the differences between C_3 and C_4 species are in line with results from most experimental studies on crops (Cure and Acock, 1986; Rogers and Dahlman, 1993). The one exception was at Arlington, where soil C under increased CO_2 alone was predicted to be slightly less than for the ambient condition. Here, in this more mesic site, corn yields were increased by about 10%, but C inputs from roots and residues were essentially unchanged. Because of the higher water use efficiency of the crop under doubled CO_2, soil moisture levels were increased slightly and thus decomposition rates during the growing season were increased. The result was slightly less soil C compared to the ambient scenario (Figure 5).

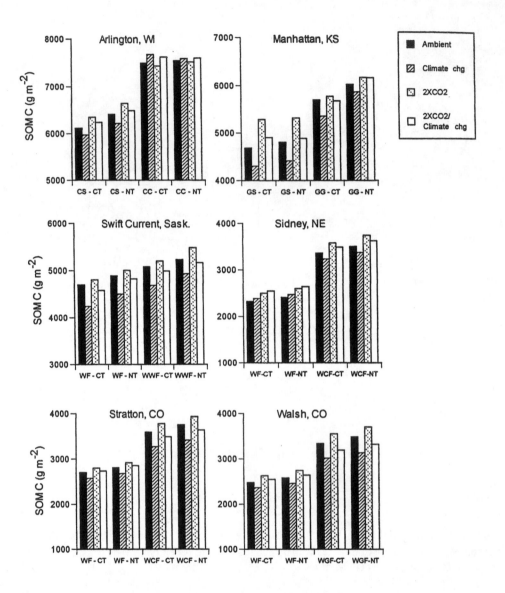

Figure 5. Soil organic C for each of the sites by management system combinations for climate change/CO_2 scenarios, i.e., i) ambient (current climate and CO2 levels), ii) climate change alone ($2XCO_2$ scenario with CCC model), iii) increased (640 µl l^{-1}) CO_2 alone, and iv) both increased CO_2 and climate change. Climate variables and CO_2 levels were ramped linearly over the course of the simulation period (1995-2055). Cropping designations are: C=corn, S=soybean, G=sorghum, W=wheat, F=summer fallow; tillage designations are CT for conventional tillage and NT for no-tillage.

Table 5. Percent change in SOM C with climate change (using the Canadian Climate Centre model) and 640 µl l^{-1} CO_2, relative to the ambient scenario (after 60 years); values represent the mean for all four management combinations at a site

Site	Mean % change
Arlington	1.4
Manhattan	2.6
Swift Current	-1.8
Sidney	6.6
Stratton	-0.8
Walsh	-1.0

The combination of increased CO_2 together with climate change usually resulted in an intermediate response compared to that for climate or CO_2 alone (Figure 5). As discussed above, climate change alone tended to decrease soil C levels at most sites, by stimulating decomposition and often reducing C inputs, whereas CO_2 alone tended to increase C levels, by enhancing crop production and rates of residue return to soil. The two factors together tended to offset one another with respect to their effects on the soil C balance. For most systems, soil C levels were predicted to be relatively close to those for the ambient scenario (Table 5).

Of all the factors included in the simulation experiments, cropping intensity had by far the greatest influence on soil C levels. In an analysis of variance across all sites (using sites as replicates), in which each of the four factors (i.e., cropping intensity, tillage, climate, and CO_2) were included, only cropping intensity showed a significant effect. In other words, if one views the entire region and considers the long-term sites as "samples" of agricultural systems within the region, then the high inherent variability in soil C levels associated with existing climate and landuse differences tends to "mask" effects of climate change and CO_2. In contrast, the effects of cropping intensity (i.e., higher C inputs and reduced fallow periods) are great enough, even at the regional level, that they exert an overriding influence, despite the background variability in soil C levels.

The other management variable, tillage, showed a consistent pattern of higher soil C levels under no-till, but simulated differences were < 10% from those under conventional tillage at all the sites. The main effects of no-tillage, as simulated in the model, are a reduced rate of decomposition due to the reduction of soil disturbance and the buildup of a surface litter layer. The model predicts very little difference in production as a function of tillage. While this may be reasonable for the two eastern sites and for the wheat-fallow systems in the four western sites, the use of no-till (with the additional water storage it makes possible) is thought to be a key component in the success of reduced fallow systems (e.g., wheat-corn-fallow) in semiarid climates (Peterson and Westfall, 1997). However, our model results show relatively little difference in soil C and production rates for these systems irrespective of tillage. One possibility is that the effects of no-till versus tillage on the water balance (i.e., effects on infiltration, surface mulching effects on bare soil evaporation, and effects of tillage operations during the fallow on soil water loss) are inadequately represented in the model.

IV. Concluding Remarks

We analyzed the direct effects and interactions of climate, CO_2, and management practices on soil organic matter C at several locations in the central U.S. and southern Canada. The selection of sites which have long-term records of soil organic matter and crop production under different management systems made it possible to evaluate the model's ability to represent differences in soil C levels as a function of climate, soil, and landuse. In general, the model was able to reasonably represent the range of soil C levels across the sites and the current trajectories of soil C that are a result of previous management practices. We believe that this approach, i.e., the use of a regionally distributed set of field experimental sites for model validation, is a crucial step in making defensible regional projections of agroecosystem responses (Paustian et al., 1995; Paustian et al., in press; Elliott and Paustian, 1996; Paul et al., 1997).

As a general conclusion, the model results suggest that for these sites, agroecosystem properties such as soil C may be more affected by changes in management than by projected climate changes. In particular, the use of more intensive cropping systems, with reduced fallow frequency and increased residue C inputs had the greatest influence on the soil C balance. Thus, there may be a strong potential to mitigate negative aspects of climate change through adaptive management. This is not surprising if one considers that management affects many factors controlling soil C, including residue inputs, soil water balance, soil temperature, soil disturbance, and soil nutrient levels.

As yet, we have not explicitly addressed questions of how management systems, and their spatial distribution, may change in response to climate change. If a changing climate affects the management decisions of the farmer, then climate change will have both direct and indirect effects on soil C levels. Moreover, we recognize that management decisions are influenced to a high degree by economic, social, and policy considerations, which have yet to be included in our analyses. We believe a better integration of environmental, management, economic, and social factors is the next step needed to improve assessments of global change impacts on agricultural ecosystems.

Acknowledgments

Support for the research presented here, from the U.S. Department of Energy-National Institute for Global Environmental Change (NIGEC)/Great Plains Region, the U.S. Department of Agriculture/Agricultural Research Service, and the U.S. Environmental Protection Agency (AERL 9101), is gratefully acknowledged.

References

Adams, R.M., C. Rosenzweig, R.M. Peart, J.T. Ritchie, B.A. McCarl, J. D. Glyer, R.B. Curry, J.W. Jones, K.J. Boote, and L. H. Allen, Jr. 1990. Global climate change and U.S. agriculture. *Nature* 345:219-224.

Armentano, T.V. and E.S. Menges. 1986. Patterns of change in the carbon balance of organic soil-wetlands of the temperate zone. *J. Ecology* 74:755-774.

Campbell, C.A. and R.P. Zentner. 1997. Crop production and soil organic matter in long-term crop rotations in the semi-arid northern Great Plains of Canada. p. 317-334. In: E.A. Paul, K. Paustian, E.T. Elliott and C.V. Cole. (eds.), *Soil Organic Matter in Temperate Agroecosystems: Long Term Experiments in North America.* CRC Press, Boca Raton, FL.

Cole, V., C. Cerri, K. Minami, A. Mosier, N. Rosenberg, D. Sauerbeck, J. Dumanski, J. Duxbury, J. Freney, R. Gupta, O. Heinemeyer, T. Kolchugina, J. Lee, K. Paustian, D. Powlson, N. Sampson, H. Tiessen, M. van Noordwijk and Q. Zhao. 1996. Chapter 23. Agricultural Options for Mitigation of Greenhouse Gas Emissions. p. 745-771. In: *Climate Change 1995. Impacts, Adaptations and Mitigation of Climate Change: Scientific-Technical Analyses.* IPCC Working Group II. Cambridge University Press, Cambridge.

Cure, J.D. and B. Acock. 1986. Crop responses to carbon dioxide doubling: a literature survey. *Agric. For. Meteorol.* 38:127-145.

Davidson, E.A. and I.L. Ackerman. 1993. Changes in soil carbon inventories following cultivation of previously untilled soils. *Biogeochemistry* 20:161-164.

Elliott, E.T. and K. Paustian. 1996. Why site networks? p. 27-3.6 In: D.S. Powlson, P. Smith, and J.U. Smith (eds.), *Evaluation of Soil Organic Matter Models Using Existing, Long-Term Datasets.* NATO ASI Series, Global Environmental Change, Vol. 38, Springer Verlag, Berlin.

Haas, H.J., C.E. Evans, and E.F. Miles. 1957. Nitrogen and carbon changes in Great Plains soils as influenced by cropping and soil treatments. Technical Bulletin No. 1164 USDA, State Agricultural Experiment Stations, 111 pp.

Havlin, J.L. and D.E. Kissel. 1997. Management effects on soil organic carbon and nitrogen in the east-central Great Plains of Kansas. p. 343-351. In: E.A. Paul, K. Paustian, E.T. Elliott, and C.V. Cole. (eds.), *Soil Organic Matter in Temperate Agroecosystems: Long Term Experiments in North America.* CRC Press, Boca Raton, FL.

IPCC. 1990. Climate change. The IPCC Scientific Assessment. In: J.T. Houghton, G.J. Jenkins, and J.U. Ephraums (eds.). *Intergovernmental Panel on Climate Change.* Cambridge University Press, Cambridge.

Kittel, T.G.F., N.A. Rosenbloom, T.H. Painter, D.S. Schimel, and VEMAP Modeling Participants. 1995. The VEMAP integrated database for modeling United States ecosystem/vegetation sensitivity to climate change. *J. Biogeogr.* 22:857-862.

Lyon, D.A., C.A. Monz, R.E. Brown, and A.K. Metherell. 1997. Soil organic matter changes over two decades of winter wheat-fallow cropping in western Nebraska. In: E.A. Paul, K. Paustian, E.T. Elliott and C.V. Cole. (eds.), *Soil Organic Matter in Temperate Agroecosystems: Long Term Experiments in North America.* CRC Press, Boca Raton, FL.

Mann, L.K. 1986. Changes in soil carbon storage after cultivation. *Soil Sci.* 142:279-288.

Metherell, A.K., L.A. Harding, C.V. Cole, and W.J. Parton. 1993. Century Soil Organic Matter Model Environment - Technical Documentation. Agroecosystem Version 4.0. USDA/ARS Great Plains System Research Unit, Technical Report No. 4. Colorado State University, Fort Collins, CO.

Parton, W.J., D.S. Schimel, C.V. Cole, and D.S. Ojima. 1987. Analysis of factors controlling soil organic matter levels in Great Plains grasslands. *Soil Sci. Soc. Am. J.* 51:1173-1179.

Parton, W.J., J.W.B. Stewart, and C.V. Cole. 1988. Dynamics of C, N, P, and S in grassland soils: a model. *Biogeochemistry* 5:109-131.

Paul, E.A., K. Paustian, E.T. Elliott, and C.V. Cole. (eds.). 1997. *Soil Organic Matter in Temperate Agroecosystems: Long Term Experiments in North America.* CRC Press, Boca Raton, FL.

Paustian, K., H.P. Collins, and E.A. Paul. 1997. Management controls on soil carbon. p. 15-49. In: E.A. Paul, K. Paustian, E.T. Elliott, and C.V. Cole. (eds.). *Soil Organic Matter in Temperate Agroecosystems: Long Term Experiments in North America.* CRC Press, Boca Raton, FL.

Paustian, K., E.T. Elliott, H.P. Collins, C.V. Cole, and E.A. Paul. 1995. Use of a network of long-term experiments for analysis of soil carbon dynamics and global change: The North America model. *Aust. J. Exp. Agr.* 35:929-939.

Paustian, K., E.T. Elliott, G.A. Peterson, and K. Killian. Modelling climate, CO_2 and management impacts on soil C in semi-arid agroecosystems. *Plant Soil* (in press).

Peterson, G.A. and D.G. Westfall. 1997. Management of dryland agroecosystems in the Central Great Plains of Colorado. p. 371-380. In: E.A. Paul, K. Paustian, E.T. Elliott, and C.V. Cole (eds.). *Soil Organic Matter in Temperate Agroecosystems: Long Term Experiments in North America.* CRC Press, Boca Raton, FL.

Rogers, H.H. and R.C. Dahlman. 1993. Crop responses to CO_2 enrichment. *Vegetatio* 104/105: 117-131.

Vanotti, M.B., L.G. Bundy, and A.E. Peterson. 1997. Nitrogen fertilizer and legume-cereal rotation effects on soil productivity and organic matter dynamics in Wisconsin. In: E.A. Paul, K. Paustian, E.T. Elliott, and C.V. Cole (eds.). *Soil Organic Matter in Temperate Agroecosystems: Long Term Experiments in North America.* CRC Press, Boca Raton, FL.

CHAPTER 32

Predicting Soil Carbon in Mollisols Using Neural Networks

Elissa R. Levine and Daniel S. Kimes

I. Introduction

In this study, the feasibility of using neural networks as a method to understand and predict was tested. This task is in response to the challenge posed to the scientific community to develop an understanding of global environmental change due to natural and anthropogenic processes (Committee on Global Change, 1988). Scientific understanding of the soil system is important in the context of the sustainability and economic viability of agriculture and forestry, as well as proper management of our natural resources. Soils strongly influence the structure and function of ecosystems and will act as a buffer to global change as well as be affected by changes. Organic carbon is a critical component of the soil because of its role in soil productivity, in hydrology, and as a potential source or sink of terrestrial carbon affecting the concentration of atmospheric carbon dioxide. Soils information is critical for modeling ecological processes and vegetation dynamics, forecasting agricultural potential, predicting climate trends (with General Circulation Models) (GCMs), determining fluxes of CO_2 and other trace gases, making interpretations of land use and socioeconomic conditions, and for better interpretation of satellite imagery (Fung and Tucker, 1986; Adams et al., 1990; Daily and Ehrlich, 1992; Huston, 1993; Levine et al., 1993; Webb et al., 1993; Dixon et al., 1994). Yet, one of the limits to adequately assessing the presence of soil carbon is the lack of suitable data which can be used to quantify existing carbon stores and used for modeling purposes.

A. Soil Organic Carbon Data and Models Presently Available

The scientific community has available a wealth of information about soils across the landscape in the form of pedon characterization data from the USDA Natural Resources Conservation Service (NRCS, formerly Soil Conservation Service), as well as from other agencies and Universities. There are many advantages to these data which allow them to be a great resource to the scientific community. There are some basic problems, however, that prevent them from directly being used as inputs to models related to many of the above issues.

Present soil characterization data represent a comprehensive suite of measurements and descriptive properties of soils as they occur across the landscape. Each individual profile described in this data set provides an accurate "snapshot" of the physical and chemical characteristics occurring at the time of sampling. With a sufficient number of cases at various sites, these data can aid in understanding the conditions under which soils form. Although samples were originally taken as discreet sampling points, with advanced computer technology, we now have the ability to assimilate this large quantity of points into a complete picture of how soils occur spatially across the landscape or region, and identify those soil properties that would be expected to occur under different conditions. Thus, individual soil profile information can be used to estimate regional, continental, or even global scale soil properties using careful analysis and modeling techniques.

The most obvious shortcoming to most soil characterization data for carbon estimates is that the original purpose for sampling the soils represented in the data base was not to estimate soil carbon or answer ecological questions. When most of the soils data available from the USDA NRCS were originally collected, the fundamental purpose was to "make predictions", specifically for engineering problems and agricultural planning (U.S. Department of Agriculture, 1993). Soils information was also collected for soil classification (Olsen, 1981) in order to describe soils based on actual properties as they exist in the field at sampling time (Smith, 1986). While both organic and inorganic forms of carbon are listed as part of the suite of analyses performed for soil classification purposes, laboratory measurements of carbon were not always made for each soil sample, and methods used to measure organic carbon content were not always consistent. In addition, soils which may represent a large sink of carbon such as forest soils, cold soils, surface horizons, or organic soils were not always sampled, since these are not generally used for agricultural purposes. In the case of organic soils, samples were generally not taken to a deep enough extent to account for the large quantity of subsurface carbon they contain.

With these sampling problems, gross underestimation of soil organic carbon contents may occur. By far, the soil organic carbon data collected is heavily weighted to the United States, requiring a major sampling effort to obtain similar data for the rest of the world. Additional sampling is under way in many locations, but data might not be immediately available, and sampling schemes and measured parameters continue to be made based on the original paradigms of soil survey. Data could be improved for missing values by reanalyzing archived soil samples for which measurements were not originally made, but would require a very concentrated effort. Thus, although the data base does contain a great deal of information about soil properties within horizons of individual soil profiles, there is also many missing or inconsistent pieces of information which makes it difficult to use in its present form to estimate organic carbon amounts and understand carbon dynamics.

Despite these shortcomings, the soil carbon budget, as well as the processes of carbon dynamics within soils have been estimated with existing data bases, or through the use of various modeling techniques. Post et al. (1992) used soil data up to 1 meter in depth from across the globe grouped by Holdridge life zone. Eswaran et al. (1993) estimated the distribution of sequestered organic and inorganic carbon in soil orders and suborders across the globe from the pedon data base of the USDA Soil Conservation Service and related organizations. Kern (1994) compared predictions of soil organic carbon made by aggregating data according to ecosystem type and those made by aggregating according to Soil Taxonomy. The taxonomy approach appeared to be more reliable because it had more complete data and provided a more realistic display of spatial distribution of carbon. Because of the hierarchical nature of Soil Taxonomy, additional information about the presence of carbon can be obtained when moving from the soil order to finer taxonomic levels (Kern, 1994). Bliss et al. (1995) performed a similar carbon inventory for the United States using the USDA NRCS State Soil Geographic (STATSGO) data base. Others have used similar approaches based on the same type of data (Bohn, 1976, 1982; Schlesinger, 1984; Batjes, 1995; Buringh, 1984; Franzmeier et al., 1985; Kimble et al., 1990).

Mathematical modeling is another approach that has been used to simulate carbon dynamics and soil carbon contents (e.g., Jenkinsen and Rayner, 1977; Juma and McGill, 1986; Parton et al., 1988; Pastor and Post, 1985; Smith, 1982; van Veen and Paul, 1981). These have the advantage of simulating complex processes involved in the formation and degradation of organic carbon and describing the relationship between various soil properties which interact simultaneously to control carbon dynamics. Empirical, stochastic, and mechanistic equations describing these interactions are represented in these mathematical models which allow the user to predict soil conditions under different scenarios and go beyond the information available in the initial input data. Results of this type of modeling approach have shown relationships between soil organic carbon and different parameters depending upon the location and type of soils. Organic carbon contents of soils of the Great Plains (Mollisols), for example, have been modeled by numerous investigators who have found similar inputs to be significantly related to organic carbon production, but not necessarily with the same importance in each study. Parton et al. (1987, 1989) stated that soil can be predicted from soil temperature, soil moisture, soil textural class, plant lignin content, and nitrogen in soil of the Great Plains. In similar soils, Burke et al. (1989) found soil organic carbon to be related to precipitation, clay content, and air temperature, while others found elevation and water regime to also be important factors (Nichols, 1984; Sims and Nielsen, 1986).

While their results are provocative, there are characteristics of soil carbon models which also limit their use in accurately assessing the distribution of carbon across the landscape. In general, these models may be overly simplistic, ignoring critical factors and non-linear relationships which may influence carbon dynamics, may be too complex and require detailed information describing the fractionation of organic components (which is not available in standard soil characterization data), or may be limited to a certain geographic region or soil type. The use of statistical or empirical models may also be hindered because strict statistical sampling schemes are not generally used in soil sampling (e.g., random or gridded sampling).

When dealing with an incomplete or imperfect soil data base, Beinroth (1990) suggested solutions of either collecting new data or making better use of existing data. While it is always important to supplement and improve data bases by collecting new data, the time and resources may not be available and would be enormous to provide the exact measurements required for modeling soil organic carbon globally. In order to make better use of the soil characterization data that have been collected to date, we need to develop new tools for data analysis. Ideally, these tools would be able to input all of the data as a whole representation of carbon across the landscape and identify the complex linear and nonlinear relationships between soil properties. They should be able to observe existing conditions which occur at the time of sampling, and use this information to learn the relationships between soil organic carbon at the site and the other soil, landscape, and climate variables. In addition, they must be insensitive to missing or "noisy" data, and must be rigorous enough to handle large quantities of data in order to create scenarios which may help in identifying key processes.

B. Neural Networks as a Tool for Understanding Soil Properties

Neural networks provide a tool to solve many of these complex data problems and allow modeling of soil carbon and soil properties using existing data. Because of the way they are organized, neural networks are able to:

1. learn patterns or relationships in data provided with a set of inputs and desired outputs,
2. generalize or abstract results from imperfect data,
3. be insensitive to minor variations in input (such as noise in the data, missing data, or a few incorrect values).

(Anderson and Rosenfeld, 1988; Fahlmann and Labriere, 1990; Zornetzer et al., 1990).

The learning ability of neural networks is based on connectionism (Fu, 1994). A network's structure is made up of a number of processing nodes with weighted connections between these nodes. The fundamental principles of many of the networks used in this study are as follows. (More details are presented by Fu, 1994 and Fahlman and Lebiere, 1990). The network consists of multiple inputs (e.g., measured or observed soil properties in numeric form), multiple hidden nodes, and a single output node. The inputs are fully connected to all hidden nodes which, in turn, are connected to the output node. In this study, the output node of the network represents soil organic carbon. All processing of signals flow from the input nodes through the hidden nodes to the output node (feed-forward). No feedback processes are allowed. Each node is a nonlinear processor of its input signals and receives signals from previous nodes or from network input. Each input signal to a node is weighted. All weighted inputs to a node are then summed and put through a nonlinear transfer function to produce the output signal of the node. Weights for each node are unique to that node and are learned (optimized) by the network. Knowledge contained in a network is encoded in the weights of each node. The result is a network that can be used to map complex nonlinear relationships between organic carbon and other soil properties.

Others have used this approach to solve environmental data problems. Lam and Pupp (1993) used neural networks to estimate missing data in an environmental database for several parameters that were occasionally not measured at monitoring stations. Their approach was to train the network with a part of the data and then use it to estimate the missing values. With this technique, they were able to predict pH values in water samples with low relative error.

In a study by Pachepsky et al. (1996), soil water retention in Ustolls, based on texture and bulk density, was predicted using neural networks, and gave better results than the same models using linear regression.

Levine and Kimes (1993) and Levine et al. (1996) predicted quantitative (cation exchange capacity) as well as descriptive (soil structure) soil properties in Ustolls using neural networks. The best network found for cation exchange capacity used inputs of clay, organic carbon, pH, and -1.5 MPa moisture/clay ratio with a prediction accuracy of 85% (r^2 of predicted vs. actual). For soil structure, the best network predicted 3 types (granular, blocky, and massive) with classification accuracies of about 79%. Inputs used by the net were % organic carbon, silt, and clay. This was an improvement over linear models which had accuracies of only 46%.

In work using multisource spatial data sets, neural networks were effectively used to classify four types of rock in Melville Peninsula, Northwest Territory, Canada (Gong, 1996). The data sources included Landsat Thematic Mapper, aeromagnetic, radiometric, and gravity data. Net results had an accuracy of 96% which was an improvement over other methods tested. Levine and Kimes (unpublished data) also used multisource, multiscale data obtained as part of the Forest Ecosystem Dynamics Project at Howland, Maine to classify soil drainage. Preliminary results from this study showed good results from neural net classification of soil drainage class using soil survey data, digital elevation data, forest cover data, and SPOT imagery.

As shown by these studies, neural networks can be used as a tool to identify parameters important for models, as well as provide a means of predicting soil and other ecosystem properties. Neural networks are used in this study because of their ability to "learn" complex relationships that may be inherent in soil data in an attempt to predict soil organic carbon.

II. Methods

A. Soil Data

The soils chosen for this study were part of the pedon database of the National Cooperative Soil Survey of the USDA, Natural Resources Conservation Service (NRCS). Profiles classified within the Mollisol taxonomic order were used. Mollisols can be described as dark colored, nutrient-rich soils formed under grassland vegetation, mainly found in middle to low latitudes (Soil Survey Staff, 1975). By limiting the analysis to Mollisols vs. a larger data set, the effects of climate and natural vegetation could be partially controlled.

Table 1. Soil parameters used as possible inputs to the neural network

Description data[1]	Analytical laboratory data
Suborder	% sand, silt, and clay
State	% total nitrogen
Year of sampling	pH
Horizon name	Total coarse fragments
Thickness of horizon	Exchangeable bases and acidity
Depth of horizon	Base saturation
Drainage class	Cation exchange capacity (CEC)
Moisture regime	CEC/clay ratio
Hue, value, and chroma	Bulk density
Physiography	Moisture retained at -1.5MPa and -0.03MPa
	-1.5MPa moisture/clay ratio

[1]Qualitative soil description data observations were assigned integer values (without ranking) in order to apply them in a numerical fashion to the neural network.

Horizons from each profile in the data set were regarded as individual points in the data. In that way, the specific properties unique to each horizon were considered in predicting the amount of carbon found there. The number of horizons used in the analysis were limited to those for which data were available (i.e., horizons which included a large range of soil parameters, and for which no missing values were found). A great number of potential parameters were available from the NRCS Pedon Data base. The subset of parameters chosen from the laboratory data and description data are given in Table 1. These were chosen because they represented parameters that may be important to soil carbon dynamics according to literature theory. Other parameters that also might be important were excluded because there were too many missing data values for these parameters in the data base. In an attempt to fill in missing values for some of the more important parameters, a neural network approach was also used to predict bulk density and % moisture at -0.03MPa (Levine and Kimes, unpublished data). With this in mind, a number of data manipulation scenarios were designed to determine the best neural network for predicting soil organic carbon. This affected the number of data elements included as well as the number of horizons that were available to train the neural nets. Model scenarios (Table 1) were tested which trained neural networks from different sets of input data including:

1. Using all analytical lab data for those horizons for which measured data (including bulk density and -0.03MPa moisture) were available (n=2,708 horizons)
2. Removing horizons from the data (listed in model scenario 1) which had a -1.5MPa moisture/clay ratio of >0.6 (indicative of poorly crystalline material) (Burt, 1995, p. 42) (n=2,100 horizons)
3. Using the data from model scenario 2 but removing % nitrogen as an input (since nitrogen is closely linked with organic carbon, this would help determine the effects of other soil parameters on soil organic carbon content) (n=2,100 horizons)
4. Predicting bulk density and -0.03MPa moisture with a neural net approach and adding these into the analytical lab data for which there were missing values (n=3,755 horizons)
5. Using a combination of qualitative soil description and analytical lab data (n=1,190 horizons)
6. Using only qualitative soil description data (n=1,190 horizons)

B. Training and Testing of Neural Networks

The "Predict"[1] software package (NeuralWare Inc.) was used to aid in processing the large data sets and in testing the numerous neural network structures. Predict automates many of the tedious data manipulation techniques required to explore and build networks. It incorporates a number of techniques including numerical transformations of the input variables, a genetic algorithm that performs input variable selection, and a cascade method of network constructions with an adaptive gradient learning rule.

Numerical transformations performed on the inputs often improve the performance of the network (Tukey, 1977; Chatterjee et al., 1991). In this study, continuous transformations were tested by Predict. In all cases, two thirds of the entire data set was chosen randomly for training, and one third chosen randomly for testing.

In this study, Predict used a genetic algorithm (Koza, 1993) to identify the most important soil inputs from a large number of potential parameters available in the data base. It searched for a set of variables that behave synergistically to cause good network performance. The algorithm started with a population of random variables of limited size and added variables according to network performance.

Predict used a cascade method of network construction (Fahlmann and Labriere, 1990) that added a hidden node one at a time while testing network performance. As each hidden node is added, it is fully connected to all previous nodes. The transfer function used for the hidden nodes was the hyperbolic tangent function. The transfer function used for the ouput node was the sigmoid function. An adaptive learning rule was applied by Predict that uses a back-propagated gradient approach (Fahlmann and Labriere, 1990). Several methods were used to avoid overfitting the training data including reducing network structure, halting training when test performance began to decline, decreasing the number of input variables, and others. In all cases, network solutions were found where the network performance was very comparable between the training and testing data (e.g., r^2 values similar to within 4%). For simplicity, the statistics reported are for the entire data sets (training and testing data combined). Both the rms (root mean error square of the predicted vs. true target value of the network) and the r^2 (multiple correlation coefficient of the predicted vs. true target value of the network) are reported.

III. Results and Discussion

Results of neural network training and testing are given in Tables 2 and 3 and Figure 1. As shown in Table 2, Model 2 gave the best results with an r^2 of 0.89 representing the relationship between actual carbon and the carbon predicted by the net. This model used analytical laboratory data for horizons which had measured values for all input parameters, but excluded horizons with -1.5MPa moisture/clay ratio >0.6 (indicating poorly crystalline clays). The r^2 for Model 2 was only slightly higher than Model 1 and Model 5, both of which had an r^2 of 0.84. Model 1 contained the same data as Model 2, but included those horizons which had poorly crystalline clays. Thus, removing the poorly crystalline clay data had a small effect on the net's predicting ability. Model 5 used a combination of analytical lab data and description data for all horizons in which both types of data were available.

Model 4 had an r^2 value of 0.80 which was slightly lower than Models 1, 2, or 5. It used the greatest number of horizons because it included all horizons available within the analytical data, as well as those for which bulk density and -0.03MPa moisture were predicted. Bulk density and

---

[1]The use of commercial products or company names is not intended as an endorsement by NASA of any products or companies.

Table 2. Results of neural net analyses

Model	n	r^{2*}	rms	Parameters used by the network
1. Laboratory data for all horizons with measured parameters	2,708	0.84	0.50	% nitrogen, CEC**/clay ratio, depth
2. Laboratory data from which horizons containing poorly crystalline materials have been excluded	2,100	0.89	0.30	% nitrogen, CEC/clay ratio
3. Same data as Model 2 but with nitrogen excluded as an input	2,100	0.60	0.55	Depth, bulk density, % silt
4. Laboratory data for all horizons including those in which bulk density and -0.03 MPa moisture were predicted	3,755	0.80	0.65	% nitrogen, CEC/clay ratio
5. Horizons for which both laboratory and description data were available	1,190	0.84	0.44	% nitrogen, CEC/clay ratio, -1.5 Mpa moisture
6. Description data with depth and % sand, silt, and clay included	1,190	0.58	0.71	Depth, % clay, moisture regime, value, chroma

*r^2 represents the relationship between the % carbon predicted by the neural network and the actual laboratory measured carbon content in each horizon.
**CEC = cation exchange capacity.

Table 3. Means and ranges of measured and predicted values of % organic carbon

	Measured		Predicted	
Model	Mean (%)	Range (%)	Mean (%)	Range (%)
1	1.30	0.01 - 12.78	1.25	0.01 - 4.40
2	1.16	0.01 - 7.51	1.14	0.06 - 3.33
3	1.16	0.01 - 7.51	1.15	0.04 - 3.04
4	1.39	0.01 - 13.66	1.34	0.01 - 4.63
5	1.39	0.01 - 10.73	1.34	0.01 - 3.71
6	1.22	0.01 - 10.73	1.18	0.01 - 3.55

-0.03MPa moisture predictions were based on previous work in which neural networks were trained and tested using Mollisol data from the NRCS Pedon Data Base (Levine and Kimes, unpublished data). The best neural network models for these parameters had r^2 values of 0.76 for bulk density and 0.81 for -0.03MPa. The lower r^2 value attributed to this model may be due to the large number of total horizons applied to the net, and thus, the greater range in carbon content in relation to other soil properties. In general, the trained neural networks were able to predict organic carbon values at the lower ranges (<4%) around which most of the points were clustered, as indicated in Table 3 and Figure 1. Fewer carbon values were measured at values >4%.

The genetic algorithm within the Predict software was able to identify a small number of soil parameters from the total given in Table 1 that performed well in predicting the desired soil carbon values. Nitrogen (%), and CEC/clay ratio were chosen as important for Models 1, 2, 4, and 5 (Table 2). Even with the addition of these data, % nitrogen and CEC/clay ratio were still chosen as the most important parameters when description data was present, as in Model 5; nitrogen and CEC/clay ratio still outweighed the descriptive parameters in the final net structure. Depth was also included as important in Model 1, and -1.5MPa moisture in Model 5.

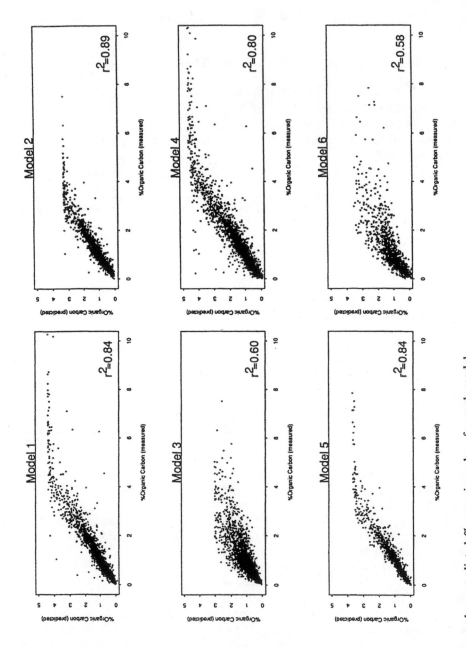

Figure 1. Measured vs. predicted % organic carbon for each model.

Model 3 used the same input data as Model 2, but omitted nitrogen. Without the influence of nitrogen, the parameters used by the network were depth, silt content, and bulk density. Overall, however, the neural network predicted organic carbon more poorly, with an r^2 value of 0.60.

When the net was restricted to using only description data, as in Model 6, depth, % clay, moisture regime, value, and chroma were chosen as the most important input variables. In this case, the genetic algorithm chose more input parameters than in the other cases, and the net had the lowest predicting ability ($r^2 = 0.58$) of the Models (Table 2). Again, quantitative data had a much greater effect and higher predicting ability than the description data as also shown in Model 5, when both description and laboratory data were combined. For cases such as this, where only description data is available, predicting soil organic carbon becomes much more difficult to accomplish. This may indicate and emphasize that additional quantitative soils information is needed to model carbon at coarser spatial scales.

Each of the parameters used by the neural networks had theoretical relationships to soil organic carbon. Nitrogen tends to parallel the organic matter content in soils under "steady-state" conditions where C/N ratios tend to be relatively constant, and can be used as an index of organic carbon (Jenny, 1980; Stevenson, 1982). The CEC/clay ratio gives an estimate of the type of clay in the horizon (Burt, 1995). Clays tend to retain organic substances using mechanisms such as: clay/organic complexes, adsorbtion to mineral surfaces (through cation exchange, bridging by polyvalent cations, hydrogen bonding, or van der Waals forces), or within the interlayers of expanding-type clays (Kononova, 1975; Stevenson, 1982), so that the type of clay will have a strong effect on the presence of carbon.

Depth made an important contribution to the neural net structure for Models 1, 3, and 6 which shows a strong relationship between the location of carbon input as well as biological activity in Mollisols to organic carbon content. Bulk density, -1.5MPa moisture, moisture regime, and value and chroma can logically be related to the influence that organic matter has on porosity, soil water holding capacity, organic decomposition, and soil color.

IV. Conclusions

Using any of the trained networks for Models 1, 2, 4, or 5 to predict soil organic carbon should produce reliable results. In addition, % nitrogen and CEC/clay ratio appear to be required as the most important soil parameters for making predictions. A caveat for using these nets, however, is that they were developed specifically with data for Mollisols. If % nitrogen, and CEC/clay ratio data are available for other Mollisols, carbon predictions can be made with reliability. With data for other soil orders or other soil conditions, it is recommended that additional neural networks be trained to determine the best model for those specific soils.

Studies such as these give preliminary evidence that the use of neural networks with soil characterization data is very promising. Additional experimentation with this method needs to be performed in order to more thoroughly make use of the soil characterization data available. The most rigorous work required for successful experimentation involves manipulating and formatting existing data before they can be presented to the nets for training. Once this formatting is done, the neural network approach can be applied to existing large soil data sets and effectively used to better understand the relationships between organic carbon and other soil properties under different scenarios and scales.

Acknowledgments

The authors would like to acknowledge and express our great appreciation to Dr. John Kimble for contributing his time and expertise to assisting us in data analysis and interpretation. This work was funded by the Director's Discretionary Fund, NASA/Goddard Space Flight Center.

References

Adams, R.M., C. Rosenzweig, R.M. Peart, J.T. Ritchie, B.A. McCarl, J.D. Glyer, R.B. Curry, J.W. Jones, K.J. Boote, and L.H. Allen, Jr. 1990. Global climate change and U.S. agriculture. *Nature*. 345:219-224.

Anderson, J.A. and E. Rosenfeld (eds.). 1988. *Neurocomputing: Foundations of Research*. The MIT Press, Cambridge, MA.

Batjes, N.H. 1995. *World Soil Carbon Stocks and Global Change*. ISRIC, Wageningen, The Netherlands.

Beinroth, F.H. 1990. Estimating missing soil data: How to cope with an incomplete data base. *Agrotechnology Transfer*. 11:10-11.

Bliss, N.B., S.W. Waltman, and G.W. Petersen. 1995. Preparing a soil carbon inventory for the United States using Geographic Information systems. In: R. Lal, J. Kimble, E. Levine, and B.A. Stewart, (eds.), *Soils and Global Change*. Adv. in Soil Sci., CRC Press, Boca Raton, FL.

Buringh, P. 1984. Organic carbon in soils of the world. *SCOPE* 23:91-109.

Bohn, H.L. 1976. Estimate of organic carbon in world soils. *Soil Sci. Soc. Am. J.* 40:468-470.

Bohn, H.L. 1982. Estimate of organic carbon in world soils, II. *Soil Sci. Soc. Am. J.* 46:1118-1119.

Burke, I.W., C.M. Yonker, W.J. Parton, C.V. Cole, K. Flach, and D.S. Schimel. 1989. Texture, climate, and cultivation effects on soil organic matter content in U.S. grassland soils. *Soil Sci. Soc. Am. J.* 53:800-805.

Burt, R. 1995. Soil Survey Laboratory Information Manual. *Soil Survey Investigations Report No. 45, version 1.0*, National Soil Survey Center, Soil Survey Lab, Lincoln, NE.

Chatterjee, S. and P. Bertram. 1991. *Regression Analysis by Example*. John Wiley & Sons.

Committee on Global Change. 1988. *Toward an Understanding of Global Change*. National Academy Press, Washington, D.C.

Daily, G.C. and P.R. Ehrlich. 1992. Population, sustainability, and Earth's carrying capacity. *Bioscience*. 42:761-771.

Dixon, R.K., S. Brown, R.A. Houghton, A.M. Solomon, M.C. Trexler, and J. Wisniewski. 1994. Carbon pools and flux of global forest ecosystems. *Science* 263:185-190.

Eswaran, H., E. Van Den Berg, and P.F. Reich. 1993. Organic carbon in soils of the world. *Soil Sci Soc. Am. J.* 57:192-194.

Fahlmann, S. and C. Lebriere. 1990. The Cascade-Correlation Learning Architecture. In: D. Touretzky (ed.), *Advances in Neural Information Processing Systems 2*. Morgan Kaufmann., San Mateo, CA.

Franzmeier, D.P., G.D. Lemme, and R.J. Miles. 1985. Organic carbon in soils of north central United States. *Soil Sci. Soc. Am. J.* 49:702-708.

Fu, L. 1994. *Neural Networks in Computer Intelligence*. McGraw-Hill, New York.

Fung, I.Y. and C.J. Tucker. 1986. Remote Sensing of the Terrestrial Biosphere. In C. Rosenzweig and R. Dickinson (eds.), Climate-Vegetation Interactions. Report OIES-2. Univ. Corp. for Atmos. Res., Boulder, CO.

Gong, P. 1996. Integrated analysis of spatial data from multiple sources: using evidential reasoning and artificial neural network techniques for geological mapping. *Photogram. Eng. and Rem. Sens.* 62:513-523.

Huston, M. 1993. Biological diversity, soils, and economics. *Science* 262:1676-1680.

Jenny, H. 1980. *The Soil Resource: Origin and Behavior*. Springer-Verlag, New York, NY.

Jenkinsen, D.S. and J.H. Rayner. 1977. The turnover of soil organic matter in some of the Rothemstad classical experiments. *Soil Sci.* 123:298-305.

Juma, N.G. and W.M. McGill. 1986. Decomposition and nutrient cycling in agro-ecosystems. p. 74-136. In: M.J. Mitchell and J.P. Nakas (eds.), *Microfloral and Faunal Interactions in Natural and Agroecosystems*. Martinus Nijhoff/Dr. W. Junk Publishers, Dordrecht.

Kern, J.S. 1994. Spatial patterns of soil organic carbon in the contiguous United States. *Soil Sci. Soc. Am. J.* 58:439-455.

Kimble, J.M., H. Eswaran, and T. Cook. 1990. Organic carbon on a volume basis in tropical and temperate soils. p. V248-V253. In: *14th Trans. Int. Congr. Soil Sci.*, Kyoto Japan.

Kononova, M.M. 1975. Humus of virgin and cultivated soils. p. 475-576. In: J.E. Gieseking (ed.), *Soil Components, Vol. 1*. Springer, N.Y.

Koza, J. 1993. *Genetic Programming*. MIT Press, Cambridge, MA.

Lam, D. and C. Pupp. 1993. Integration of GIS, expert system and modeling for state of environment reporting. Proc. of NCGIA 2nd Int. Conf. on Integrating GIS and Environ. Modeling. Breckenridge, CO.

Levine, E.R. and D.S. Kimes. 1993. Classifying Soil Structure from Soil Characterization Data Using Neural Networks. Proc. of NCGIA 2nd Int. Conf. on Integrating GIS and Environ. Modeling. Breckenridge, CO.

Levine, E., K.J. Ranson, D. Williams, H. Shugart, R. Knox, and W. Lawrence. 1993. Forest Ecosystem Dynamics: Linking Forest Succession, Soil Process, and Radiation Models. *Ecol. Modeling* 65:199-219.

Levine, E.R., D.S. Kimes, and V.G. Sigillito. 1996. Modeling soil structure using artificial neural networks. *Ecol. Modeling*, in press.

Nichols, J.D. 1984. Relation of organic carbon to soil properties and climate in the southern Great Plains. *Soil Sci. Soc. Am. J.* 48:1382-1384.

Olsen, G.W. 1981. *Soils and the Environment: A Guide to Soil Surveys and their Application*. Chapman and Hall, New York, NY.

Pachepsky, Y.A., D. Timlin, and G. Varallyay. 1996. Artificial neural networks to estimate soil water retention from easily measurable data. *Soil Sci. Soc. Am. J.* 60:727-733.

Parton, W.J., D.S. Schimel, C.V. Cole, and D.S. Ojima. 1987. Analysis of factors controlling soil organic matter levels in the Great Plains grasslands. *Soil Sci. Soc. Am. J.* 51:1173-1179.

Parton, W.J., J.W.B. Stewart, and C.V. Cole. 1988. Dynamics of C, N, P, and S in grassland soils: a model. *Biogeochemistry* 5:109-131.

Parton, W.J., C.V. Cole, J.W.B. Stewart, D.S. Ojima, and D.S. Schimel. 1989. Simulating regional patterns of soil C, N, and P dynamics in U.S. central grasslands region. p. 99-108. In M. Clarholm and L. Bergstrom (eds.), *Ecology of Arable Lands*. Kluwer Academic Publ., Dordrecht, the Netherlands.

Pastor, J. and Post, W.M. 1985. Development of a Linked Forest Productivity-Soil Process Model. Oak Ridge National Laboratory Environmental Sciences Division, Publication No. 2455. NTIS, Springfield, VA.

Post, W.M., W.R. Emanuel, P.J. Zinke, and A.G. Strangenberger. 1992. Soil carbon pools and world life zones. *Nature*. 29:156-159.

Schlesinger, W.H. 1984. Soil organic matter: A source of atmospheric CO_2. *SCOPE*. 23:111-127.

Sims, Z.R. and G.A. Nielsen. 1986. Organic carbon in Montana soils as related to clay content and climate. *Soil Sci. Soc. Am. J.* 50:1269-1271.

Soil Survey Staff. 1975. *Soil Taxonomy.* U.S. Dept. Agr. Handbook #436. U.S. Govt. Printing Office, Washington, D.C.

Smith, G.D. 1986. Rationale for Concepts in Soil Taxonomy. SMSS Tech. Monograph #11. Soil Conservation Service, Washington, D.C.

Smith, O.L. 1982. *Soil Microbiology: A Model of Decomposition and Nutrient Cycling.* CRC Press, Inc. Boca Raton. FL.

Stevenson, F.J. 1982. *Humus Chemistry.* John Wiley & Sons, NY.

Tukey, J. 1977. *Exploratory Data Analysis.* Addison Wesley.

U.S. Department of Agriculture. 1993. Soil Survey Manual. U.S. Dept. Agr. Handbook #18. U.S. Gov't Printing Office, Washington, D.C.

van Veen, J.A. and E.A. Paul. 1981. Organic C dynamics in grassland soils. 1: Background information and computer simulation. *Can J. Soil Sci.* 61:185-201.

Webb, R.S., S.E. Rosenzweig, and E.R. Levine. 1993. Specifying land surface characteristics in general circulation models: Soil profile data set and derived water holding capacities. *Global Biogeochemical Cycles* 7:97-108.

Zornetzer, S.F., J.L. Davis, and C. Lau. 1990. *An Introduction to Neural and Electronic Networks.* Academic Press, New York, NY.

CHAPTER 33

A Retrospective Modeling Assessment of Historic Changes in Soil Carbon and Impacts of Agricultural Development in Central U.S.A., 1900 to 1990

A.S. Patwardhan, A.S. Donigian, Jr., R.V. Chinnaswamy, and T.O. Barnwell

I. Introduction

Wilson (1978) stated that during the past 150 years, the Great Plains and the Corn Belt regions of the Central United States have gone under extensive ecological disturbance by the conversion of native ecosystems to agricultural lands. It has been reported in the literature that conversion of land from native vegetation into agroecosystems results in a loss of soil organic matter (Haas et al., 1957; Campbell 1978). Post et al. (1990) state that soil carbon losses have been a primary anthropogenic source of carbon dioxide, which is second only to fossil fuel combustion.

The current levels of soil carbon in agricultural soils are well below the levels observed prior to the establishment of agriculture, thus, with proper agricultural management the soil can be used as a sink of carbon, thus reducing CO_2 release to the atmosphere. In the global carbon cycle, the role of agroecosystems is very important as they contain 12% of the terrestrial soil carbon. The result of a properly managed agroecosystem could increase this pool, thus reducing the build-up of atmospheric carbon. In particular, alternative or sustainable agricultural practices could maintain or perhaps even increase soil carbon content.

Jenny (1941) stated that the factors controlling production and decomposition of soil organic matter are temperature and precipitation (climate), parent material (often represented by soil texture), relief (landscape position), organisms (particularly the plant community), and management (Elliott et al., 1993). All the factors stated above seldom act alone; however, they interact in a very complex way.

The U.S. Environmental Protection Agency (EPA) through its research laboratory in Athens, GA initiated a study in 1991 to estimate greenhouse gas emissions and carbon sequestration potential of agricultural production systems for the period 1990 to 2030 as part of a climate change assessment effort (Donigian et al., 1994). The EPA study region consists of the Corn Belt, the Great Lakes area, and a portion of the Great Plains. The CENTURY model, developed by Parton et al. (1988) and Metherell et al. (1993) at Colorado State University to simulate the dynamics of carbon (C), nitrogen (N), phosphorus (P), and sulfur (S) in cultivated and grassland soils, was used in this EPA study to estimate soil carbon levels throughout the study area in 1990. Estimates of initial soil carbon levels were required for estimating the carbon sequestration potential from agricultural production systems for the period 1990 to 2030.

In addition to the effects of variables noted by Jenny (1941), soil carbon is a direct function of the carbon inputs to the soil derived from crops, both as roots and crop residues. Thus, the impact of tillage and harvest practices that control the disposition of the crop biomass, and associated carbon, must be accurately represented in predictive models in order to mimic the carbon balance of the soil and the pattern of changes in soil carbon.

This chapter presents the approach used to simulate the historical pattern of soil organic carbon (SOC). We also discuss the assumptions that were made for historical agronomic and agricultural production practices that formed the basis of CENTURY model simulation runs. These simulations were then used to predict soil carbon inputs and resulting changes in soil carbon from 1907 to 1990. The results obtained from the historical simulations are then compared to available literature data and long-term field observations. A subsequent effort (Donigian et al., 1997) projected future changes in soil carbon, and associated carbon sequestration potential, from 1990 to 2030 under alternative scenarios of future production levels. Complete details, model results, and conclusions and recommendations are provided in the EPA final report (Donigian et al., 1994).

II. Century Model

The CENTURY model (Parton et al., 1988, Metherell et al., 1993) simulates the dynamics of C, N, P, and S in cultivated and grassland soils using a monthly time step for model simulations. The limitations and assumptions of the CENTURY model are discussed in detail by Donigian et al. (1994).

The model is composed of five organic matter pools, two of which represent litter or crop residues and the remaining three representing soil organic matter. The soil organic matter is divided into three fractions: 1) an "active" fraction which has rapid turnover rate and consists of microbial biomass and metabolites, 2) a "slow" fraction with an intermediate turnover time which represents stabilized decomposition products, and 3) a "passive" fraction which represents the highly stabilized, recalcitrant organic matter. In the model, soil moisture is calculated as a function of the ratio of monthly precipitation to monthly evapotranspiration, and the soil temperature term is calculated as a function of average monthly soil temperature at the soil surface. First-order rate constants depending on soil temperature and moisture are used to represent the decomposition rates. The other factors that affect the organic matter transformation include the lignin content of the litter. This lignin content of the litter pool influences the relative amount of the structural material which is assimilated by microorganisms (transferred to active pool) or gets incorporated into the slow pool. The soil texture influences the relative amount of decomposing organic matter that is either mineralized to CO_2 or enters the slow pool as decomposition products.

The CENTURY model has submodels for estimating plant growth and soil water balance. These submodels provide information on plant litter inputs and climatic variables for soil organic matter computations. Plant growth is expressed as a function of the growing season precipitation. The potential plant growth rate is used to calculate the crop nutrient demand. If nutrient availability is insufficient, then the crop production is reduced based on the nutrient availability.

A. CENTURY Model Data Needs

The application of the CENTURY model in this study was limited to the simulation of carbon and nitrogen dynamics in the soil profile. The CENTURY model inputs consist of meteorological data, soils data, and crop rotation along with management practices. In the following paragraphs, we describe briefly the procedures used for developing CENTURY model meteorological, soils, and crop inputs.

A collection of historical climate data for the United States (Wallis et al., 1991) was used to obtain 41 years (1948-1988) of monthly maximum and minimum temperature and total precipitation data for 589 climate stations (CS) in and near the study region. From these daily data, average monthly maximum temperature and average annual precipitation values were derived for each of the stations. A Geographic Information System (GIS) was used to create contour maps of the annual temperature and precipitation data; the temperature contours (using 2°C contour intervals) and precipitation contours (with 10-cm contour intervals) were overlayed to produce areas of similar climate defined as Climate Divisions (CD). Areas bounded by either type of contour were assumed to experience similar climate within the accuracy of the contour intervals. These values of 2°C temperature and 10-cm precipitation were derived from sensitivity analyses with CENTURY, recommendations and prior experience from the model authors (Burke et al., 1991), and prior experience in similar modeling investigations (Donigian and Mulkey, 1992).

The CENTURY model requires input data on soil texture (percent sand, silt, and clay), bulk density, field capacity, wilting point, and initial soil C and N levels. County resolution information was acquired for all of the soil data. The data were taken from three soils data bases — Data Base Analyzer and Parameter Estimator (DBAPE) (Imhoff et al., 1990), the 1982 National Resources Inventory (NRI) (Goebel and Dorsch, 1982), and the 1987 NRI/Soils-5 data base (Goebel, 1987). To develop soil texture information for the entire Study Region, the three data bases were utilized independently, meaning that DBAPE data were used for all counties contained in the data base, and data for any missing counties were obtained from either the 1982 NRI data, or 1987 NRI/Soils-5 data. Data extracted from each data base were limited to agricultural soils classes (i.e., P0, P1, and P4 categories), including agricultural soils under both irrigated and nonirrigated conditions.

All cropping, land use, and agricultural practice data for each policy option were provided by the Center for Agricultural and Rural Development (CARD) as part of the Resource Adjustment Modeling System (RAMS) output (Bouzaher et al., 1992). Essentially RAMS provided the area for each crop rotation practiced in each Production Area (PA), along with the required tillage and irrigation categories needed to accurately model the rotations. The information from RAMS is current as of the 1990-91 crop year; thus it provides an appropriate starting point for projection of soil carbon changes over the next 40 years. The most prevalent rotation is corn-soybean which makes up 25.7% of the study cropland area.

Historical crop yield data were needed as part of the procedures for estimating initial soil carbon levels in order to predict changes through the projection period; these yield values were used to calibrate the model estimated yields so that carbon inputs were properly estimated during the historical time frames. The historical yield data at a state level of resolution were obtained for corn, corn-silage, soybean, wheat, and hay for the period beginning in 1928 and ending in 1988 on 5-year intervals from United States Department of Agriculture - National Agricultural Statistics Service (NASS)(1928-88) and U.S. Department of Commerce CENSUS of Agriculture (1972-88) for all the states associated in the study region. However, for the period beginning in 1972 and ending in 1988, yearly county level yield data were obtained for all crops listed above from NASS computer tapes (USDA-NASS Tapes, 1992). The yearly county level hay yield was available only for the period 1982 to 1988.

III. Procedure for Estimating Initial Soil Organic Carbon

Significant changes in soil organic matter, either depletion or accumulation, usually occur slowly in agricultural systems with time frames on the order of decades. In estimating equilibrium soil carbon levels under native grassland and rangeland conditions, the CENTURY model is often run for 5,000 to 10,000 years in order to calculate stable values (Parton et al., 1989; Burke et al., 1990). In agricultural systems, changes in soil carbon depend on soil texture, regional climate, and carbon inputs to the soil, which are determined by cropping practices, crop yields, and management (i.e., rotations, tillage, fertilizer applications, residue/harvest practices). The challenge for this study was to develop a procedure for estimating initial soil carbon levels for each of our CDS, consistent with historical agricultural practices within each CD, as a basis for projecting future changes in SOC under both baseline (status quo) and alternative policy scenarios.

Although a number of soils data bases exist that include soil carbon, or organic matter, their utility for estimating initial soil carbon levels is limited. The primary limitation is that the data are usually associated with a specific soil series but without any definition of the variation in SOC due to the specific historical practices. Pedon data bases that provide values for defined sample sites also suffer from not having a history of the land use at the site. Moreover, the values in the data bases are derived from samples collected over the past 10 to 20 years, so they are not really current conditions. Even if appropriate soil carbon data were available, assigning values to the individual carbon pools in CENTURY would be difficult because the pools are conceptually defined based on turnover rates.

The approach developed to estimate starting SOC conditions was suggested by Dr. Vern Cole of the USDA Agricultural Research Service (Fort Collins, CO) based partly on his past work with CENTURY (Cole et al., 1989), studies by Paustian et al. (1992) on impacts of carbon inputs on soil carbon, and studies modeling historical yields and SOC on wheat-fallow systems in Colorado (V. Cole, personal communication, 1992). The underlying concept is to model a prior historic time period of agricultural activity, and calibrate to historic crop yields, so that the carbon inputs to the soil, the major factor determining soil carbon levels, are accurately represented.

To apply this general concept to the study region, we developed historical scenarios for cropping practices, tillage practices, fertilizer use, and harvest practices, and then attempted to match historical yield trends within each CD so that carbon inputs and SOC could be estimated with CENTURY. The overall steps in the methodology for estimating starting conditions are listed below, followed by a discussion of the underlying assumptions:

1. Use the IVAUTO option in CENTURY to start model runs in 1907 assuming a native grassland condition prior to the onset of agriculture. The IVAUTO option allows the user to estimate beginning (1907) SOC conditions based on regression models developed by Burke et al. (1989) that estimate SOC as a function of soil texture and regional climate. Although the data used by Burke et al. (1989) covered only the western fringes of our study region, we believe the 81-year simulation period from 1907 to 1988 will minimize the impact of any discrepancies in the 1989 starting values for our projection period.

2. Select the dominant crop rotation in each of the CDS based on the distribution and area of rotations provided by RAMS.

3. Run CENTURY for each dominant rotation in each CD for each of the six soil texture groups existing in the CD for the period 1907 to 1988; Table 1 lists the primary assumptions for weather data and historical management practices for defined time periods, which are discussed further below.

4. Calibrate the CENTURY model crop parameters to historic crop yield data for the 1907 to 1988 time period for all crops and for the dominant soil type within each rotation, so that historic carbon inputs are represented.

5. Use the 1988 year-end conditions for SOC (and all other CENTURY required state variables) for each soil texture group and rotation as the starting conditions for the 1989-2030 projections; starting in 1989 allows a 1- to 2-year period to adjust any inconsistencies between the initial conditions and the simulated scenario.

Table 1. Schedule of climate and management practices for estimating soil organic carbon

Time period	Weather data[1]	Crop varieties	Tillage practices[2]	Irrigation	Fertilizer	Grazing/manure[3]	Harvest
1907-1935	1948-1976	Low yield	Conventional	None	None	Manure return for forage crops	GS[4]
1936-1947	1977-1988	Low yield	Conventional	None	A50[5]	" "	G[6]
1948-1970	Historical	Medium yield	Conventional	Irrigated[7]	A80[5]	" "	G[6]
1971-1980	Historical	High yield	Limited conservation	Irrigated[7]	A[8]	" "	G[6]
1980-2030	1948-1988	Higher yield[9]	Policy scenario	Irrigated[7]	A[8]	" "	G[6]

[1] 1948-88 Site meteorological data was used for the periods prior to and following the historical period.
[2] See Table 2 for specific tillage operations in each time period.
[3] Grazing/manure represented manure returned from forage crops (i.e., legume and nonlegume hay and corn silage) used for animal feed stocks.
[4] At harvest, 50 percent of straw is removed along with grains.
[5] Automatic fertilizer application to maintain crop production at 50% (A50), 80%(A80) of potential maximum production, without increasing plant nutrient concentration above the minimum level.
[6] Only grains harvested.
[7] Irrigation was applied when 50% of the available water was depleted the soil water status was brought to field capacity after each irrigation. Irrigation started in 1948.
[8] Automatic fertilizer application to maintain crop production at 100% (A) of potential maximum production, and increase plant nutrient concentrations to the maximum level.
[9] Crop production parameter (PRDX) adjusted to reflect 1.5% yearly increase in yield for the projection period; impacts of alternative annual increases (i.e., 1.0% and 0.5%) are discussed by Donigian et al. (1994).

6. Run CENTURY for the 1989-2030 projection period for all rotations within the CD based on the RAMS data, using the starting conditions from 1988 (Step 5 above). See Donigian et al. (1994; 1996, this publication) for results of projections and alternative scenarios.

A. Underlying Assumptions for Starting Conditions

Table 1 lists the underlying assumptions used to approximate the evolutionary changes in agricultural practices over the 120-year simulation period, including the past 80 years and the 40-year projection period, so that we could use CENTURY to estimate changes in soil carbon. Clearly, we did not have the resources nor the data to accurately track historical changes in agricultural practices for the 489-million-acre study region. Our goal was to develop a reasonable pattern of the timing and nature of changes in agriculture that have a direct effect on soil carbon, in order to arrive at a reasonable starting point for assessing future changes.

1. Climate

The 41-year climate data base was used in simulating the actual time period of data 1948-1988, and both the prior and following time periods of 1907-1947 and 1989-2030. Although CENTURY includes an option to statistically generate monthly precipitation, testing showed that this produced much greater projected year-to-year variability in crop yields than observed. Also, since an analogous capability did not exist for monthly air temperature, use of the observed data was judged to be the best approach.

2. Tillage

The type and number of tillage operations are shown in Table 2 for each of the simulation time periods. Conventional tillage was assumed through 1970, when a gradual shift to various types of conservation tillage began. Larson and Osborne (1982) cite statistics showing that conservation tillage was only 2.3% of harvested cropland nationwide in 1965, and this increased to 16% in 1979. Citing data from the Conservation Tillage Information Center (CTIC), Mannering et al. (1987) show that conservation tillage was increasing but was still less than 10 to 15 million acres nationwide in the late 1960s. The RAMS data showed conventional tillage to still be 70% of the study region harvested cropland, i.e., 1989-1990. For the 1971-1988 simulation period, we reduced the number of tillage operations to accommodate the trend toward less tillage, but maintained conventional tillage as the dominant practice.

3. Fertilizer and Irrigation

Dramatic increases in fertilizer use in the 1940s and following WW II is cited as one of the reasons for significant yield increases for the major production crops (Tisdale et al., 1985). As shown in Table 1, we assumed the use of fertilizers began about 1936 with levels leading to 50% of maximum production for 1936-1974, 80% of maximum production for 1948-1970, and 100% of maximum production after 1970.

For irrigation applications, the Council for Agricultural Science and Technology (CAST) noted significant increases in irrigation water use in the Great Plains and the East starting in the mid-to-late 1940s (CAST, 1992). We assumed that irrigation began extensively in the time period beginning in

Table 2. Type and number of tillage operations for 1907-2030 (numbers in the column under a particular tillage practice reflect the number of times a particular tillage operation will be performed)

Time period	1907-1970	1971-1988	1989-2030				
Tillage			CTSP[1]	CTFP[2]	RT[3]	NT[4]	CC[5]
Moldboard plow (fall)	1	1	-	1	-	-	-
Moldboard plow (spring)	-	-	1	1	-	-	-
Chisel plow	-	-	-	-	1	-	-
Disk cultivator	2	1	2	2	1	-	-
Planting drill	-	1	-	-	-	-	1
Row cultivator	2	1	2	2	1	-	-
Herbicide	-	-	-	-	-	1	-
No-till drill	-	-	-	-	-	-	-

[1]Conventional till spring plow; [2]conventional till fall plow; [3]reduced till; [4]no till; [5]crop rotations with cover crops.

1948, and that irrigation water was applied when the moisture deficit reached 50% of available water holding capacity. The RAMS output identified the specific crops, rotations, and PAs where irrigation was practiced.

4. Harvest and Manure Applications

For the initial time period of 1907-35 (Table 1), we assume that 50% of the plant residue is removed along with the grain at harvest time and that, beginning in 1936, only the grain is harvested with all the plant residue remaining in the field. These assumptions were derived from Cole et al. (1989) for wheat-fallow cropping systems in the northern Great Plains to approximate the advent of harvesting combines and their greater selectivity of grain harvest. The validity of these assumptions for other crops and regions is unknown and should be further investigated.

5. Crop Varieties and Yields

As noted above, the procedure we followed to estimate initial SOC levels involved calibrating crop yields to match historic levels during the simulation period. This led to lower yielding varieties (and associated crop parameters in CENTURY) for the 1907-1947 time periods, increasing yields from 1948-70, and the highest historic yields in the latest period of 1971-1988. The crop yield data clearly showed the sometimes dramatic trends of increasing crop yields; Flach et al. (1996) cite references supporting annual increases in crop yields averaging 1.8% per year during that past 40 years.

Table 3 Comparison of CENTURY (1980-1990) SOC with values estimated from Kern and Johnson (1991)

		SOC, gCm^{-2} to 20 cm depth	
CD	Location	Kern and Johnson, 1991	Century, 1980-1990
272	Northeastern Ohio	4000-5000	4160-4450
341	Northeastern Kentucky	3000-4000	3080-3280
413	Central Iowa	5000-7000	4500-4800
473	Northern Minnesota	6000-8000	3800-3920
603	Central Missouri	3000-5000	3050-3130
632	Southwestern Kansas	3000-4000	3000-3030
643	Northwestern Arkansas	3000-4000	1820-1970

IV. Simulation Results and Comparison with Literature Values

As discussed by Donigian et al. (1996, this publication), matching historical and projected crop yield data with CENTURY yield predictions, under assumed crop residue and management practices, is the key element in estimating soil carbon levels for the historical, current, and projection time periods. The basic premise was that adjusting CENTURY model parameters to accurately portray historical crop yield data, and using reasonable assumptions for historical (and future) tillage and management practices, would result in accurate estimates of carbon inputs into the soil environment, and would thus provide a sound basis for estimating soil carbon.

Lacking long-term soil carbon databases to confirm the CENTURY model predictions, we included two consistency checks as part of the modeling effort to insure that the model results are reasonable and consistent with the general literature, observations, and impressions of carbon dynamics within the study region. These two checks included comparisons of simulated and observed crop yields, and simulated soil carbon levels in selected CDS compared to mapped estimates developed by Kern and Johnson (1991) in a recent study. In addition, in this chapter we compare the general shape of the simulated SOC curve for the study region with available historical SOC curves reported in the literature. The comparison of simulated and observed historical crop yields are not included here (due to space limitations); however, the calibrated crop yield results, reported by Donigian et al. (1994, 1995), show good agreement and year-to-year variability compared to the available observations from 1924 to 1988.

Table 3 shows a comparison of soil carbon levels for selected test CDs with corresponding values estimated by Kern and Johnson (1991), who developed a national map of agricultural soil carbon from the 1982 and 1987 NRI database (Goebel and Dorsch, 1982; Goebel, 1987), the 1982 SCS Soil Interpretation Records, and the SCS Pedon Database. The Kern and Johnson values shown in Table 3 were estimated from their published maps and adjusted to a 20-cm depth to be consistent with the CENTURY model predictions. The CENTURY values shown in the table are a range for all crop/rotations/tillage combinations in the test CD for the 1980 to 1990 time period. Except for two test CDs (CD473 and CD643), the CENTURY estimates are consistently in or near the low end of the estimated range from Kern and Johnson. The general spatial pattern of SOC from both estimates is similar, with the highest values in the central and northern portions of the study region and the lowest values in the western, southern, and eastern periphery. For the two CDs where the CENTURY values are about 40% too low, it is interesting to note that the neighboring CDs produced SOC values within the Kern and Johnson range; differences in soil texture and climate used in the two methods are possible reasons for this spatial shift and differences in SOC.

Figure 1 attempts to aggregate a portion of the study results into a timeline of agricultural soil carbon predicted by the CENTURY model for the study region for the 120-year simulation period,

A Retrospective Modeling Assessment of Historic Changes in Soil Carbon and Agricultural Impacts

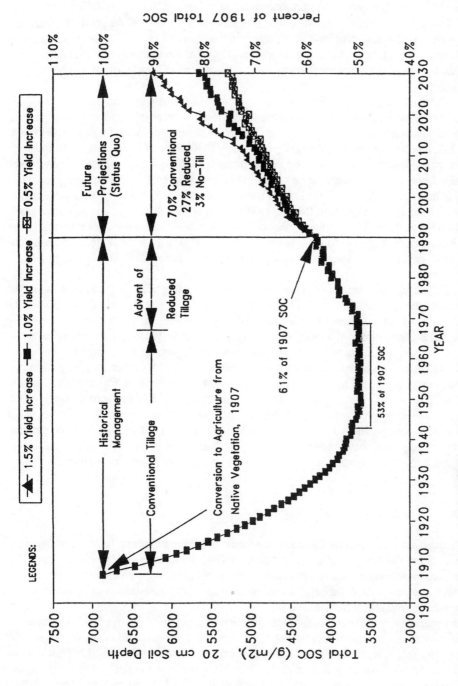

Figure 1. Simulated total soil carbon levels for the study basin region under the status quo scenario for both historical conditions (1907-1990) and future projections (1990-2030) under three alternative levels of future crop yield increases.

beginning with conversion of native vegetation to agricultural production in 1907, through current conditions (i.e., 1988-89); three alternative projections through the year 2030 are included for different levels of annual crop yield increases. These results were obtained by summing the products of the unit area changes in SOC under each crop/rotation/tillage combination and the area associated with each combination within each CD, then summing the values for all CDs in the study region, and dividing by the entire study region area; thus the values are in units of grams C per square meter (gCm^{-2}) for a 20-cm soil layer.

The total SOC values for the projection period 1990 to 2030 are discussed in detail by Donigian et al. (1994). The historical portion of the curve shows the characteristic, and well-documented, decrease in SOC following land conversion to agriculture in about 1907, a continuing drop in SOC until about 1950, a period of stable and slightly increasing SOC through 1970, and then significant SOC increases throughout the remainder of the period through 2030. A variety of studies have noted the typical steep decrease in SOC when forest and grasslands are converted to agricultural production (Mann, 1986; Schlesinger, 1984; Odell et al., 1984), but increases in SOC have also been documented when soils were initially low in carbon (Mann, 1986) or under intensive fertilization (Odell et al., 1984). Houghton et al. (1983) have shown idealized curves of SOC following conversion of native lands to agriculture that are similar to the pattern as shown in Figure 1. Similar trends showing the decrease in SOC have been reported by Brady (1990) and Johnson (1995), Figure 2a and 2b respectively.

Allison (1973) identifies a period with stable or slightly increasing SOC which is consistent with the 1950-70 simulations. Paustian et al. (1992), and others, have noted increases in SOC with increased carbon inputs due to amendments and residue management that retain most of the crop residue on the field. Jenkinson et al. (1987) report on a long term soil management experiment that shows the effect of management on SOC. Continuous spring barley was grown from 1852 under two management options, one receiving no farm yard manure (FYM) and the other receiving farm yard manure. The plots where no FYM was applied resulted in a decrease in SOC, however, the plots receiving FYM increased SOC from about 3000 gCm^{-2} to 8500 gCm^{-2} to a depth of 23 cm.

In summary, the consistency checks performed indicated reasonable model behavior when compared to selected observed data points for crop yields and expected historical soil carbon patterns. Although further validation is needed and highly recommended, these consistency checks were deemed sufficient to justify use of the modeling procedures and methodology for this assessment.

V. Closure

Historical practices, as represented by the model assumptions used in this study, have led to decreases in SOC until about the 1940s and 1950s. Since then the model predictions of SOC are increasing for most crop production systems that leave significant amounts of crop residue on the field. At this time, these predicted historical trends appear to be reasonable and consistent with the application assumptions and the available literature.

Thus, the general pattern of decreasing SOC following conversion of native land to agriculture, and subsequent increases with more intensive fertilization and increased carbon inputs, is generally consistent with the literature; however, we have no direct confirmation of these CENTURY model predictions at the scale of the study region at this time. Hopefully, the ongoing data consolidation effort by Colorado State University's Natural Resources Ecology Laboratory and Michigan State University will help to further validate these conclusions when it is available.

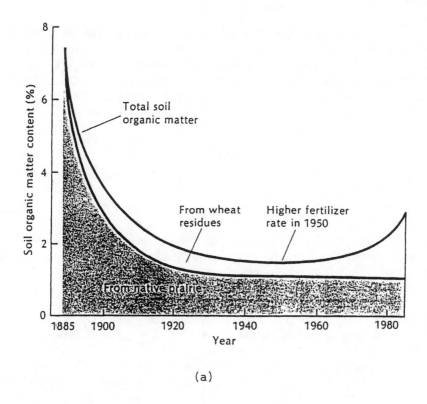

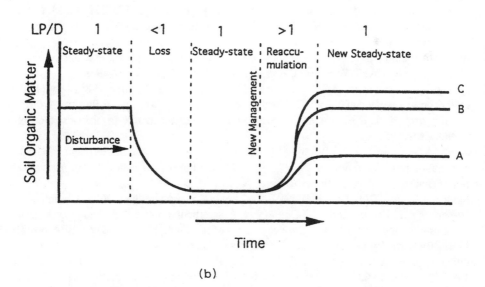

Figure 2. Changes in soil organic matter (a) from Brady, 1990; and conceptual model (b) of soil organic matter reaccumulation from Johnson, 1993. Line A indicates partial SOM reaccumulation; Line B, complete SOM reaccumulation; Line C, SOM reaccumulation in excess of original levels.

Acknowledgments

This study was funded by the Environmental Research Laboratory in Athens, GA, under EPA Contract No. 68-CO-0019 with AQUA TERRA Consultants, Mountain View, CA. Mr. Tom Barnwell was the Project Officer from the EPA Athens ERL. His support and assistance was critical to the success of the project. In addition, we would like to acknowledge the support and participation of the Natural Resources Ecology Laboratory at Colorado State University, the Center for Agricultural and Rural Development at Iowa State University, and the Kellogg Biological Center at Michigan State University.

References

Allison, F.E. 1973. *Soil Organic Matter and its Role in Crop Production*. Elsevier, Amsterdam, The Netherlands.

Bouzaher, A.B., D.J. Holtkamp, and J.F. Shogren. 1992. *The Resource Adjustment Modeling System (RAMS)*. Research Report, Center for Agricultural and Rural Development, Iowa State University, Ames, IA.

Brady, N.C. 1990. *The Nature and Properties of Soils*, 10th Edition. Macmillan Publishing Company, NY.

Burke, I.C., C.M. Yonker, W.J. Parton, C.V. Cole, K. Flach, and D.S. Schimel. 1989. Texture, climate and cultivation effects on soil organic matter contents in U.S. grassland soils. *Soil Sci. Soc. Am. J.* 53:800-805.

Burke, I.C., D.S. Schimel, C.M. Yonker, W.J. Parton, L.A. Joyce, and W.K. Lauenroth. 1990. Regional modeling of grassland biogeochemistry using GIS. *Landscape Ecology* 4:45-54.

Burke, I.C, T.G.F. Kittle, W.K. Lauenroth, P. Snook, C.M. Yonker, and W.J. Parton. 1991. Regional Analysis of the Central Great Plains - Sensitivity to climate variability. *Bioscience* 41:685-692.

Campbell, C.A. 1978. Soil organic carbon, nitrogen and fertility. p. 173-271. In: M. Schnitzer and S.U. Khan (eds.), *Soil Organic Matter*. Developments in Soil Science 8. Elsevier Science Pub. Co., Amsterdam.

Council for Agricultural Science and Technology (CAST). 1992. Preparing U.S. agriculture for global climate change. CAST Task Force Report No. 119. CAST, Ames, IA. 96p.

Cole, C.V, J.W.B. Stewart, D.S. Ojima, W.J. Parton, and D.S. Schimel. 1989. Modeling land use effects on soil organic matter in the North American Great Plains. p. 89-98. In: M. Clarholm and L. Bergstrom (eds.), *Ecology of Arable Land*. Kluwer Academic Publishers, Dordrecht, The Netherlands.

Donigian, A.S., Jr. and L.A. Mulkey. 1992. STREAM, An exposure assessment methodology for agricultural pesticide runoff. p. 297-330. In: J.L. Schnoor (ed.), *Fate of Pesticides and Chemicals in the Environment*. John Wiley & Sons, NY.

Donigian, A.S, Jr., T.O. Barnwell, R.B. Jackson, A.S. Patwardhan, R.V. Chinnaswamy, K.B. Weinrich, A.L. Rowell, and C.V. Cole. 1994. Assessment of alternative management practices and policies affecting soil carbon in agroecosystems of the Central United States. EPA/600/R-94-067. U.S. EPA, Environmental Research Laboratory, Athens, GA. 193 pp.

Donigian, A.S., Jr., A.S. Patwardhan, R.B. Jackson, IV., T.O. Barnwell, Jr., K.B. Weinrich, and A.L. Rowell. 1995. Modeling the impacts of agricultural management practices on soil carbon in the Central U.S. p. 121-135. In: R. Lal, J. Kimble, E. Levine, and B.A. Stewart (eds.), *Soil Management and Greenhouse Effect*. Advances in Soil Science. CRC Press/Lewis Publishers, Boca Raton, FL.

Donigian, A.S., Jr., A.S. Patwardhan, R.V. Chinnaswamy, and T.O. Barnwell. 1997. Modeling soil carbon and agricultural practices in the Central U.S.: An update of preliminary study results. (This publication.)

Elliott, E.T., C.A. Cambardella, and C.V. Cole. 1993. Modifications of ecosystem processes by management and the mediation of soil organic matter dynamics. p. 257-266. In: K. Mulongoy and R. Merchx (eds.), *Soil Organic Matter Dynamics and Sustainability of Tropical Agriculture.* Wiley-Sayce Co.

Flach, K.W., T.O. Barnwell, and P.R. Crosson. 1996. Impacts of agriculture on soil organic matter in the United States. In: E.A. Paul and C.V. Cole (eds.), *Soil Organic Matter in Agricultural Systems. I. Carbon Pools and Dynamics.* Lewis Publ., CRC Press, Boca Raton, FL.

Goebel, J.J. and R.K. Dorsch. 1982. National Resources Inventory: A guide for users of 1982 NRI data files. U.S. Soil Conservation Service. GPO Washington, D.C.

Goebel, J.J. 1987. National Resources Inventory. U.S. Soil Conservation Service. Resource Inventory Division. USDA Soil Conservation Service. P.O. Box 2890. Washington, D.C.

Haas, H.J., C.E. Evans, and E.R. Miles. 1957. Nitrogen and carbon changes in Great Plains soils as influenced by cropping and soil treatments. USDA Tech. Bull. 1164. United States Department of Agriculture, Washington, D.C.

Houghton, R.A., J.E. Hobbie, J.M. Melillo, B. Moore, B.J. Peterson, G.R. Shaver, and G.M. Woodwell. 1983. Changes in the carbon content of terrestrial biota and soils between 1860 and 1980: a net release of CO_2 to the atmosphere. *Ecol. Monogr.* 53:235-262.

Imhoff, J.C., R.F. Carsel, J.L. Kittle, Jr., and P.R. Hummel. 1990. Data base analyzer and parameter estimator (DBAPE) interactive program user's manual. EPA/600/3-89/083. Environmental Research Laboratory, U.S. Environmental Protection Agency, Athens, GA.

Jenkinson, D.S., P.B.S. Hart, J.H. Rayner, and L.C. Parry. 1987. Modeling the turnover of organic matter in long-term experiments at Rothamsted. In: Cooley, J.H. (ed.), Soil Organic Matter Dynamics and Soil Productivity. INTECOL Bull. 1987:15.

Jenny, H. 1941. *Factors of Soil Formation.* McGraw Hill, NY.

Johnson, M.G. 1995. The role of soil management in sequestering soil carbon. In: R. Lal, J. Kimble, E. Levine, and B.A. Stewart (eds.), *Soil Management and Greenhouse Effect.* Lewis Publishers (CRC Press), Boca Raton, FL.

Kern J.S. and M.G. Johnson. 1991. The impact of conservation tillage use on soil and atmospheric carbon in the contiguous United States. EPA/600/3-91/056. USEPA Environmental Research Laboratory, Corvallis, OR. 85 pp.

Larson, W.E. and G.J. Osborne. 1982. Tillage accomplishments and potential. p. 1-12. In: *Predicting Tillage Effects on Soil Physical Properties and Processes.* American Society of Agronomy Special Publication No. 44. Soil Sci. Soc. of Am., Madison, WI.

Mann, L.K. 1986. Changes in soil carbon storage after cultivation. *Soil Science* 142:279-288.

Mannering, J.V., D.L. Schertz, and B.A. Julian. 1987. Overview of conservation tillage. p. 3-17. In: T.J. Logan, J.M. Davidson, J.L. Baker, and M.R. Overcash (eds.), *Effects of Conservation Tillage on Groundwater Quality - Nitrates and Pesticides.* Lewis Publishers, Chelsea, MI.

Metherell, A.L., L.A. Harding, C.V. Cole, and W.J. Parton. 1993. CENTURY soil organic matter model environment, Technical Documentation, Agroecosystem Version 4.0, Great Plains System Research Unit, Technical Report No.4, USDA-ARS, Fort Collins, CO.

Odell, R.T., S.W. Melsted, and W.M. Walker. 1984. Changes in organic carbon and nitrogen of Morrow Plot soils under different treatments, 1904-1973. *Soil Science* 137:160-171.

Parton, W.J., J.W.B. Stewart, and C.V. Cole. 1988. Dynamics of C, N, P and S in grassland soils: a model. *Biogeochemistry* 2:3-27.

Parton, W.J., C.V. Cole, J.W.B. Stewart, D.S. Ojima, and D.S. Schimel. 1989. Simulating regional patterns of soil C, N, and P dynamics in the U.S. Central Grasslands Region. p. 99-108. In: M. Clarholm and L. Bergstrom (eds.), *Ecology of Arable Land.* Kluwer Academic Press.

Paustian, K., W.J. Parton, and J. Persson. 1992. Modeling Soil Organic Matter on Organic-Amended and Nitrogen-Fertilizer Long-Term Plots. *Soil Sci. Soc. Am. J.* 56:476-488.

Post, W.M., T.H. Peng, W.R. Emanuel, A.W. King, V.H. Dale, and D.L. DeAngelis. 1990. The Global Carbon Cycle. *Amer. Sci.* 78:310-326.

Schlesinger, W.H. 1984. Soil organic matter: a source of atmospheric CO_2. p 111-127. In: G.M. Woodwell (ed.), *The Role of Terrestrial Vegetation in the Global Carbon Cycle: Measurement by Remote Sensing.* John Wiley & Sons, NY.

Tisdale, S.L. W.L. Nelson, and J.D. Beaton. 1985. *Soil Fertility and Fertilizers*, 4th Edition. Macmillan Publishing Company, New York.

USDA/NASS - U.S. Department of Agriculture/National Agricultural Statistics Service. 1992. Crop estimates reel tapes ASB-101 & ASB-102. USDA Rm 5809-S Bldg., Washington, D.C. 20250.

Wallis, J.R., D.P. Lettenmaier, and E.F. Wood. 1991. A daily hydroclimatological data set for the continental United States. *Water Resources Research* 27: 1657-1663.

Wilson, A.T. 1978. The explosion of pioneer agriculture: contribution to the global CO_2 increase. *Nature* 273:40-42.

CHAPTER 34

Modeling Soil Carbon and Agricultural Practices in the Central U.S.: An Update of Preliminary Study Results

A.S. Donigian, Jr., A.S. Patwardhan, R.V. Chinnaswamy, and T.O. Barnwell

I. Introduction

As part of the U.S. Environmental Protection Agency's (EPA) Climate Change Program, selected agricultural management practices and policies were examined to evaluate their potential for sequestering additional carbon, and thereby reducing atmospheric CO_2 emissions. The research approach involved integration of physical soil carbon and econometric models and databases, and their linkage for assessment of policy impacts on agriculture and resulting soil carbon levels. The work described in this chapter focuses on the soil carbon modeling procedures used to evaluate the impacts of cropping practices, crop rotations, tillage practices, soils, and climate on soil carbon levels in the Central United States.

In April 1993, preliminary results of the study were presented at the International Symposium on Greenhouse Gas Emissions and Carbon Sequestration held in Columbus, Ohio, and subsequently published in Donigian et al. (1995). Following further investigation and review of those preliminary results, methodology and modeling assumptions were refined, further model parameterization was performed, and simulations were executed to update and improve the research approach and results. The final study results were published as an EPA Research Report entitled "Assessment of Alternative Management Practices and Policies Affecting Soil Carbon in Agroecosystems of the Central United States," (Donigian et al., 1994). This chapter summarizes the final study results, and thereby serves as an update of the preliminary results presented at the April 1993 International Symposium in Columbus, Ohio. In addition, research recommendations are provided to improve and refine the modeling procedures and extend their potential application to other regions of the U.S. and the world.

II. Study Background

In order to provide the basis for discussion of the final study results, we first describe the study background and methodology, much of which is abstracted from Donigian et al. (1994, 1995). Barnwell et al. (1992) presented the overall operational strategy for the study in terms of the high-level integration of models and databases, and their linkage for assessment of policy impacts on agriculture and resulting soil carbon levels. The objective is to estimate greenhouse gas emissions and C sequestration potential of agricultural production systems, and the impact of various policy alternatives designed to reduce emissions by increasing soil C levels. The impacts of projections reflecting increased adoption of conservation tillage and increased use of cover crops are compared to "status quo" conditions based on current practices and policies. Comparisons are made in terms of both the unit-area impact of crop rotations and practices throughout the study region, and the cumulative impacts for the entire study region to assess the potential role of agricultural management, within a national and global perspective, to sequester soil carbon for mitigation of global warming.

A. Study Methodology

In order to use models effectively to assess the soil carbon sequestration potential of agricultural regions of the U.S., a framework, or methodology, was needed to address issues of model integration, data needs and availability, temporal and spatial scales of analysis, detail of representation of agricultural practices, and associated technical model application questions. The integration of these concerns is implemented through the overall project methodology that was followed to assess both current conditions of soil carbon and soil organic matter in agricultural systems, and projected changes. These changes were evaluated under a continuation of current conditions (i.e., status quo) and selected alternative futures derived from policy changes.

The primary components of the modeling system include: the Resource Adjustment Modeling System (RAMS) (Bouzaher et al., 1992) to provide baseline production systems and cropping practices, and changes due to policy alternatives; the CENTURY soil carbon model (Parton et al., 1988) to provide unit-area (i.e., per hectare) changes in soil carbon and greenhouse gas emissions for each designated agricultural production system; the EPA Agroecosystem Carbon Pools database (E. Elliott, 1993, NREL/CSU, personal communications) to provide field data for model testing and validation; and the GIS capabilities EPA, Athens, GA to integrate the unit-area soil C values with cropping areas for display and analysis of baseline and policy alternatives. These components are discussed by Barnwell et al. (1992).

B. CENTURY and RAMS Model Overviews

A review of soil carbon models led to the selection of the CENTURY model for use in the modeling component of the study. Parton et al. (1988) developed the CENTURY model to simulate the dynamics of C, N, P, and S in cultivated and uncultivated grassland soils. The model uses monthly time steps for simulation, and model runs can be performed for time periods ranging from 10 to 10,000 years. CENTURY has submodels for estimating plant growth and soil water balance. These submodels provide information on plant litter inputs and climatic variables for soil organic matter computations. In this effort, the Agroecosystem Version 4.0 of CENTURY (Metherell et al., 1993) was used because of its capabilities for representing additional crops, complex cropping patterns, and multiyear crop rotations.

The Resource Adjustment Modeling System (RAMS) was developed in 1990 by the Center for Agricultural and Rural Development (CARD) at Iowa State University. RAMS is a regional, short-term, static, profit maximizing, linear programming model of agricultural production defined at the producing area level. RAMS provides the CENTURY model with information about production patterns and production practices (i.e., crops, rotations, tillage, and production inputs) throughout the study region.

Modeling Soil Carbon and Agricultural Practices in the Central U.S.

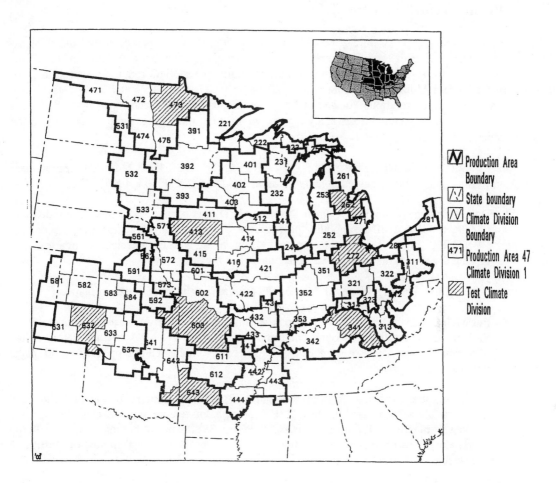

Figure 1. Study region, production areas, and climate divisions.

C. Initial Study Region and Production Areas

The study region, which corresponds to the area for which RAMS was applied, extends from the Great Lakes to eastern Colorado, and from the Canadian border to northern Arkansas, thereby including portions of the agricultural regions of the Cornbelt, Great Plains, and Great Lakes (Figure 1). Approximately 216 million acres of cropland are included in the RAMS region, representing 60% to 70% of the total U.S. cropland (USDA, 1992). In addition, the RAMS region is 25% of the continental U.S., and the 216 million cropland acres represents 44% of the entire study region. Also shown in Figure 1 are the RAMS 'Production Areas' (PAs) representing the basic spatial unit of analysis for the economic and policy analysis. PAs are hydrological areas, adjusted to county boundaries, representing aggregated subareas defined by the U.S. Water Resources Council (1970). The areas are small enough so that the assumption of homogenous production across the area can be reasonably made.

III. Project Modeling Methodology

The project modeling methodology was developed to integrate the issues of spatial detail of the required data, modeling of agricultural practices, estimation of current or initial soil carbon conditions, and procedures for assessment of alternative future conditions resulting from policy scenarios. An overriding concern throughout the work was to strike a balance between the number of combinations of soils, climate, crop rotations, tillage practices, and associated model runs needed to adequately represent the heterogeneities of the study region, within the time and resource constraints of the assessment.

Figure 2 shows the basic steps in the methodology and its application to the RAMS study region; these steps are listed below and discussed in more detail in Donigian et al. (1994, 1995):

1. Divide the study region into 'Climate Divisions' (CD) that can be represented with individual climate timeseries of the monthly precipitation and max/min air temperature needed by CENTURY. Figure 1 shows the climate divisions along with PA and state boundaries. The climate division is the basic spatial unit of analysis for this study.

2. For each CD, identify the distribution of soil textures on the agricultural cropland.

3. For each CD, identify the primary crops/rotation/tillage (C/R/T) scenarios as defined by the RAMS output for the status quo condition (i.e., current 1989-90) and alternative policy conditions. These scenarios were the acreages in each combination of crop rotation and tillage practice in each CD.

4. Establish initial conditions (Current, i.e., 1989-90) for soil carbon (SOC), and other needed model variables, to be used for each C/R/T for each soil texture within each CD. This was derived from CENTURY model runs from 1907 to 1988 with calibration to historical crop yield data; Patwardhan et al. (1996) discuss the basis for the historical simulations through a retrospective assessment.

5. For each CD, extend the historical CENTURY model runs (from Step 4, above) through the 40-year projection period of 1990-2030 for each C/R/T and each soil texture group, and output SOC (year-end values, g C/m^2) for analysis.

6. Weight the output from each soil-C/R/T (CENTURY) run by the soil texture distribution for that CD to get one time series curve of SOC for each C/R/T within each CD.

7. For each CD, multiply the weighted curve (e.g., for SOC, gC/m^2) by the area for each C/R/T combination; these areas are provided by the RAMS output for each PA, which is transformed to a CD basis. The resulting values of SOC for each C/R/T are then analyzed to determine the total SOC changes for each CD. The totals for each CD are then summed to provide totals for the entire study region for the entire projection period or any subset thereof.

8. Evaluate the status quo or baseline conditions, and compare them with each alternative scenario, based on the relative impacts on SOC changes. Alternatives that simply change the acreages of the rotations (i.e., each C/R/T combination) do not require additional model runs; Step 7 is repeated for each new set of rotation acreages, in a simple spreadsheet-type calculation. Alternatives that lead to changes in the individual C/R/T curves (e.g., changes in tillage practices, new rotations, addition of cover crops) require new model runs, so Steps 5 through 7 are all repeated, along with calculation of the total SOC changes and N emissions.

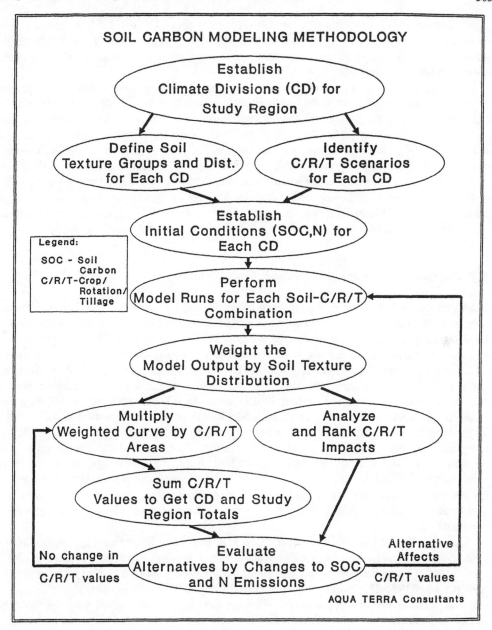

Figure 2. Soil carbon modeling methodology.

IV. Modeling Agricultural Production Systems and Scenarios

The objectives of this study and the methodology described above required development of methods to (1) represent agricultural production systems within thes study region using the CENTURY model, (2) estimate initial soil carbon (and nitrogen) values as a basis for projecting future changes, and (3) define both baseline (i.e., current) and alternative future conditions reflecting the modeling scenarios developed by RAMS. An agricultural production system is a complex combination of (1) specific crops grown in defined rotation patterns; (2) seasonally defined tillage and harvest practices using specified implements (equipment) that control the soil, crop, and residue disposition; and (3) agronomic inputs including fertilizers, manure, and/or organic amendments along with irrigation water and pesticides, as appropriate.

The project report (Donigian et al., 1994) describes the details of our approach for representing the physical aspects of agricultural production systems within the RAMS study region using the CENTURY model; the policy and market mechanisms are represented by RAMS and are also discussed by Bouzaher et al. (1993). Each of the three major components of these systems — crops and rotations, tillage and residue management, inputs — are represented within the limitations of the modeling capabilities and data available for this investigation. Clearly, as in any modeling exercise, representing agriculture on 216 million acres with 100,000 farms or more, requires some simplification of the real system to make the assessment feasible. Thus, our approach was to focus on the dominant crops, rotations, and tillage practices, and to represent the soil carbon impacts of these combinations within the spatial variability of the climate and soils conditions of the study region.

A. Status Quo/Baseline and Alternative Policy Scenarios

The status quo, or baseline, scenario represents the projection of changes in SOC from 1990 to 2030 under current cropping practices and current trends, the most critical of which is the assumed annual 1.5% increase in crop yields (see discussion below). The output of the RAMS model included the acreages for each crop rotation practiced within each PA, along with the tillage practices and areas receiving supplemental irrigation. The baseline for the RAMS model was the 1991 growing season. Three alternative policy scenarios were analyzed, including increased use of conservation tillage, increased use of cover crops, and impacts of the Conservation Reserve Program (CRP). The first two of the three scenarios were represented in RAMS by (1) targeting adoption of no-till and reduced tillage practices to levels suggested by tightening the criteria for defining highly erodible land for conservation compliance purposes, and (2) targeting planting of winter cover crops following sorghum, silage, and small grains. Due to space limitations, the CRP simulations are not discussed here; interested readers are referred to the project report. Table 1 summarizes the tillage distribution for the study region for each of the three levels of conservation tillage implementation — low, medium, high — along with the cover crop scenario. (Tillage practice abbreviations are shown at the bottom of Table 2.)

The procedure we followed to estimate initial SOC levels involved calibrating crop yields to match historic levels. This led to lower yielding varieties (and associated crop parameters in CENTURY) for the 1907-47 time periods, increasing yields from 1948-70, and the highest historic yields in the latest period of 1971-88. For the projection period 1989-2030, a 1.5% per year increase was assumed based on U.S.D.A. projections (U.S.D.A., 1990). Although the CAST report indicates that this level of yield increase may be obsolete (the projections were made in the early 1980s) and optimistic (since increases in the 1980s were considerably less), it concluded that they were plausible (CAST, 1992).

This assumption of a 1.5% per year increase in crop yields is a critical factor in projecting soil carbon levels, and associated carbon sequestration, over the next 40 years. To assess the general sensitivity of the study conclusions to this assumption, additional model runs were performed on the entire study region assuming annual yield increases of 1.0% and 0.5% (see discussion below).

Table 1. Tillage distributions for policy scenarios

Modeled scenarios	Tillage practices			
	CTFP	CTSP	RT	NT
		—%—		
Status quo/baseline	23	47	27	3
Low conservation	23	43	29	5
Medium conservation	23	41	29	8
High conservation	13	31	30	26
Cover crops	22	48	27	3

V. Revisions to the Modeling Procedures

In May 1993, the draft report on this study was prepared and distributed to the project team members, and a limited number of outside reviewers, in order to obtain comprehensive review comments on both the entire methodology and the preliminary study results. As noted above, following this review and presentation of initial study results at the 1993 Ohio Symposium, a number of methodology and operational refinements were implemented prior to performing the final model simulations. Although the changes resulted in significantly different model predictions of SOC, the general nature of the conclusions are essentially the same as reported previously by Donigian et al. (1995). The primary differences implemented in the modeling procedures for the final study results presented herein and in the Project Report (Donigian et al., 1994), are as follows:

a. The CENTURY model output variable used to assess soil carbon storages and changes was revised to include both plant root and surface residues. All analyses in this report are based on this quantity, referred to as total soil carbon (with the same acronym, Total SOC), unless otherwise specified. Total SOC is a better indicator of the carbon sequestration potential of alternative agricultural practices and policies because root and residue amounts will differ as a function of the practices and alternatives. This provides a more consistent means of assessing SOC changes for a wide range of crop rotations and policy scenarios.

b. The yield calibration procedure for the projection time period of 1989 to 2030 was adjusted to maintain a constant fraction of plant C included in the grain (i.e., CENTURY HIMAX parameter), and subsequently harvested for cash crops, based on calibrated values for the most recent historical period of 1971 to 1988. The initial results were based on calibrations for the projection period, leading to lower C content of residues and lower C inputs to the soil, especially for reduced tillage and no-till practices, resulting in lower SOC for these practices.

c. Investigations of the no-till simulations from the Draft Report indicated an over-estimation of the effect of crop residues on soil temperature, leading to lower soil temperatures and a retarding impact on crop growth and associated residues. For the final CENTURY model results, this adjustment to soil temperature was not included in the simulations in order to eliminate any bias in the comparisons of tillage alternatives.

d. As noted above, the uncertainty associated with the assumption of a 1.5% annual yield increase for all crops during the projection period, which was the basis of the preliminary results, was addressed by also evaluating the impacts of alternative annual yield increases of 1.0% and 0.5%.

Table 2. Soil carbon (gC/M²) and percent change in soil carbon for selected climatic divisions

CD	CR	Rotation sequence	Till	Soil carbon (gC/m²)						% change in soil C for selected time periods				
				1980	1990	2000	2010	2020	2030	1990/ 2000	2000/ 2010	2010/ 2020	2020/ 2030	1990/ 2030
272	100	CRN	NT	4289	4607	6043	6964	8364	9402	31.16	15.24	20.12	12.41	104.08
272	138	CRN,CRN,SOY,WHT,HLH	CTSP	4289	4560	5056	5474	6300	6761	10.87	8.26	15.08	7.31	48.26
272	186	CRN,SOY	CTSP	4289	4559	5029	5395	6076	6613	10.30	7.27	12.62	8.83	45.05
272	186	CRN,SOY	RT	4289	4569	5109	5520	6256	6834	11.81	8.04	13.33	9.23	49.57
272	186	CRN,SOY	NT	4289	4606	5364	5924	6769	7412	16.45	10.43	14.26	9.49	60.92
272	201	CRN,SOY,WWT	RT	4289	4570	5231	5689	6180	6994	14.46	8.75	8.63	13.17	53.04
272	243	CSL,CSL,OTS,HLH,HLH	CTSP	4289	4581	5546	6210	7194	7861	21.06	11.97	15.84	9.27	71.60
272	366	OTS,NLH,NLH,NLH	CTSP	4289	4592	5132	5435	6088	6653	11.75	5.90	12.01	9.28	44.88
413	100	CRN	CTSP	4677	4967	5933	6559	7510	8442	19.44	10.55	14.49	12.41	69.96
413	138	CRN,SOY	CTSP	4677	4968	5355	5658	6333	6741	7.78	5.65	11.93	6.44	35.68
413	138	CRN,SOY	RT	4677	4979	5413	5768	6513	6958	8.71	6.55	12.91	6.83	39.74
413	186	CRN,SOY	CTFP	4677	4966	5123	5286	5493	5790	3.16	3.18	3.91	5.40	16.59
413	186	CRN,SOY	CTSP	4677	4965	5193	5390	5634	5980	4.59	3.79	4.52	6.14	20.44
413	186	CRN,SOY	RT	4677	4976	5257	5491	5758	6133	5.64	4.45	4.86	6.51	23.25
413	186	CRN,SOY	NT	4677	5018	5553	5958	6356	6815	10.66	7.29	6.68	7.22	35.81
413	243	CSL,CSL,OTS,HLH,HLH	CTSP	4677	4988	5981	6588	7606	8592	19.90	10.14	15.45	12.96	72.25
413	503	HLH,HLH,HLH,HLH	CTSP	4677	5041	5216	5733	6341	7134	3.47	9.91	10.60	12.50	41.51
413	508	NLH,NLH,NLH,NLH	CTFP	4677	5054	5230	5746	6233	7114	3.48	9.86	8.47	14.13	40.75
473	002	BAR,BAR,SOY	CTSP	3895	3971	5046	5594	6327	7277	27.07	10.86	13.10	15.01	83.25
473	002	BAR,BAR,SOY	RT	3895	3977	5117	5713	6470	7508	28.66	11.64	13.25	16.04	88.78
473	100	CRN	NT	3895	3976	5435	6156	7504	8698	36.69	13.26	21.89	15.91	118.76
473	125	CRN,CRN,OTS,NLH,NLH	CTSP	3895	3958	4921	5555	6397	7265	24.33	12.88	15.15	13.56	83.55
473	145	CRN,CRN,WWT,HLH,HLH,HL	CTFP	3895	3965	4918	5618	6258	7215	24.03	14.23	11.39	15.29	81.96
473	215	H	CTSP	3895	3956	5157	5868	6859	7948	30.35	13.78	16.88	15.87	100.91
473	215	CRN,SWT,SWT	RT	3895	3967	5232	5993	7091	8317	31.88	14.54	18.32	17.28	109.65
473	215	CRN,SWT,SWT	NT	3895	3975	5428	6311	7524	9106	36.55	16.26	19.22	21.02	129.08
473	262	CRN,SWT,SWT	CTFP	3895	3975	5020	5630	6625	7597	36.28	12.15	17.67	14.67	91.11
473	463	CSL,OTS,HLH,HLH,HLH	CTSP	3895	3917	4128	4133	4345	4511	5.38	0.12	5.12	3.82	15.16
473	463	SMF,SWT	RT	3895	3917	4479	4649	5219	5758	14.34	3.79	12.26	10.32	47.00
473	508	SMT,SWT	CTFP	3895	3977	4700	5280	5674	7133	18.17	12.53	7.27	25.71	79.35

Table 2. continued--

CD	CR	Rotation sequence	Till	Soil carbon (gC/m²)						% change in soil C for selected time periods				
				1980	1990	2000	2010	2020	2030	1990/2000	2000/2010	2010/2020	2020/2030	1990/2030
643	100	CRN	NT	1787	1953	2898	3387	4097	4533	48.38	16.87	20.96	10.64	132.10
643	131	CRN,CRN,SOY	CTSP	1787	1928	2414	2630	2885	3275	25.20	8.94	9.69	13.51	69.86
643	131	CRN,CRN,SOY	RT	1787	1934	2476	2712	2987	3446	28.02	9.53	10.14	15.36	78.17
643	131	CRN,CRN,SOY	NT	1787	1953	2630	2979	3341	3811	34.66	13.26	12.15	14.06	95.13
643	218	CRN,WWT	CTFP	1787	1929	2419	2692	3105	3480	25.40	11.28	15.34	12.07	80.40
643	218	CRN,WWT	RT	1787	1935	2571	2935	3460	3919	32.86	14.15	17.88	13.26	102.53
643	366	OTS,NLH,HLH,NLH	CTSP	1787	1968	2290	2420	2968	3261	16.36	5.67	22.64	9.87	65.70
643	490	WWT	RT	1787	1894	2350	2582	3060	3448	24.07	9.87	18.51	12.67	82.04
643	503	HLH,HLH,HLH,HLH	CTSP	1787	1953	2044	2173	2509	2771	4.65	6.31	15.46	10.44	41.88

Note: Abbreviations used in the table.

CD	-	Climatic division
CTFP	-	Conventional till fall plow
RT	-	Reduced till
BAR	-	Barley
CRN	-	Corn
CSL	-	Corn silage
HLH	-	Legume hay
NLH	-	Nonlegume hay
CR	-	Crop rotation
CTSP	-	Conventional till spring plow
NT	-	No till
OTS	-	Oats
SMF	-	Summer month fallow
SOY	-	Soybean
SWT	-	Spring wheat
WWT	-	Winter wheat

e. Selected changes and refinements were made to CENTURY model parameters and/or capabilities to improve the (1) representation of tillage impacts on soil decomposition, (2) fertilizer nitrogen additions based on expected crop nitrogen content, and (3) application of irrigation amounts based on root zone soil moisture levels. Details are provided in Donigian et al. (1994).

VI. Model Simulations and Assessment Results

For this assessment we have focused primarily on soil carbon (SOC) values, and changes in SOC as impacted by the agricultural production systems and the variation in soils and climate conditions throughout the study region. We have analyzed these impacts in two ways: (1) unit area impacts, in terms of gC/m^2, to assess local impacts of the practices, and (2) aggregated impacts, in terms of total grams C, to evaluate the net change in C (i.e., sequestration potential) due to both practice and land use changes. The results presented below are brief updated study results corresponding to Those published in donigian et al. (1995). For brevity, we have included only results that have changed from the earlier paper; users should refer to the previous paper and the Project Report for more details and discussion.

A. Impacts of Crops, Rotations, and Tillage Practices

Table 2 shows model predictions of SOC for rotations and tillage combinations for selected CDs distributed throughout the study region; the locations of these CDs are shown in Figure 1. The values in Table 2 are year-end conditions at 10-year intervals from 1980 to 2030, and the percent change is calculated for each ten-year period starting in 1990 and for the entire 40-year projection period, 1990-2030. The SOC values pertain to the top 20 cm of the soil profile as this is the soil layer depth represented by CENTURY. The project report includes results for all CDs within the study region.

Figure 3 is an example of the CENTURY model predictions for SOC for the entire simulation period of 1907-2030 for CD 413 in central Iowa. These are presented to demonstrate the historical and future trends in SOC predicted by CENTURY under the model application assumptions of the study. Analysis of the model results provided the following observations:

1. Values in Table 2 are generally representative of those for all CDs and C/R/T combinations; the percent change for the 40-year projection period is typically about 30% to 80%. For all CDs and C/R/T combinations, the average increase was 55%, with an average minimum of 25% and an average maximum of 87%.

2. Historical practices, as represented by the model assumptions used in this study, have led to decreases in SOC until about the 1940s and 1950s. Since that time period, the model predictions of SOC are increasing for most crop production systems that leave significant amounts of crop residue on the field. A retrospective assessment of the general shape of this curve, the underlying modeling assumptions, and comparisons to available literature citations and field data is provided by Patwardhan et al. (1997).

3. The model results show consistent and significant increases in SOC throughout the projection period, except for a few hay or grain-based rotations in CDs located in the southern or western portions of the study region. The increases are due primarily to the assumption of a 1.5%

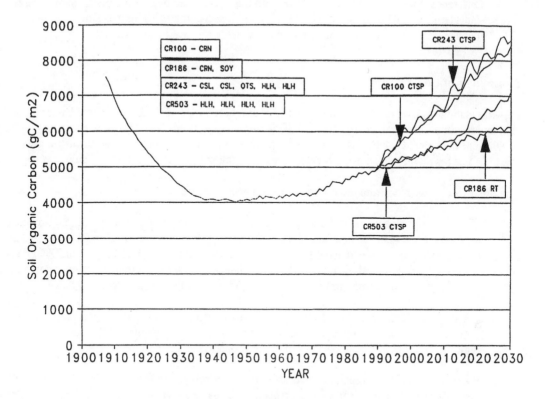

Figure 3. SOC changes and impacts of agricultural production systems on SOC in CD413, 1907-2030.

annual increase in crop yields and subsequent increase in carbon inputs. The sensitivity of the study region results to alternative annual yield increase levels is discussed further below.

4. The impacts of the tillage practices — conventional till, reduced till, no-till — are significant and highly variable across the study region. The degree of the impact is a function of the complex interactions of specific crops and rotations, climate, and soil characteristics. We generally see significant increases in SOC as one moves from CT to RT to NT. In many cases, the increase in SOC due to RT may be 10% to 15% higher, and for NT up to 50% higher, than for CT; however, in other C/R/T and CD combinations the changes are much less, and often less than generally assumed by conventional wisdom.

These results on the impacts of tillage should be considered preliminary for a number of reasons: (1) we simply do not quantitatively understand all the impacts of tillage alternatives as a function of soils and climate on crop yields, residue decomposition, and organic matter dynamics; (2) the parameter changes in current models, like CENTURY, to distinguish RT and NT are uncertain and require further investigation; and (3) some mechanisms for movement and disposition of surface residues under alternative tillage practices (e.g., earthworm activity, new tillage implements, etc.) are not well represented by the model.

5. The impact of irrigation was primarily on maintenance of crop yields; information from RAMS identified the CDs and rotations where irrigation was practiced.

6. Differences in the 1980 SOC conditions within a PA, i.e., among CDs with the same first two digits (e.g., 411, 412, 413) are due entirely to soil and climate differences since the historical management practices and dominant rotations were the same within the PA.

B. Impacts of Cover Crops

Implementation of cover crops in appropriate CDs and C/R/T combinations identified by the CARD analysis resulted in dramatic increases in SOC in many cases. Only 45 CDs out of the total number of 80 were considered conducive to cover crops due to climate, crops, and length of the growing season. As a result, these CDs were located in the central and southern portions of the study region.

The observations derived from the cover crop simulations include:

1. Use of cover crops can significantly increase SOC for many C/R/T combinations, especially for CDs located in the southern portions of the study region. For example, in CDs 603 and 612 the increases during the 40-year projection period are on the order of 100% to 150% greater with cover crops. For all CDs and C/R/T combinations with cover crops, the average increase was 63%, with an average minimum of 29% and an average maximum of 117%.

2. For the CDs located in the central portion of the study region (which corresponds to the northern tier of cover crop CDs), the SOC increase is much less than in the other crop cover regions. In these areas the benefits of cover crops would be marginal at best.

3. Even if climate conditions are favorable for cover crops, the extent of the impact on SOC depends on the specific crop rotation sequence. Thus, crop rotation 350 (CSL,CSL,SOY) can support cover crops in each winter (inter-crop) period, whereas rotation 186 (CRN,SOY) includes a cover crop only after the soybean crop, i.e., only 50% of the time. Clearly, the more often a cover crop is grown, the greater the potential impact on SOC.

C. Aggregate SOC Impacts for CDs and the Study Region

To evaluate the aggregate impact on SOC within each CD and the entire study region, the unit SOC soil storage values discussed above were converted to total SOC, i.e., gC, by multiplying times the appropriate areas associated with each C/R/T combination and summing the values for the CD. The results of this computation are shown in Figures 4 and 5, which are spatial displays of SOC by CD for 1990 and 2030, respectively, for the status quo scenario.

Observations on these model results are as follows:

1. The pattern of higher values of SOC in the central and northern portions of the study region, and lower values in the eastern, southern, and western portions, is consistent with other soil carbon mapping efforts (e.g., Kern and Johnson, 1991). The 1990 and 2030 maps of SOC show the same general spatial pattern, although with significantly higher values for 2030.

2. The regions with the greatest absolute change and percent change in SOC appear to be those CDs with dominant corn-based rotations and/or those located in the central and northern portions of the study region. The smallest absolute increases are generally in the south.

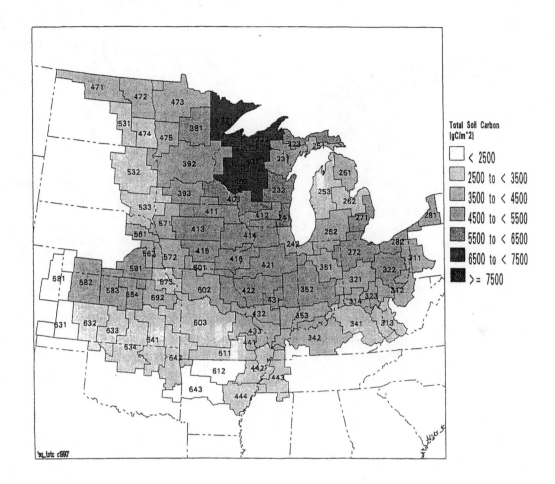

Figure 4. Simulated 1990 soil carbon (gC/m^2) distribution within the study region.

3. Figure 6 shows the 2030 SOC values weighted by cropland area in order to account for the relative distribution of cropland within the study region; this figure was generated by multiplying the SOC values in Figure 5 by the fraction of area representing cropland. Figure 6 is a more accurate spatial representation of SOC distribution within the study region since it shows the net impact of both unit area changes and cropland distribution. The greatest concentration of SOC is shown in the central area, i.e., the Cornbelt and Plains portions of study region; northern areas, such as PA 22, have large unit area values but comprise a small portion of the total SOC because of the small area of cropland.

D. Impact of Annual Yield Increase Assumptions

The critical importance of the assumption of the annual crop yield increase has been noted earlier. Using automated calibration and study region modeling capabilities developed by the EPA-Athens, the entire suite of CENTURY simulations for all 80 CDs, all C/R/T combinations, and all policy scenarios were redone for the 1990-2030 projection period for two additional levels of crop yield

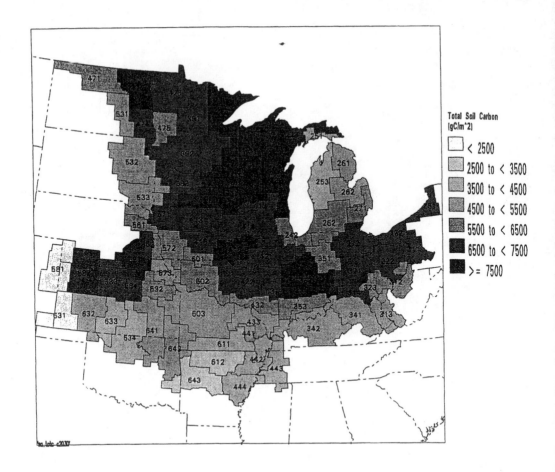

Figure 5. Simulated 2030 soil carbon (gC/m^2) distribution within the study region.

increases: 1.0% and 0.5 % per year. Table 3 shows the summary results of simulations for all three annual yield levels, in terms of both the Total SOC change (in GtC) over the 40-year projection period, and the percent change from the 1990 value and relative to the status quo scenario.

During the course of the study, thousands of simulation runs were performed as part of the model testing, calibration, and evaluation of alternative policy scenarios. Considering model runs for each of 80 climate divisions, with 4 to 6 crop rotations and multiple crops, up to 6 soil types, 3 alternative tillage practices, and multiple policy alternatives, no simple summary can adequately describe the range of conditions and impacts identified. Figure 7 attempts to aggregate a portion of the study results into a timeline of agricultural soil carbon (SOC), predicted by the CENTURY model, for the study region for the 120-year simulation period, beginning with conversion of native vegetation to agricultural production in about 1907, through current conditions (i.e., 1988-89); three alternative projections through the year 2030 are included for different levels of annual crop yield increases. These results were obtained by summing the products of the unit area changes in SOC under each crop/rotation/tillage combination and the area associated with each combination within each CD, then summing the values for all CDs in the study region, and dividing by the entire study region area; thus the values are in units of grams C per square meter (gC/m^2) for a 20-cm soil layer.

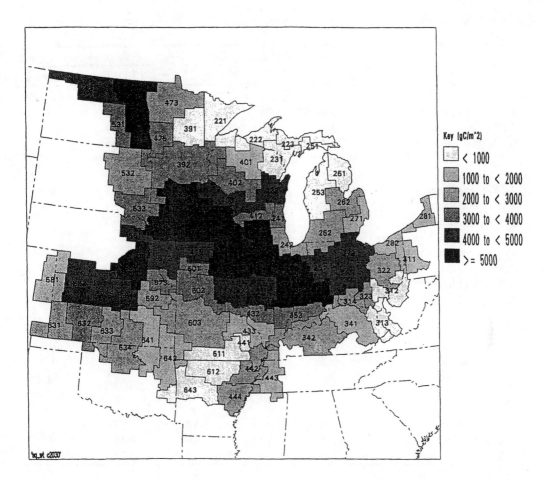

Figure 6. Simulated 2030 soil carbon (gC/m²) distribution within the study region weighted by cropland distribution.

The total SOC values for the projection period of 1990 to 2030 in Figure 7 are for the status quo condition, which represents a continuation of current 1989-90 cropping, rotation, and tillage practices through 2030, along with the impacts of the three alternative levels of annual increases in crop yields – 1.5%, 1.0%, 0.5%. The historical portion of the curve shows the characteristic, and well-documented, decrease in SOC following land conversion to agriculture in about 1907, a continuing drop in SOC until about 1950, a period of stable and slightly increasing SOC through 1970, and then significant SOC increases throughout the remainder of the period through 2030. Allison (1973) identifies a period with stable or slightly increasing SOC that is consistent with the 1950-70 simulations, but subsequent increases of the magnitude shown in Figure 7 from 1972 through 1990 have yet to be confirmed. These results reflect the impacts of increasing crop yields during this period, and modeling assumptions that represent an associated increase in retaining residues and returning them to the soil, along with a decrease in the level and intensity of tillage beginning in 1972.

For the 1990-2030 projection period, the continuing increase in SOC is based on the mix of cropping and tillage practices identified by the RAMS analysis for the current 1989-90 timeframe, along with the assumed levels of crop yield increases. The curves show steady, almost linear increases

Table 3. Impacts of alternative annual crop yield increases on study region total soil organic carbon for status quo and policy scenarios

	Total SOC, GtC			Precent increase from	
	1990	2030	1990-2030 gain	1990 value	Status quo
Status quo					
1.5% yield increase	3.66	5.45	1.80	49.1	--
1.0% yield increase	3.66	5.01	1.36	37.1	--
0.5% yield increase	3.66	4.60	0.95	26.0	--
Low conservation					
1.5% yield increase	3.66	5.47	1.81	49.4	0.7
1.0% yield increase	3.66	5.02	1.37	37.4	0.9
0.5% yield increase	3.66	4.62	0.96	26.2	1.1
Medium conservation					
1.5% yield increase	3.66	5.48	1.82	49.9	1.7
1.0% yield increase	3.66	5.04	1.39	37.9	2.2
0.5% yield increase	3.66	4.63	0.97	26.6	2.6
High conservation					
1.5% yield increase	3.66	5.57	1.90	52.0	6.1
1.0% yield increase	3.66	5.13	1.47	40.1	8.2
0.5% yield increase	3.66	4.72	1.06	28.8	11.2
Cover crops					
1.5% yield increase	3.66	5.60	1.94	53.1	8.3
1.0% yield increase	3.66	5.14	1.49	40.7	9.9
0.5% yield increase	3.66	4.74	1.08	29.7	14.2

in SOC from 1970 through 2030. For the historical period of 1970 to 1990, the increase occurs at an annual rate of about 0.6% per year, which is also about the same rate as for the 1990-2030 increase under the 0.5% crop yield increase. For the 1.5% and 1.0% yield increase levels during the 1990-2030 projection period, the total SOC increases at 1.2% and 0.9% per year, respectively. If this general pattern is accurate, agricultural SOC within the study region is making a comeback from a low of about 50% of original (i.e., native vegetation) levels in 1950-70, to about 60% of these levels in 1990. Continuing the increase would lead to 2030 total SOC levels that approach 75% to 90% of the original SOC prior to the onset of agricultural production (circa 1900).

From our modeling results presented here, with greater detail in the project report, the following observations are provided:

1. Under the status quo or baseline scenario, 1.8 Gt C will be retained or sequestered within the study region from 1990 to 2030 under the status quo assumptions described earlier. This represents a 49% increase over the 1990 level of SOC of 3.66 Gt C in the study region based on the model calculations.

2. The range in increase in SOC among the CDs goes from a minimum of 22% to a maximum of 101% during the 40-year projection period, under the status quo scenario.

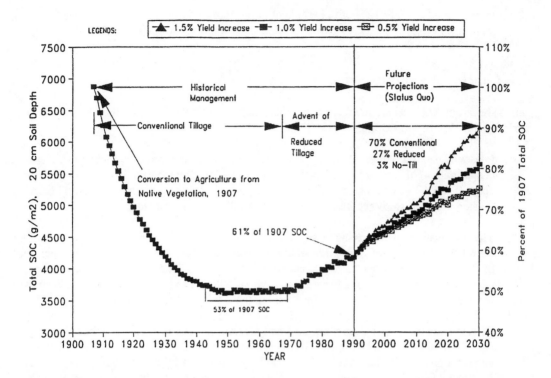

Figure 7. Simulated total soil carbon levels for the study region under the status quo scenario for three alternative levels of future crop yield increases.

3. The three conservation compliance (tillage) alternatives had a small, but significant, impact on SOC or carbon sequestration in the study region as a whole. The differences shown above, especially for the high conservation alternative, although small are considered significant because of the consistency of the increase and the relatively modest level of conservation tillage represented by the scenario. Clearly, conservation tillage has other benefits (e.g., water quality) not considered in this study.

    Moreover, as shown in Table 1, the high conservation scenario was based on a 26% level for NT, while the medium conservation scenario assumed only an 8% NT level; the corresponding levels of conservation tillage (i.e., combined RT and NT) were 37% and 56%, respectively, for medium and high scenarios. Data for 1992 from CTIC indicate that NT was practiced on 19% of cropland within the Cornbelt, and combined conservation tillage was 43%; also, dramatic increases in both practices were expected for 1993 (Dan McCain, CTIC, personal communication, 1993). Thus, the levels of RT and NT derived from the RAMS simulations are modest when compared to more recent data. Increased levels of RT and NT practices would show a greater increase on the study region SOC than indicated by the three scenarios evaluated. Although these results of tillage impacts are encouraging, further study is needed before we can confirm these conclusions. In addition, conservation tillage has other benefits (e.g., water quality improvement) not considered in this study.

4. Cover crops have a significant impact on SOC in the study region leading to a 53% increase from 1990, compared to 49% for status quo, and an additional 0.14 Gt C sequestered in the study region. Although these numbers are not large, they are the result of cover crops being

implemented on only 12% of the cropland in the study region. Among the CDs with cover crops, the average SOC increase over the status quo was 14% with a range from 0% to 132%.

VII. Conclusions and Recommendations

Below are capsule conclusions and recommendations derived from the study results, which are further expanded and discussed in Donigian et al. (1994):

1. Reasonable extrapolation of current agricultural practices and trends will lead to an increase (sequestration) of about 1 to 2 Gt C within the study region by the year 2030. This represents about a 25% to 50% increase over 1990 levels. Nationwide, the increase could be 50% greater since our study region includes 60-70% of total U.S. cropland.

2. The key assumption underlying these predictions is the projection of annual crop yield increase from 1990 to 2030; the lower range reflects an increase of 0.5% per year, while the upper limit of 50% increase reflects a 1.5% per year crop yield increase. The validity of these assumption needs to be reassessed or confirmed and, if valid, policies and research need to be promoted to support the chances of agriculture attaining these levels of yield increase.

3. Conservation tillage practices can significantly increase soil carbon, but the impacts are highly variable across the study region. The degree of impact is due to complex interactions of combinations of crops and rotations, soils, and climate. For many combinations, SOC increased 10% to 15% for reduced till and up to 50% for no-till, while much lower changes occurred in other C/R/T and CDs.

4. The overall impact of increased reduced till and no-till practices in terms of total SOC change for the study region, ranged from 2% to 3% higher than status quo under the medium conservation policy and 6% to 11% higher under the high conservation policy. These conclusions should be considered preliminary, within the capabilities of current models and our limited knowledge of the quantitative impacts of tillage practices. Moreover, current (i.e., 1992) nation-wide levels of no-till and conservation tillage (i.e., combined NT and RT) exceed the levels included under the medium conservation scenario; within the Cornbelt both NT and RT currently approach the levels included in the high conservation scenario. Thus, the conservation scenarios assume relatively modest changes in practices as compared to more current CTIC data. Further evaluation of the study procedures, conservation tillage usage, model parameter sensitivity, and more research are needed in this area of tillage impacts before these preliminary conclusions can be confirmed.

5. Cover crops can lead to significant increases in soil carbon in crop, soil and climate regimes where they are feasible and appropriate. Although only 12% of the study region cropland included cover crops (under the cover crop scenario), this increased soil carbon by 140 Mt through 2030. Since southern and eastern portions of our study region were most appropriate for cover crops, this may be an attractive alternative for promoting carbon gains in the South and Southeastern U.S.

6. The results of the CRP simulations are mixed. In many cases, 20 years of CRP leads to SOC values higher than the dominant rotation by the year 2030, and usually higher than under continuous CRP. In other CDs, the dominant rotation maintains the highest SOC throughout the projection period. The key factor is likely the relative carbon inputs of the dominant

rotation as compared to the CRP conditions. When the dominant rotation is a corn-based rotation, its 2030 SOC is usually the highest. However, this is not always true, especially if only one year of corn is in the rotation. The percent difference for the CRP simulations can range from a few percentage points to up to 20% or higher.

In addition, the study indicated a number of recommendations related to refinements in the modeling procedures for further application to the study region and other portions of the U.S. The key areas for further investigation included: further model testing and validation, especially related to historical trends and current SOC levels, corn and soybean based rotations, and impacts of tillage practices; better assessment of predictions of crop yield changes and associated biomass and residues assumptions; impacts of future climate scenarios; expansion of the study to include the impacts of erosion; improved databases and modeling procedures for animal waste applications; and improved modeling procedures for the integrated assessment of SOC with the nitrogen cycle, and associated N_2O emissions, as impacted by agricultural management alternatives.

These final study results confirm the preliminary results and provide a strong indication that agricultural trends are leading to generally improved soil fertility and increased SOC sequestration even without specific policies designed to promote these objectives. We need to continue to support those policies that will promote these current agricultural trends for continued improvements in soil fertility and productivity, since these goals are coincident with our efforts to sequester carbon and mitigate global warming and climate change impacts.

Acknowledgments

This study was funded by the U.S. Environmental Protection Agency in Athens, GA under EPA Contract No. 68-CO-0019 with AQUA TERRA Consultants, Mountain View, CA. Mr. Tom Barnwell was the EPA Project Officer. His support and assistance was critical to the success of the project. In addition, we would like to acknowledge the support and participation of the Natural Resources Ecology Laboratory at Colorado State University, the Center for Agricultural and Rural Development at Iowa State University, and the Kellogg Biological Center at Michigan State University.

References

Allison, F.E. 1973. *Soil Organic Matter and its Role in Crop Production*. Elsevier, Amsterdam, The Netherlands. 637 pp.

Barnwell, T. O. Jr., R. B. Jackson, IV, E. T. Elliott, E. A. Paul, K. Paustian, A. S. Donigian, A. S. Patwardhan, A. Rowell, and K. Weinrich. 1992. An approach to assessment of management impacts on agricultural soil carbon. *Water, Air, and Soil Pollution* 64:423-435, August 1992.

Bouzaher, A.B., D.J. Holtkamp, and J.F. Shogren. 1992. *The Resource Adjustment Modeling System (RAMS)*. Research Report, Center for Agricultural and Rural Development, Iowa State University, Ames, IA.

Bouzaher, A.B., D.J. Holtkamp, R. Reese, and J.F. Shogren. 1993. Economic and resource impacts of policies to increase organic carbon in agricultural soils. In: R. Lal, J. Kimble, E. Levine, and B.A. Stewart (eds.). *Soil Management and Greenhouse Effect, Advances in Soil Science*. CRC Press/Lewis Publishers, Boca Raton, FL. p. 309-328.

Council for Agricultural Science and Technology (CAST). 1992. *Preparing U.S. Agriculture for Global Climate Change*. CAST Task Force Report No. 119. CAST, Ames, IA. 96 pp.

Donigian, A.S. Jr. et al. (7 co-authors). 1994. *Assessment of Alternative Management Practices and Policies Affecting Soil Carbon in Agroecosystems of the Central United States.* EPA/600/R-94-067. U.S. EPA, Environmental Research Laboratory, Athens, GA. 193 pp.

Donigian, A.S. Jr., A.S. Patwardhan, R.B. Jackson, IV, T.O. Barnwell, Jr., K.B. Weinrich, and A.L. Rowell. 1995. Modeling the Impacts of Agricultural Management Practices on Soil Carbon in the Central U.S. p. 121-135. In: R. Lal., J. Kimble, E. Levine, and B.A. Stewart (eds.). *Soil Management and Greenhouse Effect*, Advances in Soil Science. CRC Press/Lewis Publishers, Boca Raton, FL.

Kern, J.S. and M.G. Johnson. 1991. *The Impact of Conservation Tillage Use on Soil and Atmospheric Carbon in the Contiguous United States.* EPA/600/3-91/056. U.S. Environmental Protection Agency, Corvallis, OR.

Metherell, A.L., L.A. Harding, C.V. Cole, and W.J. Parton. 1993. *CENTURY Soil Organic Matter Model Environment, Technical Documentation, Agroecosystem Version 4.0.* Great Plains System Research Unit, Technical Report No. 4. USDA-ARS, Fort Collins, CO.

Parton, W.J., J.W.B. Stewart, and C.V. Cole. 1988. Dynamics of C, N, P and S in grassland soils: A Model. *Biogeochemistry.* 5:109-131.

Patwardhan, A.S., A.S. Donigian, Jr., R.V. Chinnaswamy, and T.O. Barnwell. 1997. A retrospective modeling assessment of historical changes in soil carbon and impacts of agricultural development from 1900 to 1990. (This volume).

U.S. Department of Agriculture. 1992. Agricultural Resources: Cropland, Water, and Conservation - Situation and Outlook Report. AR-27. Economic Research Service, U.S. Department of Agriculture, Washington D.C. 77 pp.

U.S. Department of Agriculture. 1990. The Second RCA Appraisal. USDA Miscellaneous Publication No. 1482, Washington D.C.

U.S. Water Resources Council. 1970. *Water Resources: Regions and Sub-Regions for the National Assessment of Water and Related Land Resources.* U.S. Government Printing Office, Washington, D.C.

CHAPTER 35

Experimental Verification of Simulated Soil Organic Matter Pools

Cynthia A. Cambardella

I. Introduction

Soil organic matter (SOM) is a key component of terrestrial ecosystems and is functionally and structurally integrated into basic ecosystem processes. SOM turnover is coupled to the cycling of nutrients in soil and the agradation/degradation of soil aggregates through the activity of soil microorganisms.

Soil organic matter can be conceptually defined as a series of fractions that comprise a continuum based on decomposition rate (Schimel et al., 1985; Christensen, 1996a). These fractions have been represented in simulation models as kinetically defined pools with different turnover rates (Jansson, 1958; Jenkison and Raynor, 1977; van Veen and Paul, 1981; Parton et al., 1987). The movement of material into and out of SOM fractions can be mathematically described but, to determine how well models describe nature, one must be able to isolate biologically meaningful fractions from soil and characterize them (Cambardella and Elliott, 1994). A major limitation of current simulation models describing SOM turnover is that the conceptualized pools may not directly correspond to experimentally verifiable fractions.

Historically, chemical extractants have been used to fractionate SOM. However, chemically-defined SOM fractions aren't clearly related to the dynamics of organic matter in natural and managed systems because chemical extractants isolate SOM that may be physically protected from microorganisms and not readily available for decomposition (Oades and Ladd, 1977; Duxbury et al., 1989). Physical fractionation techniques are less destructive and more selective, and fractions isolated using these methods relate more directly to the structure and function of SOM *in situ* (Elliott and Cambardella, 1991).

The focus of this discussion is verification of simulation models that describe long-term SOM turnover. The concept of model verification can be easily confused with model validation. Model validation requires that a model which was parameterized and calibrated with one set of input variables be able to accurately predict output variables for another site or dataset. Model verification is "ground-truthing" the values for state variables and rate constants by isolation and characterization of naturally-occurring pools that correspond to conceptualized pools. A simulation model is simply an abstract representation of reality. This implies that a model is only as good as the data used to verify the variables that describe the kinetics of material transfer between pools.

II. The Problem

The rate of turnover rate of SOM varies due to the interaction of complex biological, chemical, and physical processes in soil. Biological availability of SOM is related to its chemical quality and the degree to which it is physically protected from decomposition. From an experimentalist perspective, SOM is made up of plant, animal, and microbial residues and organic decomposition products that are associated with the inorganic soil matrix. An ecosystem modeler would describe SOM as being composed of a series of fractions with different turnover rates. Ecologically, it becomes necessary to rectify these very different views into one congruent whole. In reality, SOM is a diverse mix of organic materials in various stages of decomposition. The distribution of SOM varies spatially and temporally and its turnover rate varies continuously due to the interaction of complex soil processes. These processes are controlled by stochasitc abiotic factors that are external to the system, such as landuse, management, and climate, which can affect the amount of organic matter present in soil, its turnover rate, and the degree to which that soil can sequester carbon as additional organic matter.

Simulation models can be useful as predictors of the long-term changes in SOM that occur as a result of changes in climate, landuse, and management. Societal concern over potential climate changes induced by increased inputs of greenhouse gases to the atmosphere can't be easily addressed through direct experimental evidence. Changes in climate occur very slowly, and simulation models are probably the only way to predict climate-induced changes in total SOM storage. Likewise, changes in total SOM that arise as a result of soil management and landuse changes will be verifiable experimentally only over extended time periods because it takes long periods of time for SOM levels to reach a new steady-state. A model must be relatively generalizable before it is a useful predictor of change. Unfortunately, the rigorous validation that is required before these models are generalizable is often side-stepped due to the urgency of society's expectations.

Although total SOM takes many decades to respond to change, active SOM pools are hypothesized to respond relatively quickly to changes in climate, landuse, and management. Labile SOM fractions could be valuable short-term indicators of long-term changes in total SOM. Conceptualized pools are difficult to quantify experimentally and accurate verification of model output isn't currently possible. This is because SOM exists as a continuum in nature and pool definitions require isolation of discrete fractions from soil. In addition, the pools were originally functionally defined using information from chemical fractionation procedures. Furthermore, the spatial aspects of SOM turnover were largely ignored. Potential experimental equivalents of SOM pools must integrate both the functional and structural properties of SOM and be selectively quantifiable.

III. One Solution

Despite all of this uncertainty, SOM turnover is commonly described using simplified, deterministic, multicompartmental simulations that are based primarily on empirically-derived relationships. Therefore, the concept of "SOM pools" is a working hypothesis used to explain the calibration data of simulation models (Jenkinson and Raynor, 1977; Christensen, 1996b). One way to test this hypothesis is to selectively isolate biologically meaningful fractions from soil and characterize them. A "biologically meaningful fraction" is isolated at a spatial scale that corresponds in a biologically sensible fashion to the scale of the habitat and resource base of the soil microbial community. Soil microorganisms occupy only a small percentage of the total soil pore volume. Their substrate base is primarily associated with the surfaces and crevices of soil aggregates (Elliott and Coleman, 1988). The main reason that chemically isolated SOM fractions aren't clearly related to SOM dynamics is because chemical extraction ignores the spatial compartmentalization of the organic material. Physically isolated SOM fractions obtained with fractionation procedures that take the soil apart in a hierarchical manner are more likely to be related to SOM structure and function *in situ* (Cambardella and Elliott, 1994). For example, fractionation by shaking or gentle sonication, followed by the separation of fractions based on particle size or density, provides information on where the organic matter is located in the soil. Subsequent characterization of the fractions provides information on the quality, form, and nutrient content of the organic matter associated with the fractions (Cambardella and Elliott, 1993a).

IV. The Models

Most current SOM turnover models are modifications of the Rothamsted model proposed in the late 1970s by Jenkinson and Raynor (1977). I will discuss the details of two models in addition to the original 5-compartment Rothamsted model. The Long-term Organic Carbon model (LTOC) proposed by van Veen and Paul (1981) and the CENTURY model of Parton et al. (1987) were chosen because of their similarity in structure to the Rothamsted model. All three models have a decomposable and resistant input pool, a microbial biomass pool, and several labile and stabilized SOM pools differentiated by turnover rate and C/N ratio. The mean residence times (MRT), or turnover times, for the model pools are given in Table 1.

V. Input Pool Verification

Model input pool verification can be approached by chemical analysis or by decomposition studies. Input pools are traditionally thought to be the easiest to verify. The decomposable input pool may be equivalent to either cold- or hot-water extractable SOM (Christensen, 1985). A common substitute is $0.5M$ K_2SO_4-extractable SOM which is the nonchloroform fumigated control in microbial biomass determination by fumigation and direct extraction (Brooks et al., 1985). The role of lignin and lignin/N ratio has been clearly defined for forest systems (Meentemeyer, 1978; Melillo et al., 1982). The resistant input pool is defined as having a high lignin/N ratio (Parton et al., 1987) and is commonly verified with van Soest's Acid Detergent Fiber method (Anderson and Ingram, 1989). Hemicellulose and cellulose can be similarly verified with chemical analyses. A potential problem with this reasoning is that rapid decomposition of cellulose and hemicellulose is greatly reduced *in situ* by its close association with lignin. Therefore, the functional equivalents of chemically-defined input fractions may not be straightforward.

VI. Microbial Biomass Pool Verification

The Rothamsted model (Jenkinson and Raynor, 1977) directly relates the microbial biomass compartment to biomass obtained from chloroform fumigation-incubation method (CFIM) (Jenkinson and Powlson, 1976). The active fraction of the CENTURY model (Parton et al., 1987) includes both the biomass and microbial metabolites. Partial verification is possible in this case by using CFIM biomass data. In the LTOC model (van Veen and Paul, 1981), the biomass is split into protected and unprotected fractions, and biomass protection is related to soil aggregation and texture. Since no direct method exists to quantify the protective capacity of soil structure, the LTOC biomass pool is currently not verifiable.

Table 1. Conceptualized pools and pool mean residence times (MRT) for 3 SOM turnover models

Rothamsted Model Pools (1977)*	Long-Term Org C Pools (1981)**	Century Model Pools (1987)***
Decomposable residue, 0.24 years	Decomposable residue, 1-2 weeks	Metabolic residue, 0.5 years
Resistant residue, 3.33 years	Lignified residue, 0.3 years	Structural residue, 3.0 years
Biomass, 2.44 years	Biomass, 0.5 years	Active, 1.5-10 years
Physically stabilized, 72 years	Active protected, 29 years	Slow, 25-50 years
Chemically stabilized, 2857 years	Old, >3000 years	Passive, 1000-1500 years

* Jenkinson and Raynor, 1977; ** van Veen and Paul, 1981; and *** Parton et al., 1987.

VII. Stabilized SOM Pool Verification

The stabilized SOM pools conceptualized in models are the most difficult to verify experimentally. This is partly due to the qualitative nature of the criteria used to define these pools. More importantly, verification is complicated because of the many and varied stabilizing mechanisms that are operating simultaneously in soil. The primary mechanisms for stabilization of organic matter are chemical recalcitrance, chemical stabilization, and physical protection. Chemical recalcitrance may be due to the incorporation of organic materials into heterocyclic polyaromatic compounds and also to the inherent chemical characteristics of the substrates (Christensen, 1996a). Chemical stabilization occurs via interaction of decomposable compounds with the mineral soil, primarily surface reactions with clays. Labile SOM can also be physically protected from decomposition inside soil aggregates, either by barriers that exist between substrate and decomposers, or by physical barriers that prevent faunal grazing. If SOM is physically unavailable to soil organisms, the chemical composition of that substrate is unimportant with regard to soil organic matter turnover. The microbial community must exist in the soil pore volume where there is sufficient air, water, and abundant readily utilizable substrates (Hattori and Hattori, 1976; Elliott and Coleman, 1988). The walls of the pore space are the surfaces of primary particles and/or aggregates. The accessibility of SOM to organisms or of prey to grazers, is a function of the architecture of the pores and particles. Different size pores are accessible to different size organisms, each of which may serve different functions in processing SOM (Elliott and Cambardella, 1991).

Most physical fractionation methods avoid chemical changes in SOM during the fractionation steps. The disruption energies are minimized in order to produce biologically meaningful fractions. The sequence of steps permits differentiation between SOM that is not associated with soil minerals (particulate organic matter (POM)), SOM that is incorporated into primary organomineral complexes (sand-, silt-, and clay-size complexes), and SOM that is trapped inside aggregates (Christensen, 1996a). Isolation of sand-, silt-, and clay-size complexes requires complete soil dispersion and produces fractions that are functionally analogous to soil textural classes. Wet sieving and limited soil dispersion are used to isolate aggregate size fractions. The methods employed to isolate aggregate fractions are based on the concept of aggregate hierarchy (Tisdall and Oades, 1982) and related to the processes of aggregate formation and stability.

Utilization of one or more physical separation methods, followed by chemical extraction of the physically isolated fractions and subsequent characterization have been used to verify SOM pools (Turchenek and Oades, 1979; Anderson et al., 1981; Dormaar, 1983; Tiessen and Stewart, 1983; Skjemstad et al., 1990; Cambardella and Elliott, 1992 and 1994; Beare et al., 1994; Golchin et al., 1994; Buyanovsky et al., 1994; Gregorich et al., 1995; Jastrow et al., 1996; Wander and Traina., 1996). For example, the specific combination of wet sieving to obtain macroaggregate size fractions, sonication to produce the constituent parts, followed by density floatation of the organic material associated with the constituent parts, yields discrete particle size/density fractions all of which were originally contained within the structure of the large soil aggregates (Cambardella and Elliott, 1993a). The SOM associated with these fractions can be characterized biologically through laboratory incubations that assess C and N mineralization potential, and chemically, through humic/fulvic separations or nutrient analysis.

A number of studies suggest that the POM is a dynamic soil property and an important pool of available C. POM responds selectively and rapidly to changes in landuse and soil management (Greenland and Ford, 1964; Tiessen and Stewart, 1983; Cambardella and Elliott, 1992; Wander and Traina, 1996). The majority of the POM fraction is physically stabilized inside soil macroaggregates and hypothesized to play a critical role in the formation and stabilization of aggregates (Cambardella and Elliott, 1993b). The stabilization of POM through encrustation and occlusion inside aggregates has been postulated to be the primary mechanism involved in the physical protection of bioavailable SOM (Oades, 1984; Elliott and Coleman, 1988).

POM can account for over 40% of the total SOM in native soils and is often less than 10% of the total SOM in soils that have been cultivated for a long time (Tiessen and Stewart, 1983; Cambardella and Elliott, 1992). Losses from the POM fraction account for the majority of the initial drop in SOM when native soils are brought under cultivation (Greenland and Ford, 1964; Balesdent et al., 1988). SOM associated with coarse-clay and fine-silt size fractions can account for up to 90% of the total SOM in soils that have been cultivated a very long time (Tiessen and Stewart, 1983). Accumulation of SOM in these fractions is associated with the stability of coarse-clay and silt-size aggregates (Turchenek and Oades, 1979). Cultivation also destroys the macroaggregate structure of soil (Chaney and Swift, 1984) with a concomitant reduction in SOM (Jenny, 1941). The reduction in SOM has been directly related to losses from the POM fraction (Tiessen and Stewart, 1983; Cambardella and Elliott, 1992 and 1993b). Therefore, the loss of structural stability associated with cultivation is related to losses from the POM fraction (Cambardella and Elliott, 1993b).

This information emphasizes the importance of identifying potential experimental equivalents to conceptualized fractions that integrate both structural and functional properties of SOM.

VIII. Input Pool Rate Constant Verification

In order to simplify the verification of rate constants parameterized in simulation models describing SOM turnover, it is generally assumed that all of the transformations follow first-order kinetics. Labeling plant residues with ^{13}C or ^{14}C is a commonly applied method to study the transfer kinetics of C from residue to SOM pools. These types of studies are most appropriate for examining short-term transfer kinetics and are most accurate for estimating turnover times for input pools and the most active SOM pools. Buyanovsky et al. (1994) describe the partitioning of ^{14}C originating from labeled soybean residues, among free POM, aggregates, and mineral-associated SOM. They report that free, undecomposed POM has the shortest MRT and that the MRT is inversely related to size. Larger pieces of free POM turn over every 1-2 years, and smaller pieces, every 2-3 years. The turnover times for free POM are similar to the MRT predicted for the resistant residue input pool of simulation models (Table 1). The calculated MRT for macroaggregates was 1-4 years, which is very similar to the MRT calculated for free POM. This isn't surprising, since macroaggregates form around fragments of decomposing plant residue.

IX. Stabilized SOM Pool Rate Constant Verification

Many recent studies have exploited shifts in ^{13}C natural abundance ratios that occur when the photosynthetic pathway of the original vegetation has changed to estimate long-term SOM turnover (Balesdent, et al. 1987 and 1988; Martin et al., 1990; Skjemstad et al., 1990; Cambardella and Elliott, 1992; Golchin et al., 1994; Jastrow et al., 1996). Since the isotopic composition of soil organic C reflects the material from which it was derived, the introduction of vegetation with a different photosynthetic pathway provides an *in situ* label enabling quantification of both the loss rate of the original SOM and the net input rate from the new source. Jastrow et al. (1996) report that after removal of free POM from aggregate size fractions, the turnover time of old C was 412 years for microaggregates (53-212 µm) and 140 years for macroaggregates (>212 µm). The turnover time for the macroaggregates roughly corresponds to that predicted by simulation models for the slow organic matter pool (Table 1) where physical protection is the dominant mechanism for SOM stabilization. The microaggregates are turning over at rates that more closely approximate those for the passive organic matter pool, where both biochemical recalcitrance as well as physical protection stabilize the organic matter. They also calculated that net input rates of new C increased with aggregate size, supporting the hierarchical aggregate model of Tisdall and Oades (1982).

Skjemstad et al. (1990) calculated turnover times for macro- and microaggregates from an Australian grassland soil that had developed under subtropical rainforest. They found that without removal of free POM, turnover times were 60 and 75 years for macroaggregates and microaggregates, respectively. Using soil from the same field site, Golchin et al. (1994) calculated turnover times for free and aggregate-occluded POM. They report that free POM turned over every 27 years, and occluded POM every 49 years. Using ^{13}C natural abundance ratios, Cambardella and Elliott (1992) calculated a turnover time of 19 years for POM isolated from a western Nebraska grassland-derived soil. For a forest-derived soil in eastern Ontario, Gregorich et al. (1995) report a turnover time of 11 years for free POM. All of these calculated turnover times are very similar to the 25-75 years predicted by simulation models for the slow or intermediately-labile SOM pool, suggesting that POM and/or macroaggregates may function as equivalents for this pool. Although the data are less abundant, there is some evidence to support the hypothesis that microaggregates and/or mineral-associated SOM are adequate equivalents to the passive or old SOM pool.

X. Conclusions

Verification of SOM pools conceptualized in simulation models has been hindered by our inability to isolate experimental equivalents to these pools. Data collected in the past 15 years or so suggests physically isolated SOM fractions obtained with fractionation procedures that take the soil apart in a hierarchical manner are most likly to be related to SOM structure and function *in situ*. Physically isolated SOM fractions have been related to conceptualized pools using calculated turnover times and location within the soil aggregate structure as guidelines for comparison. It is anticipated that model verification using physically isolated SOM fractions will undoubtedly lead to a redefinition of conceptualized pools and the rate constants that control the transfer of material between these pools (Christensen, 1996b). The search for quantifiable equivalents to current conceptualized pools should continue, especially taking into account the spatial arrangement of SOM in the soil matrix. These efforts will require closer cooperation between system modelers and experimental scientists collecting the data that will be used to calibrate and validate the models. Ideally, this will require a unified effort centered around the use of standardized methods that still permit the flexibility necessary for creative thought.

References

Anderson, D.W., S. Sagger, J. R. Bettany, and J.W.B. Stewart. 1981. Particle-size fractions and their use in studies of soil organic matter. I. The nature and distribution of forms of carbon, nitrogen and sulfur. *Soil Sci. Soc. Am. J.*. 45:767-772.

Anderson, J. M. and J.S.I. Ingram (eds.). 1989. *Tropical Soil Biology and Fertility: A Handbook of Methods.* C.A.B. International. Wallingford, U.K.

Balesdent, J.A., A. Mariotti, and B. Guillet. 1987. Natural ^{13}C abundance as a tracer for studies of soil organic matter dynamics. *Soil Biology Biochemistry* 19:25-30.

Balesdent, J.A., G.H. Wagner, and A. Mariotti. 1988. Soil organic matter turnover in long-term field experiments as revealed by carbon-13 natural abundance. *Soil Sci. Soc. Am. J.*. 52:118-124.

Beare, M.H., P.F. Hendrix, and D. C. Coleman. 1994. Aggregate-protected and unprotected organic matter pools in conventional- and no-tillage soils. *Soil Sci. Soc. Am. J.* 58:787-795.

Brooks, P.C., A. Landman, G. Pruden, and D. S. Jenkinson. 1985. Chloroform fumigation and the release of soil nitrogen: A rapid direct extraction method to measure microbial biomass nitrogen in soil. *Soil Biology and Biochemistry* 17:837-842.

Buyanovsky, G.A., M. Aslam, and G. H. Wagner. 1994. Carbon turnover in soil physical fractions. *Soil Sci. Soc. Am. J.* 58:1167-1173.

Cambardella, C.A. and E.T. Elliott. 1992. Particulate soil organic matter changes across a grassland cultivation sequence. *Soil Sci. Soc. Am. J.* 56:777-783.

Cambardella, C.A. and E.T. Elliott. 1993a. Methods for physical separation and characterization of soil organic matter fractions. *Geoderma* 56:449-457.

Cambardella, C.A. and E.T. Elliott. 1993b. Carbon and nitrogen distribution in aggregates from cultivated and native grassland soils. *Soil Sci. Soc. Am. J.* 57:1071-1076.

Cambardella, C.A. and E.T. Elliott. 1994. Carbon and nitrogen dynamics of soil organic matter fractions from cultivated grassland soils. *Soil Sci. Soc. Am. J.* 58:123-130.

Chaney, K. and R.S. Swift. 1984. The influence of organic matter on aggregate stability in some British soils. *J. Soil Sci.* 35:223-230.

Christensen, B.T. 1985. Decomposability of barley straw: effect of cold-water extraction on dry-weight and nutrient content. *Soil Biology Biochemistry* 17:93-97.

Christensen, B.T. 1996a. Carbon in primary and secondary organomineral complexes. p. 97-165. In: M.R. Carter and B.A. Stewart (eds.). *Structure and Organic Matter Storage in Agricultural Soils.* Advances in Soil Science. CRC Press, Inc., Boca Raton, FL.

Christensen, B.T. 1996b. Matching measurable soil organic matter fractions with conceptual pools in simulation models of carbon turnover: revision of model structure. p. 143-159. In: D.S. Powlson et al. (eds.). *Evaluation of Soil Organic Matter Models.* Springer Verlag. Berlin, Heidelberg.

Dormaar, J.F. 1983. Chemical properties and water-stable aggregates after sixty-seven years of cropping to spring wheat. *Plant Soil* 75:51-61.

Duxbury, J.M., M.S. Smith, and J.W. Doran. 1989. Soil organic matter as a source and a sink of plant nutrients. p. 33-67. In: D.C. Coleman et al. (eds.). *Tropical Soil Organic Matter.* University of Hawaii Press. Honolulu, HI.

Elliott, E.T. and D.C. Coleman. 1988. Let the soil work for us. *Ecological Bulletins* 39:1-10.

Elliott, E.T. and C.A. Cambardella. 1991. Physical separation of soil organic matter. *Agriculture, Ecosystems, Environment* 34:407-419.

Golchin, A., J.M. Oades, J.O. Skjemstad, and P. Clarke. 1994. Soil structure and carbon cycling. *Australian J. Soil Res.* 32:1043-1068.

Greenland, D.J. and G.W. Ford. 1964. Separation of partially humified organic materials from soils by ultrasonic dispersion. p. 137-148. In: Transactions International Congress Soil Science 8th, Volume 3. Bucharest.

Gregorich, E.G., B.H. Ellert, and C.M. Monreal. 1995. Turnover of soil organic matter and storage of corn residue carbon estimated from natural ^{13}C abundance. *Canadian J. Soil Sci.* 75:161-167.

Hattori, T. and R. Hattori. 1976. The physical environment in soil microbiology: an attempt to extend principles of microbiology to soil microorganisms. *CRC Critical Reviews in Microbiology* 4:423-461.

Jansson, S.L. 1958. Tracer studies on nitrogen transformations in soil with special attention to mineralization-immobilization relationships. *Lanterukshoegsk. Ann.* 24:101-361.

Jastrow, J.D., T.W. Boutton, and R.M. Miller. 1996. Carbon dynamics of aggregate-associated organic matter estimated by carbon-13 natural abundance. *Soil Sci. Soc. Am. J.* 60:801-807.

Jenkinson, D.S. and D.S. Powlson. 1976. The effects of biocidal treatments on metabolism in soil. V. A method for measuring soil biomass. *Soil Biology Biochemistry.* 8:209-213.

Jenkinson, D.S. and J.H. Raynor. 1977. The turnover of soil organic matter in some of the Rothamsted classical experiments. *Soil Science* 123:298-305.

Jenny, H. 1941. *Factors of Soil Formation.* McGraw Hill, New York.

Oades, J.M. 1984. Soil organic matter and structural stability: mechanisms and implications for management. *Plant Soil* 76:319-337.

Oades, J.M. and J.N. Ladd. 1977. Biochemical properties: carbon and nitrogen metabolism. p. 127-162. In: J.S. Russell and E.L. Greacen (eds.). *Soil Factors in Crop Production in a Semi-arid Environment.* Univ. Queensland Press. St. Lucia.

Martin, A., A. Mariotti, J. Balesdent, P. Lavelle, and R. Vuattoux. 1990. Estimate of organic matter turnover rate in a savanna soil by ^{13}C natural abundance measurements. *Soil Biology Biochemistry* 22:517-523.

Meentemeyer, V. 1978. Macroclimate and lignin control of litter decomposition rates. *Ecology* 59:465-472.

Melillo, J.M., J.D. Aber, and J.F. Muratore. 1982. Nitrogen and lignin control of hardwood leaf litter decomposition dynamics. *Ecology* 63:621-626.

Parton, W.J., D.S. Schimel, C.V. Cole, and D.S. Ojima. 1987. Analysis of factors controlling soil organic matter levels in Great Plains grasslands. *Soil Sci. Soc. Am. J.* 51:1173-1179.

Schimel, D.S., D.C. Coleman, and K.A. Horton. 1985. Soil organic matter dynamics in paired rangeland and cropland toposequences in North Dakota. *Geoderma* 36:201-214.

Skjemstad, J.O., R.P. Le Feuvre, and R. E. Prebble. 1990. Turnover of soil organic matter under pasture as determined by ^{13}C natural abundance. *Australian J. Soil Res.* 28:267-276.

Tiessen, J. and J.W.B. Stewart. 1983. Particle-size fractions and their use in studies of soil organic matter. II. Cultivation effects on organic matter composition in size fractions. *Soil Sci. Soc. Am. J.* 47:509-514.

Tisdall, J.M. and J.M. Oades. 1982. Organic matter and water-stable aggregates in soils. *J. Soil Sci.* 33:141-163.

Turchenek, L.W. and J.M. Oades. 1979. Fractionation of organo-mineral complexes by sedimentation and density techniques. *Geoderma* 21:311-343.

van Veen, J.A. and E.A. Paul. 1981. Organic carbon dynamics in grassland soils. 1. Background information and computer simulation. *Canadian J. Soil Sci.* 61:185-201.

Wander, M.M. and S.J. Traina. 1996. Organic matter fractions from organically and conventionally managed soils: I. Carbon and nitrogen distribution. *Soil Sci. Soc. Am. J.* 60:1081-1086.

CHAPTER 36

Modeling Tillage and Surface Residue Effects on Soil C Storage under Ambient vs. Elevated CO$_2$ and Temperature in *Ecosys*

R.F. Grant, R.C. Izaurralde, M. Nyborg, S.S. Malhi, E.D. Solberg, and D. Jans-Hammermeister

I. Introduction

Soil cultivation and consequent exposure has caused widespread declines in soil organic matter contents. Reduced cultivation practices prevent and eventually reverse these declines, due largely to the accumulation of plant residues on the soil surface (Edwards et al., 1992; Mielke et al., 1986). Surface residues decompose more slowly than those that are incorporated by tillage because of reduced contact with soil micro-organisms (Reicosky et al., 1995). Surface residues also reduce evaporation (Langdale et al., 1992), soil temperature (Grant et al., 1995a), and gas exchange from biological oxidation of organic matter (Reicosky and Lindstrom 1993, 1995), thereby reducing the oxidation of subsurface residues. Organic matter quality also improves under reduced tillage due to increased retention of microbial products (Angers et al., 1993; Biederbeck et al., 1994; Cambardella and Elliott, 1994).

Increases in atmospheric CO$_2$ concentration (C$_a$) and consequent increases in temperature have raised concerns about declines in soil organic matter beyond those due to cultivation that may contribute to further increases in C$_a$ (Jenkinson et al., 1991; Kirschbaum, 1995; Parton et al., 1995). The magnitude of such declines, if any, depends upon comparative increases in primary production caused by increased C$_a$ and air temperature, and those in soil respiration caused by increased soil temperature. Primary production and respiration may be affected differently by changes in precipitation and evapotranspiration rates under increased air temperature. Because more organic matter is retained under reduced vs. conventional tillage, the expanded adoption of reduced tillage has been proposed to reduce declines in soil organic matter and consequent increases in C$_a$. However, there is uncertainty about the rates at which C accumulates under reduced vs. conventional tillage, and about how these rates would change under elevated C$_a$ and temperature.

The complexity of interactions between C production and respiration has led to the use of mathematical modeling to study changes in soil organic matter under different climates and managements. However, in earlier studies, important effects of soil and climate on production and respiration were omitted (Jenkinson et al., 1991) or simulated at a temporal resolution much lower than that at which these effects are known to take place (Parton and Rasmusson, 1994; Parton et al., 1995). To our knowledge, the effects of tillage and surface residue management on soil C have not yet been modeled. We hypothesize that increased surface residue accumulation under reduced vs. conventional tillage causes reduced mass and energy exchange between soil surfaces and the atmosphere that in turn cause changes in soil temperature and water content that favor increased C storage. We propose to test this hypothesis using the mathematical model *ecosys* (Grant, 1993a) in which energy exchange among the atmosphere, plant canopies, and both soil and residue surfaces are explicitly simulated at high temporal resolution. These exchanges are fully coupled to energy transfers through soil profiles and plant canopies which have been used to simulate seasonal changes of soil temperature and water content under different plant and residue covers (Grant et al., 1995a). Our hypothesis will be supported if the effects of these changes on microbial (Grant, 1994; Grant et al., 1993a,b,c,d; Grant and Rochette, 1994) and plant (Grant et al., 1993e, 1995b,c) activity in *ecosys* allow the effects of surface tillage on soil C transformations to be simulated.

II. Methods

A. Model Development

1. General

The mathematical model *ecosys* is a general-purpose, research-level model of the transformations and transfers of water, heat, carbon, oxygen, nitrogen, phosphorus, and salts within a complex plant-microbe biome under defined conditions of soil, climate, and management. The objectives of the model are to provide a predictive capability for changes in ecosystem behavior under defined changes in these conditions. The model algorithms of particular relevance to the hypothesis proposed above concern the determination of surface residue mass, the effect of tillage on surface residue mass, and the effects of surface residue mass on gas and energy exchange between the soil and the atmosphere. These algorithms are described below, with reference to related algorithms published elsewhere.

2. The Effect of Surface Residue on Gas and Energy Exchange

Surface residue affects mass and energy exchange in *ecosys* by causing shortwave and longwave radiation to be partitioned at the ground surface between fractions of exposed soil and surface residue. These fractions are estimated from surface residue dry mass according to Sloneker and Moldenhauer (1977). Radiation absorbed by each fraction is used to drive energy exchanges with the atmosphere as described in Eqs. [1-11] of Grant et al. (1995a). Latent and sensible heat fluxes between the surface residue and the atmosphere are calculated from vapor density and temperature differences (Eqs. [8-9] of Grant et al., 1995a) and from residue boundary layer resistance r_{br} calculated from Lascano and van Bavel (1983) with a nonisothermal stability correction factor from van Bavel and Hillel (1976). Vapor density in the surface residue ρ_r is calculated from saturated vapor density at current residue temperature T_r and from relative humidity at current residue water content θ_r:

$$\rho_r = \rho_r^* (1 - e^{-8.6\theta_r^{0.965}}) \qquad [1]$$

(Tanner and Shen, 1990). Because vapor density in the atmosphere controls that in the residue, it also controls θ_r during evaporation. The definition and units of all variables are given in the Appendix below. Values T_r and θ_r are calculated from the residue energy balance (Eqs. [6-11] of Grant et al., 1995a) in which Eq. [1] is incorporated. These values are also determined by absorbed precipitation calculated from residue dry mass according to Shaffer and Larson (1987).

Latent and sensible heat fluxes between the surface soil and the atmosphere are also calculated from gradients of vapor density and temperature (Eqs. [3-4] of Grant et al., 1995a) and from soil boundary layer resistance r_{bs}. In *ecosys*, the soil surface is represented as a soil layer 0.01 m deep, although shallower depths may also be used. Vapor density at the soil surface ρ_s is calculated from saturated vapor density at current surface temperature T_s and from relative humidity at current surface water potential ψ_s (Eq. [12] of Grant, 1992). Values of T_s and ψ_s are calculated from the soil surface energy balance (Eqs. [1-5] and [11] of Grant et al., 1995a). Boundary layer resistance at the soil surface is calculated from r_{br} plus a diffusive resistance to vapor and heat (assumed the same as that of vapor) transport imposed by the residue layer:

$$r_{bs} = r_{br} + d_r / (D_r P_r^{2.33} c_r) \qquad [2]$$

where surface residue depth d_r is a linear function of residue dry mass derived from data of Tanner and Shen (1990) and Hares and Novak (1992b), diffusivity in air D_r is corrected for T_r according to Campbell (1985), and P_r is calculated from d_r, residue dry mass and residue specific density. The term c_r in Eq. [2] is a correction term for wind-driven dispersion within the surface residue (Hares and Novak, 1992a) calculated from wind speed according to Tanner and Shen (1990). Values for diffusive resistance calculated from Eq. [2] reproduce those measured in wind tunnels by Tanner and Shen (1990) using values for d_r, residue dry mass and wind speeds from their experiments.

Gas and energy exchanges between the atmosphere and the soil-residue surfaces are coupled to subsurface water, gas and energy transfers as described in Eqs. [12-17] of Grant et al. (1995a). Values of r_{bs} are also used to calculate surface fluxes Q_G of ecosystem gases G (including O_2, CO_2, N_2, N_2O, and NH_3) from differences between atmospheric (G_e) and surface (G_s) concentrations. These fluxes are calculated for both the gaseous g and aqueous l phases of the soil surface:

$$Q_{Gg} = (G_e - G_{sg}) / r_{bs} \qquad [3]$$

$$Q_{Gl} = (G_e - G_{sl}) / r_{bs} \qquad [4]$$

The terms G_{sg} and G_{sl} represent the surface concentrations of each gas G at which the flux between the atmosphere and the soil surface equals that between the surface and the midpoint of the surface layer through the gaseous and aqueous phases, respectively. Diffusivity and tortuosity effects on the transfer of G through each phase is accounted for in the calculation of these concentrations. The flux through the aqueous phase usually contributes comparatively little to the total, except when the soil surface approaches saturation. Surface fluxes of each G are coupled to subsurface and interphase fluxes as described in Eqs. [27-35] of Grant (1993b) and in Eqs. [18-22] of Grant et al. (1993c). The surface residue also exchanges heat with the soil surface beneath it through conduction (Eq. [11] of Grant et al., 1995a).

3. Surface Residue and Tillage

In *ecosys*, tillage is simulated as the homogenization of all soil variables to the degree and depth indicated for a specified tillage implement, values of which are taken from Williams et al. (1989). For example, homogenization of soil water during a tillage event is calculated as:

$$\theta_{zt} = z_t \theta_{zt-1} + F \left(\sum_{z=1}^{Z} (z_t \theta_{zt-1}) - z_z \theta_{zt-1} \right) + (1 - z_z) \theta_{zt-1} + Z_{zF} \theta_{zt-1} \qquad [5]$$

such that as F approaches 1, all θ_{zt} as a fraction of soil volume approach a common value through the tillage zone. Homogenization in Eq. [5] allows redistribution of all forms of water, heat, carbon, oxygen, nitrogen, phosphorus, and salts among soil layers within the tillage zone. All surface residue variables during a tillage event are reduced to a fraction 1 - F of their pretillage values, and a fraction F of these values is distributed among subsurface residues in the soil layers within the tillage zone according to Z_z.

4. Surface Residue Dry Mass

The entire set of variables for carbon, nitrogen, and phosphorus used to represent plant residues in the soil (Grant et al., 1993a,b) are also used in *ecosys* to represent those on the surface. Variables for carbon are used to calculate residue dry mass for gas and energy exchange calculations described above. Surface residue variables are increased by deposition of plant material during simulated senescence and harvest of surface vegetation (Grant et al., 1995b,c) and decreased through decomposition by simulated microbial populations (Grant et al., 1993a). These populations are shared with those of the residue already incorporated in the soil surface layer such that microbial concentrations (Eq. [2] of Grant and Rochette, 1994) in surface soil water (θ_s) and in residue water (θ_r) are equal. Because θ_r is normally low (Eq. [1]), microbial biomass associated with the surface residue is usually limited and its decomposition is slow. When θ_r increases due to increased atmospheric humidity or to precipitation, microbial access to surface residue increases, and, hence, residue decomposition increases. Products of microbial decomposition of the surface residue are added to those of the surface soil (Eq. [1] in Grant et al., 1993a) where they contribute to microbial growth (Eqs. [11-15] of Grant et al., 1993a) and are eventually stabilized as microbial residue and soil organic matter (Eqs. [26 - 27] in Grant et al., 1993a). The rate of microbial decomposition of the surface residue is controlled by θ_r and T_r, as is that of soil organic matter by θ_z and T_z in Eqs. [2] and [3], respectively, of Grant and Rochette (1994). Values of z_r and T_r are determined by the effects of surface residue on atmospheric gas and energy exchange as described above.

Subsurface residues in *ecosys* are increased by root exudation (Grant, 1993c) and senescence (Grant, 1993b; Grant and Robertson, 1997), and by surface residues during tillage (Eq. [5]). Residues are decreased through decomposition by simulated microbial populations (Eq. [1] in Grant et al., 1993a) and the subsequent humification of microbial products (Eqs. [26 - 27] in Grant et al., 1993a). Decomposition of each organic fraction (plant and animal residues, active and passive organic matter) is driven by part of the active microbial biomass associated with the fraction itself (Eq. [24] of Grant et al., 1993a) and by part of that associated with each of the other organic fractions, rather than by the total active microbial biomass associated with all fractions as proposed earlier. This change reflects the partial mixing of residues with organic matter under field conditions, rather than the complete mixing under laboratory conditions where the model was originally tested, but still preserves the simulated priming effect of residue addition on organic matter decomposition (Grant et al., 1993b).

B. Field Experiment

Field plots (2.8 m x 6.9 m) were established on a Breton loam (Typic Cryoboralf) at Breton, Alberta (110 km SW of Edmonton) in 1979 to examine the effects of tillage (conventional CT vs. minimum MT), surface residue management (removed or retained) and N fertilizer (0 or 56 kg N ha^{-1} yr^{-1} as urea) on barley (*Hordeum vulgare* L.) growth. The MT treatment included only a hoe-opener seed drill in late May each year, while the CT treatment also included a pass in early May and early October with a small sweep-chisel cultivator followed by a coil packer. At harvest in mid-September, grain and straw yields were determined from 2.3 m² samples following drying at 65°C. All harvested straw was returned to those plots on which surface residue was retained.

In May 1980, organic C concentrations were measured using an ignition technique on composite samples taken at 0.05 m depth increments from the upper 0.15 m of the field site. In October 1990, bulk densities were measured on ten soil cores taken from each plot in 0.05 m increments to a depth of 0.15 m. Composite samples from each increment in each plot were air dried and sieved (2 mm mesh). Total C (LECO Carbon Determinator, model CR12, St. Joseph, MI)[1] and N (Technicon Industrial Systems, 1977 colorimetric analysis) contents were then measured. None of the samples contained significant amounts of free carbonates. Further details of the field experiment are reported in Nyborg et al. (1995).

C. Simulation Experiment

The ecosystem simulation model *ecosys* was run between simulated dates of Jan. 1, 1980 and Dec. 31, 1994 for both CT and MT under current C_a (340 µmol mol^{-1}) and air temperature, and under doubled C_a and current air temperature +3°C or +6°C (the range of increases projected for central Alberta by 2050 in recent climate change studies) using physical and chemical properties of the Breton loam (Table 1), biological properties of barley (assumed to be the same as those of wheat in Grant et al. (1995b,c) except earlier maturing), and daily values of solar radiation, maximum and minimum temperatures, relative humidity, wind travel and precipitation recorded at the experimental site. These values were used by *ecosys* to generate hourly values of radiation and temperature from sine functions, of humidity from the assumption that dewpoint temperatures were equal to minimum daily temperatures, of wind speed assuming constant values during each day, and of precipitation assuming daily events occurred uniformly between 1300 and 1700 h. A concentration of 0.8 µg g^{-1} NO$_3^-$ N was added to the simulated precipitation based on local measurements of rainwater quality. All biological transformation rates in the simulated plant-microbe biome were calculated on an hourly time step. These rates were coupled to mass and energy fluxes through the soil-residue-snowpack profile, solved on a time step of 120 s, by assuming that transformation rates remained constant during each hour.

During each model run, fertilizer N was applied as urea on May 13 of each year at a depth of 0.05 m. During the simulation of CT, depth and mixing factors for a sweep-chisel cultivator and a coil packer (Williams et al., 1989) were used on 1 May and 7 October of each year to incorporate surface residue and redistribute soil (Eq. [5]). During the simulation of both CT and MT, depth and mixing factors for a seed drill (Williams et al., 1989) were used on May 21 of each year, accompanying the planting of barley at a depth of 0.015 m and a density of 200 plants m^{-2}. Harvesting was simulated on September 17 of each year, at which time grain yield was removed from the simulation, all remaining shoot phytomass was transferred to surface residue, and all remaining root phytomass was transferred to residue in the soil layers in which it resided at the time of harvest. The lower boundary of the simulated soil profile was maintained close to the annual mean temperature of the field site, and was assumed to be impermeable to water, based on drainage problems observed in the Breton soil.

III. Results

Simulated tillage and incorporation of surface residues (Eq. [5]) caused changes in the surface energy balance that, in turn, caused changes in soil temperature and respiration rates. An example of these changes is shown for the fall tillage of 1988, almost 9 simulated years after model initialization, using

---

[1] Mention of a commercial product by name does not imply endorsement by the Univ. of Alberta or Agriculture and Agri-Food Canada.

Table 1. Physical and chemical properties of the Breton loam (Typic Cryoboralf)

	Soil depth (m)							
Property (unit)	0.05	0.10	0.15	0.30	0.76	1.12	1.50	1.70
Bulk density (Mg m^{-3})	1.35	1.35	1.35	1.40	1.50	1.50	1.50	1.50
$\theta_{-0.03}$ (m^3 m^{-3})	0.251	0.251	0.251	0.286	0.317	0.296	0.268	0.272
$\theta_{-1.5Mpa}$ (m^3 m^{-3})	0.95	0.95	0.95	0.158	0.208	0.190	0.154	0.159
K_{sat} (mm h^{-1})	11.5	11.5	11.5	6.40	5.08	5.08	7.10	6.74
Sand (g kg^{-1})	260	260	260	340	320	310	350	320
Silt (g kg^{-1})	620	620	620	370	350	360	380	400
pH	5.9	5.9	5.9	5.3	5.2	5.0	5.4	6.7
CEC (cmol$^+$ kg^{-1})	15.0	15.0	15.0	18.0	21.0	21.0	20.0	18.0
Organic C (g kg^{-1})[a]	13.8	13.8	13.8	3.0	5.0	3.0	2.0	1.7
Total N (g Mg^{-1})[a]	1200	1200	1200	300	500	300	205	168
NO$_3$-N (g Mg^{-1})[a]	10	10	10	9	6	4	3	2

[a] Initial values on January 1, 1980.

output recorded during the model runs under current C_a (Figure 1). At this time, simulated surface residue was 467 and 564 g m^{-2} for the CT and MT treatments. During the period from 2 days before to 10 days after tillage, solar radiation averaged 9 MJ m^{-2} d^{-1} (Figure 1a). Maximum daily air temperatures remained above 20°C until a 3 mm rainfall during day 288, after which temperatures declined. The simulated tillage event at the start of day 281 caused upward latent heat fluxes (LE), represented as negative values in Figure 1b, to become larger under CT than under MT during the following 2 d because wetter soil was redistributed to the soil surface by tillage. After the surface soil dried again, LE under CT remained low, following diurnal cycles characteristic of soil-limited evaporation in which maximum rates occur in the morning and declining rates during the afternoon (Idso et al., 1974). During dry periods, water content at the soil surface stabilized at higher values under MT than under CT, so that upward LE was greater. This increase is consistent with the observation by Steiner (1989) that soil-limited evaporation rates increase with surface residue. Fluxes under both tillage treatments increased briefly following rainfall on day 288. The initial effect of simulated tillage on sensible heat (S) was to increase downward (positive) and reduce upward (negative) fluxes (Figure 1c), indicating greater advection of heat into the soil under CT than under MT. After the soil surface had dried, upward S under CT increased over that under MT, indicating that soil warming was causing greater sensible heat loss. Soil heat flux (H) after simulated tillage became more positive under CT vs. MT during the day, indicating greater soil warming, and more negative at night, indicating greater soil cooling (Figure 1d). Diurnal variation of H under MT was less than that under CT due to the insulating effect of greater surface residue, and values were generally lower, indicating less soil warming. Changes in the surface energy balance following simulated tillage and partial residue incorporation caused temperatures under CT to increase over those under MT by more than 5°C in the 0.05 - 0.10 m soil layer (Figure 1e). These increases declined with depth in the soil.

Simulated tillage and residue incorporation caused an increase in CO_2 evolution (Figure 1f) through an increase in specific microbial respiration. This increase in respiration was caused by (a) an increase in soil temperature (Figure 1e) (Eq. [3] of Grant and Rochette, 1994), (b) an increase in the ratio of water to substrate for incorporated vs. surface residue, because surface residue does not hold much water (Eq. [1] above), and hence an increase in microbial access to incorporated vs. surface residue (Eq. [2] of Grant and Rochette, 1994), and (c) a decrease in r_{bs} with decreased surface residue (Eq. [2]), and, hence, an increase in surface (Eqs. [3-4]) and subsurface (Eqs. [18-22] of Grant et al., 1993c); O_2 fluxes and in microbial O_2 uptake (Eqs. [8-9] of Grant and Rochette, 1994) and growth (Eqs. [17-18] of Grant and Rochette, 1994). The increase in specific microbial respiration under CT

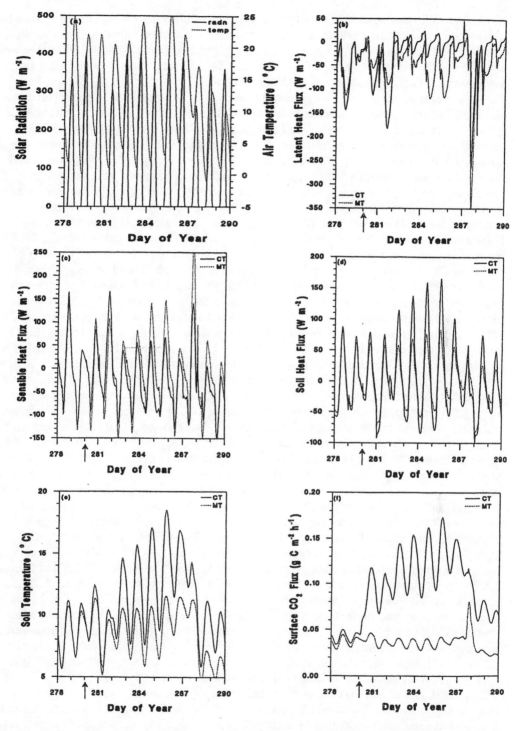

Figure 1. (a) Solar radiation and air temperature, (b) latent, (c) sensible, and (d) soil heat fluxes, (e) soil temperature at 0.05 - 0.10 m depth, and (f) surface CO_2 flux simulated under conventional (CT) vs. minimum (MT) tillage from two days before to ten days after tillage (CT only) in 1988. Arrow indicates date of tillage.

is consistent with that observed by Saffigna et al. (1989) from incubated soil samples taken from conventional- vs. zero-tilled field plots. The increase in CO_2 evolution following simulated tillage is consistent with that observed following field tillage by Reicosky and Lindstrom (1993). A transient increase in CO_2 evolution from the MT treatment on day 288 was caused by an increase in θ, and, hence, surface residue microbial activity following rainfall.

The increase in soil temperature simulated under CT vs. MT in Figure 1e was apparent during much of the tillage experiment. Maximum daily temperatures at 0.05 - 0.10 m increased earlier in the spring, and remained 2 - 5°C higher until autumn under CT vs. MT (Figure 2b). Minimum daily temperatures under the two tillage treatments were similar (Figure 2b). Simulated soil water contents were higher under MT vs. CT (Figure 2c), especially during spring following snowmelt and soil thawing. Low water contents simulated during winter indicate ice formation during soil freezing (Eqs. [26-28] of Grant, 1992). Increases in water contents under MT vs. CT during some winters indicate thawing events. These results are consistent with those reported from several field experiments in which increased residue cover from reduced tillage caused decreased frost penetration during winter (Benoit and Mostaghimi, 1985; Flerchinger and Saxton, 1989; Thunholm and Håkansson, 1988) and increased soil water content during spring (Gauer et al., 1982; Maulé and Chanasyk, 1989).

Under CT, organic C simulated in the upper 0.05 m of the soil oscillated between 1200 and 1300 g m^{-2} from 1981 to 1994 (Figure 3a). Each year values of organic C increased during crop senescence and harvest in late summer, then declined after tillage in the fall, remained stable during winter, and then declined further during the following spring and summer. These declines are the net result of C oxidation (Eqs. [1-5] in Grant et al., 1993a) offset by C addition through root senescence (Eq. [12] in Grant, 1993b) and exudation (Eq. [1] in Grant, 1993c), both of which are driven by C transfers from shoots (Eq. [11] in Grant and Robertson, 1997). Declines simulated during summers are consistent with those of 0.2 - 0.4% in organic C reported by Leinweber et al. (1994) from long-term field experiments. Under MT, organic C in the upper 0.05 m of the soil increased during the experiment to about 1600 g m^{-2} due largely to the absence of tillage in the fall. Seasonal oscillations in organic C under MT were less than those under CT because less C was added to (Table 2a) and respired from (Figure 1f) the cooler soils (Figure 2a) under MT. Organic C in the 0.05 - 0.10 soil zone under CT oscillated between 1075 and 1125 g m^{-2} due to residue incorporation from fall and, to a small extent, spring tillage (Figure 3b). These incorporations offset small declines in organic C during winter, and larger declines during summer. Organic C in this soil zone was not replenished by surface residue under MT, and, hence, declined gradually to about 950 g m^{-2} by the end of 1994. Organic C in the 0.10 - 0.15 m soil zone declined gradually under both tillage treatments, but slightly more so under CT (Figure 3c). Total organic C in the upper 0.15 m of the soil under MT increased over that under CT by an average of 15 g m^{-2} y^{-1} from 1980 through 1994 (Figure 3d). This increase was more rapid during the spring and summer when respiration under MT was less (Figure 1f), but was frequently reduced by increased residue additions from CT vs. MT at harvest in September (Table 2a). Seasonal variation in this increase should therefore be accounted for when designing field sampling protocols.

In *ecosys*, organic matter transformations are driven by active microbial biomass, which is a fraction of total microbial biomass (Eq. [24] of Grant et al., 1993a). Total simulated biomass increased more under MT than under CT in the upper 0.05 m of the soil profile (Figure 4a), but less in the 0.05 - 0.10 soil zone (Figure 4b), reflecting tillage-induced changes in access to C substrate (Figs. 3a and b). Higher values were also found for microbial residues in the upper 0.05 m under MT vs. CT (Eq. [27 of Grant et al, 1993a) which is consistent with the findings of Cambardella and Elliott (1994) that the "enriched labile fraction," postulated to be a product of microbial activity, was increased under MT. Biomass remained stable in the 0.10 - 0.15 m soil zone in which no incorporation of surface residue occurred in either treatment (Figure 4c), again reflecting changes in organic C (Figure 3c). Seasonal changes in biomass reflect growth during soil heating and following tillage events (Figure 1f), and declines during soil cooling. Increases in biomass simulated during the earlier part of the experiment followed by more constant values during the later part (e.g., Figure 4a and b) indicate the gradual

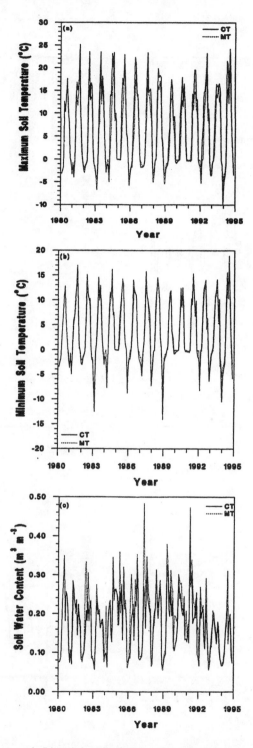

Figure 2. (a) Maximum and (b) minimum daily soil temperatures, and (c) soil water contents simulated at 0.05 - 0.10 m depth under conventional (CT) vs. minimum (MT) tillage from 1980 through 1994.

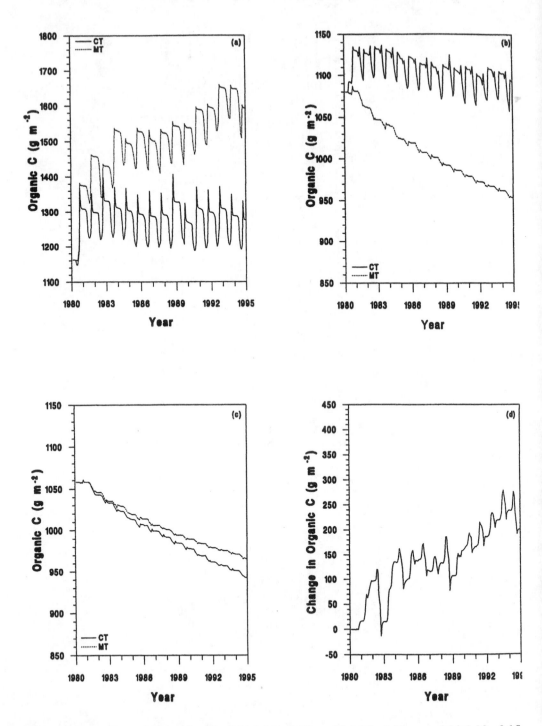

Figure 3. Organic C simulated at depths of (a) 0.00 - 0.05 m, (b) 0.05 - 0.10 m, and (c) 0.10 - 0.15 m under conventional (CT) vs. minimum (MT) tillage, and (d) increase in organic C simulated at 0.00 - 0.15 m under MT vs. CT from 1980 through 1994.

Table 2. Average (± s.d.) annual C inputs to soil from 1984 to 1991 measured and simulated under conventional (CT) or minimum (MT) tillage and (a) current C_a and temperature or double current C_a and current temperature (b) + 3°C or (c) + 6°C

	CT		MT	
Input	Measured	Simulated	Measured	Simulated
		g m^{-2} DM		
		(a) current CO_2 and temperature		
Straw	313 ± 115	349 ± 50	248 ± 87	287 ± 38
Roots[a]		88 ± 21		55 ± 14
Exudates		51 ± 11		34 ± 8
		(b) double current CO_2 and temperature + 3°C		
Straw		402 ± 59		313 ± 63
Roots[a]		91 ± 30		74 ± 20
Exudates		50 ± 18		44 ± 12
Straw		442 ± 80		310 ± 121
Roots[a]		125 ± 19		82 ± 22
Exudates		64 ± 15		48 ± 14

[a] Excluding root senescence.

equilibration of biomass with soil organic C in the model under different environmental conditions. Total biomass simulated in the upper 0.15 m of the soil initially declined under MT vs. CT, but gradually increased to values comparable to those under CT by the end of the experiment (Figure 4d).

Average annual inputs to surface residue were closely reproduced in *ecosys* (Table 2a), although interannual variation was underestimated. The same biological properties used to calculate phytomass at this site were used to calculate phytomasses greater than 2000 g m^{-2} under more favorable conditions elsewhere (Grant et al., 1995b,c), indicating that the simulation of primary productivity in *ecosys* is robust. Both measured and simulated inputs were higher under CT than under MT. In the model, higher inputs were caused by higher rates of N mineralization (Eq. [23] of Grant et al., 1993a) and uptake (Eqs. [3-4] of Grant, 1991), both of which are strongly dependent upon soil temperature. Simulated inputs below the soil surface consisted of roots and exudates, average annual values of which were also higher under CT, reflecting the effects of warmer, drier soil (Figure 1) on C partitioning in the simulated plant. Inputs from root senescence are simulated in *ecosys*, but not included in Table 2. Values of root mass are consistent with those reported for barley at the Breton field site by Izaurralde et al. (1993). Values of root exudate are consistent with root:exudate C ratios of 1.3 - 1.5 reported for barley by Xu (1993). Values of straw:root + exudate ratios are consistent with that used by Campbell et al. (1995) to estimate below-ground C inputs from straw mass.

Between 1980 and 1990, organic C simulated and measured under CT increased in the upper 0.10 m of the soil profile, but declined slightly below 0.10 m (Table 3). The comparability of values measured in 1980 and 1990 is limited by changes in sampling and measurement protocols. The total increase simulated in the upper 0.15 m under CT was less than that measured, mostly because of differences in the 0.05 - 0.10 m soil zone. Organic C simulated and measured in the upper 0.05 m of the soil profile in 1990 under MT increased from initial values more than did that under CT. However, organic C simulated below 0.05 m declined slightly from initial values while C measured below this depth increased strongly. The total increase in organic C from 1980 to 1990 simulated in the upper 0.15 m was 161 g m^{-2} greater under MT than under CT, while that measured was 749 g m^{-2} greater.

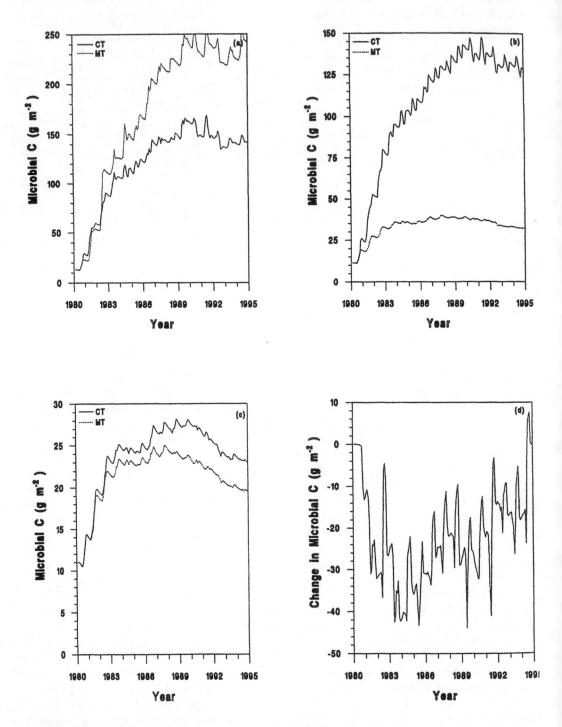

Figure 4. Microbial biomass simulated at depths of (a) 0.00 - 0.05 m, (b) 0.05 - 0.10 m, and (c) 0.10 - 0.15 m under conventional (CT) vs. minimum (MT) tillage, and (d) increase in microbial biomass simulated at 0.00 - 0.15 m under MT vs. CT from 1980 through 1994.

Table 3. Changes in (a) organic C and (b) total N simulated and measured under continuous conventional (CT) and minimum (MT) tillage of barley from the beginning of 1980 to October, 1990

Soil depth (m)	1980 Initial	CT Simulated	CT Measured	MT Simulated	MT Measured
(a) Organic C (g m^{-3})					
0.00-0.05	1162	1314	1156 ± 95	1591	1537 ± 144
0.05-0.10	1080	1111	1348 ± 134	980	1417 ± 249
0.10-0.15	1058	970	949 ± 297	985	1248 ± 233
Total <0.15	3300	3395	3453	3556	4202
>0.15	8120	8183		8160	
(b) Total N (g m^{-3})					
0.00-0.05	94	98	102 ± 7	115	134 ± 11
0.05-0.10	86	93	110 ± 6	79	125 ± 17
0.10-0.15	84	78	90 ± 23	79	115 ± 21
Total <0.15	264	269	302	273	374
>0.15	740	744		742	

In the model, the greater increase under MT consisted of a large increase in the upper 0.05 m of the soil profile (Figure 3a), partially offset by a smaller reduction in the 0.05 - 0.10 m soil zone (Figure 3b). Average annual increases in organic C under MT vs. CT during the entire experiment were about 15 g m^{-2} y^{-1} (Figure 3d). These increases occurred because reduced temperature under MT vs. CT (Figure 2) caused CO_2 evolution (including root respiration) between 1980 and 1990 to be reduced by an average of 76 g C m^{-2} y^{-1} (e.g., Figure 1e) but caused C inputs during this period to be reduced by an average of only 61 g C m^{-2} y^{-1} (Table 2). Organic C simulated and measured below 0.15 m changed little (Table 3).

Changes in total N simulated under CT and MT reflect differences in mineral N inputs between 1980 and 1990 (13, 62 and 7 g m^{-2} from initial soil, fertilizer, and precipitation, respectively, for both treatments) and removals (62 and 57 g m^{-2} in grain, 6 and 7 g m^{-2} in runoff, and 5 and 7 g m^{-2} as N_2O, N_2 and NH_3 from CT and MT, respectively) (Table 3). Greater increases in total N from initial values simulated under MT vs. CT mainly reflect lower grain N removals. These increases were less than those measured, although as for organic C, the comparability between 1980 and 1990 of the measured results is limited.

Increases in air temperature of 3 and 6°C caused increases in soil temperature (Figure 5a,c) and reductions in soil water content (Figure 5b,d) due to increased vapor pressure deficits and, hence, evapotranspiration rates. Under doubled C_a and a 3°C increase in temperature, annual C inputs increased by about 30 g m^{-2} y^{-1} under CT and MT (Table 2b). These increases included those due to increased CO_2 fixation (Grant et al., 1995b) and improved water relations (Grant et al., 1995b,c) under increased C_a, offset by reductions due to shortened plant growth cycles, lower soil water contents (Figure 5b), and increased N constraints under increased temperatures. Increases in C inputs were allocated less to the shoot and more to the root under MT than CT. Under a 3°C increase in temperature, further increases in annual C inputs were simulated under CT but not under MT (Table 2c). Differences in response of C inputs to increased temperatures under CT and MT were caused by the greater interception and reevaporation of precipitation from surface residue under MT which caused soil water to become more limiting to plant growth (Figure 5b and d). Increases in soil C evolution under doubled C_a and a 3 or 6°C increase in temperature were about 35 and 75 g m^{-2} y^{-1} respectively (some of which was increased root respiration, and, hence, not included as C input) caused

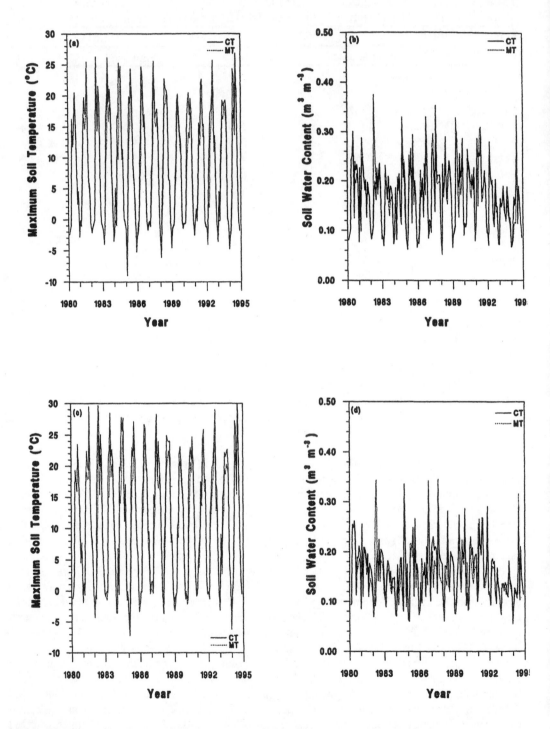

Figure 5. (a,c) Maximum soil temperature and (b,d) soil water content simulated at 0.05 - 0.10 m under conventional (CT) vs. minimum (MT) tillage, doubled atmospheric CO_2 concentration, and current temperature + 3°C (a,b) or + 6°C (c,d).

by higher soil temperatures (Figure 5a and c). These increases in soil C evolution were greater than those in residue C inputs (Table 2) so that organic C in October 1990 under increased C_a and temperature was reduced from that under current C_a and temperature in the upper 0.15 m of the soil profile (Figure 6a,b,c) (Table 4a). These reductions were partially offset by increases below 0.15 m, caused by deeper rooting patterns simulated in *ecosys* under warmer, drier soil conditions. Reductions of total N simulated in the upper 0.15 m under increased vs. current C_a and temperature (Table 4b) reflected increased removal of N in grain. Increases in organic C under MT vs. CT in the upper 0.15 m of the soil profile were greater under increased C_a and temperature than under current (Table 4a and Figure 6d vs. Figure 3d), with average rates of 20 vs. 15 g m^{-2} y^{-1}.

IV. Discussion

In *ecosys*, the effects of tillage on soil organic C are determined by changes in amounts of surface residue, and, hence, in transfers of heat, water, and gas between the atmosphere and the soil profile beneath. These transfers are driven by a physically-based exchange of radiative, latent, and sensible energy between the atmosphere and both soil and residue surfaces. However, the simulated effect of residue in the partitioning of radiative energy between these surfaces remains empirical, and its accuracy needs to be confirmed by further testing. Partitioning will depend not only on residue mass as currently assumed, but also on residue dispersal, and so may be site specific. The simulated effect of residue on latent and sensible energy transfers (Eqs. [1 - 2]; Figure 1b,c) is also physically-based, and has undergone rigorous testing. However, it depends upon an empirical estimate of residue depth to calculate a diffusive path length, although the uncertainty caused by this is likely less than that caused by radiative energy partitioning. There is also uncertainty about the amount of precipitation that is absorbed and reevaporated by residue. The existing technique for calculating this amount, taken from Shaffer and Larson (1987), should be confirmed in further studies.

The simulated effects of increased residue under MT vs. CT on surface mass and energy transfer (Figure 1b, c,d) caused decreases in soil temperature and increases in soil water content (Figure 1e, Figure 2) that are consistent with field observations under a wide range of conditions. These changes caused reductions in soil C respiration under MT vs. CT (e.g., Figure 1f) through reductions in simulated microbial activity and microbe-substrate contact that reproduce the mechanisms proposed for these reductions by Reicosky et al. (1995). However, increases in bulk density and aggregate size under MT vs. CT and associated decreases in gas exchange and microbe-substrate contact are not currently represented in *ecosys* because of uncertainty in their calculation. Although aggregate disruption during tillage may not itself greatly increase soil respiration (Rovira and Greacen, 1957), interactions between disruption and residue incorporation following tillage may contribute to tillage effects on soil respiration (Reicosky et al., 1995). Reductions in soil C respiration simulated under MT vs. CT were larger than those in soil C inputs (Table 2) because inputs are determined comparatively more by atmospheric conditions than is respiration. The consequent increase in soil organic C simulated by *ecosys* under MT vs. CT is consistent with that reported from other studies with similar temperatures and primary productivity (e.g., Campbell et al., 1995) but is less than that measured in this study (Table 3a).

The simulated increase in soil C under MT vs. CT lends support to our hypothesis that reductions in gas and energy exchange between soil surfaces and the atmosphere cause changes in soil temperature and water content that favor increased C storage. However, the cause of the difference between this increase and that measured in the field experiment (Table 3a) remains unclear. Clarification of this difference would require that the source of the increased N (Table 3b) associated with the increase in C measured under MT vs. CT be identified. Differences in organic C and N lost as sediment in runoff (simulated values of which were 44 and 54 mm y^{-1} under MT vs. CT), if present, were not measured in the field experiment and were not accounted for in the model.

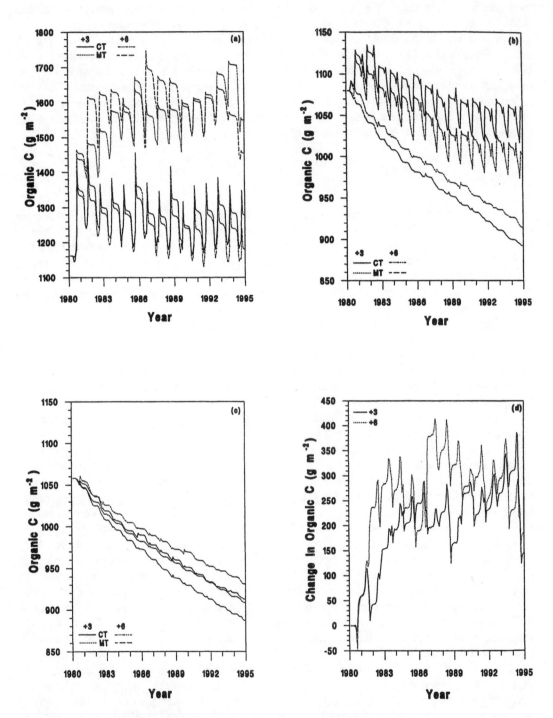

Figure 6. Organic C simulated at depths of (a) 0.00 - 0.05 m, (b) 0.05 - 0.10 m, and (c) 0.10 - 0.15 m under conventional (CT) vs. minimum (MT) tillage, and (d) increase in organic C simulated at 0.00 - 0.15 m under MT vs. CT, doubled atmospheric CO_2 concentration and current temperature + 3°C or + 6°C.

Table 4. Changes in (a) organic C and (b) total N simulated and measured under continuous conventional (CT) and minimum (MT) tillage of barley from the beginning of 1980 to the end of 1990 under current C_a and temperature (current) or double current C_a and current temperature + 3°C or + 6°C

Soil depth (m)	1980 Initial	CT			MT		
		Current	+3°C	+6°C	Current	+3°C	+6°C
(a) Organic C (g m^{-3})							
0.00-0.05	1162	1314	1296	1243	1591	1610	1604
0.05-0.10	1080	1111	1071	1025	980	953	934
0.10-0.15	1058	970	943	921	985	960	941
Total <0.15	3300	3395	3310	3189	3556	3523	3479
>0.15	8120	8183	8228	8337	8160	8191	8288
(b) Total N (g m^{-3})							
0.00-0.05	94	98	94	91	115	113	112
0.05-0.10	86	93	89	85	79	77	75
0.10-0.15	84	78	76	73	79	77	75
Total <0.15	264	269	259	249	273	267	262
>0.15	740	744	747	752	742	746	759

Doubling C_a and increasing air temperatures by 3 or 6°C caused organic C simulated above a depth of 0.15 m to decline but that below to increase such that reductions through the entire profile were small. These reductions suggest that, under the climatic change simulated in this study, the soil at Breton would not become a major source of CO_2 within the time period of this study, especially if fertilizer N additions were increased. However, more complex changes including those in humidity and precipitation should be evaluated in a more rigorous analysis of climate change effects on soil quality. The increase in C sequestration simulated under MT vs. CT was greater under increased C_a and temperature than under current (Figure 5d vs. Figure 3d), indicating that tillage effects upon soil organic C are site and climate specific. Regional estimates of changes in soil-atmosphere C distributions under different land management practices should, therefore, be based upon a comprehensive range of soil and climate conditions.

All parameters used in the simulation of surface residue mass, and of its effects on soil heat, water, and organic C were derived independently of the model, and function at temporal and spatial resolutions higher than those at which the prediction of these effects occurs. No site-specific alteration of any model parameters is permitted during testing of *ecosys*. These parameter attributes are intended to insure that *ecosys* is capable of representing management effects upon soil organic C and N under diverse conditions of soil and climate. Determining the extent of this capability is the subject of ongoing research.

V. Summary and Conclusions

Increased C storage under minimum (MT) vs. conventional (CT) tillage is believed to be caused by increased surface residue accumulation that reduces gas and energy exchange between the soil surface and the atmosphere. These reductions decrease soil temperature and increase soil water content, thereby favoring C storage. In the mathematical model *ecosys*, the effects of surface residue accumulation on soil temperature and water content caused C storage in the upper 0.15 m of the soil profile at Breton to increase by about 15 g m^{-2} y^{-1} under MT vs. CT. This increase is less than that measured at this site, but is consistent with that measured in similar studies elsewhere. This increase varies seasonally, suggesting that field detection methods should be standardized for time of year. Doubling C_a and increasing air temperatures by 3 or 6°C at Breton caused declines in organic C in the upper 0.15 m of soil, but smaller increases below 0.15m. Increases in C storage under MT vs. CT when C_a and temperature were increased was about 20 g m^{-2} y^{-1}, which was 5 g m^{-2} y^{-1} higher than that when C_a and temperature were maintained at current levels. Differences in these increases suggests that they are climate-specific for any soil and management regime, requiring that regional assessments of changes in C storage under different land managements account for changes in climate.

Acknowledgments

This research was carried out as part of the Global Change and Terrestrial Ecosystems project in the International Geosphere-Biosphere Program. It was partially supported by a grant from the National Science Foundation for use of the SGI Power Challenge facilities in the National Center for Supercomputing Applications at the University of Illinois in Urbana-Champaign.

Appendix (square brackets refer to equation in which variable is used)

c_r correction term for wind-driven dispersion of vapor and heat within the surface residue [2]

D_r diffusivity of water vapor in air (m^2 s^{-1}) [2]

d_r depth of residue (m) [2]

F incorporation factor of tillage implement [5]

G_e concentration of gas G in the atmosphere (g m^{-3}) [3,4]

G_{sg} concentration of gas G in the gaseous phase at the soil surface (g m^{-3}) [3]

G_{sl} concentration of gas G in equilibrium with the aqueous phase at the soil surface (g m^{-3}) [4]

P_r porosity of surface residue (m^3 m^{-3}) [2]

Q_{Gg} flux of gas G between the atmosphere and the gaseous phase at the soil surface (g m^{-2} s^{-1}) [3]

Q_{Gl} flux of gas G between the atmosphere and the aqueous phase at the soil surface (g m^{-2} s^{-1}) [4]

r_{br} boundary layer resistance between surface residue and atmosphere (s m^{-1}) [2]

r_{bs} boundary layer resistance between soil surface and atmosphere (s m^{-1}) [2,3,4]

t current time step [5]

$t-1$ previous time step [5]

Z number of soil layers in which homogenization of soil from tillage is occurring [5]

z_z depth of tillage in layer l as fraction of depth of tillage [5]

z counter for soil layer [5]

z_z depth of tillage in layer l as fraction of depth of layer l [5]

β_r residue water content (m^3 layer^{-1}) [1,5]

θ_z soil water content in layer z (m^3 layer^{-1}) [5]

ρ_r vapor density in the surface residue (m^3 m^{-3}) [1]

ρ_r^* saturated vapor density in the surface residue (m^3 m^{-3}) [1]

References

Angers, D.A., A. N'dayegamiye, and D. C. Côté. 1993. Tillage induced differences in organic matter of particle-sized fractions and microbial biomass. *Soil Sci. Soc. Amer. J.* 57:512-516.

Benoit, G.R. and S. Mostaghimi. 1985. Modeling soil frost depth under three tillage systems. *Trans ASAE* 28:1499-1505.

Biederbeck, V.O., H.H. Janzen, C.A. Campbell, and R.P. Zentner. 1994. Labile soil organic matter as influenced by cropping practices in an arid environment. *Soil Biol. Biochem.* 26: 1647-1656.

Cambardella, C.A. and E.T. Elliott. 1994. Carbon and nitrogen dynamics of soil organic matter fractions from cultivated grassland soils. *Soil Sci. Soc. Amer. J.* 58:123-130.

Campbell, G.S. 1985. *Soil Physics with BASIC.* Elsevier, Amsterdam.

Campbell, C.A., B.G. McConkey, R.P. Zentner, F.B. Dyck, F. Selles, and D. Curtin. 1995. Carbon sequestration in a brown Chernozem as affected by tillage and rotation. *Can. J. Soil Sci.* 75:449-458.

Edwards, J.H., C.W. Wood, D.L. Turlow, and M.E. Ruf. 1992. Tillage and crop rotation effects on fertility status of a Hapludult soil. *Soil Sci. Soc. Amer. J.* 56:1577-1582.

Flerchinger, G.N. and K.E. Saxton. 1989. Simultaneous heat and water model of a freezing snow-soil-residue system. II. Field verification. *Trans. ASAE* 32:573-578.

Gauer, E., C.F. Shaykewich, and E.H. Stobbe. 1982. Soil temperature and soil water under zero tillage in Manitoba. *Can. J. Soil Sci.* 62:311-325.

Grant, R.F. 1991. The distribution of water and nitrogen in the soil-crop system: a simulation study with validation from a winter wheat field trial. *Fert. Res.* 27:199-213.

Grant, R.F. 1992. Dynamic simulation of phase changes in snowpacks and soils. *Soil Sci. Soc. Amer. J.* 56:1051-1062.

Grant, R.F. 1993a. *ecosys*. p. 13-23. In: Global Change and Terrestrial Ecosystems Focus 3 Wheat Network. Model and Experimental Meta-Data. International Geosphere-Biosphere Programme.

Grant, R.F. 1993b. Simulation model of soil compaction and root growth. I. Model development. *Plant and Soil.* 150:1-14.

Grant, R.F. 1993c. Rhizodeposition by crop plants and its relationship to microbial activity and nitrogen distribution. *Model. Geo-Biosph. Proc.* 2:193-209.

Grant, R.F. 1994. Simulation of ecological controls on nitrification. *Soil Biol. Biochem.* 26:305-315.

Grant, R.F., N.G. Juma, and W.B. McGill. 1993a. Simulation of carbon and nitrogen transformations in soils. I. Mineralization. *Soil Biol. Biochem.* 27:1317-1329.

Grant, R.F., N.G. Juma, and W.B. McGill. 1993b. Simulation of carbon and nitrogen transformations in soils. II. Microbial biomass and metabolic products. *Soil Biol. Biochem.* 27:1331-1338.

Grant, R.F., M. Nyborg, and J. Laidlaw. 1993c. Evolution of nitrous oxide from soil: I. Model development. *Soil Sci.* 156:259-265.

Grant, R.F., M. Nyborg, and J. Laidlaw. 1993d. Evolution of nitrous oxide from soil: II. Experimental results and model testing. *Soil Sci.* 156:266-277.

Grant, R.F., P. Rochette, and R.L. Desjardins. 1993e. Energy exchange and water use efficiency of crops in the field: Validation of a simulation model. *Agron. J.* 85:916-928.

Grant, R.F. and P. Rochette. 1994. Soil microbial respiration at different temperatures and water potentials: Theory and mathematical modelling. *Soil Sci. Soc. Amer. J.* 58:1681-1690.

Grant, R.F., R.C. Izaurralde, and D.S. Chanasyk. 1995a. Soil temperature under different surface managements: testing a simulation model. *Agric. and For. Meteorol.* 73:89-113.

Grant, R.F., R.L. Garcia, P.J. Pinter Jr., D. Hunsaker, G.W. Wall, B.A. Kimball, and R.L. LaMorte. 1995b. Interaction between atmospheric CO_2 concentration and water deficit on gas exchange and crop growth: Testing of *ecosys* with data from the Free Air CO_2 Enrichment (FACE) experiment. *Global Change Biology* 1:443-454.

Grant, R.F., B.A. Kimball, P.J. Pinter Jr., G.W. Wall, R.L. Garcia, R.L. LaMorte, and D.J. Hunsaker. 1995c. CO_2 effects on crop energy balance: testing *ecosys* with a free air CO_2 enrichment (FACE) experiment. *Agron. J.* 87:446-457.

Grant, R.F. and J.A. Robertson. 1997. Phosphorus uptake by root systems: mathematical modelling in *ecosys*. *Plant and Soil.*

Hares, M.A. and M.D. Novak. 1992a. Simulation of surface energy balance and soil temperature under strip tillage: I. model description. *Soil Sci. Soc. Amer. J.* 56:22-29.

Hares, M.A. and M.D. Novak. 1992b. Simulation of surface energy balance and soil temperature under strip tillage: II. field test. *Soil Sci. Soc. Amer. J.* 56:29-36.

Idso, S.B., R.J. Reginato, R.D. Jackson, B.A. Kimball, and F.S. Nakayama. 1974. The three stages of drying of a field soil. *Soil Sci. Soc. Amer. Proc.* 38:831-837.

Izaurralde, R.C., N.G. Juma, W.B. McGill, D.S. Chanasyk, S. Pawluk and M.J. Dudas. 1993. Performance of conventional and alternative cropping systems in cryoboreal subhumid central Alberta. *J. Agric. Sci.* 120:33-41.

Jenkinson, D.S., D.E. Adams, and A. Wild. 1991. Model estimates of CO2 emissions from soils in response to global warming. *Nature* 351:304-306.

Kirschbaum, M.U.F. 1995. The temperature dependence of soil organic matter decomposition and the effect of global warming on soil organic C storage. *Soil Biol. Biochem.* 27:753-760.

Langdale, G.W., L.T. West, R.R. Bruce, W.T. Miller, and A.W. Thomas. 1992. Restoration of eroded soil with conservation tillage. *Soil Tech.* 5:81-90.

Lascano, R.J. and C.H.M. van Bavel. 1983. Experimental verification of a model to predict soil moisture and temperature profiles. *Soil Sci. Soc. Amer. J.* 47:441-448.

Leinweber, P., H-R. Schulten and M. Körschens. 1994. Seasonal variations in soil organic matter content in a long term agricultural experiment. *Plant Soil* 160:225-235.

Maulé, C.P. and D.S. Chanasyk. 1989. The effects of tillage upon snow cover and spring soil water. *Can. Agric. Eng.* 32:25-31.

Mielke, L.N., J.W. Doran and K.A. Richards. 1986. Physical environment near the surface of plowed and no-tilled soils. *Soil Till. Res.* 7:355-366.

Nyborg, M., E.D. Solberg, S.S. Malhi, and R.C. Izaurralde. 1995. Fertilizer N, crop residue, and tillage alter soil C and N content in a decade. p. 93-99. In: R. Lal, J. Kimble, E. Levine and B.A. Stewart (eds.), *Soil Management and Greenhouse Effect.* CRC Press, Boca Raton, FL.

Parton, W.J. and P.E. Rasmussen. 1994. Long term effects of crop management in wheat-fallow: II. CENTURY model simulations. *Soil Sci. Soc. Amer. J.* 58:530-536.

Parton, W.J., J.M.O. Scurlock, D.S. Ojima, D.S. Schimel, D.O. Hall, and SCOPEGRAM group members. 1995. Impact of climate change on grassland production and soil carbon worldwide. *Global Change Biology.* 1:13-22.

Reicosky, D.C., W.D. Kemper, G.W. Langdale, C.L. Douglas Jr., and P.E. Rasmussen. 1995. Soil organic matter changes resulting from tillage and biomass production. *J. Soil Water Cons.* 50: 253-261.

Reicosky, D.C. and M.J. Lindstrom. 1993. Effect of fall tillage method on short term carbon dioxide flux from soil. *Agron. J.* 85:1237-1243.

Reicosky, D.C. and M.J. Lindstrom. 1995. Impact of fall tillage on short term carbon dioxide flux. p. 177-187. In: R. Lal, J. Kimble, E Levine, and B.A. Stewart (eds.), *Soils and Global Change.* Lewis Publishers, Chelsea, MI.

Rovira, A.D. and E.L. Greacen. 1957. The effect of aggregate disruption on the activity of micro-organisms in the soil. *Aust. J. Agric. Res.* 8:659-673.

Saffigna, P.G., D.S. Powlson, P.C. Brookes, and G.A. Thomas. 1989. Influence of sorghum residues and tillage on soil organic matter and soil microbial biomass in an Australian soil. *Soil Biol. Biochem.* 21:759-765.

Shaffer, M.J. and W.E. Larson. 1987. NTRM, a soil-crop simulation model for nitrogen, tillage and crop-residue management. USDA Conserv. Res. Rep. 34-1. National Technical Information Service, Springfield, VA.

Sloneker, L.L. and W.C. Moldenhauer. 1977. Measuring the amounts of crop residue remaining after tillage. *J. Soil Water Conserv.* 32:231-236.

Steiner, J.L. 1989. Tillage and surface residue effects on evaporation from soils. *Soil Sci. Soc. Amer. J.* 53: 911-916.

Tanner, C.B. and Y. Shen. 1990. Water vapor transport through a flail-chopped corn residue. *Soil Sci. Soc. Amer. J.* 54:945-951.

Thunholm, B. and I. Håkansson. 1988. Influence of tillage on frost depth in a heavy clay soil. *Swed. J. Agric. Res.* 18:61-65.

van Bavel, C.H.M. and D.I. Hillel. 1976. Calculating potential and actual evaporation from a bare soil surface by simulation of concurrent flow of water and heat. *Agric. Meteorol.* 17:453-476.

Williams, J.R., P.T. Dyke, W.W. Fuchs, V.W. Benson, O.W. Rice, and E.D. Taylor. 1989. *EPIC - Erosion Productivity Impact Calculator.* USDA Tech. Bull. 17.

Xu, J.G. 1993. Transformation and stabilization of carbon in two barley soil ecosystems. Ph. D. thesis. Univ. of Alberta. 104 pp.

CHAPTER 37

Using Bulk Soil Radiocarbon Measurements to Estimate Soil Organic Matter Turnover Times

Kevin G. Harrison

I. Introduction

This chapter outlines a strategy for using bulk soil radiocarbon measurements to estimate soil organic matter turnover in native, cultivated, and recovering soil. The turnover of soil organic nitrogen can also be inferred from these measurements. Knowing soil carbon turnover times allows a first step toward understanding how soil carbon, which contains three times the amount of carbon present in the preindustrial atmosphere, will respond to anthropogenic perturbations, including changing land use, anthropogenic nitrogen deposition, changing climate, and CO_2 fertilization. The radiocarbon results suggest that soil carbon (exclusive of litter) exchanges significant amounts of carbon with the atmosphere (~20-25 Gt C/year), thus having the potential to respond to perturbations. The increase in soil carbon storage due to CO_2 fertilization may potentially explain most of the so-called "missing sink."

Many workers have estimated the global inventory of soil carbon using a variety of techniques: Schlesinger (1977) used vegetation types to estimate an inventory of 1456 Gt. C; Post et al. (1982) estimated soil humus to hold 1395 Gt C using climatic life zones; Eswaran et al. (1993) used soil orders to estimate an inventory of 1576 Gt C; and Batjes (1996) found 1462-1548 Gt C using data from 4,353 soil profiles. Although there are differences between these techniques, the results generally agree. How this large pool of carbon influences atmospheric CO_2 is uncertain, because the individual turnover times of the multitude of soil compounds are not well known. These turnover times may range from days to millennia. The purpose of this chapter is to introduce a strategy for estimating the turnover time of soil organic matter using bulk soil radiocarbon measurements.

The technique builds on the approaches of others: The Century and Rothamsted models, mass balance studies, and fractionation of soil humus. The Century and Rothamsted models use measurements of soil carbon decomposition as the foundation for sophisticated ecosystem models. The model structures are similar, having soil organic material consisting of fast, active, and passive fractions. The Century model (Parton et al., 1987, 1989, 1993; Schimel et al., 1994) has an active carbon turnover time ranging from 20 to 50 years, and, it assigns a turnover of 800 to 1200 years to passive carbon. The Rothamsted model (Jenkinson, 1990) uses a 20-year turnover time for active carbon and a near infinite turnover time for passive carbon.

O'Brien and Stout (1978) use a sophisticated model to interpret their New Zealand soil radiocarbon measurements. Their model includes carbon input, decomposition rates, and soil diffusivity, which is constrained by the depth distribution of radiocarbon and total carbon, and it

Table 1. Prebomb soil radiocarbon values for uncultivated soil

% modern	Depth	Year	Reference	Soil type	Location
96	1-8 cm	1959	O'Brien, 1986	Mollisol	New Zealand
96	0-23 cm	1959	Trumbore, 1993	T. forest	California
82	0-12 cm	1927	Trumbore et al., 1990	Spodosol	USSR
96	A-horizon	1962	Campbell et al., 1967	Chernozemic	Canada
90	A-horizon	1962	Campbell et al., 1967	Mollisol	Canada
94	0-2 cm	1959	Vogel, 1970	Forest	Germany

includes carbon input, decomposition, and soil diffusivity. They assign a 50-year turnover time for active carbon and a near infinite time for passive carbon.

Researchers have tried to separate active and passive components using physical and chemical fractionation techniques (Paul et al., 1964; Cambell et al., 1967; Martel and Paul, 1974b; Goh et al., 1976, 1977, 1984; Scharpenseel et al., 1968a,b; Trumbore et al., 1989, 1990). Trumbore (1993) summarizes the results of various fractionation techniques. One way to test the effectiveness of fractionation is to see if the amount of bomb radiocarbon is distributed in the soil as predicted by estimates of the residence time for soil organic matter. To date, none of the available fractionation schemes reproduce the expanded distribution of bomb carbon (Trumbore, 1993).

II. Estimating Soil Carbon Turnover Times Using Bulk Radiocarbon Measurements

My research uses a time-step one-box model and bulk soil radiocarbon measurements to estimate turnover times and inventories of active and passive carbon. The model has atmospheric C-14 values and CO_2 concentrations for every year from 1800 until the present. The user selects the carbon inventory and the turnover time. The turnover time equals the carbon inventory divided by the exchange flux. The exchange flux equals the amount of carbon that is added to the box (from photosynthesis) or lost from the box (respiration). Losses through erosion and dissolution are thought to be small (Schlesinger, 1986) and are not considered. The model can be run in either a steady state mode (where the flux in equals the flux out) or in a nonsteady state mode (in which carbon is either accumulating or decreasing). My research uses this model and soil radiocarbon data to show that soil carbon has more than one component and to estimate the turnover time of the passive fraction, the proportions of active and passive carbon in surface soil, and the active soil carbon residence time.

Many researchers have concluded that soil consists of a complex mix of organic molecules whose turnover times range from a few years to thousands of years. This pool cannot be characterized by a single turnover time (O'Brien and Stout, 1984; Balesdent, 1988; Parton, 1987, 1989, 1993; Jenkinson and Raynor, 1977). For example, six published prebomb values for the average radiocarbon content of surface soil are 92% modern (Table 1). "Modern" means the amount of radiocarbon relative to 1850 wood. This 92% modern value indicates a 650-year turnover time, which would show very little increase in bomb radiocarbon with time. Further, different types of vegetation and soil types within the same climate will often have different values. In reality, the soil radiocarbon values increase in the 1960s and then level off (Figure 1). This increase suggests that soil organic material contains an active component with a turnover time significantly less than 650 years. This active component must be diluted with a passive component having a turnover time of greater than 650 years.

The turnover time of passive soil carbon can be estimated from soil radiocarbon measurements made at depths where little or no active soil carbon is present. Some tropical soils have active soil carbon several meters below the surface (Nepsted et al., 1994 and Fisher et al., 1994); however,

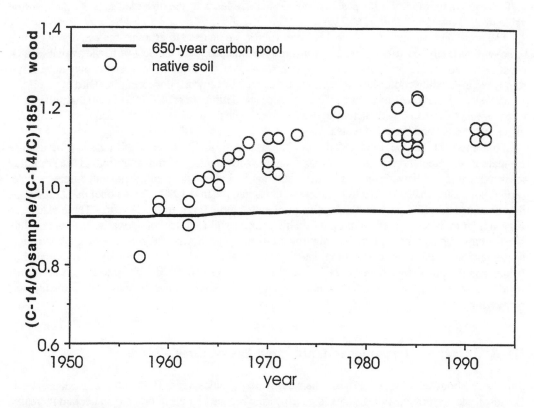

Figure 1. Soil radiocarbon values vs. time. Measured soil radiocarbon values for noncultivated soils are plotted against time from 1950 to 1991. The values tend to increase during the 1960s and then level off. Model results for a theoretical carbon pool having a 650-year residence time are shown by the line. Data for these figures are from Tables 1 and 2, and Harrison (1994).

Table 2. Deep soil radiocarbon values for cultivated and uncultivated soil

% modern	Depth (cm)	Reference	Soil type	Location
49	65-109	Becker-Heidmann et al., 88	Mollisol	China
43	74-94	O'Brien and Stout, 78	Mollisol	New Zealand
60	60-140	Scharpenseel and Becker-Heidmann, 89	Vertisol	Israel
60	60-140	Becker-Hiedmann, 89	Udic	India
62	85-110	Tsutsuki et al., 88	Mollisol	Germany

radiocarbon values decrease with depth, which shows a decrease in the proportion of active to passive carbon (Harrison et al., 1993a). At some depth, the soil radiocarbon measurements approach a minimum, where values tend to decrease very slowly. These values and depths vary for different locations (see Table 2). The average value for the sites listed in Table 2 is 55% modern, which corresponds to a 4700-year turnover time for passive soil carbon.

For the prebomb condition, one can estimate the proportions of active and passive components in surface soil using the 55% modern passive soil radiocarbon value and the 92% modern value measured for the bulk soil (Table 1). If the active component turns over quickly (<100 years) enough so that we can assume that its radiocarbon value is almost 100 percent modern (radiocarbon has a half life of

5,700 years), a mixture of 17% passive and 83% active leads to the observed average radiocarbon value of 92% modern (Figure 2).

The post-bomb increase in soil radiocarbon values can be used to estimate the active soil carbon turnover time (Harrison, 1993a). A 25-year turnover time produces the best fit to the available data. Most of the points are for temperate ecosystems, and warmer tropical ecosystems may have faster turnover times, while cooler boreal turnover times may be slower. For example, Bird et al. (1996) found faster soil carbon turnover times in the tropics and Trumbore et al. (1996) found that soil carbon turnover times decreased with increasing temperature. The proportions of active and passive soil may differ for tropical and boreal climates.

This approach can be validated by looking at a specific site where field data can be used to compare model predictions with soil radiocarbon measurements. O'Brien and Stout (1978) reported measurements for a New Zealand grassland soil that included a deep soil value and a time series of surface soil values that extended from prebomb times into the mid-1970s. The model that best fit, the data consisted of a 12% passive, 88% active portion that turned over every 25 years. These values are very similar to those derived from the available data for soil radiocarbon globally, with the slightly shorter turnover time being the most significant difference. Further, Figure 3 shows the excellent agreement between the model and the data. This model reproduces the prebomb soil radiocarbon values and the post-bomb increase in radiocarbon values in native soil. The model can be further validated by seeing if derivatives could explain radiocarbon measurements in cultivated and recovering soil.

III. Soil Carbon Dynamics in Cultivated Ecosystems

Soil loses about 25% of its carbon when cultivated (Schlesinger, 1986; Post and Mann, 1990; Davidson and Ackermen, 1993). This loss stems from reduced inputs of organic matter and increased rates of organic matter mineralization. Cultivated soil generally has lower radiocarbon values than native soil that has been sampled at the same time (Figure 4; Martel and Paul, 1974). Hsieh (1992 and 1993) has developed a two-component model that reproduces temporal changes in radiocarbon values in cultivated soil. Using a similar approach, Harrison et al. (1993b) assumed that carbon lost from soil due to cultivation would be lost from the active carbon pool. However, the oxidation and loss of a fraction of the active-soil carbon could only explain about half the observed radiocarbon depletion. Mixing subsurface soil with the shallow surface soil through cultivation (i.e., the plow mixes up the soil) can account for the remaining depletion in radiocarbon values. The model, which included mixing and oxidation, produced good agreement with the available data for changes in soil radiocarbon upon cultivation (Figure 4).

IV. Soil Carbon Turnover in Recovering Ecosystems

To further test the model, the baseline model was modified to explore the carbon dynamics of a recovering soil that was increasing its carbon stores. One example includes a recovering temperate forest located in the Calhoun National Forest of South Carolina that is described by Richter et al. (1994, 1995), Harrison (1994), and Harrison et al. (1995). This site contains Loblolly pine (*Pinus taeda*) that was planted in 1959 on land that had been cultivated for 150 years. From 1962 to 1968, the surface carbon concentration increased from 5.9 Mg/ha to 8.0 Mg/ha. The native soil carbon model was modified to take into account this carbon accumulation by increasing the flux of carbon into the active-soil carbon pool. The turnover time that best reproduced the observed radiocarbon measurements was 12 years, which is about twice as fast as carbon turnover in native ecosystems. Figure 5 shows the agreement between the model and the data for this accumulating ecosystem.

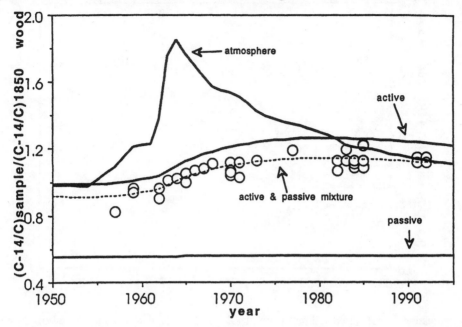

Figure 2. Soil radiocarbon model predictions and soil radiocarbon observations vs. time. A mixture of 83% active and 17% passive soil carbon produces the best visual fit to the native soil radiocarbon measurement (open circles). The concentration of atmospheric radiocarbon almost doubled because of nuclear bomb testing.

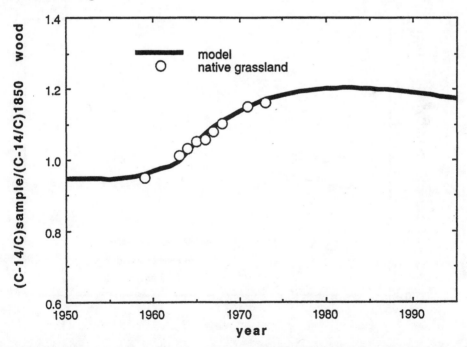

Figure 3. New Zealand test case. O'Brien and Stout (1978) published radiocarbon data for a New Zealand grassland site comprised of a time series of surface soil, including one prebomb and one deep soil radiocarbon value. This information was used to to attempt to validate the model for a specific site.

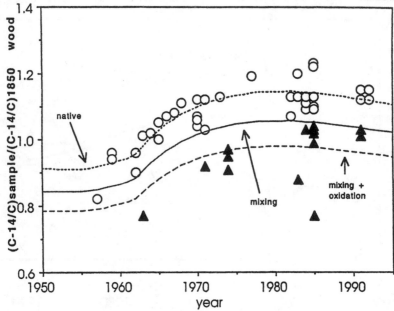

Figure 4. Native vs. cultivated soil radiocarbon values. Cultivated soil has lower radiocarbon values than native soil. This difference is caused by mixing and oxidation. The plow mixes deeper radiocarbon depleted soil with radiocarbon rich surface soil, diluting the amount of active soil carbon in the surface. Increased soil organic matter oxidation further reduced the inventory of active soil carbon.

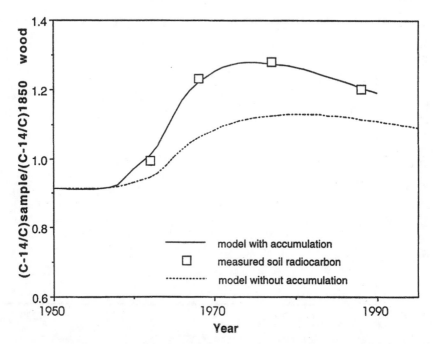

Figure 5. Radiocarbon measurements and model results for South Carolina. The model results that best fit the surface soil radiocarbon measurements were those from an accumulation model that took into account the increase in carbon inventory. The carbon increased from 5.9 Mg ha^{-1} to 8.0 Mg ha^{-1}. The active reservoir turnover time is about twice as fast as the global average for native soils.

Table 3. Soil carbon sequestration due to CO_2 fertilization

	Schlesinger 1977, 1991	Post et al., 1982	Eswaran et al., 1993
Total soil carbon (Gt C)	1500	1400	1600
Nonwetland soil carbon (Gt C)	1000	1200	1250
Active soil carbon (Gt C)	500	600	625
Exchange flux (Gt C yr^{-1})	20	24	25
Potential C sequestration due to CO_2 fertilization (Gt C yr^{-1})	0.5	0.6	0.7
% "missing sink"	45	55	65

This table uses estimates of the inventory of active soil carbon and a simple greening model to estimate the potential carbon sequestration in soil due to CO_2 fertilization. Only nonwetland soil was included in the calculation. The inventory of active soil was esimtated assuming that a soil profile contains about 50% active carbon; a CO_2 fertilization factor of 0.35 was used in the CO_2 fertilization model to calculate the amount of carbon sequestered during an average year in the 1980s (Harrison, 1993a). Dixon et al.'s (1994) missing sink estimate for the 1980s of 1.1 Gt C/year was used.

V. Determining the Global Inventory of Active Soil Organic Matter

Table 3 lists the estimates for the global inventory of soil organic matter in nonwetland ecosystems. These values can be used to estimate the global inventory of active soil carbon. The active-to-passive proportions found in surface soil cannot be applied to all carbon present in nonwetland soil because the proportion of active to passive carbon decreases with increasing depth. The integrated inventories suggest that the global pool is about 50% passive and 50% active (Harrison, 1993a). While extrapolating these distributions involves uncertainty (i.e., the proportions are likely to differ for other climates and types of vegetation), the global inventory of active carbon ranges from 500 to 625 Gt C.

A way to confirm the model is to see if measured fluxes from the soil agree with values predicted by the model. A 500 Gt. C pool turning over every 25 years emits 20 Gt. C from the soil annually. This translates into a 150 g C/m²/yr from the world's uplands surface. The 600 and 625 Gt. C. values result in 24 Gt C/year (180 g C/m²/yr) and 25 Gt C/year (190 g C/m²/yr). All are still significantly lower than the observed flux from a temperate forest soil ranging from 400 to 500 g C/m²/year (Raich and Schlesinger, 1992). Yet, the measured values include sources of carbon dioxide besides microbial respiration of soil organic matter, such as root respiration and oxidation of litter and fine roots (perhaps as much as 50% of NPP). It is impossible to separate these CO_2 sources, but it is unlikely that they are greater than 50% of the total. Also, land having low organic carbon contents, such as desert soil, make it difficult to compare global and regional values. It would be impossible to get better agreement between the fluxes predicted by the model and measured fluxes for a temperate forest ecosystem because of these differences. However, if one considers the measured fluxes to be an upper limit, the predicted values fall well below it.

VI. Carbon Dioxide Fertilization

Having estimated the turnover time and inventory of fast cycling soil carbon, it is possible to estimate the amount of carbon potentially stored in soil because of CO_2 fertilization. CO_2 fertilization occurs when plants increase their growth when exposed to elevated carbon dioxide levels (Strain and Cure, 1985; Bazzaz and Fajer, 1992). A convenient way of expressing CO_2 fertilization is with a CO_2 fertilization factor (i.e., the percentage increase in growth for a doubling of CO_2 concentration).

Although many indoor CO_2 fertilization experiments have shown increased growth at elevated CO_2 levels (Strain and Cure, 1985), extrapolating these results to natural vegetation is highly controversial (Bazzaz and Fajer, 1992). Further, applying these CO_2 fertilization factors to the carbon flux going into soil is speculation. Still, Zak et al. (1993) have shown that the soil carbon under plants grown in doubled CO_2 in open-top chambers was greater than in their nonelevated paired counterparts, although their results were not statistically significant. Also, Norby et al. (1992) found evidence of increased fine root density for trees grown in elevated CO_2 concentrations, lending credibility to the belief that if plant growth is stimulated, so will soil carbon storage.

For this study, a CO_2 fertilization model has been developed to estimate the additional amount of carbon stored in soil because of CO_2 fertilization. The flux of carbon into soil organic matter can be increased by adding the term $\beta * $ delta $pCO_2 *$ EF. β is the CO_2 fertilization factor (0.35 after Harrison, 1993a), delta pCO_2 is the fractional change in carbon dioxide, and EF is the exchange flux. As the flux of carbon into the box increases, the decay flux (i.e., the decay constant times the amount of carbon in the box) also increases. If the level of atmospheric carbon dioxide stops increasing, the soil will attain a higher steady state carbon content with an e-folding time of 25 years.

Table 3 lists the model predictions. The amount of carbon sequestered in soil because of CO_2 fertilization ranges from 0.5 to 0.7 Gt C/year for the 1980s. Dixon et al. (1994) estimate that the "missing sink" is 1.1 Gt C/year for this time period. Thus, carbon dioxide fertilization can be storing much of the "missing sink" in soil.

VII. Conclusion and Future Research

This chapter presents a strategy for estimating the global inventory and turnover time for nonwetland soil. The available radiocarbon data suggests that active soil carbon has a 25-year turnover time and a 500 to 625 Gt. C inventory. Therefore, active soil carbon may respond significantly to perturbations such as CO_2 fertilization, changing climate, and anthropogenic nitrogen deposition. CO_2 fertilization can potentially store 0.5 to 0.7 Gt C/year in soil, thus explaining a major portion of the "missing sink."

Acknowledgments

I thank Bill Schlesinger for helpful comments for improving the manuscript and many discussions that provided the context for this study and Dan Richter for access to Calhoun data and providing Calhoun soil for radiocarbon measurements. I thank Beth Ann Zambella for her support and encouragement. The National Science Foundation supported this research.

References

Balesdent, J., G.H. Wagner, and A. Mariotti. 1988. Soil organic matter turnover in long-term field experiments as revealed by C-13 natural abundance. *Soil Sci. Soc. Am. J.* 52:118-124.

Batjes, N. H. 1996. Total carbon and nitrogen in the soils of the world. *European J. Soil Sci.* 47:151-163.

Bazzaz, F.A. and E.D. Fajer. 1992. Plant life in a CO_2-rich world. *Sci. Amer.* 266:68-74.

Bird, M. I., A. R. Chivas, and J. Head. 1996. A latitudinal gradient in carbon turnover times in forest soils. *Nature* 381:143-146.

Campbell, C.A., E.A. Paul, D.A. Rennie, and K.J. McCallum. 1967. Applicability of the carbon-dating method of analysis to soil humus studies. *Soil Sci.* 104:217-223.

Davidson, E. A. and Ackerman, I.L. 1993. Changes in soil carbon inventories following cultivation of previously untilled soils. *Biogeochemistry* 20:161-193.

Dixon, R.K., S. Brown, R.A. Houghton, A.M. Solomon, M.C. Trexler, and J. Wisniewski. 1994. Carbon pools and flux of global forest ecosystems. *Science* 263:185-190.

Eswaran, H., E.V. Den Berg, and P. Reich. 1993. Organic carbon in soils of the world. *Soil Sci. Soc. Am. J.* 57:192-194.

Fisher, M. J., I.M. Rao, M.A. Ayarza, C.E. Lascano, J.I. Sanz, R.J. Thomas, and R.R. Vera. 1994. Carbon storage by introduced deep-rooted grasses in the South American savannas. *Nature* 371:236-238.

Goh, K.M, T.A. Rafter, J.D. Stout, and T.W. Walker. 1976. Accumulation of soil organic matter and its carbon isotope content in a chronosequence of soils developed on aeolian sand in New Zealand. *J. Soil Sci.* 27:89-100.

Goh, K.M., J.D. Stout, and T.A. Rafter. 1977. Radiocarbon enrichment of soil organic fractions in New Zealand soils. *Soil Sci.* 123:385-390.

Goh, K.M. J.D. Stout, and J.O'Brien. 1984. The significance of fractionation dating the age and turnover of soil organic matter. *New Zealand J. Soil Sci.* 35:69-72.

Harrison, K. G. 1994. The impact of CO_2 fertilization, changing land use, and N-deposition on soil carbon storage. Columbia University Ph.D. Thesis.

Harrison, K. G., W. M. Post, and D. D. Richter. 1995. Soil carbon turnover in a recovering temperate forest. *Global Biogeochemical Cycles* 9:449-454.

Harrison, K. G., W. S. Broecker, and G. Bonani. 1993a A strategy for estimating the impact of CO_2 fertilization on soil carbon storage. *Global Biogeochem. Cycles* 7:69-80.

Harrison, K.G., W.S. Broecker, and G. Bonani. 1993b. The effect of changing land use on soil radiocarbon. *Science* 262:725-726.

Becker-Heidmann, P. 1989. Die Teifenfunktionen der naturlichen Kohlenstoff-Isotopengehalte von vollstandig dunnschichtweise beprobten Parabraunerde und ihre Relation zur Dynamic der organischen Substanz in diesen Boden." Ph.D. Thesis, Hamburg University.

Becker-Heidmann, P., Liu Liang-wu, and H.W. Scharpenseel. 1988. Radiocarbon dating of organic matter fractions of a Chinese mollisol. *Z. Pflanzenernahr. Bodenk.* 151:37-39.

Hsieh, Y-P., 1992. Pool size and mean age of stable soil organic carbon in cropland, *Sol. Sci. Soc. Am. J.*, 56:460-464.

Hsieh, Y.P. 1993. Radiocarbon signatures of turnover rates in active soil organic carbon pools. *Soil Sci. Soc. Am. J.* 57:1020-1022.

Jenkinson, D.S. and J.H. Raynor. 1977. The turnover of organic matter in some of the Rothamsted classical experiments. *Soil Sci.* 123:298-305.

Jenkinson, D.S. 1990. The turnover of organic carbon and nitrogen in soil. *Phil. Trans. R. Soc. Lon.* B, 329, 361-368.

Martel, Y.A. and E.A. Paul. 1974a. Effects of cultivation on the organic matter of grassland soils as determined by fractionation and radiocarbon dating. *Can. J. Soil Sci.* 54:419-426.

Martel, Y.A. and E.A. Paul. 1974b. Use of radiocarbon dating of organic matter in the study of soil genesis. *Soil Sci. Soc. Amer. Proc.* 38:501-506.

Nepsted, D.C., D.R. de Carvalho, E.A. Davidson, P.H. Jipp, P.A. Lefebvre, G.H. Negreiros, E.D. da Silva, T.A. Stone, S.E. Trumbore, and S. Vieira. 1994. The role of deep roots in the hydrological and carbon cycles of Amazon forests and pastures. *Nature* 372:666-669.

Norby, R.J., C.A. Gunderson, S.D. Wullschleger, E.G. O'Neill, and M.K. McCracken. 1992. Productivity and compensatory responses of yellow-poplar trees in elevated CO_2. *Nature* 357:322-324.

O'Brien, B.J. and J.D. Stout. 1978. Movement and turnover of soil organic matter as indicated by carbon isotope measurements. *Soil Biol. Biochem.* 10:309-317.

O'Brien, B.J. 1984. Soil organic carbon fluxes and turnover rates estimated from radiocarbon enrichments. *Soil Biol. Biochem.* 16:115-120.

O'Brien, B.J. 1986. The use of natural and anthropogenic C-14 to investigate the dynamics of soil organic carbon. *Radiocarbon* 28:358-362.

Parton, W.J., C.V. Cole, J.W.B. Stewart, D.S. Ojima, and D.S. Schimel. 1989. Stimulating regional patterns of soil, C,N and P dynamics in the U.S. central grasslands region. p. 99-108. In: M. Clarholm and L. Bergström (eds.), *Ecology of Arable Land.* Kluwer Academic Publishers. Norwell, MA.

Parton, W.J., D.S. Schimel, C.V. Cole, and D.S. Ojima. 1987. Analysis of factors controlling soil organic matter levels in Great Plains Grasslands. *Soil Sci. Soc. Am. J.* 51:1173-1179.

Parton, W.J., M.O. Scurlock, D.S. Ojima, T.G. Gilmanov, R.J. Scholes, D.S. Schimel, T. Kirchner, J-C. Menaut, T. Seastedt, E. Garcia Moya, A. Kamnalrut, and J. I. Kinyamario. 1993. Observations and modeling of biomass and soil organic matter dynamics for the grassland biome wordwide. *Global Biogeochemical Cycles* 7:785-809.

Paul, E.A., C.A. Campbell, D.A. Rennie, and K.J. McCallum. 1964. Investigations of the dynamics of soil humus utilizing carbon dating techniques. In: Transactions 8th Int. Soil Sci. Soc., Bucharest, Romania, 201-208.

Post, W.M. and Mann, L.K. 1990. Changes in soil organic carbon and nitrogen as a result of cultivation. p. 401-407. In: *Soils and the Greenhouse Effect.* John Wiley & Sons, NY.

Post, W.M., W.R. Emanuel, P.J. Zinke, and A.G. Stangenverger. 1982. Soil carbon pools and world life zones. *Nature* 298:156-159.

Raich, J.W. and W.H. Schlesinger. 1992. The global carbon dioxide flux in soil respiration and its relationship to vegetation and climate. *Tellus* B, 44,81-89.

Richter, D.D., D. Markewitz, C.G. Wells, H.L. Allen, J. Dunscombe, K. Harrison, P.R. Heine, A. Stuanes, B. Urrego, and G. Bonani. 1995. Carbon cycling in a Loblolly pine forest: implications for the missing carbon sink and for the concept of soil. p. 233-251. In: W. MacFee (ed.), *Proceedings, Eighth North American Forest Soils Conference*, Univ. of Florida, Gainesville.

Richter, D.D., Markewitz, D., Wells, C.G., Allen, H.L., April, R., Heine, P.R., and B. Urrego. 1994. Soil chemical change during three decades in an old-field loblolly pine ecosystem. *Ecology* 75:1463-1473.

Scharpenseel, H.W., C. Ronzani, and F. Pietig. 1968a. Comparative age determinations on different humic-matter fractions. p. 67-74. In: Proc. Symposium on the use of isoptopes and radiation in soil organic matter studies, Vienna. International Atomic Energy Commission.

Scharpenseel, H.W., M.A. Tamers, and F. Pietig. 1968b. Altersbestimmun bon Boden durch die Radiokohlenstoffdatierungsmethode. *Zeitschr. Pflanzenemernahr Bodenkunde* 119:34-52.

Scharpenseel, H.W. and P. Becker-Heidmann. 1989. Shifts in C-14 patterns of soil profiles due to bomb carbon. *Radiocarbon* 31:627-636.

Schimel, D.S., Braswell, B.H., Holland, E.A., McKeown, R., Ojima, D.S., Painter, T.H., Parton, W.J., and Townsend, A.R. 1994. Climatic, edaphic, and biotic controls over storage and turnover of carbon in soils. *Global Biogeochemical Cycles* 8:279-273.

Schlesinger, W. H. 1991. *Biogeochemistry: An Analysis of Global Change.* New York: Academic Press, 433 pp.

Schlesinger, W. H. 1977. Carbon Balance in Terrestrial Detritus. *Ann. Rev. Ecol. Syst.* 8:51-81.

Schlesinger, W.H. 1986. Changes in soil carbon storage and associated properties with disturbance and recovery, p. 194-220. In: J.R. Trabalka and D.E. Reichle (eds.), *The Changing Carbon Cycle – A Global Analysis*, Springer-Verlag, NY.

Strain, B.R. and J.D. Cure. 1985. Direct effects of increasing carbon dioxide on vegetation. *DOE/ER-0238*.

Trumbore, S. E. 1993. Comparison of carbon dynamics in tropical and temperate soils using radiocarbon measurement. *Global Biogeochemical Cycles* 7:275-290.

Trumbore, S.E., G. Bonani, and W. Wolfi. 1990. The rates of carbon cycling in several soils from AMS C-14 measurements of fractionated soil organic matter. p. 407-414. In: A.F. Bouwman (ed.), *Soils and the Greenhouse Effect*, John Wiley & Sons, NY.

Trumbore, S.E., J.S. Vogel, and J.R. Southon. 1989. AMS C-14 measurments of fractionated soil organic mater. *Radiocarbon*, 31:644-654.

Trumbore, S. E., O. A. Chadwick, and R. Amundson. 1996. Rapid exchange between soil carbon and atmospheric carbon dioxide driven by temperature change. *Science* 272:393-396.

Tsutsuki, K., C. Suzuki, S. Kuwatsuka, P. Becker-Heidmann, and H.W. Scharpenseel. 1988. Investigation of the stabilization of the humus in mollisols. *S. Pflanzenernahr. Bodenk.* 151:87-90.

Vogel, J. G. 1970. Carbon-14 Dating of Groundwater. *Isotope Hydrology* IAEA-SM-129/15:225-237.

Zak, D.R., K.S. Pregitzer, P.S. Curtis, J.A. Teeri, R. Fogel, and D.L. Randlett. 1993. Elevated atmospheric CO_2 feedback between carbon and nitrogen cycles. *Plant and Soil* 151:105-117.

CHAPTER **38**

Impacts of Climatic Change on Carbon Storage Variations in African and Asian Deserts

E. Lioubimtseva

I. Introduction

The carbon reservoir in the continental biosphere is still poorly known and its variability in long-time series is quite conjectural. Knowledge of the preindustrial concentrations of carbon in atmosphere and terrestrial biota is important for several reasons: for identifying the sources and sinks controlling the atmosphere's composition, for the realistic assessment of man's contribution to these increases, and for estimating the associated trends of change in greenhouse effect due to natural and anthropogenic factors.

Continuous desert belt, stretching for more than 15,000 km from the Atlantic coast to northern China, includes the Sahara, deserts of the Arabian peninsula and the Thar, the Iranian highlands, the Karakumi and Kizilkumi deserts of the CIS Central Asia, the Takhla Makhan and the Gobi desert in Mongolia and China. It includes extra-arid, arid, and semiarid zones of the African and Asian continents and forms the most extensive arid area in the world (the so-called Sahara-Gobi desert belt).

Although it is evident that the changes and shifts of vegetation zones respond to global climatic fluctuations, these relationships are not always clear and direct. Among the most complicated key questions of the origin and evolution of the desert landscapes are the role of global climatic change and biomass variations during the Late Pleistocene and Holocene.

II. Subject and Objectives of Study

A. The Present-Day Climate and Carbon Storage in Desert Landscapes

Deserts occupy about 18,000,000 km² in Africa northward to equator and in Asia. The present-day climatic and geoecological conditions of this macroregion are rather diverse. The Sahara, deserts of the Arabian peninsula (Ar Rub al'Khali, Dekhna, and Nefud deserts), and the Thar desert in Radjasthan represent the zone of tropical deserts with high mean annual temperatures and precipitation far below 100 mm (below 10 mm in the Libyan desert) throughout the year with some variations due to the local topography.

In contrast, the Irano-Turanian deserts (Deshte Kebir, Karakumi, Kizilkumi, and Moyankumi), which belong to the temperate climatic zone, receive relatively high winter rainfall (up to 200 to 250 mm) and have a distinctive seasonal curve of annual temperatures which falls below zero in winter. Distinctive hot and cold seasons are also typical for vast closed intermountain depressions of Central

Asia (the Takhla Makan and Dzunghar deserts) and the Gobi desert, while the mean annual precipitation is very low (e.g., below 25 mm in the Lake Lob Nur area, Xingsang province). In its Asian part, the belt does not constitute a homogenous desert zone, being interrupted by numerous mountain systems, whose altitude and aspect mainly determine the pattern of the landscape mosaic. The distribution of the aridity index (P/PET) is the main factor controlling patterns of vegetation zones in the Sahara-Gobi belt both in the past and present. Most of the Sahara, the Arabian peninsula, the Takhla Makan and Gobi deserts have hyper-arid climate (<0.05). Arid (0.05 to 0.20), semi-arid (0.20 to 0.50), and dry subhumid (0.50 to 0.65) landscapes form marginal zones around this chain of hyperarid desert areas.

Although most of the present-day landscapes of the Sahara-Gobi belt generally correspond to the zone of deserts (tropical and temperate), their climatic and topographic diversity effects in a spatial variety of the biomass distribution. The latter gives a key to understanding the spatial distribution of carbon stocks in ecosystems within the area.

Let us remember that biomass is the total mass of vegetation, soils, and litter within a given area. Carbon makes up more than 45% of this mass. According to estimations of different authors (Bazilevich et al., 1986; Olson et al., 1985; Zinke et al., 1985; Adams et al., 1990), carbon density in tropical deserts does not exceed 0.01 kg/m^2, in extreme temperate desert it is about 0.6 kg/m2, while in temperate semideserts it amounts to 0.9-1 kg/m^2 (Table 1).

B. Climatic and Environmental Changes

Temporal synchronism of climatic changes in the Sahara, southwestern Asia, and northern China during the last glacial cycle and its correlation with monsoon activity were recently clearly shown by Petit-Maire and colleagues (Yan and Petit-Maire 1994; Petit-Maire et al., 1995). In this study, I tried to analyze environmental changes throughout this vast belt of desert landscape (including the CIS Central Asia, whose paleogeography is still poorly known to the international scientific community), and to estimate the associated quantitative changes in biomass and carbon storage and their role in global carbon cycle variations.

The time of the initial formation of desert landscapes in different parts of the Sahara-Gobi area is a matter of discussions. Presently there is no scientific consensus on the age of the Sahara, or that of the deserts of Western and Central Asia. It is known that arid conditions dominated in Northern Africa in the Miocene (Goudie, 1991). There is geological evidence of dust transport by northwestern winds from the Sahara over the Atlantic in the Pliocene (Coudé-Gaussen, 1992). In the case of Central Asia, it was the dramatic lifting of the Tibetan plateau and the formation of the Himalayan mountains which had a crucial impact on monsoon circulation and which caused aridization of climate. However, a strong general trend towards more arid conditions started during the Late Pliocene, being caused by a series of global cooling intervals.

During the Pleistocene, this trend became stronger and the global cooling increased in the Late Pleistocene, first during isotopic stage 4 (70 to 50 ka) and then reaching its maximum around 20 to 18 ka BP (isotopic stage 2). Much paleogeographical, paleoclimatic and archaeological evidence found in different parts of the study area show that the Pleistocene global coolings (of which the last glacial cycle is the best known period) caused increase in aridity throughout the area (Varuschenko et al., 1987; Kes et al., 1993; Mamedov, 1990; Petit-Maire, 1989, 1992; Petit-Maire, 1991; Milanova and Lioubimtseva, 1992; Petit-Maire et al., 1995; Sanlaville, 1991; Singh et al., 1990; Yan and Petit-Maire, 1994). By contrast, global climatic optima of isotopic substage 5 e and stage 1 were characterized by much more humid conditions compared to the present.

Available paleohydrological (lake-level, run-off), paleozoological, pollen, and paleopedological data from various parts of the Sahara-Gobi region suggest that climatic conditions of the LI (roughly around 125 to 120 ka according to Th/U dating) were much wetter than now or even than at the HCO.

Table 1. Values of carbon storage in the present-day natural (potential) landscapes of the Sahara-Gobi belt

Type of landscape	Aridity index P/PET	Vegetation C density (kg/m^2)	Authors	Soil C density (kg/m^2)	Authors	Carbon storage totals (kg/m^2)
Extra-arid tropical desert on leptosols and xerosols	<0.05	0.01	Olson et al., 1985; Ajtai et al., 1979	0.2 to 0.5	Bolin et al., 1979; Post et al., 1982	0.21 to 0.51
Acid tropical desert on arenosols, desert leptosols and xerosols	0.05 to 0.2	0.1 to 0.2	Olson et al., 1985; Ajtai et al., 1979	0.5 to 1	Post et al., 1982; Zinke et al., 1985	0.6 to 1.2
Acid and semiarid tropical semidesert on chromic xerosol and eutric leptosols	0.05 to 0.5	0.6 to 1.5	Olson et al., 1985	5.5 to 7	Zinke et al., 1985	6.1 to 8.5
Dry mediterranean woodland or shrub	0.5 to 1	4 to 6	Olson et al., 1985; Bazilevich et al., 1986	8 to 10	Zinke et al., 1985; Post et al., 1982	12 to 16
Dry savanna and tropical steppe on rhodic nitosols and chromic xerosols	0.2 to 0.65	2.5 to 3.5	Olson et al., 1985; Ajtai et al., 1979	10 to 12	Zinke et al., 1985; Post et al., 1982	12.5 to 14.5
Extra arid temperate desert on vermosols	<0.05	0.05 to 0.1	Bazilevich et al., 1986; Olson et al., 1985	1 to 1.5	Bolin et al., 1979; Post et al., 1982	1.05 to 1.6
Arid temperate desert on yermosols and haplic xerosols	0.05 to 0.2	0.1	Bazilevich et al., 1986; Olson et al., 1985	3 to 3.5	Post et al., 1982	3.1 to 3.6
Semiarid temperate semidesert on haplic xerosols and yermosols	0.2 to 0.5	2 to 2.5	Olson et al., 1985; Ajtai et al., 1979	10 to 15	Post et al., 1982; Zinke et al., 1985	12 to 17.5
Dry temperate steppe on luvid kastanozems and solonetz-mollic calcisols	0.5 to 1	2.5 to 3	Bazilevich et al., 1986; Ajtai et al., 1979; Olson et al., 1985	20 to 25	Bolin et al., 1979; Post et al., 1982; Zinke et al., 1985	22.5 to 28

Considerable increase of humidity at those times is supported by evidence of the transgressive lacustrine phase throughout the area (e.g., transgression of the Lake Tchad and the Lake Khazar transgression of the Caspian sea and such numerous smaller lakes as the Chemchen in the western Sahara, Shati in Fezzan, Mundafan in the Rub al'Khali, Didwana in the Thar desert, and the so-called Great Lakes of central Asia). Pollen spectra of this period show a complete absence of desert species and considerable increase of mesophillic vegetation in what are currently desert and semidesert zones.

A severe arid phase occurred everywhere during the LGM (ca. 20-18 ka BP). In the tropics (the Sahelian zone, southern Arabia, and Radjasthan) mean annual precipitation decreased by at least 300 mm compared to the present and the 100 mm isohyete shifted to 13-14° N in Africa (Petit-Maire et al., 1995). Aridization resulted in the spread of the extra-arid desert zone by 300 to 400 km towards the south in northern Africa compared to the present situation (Petit-Maire, 1989). During that period semidesert and dry savanna landscapes spread throughout the present-day zone of humid tropical forests at that period (Talbot and Johanessen, 1992).

Cold and extremely dry conditions dominated at those times in the temperate zone. In Kazakhstan, annual rainfall decreased by not less than 150-200 mm, mean July temperature was 2°C lower than at present, while the mean January temperature dropped by 12°C compared to the present (Aubekerov et al., 1993). According to pollen data in western Turkmenistan, the mean annual temperature decreased by 4.5°C (Kes et al., 1993). A significant decrease of temperature (by at least 12°C) and of precipitation (by 100 to 200 mm) occurred in Mongolia and northern China (Liu et al., 1995; Petit-Maire et al., 1995).

Such climatic conditions caused a considerable spread of cold (and cool) deserts and semideserts in the Asian part of the belt, both northward and southward of the present-day desert zone. On the Loess plateau of China, subtropical subhumid forests gave way to cold dry steppes with considerable aeolian loess accumulation (Liu et al., 1995). Intensive processes of loess accumulation (more than 0.3 mm per year) also occurred in southern Uzbekistan and Tadjikistan (Dodonov, pers. comm., 1995). Cool desert conditions were also accompanied by low levels of the Caspian and Aral seas. In the history of the Caspian sea, the interval between 25 and 16 ka is known as the Yenotaevskaja regression when the sea level dropped to −64 m and, for some short period, even to −135 m (Varuschenko et al., 1987).

It has been shown with evidence from many regions that deglaciation was interrupted by the Younger Dryas event around 11 ka in northern Africa, and the CIS central Asia and China, although this signal was not recorded everywhere.

The Holocene global warming affected everywhere in the Sahara-Gobi desert belt in significant increase of precipitation (and P/PET ratio). Despite some temporal variability of the start and pick of the optimum climatic conditions in the Holocene (the Holocene Climatic Optimum - HCO), a significant increase of humidity generally occurred between 9 to 9.5 ka and 6.5 to 5.5 ka, varying according to regional geographical conditions. The HCO increase of humidity caused by reactivating of monsoons led to almost total disappearance of the desert landscapes both in tropical and temperate zones. In the Sahara, climatic optimum of 8.5 to 6.5 ka resulted in an almost 50-times increase of precipitation (by 200-300 mm) compared to the present (Petit-Maire, 1989). The Saharo-Sahelian boundary shifted at those times up to 23 to 23° N, that meant by 500 km to the north compared to its present-day position and by 1000 km compared to the LGM situation (Petit-Maire, 1989; Petit-Maire et al., 1995). Considerable increase of precipitation occurred on the Arabian peninsula and in Radjasthan resulting in total disappearance of arid landscapes and in spread of savanna vegetation in Ar Rub al'Khali and Thar deserts (Sanlaville, 1992; Singh et al., 1990).

In the deserts of the CIS central Asia, climatic and hydrological optimum of the first half of the Holocene is known as the Lavliakan pluvial and was dated by radiocarbon in different archaeological sites from 8 to 4 ka with its maximum around 6.5 ka (Mamedov, 1990). In western Turkmenistan and Uzbekistan, this interval was characterized by the mean summer temperatures 2-4°C lower than today and by the annual precipitation not less than 175-300 mm (Varuschenko et al., 1987). Such climatic conditions favored development of *Artemisia* and *Gramineae* steppes on the currently desert Caspian-Aral watershed. In the Holocene history of the Caspian Sea, three transgressive intervals mainly correspond to the optima climatic conditions in its basin: 8 to 6 ka, 6.4 to 5.4 ka, and 4.3 to 4.1 ka with maxima sea levels −16.5 to −22 m, −18.5 to −28 m, and −18 to −22 m, respectively (Varuschenko et al., 1987).

According to available pollen data in northern and eastern Kazakhstan, dry cold steppes of the Younger Dryas were replaced in the Holocene by mesophilic forest-steppe vegetation with the maximum increase of arborescent species (up to 40%) and presence of broad-leaved species around 6 ka (Aubekerov, 1989; Tarasov, 1992).

In Mongolia and on the northwest of China, the HCO occurred between 8.5 ka and 6.5 ka and caused the increase of precipitation by more than 100 mm and the mean annual temperature −1.5° higher than at present (Shi et al., 1993; Yan and Petit-Maire, 1994). On the Loess plateau of China, desert landscapes with intensive aeolian loess accumulation gave way to temperate steppes on Luvic Phaenozems (Liu et al., 1995).

After 2.5 to 3 ka, climatic deterioration toward the present-day conditions occurred everywhere from the Atlantic coast of the Sahara to Xingsang.

III. Methods of Macroregional Paleoreconstructions and Assessment of Carbon Storage

A. Paleolandscape Reconstructions

Reconstruction of paleoscenarios helps to understand the relationships between climatic and biogeochemical factors of environmental change at global and regional scales.

The methodology proposed in this chapter for assessments of biomass and carbon storage variations in paleolandscapes includes several successive stages of research. Four time-slices with distinctively different climatic conditions were selected to reconstruct paleolandscapes of the Sahara-Gobi belt: the LI (ca. 125 to 120 ka), the LGM (20 to 18 ka), the HCO (9 to 8 ka), and the present (with an assumption of absent human impact) (Figures 1, 2, 3, Tables 2, 3, 4).

Palaeoreconstructions were based on the analysis of paleohydrological, pollen, faunal, pedological, and geomorphological data, and extrapolation to the past of the existing known relationships between various parameters of landscape components, such as zonal vegetation and soils, pollen spectra composition, granulometric composition of deposits, precipitation, runoff, lake levels, dune movement velocity, etc. Proceeding from the principal of actualism broadly used in paleogeography and extrapolating the actual natural relationships into the past, one has to assume that paleolandscapes analogous to the present potential landscapes (without human impact) existed in the past under climatic conditions analogous to the present ones. That is why paleolandscape maps are not free of unavoidable mistakes and uncertainties due to possibility of existence in the past of landscapes with interrelationships and parameters different from those existing at the present. Therefore, our reconstructions do not claim to provide on exact and detailed mapping of paleolandscapes, representing only an approximate pattern of the landscape structure during the selected intervals.

The reconstruction of the paleolandscapes of the selected intervals was based upon the analyses of paleogeographical data from numerous sites described in literature and generalizations of regional and global reconstructions of many authors (Varuschenko et al., 1987; Velichko, 1993; Street-Perrott, et al., 1989; Petit-Maire, 1991; Petit-Maire et al., 1995; Lioubimtseva, 1995).

Compared to the interval 20 to 18 ka, the Holocene time yields much more paleobotanical and pedological data, which appear to be the most exact and reliable for reconstructing paleoecosystems (Figures 1, 2). Therefore, reconstructions of the Holocene landscapes are based mainly on the

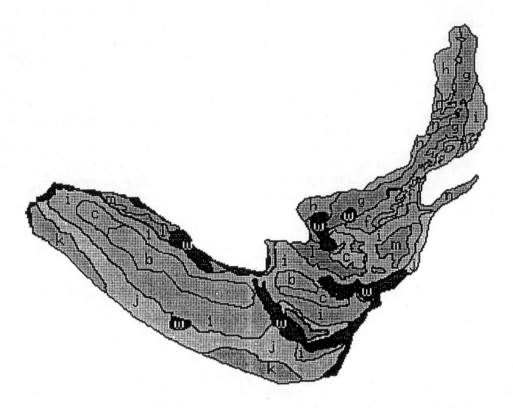

Figure 1. Landscapes of the Sahara-Gobi desert belt 9-8 ka (a: extra-arid tropical desert; b: arid tropical desert; c: tropical semidesert; d: arid temperate desert; e: extra-arid temperate desert; f: temperate semidesert; g: temperate steppe; h: steppe-forest; i: dry savanna and subtropical steppe; j: typical savanna; k: tropical woodland; l: mediterranean shrubland; m: tropical and subtropical forest; n: temperate or mountain forest; w: surface water).

analyses of correlations between climatic parameters and pollen spectra composition as well as of some paleosols information..

Paleolandscape reconstruction of the last interglacial (ca. 125 to 120 ka) was based upon the same methodological approach (Figure 3). This time-slice, however, yields much less credible paleogeographical data while a number of test sites and the quality of dating are still rather poor compared to the later time periods. Although paleoreconstruction of this period is surely the most hypothetical and uncertain, it seems feasible to assess at least rough patterns of vegetation zones, that would allow assessment of dynamics of carbon cycle and analyses of its relations with global climatic variations during the complete glacial cycle.

The estimations of carbon storage during the LI are also more difficult compared to the later intervals. They require the most thorough analyses of all available information on paleolandscapes and their functioning, and should not be based only on a direct extrapolation of the present-day analogue relationships into the past.

The present-day landscapes of the Sahara-Gobi area differ remarkably from those which were reconstructed for the previous intervals. The present status of the landscape boundaries seems to be intermediate between their patterns in the LGM and during the optima. The present mean global temperature is 5°C higher than during the LGM, 1.5°C lower than during the HCO, and 3 to 4°C

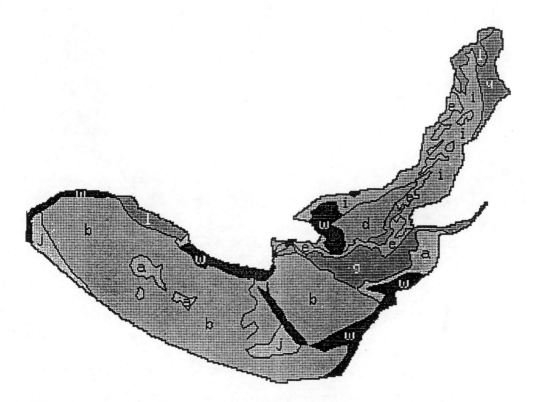

Figure 2. Landscapes of the Sahara-Gobi desert belt 18 ka (a: extra-arid tropical desert; b: arid tropical desert; c: tropical semidesert; d: arid temperate desert; e: extra-arid temperate desert; f: temperate semidesert; g: temperate steppe; h: steppe-forest; i: dry savanna and subtropical steppe; j: typical savanna; k: tropical woodland; l: mediterranean shrubland; m: tropical and subtropical forest; n: temperate or mountain forest; w: surface water).

than during the LI. Therefore, the present pattern of landscape zonality in the Sahara-Gobi area seems to be a good macroregional indicator of the global change trends.

Changes in vegetation and landscape cover caused by climatic variations were interpreted and studied in terms of qualitative changes in carbon storage. Evaluation of the latter is necessary for the better understanding of the global carbon cycle and its role in climatic and biogeochemical planetary changes. Estimations of carbon storage in paleoanalogues landscapes can also be extremely useful for validation of data simulated by ecological models. Combination of both paleogeographical mapping and mathematical modeling approaches could provide more reliable scenarios of carbon storage in landscapes under different climatic conditions.

B. Carbon Storage Calculations and Computer Data Base

There are several possible approaches to assess biomass in palaeoecosystems, including collection of direct and indirect bibliographical data, mathematical evaluation of biomass as a function of precipitation and temperature values calculated from paleogeographical evidences and pollen spectra, and modeling combined with extrapolation of the present-day remote sensing data into the past

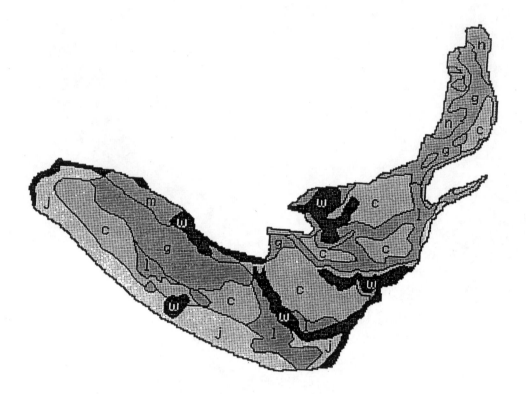

Figure 3. Landscapes of the Sahara-Gobi desert belt 125 ka (a: extra-arid tropical desert; b: arid tropical desert; c: tropical semidesert; d: arid temperate desert; e: extra-arid temperate desert; f: temperate semidesert; g: temperate steppe; h: steppe-forest; i: dry savanna and subtropical steppe; j: typical savanna; k: tropical woodland; l: mediterranean shrubland; m: tropical and subtropical forest; n: temperate or mountain forest; w: surface water).

(Branchu et al., 1992). Combination of these approaches could give the most exact results and should allow us to diminish the disadvantages of each method.

The most direct indications that we have about carbon storage of paleolandscapes can be gathered from the present-day world (Faure, 1989; Adams et al., 1990; Branchu et al., 1992). Although there are certain considerable differences in the landscape status and functioning between the present and recent geological past, at the moment there is no other way to assess biomass of paleolandscapes than to extrapolate our knowledge on their present-day analogues. This assumption allows us to estimate carbon storage throughout a given area during a given period of time by reconstructing and measuring the surface of each landscape zone that existed at that period, and multiplying it by the value of carbon density in its present-day analogues. An approach to biomass/carbon density evaluation that is suggested here is based on the surface measurements of reconstructed paleolandscapes, and could be applied elsewhere in the regions with sufficient paleoecological data sets.

Creation of a computer data base of carbon storage in paleolandscapes of the Northern Hemisphere and more detailed data base on the landscapes of the Sahara-Gobi belt (Lioubimtseva and Simon, 1994, 1995, unpubl.) allowed us to select the most credible scenarios of carbon stocks in palaeoecosystems during the LI, LGM, and HCO, and to calculate variations during the last 125 thousand years. The data base consists of the following main sections: digitized original reference

Table 2. The LI palaeoeolandscapes of the Sahara-Gobi belt

Type of landscape	Aridity index P/PET	Vegetation C (kg/m²)	Soil[a] C (kg/m²)	Area (km²x 10⁶)	Totals of C storage in landscapes (Gt)
Semiarid tropical semidesert	0.5 to 0.65	0.6 to 1	5.5 to 7	6.6	40.26 to 52.8
Subhumid tree savanna and open woodlands	0.5 to 1	5.5 to 6.5	7 to 9	2.82	35.25 to 43.71
Semiarid mediterranean woodland or shrub	0.5 to 1	4 to 7	6 to 8	2.42	24.2 to 36.3
Subhumid tropical forest	0.6 to 1.15	12 to 15	8 to 10	1.1	22 to 25
Temperate steppe	0.5 to 0.65	0.6 to 1	20 to 25	3.9	80.34 to 101.4
Tropical and subtropical forest	1 to 1.15	18 to 20	12 to 13	1.09	32.7 to 35.97
Temperate and mountain mixed forest	1 to 1.5	17 to 22	13 to 14	0.77 to 9	23.7 to 28.44
Totals (Gt)				18.03	Min 256.45 Max 359.59

[a]Values of carbon storage in soils are given as hypothetical ones according to the present-day landscape analogues.

Table 3. The LGM palaeoelandscapes of the Sahara-Gobi belt

Type of landscape	Aridity index P/PET	Vegetation C (kg/m²)	Soil[a] C (kg/m²)	Area (km²× 10⁶)	Totals of C storage in landscapes (Gt)
Extra-arid tropical desert on desert leptosols or calcic xerosols	<0.05	0.01	0.2 to 0.5	0.74	0.13 to 0.38
Arid tropical desert on arenosols, desert leptosols, or xerosols	0.05 to 0.2	0.1 to 0.2	0.5 to 1	9.97	6.0 to 11.96
Arid and semiarid tropical semi-desert on chromic xerosols and eutric leptosols	0.05 to 0.5	0.6 to 1	5.5 to 7	1.07	6.52 to 7.49
Dry savanna and tropical steppe on rhodic nitosols and chromic xerosols	0.2 to 0.65	2.5 to 3.5	10 to 12	1.14	14.25 to 17.67
Dry mediterranean woodland or shrub	0.5 to 1	4 to 6	8 to 10	1.05	12.6 to 16.8
Arid temperate desert on yermosols and haplic xerosols	0.05 to 0.2	0.1	3 to 3.5	1.83	5.67 to 6.59
Temperate dry steppe on luvic kastanozems and solonetz-mollic calcisols	0.6 to 0.7	2.5 to 3	20 to 25	0.93	20.93 to 26.04
Mountain subhumid tundra-steppe and steppe	0.9 to 1.1	0.34 to 0.95	15 to 20	0.29 to 4	4.51 to 6.15
Totals (Gt)				17.5	Min 70.61 Max 93.08

[a]Values of carbon storage in soils are given as hypothetical ones according to the present-day landscape analogues.

Table 4. The HCO palaeoeolandscapes of the Sahara-Gobi belt

Type of landscape	Aridity index P/PET	Vegetation C (kg/m²)	Soil[a] C (kg/m²)	Area (km²× 10⁶)	Totals of C storage in landscapes (Gt)
Arid tropical desert on arenosols, desert leptosols, and xerosols	<0.2	0.1 to 0.2	0.5 to 1	2.1	1.26 to 2.52
Arid and semiarid tropical semi-desert on chromic xerosols and eutric leptosols	0.05 to 0.5	0.6 to 1	5.5 to 7	1.1	6.16 to 8.8
Dry savanna and tropical steppe on rhodic nitosols and chromic xerosols	0.2 to 0.65	2.5 to 3.5	10 to 12	6.9	86.25 to 106.95
Subhumid typical savanna and open woodlands on rhodic and calcic-chromic nitosols	0.5 to 1	5.5 to 6.5	7 to 9	3.05	38.13 to 47.28
Dry mediterranean woodland or shrub	0.5 to 1	4 to 6	8 to 10	0.76	9.12 to 12.16
Acid temperate desert on yermosols and haplic xerosols	0.05 to 0.2	0.1	3 to 3.5	0.04	0.12 to 0.14
Semiarid temperate semidesert on haplic xerosols and yermosols	0.2 to 0.5	2 to 2.5	10 to 15	0.93	11.6 to 16.2
Temperate steppe on luvic kastanozems and solonetz-mollic calcisols	0.6 to 0.7	2.5 to 3	20 to 25	1.85	37 to 51.8
Subhumid temperate forest-steppe on orthic greyzems, haplic and luvic phaenozems, chernozems, and kastanozems	0.65 to 1	5 to 8	18 to 22	1.25	28.75 to 37.5
Totals of carbon storage (Gt)					
Minimum				17.8	218.39
Maximum				9	283.35

[a] Values of carbon storage in soils are given as hypothetical ones according to the present-day landscape analogues.

maps of the LI, LGM, HCO and present-day landscapes, legends, and tables with carbon density values (vegetation and soils) in landscapes present at all chosen time-slices; operational system allowing calculations of land surfaces and carbon storage.

In order to facilitate superimposing of maps, which correspond to different time periods, and compatibility with computerized maps of other macroregions, and, in view of future development of the system, the map grid was acquired and stored as a separate file. Digitized maps were transferred as 1024x1024 files (ASCII codes,1 pixel corresponds to 1 byte) to IBM calculator 3090-600/VF. Operational system comprises high resolution IBM 5080 post graphic, IAX program, calculator 37 xx operating with VM/HPO exploitation system. Image Assistant program was used for data visualization and treatment.

IV. Results and Discussion

Reconstruction of paleolandscapes of the Sahara-Gobi desert belt for the LI, LGM, and HCO periods allowed us to approximately estimate variations of landscapes and carbon storage, which were induced by climatic variations during the climatic macrocycle. Comparison of a set of palaeogeographic maps of this extent region for the chosen time slices (125 to 120 ka, 20 to 18 ka, 9 to 8 ka, and the present) let us discuss the contribution of palaeocontinental proxy-data in palaeobiomass calculations and their accuracy.

Warm and humid conditions of the isotopic stage 5e effected an almost complete disappearance of arid zones and development of rich in biomass ecosystems. Our preliminary estimations let us assume that total carbon storage in the Sahara-Gobi belt around 120 to 125 ka could be between 256.5 and 359.6 Gt.

During the LGM, arid and extra-arid zones were widely expanded all over the Sahara-Gobi area and its margins, while semiarid and subhumid areas were considerably reduced (both in tropical and temperate belts). According to our estimations, the reduction of savanna zone along the southern Sahara margin amounted to about 3.3×10^6 km^2, while the areas of semideserts and deserts spread for 2.2×10^6 km^2 and 5.3×10^6 km^2, respectively, both northward and southward compared to their present day position.

On the Asian continent, arid landscapes also shifted throughout the present-day steppe and steppe-forest zones. Here, considerable changes in the landscape pattern were related not only to the latitudinal expansion of deserts and semideserts on the plains but also with the shift of arid altitudinal zones up over the mountain slopes. According to our assessments in the CIS central Asia, for instance, temperate desert zone expanded during the LGM by ca. 1.8×10^6 km^2, mainly due to the northward shift of its northern margin into the periglacial area.

In contrast, climatic changes during the HCO led to considerable increase in humidity. In the Sahara, the surface of desert zone at those times did not exceed 1.7×10^6 km^2 and its reduction was about 4.2×10^6 km^2 compared to the present and 6.4×10^6 km^2 compared to the LGM. On the Arabian peninsula and in Radjasthan, tropical deserts disappeared giving way to savanna. In the CIS central Asia, in China and Mongolia, warm temperate steppes and forest-steppes occurred in what is currently desert and semidesert zones, while the total surface of temperate desert zone did not exceed 7×10^4 km^2.

The calculations show that the surface of area, whose landscapes were affected by considerable zonal shifts due to changes in aridity from the LGM to HCO, could amount in the Sahara-Gobi area to ca. $16-16.5 \times 10^6$ km^2. This result is in a good agreement with the estimations of Petit-Maire et al. (1995), whose assessments of zonal shifts over this area (excluding the CIS central Asia) yields ca. 14×10^6 km^2.

At the maximum of the glaciation, around 18 ka total carbon storage in the landscapes of the Sahara-Gobi belt could be between 70.61 Gt and 93.08 Gt. The total carbon storage in the landscapes of the Sahara-Gobi belt during the HCO was between 218.4 Gt and 283.35 Gt. The present-day total carbon content in landscapes of the Sahara-Gobi belt is about 120 to 160 Gt in the phytomass and soils.

If we regard paleoenvironmental changes in terms of carbon storage variations, they will mean that the increase in biota carbon from the LGM to the HCO amounted to about 200 Gt (almost 3 times). Its decrease from the LI to LGM was even higher: about 240 to 270 Gt. The carbon reservoir of the Sahara-Gobi area has considerably increased from the LGM to the HCO. After the last humid phase, a rapid decrease seems to cumulate natural aridification and anthropogenic desertification. Careful studies are still needed to assess the biomass decrease from the Holocene optimum to present with more accuracy, but it is assumed here that it could decrease by about 1/2 or more.

The calculations presented here are not free of inevitable errors, the most important of which is due to insufficiently precise mapping of ecosystems limits, because the hypotheses on palaeoecosystems distribution vary greatly from one author to another. There are also less significant uncertainties associated with insufficient accuracy of surface measurements and projection conversion, although the last ones can be neglected at the scale of the present study.

The results compiled about the order of magnitude of biomass changes in arid zones are important to understand the behavior of the carbon cycle since the last interglacial. The general increase in the extent of deserts at the expense of arborescent and tall-grass vegetation during the LGM in tropical areas, may have been driven not only by cooling or by dryer conditions, but also partly by the lowering of the atmospheric CO_2.

V. Conclusion

Our calculations show that climatic change during the last glacial-interglacial cycle had a considerable effect on carbon cycle in arid ecosystems of both temperate and tropical zones. The change of biomes distribution pattern and latitudinal shifts of vegetation zones resulted in significant variations of biomass and carbon storage order of 3 times. The computation of carbon content in the main landscape zones from a reconstruction of successive "snap shots" of the study area has shown that deserts turn to be a sink or source of carbon during rapid climatic and vegetation change. The maxima carbon storage in this macroregion was reached under the relatively moist climatic conditions around 120 to 125 ka and 9 to 6 ka. Since the Mid-Holocene, increasing aridity has caused the release of about 110 to 140 Gt of vegetation and soil carbon from the Sahara and of about 120 Gt from Asian deserts under 5 to 6 ka. The increase of carbon flux to the atmosphere could occur both because of aridization during the second half of the Holocene and acceleration of anthropogenic disturbance.

Currently, arid and semiarid zones of Africa and Asia were a sink for 256.5 to 359.6 Gt of carbon during the last interglacial and about 218.4 Gt to 283.35 Gt in the HCO. But after 4-3 ka, deserts become a source of carbon to the atmosphere. A more accurate estimate of the Sahara-Gobi belt and desert zones role as a whole in variations of global carbon cycle can be achieved only by more detailed paleolandscape mapping for several key dates, based on well distributed and dated new records of all measurable proxy-data. Recent anthropogenic impact on arid and semiarid ecosystem could make an important contribution to the increase of the natural climatic induced carbon flux. As a general rule, in studies of global changes, particular attention should be paid to continental changes through proxy-data but also isotopic data of carbonates and organic matter, soils, peats, etc. in order to better understand the correlation with the results of other global and regional studies and with paleo-oceanic data in particular.

Acknowledgments

This work is a part of the INTAS Project N° 93-62037 "Dynamics of terrestrial biota and its role in the global carbon cycle" funded by European Community. I am indebted to Dr. B.Simon (CEREGE) and Prof. Hugues Faure (LGQ CNRS) for collaboration.

References

Adams, J.M., H. Faure, L. Denard, J.M. Mc Glade, and F.J. Woodward. 1990. Increases in terrestrial carbon storage from the Last Glacial Maximum to the Present. *Nature* 348:711-714.

Ajtay, G.L., P. Ketner, and P. Duvigneaud. 1979. Terrestrial Primary Production and Phytomass. p. 129-182. In: B. Bolin, E.T. Degens, S. Kempe, and P. Ketner (eds.), *The Global Carbon Cycle*, Surrey: Gresham Press, SCOPE 13.

Aubekerov, B.Zh., E.V. Chalihjan, and Sh.A. Zakupova. 1989. Izmenenija klimata i paleogeograficheshih uslovij Centralnogo Kazakhstana v pozdnelednikovie i golocene (in Russ.) (Variations of climate and palaeogeographical conditions in Central Kazakhstan during the Late Glaciation and Holocene). p. 98-102. In: *Paleoclimati Pozdnelednikovia i Golocena*, Moscow, Nauka.

Aubekerov, B.Zh. 1993. Stratigrafija i paleogeografija ravninnih oblastej Kazakhstana v pozdnem pleisticene i golocene (in Russ.) (Central and Northern Kazakhstan during the Late Pleistocene and Holocene). p.88-90. In: A.A. Velichko (ed.), *Evolution of Landscapes and Climates of the Northern Eurasia*, Moscow, Nauka.

Bazilevich N.I. 1986 Biologicheskaja produktivnost pochvenno-rastitelnih formatsij SSSR (in Russ.) (Biological productivity of soil-vegetation formations of the USSR). *Izvestia Acad. Science USSR, ser. geogr.* 2:49-62

Bolin B., E.T.Degens, P. Duvigneaud, and S. Kempe. 1979. The global biogeochemical carbon cycle. p. 129-182. In: B. Bolin, E.T. Degens, S. Kempe, and P. Ketner (eds.), *The Global Carbon Cycle,* Surrey: Gresham Press, SCOPE 13.

Branchu, P. 1991. Cycle Biogeochemique du Carbon: Variations du Reservoir Biomasse Continentale Depuis Le Dernier Maximum Glaciaire. Exemple de l'Afrique. These D.E.A. Univ. d'Aix-Marseille III. 60 pp.

Coude-Goussen, G. 1991. Les Poussieres Sahariennes (cycle sedimentaire et place dans les environnements desertiques). Paris, Universites Francophones. 480 pp.

Faure, H., E. Manguin and Nydal R. 1963 Formations lacustres du quaternaire superérieur du Niger oriental: Diatomites et âges absolus. *Bull. B.R.G.M.* 3:41-63

Frenzel, B., M. Pecsi, and A.A. Velichko (eds.). 1992. *Atlas of Palaeoclimates and Palaeo-Environments of the Northern Hemisphere (Late Pleistocene - Holocene).* Budapest - Stuttgart. 143 pp.

Goudie, A.S. 1992 *Environmental Change.* Oxford, Clarendon Pres. 329 pp.

Kes, A.S., E.D. Mamedov, S.O. Khondkaryan, G.N. Trofimov, and K.V. Kremenetsky. 1993. Stratigrafija i paleografija ravninnih oblastej Srednej Azii v pozdnem pleistocene i golocene (in Russ.) (Plains of Northern central Asia during the Late Pleistocene and Holocene: stratigraphy and palaeogeography.) p. 82-87. In: A.A. Velichko (ed.), *Evolution of Landscapes and Climates of the Northern Eurasia,* Moscow, Nauka.

Lioubimtseva E. 1995. Landscape evolution of the Saharo-Arabian area during the last glacial cycle. *J. Arid Environ.* 30:1-17

Lioubimtseva, E. and B. Simon. 1994. Image de carbone dans l'environnement. *Rep. internes de STSC* i:48.

Lioubimtseva, E. and B. Simon. 1995. Image de carbone dans l'environnement. Rep. internes de STSC ii:45.

Liu, T., Z. Guo, J. Liou , J. Han, Z. Ding, Z. Gu, and Wu N. 1995. Variations of Eastern Asian Monsoon over the last climatic cycle. *Bull. Soc. Geol. France 166* 2:221-230.

Mamedov, E.D. 1990. Raschetnije gidroclimaticheskije harakteristiki aridnih i pluvialnih faz pozdnego pleistocena i golocena (in Russ.) (Estimated hydroclimatic characteristics of arid and pluvial phases of the late Pleistocene and Holocene). p. 148-149. In: Theses of VII All-Union conference Quaternary period: methods of research, stratigraphy and ecology, Tallin.

Milanova, E. and E. Lioubimtseva. 1992. Osnovnije trendi landshaftnoi evolutsii Saharo-Gobijskoi oblasti. (The major trends of the landscape evolution in the Sahara-Gobi desert belt during the Late Pleistocene and Holocene). *Problems Desert Development* (in Russ., abstract in English) 5:48-51.

Olson, J.S., H.A. Pfunder, and Y. Chan. 1983. Carbon in Life Vegetation of Major World Ecosystems. Oak Ridge Natl. Lab., ORNL 5862. 152 pp.

Petit-Maire, N. 1989 Interglacial environments in presently hyperarid Sahara: palaeoclimatic implications. p. 637-661 In: M. Leinen and M. Sarnthein (eds.), Palaeoclimatology and Palaeometeorology: *Modern and Past Patter of Global Atmospheric Transport.* Kluwer Acad. Publ.

Petit-Maire, N. (ed.). 1991. Palaeoenvironments du Sahara. Lacs Holocenes a Taoudenni (Mali). CNRS, Marseille, Paris. 239 pp.

Petit-Maire, N. 1992. Environnements et climats de la ceinture tropicale nord africaine depuis 140,000 ans. p. 27-34. In: J.C. Miskovsky. Les Applications De La Geologie a La Connaissance De L'environnement De L'homme Fossile. (Dir. Publ.). Mem. Soc. Geol. Fr. 160.

Petit-Maire, N, P. Sanlaville, and Z. Yan. 1995 Oscillations de la limite nord du domaine des moussons africaine, indienne et asiatique, au cours du dernier cycle climatique. *Bull. Soc. Geol. france, 166* 2:213-220.

Post, W.M., W.R. Emmanuel, P.J. Zinke, and A.G. Stangenberg. 1982. Soil carbon pools and world life zones. *Nature* 346:48-51

Sanlaville, P. 1992. Changements climatiques dans la peninsule Arabique durant le Pleistocene superieur et l'Holocene. Paleorient.

Singh, G., R.J. Wasson, and D.P. Agrawal. 1990. Vegetation and seasonal climatic changes since the last full glacial in the Thar desert, NW India. *Rev. Paleobotany and Palynology* 64:351-358.

Shi, Y., Z.H. Kong, S. Wang et al. 1993. Mid-Holocene climates and environments in China. *Global and Planetary Change* 7:219-293.

Velichko, A.A. (ed.). 1993. Razvitije Landshaftov I Climata Severnoj Evrazii (in Russ.) (Evolution of Landscapes and Climates of the Northern Eurasia). Moscow, Nauka. 101 pp.

Street-Perrott, F.A., D.S. Marchand, N. Robert, and S.P. Harrison. 1989. Global Lake-Level Variations from 18,000 to 0 Years Ago: a Palaeoclimatic Analysis. U.S. Dept. Energy, Washington D.C. 213 pp.

Talbot, M.R. and T. Johanessen. 1992. A high resolution palaeoclimatic record for the last 27,500 years in tropical West Africa from the carbon and nitrogen isotopic composition of lacustrine organic matter. *Earth Planet Scient. Letters* 110:23-37.

Tarasov. 1992. Evlutsia klimata i landshaftov Severnogo i centralnogo Kazakhstana (in Russ.) (Climatic and landscape evolution of northern and central Kazakhstan) Ph.D. theses, Moscow State Univ. 120 pp.

Varuschenko, S.I., A.N. Varuschenko, and R.K. Klige. 1987. Izmenenija Rezima Kaspijskogo Morja i Besstochnih Vodoemov V Paleovremeni (in Russ.) (Variations of the Caspian Sea Regime and of Closed Lakes in Palaeotimes.) Moscow: Nauka. 239 pp.

Yan, Z.W. and N. Petit-Maire. 1994. The last 140 ka in the Afro-Asian climatic transitional zone. *Palaeogeogr., Palaeoclimatol., Palaeoecol.* 110:217-233.

Zinke, P.J., A.G. Stangenberger, W.M. Post, W.R. Emmanuel, and J.S. Olson. 1984. Worldwide organic soil carbon and nitrogen data - Soil data file. Environmental Sciences Division. Oak Ridge National Laboratory.

CHAPTER 39

Carbon Turnover in Different Climates and Environments

H.W. Scharpenseel and E.M. Pfeiffer

I. Introduction

Due to its exceptional importance for fertility and potential production, the carbon turnover in soils of different climates and environments has been assessed rather judiciously. Diverse methods of land survey, of balancing sustained cropland-, grassland-, woodland-, and wetland experiments over longer periods and the application of tracers have been involved. Yet, many questions still remain unanswered.

C-turnover in soils is involved in basic processes — CO_2 emission to the atmosphere; in mitigation of CO_2 release by C sequestration in SOM; nutrient release, supportive to new CO_2 transfer from the atmosphere into new biomass and SOM; and as precondition of the Second Paradigm for promotion of recycling vs. rationalizing use of agrochemistry.

This chapter first presents a short overview of C turnover behavior of SOM types and fractions, of C in isothermic and nonisothermic soils, in relation to the lignine/N ratio, and concerning the C pools in different climates and environments. It then presents results of radioactive and stable isotope tracer work. This comprises turnover studies, based on uniformly ^{14}C labelled plant substances, assessing intensities of biotic, abiotic, and photochemical types of turnover, C-residence time scanning by radiocarbon dating (^{14}C), and delta ^{13}C measurements in (thin) layerwise sampled soil profiles of different soil orders, whereby the delta ^{13}C reveals changes in vegetation at different phases of C3, C4, and CAM (mainly succulents) plant changes or can be used to indicate C-fixation or release in the process of methanogenesis. It also shows the approximate rate of biological/respiratory C fixed in caliches and other secondary soil carbonates.

Turnover studies with radioactive or stable isotope labeled biomass of agroforestry vegetation are still required. The soil transect program in the "Global Change in Terrestrial Ecosystems" core project of the IGBP (International Geosphere-Biosphere Programme) would offer a unique chance to expand our knowledge of C turnover, C residence time, and extent of historic vegetation changes by including (thin) layer soil profile scans regarding ^{14}C and delta ^{13}C as a regular routine procedure.

II. Basic and Regional Orientation

In a classic approach, Schlesinger (1977) studied the entire regime of processes and reactions in grassland, from litterfall and root biomass via pools of undecomposed litter. He dealt with the formation of fulvic acids, humuc acids, and humins and with the different fraction rates of turnover, mainly via soil respiration, expressed as kg C (of CO_2) per $m^2 y^{-1}$.

Equally revealing in the same report is a latitude-related overview of C inputs by litter vs. CO_2 emission, this time in the woodlands of the world, where C input and loss measurements are very much dominated by data derived from latitude 33-66°.

Buol et al. (1990) estimated the organic C contents of isothermic soils (<5°C difference from winter to summer) and nonisothermic soils (>5°C difference from winter to summer), where the remaining organic carbon in the upper 30 cm in kg m^{-2} is 2 to 3 times as high in the isothermic soils, relating to an interval between 4.5 and 25.5°C mean annual temperature.

Mellilo et al. (1982) and Haider (1992) demonstrate that the lignin/N ratio of the litter-derived organic matter strongly relates to the decomposition in the first year, whereby a high lignine/N ratio is drastically conservationist.

Paul and Clark (1989) attempted a background overview on the global distribution of C in plant biomass and humus (Table 1).

Extracts of chronosequence studies (Schlesinger, 1990) indicate accumulation rates in g C $m^{-2} y^{-1}$ for different accumulation intervals and ecosystems (Table 2).

Directly related to organic matter decomposition in the total soil organic C pool of ca. 1,550 Pg C is the gaseous emission as CO_2, the aquatic transport in solution, the reprecipitation to carbonate-C, and the sequestration into stable SOM-humic substances.

Table 3 reveals the C percentages of the SOM-C pool in different ecosystems (Post et al.,1982). Lal and Kimble (1994) estimate soil-C in the northern permafrost ecozone as having a level of 350-455 Pg C, about 22.5-29.4%, with the arctic tundra alone comprising a share of about 192 Pg of soil C, about 12.4%, both percentages related to the total world soil C pool.

In total, the forest soils are the most important C reserves, accounting for about 60% of the terrestrial C-pool, according to Dixon et al. (1994), standing for 787 Pg C as SOM-C, plus 359 Pg C as vegetation/biomass-C. Thus, together, SOM and biomass-C in the 4.1 bil ha forest systems (37% in low latitudes, 14 % in mid latitudes, 49 % in high latitudes) comprise 1,146 Pg C (SOM + biomass), with presently a deforestation related net flux of 0.9 +/- 0.4 Pg C y^{-1} to the atmosphere.

Brown et al. (1992) estimate that the tropical forests alone could, without further human pressure, have a potential to sequester up to 2.5 Pg C y^{-1}.

Post (1992) arrives at a net terrestrial flux of 1-3 Pg C y^{-1}. The corresponding terrestrial C-bio-sink amounts, according to Blanis (1995), decrease to 0.9 Pg C for 1990.

Degens (1989) made the assessment that land use changes in the more recent past caused a loss of ca. 260 Pg biospheric carbon.

Cole (1995), in the IPCC Report, estimates a loss of C by anthropogenic land use intervention of ca 55 Pg C.

Thiessen et al. (1994) estimate the economically feasible span of land use without fertilization, mainly due to C-turnover, as ca. 65 yrs in temperate prairie soils and only about 6 yrs in tropical semiarid forest-dominated ecosystems. Abrupt changes in drought and rainfall have a strong impact; for example, in the Sahel, ENSO (El Nino Southern Oscillation)-related extreme droughts and floods were most severe in many semiarid ecozones from 1990 -1995.

Glenn and Squires (1993) note that, of the worldwide 43% of nonforested dryland, about 70% are affected by desertification, with a 3.5% annual loss for economic production, but with a potential for 0.5-1.0 Pg C y^{-1} net C sequestration by revegetation and appropriate rangeland farming.

Cole et al. (1993) also opt for a C sequestration potential in semiarid soils of near to 0.5 kg C $m^{-2} y^{-1}$. The transfer of mainly respiratory CO_2 into secondary soil carbonates is described by Schlesinger (1985) with 10 Tg C y^{-1}.

While in grassland soils, the C residence time is mostly higher than in adjacent woodland soils (with the same climate, geomorphology, and parent material), grassland soils are most sensitive to changes in rainfall and N mineralization rates (Whittaker and Likens,1973). Principally, tropical

Table 1. Global distribution of carbon in vegetative biomass and SOM

Area ecosystem	Surface (billion ha)	NPP (thousand kg C ha^{-1} yr^{-1})	Humus-C (Pg)
Tropical forests	2.5	19	250
Temperate and northern forests	2.4	12	320
Bushland, savanna	2.3	8	120
Temperate grassland	0.9	6	170
Tundra	0.8	1	180
Bush desert	1.8	1	111
Mountainous deserts	2.4	0.03	20
Cropland	1.4	3	180
Swamps and marshes	0.2	30	140
Total	14.7		1,490

(Adapted from Paul and Clark, 1989.)

Table 2. Long-term rates of enrichment of organic carbon in soils since Holocene time

Ecosystem	Accumulation phase (yr)	Accumulation rate g Cm^{2}yr^{-1}
Polar desert	8,000 to 9,000	0.2
Tundra	9,000	1.1 to 2.4
Boreal forests	3,000 to 5,000	5.7 to 11.7
Temperate forests	2,000 to 10,000	0.7 to 2.5
Tropical forests	4,000 to 8,000	2.3 to 2.5
Temperate grassland	9,000	2.2

(According to Schlesinger, 1990.)

Table 3. Carbon pools in major ecosystems and zones

Estimate from assessed global reservoir of SOM-C equal to 1,400 Pg C, based on 2,700 soil profiles

Wetlands	14.5% of C
Tundra	13.7% of C
Croplands (agricultural areas)	12.0% of C
Boreal forests	9.5% of C
Tropical woodland and savanna	9.6% of C
Cool temperate steppe	8.6% of C

(According to Schlesinger, 1990.)

grassland and savannas seem to be affected less by natural and anthropogenic impacts than those of the higher latitudes, such as the European grasslands.

But high management pressures from overgrazing and overfrequent burning make it difficult to assess the fate of C stocks in the pulse of enhanced biomass production at higher CO_2 levels, provided the mineral nutrient supply is adequate and the more periodically heterogenous state of soil moisture allows such rise in NPP (Hall et al., 1994).

Alexandrov and Oikawa (1995) see a sink of C in global savannas of about 0.2 Pg C y^{-1}, sustained since the last century and probably related to the rise of atmospheric CO_2 concentration.

In the wetlands, peatlands and ricelands are probably best explored. According to Gorham (1991), in boreal and subarctic peatlands about 455 Pg C are accumulated, with 96 Tg C y^{-1} new sequestration in course of the Holocene. But with continued land use and related drainage, oxidation to CO_2 can be expected at a rate of 8.5 Tg C y^{-1}, plus about 26 Tg C y^{-1} release due to combustion of peat as fuel, plus an emission of 46 Tg C y^{-1} as CH_4 by methanogenesis. From ca. 500 Tg C y^{-1}, originally released by methanogenesis, only about 45 Tg remain unoxidized and are added to the atmosphere, the ca. 115 Tg C as methane from wetlands and about 60 Tg C y^{-1} especially from riceland contributing through CH_4 and CO_2 emission (Neue, Gaunt et al., 1994; Neue and Scharpenseel, 1987).

The C sinks in temperate and boreal wetlands decreased according to Downing et al. (1993) since 1850 by about 50%, from 0.2 to 0.1 Pg C y^{-1}. It may be similar in the tropical wetlands, but valid data are rare.

Maybeck (1981) estimated the annual rate of terrestrial C, lost to the oceans by riverine transport, as amounting to ca. 0.45 Pg C y^{-1}.

III. Radiometric and Stable Isotope Approaches to Assess Carbon Turnover in Soils

In order to make biomass decomposition and C turnover in specific systems visible and measurable, labeling procedures have been applied with preference.

A. Stable or Radio Isotope Lebeled Biomass Turnover in Soils

Uniformly ^{14}C (or ^{13}C or ^{3}H or ^{15}N) labeled plant substance, typical for the examined ecosystem, was produced in special growth chambers. Its decomposition in the soil was measured by counting the released gaseous decomposition products (mainly $^{14}CO_2$) or the remainder of labeled substance in the soil after time intervals. Variables, such as soil depth, different season, rainfall, temperature, exposure to sunshine, being shaded by a cover crop, fallow years, or phase within crop rotation require substancial experimental effort to assure results of high accuracy. Major reports with results for different climates and environments have been presented (Mayaudon and Simonart, 1958; Snolnoki and Vago, 1959; Jenkinson, 1964; Jenkinson, 1966; Smith, 1966; Sauerbeck, 1966; Oberländer and Roth, 1968; Shields and Paul, 1973; Oberländer and Roth, 1974; Martel and Paul, 1974; Oberländer and Roth, 1975; Jenkinson and Ayanaba, 1977; Sauerbeck and Gonzales, 1977; Jenkinson and Raina, 1977; Paul and Van Veen, 1978; Oberländer and Roth, 1980; Tsutsuki and Ponamperuma, 1982; Martin et al., 1983; Van Veen et al., 1985; Ladd et al., 1985; Neue and Scharpenseel, 1987; Rainer and Goswami, 1988; Neue et al., 1990; Singer and Scharpenseel, 1993; Scharpenseel et al., 1992).

As an example, Figure 1 integrates decomposition curves from similar studies in Europe, Africa, S-Asia, and SE-Asia by several authors. Table 4 reflects ^{14}C remainder rates after 1 year. The results of Jenkinson and Ayanaba (1977) and of Trumbore et al. (1996) point adequately at temperature as a main determinant of turnover rates. Also slope/altitude are related and, in the other turnover studies of uniformly labeled biomass (Figure 1), pH, moisture, albedo, and type of clay and minerals matter.

B. Measurement of C Residence Time in Soils, Based on (Thin) Layer Wise ^{14}C-Dating

(Thin) layer wise soil profile scan for C-residence time by the ^{14}C-dating procedure. The age vs. depth curves indicate, as well, the apparent mean residence time of C in various profile depth and genetic soil horizons. Via the factor of the X-member in the regression equation, the steepness of the age increases with depth.

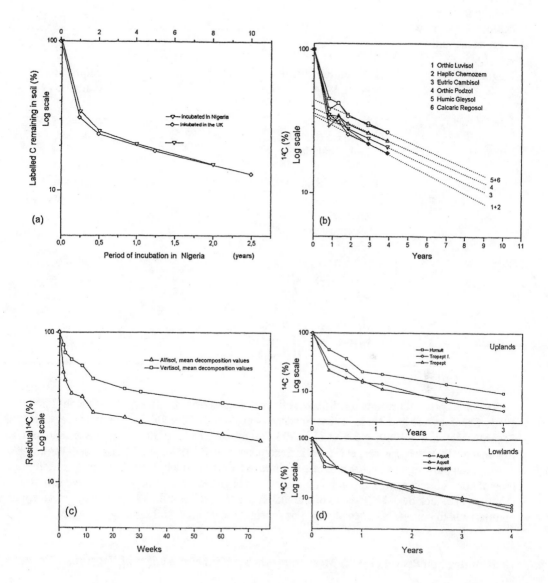

Figure 1. Biomass turnover in different climate and ecosystems; dependence on temperature (T), moisture (M), pH, albedo (Alb), and type of clay minerals (HAC = high activity clays or LAC = low activity clays).

Table 4. Decomposition speed of ^{14}C labeled plant material

Labeled substrate	Country, Soils		Approx. C remaining after 1 yr (%)	Reference
Straw + urine	Austria, Hapludoll		64	Oberländer and
Stable manure 1			62	Roth, 1980
Stable manure 2			56	
Straw			40 to 42	
Green manure			33	
Stable manure:straw: green manure			after 8 yr, remainder = 100:68:62	
Ryegrass	England, Rothamsted		31	Jenkinson and
	Nigeria, IITA		21	Ayanaba, 1977
Wheat straw	Germany, Bonn		35	Sauerbeck and
	Costa Rica (fallow plot), both Hapludalfs		23 to 36	Gonzalez, 1977
Peanut straw	India, ICRISAT, both plots under shade			Singer and Scharpenseel, 1993
	Rhodustalf		21	
	Chromustert		36	
Rice straw	Philippines			Neue and
	Rice paddy	Humult	20.7	Scharpenseel, 1987
	Lowland	Aquoll	19.3	
		Aquept	17.4	
	Upland	Humult	18.6	
	Rice soils	Tropept 1	12.9	
		Tropept 2	11.1	

Major reports with results for different climates and environments have been made (Paul et al., 1964; Scharpenseel and Neue, 1984; Scharpenseel et al., 1986; Becker-Heidmann et al., 1988; Scharpenseel et al., 1989; Neue et al., 1990; Trumbore et al., 1990; Scharpenseel and Becker-Heidmann, 1992; Scharpenseel et al., 1992; Scharpenseel et al., 1994; and Scharpenseel et al., 1996).

As examples, Figure 2(a-d) presents the integral mean residence time curves of different soil orders. For a correct age vs. depth relation, reflecting the C dynamics of the respective soil order, it is important to include only ^{14}C-dates of completely scanned soil profiles (2 cm thin layer scan appears optimal) and to ommit ^{14}C-dates of single or several, just spotted, soil sampling ventures.

C. Scanning Soil Profiles by ^{13}C Measurements for Historic Changes of Vegetation Cover

In some cases, the landscape history is sufficiently known, i.e., the date/age of changes in photosynthetic mechanism of the vegetation cover is known (forest C3, savanna grasses C4, succulent plants CAM). Such successions are leading to shifts in the stable isotope ratio (Figure 3). From the extent of shifts, one can, against the background of total soil-C, roughly calculate the amount of C input of the new delta ^{13}C-source during the time since the vegetation change in landscape history began. This also can be expressed in percentage loss of the old resident carbon.

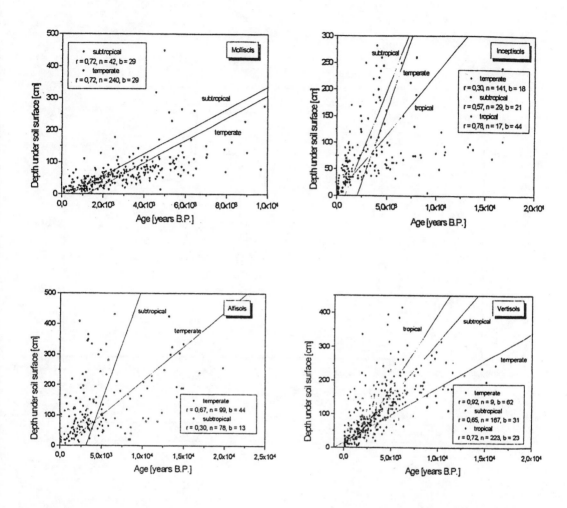

Figure 2. Carbon residence time according to ^{14}C-dates of layers of scanned soil in different soil profiles.

Principally, but restrained by the low precision of knowledge regarding the caliche age, the shift of delta ^{13}C in calcareous crusts/caliches is intriguing regarding input of respiratory biotic C during the transporting bicarbonate phase and regarding which amounts vs. time (Schleser et al., 1983; Schleser and Jayasekera, 1985; Freytag, 1985; Scharpenseel et al., 1986; Martin et al., 1990; Wagner, 1990; and Gerzabek et al.,1990).

D. Approaches to Assess Biotic and Abiotic Biomass Decomposition in Soils

Specific studies on biotic, abiotic (also protolytic), photochemical decomposition of organic matter in soils were carried out for more distinctive exploration of their contribution under various background conditions. An early basic and intensive study on biomass input and decomposition rate

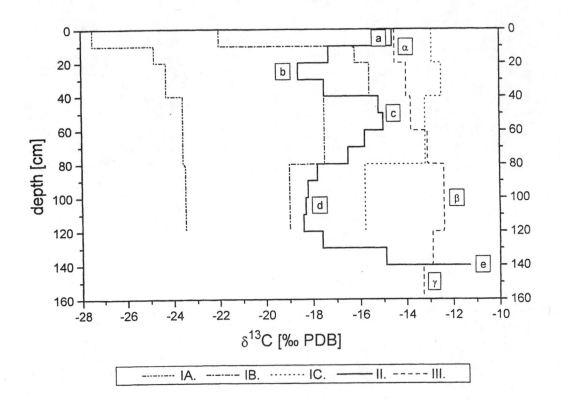

Figure 3. ^{13}C-depth profiles of different soil-plant systems:
  (1) Lamto, Ivory Coast;
    I.[a] A = Gallery forest (mean of 1)
         B = Savanna, protected from fire (mean of 4)
         C = Shrub + grass savanna (mean of 3)
  (2) ICRISAT, Hyderabad, India:
    II.[b] a = sorghum (C4); b = crops (more C3); c = savanna grasses (C4); d = forest (C3);
          e = free carbonates in soil
  (3) Gezira, Sudan:
    III.[b] α = cotton (C3); β = durra (C4); δ = towards Nile alluvium (C3 + C4-mix)

I. = Ferralsol; II. = Rhodustalf; III. = Pellustert
[a]Martin et al., 1990; [b]Scharpenseel et al., 1989.

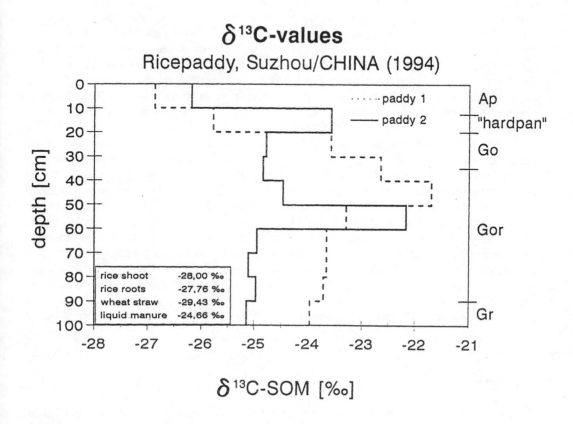

Figure 4. ^{13}C-values of SOM in rice paddy soils in Suzhou, China: paddy 1 with rice-wheat cropping system and fallow period; paddy 2 with irrigated rice.
(According to unpublished data by Pfeiffer and Wagner, 1995.)

was carried out by Broadbent and Bartholomew (1948). Steril conditions for abiotic decomposition tests were achieved either by biocidal treatment or by gamma irradiation, or by autoclaving or fumigation (Powlson and Jenkinson,1976), and also by addition of 0.5 g Na-ethylmercurythiosalicilate per 100 g of soil (Scharpenseel,Wurzer, Freytag and Neue,1984), where CO_2 release over length of the experiment was: biotic : abiotic : biotic + UV irradiation : abiotic + UV irradiation = 100 : 20 : 70 : 50.

Laura (1975, 1976) tested especially CO_2 contributions by protolytic action. Martin et al. (1981) distinguished between crop biomass-, lignine-, and polysaccharide-carbon fractions in soil biodegradation. Later, Martin and Haider (1986) studied the influence of mineral colloids on SOM turnover rates. Tsutsuki and Ponamperuma (1982) focused on conventional methods, studying decomposition of organic matter in anaerobic soils.

E. Anoxic Decomposition Related Isotope Discrimination, Revealing Methanogenesis

The bacterial decomposition processes tend to fractionate the C-isotopes of SOM by favoring the lighter ^{12}C-carbon over the heavier ^{13}C-compounds. Under anoxic conditions, the C-decay is dominated by the processes of methanogenesis and leads to an enrichment of ^{13}C in the residual soil organic matter – e.g., SOM with -22‰ – while the product of decay (CH$_4$ with ^{13}C-values of about -60‰) is strongly depleted. Typical examples are given in Figure 4, which shows the deep distribution of the ^{13}C-values of the SOM in two rice paddies in Suzhou, China. The highest values are measured in the anoxic soil zones, which are characterized by an annual methane production. Paddy 1 is an example for a rice-wheat-cropping system, including a fallow period, and shows a zone of ^{13}C-enrichment (-21.6‰) in the deeper soil horizons which is influenced by ground water level (Gor, the gleyic horizons) and the availability of the methanogenic substrates (acetate, CO$_2$/H$_2$). Paddy 2 presently is planted by irrigated rice and has two distinct ^{13}C-enriched zones: one is located in the hardpan layer (-23.7‰ ^{13}C), which is characterized by high water saturation and by high CH$_4$ production and is not mixed by puddling. The second zone with ^{13}C-values of about -22.1‰, is situated in the deeper groundwater horizons. In comparison to the plant sources, rice and wheat with ^{13}C-values of -28.7‰ (see Figure 4), the methanogenesis has led to an ^{13}C-enrichment of about 5‰.

Isotope-related and other studies of C turnover have been successfully conducted in different climates and environments. The clear and unmistakable results on percentage decomposition or rate of remainder by the use of uniformly labeled plant substance (mostly straw) suggest continuation, especially with typical species of savanna and agroforestry environments, where knowledge gaps exist. For an understanding of SOM-C, with 1,550 Pg the largest accessible organic C pool, residence time and replacement rates after alternation of vegetation with different photosynthesis mechanisms, including thin layer soil profile sampling with subsequent D ^{14}C as well as delta ^{13}C scan of typic (benchmark) soil profiles, is essential. It should be included in standard methods and treatments of the forthcoming IGBP/GCTE landscape transects to elucidate the diverse trends of C sequestration stability or turnover in the various regions and ecosystems.

References

Alexandrov, G.A. and T. Oikawa. 1995. Net ecosystem production resulted from CO$_2$ enrichment: Evaluation of potential response of a savannah ecosystem to global changes in atmospheric composition. p. 117-119. In: Proc. of the Tsukuba Global Carbon Cycle Workshop. Tsukuba, Japan.

Becker-Heidmann, P., Liu Liang-Wu, and H.W. Scharpenseel. 1988. Radiocarbon dating of organic matter fractions of a Chinese Mollisol. *Z. Pflanzenernähr. Bodenkunde* 151:37-39.

Blanis, D., Y. Matsuoka, and M.M. Kainuma. 1995. Carbon fertilization of the terrestrial vegetation. p. 126-133. In: Proc. of the Tsukuba Global Carbon Cycle Workshop. Tsukuba, Japan.

Broadbent, F.E. and W.V. Bartholomew. 1948. The effect of quantity of plant material added to a soil on its rate of decomposition. *Proc. Soil Sci. Soc. Amer.* 13:271-272.

Brown, S., A.E. Lugo, and L.R. Iverson. 1992. Processes and lands for sequestering carbon in the tropical forest landscape. *Water, Air and Soil Pollution* 64:139-155.

Buol, S.W., P.A. Sanchez, J.M. Kimble, and S.B. Weed. 1990. Predicted Impact of Climatic Warming on Soil Properties and Use. American Soc. Agron. Special Publ.

Cole, C.F., K. Flach, J. Lee, D. Sauerbeck, and B. Stewart. 1993. Agricultural sources and sinks of carbon. *Water, Air and Soil Pollution* 70:111-122.

Cole, C.F. 1995. Soils Chapter. IPCC-Report.

Degens, E. 1989. *Perspectives on Biogeochemistry*. Springer Verlag, NY. 208 pp.

Dixon, R.K., S. Brown, R.A. Houghton, A.M. Solomon, M.C. Trexler, and J. Wisniewski. 1994. Carbon pools and flux of global forest ecosystems. *Science* 263:185-190.

Downing, J.P., M. Meybeck, J.C. Orr, R.R. Twilley, and H.W. Scharpenseel. 1993. Land and water interface zones. *Water, Air and Soil Pollution* 70:123-137.

Freytag, J. 1985. Das $^{13}C/^{12}C$-Isotopenverhältnis als aussagefähiger Bodenparameter, untersucht an tunesischen Kalkkrusten und sudanesischen Vertisolprofilen. *Hamburger Bodenkundliche Arbeiten* 3:1-265.

Gerzabek, M.H., F. Pichlmeyer, K. Blochberger, and K. Schaffer. 1990. Use of ^{13}C measurements in humus dynamics studies. Intern. Symposium on Use of Stable Isotopes in Plant Nutrition, Soil Fertility and Environmental Studies. IAEA, Vienna, Oct. 1990. SM 313/42.

Glenn, E., V. Squires, M. Olsen, and R. Freye. 1993. Potential for carbon sequestration in the drylands. *Water, Air and Soil Pollution* 70:341-355.

Gorham, E. 1991. Northern peatlands: role in the carbon cycle and probable responses to climate warming. *Ecological Applications* 2: 182-195.

Haider, K. 1992. Auswirkungen von Klimaänderungen auf die Landwirtschaft. Deutscher Bundestag, Enquete Commission, Protection of the Earth Atmosphere. Commission Document 12/5-b. 21 pp.

Hall, D.O., J.M.O. Scurlock, D.S. Ojima, and W.J. Parton. 1994. Grassland and the global carbon cycle: Modelling the effects of climate change. In: *The Carbon Cycle*, 1993 OIES Global Institute Proceedings, Cambridge University Press.

Jenkinson, D.S. 1964. Decomposition of labelled plant material in soil. p. 199-207. In: *Experimental Pedology*. Butterworth, London.

Jenkinson, D.S. 1966. The turnover of organic matter in soil. p. 187-197. In: The Use of Isotopes in Soil Organic Matter Studies. Rep. FAO/IAEA Techn. Meeting, Pergamon, Oxford.

Jenkinson, D.S. and A. Ayanaba. 1977. Decomposition of carbon-14 labelled plant material under tropical conditions. *Soil Sci. Soc. Amer. J.* 41:912-915.

Jenkinson, D.S. and J.H. Rainer. 1977. The turnover of organic matter in some of the Rothamsted classical experiments. *Soil Sci.* 123:298-305.

Ladd, J.N., M. Amato, and J.M. Oates. 1985. Decomposition of plant material in Australian soils. III: Residual organic and microbial biomass C and N from isotope labelled legume material and SOM decomposing under field conditions. *Austr. J. Soil Res.* 23:603-611.

Lal, R. and J.M. Kimble. 1994. Soil management and the greenhouse effect. p.1-5. In: R. Lal, J.M. Kimble and E. Levine (eds.), *Soil Resources and the Greenhouse Effect*, USDA-SCS, Washington DC.

Laura, R.D. 1975. The role of protolytic action of water in the chemical decomposition of organic matter in the soil. *Pedologie* XXV, 3:157-170.

Laura, R.D. 1976. On decomposition and "primed decomposition" of organic materials in soils. Letter to the editor. *Can. J. Soil Sci.* 56:379-382.

Martel, Y.A. and E.A. Paul. 1974. The use of radiocarbon dating of organic matter in the study of soil genesis. *Soil Sci. Soc. Amer. Proc.* 38:501-506.

Martin, U., H.U. Neue, H.W. Scharpenseel, and P. Becker. 1983. Anaerobe Zersetzung von Reisstroh in einem gefluteten Reisboden auf den Philippinen. *Mitt. Dt. Bodenkdl. Ges.* 38:245-250.

Martin, J.P. and K. Haider. 1986. Influence of mineral colloids on turnover rates of soil organic carbon. p. 283-304. In: Interactions of Soil Minerals with Natural Organics and Microbes. Soil Sci Soc. America Special Publ. 17, Madison.

Martin, J.P., K. Haider, and G. Kassim. 1981. Biodegradation and stabilization after 2 years of specific crop, lignine and polysaccharide carbons in soils. *Soil Sci. Soc. Amer. J.* 44:1250-1255.

Martin, A., A. Mariotti, J. Balesdent, P. Lavelle, and R. Vuattoux. 1990. Estimate of organic matter turnover rate in a savanna soil by delta ^{13}C natural abundance measurements. *Soil Biol. Biochem.* 22:517-523.

Maybeck, M. 1981. Flux of organic carbon by rivers to the oceans. p. 219-269. In: National Technical Information Service, Springfield, Virginia, U.S.

Mayaudon, J. and P. Simonart. 1958. Etude de la decomposition de la matiere organique dans les sols au moyen de carbon radioactif, I et II. *Plant Soil* 9:376-384.

Mellilo, J.M., J. Aber, and J.F. Muratore. 1982. Nitrogen and lignine control of hardwood lead litter decomposition dynamics. *Ecology* 63: 621-626.

Neue, H.U. and H.W. Scharpenseel. 1987. Decomposition pattern of ^{14}C-labelled rice straw in aerobic and submerged rice soils of the Philippines. *The Sci. Total Environ.* 62:431-434.

Neue, H.U., P. Becker-Heidmann, and H.W. Scharpenseel. 1990. Organic matter dynamics, soil properties and cultural practices in rice lands and their relationship to methane production, pp 457-466. In: A.F.Bouwman (ed.), *Soils and the Greenhouse Effect.* John Wiley & Sons, NY.

Neue, H.U., J.L. Gaunt, Z.P. Wang, P. Becker-Heidmann, and C. Quijano. 1994. Carbon in tropical wetlands. p. 201-220. In: Transactions 15th World Congress Soil Science, Acapulco, Vol 9, Supplement.

Oberländer, H.E. and K. Roth. 1968. Transformation of ^{14}C labeled plant material in soils under field conditions. p. 351-361. In: Isotopes and Radiation in Soil Organic Matter Studies. Technical Meeting FAO/IAEA, Vienna.

Oberländer, H.E. and K. Roth. 1974. Ein Kleinfeldversuch über den Anbau und die Humifizierung von ^{14}C-markiertem Stroh und Stallmist. *Bodenkunde* 25:139.

Oberländer, H.E. and K. Roth. 1975. Die Umwandlung eines ^{14}C-markierten Düngers aus Gülle und Stroh im Boden. *Bodenkunde* 26:139.

Oberländer, H.E. and K. Roth. 1980. Der Umsatz ^{14}C-markierter Wirtschaftsdünger im Boden. *Landwirtschaftliche Forschung* 33:179-188.

Paul, E.A., C.A. Campbell, D.A. Rennie, and K.J. McCallum. 1964. Investigations on the dynamics of soil humus utilizing carbon dating techniques. p. 201-208. In: Int. Congr. Soil Sci. Transact. 8th.

Paul, E.A. and J.A. Van Veen. 1978. The use of tracers to determine the dynamic nature of organic matter. p. 61-102. In: Dep. Soil Sci. Univ. of Saskatchewan, Saskatoon Transaction of 11th ISSS-Congress, Edmonton, Vol.3, Symposia Papers.

Paul, E.A. and E.E. Clark. 1989. *Soil Microbiology and Biochemistry.* Academic Press San Diego, 273 pp.

Pfeiffer, E.M. and D. Wagner. 1995. Isotopenanalytische Untersuchung zum Einfluss einer Reis-Weizen-Fruchtfolge auf die Umsetzung der organischen Substanz im Boden und die Methanfreisetzung. *Report to BMBF, Joint Venture on Trace Elements Cycles* (in press).

Post, W.M., W.R. Emanuel, P.J. Zinke, and A.G. Stangenberger. 1982. Soil carbon pools and world life zones. *Nature* 298:156-159.

Post, W.M., J. Pastor, A.W. King, and W.R. Emanuel. 1992. Aspects of the interaction between vegetation and soil under global change. *Water, Air and Soil Pollution* 64:345-363.

Powlson, D.S. and D.S. Jenkinson. 1976. The effects of biocidal treatments on metabolism in soils. II. Gamma radiation, autoclaving, air drying and fumigation. *Soil Biol. Biochem.* 8:180-188.

Raina, J.N. and K.P. Goswami. 1988. Effect of added ^{14}C labelled organic materials on the decomposition of native soil organic matter. *J. Indian Soc. Soil Sci.* 36:646-651.

Sauerbeck, D. 1966. Über den Abbau ^{14}C-markierter Substanzen in Böden und ihren Einfluß auf den Humusgehalt. Habilitation Thesis, University Bonn, Germany

Sauerbeck, D. and M.A. Gonzalez. 1977. Decomposition of carbon 14 labelled plant residues in various soils of the Federal Republic of Germany and Costa Rica. In: Soil Organic Matter Studies. Proc. FAO/IAEA Symposium, Brunswig, FRG.

Scharpenseel, H.W., M. Wurzer, J. Freytag, and H.U. Neue. 1984. Biotisch und abiotisch gesteuerter Abbau von organischer Substanz im Boden. *Z. Pflanzenern. Bodenkunde* 147: 502-516.

Scharpenseel, H.W., H.U. Neue, and S. Singer. 1992. Biotransformations in different climate belts: source-sink relationships. p. 91-105. In: J. Kubat (ed.), *Humus, Structure and Role in Agriculture and Environment*, Elsevier Publishers.

Scharpenseel, H.W. and H.U. Neue. 1984. Use of isotopes in studying the dynamics of organic matter in soils. p. 275-309. In: Organic Matter and Rice. IRRI Publications, Los Banos, Philippines.

Scharpenseel, H.W., K. Tsutsuki, P. Becker-Heidmann and J. Freytag. 1986. Untersuchungen zur Kohlenstoffdynamik und Bioturbation von Mollisolen. *Z. Pflanzenernähr. Bodenkunde* 129:582-597.

Scharpenseel, H.W., P. Becker-Heidmann, H.U. Neue, and K. Tsutsuki. 1989. Bomb-carbon, ^{14}C-dating and delta ^{13}C measurements as tracers of organic matter dynamics as well as of morphogenetic and turbation processes. *The Sci. Total Environm.* 81/82:99-110.

Scharpenseel, H.W. and P. Becker-Heidmann. 1992. Twenty-five years radiocarbon dating of soils, paradigm of erring and learning. *Radiocarbon* 34: 541-549.

Scharpenseel. H.W., E.M. Pfeiffer, and P. Becker-Heidmann. 1994. Soil organic matter studies and nutrient cycling. p. 285-305. In: *Intern. Symp. Nuclear and Related Techniques in Soil/Plant Studies on Sustainable Agriculture and Environmental Preservation*, IAEA, Vienna.

Scharpenseel, H.W., E.M. Pfeiffer, and P. Becker-Heidmann. 1996. Alter der Humusstoffe. In: W.R. Fischer (ed.), *Handbuch für Bodenkunde*, Ecomed Publ., Landsberg/Lech (in preparation).

Schleser, G.H., H.G. Bertram, H.W. Scharpenseel, and W. Kerpen. 1983. Aussagen über Bildungsprozesse tunesischer Kalkkrusten mittels ^{13}C/^{12}C Isotopenanalysen. *Mitt. Deutsch. Bodenkundl. Gesellsch.* 38:573-578.

Schleser, G.H. and R. Jayasekara. 1985. Delta ^{13}C-variations of leaves in forests as an indication of reassimilated CO_2 from the soil. *Oecologia* 65:536-542.

Schlesinger, W.H. 1977. Carbon balance in terrestrial detritus. *Annual Rev. of Ecology and Systematics* 8:51-81.

Schlesinger, W.H. 1985. The formation of caliche in soils of the Mojave desert, California. *Geochim. Cosmochim. Acta.* 49:57-66.

Schlesinger, W.H. 1990. Evidence from chronosequence studies for a low carbon storage potential of soils. *Nature* 348:232-239.

Shields, J.A. and E.A. Paul. 1973. Decomposition of ^{14}C labelled plant material under field conditions. *Canad. J. Soil Sci.* 53:297.

Singer, S. and H.W. Scharpenseel. 1993. Losses of uniformly ^{14}C-labelled groundnut straw and tissues in soils of semi-arid tropical (SAT) India. *Agrokemia es Talajtan, Tom* 42, No 1-2:147-156.

Smith, J.H. 1966. Some interrelationships between decomposition of various plant residues and loss of soil organic matter as measured by carbon-14 labelling. p. 223-233. In: FAO-IAEA Conf., Spec. Supplem. *Int. J. Appl. Radiat. Isotopes.*

Szolnoki, J. and E.T. Vago. 1959. Abbau und Humifikation von mit dem Isotop C-14 markiertem Stroh im Boden. *Acta Agron. Acad. Sci. Hung.* 9:407-414.

Tiessen, H., E. Cuevas, and P. Chacon. 1994. The role of soil organic matter in sustaining soil fertility. *Nature* 371:783-787.

Trumbore, S.E., G. Bonani, and W. Wölfli. 1990. The rates of carbon cycling in several soils from AMS ^{14}C measurements of fractionated soil organic matter. p. 407-414. In: A.F. Bouwman (ed.), *Soils and the Greenhouse Effect*. John Wiley & Sons, NY.

Trumbore, S.E., O.A. Chadwick, and R. Amundson. 1996. Rapid exchange between soil carbon and atmospheric carbon dioxide driven by temperature change. *Science* 272:393-395.

Tsutsuki, K. and F.N. Ponamperuma. 1982. Decomposition of organic matter in anaerobic soils. Proc. 13th Ann. Scient. Conference, Crop Sci. Soc. of the Philippines, Cebu.

Van Veen, J.A., J.N. Ladd, and M. Amato. 1985. Turnover of C and N through the microbial biomass in a sandy loam and a clay soil, incubated with ^{14}C(U) glucose and (^{15}NH$_4$)$_2$SO$_4$ under different moisture regimes. *Soil Biol. Biochem.* 17:747-756.

Wagner, G.H. 1990. Using natural abundance of ^{13}C and ^{15}N to examine soil organic matter accumulated during 100 years of cropping. *Proc. Intern. Sympos. on the Use of Stable Isotopes in Plant Nutrition, Soil Fertility and Environmental Studies.* IAEA-SM-313. Vienna.

Whittaker, R.H. and G.E. Likens. 1973. The primary production of the biosphere. *Human Ecology* 1:299-369.

CHAPTER 40

Carbon Sequestration in Soil: Knowledge Gaps Indicated by the Symposium Presentations

D.J. Greenland

I. Introduction

There is no more important factor relating to long-term soil productivity than soil organic matter content, and no greater potential source for the sequestration for global carbon than the soil. Thus, by taking steps to increase soil carbon, it is possible not only to reduce the effects of global atmospheric pollution by carbon dioxide, but also to improve the prospects for a long-term solution to the world food problem. Measures to increase sequestration of carbon by soils create a genuine win-win situation, in which solutions to an environmental problem and a production problem can be solved by the same measures. As was made clear during the opening session of the Symposium, this opportunity merits far more public interest and government support than it at present receives.

There are, however, several questions which scientists have to resolve relating to how the potential benefits may be best quantified, how the true extent of the potential of soils to sequester carbon may best be established, and how practical methods to implement measures to sequester more carbon in the soil can best be promoted.

This volume and two others on the symposium present the state of knowledge regarding the role of world soils in the global carbon cycle, the processes and practices which determine the balance between sequestration and release of carbon from soils, and the adequacy of the present world database regarding the carbon balance of the soils of the world.

The relevant chapters can be grouped into the following:

1. Soil carbon dynamics and processes, which include papers concerned with the changes in the carbon content of soils in different regions of the world, and the processes which affect the gains and losses from different soils;
2. Aggregation and soil carbon, in which the protection of carbon by sequestration in soil aggregates was reviewed;
3. Soil quality indicators, in which factors determining soil carbon levels, and their relation to the economic and environmental value of soils, are discussed;
4. Soil management for carbon sequestration, in which different agricultural methods of soil management are discussed in relation to their effects on carbon sequestration;
5. Management of grasslands and woodlands, in relation to carbon sequestration;

6. Modeling of carbon dynamics, in which the effectiveness and accuracy of models at present available are reviewed, and the additional data needed to make the predictive value of such models more reliably assessed;
7. Methods of carbon measurement, when some possible changes to improve present methodology are presented; and
8. Carbon pools, in which the forms of carbon, and their effects on carbon sequestration and release are assessed.

A wealth of new research information from all parts of the globe has been presented on each of these topics.

1. Soil carbon dynamics and processes. Gains and losses of carbon from soils have been described from sites in North and South America, Africa, Asia, and Europe, and from the tropics to the Arctic. Most were derived from long-term studies of soils, some from experiments, and others from determinations made on a national or regional level. Most related to time scales of decades, although one included reference to the changes which had occurred since the land was first farmed by the Incas in 800 AD, and another reviewed the information available on carbon dioxide from the time of the earliest geological records. Although some gave data relating to net losses of carbon from the soil to the atmosphere, several showed that by good soil management practices soil sequestration of carbon could be increased under arable crop management, and that even greater sequestration of carbon could be achieved by afforestation or the improvement of the productivity of grasslands. Processes of gain and loss have been described, and the effects of tillage and zero tillage on the balance of gains and losses. The importance of interdisciplinary studies is highlighted in one chapter describing changes which had occurred since the initial plant colonization of soils.

2. Several authors have shown that in many different soils the interaction between the clay fraction of soils and the organic materials in soils provided an important mechanisms by which the carbon was protected from those soil organisms which convert the carbon to carbon dioxide. The protection was further enhanced by the adsorption of carbon compounds on the surfaces of clay minerals, and by the formation of microaggregates in which the soil pores are too small to allow the organisms to reach the organic compounds. These microaggregates are stabilized by the interaction between components of soil organic matter, and often by fungal hyphae. They are formed by the action of soil fauna. Although found in many soils of the temperate zone, information regarding their presence and significance in tropical soils is lacking.

3. Recognition of the importance of soil quality to environmental management has led to renewed and enhanced efforts to arrive at a simple index of the many factors which determine the response of soils to different management. Again there exists a wide geographical spread in the sources of information about soil quality and its relation to carbon sequestration. Among the problems recognized in defining soil quality is the need for uniform methods in the determination of the various indicators, so that results from different countries and within countries might be compared. An important development reported is the establishment of an international network, SOMNET, the International Soil Organic Matter Network, in which scientists from many countries are collaborating in the development of methodologies for the establishment of national soil carbon inventories. Participants in SOMNET use long-term experiments to evaluate models of the dynamics of soil organic matter change. The value of the models lies in the evaluation of methods proposed for changes in soil management to increase carbon sequestration.

4. Soil management and carbon sequestration. Management to increase carbon sequestration in soils may involve increasing the quantity of organic matter returned or added to the soil, or reducing the rate at which carbon in the soil is lost, by oxidation or erosion, or a combination of both. Although some methodological problems exist due to the difficulties of establishing the quantitative significance of relatively small changes occurring over periods of only a few years, it is clear that the Conservation Reserve Program (CRP), and afforestation or agroforestry practices, and production of improved pastures, even on formerly forested land, can make a highly significant contribution to carbon sequestration. Changes in tillage practices, particularly the use of minimum tillage, also reduce the rate at which soil carbon is converted to carbon dioxide and lost to the atmosphere. More work is needed to quantify the extent to which such changes in practice can reduce carbon loss. These results highlight the need to ensure that organic residues are incorporated in the soil as deeply as possible. Minimum tillage slows down the rate of decomposition, but leaves organic matter exposed close to the soil surface. There is a need to determine whether cultivation of the soil to bury the organic matter which accumulates at the soil surface under minimum tillage practices can make a significant further contribution to long-term carbon sequestration. Inorganic fertilizers generally increase plant growth and, hence, the amount of carbon returned to the soil from crop residues, but its association with intensive tillage has often meant that its beneficial effects are confounded by the enhanced rate of loss due to cultivation. Addition of biologically-fixed nitrogen through production of soybeans and other legumes appeared to have a clear advantage in contributing to greater carbon sequestration.

5. It is now generally agreed that the most rapid sequestration of carbon occurred when cultivated soils which were formerly forested or under prairie vegetation were returned to forest or improved pasture. More work is needed to provide a quantitative assessment of the rates of change and of the economics of making such changes. The potential to increase the quantities of carbon sequestered in the deep soils of many tropical savannas, which currently have low carbon contents, has been appropriately highlighted. The importance of inclusion of a legume in the pasture to add the nitrogen which forms an intrinsic component of soil organic matter is also demonstrated. The magnitude of this potentially very large sink for soil carbon in the tropics, and the economic possibilities to improve the productivity of the tropical grasslands is one area which undoubtedly merits much further attention.

   Agroforestry practices also have an important potential to increase the soil content of sequestered carbon, in both temperate and tropical regions. The magnitude of the changes which can be attained depends on several soil factors, again including the nitrogen availability, and probably also the phosphorus and sulfur status of the soil.

6. Several models of the factors determining soil carbon levels are described, and it is shown that these are now able to predict the past organic matter changes for large regions within the United States of America and, hence, could be used with confidence to predict future changes in the amounts of carbon sequestered in the soil, under current climatic conditions, and agricultural practices, as well as how changes of climate might affect the amounts sequestered.

7. Methods to improve the quality of the information regarding the quantities of carbon presently contained in the soil, and how this carbon may be protected from conversion to carbon dioxide, have been discussed. The need to standardize methods so that comparisons can safely be made between data coming from different countries, and different laboratories within a country, is generally agreed, although it is recognized that considerable discussion will be needed to reach agreement on the methods to be used.

8. There is general agreement that the rate of conversion of soil carbon to carbon dioxide, and the rate of sequestration of carbon in the soil, differs considerably depending on the form of carbon in the soil. There are pools of highly labile carbon, readily mineralized carbon, and highly resistant carbon. Advances in isotopic techniques to determine the magnitudes of the pools, and how they are influenced by soil management practices, now enable these pools to be measured successfully and, hence, in conjunction with the models discussed above, the contribution that soils can make to amelioration of the greenhouse effect to be predicted, wherever the necessary data exist to construct the models.

It is also necessary that measures to ensure maximum carbon sequestration are economic. The need for this aspect to be considered in relation to other costs of minimizing greenhouse gas emissions, for instance, by reducing fossil fuel use, need to be emphasized.

While there is undoubtedly a great need for further research to provide a better quantification of the information we have regarding the role of the soil in ameliorating the greenhouse effect, there is more than adequate data available to establish that it would cost much less to increase the soil store of carbon by promoting appropriate agricultural practices, than it would to reduce the use of fossil fuels.

CHAPTER 41

Knowledge Gaps and Researchable Priorities

R. Lal, J. Kimble, and R. Follett

I. Introduction

Pedospheric processes are vital to the production of food, fiber, fodder, and fuel to meet the basic needs of present and future generations. Pedospheric processes play a crucial role in the global C cycle. Principal pedospheric processes that influence the biosphere, the hydrosphere, and the atmosphere include leaching, erosion, and gaseous fluxes. The pedosphere affects the global C cycle by influencing the magnitude of global C pools, their dynamics, and the fluxes between them. Through the dynamics and the interaction of soil properties and processes, properties of the atmosphere, geosphere, lithosphere, hydrosphere, and biosphere are influenced by the pedosphere. Properties of the atmosphere that are affected by the pedosphere can result from the fluxes of radiatively-active greenhouse gases (e.g., CO_2, CH_4, N_2O), relative humidity and vapor pressure, and temperature. Pedospheric influences on air temperature are due to direct and indirect effects. Direct effects on air temperature result from heat exchange between the soil and the atmosphere, especially in the vicinity of the ground surface. Indirect effects of the pedosphere include the heat/energy balance of the pedosphere, hydrosphere, biosphere, and the atmosphere. These effects are moderated through the dynamics of the radiation balance of the atmosphere as influenced by its gaseous composition, principally those of the greenhouse gases (H_2O, CO_2, CH_4, N_2O, NO_x, and others).

II. Interaction Between the Biosphere and Atmosphere

The biosphere and atmosphere interact through photosynthesis and respiration. The global rate of C fixation through photosynthesis is 120 Pg C/yr and that of the respiration by plants and in soil is 60 Pg C/yr (Schlesinger, 1995). Out of the total biomass produced, the fraction retained in the soil as humus plays an important role in the global C cycle by immobilizing C within the pedosphere. Principal pathways of C returned to the soil are depicted in Figure 1. Two pathways through which biomass C is returned to the atmosphere are: (i) decomposition and volatilization, and (ii) biotic respiration. Pedospheric C is redistributed by the processes of soil erosion and leaching. The fate of C redistributed over the landscape by erosion and leached out of the pedosphere as dissolved organic carbon (DOC) or particulate organic carbon (POC) are not clearly understood. The impact of C deposited or buried in depressional sites and water bodies is unclear.

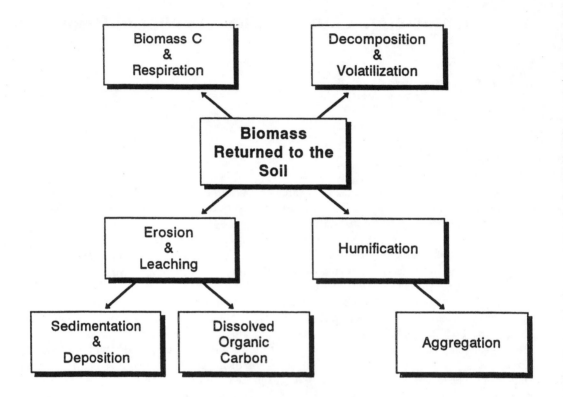

Figure 1. Soil specific processes that affect the fate of biomass carbon returned to the soil.

III. Soil Types and the Global Carbon Cycle

Some soil types influence the global C cycle more than others because of their areal extent, high C concentration, or both. Among the two broad categories of soils shown in Figure 2, the epipedons of organic soils contain more organic C than those of inorganic soils. Most organic soils are formed in flood plains, wetlands, and hydromorphic environments under complete or partial anaerobiosis. The dynamics of C in organic soils (Histosols) can have an important impact on the magnitude of C efflux from soil to the atmosphere. Change from anaerobic to aerobic environments, through drainage of wetlands due to natural or anthropogenic factors, can drastically influence the C balance of organic soils and the global C cycle. Change in land use, deforestation, drainage, and cultivation lead to a rapid loss of C from organic soils by oxidation, volatilization, and mineralization. Dried peat and muck soils are also prone to burning and wind erosion. Cultivation of organic soils leads to a rapid loss in the thickness of the organic horizon (Plate 1).

On the basis of their position within the landscape, inorganic soils are also of two broad categories: uplands and wetlands (Figure 2). Important upland soil orders in relation to the global C cycle are Mollisols, Andisols, and Aridisols. Andisols, soils of volcanic origin, are young and inherently fertile soils (Plate 2). Because of their high biomass productivity, these soils have a large potential to sequester C. Aridisols, especially those derived from calcareous parent materials, have low SOC but

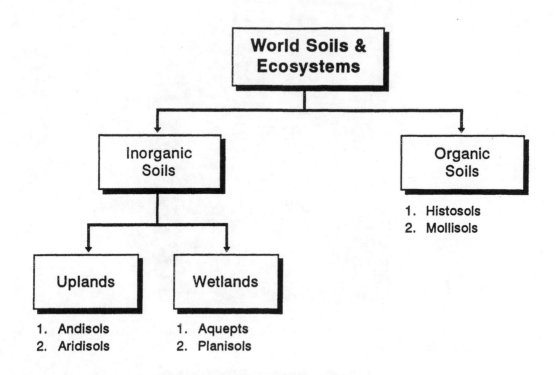

Figure 2. Predominant soils of the world with potentially significant impact on global cycle.

high soil inorganic carbon (SIC) content (Plate 3). Exact estimates of SIC content and their dynamics in relation to land use, farming/cropping systems, and soil and crop management techniques are not known. The control concept of Mollisols are the grasslands of the world. These areas are naturally fertile and have been the major granaries of the world. Their cultivation has greatly reduced the carbon storage, in many cases by 50%. Through conservation tillage and increased biomass return to the soils, they have the ability to be a major sink for carbon storage.

Another group of soils important in global C cycle are wetland soils, comprising Aquepts (Plate 4) and other hydromorphic soils (Plate 5). These soils are inundated or in the vicinity of their saturated soil moisture content most of the time, and their surface horizon is subject to anaerobiosis and rich in SOC. Hydromorphic soils are widely used for rice cultivation in the tropics and subtropics. However, the large amount of biomass-C (rice straw, sugarcane, and their residues) produced on these soils is burnt, and CO_2-C returned to the atmosphere (Plate 6). Wetlands and soils subject to anaerobiosis are also a major source of CH_4 emission into the atmosphere (IPCC, 1995).

Tropical ecosystems and soils of the tropics play a dominant role in the greenhouse effect. Principal soils of these ecoregions are Oxisols, Ultisols, and Alfisols. Deforestation of the tropical rainforest (Plate 7) affects SOC dynamics directly and indirectly. The impact of biomass burning, intensive cultivation for seasonals or annuals, and conversion to pastures have a major impact on the SOC budget which is not known. Tropical savannas are being rapidly transformed into intensively managed farmland (Plate 8). The impact of fertilizer application and high rates of biomass production on pedosphere processes is not known.

Plate 1. Decomposition of organic soils can lead to a drop of surface layer by several meters, as is the case in organic soils in Florida.

Plate 2. Andisols are young soils, have high inherent fertility and high SOC content; this soil from Japan shows stratification and buried A horizons with high levels of SOC to more than a meter depth.

Plate 3. Some aridisols have a calciferous horizon close to the soil surface; an aridisol with a petrocalcic horizon in New Mexico.

Plate 4. Wetlands are important in global C dynamics; plate shows a wetland in southern Louisiana..

Plate 5. Hydromorphic soils are intensively cultivated for rice production in Asia and elsewhere in the tropics, as in this rice field in southern India.

Plate 6. Land clearing by slash and burn in a tropical rainforest in southern Venezuela.

Plate 7. Deforestation of tropical rainforest impacts global carbon cycle, as in this deforestation of rainforest in Sumatra, Indonesia.

Plate 8. Rapid conversion of tropical savannas to row crop farming and pastures in South America and Africa affects emissions of radiatively-active gases from terrestrial ecosystems to the atmosphere.

IV. State of the Knowledge

Information presented in this and the companion volumes shows tremendous progress, especially in the following areas:

(i) **National and regional soil C budgets:** SOC pools and their dynamics have been quantified for North and South America, Africa, Asia, Australia, and Europe. As a result, statistics of the C balance is greatly improved.

(ii) **Soil processes and C dynamics:** Our knowledge of processes and mechanisms influencing C dynamics for different land use and management systems is also strengthened. We now understand more about many of the processes relevant to gains and losses of SOC, and indicators of how soil processes are affected by management changes. The importance of soil biota in the C cycle is justifiably highlighted, especially in relation to aggregation, soil structure, soil respiration, temporal and spatial changes, and effects of management systems.

(iii) **Modeling C dynamics:** Considerable improvements in models have taken place to more accurately represent temporal and spatial changes in SOC.

(iv) **Land use and soil management:** Data from long-term experiments in different ecoregions have demonstrated changes in SOC in relation to management systems. These data also highlight the significance of C accumulation within the entire soil profile and in the ecosystem.

(v) **Soil quality and atmospheric CO_2 concentration:** There exists a strong relation between SOC and soil quality. Soil can be a significant sink for CO_2 and the C thus accumulated can enhance soil quality and productivity.

V. Knowledge Gaps

Despite the tremendous progress reported, several major gaps in our knowledge exist. The following ecoregions and topics are of a high priority for future research:

(i) **Tropical ecosystems:** Knowledge about processes and properties in relation to C dynamics is scanty. Because of continuous high temperatures, losses as well as gains of C can occur more rapidly in the tropics. Understanding these processes will help realize the potential that exists for C sequestration in tropical ecosystems.

(ii) **Frozen soils:** Information on C pools and changes in C dynamics are not extensively available for soils with permafrost (Plate 9). These soils have large C reserves. Yet, potential climate change is likely to influence this large C pool as this area is predicted to have the greatest increase in temperature.

(iii) **Wetlands, Histosols, Andisols, and Aridisols:** There is insufficient information on C dynamics in wetlands and Histosols, especially with regard to historical losses of organic C. There is also a lack of information on greenhouse gas fluxes from Histosols and wetlands. Little is known abut the C dynamics in Andisols and the processes governing stabilization of organic compounds. Inorganic C (carbonates and bicarbonates) is a major pool in soils or arid regions and its magnitude and dynamics in relation to land use are not known. The total carbon pool in Histosols is poorly understood. Many of our databases look at C only in the top meter and many Histosols may be much deeper.

(iv) **C sequestration in subsoils:** More information is needed about changes in SOC with depth and with land use.

(v) **Soil erosion and C dynamics:** Information about the impact of erosion on the C balance is limited at all scales. It remains an open question whether eroded soil and its deposition as deep sediments and burial of other soils is a sink for C rather than a source of loss.

(vi) **Plant nutrients and their interaction with soil C:** More information is needed about the interaction of N, P, and S on C sequestration in soils.

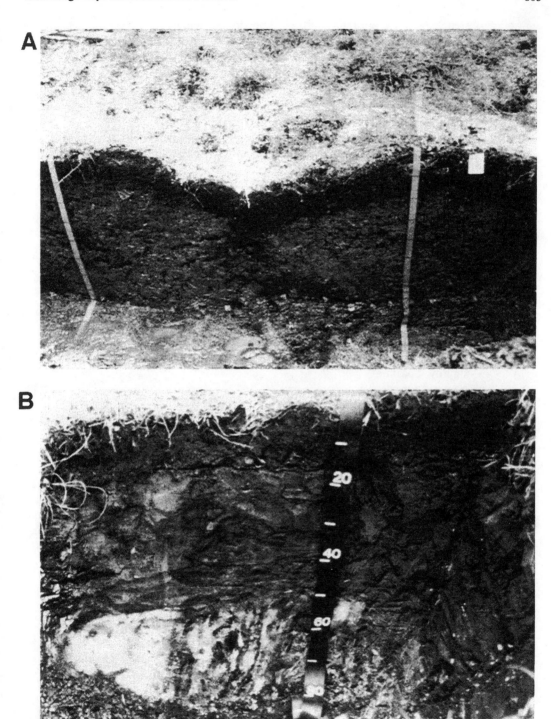

Plate 9. Soils with permafrost can play a major role in the global carbon cycle, especially with the potential greenhouse effect; (top) permafrost soil south of Inukiv in the Northwest Territory of Canada, and (bottom) permafrost soil on the north slope of Alaska.

(vii) **Soil structure and soil quality indices:** Knowledge about the processes and dynamics of soil structure, and relationship between soil structure and soil quality for developing appropriate indices and management systems that enhance soil capacity as a C sink are needed.

VI. Researchable Priorities

There are several researchable issues that need to be addressed:
(i) What is the potential for additional C sequestration in world soils?
(ii) How much of the 80 ppmv increase in atmospheric CO_2 concentration over the last 150 years is due to depletion of soil organic C?
(iii) How much of the missing C sink in the global C balance, estimated at 1.8 Pg C/yr, is due to pedosphere processes?
(iv) What are the principal processes for of C sequestration into soils, and what is the relative importance of:
- soil aggregation,
- leaching of dissolved organic C,
- soil quality enhancement, and
- C storage in the subsoil by biotic and pedosphere processes?

(v) What is the relative significance of agricultural, agroforestry, and pasture management practices to C sequestration into soils, especially with regard to:
- conservation tillage and crop residues management,
- forestry and tree plantations,
- improved pastures,
- conservation reserves program, and
- restoration of degraded ecosystems?

(vi) What are the most important policy considerations required to enhance the role of soils in ameliorating the adverse effects of CO_2 emissions?
(vii) What methods of SOC determination need to be standardized so that results that describe soil-C sequestration are comparable?
(viii) Which ecosystems play the predominant roles in the global C cycle, e.g., frozen soils, soils of the tropics, grasslands and/or forests,
(ix) What is the role of the pedosphere and world soils in biodiversity and as a genebank, and
(x) How can the data base on C pools in world soils be strengthened and information exchange improved? Could the missing carbon be just a result of bad database?

References

IPCC. 1995. Technical Summary. Inter-Governmental Panel on Climate Change, WMO, Geneva, Switzerland. 44 pp.

Schlesinger, W.H. 1995. An overview of the carbon cycle. p. 9-26. In: R. Lal, J. Kimble, E. Levine, and B.A. Stewart (eds.), *Soils and Global Change*. CRC/Lewis Publishers, Boca Raton, FL.

Index

active soil carbon 550, 552, 554-556
aggregate binding agents 208, 211
aggregate cycling 208
aggregate fractions 200, 201, 356, 364, 522
aggregate size fractions 200, 216, 267, 269-280, 443, 522, 524
aggregate stability 170, 172, 174, 175, 188, 189, 191, 192, 208, 211, 212, 218, 219, 233-235, 239, 267, 365, 392, 415, 417, 428-432, 435, 436, 439-442, 447-449, 452, 453
aggregate stabilization 202, 207, 208
aggregate strength 184
aggregating agents 245, 246
aggregation 3, 6, 7, 72, 73, 121, 124, 184, 189, 199, 201-204, 209, 215-217, 220, 225, 226, 237, 240, 241, 245-247, 257, 259, 261-263, 267, 365, 521, 591, 602, 604
agricultural management practices 499
agricultural production systems 485, 501, 506, 513, 515
agroecosystems 57, 62, 71, 73, 74, 110, 427, 485, 499
Aquic Cryoboroll 226
arable cropland 57, 58
Arctic 81, 82, 87-91, 127-130, 133, 136-139, 143, 144, 152, 153, 157, 158, 161, 162, 164, 165, 578, 592
atmosphere 1-4, 6, 7, 9, 10, 53, 63, 75, 81, 123, 143, 281, 288, 293-299, 301-303, 305-309, 335, 354, 442, 463, 485, 520, 528, 529, 541, 543, 549, 561, 573, 577, 578, 580, 592, 593, 595-597, 601
autoclaving 585
available water content 175, 181, 192
binding agents 202, 208-212, 216-218, 220, 247, 249, 251, 257, 411, 447, 448
biodiversity 3, 353, 407, 452, 604
biological indicators 335, 390
biosphere 1, 3, 4, 7, 294-299, 301-303, 305-309, 388, 561, 577, 595
biota 3, 225, 241, 295, 301, 305-308, 411, 415, 561, 573, 602
bison 416
Boreal forest 81, 89, 283
buffer capacity 390

bulk density 9, 12-15, 30, 48-51, 61, 67, 84, 90, 96-98, 111, 112, 115, 121, 122, 136, 137, 146, 160, 161, 178-181, 186-188, 355, 369, 371, 372, 374, 376, 390, 395, 412, 435, 440, 476-479, 481, 487, 532, 541
C:N ratio 23, 110, 113, 114, 162, 212, 262, 269, 270, 272, 273, 279-281
carbonate-bound sediments 297, 299, 307
carbonate content 438
carbonate equilibrium 297, 305
carbon balance 140, 307, 459, 486, 591
carbon cycling 6, 111, 294-297, 299, 301, 305, 308, 309, 345
carbon-dating 62
carbon dioxide 258, 285, 293, 296-298, 303, 307, 309, 436, 473, 485, 555, 556, 591-593
carbon encapsulation 350
carbon geochemistry 295, 299, 305
carbon mass 62, 82, 84, 86, 88-91, 95, 97, 98, 409
carbon pools 3, 4, 20, 110, 488, 501, 579, 592
carbon sequestration potential 485, 486, 501
cascade method 478
cation exchange capacity 13, 410, 415, 428, 432, 476, 477, 479
cattle 9, 378, 380, 412, 413, 416
chemical quality 23, 443, 520
clay content 13, 16, 29, 30, 32, 39, 68, 74, 171, 174, 175, 177-183, 185, 186, 191, 235, 336, 339, 475
clay dispersion 246
CO_2 fertilization 464, 549, 555, 556
community structure 438, 443
compactibility 188, 407
composted manure 68, 380, 382
compressibility 170
cone index 395
Conservation Reserve Program 417, 507, 593
conservation tillage 114, 123, 490, 501, 506, 507, 524, 525, 597, 604
continuous wheat 66, 67, 316, 317, 436, 440, 441, 445
conventional tillage 65, 178, 336, 354, 436, 448, 449, 465, 468, 469, 490, 527

cover crops 114, 121-123, 417, 491, 501, 504, 506, 507, 516, 522, 524, 525
crop residues 68, 115, 118, 119, 122, 201, 347, 350, 448, 486, 508, 593, 604
Cryosols 81, 84, 86, 89, 90
cultivated ecosystems 552
cultivation 9, 10, 49, 59, 62-64, 75, 110, 112, 114-116, 119, 120, 123, 189, 203, 210, 212, 215, 247, 249, 251, 267, 279, 280, 335, 337, 353, 354, 356, 415, 417, 435, 440, 450, 452, 459-461, 463, 523, 527, 552, 593, 596, 597
database 7, 30-32, 42, 43, 81, 82, 84, 89, 90, 93-95, 98, 121, 132, 463, 464, 476, 492, 501, 591, 604
decomposition 3, 6, 42, 53, 58, 59, 62, 64, 68, 70-73, 89, 103, 109-111, 118, 123, 124, 129, 136, 137, 157, 162, 164, 173, 175, 179, 184, 199, 201, 203, 204, 207-213, 219, 225, 240, 245-249, 251, 254, 257-259, 262, 263, 280, 283, 289, 290, 297, 315, 316, 335, 337, 364, 382, 391, 412, 442, 448, 453, 459, 465-467, 469, 481, 485, 486, 513, 515, 519-522, 530, 549, 550, 578, 580, 582, 583, 585, 586, 593, 595, 598
deep alluvium 45
deforestation 9, 10, 23, 110-112, 115, 120, 123, 261, 578, 596, 597, 601
direct drilling 354
disturbance regime 415, 416
earthworm 236, 238, 240, 345-348, 350, 515
earthworms 118, 122, 172, 173, 179, 183, 345-347, 350, 393, 415
ecoclimatic provinces 81-83, 87-91, 93, 98
ecological functions 438, 439, 442, 452
economic land values 414
electron microscopy 213, 249, 257, 261
enrichment 121, 296, 364, 459, 579, 586
environmental degradation 452
environmental filter 388, 449, 452, 453
environmental quality 5, 169, 190, 389, 390, 395, 405, 407-409, 418, 449
erosion 3, 46, 47, 58, 61, 64, 109, 111, 114, 119, 121-124, 129, 138, 169, 188, 281, 307, 353-365, 369-372, 374-379, 389, 390, 393, 395, 400, 407, 409, 410, 413, 414, 416, 417, 435-438, 441, 443, 448, 452, 526, 550, 592, 595, 596, 602

evapotranspiration 29, 32, 38-42, 109, 171, 172, 290, 486, 527, 539
extracted organic carbon 164
failure zones 170, 172-175, 182,186, 189-191, 199
farming system 388, 389, 394, 398
fauna 3, 115, 118, 170, 172, 173, 179, 182, 183, 211, 225, 237, 241, 251, 337, 438, 443, 592
fecal pellets 211, 345
feedbacks 207-209, 215, 218, 220
fractionation 42, 43, 161, 162, 199, 213, 355, 372, 475, 519-522, 524, 549, 550
freezing 129, 170-173, 183, 190, 534
friability 183, 184
fulvic acids 104, 105, 163, 164, 226, 239, 578
fumigation 428, 521, 585
fungal hyphae 136, 174, 183, 189, 192, 209, 210, 246, 247, 249, 411, 435, 592
geochemical role of carbon 294
GIS 81, 90, 93, 487, 501
glacial till 158
global warming 57, 143, 152, 158, 293, 294, 501, 526, 564
grassland 39, 62, 63, 69, 87, 88, 90, 219, 226, 237, 353, 415, 416, 460, 461, 476, 485-488, 502, 524, 552, 553, 577-579
grazing 9, 16, 24, 123, 136, 138, 268, 416, 489, 522
greenhouse effect 110, 143, 305, 561, 594, 597, 603
heavy metals 393, 438, 449, 450, 452, 453
human health 388
humic acids 104, 105, 163, 164, 226
humification 3, 7, 23, 103, 105, 122, 129, 137, 162, 164, 225, 228, 238, 530
humins 164, 578
humus 6, 103-106, 119, 121, 158, 162, 237, 240, 293, 308, 309, 315-319, 337, 389, 549, 578, 579, 595
hydrosphere 1, 3, 4, 295-297, 301, 303, 305-307, 595
indicators 73, 129, 133, 335, 387, 389-398, 400-402, 408, 414, 427-432, 435, 436, 438-440, 442, 443, 445, 447, 449, 452, 453, 520, 591, 592, 602
indicator reference values 399, 401
infiltration 14, 111, 171, 390, 393, 395, 405-407, 413, 435, 448, 469

initial cultivation 63, 64, 75, 460, 463
intercropped 370
ionization mass spectrometry 437
isotope signatures 345
isotopic composition of carbon 296
landcover 143, 144, 146, 147, 153
land use 1, 3, 9, 10, 12, 13, 15, 16, 21, 23,
    24, 31, 81, 109-112, 114, 115, 117, 120,
    121, 123, 169, 190, 267, 270, 271, 278,
    280, 290, 335, 354, 387-389, 393, 395,
    398-402, 440, 473, 487, 488, 513, 549,
    578, 580, 596, 597, 602
least limiting water range 3, 186, 187
light fraction 62, 66, 68, 69, 71-73, 173,
    174, 177, 199, 203, 220, 251, 369, 372,
    376, 378, 379, 382, 383, 389, 410
lignin 10, 13, 22, 29, 163, 164, 207, 213,
    258, 259, 263, 267, 442, 445, 447, 450,
    453, 475, 486, 521, 578
lithosphere 1, 3, 4, 295, 297, 299, 301,
    306-309, 595
living habitats 443
living matter 295, 297-299, 305, 308
long-term experiments 335-343, 370, 592,
    602
macroaggregates 71, 72, 174, 191, 200, 203,
    209-212, 215-220, 246, 247, 249, 251,
    257, 262, 263, 354, 355, 364, 365, 411,
    412, 437, 439, 440, 523, 524
manure 68, 179, 181, 182, 204, 269, 316,
    336, 343, 369, 370, 372-374, 376, 378-
    382, 390, 412, 489, 491, 494, 506, 582
meadow soils 103, 104
mesopores 180, 182, 183
metabolic quotients 343
metal ions 438, 449, 450, 453
metals 281, 393, 409, 438, 449, 450, 452,
    453
microaggregates 72, 173, 174, 176, 191,
    200, 202, 203, 208-213, 217-220, 246,
    247, 249, 251, 254, 256, 257, 261-263,
    353, 364, 365, 410-412, 524, 592
microbial biomass 66, 71, 73, 162, 174, 175,
    213, 215-217, 220, 247, 257, 335, 339,
    356, 389-392, 408, 427-433, 443, 486,
    521, 530, 534, 538
microbial colonization 247
microbial respiration 345, 350, 532, 555

microorganisms 173-175, 182, 201-203,
    213, 225, 246, 254, 257-259, 262, 263,
    337, 408, 438, 443, 446, 486, 519, 520
mineralization 6, 61, 73, 103, 104, 114, 122,
    173, 175, 189, 203, 212, 258, 261, 267,
    280, 309, 315, 337, 341, 343, 390, 408,
    410, 438, 442, 443, 446, 447, 452, 453,
    523, 537, 552, 578, 596
minimum tillage 66, 68, 593
missing carbon 290, 604
MLRA 45, 46, 48, 49, 51, 53
models 16, 21, 42, 64, 89, 98, 143, 180, 212,
    220, 225, 267, 299, 305, 306, 372, 381,
    390, 392, 395, 401, 402, 440, 459, 463-
    466, 473-476, 478, 479, 481, 486, 488,
    499-502, 515, 525, 519-524, 549, 567,
    591-594, 602
moisture regime 6, 46, 110, 477, 481
mountain forest 103, 566-568
mulch farming 115, 123
mycorrhizal fungi 211, 216
net primary productivity 280, 408
nitrifiers 436, 442, 443, 447, 452
nutrient cycling 119, 124, 289, 429, 435,
    438, 442, 443, 452
nutrient management 118
occluded organic carbon 173, 174, 184
organic carbon 1, 3, 5, 10, 29, 45, 47-54, 57,
    59, 61, 65, 81, 82, 84, 86, 88-90, 93-100,
    103-106, 110-113, 117, 119-122, 146,
    161-164, 169, 171, 173-175, 177-192,
    215, 216, 230, 259, 261, 267, 271, 278,
    281, 285, 287, 288, 290, 295, 299, 316,
    337, 343, 353, 356, 360-362, 364, 405,
    406, 413, 417, 428-432, 436, 437, 440,
    441, 459, 473-477, 479-481, 486, 487,
    489, 521, 555, 578, 579, 595
organic matter decomposition 3, 245, 248,
    262, 382, 391, 530, 578
organic matter transformation 239, 240, 486
organic matter turnover rates 415
organo-clay edge complexes 232
organo-mineral interactions 245
paleoecology 127-129, 139
particulate organic carbon 173, 174, 183,
    281, 595
passive soil carbon 550, 551, 553
pasture systems 267
pedon database 43, 476, 492

pedons 31-33, 37, 39, 42, 48, 49, 54, 62, 81, 89, 144, 146, 151, 161, 178, 227
pedosphere 1, 3, 4, 7, 283, 284, 290, 595, 597, 604
pedotransfer functions 177-182, 186, 187, 191, 395, 401, 402, 407
permafrost 89, 129, 130, 137, 138, 140, 143, 144, 146, 151, 153, 157, 158, 161, 162, 578, 602, 603
pesticides 53, 390, 449, 506
physically protected 175, 207, 208, 212, 217, 220, 261, 409, 410, 416, 519, 520, 522
physical protection 6, 72, 182, 199, 203, 207, 208, 217, 262, 522-524
plant nutrients 315, 319, 407, 415, 438, 602
plant residues 58, 72, 104, 174, 175, 199-204, 247, 251, 254, 316, 345, 369, 380, 523, 527, 530
plant roots 210, 211, 225, 407
plastic index 395
pollen 127-140, 562, 564-567
polysaccharides 164, 174, 175, 180, 185, 191, 192, 203, 209, 211, 215, 216, 257, 410
polyvalent cations 225, 365, 481
polyvalent cation bridging 257
pools of carbon 293, 294, 305, 306
porosity 14, 50, 170, 174, 178-180, 182-186, 188, 210, 216, 435, 440, 481
prairie 45, 211, 215, 217, 219, 370, 382, 438, 440, 460, 578, 593
precipitation 3, 16, 17, 19, 20, 22, 29, 30, 32, 38-41, 46, 57, 128, 138, 140, 144, 157, 207, 281-284, 288-290, 297, 307, 336, 337, 405, 407, 409, 437, 460, 463-466, 475, 485-487, 490, 504, 527, 528, 530, 531, 539, 541, 543, 561, 562, 564, 565, 567
rangelands 413, 414
rangeland soils 405
recovering ecosystems 552
resilience 405-407, 415-418
resistance 3, 49, 170, 172, 185-189, 211, 218, 225, 258, 365, 394, 405-408, 415-418, 448, 449, 528, 529
resistance to penetration 173, 185-187
respiration 3, 13, 21, 22, 58, 60, 73, 109, 111, 174, 261, 281, 306, 308, 316, 337, 341, 345, 350, 379-383, 390-392, 527, 531, 532, 534, 539, 541, 550, 555, 578, 595, 602
rooting depth 3, 96, 400, 438, 442, 452
root biomass 16, 110, 114, 122, 267, 279, 577
root development 169, 170, 172, 173, 180
root physiology 408
root scale 411-413
rotations 66, 73, 227, 410, 417, 427, 431, 436, 440, 441, 443, 445, 448, 452, 460, 463, 487, 488, 490-492, 499, 501-504, 506, 508, 513-515, 517, 519, 525, 526
salt crusting 391
sand dunes 391
savannas 9, 10, 16, 110, 114, 267, 279-281, 579, 593, 597, 601
scoring function 395
shrinking 172, 190
shrub 39, 132, 133, 144, 413, 414, 416, 563, 569-571, 584
simulated biomass 534
simulation models 64, 220, 225, 392, 440, 459, 519, 520, 523, 524
soil biological activity 104
soil biomarkers 443
soil biota 241, 411, 415, 602
soil carbon 9, 15, 22, 57, 59, 81, 82, 84-91, 93, 97, 98, 111, 127, 139, 140, 143, 147, 153, 157, 158, 161, 173, 183, 281, 293, 294, 307-309, 316, 350, 408, 412, 416, 418, 459, 473-475, 477, 479, 485-488, 490, 492-494, 499, 501-504, 506-513, 517-520, 523, 525, 549-556, 573, 591-593
soil characterization data 30, 473-475, 481
soil color 389, 393, 410, 481
soil crusting 391
soil enzymes 443, 453
soil erodibility 407
soil erosion 3, 111, 119, 121-123, 169, 353, 369, 390, 395, 414, 437, 438, 441, 595, 602
soil fertility 14, 109, 110, 122, 123, 390, 407, 526
soil flora 172, 173, 337
soil food webs 406, 407, 415, 418
soil function 388, 389, 395-397, 406, 407, 409, 410, 416, 418
soil functioning 308, 397, 399, 402
soil health 388, 394, 405

soil lipids 437, 449
soil matrix 128, 170, 171, 174, 179, 183, 203, 227, 245-247, 249, 261, 262, 412, 520, 524
soil organic matter 3, 9, 10, 16, 17, 20-23, 29, 70, 109, 157, 160-162, 199, 203, 207, 208, 225, 233, 238-240, 245, 261, 283, 309, 315, 316, 319, 335, 339, 345, 369, 370, 375, 377, 381-383, 389, 402, 405, 407, 409-411, 413, 415, 416, 435, 437, 453, 459, 460, 464, 465, 469, 485-487, 495, 501, 502, 519, 522, 527, 530, 549, 550, 554-556, 586, 591-593
soil pH 13, 14, 146, 226, 268, 432
soil properties 9, 16, 31, 43, 121, 127, 128, 144, 146, 153, 161, 173, 175, 177, 178, 180, 185, 188, 191, 228, 239, 315, 335, 339, 341, 343, 381, 389, 391, 395-397, 405, 407, 411-416, 435, 474-476, 479, 481, 595
soil quality assessment 389, 392, 428
soil quality index 394, 395, 428, 431-433, 453
soil radiocarbon measurements 549-552, 554
soil resources 353, 387, 418
soil structure 3, 6, 9, 19, 68, 109, 124, 169-173, 175, 177, 186, 190-192, 199, 203, 207, 208, 225, 245-247, 262, 393, 407, 410, 411, 414, 415, 417, 435, 438, 447, 448, 453, 476, 521, 602, 604
Soil Taxonomy 13, 30, 31, 33, 43, 146, 147, 178, 401, 474
solum depth 438, 440, 442, 452
spatial extrapolation 31, 38
spatial heterogeneity 411, 414
spatial patterns 29, 30, 38, 43, 413, 414
spatial scales 208, 210, 216, 219, 225, 481, 501
spores 127-129, 133, 136
stabilization 173, 182, 189, 199, 201, 202, 204, 207-210, 212, 215, 216, 225, 233, 240, 241, 245-247, 251, 257, 261, 262, 365, 407, 410, 412, 448, 449, 522-524, 602
structural form 170, 171, 174, 178, 184, 186, 188, 190-192
structural stability 170, 174, 183, 188, 262, 353, 354, 448, 523
structural vulnerability 170, 190, 192
stubble 67, 68, 354

substrate induced respiration 337, 341
surface residue 527-532, 534, 537, 539, 541, 543
sustainability 301, 353, 388, 391, 418, 436, 473
sustainable development 293, 295, 301-303, 309
swelling 172, 180, 181, 183, 189-191
tensile strength 172, 184, 185
termites 118, 122, 415
terrestrial ecosystems 1, 9, 121, 299, 307, 408, 459, 519, 577, 601
tillage 47, 61, 65-68, 72, 74, 106, 110-112, 114-116, 118, 119, 123, 169-171, 173, 174, 178, 180, 182, 183, 185, 186, 188, 190, 200, 203, 204, 247, 315, 336, 354, 356, 370, 381, 382, 391, 393, 395, 408, 414, 416, 417, 427, 429-431, 435, 436, 439, 443, 448, 449, 452, 460, 462-465, 468, 469, 486-492, 494, 499, 501-504, 506-508, 513, 515, 519-521, 523-543, 592, 593, 597, 604
traffic 169-171, 173, 180, 183, 188
tundra 81, 127-129, 132, 133, 136, 138, 140, 143, 144, 146-153, 157-159, 161, 162, 164, 281, 570, 578, 579
turnover characteristics 409, 410
turnover times 21, 23, 42, 71, 217-219, 262, 521, 523, 524, 549, 550, 552
urease 443
van der Waals forces 213, 257, 481
vegetation index 143
volcanic activity 137, 298, 299, 301, 303, 308, 309
water erosion 355, 393, 413, 435
water holding capacity 29, 32, 38-42, 267, 353, 412, 481, 491
water quality 3, 387, 393, 395, 405, 408, 524
weather 171, 172, 190, 488, 489
wheat-fallow 258-260, 436, 440, 441, 445, 447, 462, 463, 466, 469, 488, 491
wind erosion 393, 414, 596
zero tillage 178, 592